ANNOTATED INSTRUCTOR'S EDITION

Basic College Mathematics

FOURTH EDITION

ANNOTATED INSTRUCTOR'S EDITION

Basic College Mathematics

FOURTH EDITION

CHARLES D. MILLER

STANLEY A. SALZMAN
American River College

DIANA L. HESTWOOD
Minneapolis Community College

HarperCollinsCollegePublishers

Sponsoring Editor: Karin E. Wagner
Developmental Editor: Sandi Goldstein
Project Editor: Ann-Marie Buesing
Design Administrator: Jess Schaal
Text Design: Lesiak/Crampton Design: Lucy Lesiak
Cover Design: Lesiak/Crampton Design: Lucy Lesiak
Cover Photo: Joyce P. Lopez
Photo Researcher: Nina Page
Production Administrator: Randee Wire
Compositor: Interactive Composition Corporation
Printer and Binder: R. R. Donnelley & Sons Company
Cover Printer: The Lehigh Press, Inc.

Basic College Mathematics Fourth Edition
Annotated Instructor's Edition

ISBN 0-673-46950-6

95 96 97 9 8 7 6 5 4 3 2

Content Overview

Contents

Preface

This fourth edition of *Basic College Mathematics* is designed to help students achieve success in a developmental mathematics program. This book is comprehensive, providing the necessary background and review in whole numbers, fractions, decimals, ratio and proportion, percent, and measurement, as well as an introduction to algebra and geometry and a preview of statistics.

This text retains the successful features of previous editions: learning objectives for each section, careful exposition, fully developed examples, margin problems, cautions and notes, and boxes that set off important definitions, formulas, rules, and procedures. In this new edition, we have made several changes in content. Many of these changes follow the guidelines set forth in the *Curriculum and Evaluation Standards for School Mathematics,* published by the National Council of Teachers of Mathematics.

CHANGES IN CONTENT

- Greater emphasis has been placed on estimation skills, particularly in Chapter 1 and Chapter 4, and more exercises requiring estimation have been added throughout.
- Every exercise set has been extensively revised. Eleven accuracy checkers were hired to ensure that we maintained the highest level of accuracy possible.
- Where appropriate, students are encouraged to solve problems using a calculator. Special calculator exercises are included in selected sections as appropriate and are identified with a calculator symbol.
- In Chapter 6, "Percent," we have included compound interest as an additional application of percent.
- Chapter 7, "Measurement," has been rewritten substantially to better prepare students for a world where the metric system is now the only system used in science courses, health occupations, and international business. Section 7.2, "Denominate Numbers," has been eliminated to provide space for more background material to help students "think metric." The chapter now includes a new section on applications of metric measurement, and more discussion of the metric (Celsius) temperature scale.
- Chapter 11, "Consumer Mathematics," has been eliminated from the text, and the material contained within that chapter has been integrated into the rest of the chapters where appropriate.
- Additional visual elements, figures, graphs, and tables have been incorporated throughout the text to promote student visualization and understanding.
- The number of application exercises has been increased and they have been updated to include current data and topics relevant to students. In addition, more applications of geometry have been added throughout the text where appropriate.
- Cumulative reviews now appear after each chapter, beginning with Chapter 2, and cover material learned up to that point.

FEATURES

The following three pages illustrate important features, which are designed to assist students in the learning process. Pedagogical features have been enhanced to increase interest and accessibility.

QUEST FOR NUMERACY

Watts It Worth?

The amount of electricity used by different appliances is measured in watt-hours. You are charged a certain amount to use 1000 watt-hours (a *kilo* watt-hour, abbreviated kwh). Complete this table showing one family's yearly cost to run various appliances. They pay $0.08 per kwh. The cost in your area may be different. Round the yearly cost to the nearest dollar.

Item	*Hours in use per day*	*Hours in use per year (365 days)*	*Watts*	*Watt-hours per year*[1]	*Cost per year at $0.08 per kwh*[2]
Refrigerator[3]	12		350		
Color television	5		300		
Microwave oven	0.25		600		
Clock	24		2		
Radio	3		50		
Electric stove top	0.5		6000		
Toaster	0.1		1000		
Table lamp	6		100		
Computer	1		120		

[1] To find watt-hours, multiply watts times number of hours appliance is in use.
[2] To find number of kwh, divide watt-hours by 1000. "Kilo" means "thousand" and there are 1000 watt-hours in one kwh. Or, using unit fractions, multiply watt-hours by $\frac{1 \text{ kwh}}{1000 \text{ watt-hours}}$.
[3] Although the refrigerator is plugged in 24 hours per day, it is assumed that the motor is running only half the time.

440

Quest for Numeracy
In these pages, we address various concerns of the NCTM standards: problem solving, use of calculators, reading and using graphs and tables, estimation skills, statistical techniques, and group discussions.

QUEST FOR NUMERACY

Is the Answer Reasonable?

Even with most of the problem missing, you can tell that these answers are *not* reasonable. (They are actual student answers to test questions.) Write a sentence explaining why each answer does not make sense. Then give a reasonable range for the correct answer.

What is the teacher's monthly take-home pay?
Answer: $19.57
Unreasonable because:
Answer should be between
______ and ______.

How many hours did you work each day?
Answer: $24\frac{3}{4}$ hours
Unreasonable because:
Answer should be between
______ and ______.

Joe had $2\frac{2}{3}$ cans of paint. How much paint did he use on each window?
Answer: $4\frac{1}{2}$ cans
Unreasonable because:
Answer should be between
______ and ______.

How far can the car travel on one gallon of gas?
Answer: 3703.86 miles
Unreasonable because:
Answer should be between
______ and ______.

A $36\frac{1}{4}$ mile hike is planned. How many miles are left to travel?
Answer: $56\frac{3}{8}$ miles
Unreasonable because:
Answer should be between
______ and ______.

How much do you pay for rent?
Answer: $2.70
Unreasonable because:
Answer should be between
______ and ______.

262

Notes *Important comments are highlighted graphically and identified with the heading "Note."*

5.5 APPLICATIONS OF PROPORTIONS

1 Proportions can be used to solve a wide variety of problems. Watch for problems in which you are given a ratio or rate and then asked to find part of a corresponding ratio or rate. Remember that a ratio or rate compares two quantities and often includes one of these indicator words:

in for on per from to

When setting up the proportion, use a letter to represent the unknown number. We have used the letter x, but you may use any letter you like.

■ **EXAMPLE 1** *Using a Proportion*

Mike's car can go 163 miles **on** 6.4 **gallons** of gas. How far can it go on a full tank of 14 **gallons** of gas? Round to the nearest whole mile.

Approach Decide what is being compared. This example compares miles to **gallons**. Write the two rates described in the example. Be sure that *both* rates compare miles to gallons in the same order. In other words, miles is in both numerators and gallons is in both denominators. Use a letter to represent the missing number.

compares miles to gallons $\left\{\frac{163 \text{ miles}}{6.4 \text{ gallons}} \quad \frac{x \text{ miles}}{14 \text{ gallons}}\right\}$ compares miles to gallons

Solution Both rates compare miles to **gallons**, so you can set them up as a proportion.

> *Note* Do **not** mix up the units in the rates.
>
> compares miles to gallons $\left\{\frac{163 \text{ miles}}{6.4 \text{ gallons}} \quad \frac{14 \text{ gallons}}{x \text{ miles}}\right\}$ compares gallons to miles
>
> These rates do **not** compare things in the same order and **cannot** be set up as a proportion.

With the proportion set up correctly, solve for the missing number.

$$\frac{163 \text{ miles}}{6.4 \text{ gallons}} = \frac{x \text{ miles}}{14 \text{ gallons}}$$

matching units

matching units

Ignore the units while finding the cross products and dividing both sides by 6.4.

$6.4 \cdot x = 163 \cdot 14$ Show that cross products are equivalent.

$6.4 \cdot x = 2282$

$\frac{6.4 \cdot x}{6.4} = \frac{2282}{6.4}$ Divide both sides by 6.4.

$x = 356.5625$

Rounded to the nearest mile, the car can go 357 *miles* on a full tank of gas. Be sure to *include the units* in your answer. ■

WORK PROBLEM 1 AT THE SIDE.

OBJECTIVE

1 Use proportions to solve word problems.

FOR EXTRA HELP

Tape 7 | SSM pp. 142–145 | MAC: A IBM: A

1. Set up and solve a proportion for each problem.

(a) If 2 pounds of fertilizer will cover 50 square feet of garden, how many pounds are needed for 225 square feet?

(b) A U.S. map has a scale of 1 inch to 75 miles. Lake Superior is 4.75 inches long on the map. What is the lake's actual length in miles?

(c) Cough syrup is to be given at the rate of 30 milliliters for each 100 pounds of body weight. How much should be given to a 34-pound child? Round to the nearest whole milliliter.

ANSWERS

1. (a) $\frac{2 \text{ pounds}}{50 \text{ sq feet}} = \frac{x \text{ pounds}}{225 \text{ sq feet}}$ $x = 9$ pounds

(b) $\frac{1 \text{ inch}}{75 \text{ miles}} = \frac{4.75 \text{ inches}}{x \text{ miles}}$ $x = 356.25$ miles

(c) $\frac{30 \text{ milliliters}}{100 \text{ pounds}} = \frac{x \text{ milliliters}}{34 \text{ pounds}}$ $x = 10$ milliliters (rounded)

311

Percent 6

$13 is 25% of $52

46% = 0.46

ratio

term 0.75 = 75%

6.1 BASICS OF PERCENT

Notice that the figure below has one hundred squares of equal size. Eleven of the squares are shaded. The shaded portion is $\frac{11}{100}$, or 0.11, of the total figure.

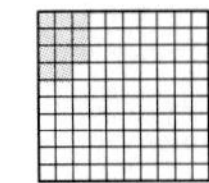

The shaded portion is also 11% of the total, or "eleven parts out of 100 parts." Read **11%** as "eleven percent."

1 As shown above, a percent is a ratio with a denominator of 100.

THE MEANING OF PERCENT

Percent means *per one hundred.* The "%" sign is used to show the number of parts per one hundred parts.

■ **EXAMPLE 1** *Understanding Percent*

(a) If 43 out of 100 students are men, then 43 per (out of) 100 or $\frac{43}{100}$ or **43%** of the students are men.

(b) If a person pays a tax of $7 per $100, then the tax rate is $7 per $100. The ratio is $\frac{7}{100}$ and the percent is 7%. ■

WORK PROBLEM 1 AT THE SIDE.

OBJECTIVES

1 Learn the meaning of percent.

2 Write percents as decimals.

3 Write decimals as percents.

FOR EXTRA HELP

Tape 7 | SSM pp. 160–162 | MAC: A IBM: A

1. Write as percents.

(a) In a group of 100 people, 63 are unmarried.

(b) The tax is $14 per $100.

(c) Out of 100 students, 36 are attending school full time.

ANSWERS

1. (a) 63% (b) 14% (c) 36%

331

Student Resources *This new feature cross-references relevant student supplements to the respective text section. They are located at the head of each section.*

Margin Exercises *Practice exercises appear in the margins throughout the text. They parallel each example and keep students involved with the presentation by allowing them to practice new concepts immediately. Answers conveniently appear at the bottom of the page.*

446 CHAPTER 7 MEASUREMENT

5. Write the most reasonable metric unit in each blank. Choose from kg, g, and mg.

(a) A thumbtack weighs 800 ____.

(b) A teenager weighs 50 ____.

(c) This large cast-iron frying pan weighs 1 ____.

(d) Jerry's basketball weighed 600 ____.

(e) Tamlyn takes a 500 ____ calcium tablet every morning.

(f) On his diet, Greg can eat 90 ____ of meat for lunch.

(g) One strand of hair weighs 2 ____.

(h) One banana might weigh 150 ____.

ANSWERS
5. (a) mg (b) kg (c) kg (d) g (e) mg (f) g (g) mg (h) g

To make larger or smaller weight units, we use the same **prefixes** as we did with length and capacity units. For example, *kilo* means 1000 so a *kilo*meter is 1000 meters, a *kilo*liter is 1000 liters, and a *kilo*gram is 1000 grams.

prefix	*kilo-* gram	*hecto-* gram	*deka-* gram	gram	*deci-* gram	*centi-* gram	*milli-* gram
meaning	1000 grams	100 grams	10 grams	1 gram	$\frac{1}{10}$ of a gram	$\frac{1}{100}$ of a gram	$\frac{1}{1000}$ of a gram
symbol	**kg**	**hg**	**dag**	**g**	**dg**	**cg**	**mg**

The units you will use most often in daily life are kilograms (kg), grams (g), and milligrams (mg). *Kilo*grams will be used instead of pounds. A kilogram is 1000 grams. It is about 2.2 pounds. This textbook weighs about 1.7 kg. An average newborn baby weighs 3 to 4 kg; a college football player might weigh 100 to 110 kg.

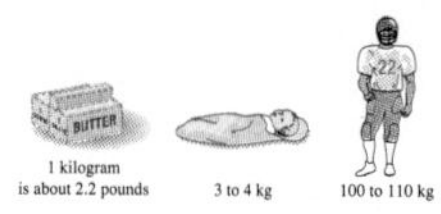

Extremely small weights are measured in *milli*grams. It takes 1000 mg to make 1 g. Recall that a dollar bill weighs about 1 g. Think of cutting it into 1000 pieces; the weight of one tiny piece would be 1 mg. Dosages of medicine and vitamins are given in milligrams. You will also use milligrams in science classes.

EXAMPLE 3 *Using Metric Weight Units*

Write the most reasonable metric unit in each blank. Choose from kg, g, and mg.

(a) Ramon's suitcase weighed 20 ____.

(b) LeTia took a 350 ____ aspirin tablet.

(c) Jenny mailed a letter that weighed 30 ____.

(a) 20 kg because kilograms are used instead of pounds. 20 kg is about 44 pounds.

(b) 350 mg because 350 g would be more than the weight of a hamburger, which is too much.

(c) 30 g because 30 kg would be much too heavy and 30 mg is less than the weight of a dollar bill. ■

WORK PROBLEM 5 AT THE SIDE.

Examples *More than 570 examples include detailed, step-by-step solutions and descriptive side comments in color. Each example includes a brief descriptive title to help students undertand the purpose of the example and to aid in studying for examinations. Many examples have been rewritten and updated to include current data and topics relevant to students.*

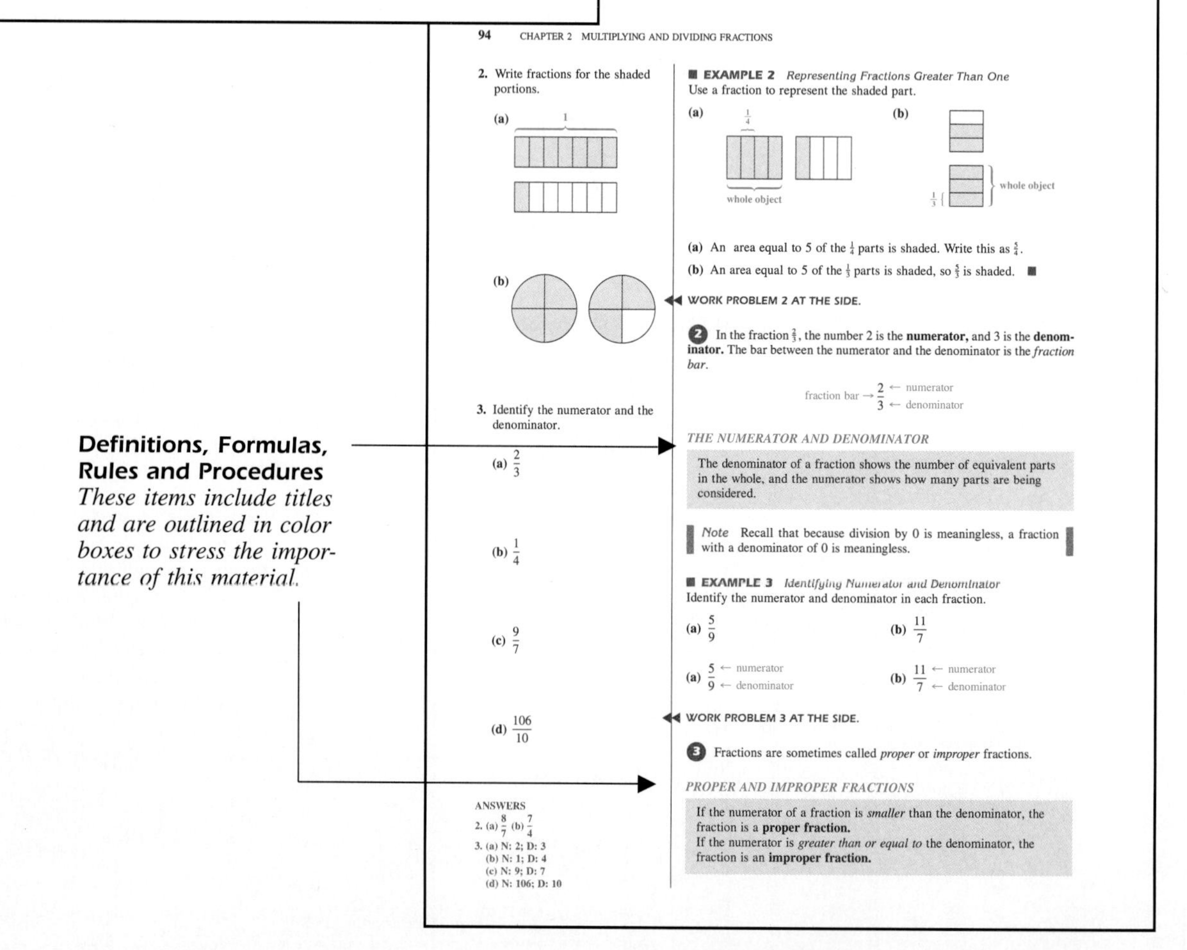

EXERCISES

As a key feature of the text, approximately 5900 exercises are provided—including 1565 review exercises, 145 conceptual and writing exercises, and over 1000 margin problems. Every exercise set has been extensively revised. Care has been taken to pair exercises (evens with odds) and to grade the exercises with regard to increasing difficulty. Exercises now include a number of special types:

Estimation exercises encourage students to estimate their answers before solving problems and to help determine whether or not their answer is reasonable. These exercises are identified with ≈.

Conceptual and writing exercises are designed to require a deeper understanding of concepts. Nearly all of these 145 exercises require the student to respond by writing a few sentences. Answers are not given for the writing exercises because they are open-ended and instructors may use them in different ways.

Challenging exercises require the student to go beyond the examples in the text.

Applications now include more realistic and interesting examples, in many cases using actual data from current events, science, sports, business, etc.

Calculator exercises are provided in selected sections and encourage the student to solve problems using the calculator. The first in a group of such exercises is identified with ▦. Subsequent exercises are identified with a colored exercise number.

Cumulative reviews end each chapter after Chapter 1 with a set of exercises that tests students on the topics covered from the beginning of the text up to that point.

Preview Exercises are intended to review the basic skills needed to do the work in the next section. These exercises also help to show how earlier material connects with and is needed for later topics.

Quick Reviews end each chapter with a list of key terms and new symbols followed by a chart pairing each of the important ideas of the chapter with a corresponding example.

SUPPLEMENTS

An extensive supplemental package is available for use with *Basic College Mathematics,* Fourth Edition.

For the Instructor

Annotated Instructor's Edition

This edition provides instructors with immediate access to the answers to every exercise in the text, with the exception of writing exercises. Each answer is printed in color next to the corresponding text exercise. Symbols are used to identify the conceptual ● and writing ✎ exercises to assist in making homework assignments. Calculator ▦ and challenging ▲ exercises are also marked for the instructor for this purpose. Suggestions for extending the *Quest for Numeracy* activities are included on those pages.

Instructor's Resource Manual

The Instructor's Resource Manual includes suggestions for using the textbook in a mathematics laboratory; short-answer and multiple-choice versions of a placement test; six forms of chapter tests for each chapter, including four open-response and two multiple-choice forms; short-answer and multiple-choice forms of a final examination; and an extensive set of additional exercises, providing 10 to 20 exercises for each textbook objective, which instructors can use as an additional source of questions for tests, quizzes, or student review of difficult topics. This manual also includes a list of all conceptual, writing, challenging, calculator, and estimation exercises.

Instructor's Solution Manual
This book includes solutions to the even-numbered section exercises and the even-numbered appendix exercises. The two solution manuals plus the solutions given at the back of the textbook provide detailed, worked-out solutions to each exercise and margin problem in the book. This manual also includes solutions to selected conceptual and writing exercises as well as a list of all conceptual, writing, challenging, calculator, and estimation exercises.

Instructor's Answer Manual
This manual includes answers to all exercises and a list of conceptual, writing, challenging, calculator, and estimation exercises.

Hands-On Activities for Basic College Mathematics
A special group of classroom activities for individual and/or small groups is included in this reference. Activity explanation, a list of necessary tools for implementation, blackline masters, suggestions for classroom uses, and textbook correlations are provided. These activities are especially useful for instructors who wish to implement the NCTM standards.

HarperCollins Test Generator/Editor for Mathematics with QuizMaster
Available in IBM and Macintosh versions, the test generator is fully networkable. The test generator enables instructors to select questions by objective, section, or chapter, or to use a ready-made test for each chapter. The editor enables instructors to edit any preexisting data or to easily create their own questions. The software is algorithm driven, allowing the instructor to regenerate constants while maintaining problem type, providing a nearly unlimited number of available test or quiz items in multiple-choice and/or open-response formats for one or more test forms. The system features printed graphics and accurate mathematics symbols. **QuizMaster** enables instructors to create tests and quizzes using the Test Generator/Editor and save them to disk so students can take the test or quiz on a stand-alone computer or network. **QuizMaster** then grades the test or quiz and allows the instructor to create reports on individual students or entire classes. CLAST and TASP versions of this package are also available for IBM and Mac machines.

For the Student

Student's Solution Manual
This book contains solutions to every other odd-numbered section exercise (those not included at the back of the textbook) as well as solutions to the odd-numbered appendix exercises. This manual also includes solutions to all margin problems, chapter review exercises, chapter tests, and cumulative review exercises. (ISBN 0-673-99064-8)

Interactive Mathematics Tutorial Software with Management System
This innovative package is available in DOS, Windows. and Macintosh versions and is fully networkable. As with the Test Generator/Editor, this software is algorithm driven, which automatically regenerates constants, so students will not see the numbers repeat in a problem type if they revisit any particular section. The tutorial is objective-based, self-paced, and provides unlimited opportunities to review lessons and to practice problem solving. If students give a wrong answer, they can ask to see the problem worked out and get a textbook page reference. The program is menu-driven for ease of use, and on-screen help can be obtained at any time with a single keystroke. Students' scores are automatically recorded and can be printed for a permanent record. The optional **Management System** allows instructors to record student scores on disk and print diagnostic reports for individual students or classes. CLAST and TASP versions of this tutorial are also available for both IBM and Mac machines. This software may also be purchased by students for use outside the classroom or lab.

Videotapes
A new videotape series has been developed to accompany *Basic College Mathematics,* Fourth Edition. In a separate lesson for each section in the book, the series covers all objectives, topics, and problem-solving techniques discussed within the text.

Overcoming Math Anxiety
This book, written by Randy Davidson and Ellen Levitov, includes step-by-step guides to problem solving, note taking, and word problems. Students can discover the reasons behind math anxiety and ways to overcome those obstacles. The book also will help them learn relaxation techniques, build better math skills, and improve study habits. (ISBN 0-06-501651-3)

ACKNOWLEDGMENTS

We appreciate the many contributions of users of the third edition of the book. We also wish to thank our reviewers for their insightful comments and suggestions:

Carla K. Ainsworth, *Salt Lake Community College*
Karen Anglin, *Blinn College*
Joyce Baker, *Blinn College*
Solveig R. Bender, *William Rainey Harper College*
Linda Kay Buchanan, *Howard College*
Laurence Chernoff, *Miami Dade Community College–South*
John W. Coburn, *St. Louis Community College at Florissant Valley*
Roger Contreras, *University of Texas at Brownsville*
Pat C. Cook, *Weatherford College*
Susan Davenport, *Blinn College*
Lucy Dechéne, *Fitchburg State College*
Diane Fariss, *McLennan Community College*
Virginia Hamilton, *Shawnee State University*
Mary Lou Hart, *Brevard Community College*
Constance C. Holden, *University College of the University of Maine*
Barbara Jordan, *Barton County Community College*
Jeff A. Koleno, *Lorain County Community College*
Jeanette Lukeman, *Blinn College*
Debra Madrid-Doyle, *Santa Fe Community College*
Grace Malaney, *Donnelly College*
Gael T. Mericle, *Mankato State University*
Julienne K. Pendleton, *Brookhaven College*
Peggy N. Phillips, *Brevard Community College*
James Price, *Tulsa Junior College*
Anne Smith, *Blinn College*
Donna Souza, *Southern Vermont College*
William J. Thieman, *Ventura College*
Priscilla R. Ware, *Franklin University*
Patti Wells, *Blinn College*
Beverly Williams, *Blinn College*

As always, Paul Eldersveld, *College of DuPage,* has done an outstanding job of coordinating all the print supplements for us.

We wish to thank all of those who did an excellent job checking all of the answers for us: Solveig R. Bender, *William Rainey Harper College;* Jean Bolyard, *Fairmont State College;* John W. Coburn, *St. Louis Community College at Florissant Valley;* Barbara Cribbs, *Stark Technical College;* Walter Faucette, *Central Arizona College;* Maria Gushanas, *Seton Hall University;* Charlene Hallermann, *St. Louis Community College at Forest Park;* Sue Jolin, *Fox Valley Technical College;* Nancy Ketchum, *Moberly Area Community College;* Linda Padilla, *Joliet Junior College;* and Peggy N. Phillip, *Brevard Community College.* We also wish to thank Irene Doo, *Austin Community College,* who wrote the solutions at the back of the text, and Jane Haasch who edited those solutions.

Our appreciation also goes to our typists, Sheri Minkner and Judy Martinez, for their fine work, to Jennifer Salzman for her help and inspiration, and to Tommy Thompson, *Cedar Valley College,* for his suggestions for the feature "To the Student: Success in Mathematics."

Special thanks go to the dedicated staff at HarperCollins who have worked so long and hard to make this book a success: Karin Wagner, Ann Shaffer, Sandi Goldstein, Ann-Marie Buesing, Anne Kelly, Linda Youngman, Kevin Connors, and Ed Moura.

In appreciation of your lasting support and never-ending enthusiasm: family, colleagues, and more than a generation of motivated students.

Stan Salzman

This book is dedicated to my dad, who always told me when I was young that girls could learn math, and to my students at Minneapolis Community College, who keep me in touch with the real world.

Diana L. Hestwood

To the Student: Success in Mathematics

The main reason students have difficulty with mathematics is that they don't know how to study it. Studying mathematics *is* different from studying subjects like English or history. The key to success is regular practice. This should not be surprising. After all, can you learn to ski or play a musical instrument without a lot of regular practice? The same thing is true for learning mathematics. Working problems nearly every day is the key to becoming successful. Here are suggestions to help you succeed in studying mathematics.

1. Pay attention in class to what your instructor says and does, and make careful notes. Note the problems the instructor works on the board and copy the complete solutions. Keep these notes separate from your homework to avoid confusion when you read them later.
2. Feel free to ask questions in class. It is not a sign of weakness, but of strength. There are always other students with the same question who are too shy to ask.
3. Determine whether tutoring is available and know how to get help when needed. Use the instructor's office hours and contact the instructor for suggestions and direction if necessary.
4. Before you start on your homework assignment, rework the problems the instructor worked in class. This will reinforce what you have learned. Many students say, "I understand it perfectly when you do it, but I get stuck when I try to work the problem myself."
5. *Read your text carefully.* Many students read only enough to get by, usually only the examples. Reading the complete section will help you be successful with the homework problems. As a bonus you will be able to do the problems more quickly if you have read the text first. As you read the text, work the example problems and check the answers. This will test your understanding of what you have read. Pay special attention to highlighted statements and those labeled "Note."
6. Do your homework assignment only after reading the text and reviewing your notes from class. Estimate the answer before you begin working the problem in the worktext. Check your work before checking with the answers in the back of the book. If you get a problem wrong and are unable to understand why, mark that problem and ask your instructor about it.
7. Work as neatly as you can using a *pencil* and organize your work carefully. Write your symbols clearly, and make sure the problems are clearly separated from each other. Working neatly will help you to think clearly and also make it easier to review the homework before a test.
8. After you have completed a homework assignment, look over the text again. Try to decide what the main ideas are in the lesson. Often they are clearly highlighted or boxed in the text.

9. Keep any quizzes and tests that are returned to you for studying for future tests and the final exam. These quizzes and tests indicate what your instructor considers most important. Be sure to correct any problems on these tests that you missed, so you will have the corrected work to study. Write all quiz and test scores on the front page of your notebook.
10. Don't worry if you do not understand a new topic right away. As you read more about it and work through the problems, you will gain understanding. Each time you review a topic you will understand it a little better. No one understands each topic completely right from the start.

USING THE ANNOTATED INSTRUCTOR'S EDITION

Answers *Answers to every exercise in the text, excluding writing exercises, are printed in color next to the problem on the page.*

Challenging Exercises *Challenging exercises, which will require most students to stretch beyond the concepts discussed in the text, are marked with the symbol ▲ in the Annotated Instructor's Edition.*

406 CHAPTER 6 PERCENT

37. Al Granard lends $7500 to the owner of Rick's Limousine Service. He will be repaid at the end of 6 years at 8% interest compounded annually. Find (a) the total amount that he should be repaid and (b) the amount of interest earned.
(a) $11,901.75 (b) $4401.75

38. Sadie Simms has $28,500 in an Individual Retirement Account (IRA) that pays 6% interest compounded semiannually. Find (a) the total amount she will have at the end of 5 years and (b) the amount of interest earned.
(a) $38,301.15 (b) $9801.15

▲ 39. There are two banks in Citrus Heights. One pays 8% interest compounded annually, and the other pays 8% compounded quarterly.
(a) If Bobbi deposits $10,000 in each bank, how much will she have in each bank at the end of 6 years?
(a) $15,869; $16,084
(b) How much more will she have in the bank that pays more interest?
(b) $215

▲ 40. Which yields more interest for Barker Aluminum: $5000 deposited for 7 years at 10% simple interest or $4000 deposited for 7 years at 10% interest compounded semiannually? What is the difference in interest?
$4000 for 7 years at 10% semiannually; $419.60

▲ 41. Jennifer Del Campo deposits $10,000 at 8% compounded quarterly. Two years after she makes the first deposit, she adds another $20,000, also at 8% compounded quarterly.
(a) What total amount will she have five years after her first deposit?
(b) What amount of interest will she have earned?
(a) $40,223.50 (b) $10,223.50

▲ 42. Scott Striver invests $9000 at 10% compounded quarterly. Three years after he makes the first deposit, he adds another $15,000, also at 10% compounded quarterly.
(a) What total amount will he have five years after his first deposit?
(b) What amount of interest will he have earned?
(a) $33,023.64 (b) $9023.64

PREVIEW EXERCISES

*Simplify by using the order of operations. (For help, see **Section 4.6.**)*

43. $6.3 \div 4.2 \cdot 3.1$
4.65

44. $18.304 \div 8.32 \cdot 3$
6.6

45. $19.3 + (6.7 - 5.2) \cdot 58$
106.3

46. $2.12 + (9.7 - 7.9) \cdot 4.5$
10.22

47. $5.34 - 2.6 \cdot 5.2 \div 2.6$
0.14

48. $61.5 - 22.8 \cdot 15 \div 5.7$
1.5

7.5 EXERCISES NAME DATE HOUR

Use the table on page 457 to make approximate conversions from metric to English or English to metric. Round your answers to the nearest tenth.

Example: 36 meters to yards

Solution: Look in the "Metric to English" part of the table.
36 meters ≈ 36 · 1.09 ≈ **39.2 yards** (rounded)

1. 20 meters to yards
21.8 yards

2. 8 kilometers to miles
5.0 miles

3. 80 meters to feet
262.4 feet

4. 85 centimeters to inches
33.2 inches

5. 16 feet to meters
4.8 m

6. 3.2 yards to meters
2.9 m

7. 150 grams to ounces
5.3 ounces

8. 2.5 ounces to grams
70.9 g

9. 248 pounds to kilograms
111.6 kg

10. 7.68 kilograms to pounds
16.9 pounds

11. 28.6 liters to quarts
30.3 quarts

12. 15.75 liters to gallons
4.1 gallons

13. Jeanette bought 18 gallons of gas for her car. How many liters of gas did she buy?
68.0 L

14. The PTA used 16 quart-size containers of orange juice at the pancake breakfast. How many liters of juice were used?
15.2 L

Circle the more reasonable temperature for each of the following.

15. A snowy day
28°C (28°F)

16. Brewing coffee
(80°C) 80°F

17. A high fever
(40°C) 40°F

18. Swimming pool water
78°C (78°F)

19. Oven temperature
(150°C) 150°F

20. Light jacket weather
(10°C) 10°F

21. Would a drop of 20 Celsius degrees be more or less than a drop of 20 Fahrenheit degrees? Explain your answer.
More; explanation varies.

22. Describe one advantage of switching from the Fahrenheit temperature scale to the Celsius scale. Describe one disadvantage.
Answer varies.

Writing · Conceptual · ▲ Challenging · ≈ Estimation

461

Conceptual and Writing Exercises *The conceptual and writing exercises are marked in the Annotated Instructor's Edition so instructors may assign these problems at their discretion. Conceptual exercises are marked with the ◉ symbol. Writing exercises are marked with this symbol, ✎.*

Symbol Key *A key to all symbols used in the exercises is provided on the first page of every exercise set in the Annotated Instructor's Edition.*

Whole Numbers

1.1 READING AND WRITING WHOLE NUMBERS

1 The **decimal system** of writing numbers uses the ten digits

0, 1, 2, 3, 4, 5, 6, 7, 8, 9

to write any number. For example, these digits can be used to write the **whole numbers:**

0, 1, 2, 3, 4, 5, 6, 7, 8, 9, 10, 11, 12, 13

and so on.

2 Each digit in a whole number has a **place value.** The following place value chart shows the names of the different places used most often.

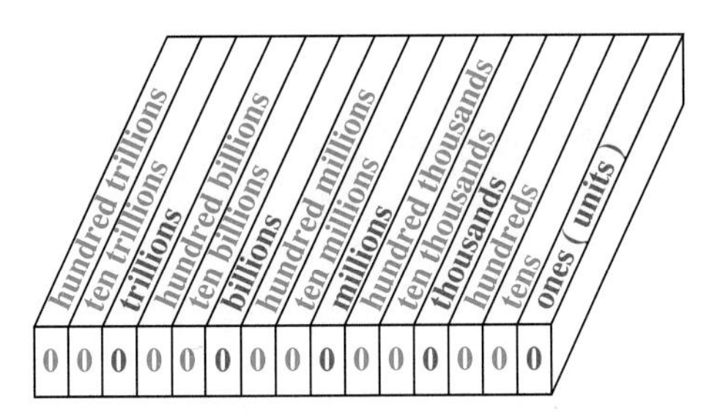

EXAMPLE 1 *Identifying Whole Numbers*

(a) In the whole number 42, the 2 is in the ones place and has a value of 2 *ones*.

(b) In 29, the 2 is in the tens place and has a value of 2 *tens*.

(c) In 281, the 2 is in the hundreds place and has a value of 2 *hundreds.*

The value of 2 in each number is different; depending on its location (place) in the number.

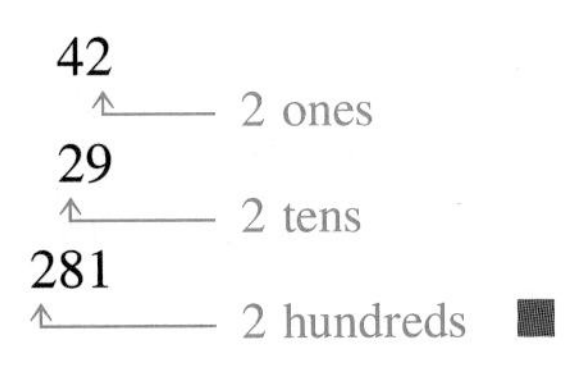

WORK PROBLEM 1 AT THE SIDE.

OBJECTIVES

1. Identify whole numbers.
2. Give the place value of a digit.
3. Write a number in words or digits.

FOR EXTRA HELP

Tape 1 | SSM pp. 1–2 | MAC: A IBM: A

1. Identify the place value of the 8 in each whole number.

(a) 582

(b) 308

(c) 896

ANSWERS

1. (a) tens (b) ones (c) hundreds

2. Identify the place value of each digit.

   (a) 24,386

   (b) 371,942

3. In the number 3,251,609,328 identify the digits in each of the following periods (groupings).

   (a) billions period

   (b) millions period

   (c) thousands period

   (d) ones period

ANSWERS

2. (a) 2: ten thousands
4: thousands
3: hundreds
8: tens
6: ones
(b) 3: hundred thousands
7: ten thousands
1: thousands
9: hundreds
4: tens
2: ones
3. (a) 3 (b) 251 (c) 609 (d) 328

■ **EXAMPLE 2** *Identifying Place Values*

Find the place value of each digit in the number 725,283.

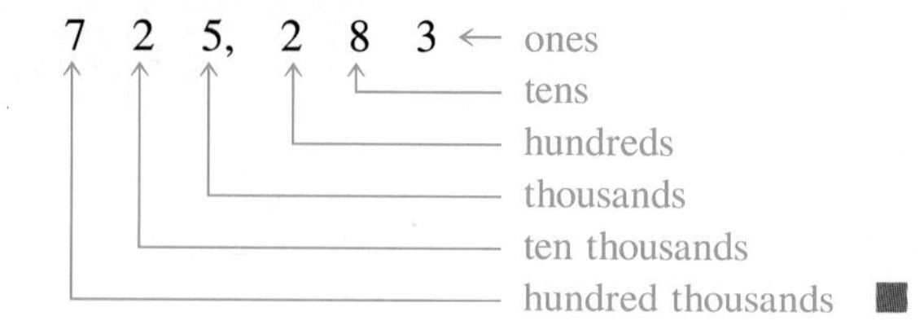

■

◀ **WORK PROBLEM 2 AT THE SIDE.**

Notice the comma between the hundreds and thousands position in the number 725,283 above.

USING COMMAS

Commas are used to separate each group of three digits, starting from the right. This makes numbers easier to read. (An exception: commas are frequently omitted in four-digit numbers such as 9748 or 1329.) Each three-digit group is called a **period.**

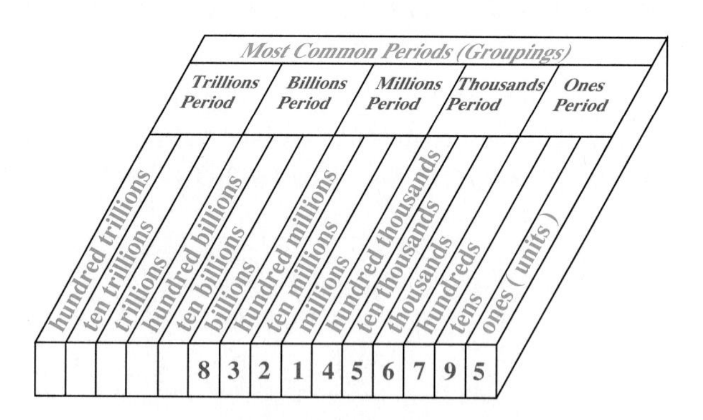

■ **EXAMPLE 3** *Knowing the Period Names (Groupings)*

Write the digits in each period of 8,321,456,795.

billions period	8	← 8,321,456,795
millions period	321	
thousands period	456	
ones period	795	■

◀ **WORK PROBLEM 3 AT THE SIDE.**

Use the following rule to read a number with more than three digits.

READING NUMBERS

Start at the left when reading a number. Read the digits in each period (grouping), followed by the name of the period, except for the period name "ones," which is *not* used.

3 The following examples show how to write names for whole numbers.

EXAMPLE 4 *Writing Numbers in Words*
Write each number in words.

(a) 57
This number means 5 tens and 7 ones, or 50 ones and 7 ones. Write the number as

fifty-seven.

(b) 94

ninety-four

(c) 874

eight hundred seventy-four

(d) 601

six hundred one ■

Note Do not use the word "and" when writing whole numbers. You will use the word "and" later to indicate the decimal position.

WORK PROBLEM 4 AT THE SIDE.

EXAMPLE 5 *Writing Numbers in Words by Using Period Names*
Write each number in words.

(a) 725,283

seven hundred twenty-five thousand, two hundred eighty-three

number in period (seven hundred twenty-five); name of period (thousand); number in period (not necessary to write "ones") (two hundred eighty-three)

(b) 7835

seven thousand, eight hundred thirty-five

name of period (thousand); no period name needed (eight hundred thirty-five)

(c) 111,356,075

one hundred eleven million, three hundred fifty-six thousand, seventy-five

(d) 6,000,005,000

six billion, five thousand ■

WORK PROBLEM 5 AT THE SIDE.

4. Write each number in words.

(a) 27

(b) 68

(c) 293

(d) 205

5. Write each number in words.

(a) 7309

(b) 95,372

(c) 100,075,002

(d) 17,022,040,000

ANSWERS

4. (a) twenty-seven **(b)** sixty-eight **(c)** two hundred ninety-three **(d)** two hundred five

5. (a) seven thousand, three hundred nine **(b)** ninety-five thousand, three hundred seventy-two **(c)** one hundred million, seventy-five thousand, two **(d)** seventeen billion, twenty-two million, forty thousand

EXAMPLE 6 *Writing Numbers in Digits*
Rewrite each of the following numbers using digits.

(a) seven thousand, eighty-five

7085

(b) two hundred fifty-six thousand, six hundred twelve

256,612

(c) nine million, five hundred fifty-nine

9,000,559

zeros indicate there are no thousands ■

WORK PROBLEM 6 AT THE SIDE.

6. Rewrite each of the following numbers using digits.

(a) five thousand, six hundred twenty-three

(b) nine hundred seventy-one thousand, six

(c) eighty-two million, three hundred twenty-five

ANSWERS
6. (a) 5623 (b) 971,006 (c) 82,000,325

1.1 EXERCISES

NAME DATE HOUR

Fill in the digit for the given **place value** *in each of the following whole numbers.*

Example:		**Solution:**
782	hundreds	**7**
	ones	**2**

1. 2615
thousands 2
tens 1

2. 8113
thousands 8
ones 3

3. 18,015
ten thousands 1
hundreds 0

4. 75,229
ten thousands 7
ones 9

5. 7,628,592,183
millions 8
thousands 2

6. 1,700,225,016
billions 1
millions 0

Fill in the number for the given **period** *(grouping) in each of the following whole numbers.*

Example:		**Solution:**
58,618	thousands	**58**
	ones	**618**

7. 10,678,286
millions 10
thousands 678
ones 286

8. 28,785,203
millions 28
thousands 785
ones 203

9. 60,000,502,109
billions 60
millions 0
thousands 502
ones 109

10. 100,258,100,006
billions 100
millions 258
thousands 100
ones 6

11. Do you think the fact that humans have four fingers and a thumb on each hand explains why we use a number system based on ten digits? Explain. Answer varies.

12. The decimal system uses ten digits. Fingers and toes are often referred to as digits. In your opinion, is there a relationship here? Explain. Answer varies.

Rewrite the following numbers in words.

Example: 1,630,254 **Solution: one million, six hundred thirty thousand, two hundred fifty-four**

13. 79,613 seventy-nine thousand, six hundred thirteen

14. 37,886 thirty-seven thousand, eight hundred eighty-six

15. 725,009 seven hundred twenty-five thousand, nine

16. 218,033 two hundred eighteen thousand, thirty-three

17. 25,756,665 twenty-five million, seven hundred fifty-six thousand, six hundred sixty-five

18. 999,993,000 nine hundred ninety-nine million, nine hundred ninety-three thousand

Rewrite each of the following numbers using digits.

Example: three thousand, four hundred twenty **Solution: 3420**

19. eighty-three thousand, one hundred thirty-five 83,135

20. fifty-eight thousand, six hundred eight 58,608

21. ten million, two hundred twenty-three 10,000,223

22. one hundred million, two hundred 100,000,200

Rewrite the numbers from the following sentences using digits.

Example: Her income tax refund was one thousand, five hundred twelve dollars.

Solution: In digits, the number is **1512.**

23. Center School District used three thousand, fifty bandage strips last year.

3050

24. Every year, nine hundred seventy-two thousand, six hundred fifty people visit a certain historical area.

972,650

25. A supermarket has thirteen thousand, one hundred twelve different items for sale.

13,112

26. Yosemite National Park had three million, four hundred eighty-five thousand, five hundred visitors last year.

3,485,500

27. An electronics firm produces forty-eight million, six hundred five computer chips.

48,000,605

28. The number of registered motor vehicles is one million, one thousand, one.

1,001,001

▲ **29.** Rewrite eight hundred billion, six hundred twenty-one million, twenty thousand, two hundred fifteen by using digits.

800,621,020,215

▲ **30.** Rewrite 306,735,002,102 in words.

three hundred six billion, seven hundred thirty-five million, two thousand, one hundred two

1.2 ADDITION OF WHOLE NUMBERS

OBJECTIVES

1. Add two single-digit numbers.
2. Add more than two numbers.
3. Add when carrying is not required.
4. Add with carrying.
5. Solve word problems with carrying.
6. Check the answer in addition.

FOR EXTRA HELP

Tape 1

SSM pp. 2–4

MAC: A
IBM: A

There are 4 triangles at the left and 2 at the right. In all, there are 6 triangles.

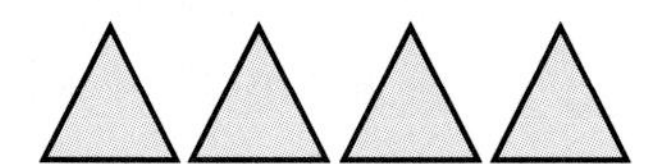 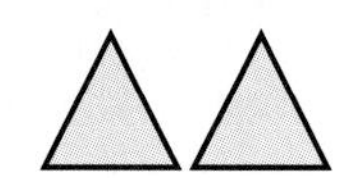

The process of finding the total is called **addition.** Here 4 and 2 were added to get 6. Addition is written with a + sign, so that

$$4 + 2 = 6.$$

1 In addition, the numbers being added are called **addends,** and the resulting answer is called the **sum** or **total.**

$$\begin{array}{r l} 4 & \leftarrow \text{addend} \\ +\ 2 & \leftarrow \text{addend} \\ \hline 6 & \leftarrow \text{sum (answer)} \end{array}$$

Addition problems can also be written horizontally as follows.

$$\begin{array}{ccccc} 4 & + & 2 & = & 6 \\ \uparrow & & \uparrow & & \uparrow \\ \text{addend} & & \text{addend} & & \text{sum} \end{array}$$

COMMUTATIVE PROPERTY OF ADDITION

> To change the order of the numbers in an addition problem we use the **commutative property of addition.** This does not change the sum.

For example, the sum 4 + 2 is the same as the sum 2 + 4. This allows the addition of the same numbers in a different order.

■ **EXAMPLE 1** *Adding Two Single-Digit Numbers*
Add the following.

(a) 6 + 2 = 8

(b) 5 + 9 = 14

(c) 8 + 3 = 11

(d) 8 + 8 = 16 ■

WORK PROBLEM 1 AT THE SIDE. ▶

ASSOCIATIVE PROPERTY OF ADDITION

> By the **associative property of addition,** grouping the addition of numbers in any order does not change the sum.

1. Add.

(a) 2 + 5

(b) 9 + 9

(c) 7 + 8

(d) 4 + 8

ANSWERS
1. (a) 7 (b) 18 (c) 15 (d) 12

For example, the sum of 3 + 5 + 6 may be found as follows.

$$(3 + 5) + 6 = 8 + 6 = 14$$ Parentheses tell what to do first.

Another way to add the same numbers is shown below.

$$3 + (5 + 6) = 3 + 11 = 14$$

Either method gives the answer 14.

2 To add several numbers, first write them in a column. Add the first number to the second. Add this sum to the third digit; continue until all the digits are used.

■ EXAMPLE 2 *Adding More Than Two Numbers*

Add 2, 5, 6, 1, and 4.

```
  2 ┐
  5 ┘─ 2 + 5 = 7
 (6)──── 7 + 6 = 13
 (1)──── 13 + 1 = 14
+(4)──── 14 + 4 = 18
 18
```
■

> *Note* By the commutative property of addition, numbers may be added starting at the bottom of a column.

◀◀ WORK PROBLEM 2 AT THE SIDE.

3 If numbers have two or more digits, first you must arrange the numbers in columns so that the ones digits are in the same column, tens are in the same column, hundreds are in the same column, and so on. Next, you add column by column.

■ EXAMPLE 3 *Adding Without Carrying*

Add 511 + 23 + 154 + 10.

First line up the numbers in columns, with the ones column at the right.

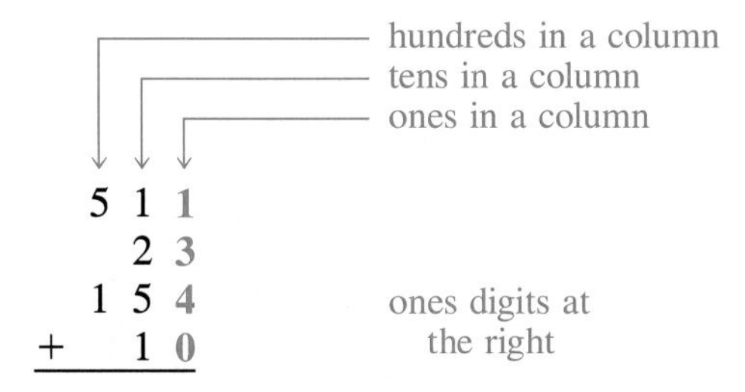

Now start at the right and add the ones digits. Add the tens digits next, and finally, the hundreds digits.

```
  5 1 1
    2 3
  1 5 4
+   1 0
  6 9 8
  ↑ ↑ ↑
  │ │ └── sum of ones
  │ └──── sum of tens
  └────── sum of hundreds
```

The sum of the four numbers is 698. ■

2. Add the following columns of numbers.

(a)
```
  2
  9
  6
  4
+ 5
```

(b)
```
  6
  2
  1
  5
+ 7
```

(c)
```
  9
  2
  1
  3
+ 4
```

(d)
```
  3
  8
  6
  4
+ 8
```

ANSWERS

2. (a) 26 (b) 21 (c) 19 (d) 29

WORK PROBLEM 3 AT THE SIDE.

4 If the sum of the digits in a column is more than 9, use **carrying.**

■ **EXAMPLE 4** *Adding with Carrying*
Add 47 and 29.

Add ones.

$$\begin{array}{r} 47 \\ +\ 29 \\ \hline \end{array}$$

Sum of ones is 16.

Because 16 is 1 ten plus 6 ones, place 6 in the ones column and carry 1 to the tens column.

$$\begin{array}{r} 1 \\ 47 \\ +\ 29 \\ \hline 6 \end{array} \quad 7 + 9 = 16$$

Add in the tens column.

$$\begin{array}{r} 1 \\ 47 \\ +\ 29 \\ \hline 76 \end{array}$$

sum of digits in tens column ■

WORK PROBLEM 4 AT THE SIDE.

■ **EXAMPLE 5** *Adding with Carrying*
Add 324 + 7855 + 23 + 7 + 86.

Step 1 Add the digits in the ones column.

$$\begin{array}{r} 2 \\ 324 \\ 7855 \\ 23 \\ 7 \\ +\ 86 \\ \hline 5 \end{array}$$

Carry 2 to the tens column.

Sum of the ones column is 25.

Write 5 in the ones column.

In 25, the 5 represents 5 ones and is written in the ones column, while 2 represents 2 tens and is carried to the tens column.

Step 2 Now add the digits in the tens column, including the carried 2.

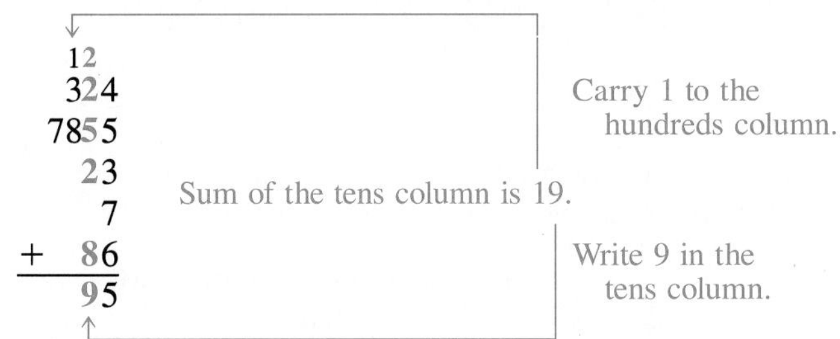

3. Add.

(a) $\begin{array}{r} 14 \\ +\ 75 \\ \hline \end{array}$

(b) $\begin{array}{r} 426 \\ +\ 572 \\ \hline \end{array}$

(c) $\begin{array}{r} 42{,}305 \\ +\ 11{,}563 \\ \hline \end{array}$

4. Add by using carrying.

(a) $\begin{array}{r} 58 \\ +\ 24 \\ \hline \end{array}$

(b) $\begin{array}{r} 76 \\ +\ 18 \\ \hline \end{array}$

(c) $\begin{array}{r} 67 \\ +\ 29 \\ \hline \end{array}$

(d) $\begin{array}{r} 34 \\ +\ 49 \\ \hline \end{array}$

ANSWERS
3. (a) 89 (b) 998 (c) 53,868
4. (a) 82 (b) 94 (c) 96 (d) 83

5. Add by carrying as necessary.

(a)
```
   576
    92
    43
+  274
```

(b)
```
  4271
   372
  8976
+  162
```

(c)
```
    57
     4
   392
   804
    51
+   27
```

(d)
```
  5318
   629
    35
   407
+ 1382
```

6. Add with mental carrying.

(a)
```
   278
   825
    14
     3
     7
+ 9275
```

(b)
```
  5280
   407
   805
    31
    20
     4
+    2
```

(c)
```
  15,829
     765
      78
      15
       9
       7
+ 13,179
```

ANSWERS
5. (a) 985 (b) 13,781 (c) 1335 (d) 7771
6. (a) 10,402 (b) 6549 (c) 29,882

Step 3
```
 112
  324
 7855
   23
    7
+  86
  295
```
Carry 1 to the thousands column.
Sum of the hundreds column is 12.
Write 2 in the hundreds column.

Step 4
```
 112
  324
 7855
   23
    7
+  86
 8295
```
Sum of the thousands column is 8.

Finally, $324 + 7855 + 23 + 7 + 86 = 8295$. ■

◀ **WORK PROBLEM 5 AT THE SIDE.**

Note For additional speed, try to carry mentally. Do not write the number carried, but just carry the number mentally to the top of the next column being added. Try this method. If it works for you, use it.

◀ **WORK PROBLEM 6 AT THE SIDE.**

5 In Section 1.9 we will describe how to solve word problems in more detail. The next two examples have word problems that require adding.

■ **EXAMPLE 6** *Applying Addition Skills*
On this map, the distance in miles from one location to another is written alongside the road. Find the shortest distance from Altamonte Springs to Clear Lake.

Approach Add the mileage along various routes to determine the distances from Altamonte Springs to Clear Lake. Then select the shortest route.

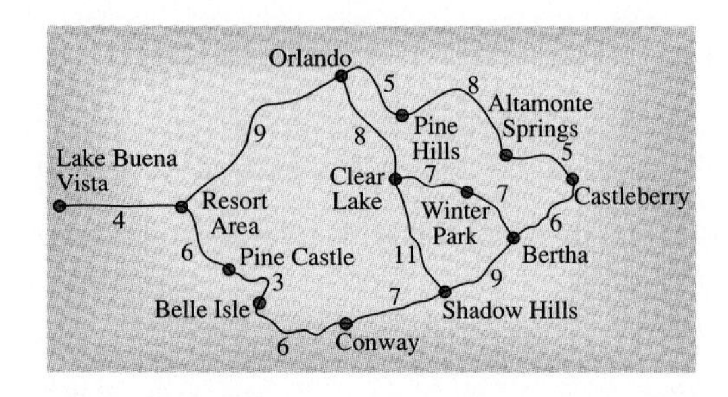

Solution One way from Altamonte Springs to Clear Lake is through Orlando. Add the mileage numbers along this route.

8	Altamonte Springs to Pine Hills
5	Pine Hills to Orlando
+ 8	Orlando to Clear Lake
21	miles from Altamonte Springs to Clear Lake, going through Orlando

Another way is through Bertha and Winter Park. Add the mileage numbers along this route.

5	Altamonte Springs to Castleberry
6	Castleberry to Bertha
7	Bertha to Winter Park
+ 7	Winter Park to Clear Lake
25	miles from Altamonte Springs to Clear Lake through Bertha and Winter Park

The shortest way from Altamonte Springs to Clear Lake is through Orlando. ■

WORK PROBLEM 7 AT THE SIDE. ▶▶

■ **EXAMPLE 7** *Finding a Total*
Find the total distance from Shadow Hills to Castleberry to Orlando and back to Shadow Hills.

Approach Add the mileage from Shadow Hills to Castleberry to Orlando and back to Shadow Hills to find the total distance.

Solution Use the numbers from the map.

9	Shadow Hills to Bertha
6	Bertha to Castleberry
5	Castleberry to Altamonte Springs
8	Altamonte Springs to Pine Hills
5	Pine Hills to Orlando
8	Orlando to Clear Lake
+ 11	Clear Lake to Shadow Hills
52	miles from Shadow Hills to Castleberry to Orlando and back to Shadow Hills ■

WORK PROBLEM 8 AT THE SIDE. ▶▶

6 Checking the answer is an important part of problem solving. A common method for checking addition is to re-add from bottom to top.

■ **EXAMPLE 8** *Checking Addition*
Check the following addition.

```
                1428 ↑
(add down)       738 |  Adding down and
    |             63 |    adding up should give
    |            125 |    the same answer.
    |             17 |
    |          + 485 ┘  (add up)
    ↓           1428     check
```

Here the answers agree, so the sum is probably correct. ■

7. Use the map to find the shortest distance from Lake Buena Vista to Winter Park.

8. The road is closed between Orlando and Clear Lake, so this route cannot be used. Use the map to find the next shortest distance from Orlando to Clear Lake.

ANSWERS
7. 28 miles
8.

Orlando to Pine Hills	**5**
Pine Hills to Altamonte Springs	**8**
Altamonte Springs to Castleberry	**5**
Castleberry to Bertha	**6**
Bertha to Winter Park	**7**
Winter Park to Clear Lake	**+ 7**
	38 miles

9. Check the following addition. If an answer is incorrect, give the correct answer.

(a)

```
   16
    3
    5
+  27
   51
```

(b)

```
  715
  622
   38
+ 198
 1573
```

(c)

```
   79
  218
    7
+ 639
  953
```

(d)

```
  21,892
  11,746
+ 43,925
  79,563
```

ANSWERS

9. (a) correct (b) correct
(c) incorrect, should be 943
(d) incorrect, should be 77,563

■ **EXAMPLE 9** *Checking Addition*

Check the following additions.

(a)

```
                 1033 ↑   correct, because both
                 ----        answers are the same
   785            785
    63             63     (add up)
+  185          + 185  └─ check
  1033           1033
```

(b)

```
                 2454 ↑   error, because answers
                 ----        are different
   635            635
    73             73
   831            831     (add up)
+  915          + 915  └─ check
  2444           2444
```

Re-add to find that the correct answer is 2454. ■

◀◀ **WORK PROBLEM 9 AT THE SIDE.**

EXERCISES

Add.

Examples:

$$\begin{array}{r} 57 \\ +\ 42 \\ \hline \end{array} \qquad 23 + 721 + 834$$

Solutions:
Line up the numbers in columns.

$$\begin{array}{r} 57 \\ +\ 42 \\ \hline \mathbf{99} \end{array} \qquad \begin{array}{r} 23 \\ 721 \\ +\ 834 \\ \hline \mathbf{1578} \end{array}$$

ones added
tens added

1. $\begin{array}{r} 13 \\ +\ 36 \\ \hline \end{array}$ 49

2. $\begin{array}{r} 12 \\ +\ 14 \\ \hline \end{array}$ 26

3. $\begin{array}{r} 26 \\ +\ 52 \\ \hline \end{array}$ 78

4. $\begin{array}{r} 83 \\ +\ 15 \\ \hline \end{array}$ 98

5. $\begin{array}{r} 317 \\ +\ 572 \\ \hline \end{array}$ 889

6. $\begin{array}{r} 651 \\ +\ 228 \\ \hline \end{array}$ 879

7. $\begin{array}{r} 258 \\ 421 \\ +\ 320 \\ \hline \end{array}$ 999

8. $\begin{array}{r} 135 \\ 253 \\ +\ 410 \\ \hline \end{array}$ 798

9. $\begin{array}{r} 6310 \\ 252 \\ +\ 1223 \\ \hline \end{array}$ 7785

10. $\begin{array}{r} 121 \\ 5705 \\ +\ 3163 \\ \hline \end{array}$ 8989

11. 321 + 175 + 403
899

12. 251 + 120 + 618
989

13. 1251 + 4311 + 2114
7676

14. 3241 + 1513 + 2014
6768

15. 12,142 + 43,201 + 23,103
78,446

16. 41,124 + 12,302 + 23,500
76,926

17. 9213 + 6715
15,928

18. 7651 + 4283
11,934

19. 38,204 + 91,020
129,224

20. 53,251 + 96,305
149,556

Writing Conceptual Challenging Estimation

Add the following numbers by carrying as necessary.

Example:	**Solution:**
$\begin{array}{r} 185 \\ +\ 769 \\ \hline \end{array}$	$\begin{array}{r} {\scriptstyle 11} \\ 185 \\ +\ 769 \\ \hline \mathbf{954} \end{array}$

21. $\begin{array}{r} 78 \\ +\ 65 \\ \hline 143 \end{array}$ **22.** $\begin{array}{r} 82 \\ +\ 49 \\ \hline 131 \end{array}$ **23.** $\begin{array}{r} 65 \\ +\ 77 \\ \hline 142 \end{array}$ **24.** $\begin{array}{r} 29 \\ +\ 98 \\ \hline 127 \end{array}$ **25.** $\begin{array}{r} 47 \\ +\ 74 \\ \hline 121 \end{array}$

26. $\begin{array}{r} 99 \\ +\ 99 \\ \hline 198 \end{array}$ **27.** $\begin{array}{r} 73 \\ +\ 89 \\ \hline 162 \end{array}$ **28.** $\begin{array}{r} 96 \\ +\ 47 \\ \hline 143 \end{array}$ **29.** $\begin{array}{r} 73 \\ +\ 29 \\ \hline 102 \end{array}$ **30.** $\begin{array}{r} 68 \\ +\ 37 \\ \hline 105 \end{array}$

31. $\begin{array}{r} 906 \\ +\ 875 \\ \hline 1781 \end{array}$ **32.** $\begin{array}{r} 621 \\ +\ 359 \\ \hline 980 \end{array}$ **33.** $\begin{array}{r} 306 \\ +\ 848 \\ \hline 1154 \end{array}$ **34.** $\begin{array}{r} 798 \\ +\ 206 \\ \hline 1004 \end{array}$ **35.** $\begin{array}{r} 278 \\ +\ 135 \\ \hline 413 \end{array}$

36. $\begin{array}{r} 172 \\ +\ 156 \\ \hline 328 \end{array}$ **37.** $\begin{array}{r} 928 \\ +\ 843 \\ \hline 1771 \end{array}$ **38.** $\begin{array}{r} 686 \\ +\ 726 \\ \hline 1412 \end{array}$ **39.** $\begin{array}{r} 526 \\ +\ 884 \\ \hline 1410 \end{array}$ **40.** $\begin{array}{r} 116 \\ +\ 897 \\ \hline 1013 \end{array}$

41. $\begin{array}{r} 7968 \\ +\ 1285 \\ \hline 9253 \end{array}$ **42.** $\begin{array}{r} 1768 \\ +\ 8275 \\ \hline 10{,}043 \end{array}$ **43.** $\begin{array}{r} 7896 \\ +\ 3728 \\ \hline 11{,}624 \end{array}$ **44.** $\begin{array}{r} 9382 \\ +\ 7586 \\ \hline 16{,}968 \end{array}$ **45.** $\begin{array}{r} 9625 \\ +\ 7986 \\ \hline 17{,}611 \end{array}$

46. $\begin{array}{r} 6829 \\ 6076 \\ +\ 8218 \\ \hline 21{,}123 \end{array}$ **47.** $\begin{array}{r} 9056 \\ 78 \\ 6089 \\ +\ 731 \\ \hline 15{,}954 \end{array}$ **48.** $\begin{array}{r} 4022 \\ 709 \\ 8621 \\ +\ 37 \\ \hline 13{,}389 \end{array}$ **49.** $\begin{array}{r} 18 \\ 708 \\ 9286 \\ +\ 636 \\ \hline 10{,}648 \end{array}$ **50.** $\begin{array}{r} 1708 \\ 321 \\ 61 \\ +\ 8926 \\ \hline 11{,}016 \end{array}$

51. $\begin{array}{r} 218 \\ 7022 \\ 335 \\ +\ 9283 \\ \hline 16{,}858 \end{array}$ **52.** $\begin{array}{r} 6505 \\ 173 \\ 7044 \\ +\ 168 \\ \hline 13{,}890 \end{array}$ **53.** $\begin{array}{r} 321 \\ 9603 \\ 8 \\ 21 \\ +\ 1604 \\ \hline 11{,}557 \end{array}$ **54.** $\begin{array}{r} 7631 \\ 5983 \\ 7 \\ 36 \\ +\ 505 \\ \hline 14{,}162 \end{array}$ **55.** $\begin{array}{r} 2109 \\ 63 \\ 16 \\ 3 \\ +\ 9887 \\ \hline 12{,}078 \end{array}$

56.
```
    244
     67
   7076
     13
    618
 + 3005
 ------
 11,023
```

57.
```
    553
     97
   2772
    437
     63
 +  328
 ------
   4250
```

58.
```
   3187
    810
    527
     76
   2665
 +  317
 ------
   7582
```

59.
```
    413
     85
   9919
    602
     31
 + 1218
 ------
 12,268
```

60.
```
    576
   7934
     60
    781
   5968
 +  371
 ------
 15,690
```

Check the following additions. If an answer is incorrect, give the correct answer.

Example:
```
    835
    278
 +  422
 ------
   1535
```

Solution:
```
   1535
    835
    278
 +  422
 ------
   1535
```
Correct

61.
```
    628
    265
 +  128
 ------
   1021
```
correct

62.
```
    483
    918
 +  754
 ------
   2155
```
correct

63.
```
    179
    214
 +  376
 ------
    759
```
incorrect; should be 769

64.
```
     17
    296
    713
 +   94
 ------
   1220
```
incorrect; should be 1120

65.
```
   4713
     28
    615
 +   64
 ------
   5420
```
correct

66.
```
  6 215
    744
     36
+ 4 284
-------
 11,279
```
correct

67.
```
    678
  7 952
     56
    718
+ 2 173
-------
 11,377
```
incorrect; should be 11,577

68.
```
    516
  8 760
     24
    189
+ 1 723
-------
 11,212
```
correct

69.
```
  4 714
     27
     77
  8 878
+   636
-------
 14,332
```
correct

70.
```
  6 715
    283
  9 617
     13
+    81
-------
 16,719
```
incorrect; should be 16,709

71. Explain the commutative property of addition. How is this used when checking an addition problem? Answer varies.

72. Explain in your own words the associative property of addition. How can this be used when adding columns of numbers? Answer varies.

Using the map below find the shortest distance between the following cities.

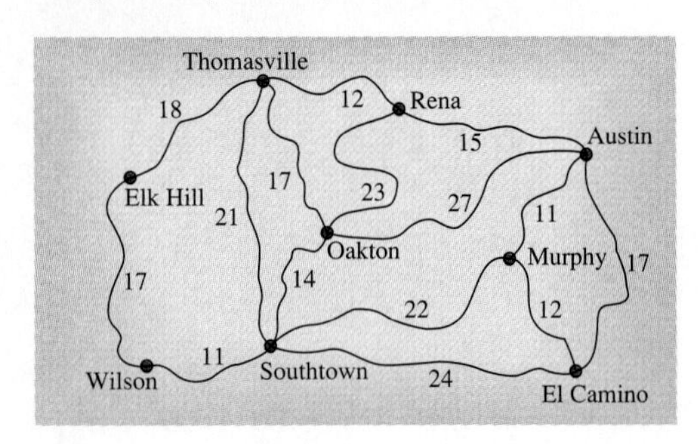

73. Wilson and El Camino 35 miles

74. Elk Hill and Oakton 35 miles

75. Thomasville and Murphy 38 miles

76. Murphy and Thomasville 38 miles

Solve the following word problems by using addition.

77. A history book costs \$47 and a math book \$43. Find the total cost for both books.

\$90

78. Jack Pritchard has 75 lock washers and 38 standard washers. How many washers does he have altogether?

113 washers

79. There are 413 women and 286 men on the sales staff. How many people are on the sales staff?

699 people

80. One department in an office building has 283 employees while another department has 218 employees. How many employees are in the two departments?

501 employees

81. At a charity bazaar, a library has a total of 9792 books for sale, while a book dealer has 3259 books for sale. How many books are for sale?

13,051 books

82. A plane is flying at an altitude of 5924 feet. It then increases its altitude by 7284 feet. Find its new altitude.

13,208 feet

Find the perimeter or total distance around each of the following figures.

83.

72 inches

58 inches 58 inches

72 inches

260 inches

84.

65 meters

73 meters 73 meters

98 meters

309 meters

▲ **85.**

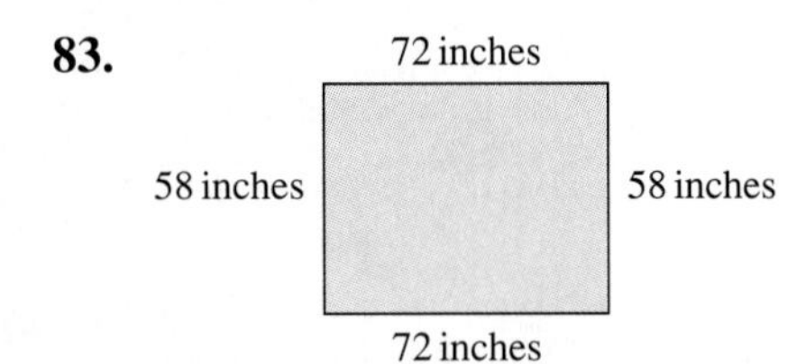

708 feet

▲ **86.**

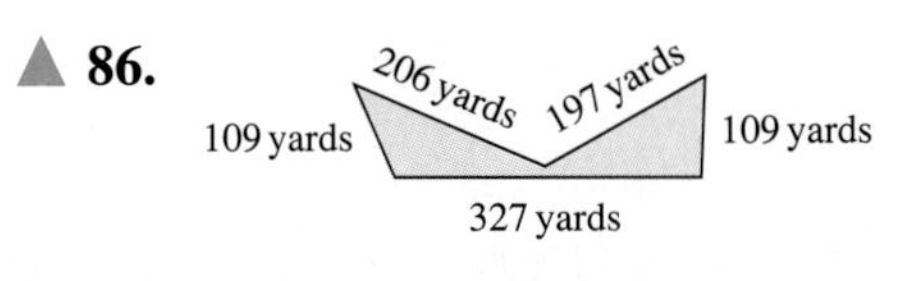

948 yards

1.3 SUBTRACTION OF WHOLE NUMBERS

Suppose you have \$8, and you spend \$5 for gasoline. You then have \$3 left. There are two different ways of looking at these numbers:

As an addition problem:

\$5	+	\$3	=	\$8
↑		↑		↑
amount spent		amount left		original amount

As a subtraction problem:

\$8	−	\$5	=	\$3
↑	↑	↑		↑
original amount	subtraction symbol	amount spent		amount left

1 As this example shows, an addition problem can be changed to a subtraction problem and a subtraction problem can be changed to an addition problem.

EXAMPLE 1 *Changing Addition Problems to Subtraction*
Change each addition problem to a subtraction problem.

(a) $4 + 1 = 5$
Two subtraction problems are possible:

$$5 - 1 = 4 \quad \text{or} \quad 5 - 4 = 1$$

These figures show each subtraction problem.

○ ○ ○ ○ ○ ○ ○ ○ ○ ○

$5 - 1 = 4$ $5 - 4 = 1$

(b) $7 + 9 = 16$

$$16 - 9 = 7 \quad \text{or} \quad 16 - 7 = 9$$ ■

WORK PROBLEM 1 AT THE SIDE. ▶

EXAMPLE 2 *Changing Subtraction Problems to Addition*
Change each subtraction problem to an addition problem.

(a) $8 - 3 = 5$
↓ ↓ ↓
$8 = 3 + 5$

It is also correct to write $8 = 5 + 3$.

OBJECTIVES

1. Change addition problems to subtraction.
2. Identify the minuend, subtrahend, and difference.
3. Subtract when no borrowing is needed.
4. Check answers.
5. Subtract by borrowing.
6. Solve word problems with subtraction.

FOR EXTRA HELP

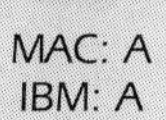

Tape 1	SSM pp. 4–6	MAC: A IBM: A

1. Write two subtraction problems for each addition problem.

(a) $3 + 2 = 5$

(b) $6 + 5 = 11$

(c) $15 + 22 = 37$

(d) $36 + 44 = 80$

ANSWERS
1. (a) $5 - 2 = 3$ or $5 - 3 = 2$
(b) $11 - 5 = 6$ or $11 - 6 = 5$
(c) $37 - 22 = 15$ or $37 - 15 = 22$
(d) $80 - 44 = 36$ or $80 - 36 = 44$

2. Write an addition problem for each subtraction problem.

(a) $6 - 2 = 4$

(b) $8 - 3 = 5$

(c) $21 - 15 = 6$

(d) $58 - 42 = 16$

3. Subtract.

(a) $\begin{array}{r} 78 \\ -\ 52 \\ \hline \end{array}$

(b) $\begin{array}{r} 47 \\ -\ 15 \\ \hline \end{array}$

(c) $\begin{array}{r} 378 \\ -\ 235 \\ \hline \end{array}$

(d) $\begin{array}{r} 3927 \\ -\ 2614 \\ \hline \end{array}$

(e) $\begin{array}{r} 5464 \\ -\ 324 \\ \hline \end{array}$

ANSWERS
2. (a) $6 = 2 + 4$ (b) $8 = 3 + 5$
(c) $21 = 15 + 6$ (d) $58 = 42 + 16$
3. (a) 26 (b) 32 (c) 143
(d) 1313 (e) 5140

(b) $19 - 14 = 5$
$19 = 14 + 5$

(c) $29 - 13 = 16$
$29 = 13 + 16$ ■

◀ WORK PROBLEM 2 AT THE SIDE.

2 In subtraction, as in addition, the numbers in a problem have names. For example, in the problem, $8 - 5 = 3$, the number 8 is the **minuend,** 5 is the **subtrahend,** and 3 is the **difference** or answer.

$$\underset{\text{minuend}}{\underset{\uparrow}{8}} \quad - \quad \underset{\text{subtrahend}}{\underset{\uparrow}{5}} \quad = 3 \leftarrow \text{difference}$$

$$\begin{array}{rl} 8 & \leftarrow \text{minuend} \\ -\ 5 & \leftarrow \text{subtrahend} \\ \hline 3 & \leftarrow \text{difference} \end{array}$$

3 Subtract two numbers by lining up the numbers in columns so the digits in the ones place are in the same column. Next subtract by columns, starting at the right with the ones column.

■ **EXAMPLE 3** *Subtracting Two Numbers*

Subtract.

(a) Ones digits are lined up in the same column.

$$\begin{array}{r} 53 \\ -\ 21 \\ \hline 32 \end{array}$$

$3 - 1 = 2$
$5 - 2 = 3$

(b) Ones digits are lined up.

$$\begin{array}{r} 385 \\ -\ 161 \\ \hline 224 \end{array}$$

$5 - 1 = 4$
$8 - 6 = 2$
$3 - 1 = 2$

(c)

$$\begin{array}{r} 9431 \\ -\ 210 \\ \hline 9221 \end{array}$$

$1 - 0 = 1$
$3 - 1 = 2$
$4 - 2 = 2$
$9 - 0 = 9$ ■

◀ WORK PROBLEM 3 AT THE SIDE.

4 Use addition to check your answer to a subtraction problem. For example, check $8 - 3 = 5$ by *adding* 3 and 5:

$$3 + 5 = 8, \quad \text{so} \quad 8 - 3 = 5 \quad \text{is correct.}$$

EXAMPLE 4 *Checking Subtraction*

Check each answer.

(a)
$$\begin{array}{r} 89 \\ -\ 47 \\ \hline 42 \end{array}$$

Rewrite as an addition problem, as shown in Example 2.

subtraction problem
$$\begin{array}{r} 89 \\ -\ 47 \\ \hline 42 \\ 89 \end{array}$$
addition problem
$$\begin{array}{r} 47 \\ +\ 42 \\ \hline 89 \end{array}$$

Because 47 + 42 = 89, the subtraction was done correctly.

(b) 72 − 41 = 21

Rewrite as an addition problem.

$$72 = 41 + 21$$

But, 41 + 21 = 62, not 72, so the subtraction was done incorrectly. We rework the original subtraction to get the correct answer, 31.

(c)
$$\begin{array}{r} 374 \\ -\ 141 \\ \hline 233 \end{array}$$
374 ← match → 141 + 233 = 374

The answer checks. ■

WORK PROBLEM 4 AT THE SIDE. ▶▶

5 If a digit in the minuend is smaller than the one directly below, **borrowing** will be necessary.

EXAMPLE 5 *Subtracting with Borrowing*

Subtract 19 from 57.

Write the problem.
$$\begin{array}{r} 57 \\ -\ 19 \\ \hline \end{array}$$

In the ones column, 7 is smaller than 9, so borrow a 10 from the 5 (which represents 5 tens, or 50).

50 − 10 = 40 → 4 17 ← 10 + 7 = 17
$$\begin{array}{r} 4\ \ 17 \\ \not{5}\ \ \not{7} \\ -\ 1\ \ 9 \\ \hline \end{array}$$

Now we can subtract 17 − 9 in the ones column and then 4 − 1 in the tens column,

$$\begin{array}{r} 4\ \ 17 \\ \not{5}\ \ \not{7} \\ -\ 1\ \ 9 \\ \hline 3\ \ 8 \end{array}$$
difference

Finally, 57 − 19 = 38. Check by adding 19 and 38. ■

WORK PROBLEM 5 AT THE SIDE. ▶▶

4. Decide whether these answers are correct. If incorrect, what should they be?

(a)
$$\begin{array}{r} 73 \\ -\ 21 \\ \hline 52 \end{array}$$

(b)
$$\begin{array}{r} 46 \\ -\ 32 \\ \hline 24 \end{array}$$

(c)
$$\begin{array}{r} 587 \\ -\ 342 \\ \hline 345 \end{array}$$

(d)
$$\begin{array}{r} 7531 \\ -\ 4301 \\ \hline 3230 \end{array}$$

5. Subtract.

(a)
$$\begin{array}{r} 82 \\ -\ 48 \\ \hline \end{array}$$

(b)
$$\begin{array}{r} 78 \\ -\ 29 \\ \hline \end{array}$$

(c)
$$\begin{array}{r} 31 \\ -\ 17 \\ \hline \end{array}$$

(d)
$$\begin{array}{r} 863 \\ -\ 47 \\ \hline \end{array}$$

(e)
$$\begin{array}{r} 951 \\ -\ 39 \\ \hline \end{array}$$

ANSWERS

4. (a) correct (b) incorrect, should be 14 (c) incorrect, should be 245 (d) correct

5. (a) 34 (b) 49 (c) 14 (d) 816 (e) 912

6. Subtract.

(a) $\begin{array}{r} 249 \\ -\ 73 \\ \hline \end{array}$

(b) $\begin{array}{r} 457 \\ -\ 68 \\ \hline \end{array}$

(c) $\begin{array}{r} 653 \\ -\ 379 \\ \hline \end{array}$

(d) $\begin{array}{r} 1437 \\ -\ 988 \\ \hline \end{array}$

(e) $\begin{array}{r} 8739 \\ -\ 3892 \\ \hline \end{array}$

ANSWERS
6. (a) 176 (b) 389 (c) 274 (d) 449 (e) 4847

■ **EXAMPLE 6** *Subtracting with Borrowing*
Subtract by borrowing as necessary.

(a) $\begin{array}{r} 7856 \\ -\ 137 \\ \hline \end{array}$

There is no need to borrow, as 4 is greater than 3.

10 + 6 = 16

$$\begin{array}{r} 4\ 16 \\ 7\ 8\ \not{5}\ \not{6} \\ -\ \ \ 1\ 3\ 7 \\ \hline 7\ 7\ 1\ 9 \end{array} \quad \text{difference}$$

(b) $\begin{array}{r} 635 \\ -\ 546 \\ \hline \end{array}$

100 + 20 = 120

600 − 100 = 500 10 + 5 = 15

$$\begin{array}{r} 5\ 12\ 15 \\ \not{6}\ \not{3}\ \not{5} \\ -\ 5\ 4\ 6 \\ \hline 8\ 9 \end{array} \quad \text{difference}$$

(c) $\begin{array}{r} 412 \\ -\ 225 \\ \hline \end{array}$

$$\begin{array}{r} 3\ 10\ 12 \\ \not{4}\ \not{1}\ \not{2} \\ -\ 2\ 2\ 5 \\ \hline 1\ 8\ 7 \end{array}$$ ■

◀◀ **WORK PROBLEM 6 AT THE SIDE.**

Sometimes a minuend has zeros in some of the positions. In such cases, borrowing may be a little more complicated than what we have shown so far.

■ **EXAMPLE 7** *Borrowing with Zeros*
Subtract.

$$\begin{array}{r} 4607 \\ -\ 3168 \\ \hline \end{array}$$

It is not possible to borrow from the tens position. Instead we must first borrow from the hundreds position.

600 − 100 = 500 100 + 0 = 100

$$\begin{array}{r} 5\ 10\ \ \\ 4\ \not{6}\ \not{0}\ 7 \\ -\ 3\ 1\ 6\ 8 \\ \hline \end{array}$$

Now we may borrow from the tens position.

$$\begin{array}{r} 9\ \ \ \\ 5\ 10\ 17 \\ 4\ \not{6}\ \not{0}\ 7 \\ -\ 3\ 1\ 6\ 8 \\ \hline 9 \end{array}$$

9 ← 100 − 10 = 90
10 + 7 = 17

Complete the problem.

```
        9
      5 10 17
    4 6  0  7
  - 3 1  6  8
  -----------
    1 4  3  9    difference
```

As above, check by adding 1439 and 3168; you should get 4607. ■

WORK PROBLEM 7 AT THE SIDE. ▶▶

■ **EXAMPLE 8** *Borrowing with Zeros*

Subtract.

(a)
```
   708
 - 149
```

```
 100 + 0 = 100        100 - 10 = 90
 700 - 100 = 600      10 + 8 = 18
          9
        6 10 18
        7  0  8
      - 1  4  9
      ---------
        5  5  9
```

(b)
```
   380
 - 276
```

```
 80 - 10 = 70          10 + 0 = 10
        7 10
      3 8  0
    - 2 7  6
    --------
      1 0  4
```

(c)
```
   9000
 - 6999
```

```
          9  9
      8 10 10 10
      9  0  0  0
    - 6  9  9  9
    ------------
      2  0  0  1   ■
```

WORK PROBLEM 8 AT THE SIDE. ▶▶

As explained above, an answer to a subtraction problem can be checked by adding.

■ **EXAMPLE 9** *Checking Subtraction*

Check the following answers.

(a)
```
                 check
   613             275
 - 275   match   + 338
   338             613   correct
```

7. Subtract.

(a)
```
   607
 - 463
```

(b)
```
   206
 - 148
```

(c)
```
   5073
 - 1632
```

8. Subtract.

(a)
```
   202
 - 178
```

(b)
```
   370
 - 163
```

(c)
```
   1570
 -  983
```

(d)
```
   7001
 - 5193
```

(e)
```
   6000
 - 2768
```

ANSWERS

7. (a) 144 (b) 58 (c) 3441

8. (a) 24 (b) 207 (c) 587 (d) 1808 (e) 3232

9. Check the answers in the following problems. If the answer is incorrect, give the correct anwer.

(a)
$$\begin{array}{r} 318 \\ -\ 275 \\ \hline 43 \end{array}$$

(b)
$$\begin{array}{r} 670 \\ -\ 439 \\ \hline 241 \end{array}$$

(c)
$$\begin{array}{r} 12{,}315 \\ -\ 9\ 647 \\ \hline 2\ 668 \end{array}$$

10. Using the table from Example 10, how many more packages did Jennifer deliver on

(a) Thursday than on Tuesday?

(b) Monday than on Saturday?

ANSWERS
9. (a) correct
(b) incorrect, should be 231
(c) correct
10. (a) 168 (b) 110

(b)
$$\begin{array}{r} 1915 \\ -\ 1635 \\ \hline 280 \end{array} \quad \text{match} \quad \begin{array}{r} \text{check} \\ 1635 \\ +\ 280 \\ \hline 1915 \end{array} \quad \text{correct}$$

(c)
$$\begin{array}{r} 15{,}803 \\ -\ 7\ 325 \\ \hline 8\ 578 \end{array} \quad \text{no match} \quad \begin{array}{r} \text{check} \\ 7\ 325 \\ +\ 8\ 578 \\ \hline 15{,}903 \end{array} \quad \text{error}$$

Rework the original problem to get the correct answer, 8478. ■

◀ **WORK PROBLEM 9 AT THE SIDE.**

6 As shown in the next example, subtraction can be used to solve a word problem.

■ **EXAMPLE 10** *Applying Subtraction Skills*
Using the table below, decide how many more packages were delivered by Jennifer on Friday than on Saturday.

PACKAGE DELIVERY (Jennifer)

Day	*Number of Packages Delivered*
Sunday	0
Monday	385
Tuesday	278
Wednesday	0
Thursday	446
Friday	398
Saturday	275

Jennifer delivered 398 packages on Friday, but she delivered only 275 packages on Saturday. Find how many more packages were delivered on Friday than on Saturday by subtracting 275 from 398.

$$\begin{array}{rl} 398 & \text{packages on Friday} \\ -\ 275 & \text{packages on Saturday} \\ \hline 123 & \end{array}$$

Jennifer delivered 123 more packages on Friday than on Saturday. ■

◀ **WORK PROBLEM 10 AT THE SIDE.**

1.3 EXERCISES

NAME DATE HOUR

Solve the following subtraction problems. Check each answer.

Example:

$$\begin{array}{r} 3722 \\ -\ 1610 \\ \hline \end{array}$$

Solution:

$$\begin{array}{r} 3722 \\ -\ 1610 \\ \hline \mathbf{2112} \end{array}$$

Check

$$\begin{array}{r} 1610 \\ +\ 2112 \\ \hline \mathbf{3722} \end{array}$$

match

The answer, 2112, checks.

1. $\begin{array}{r} 27 \\ -\ 15 \\ \hline 12 \end{array}$ **2.** $\begin{array}{r} 19 \\ -\ 14 \\ \hline 5 \end{array}$ **3.** $\begin{array}{r} 84 \\ -\ 33 \\ \hline 51 \end{array}$ **4.** $\begin{array}{r} 38 \\ -\ 17 \\ \hline 21 \end{array}$ **5.** $\begin{array}{r} 77 \\ -\ 60 \\ \hline 17 \end{array}$

6. $\begin{array}{r} 87 \\ -\ 63 \\ \hline 24 \end{array}$ **7.** $\begin{array}{r} 335 \\ -\ 122 \\ \hline 213 \end{array}$ **8.** $\begin{array}{r} 602 \\ -\ 301 \\ \hline 301 \end{array}$ **9.** $\begin{array}{r} 552 \\ -\ 451 \\ \hline 101 \end{array}$ **10.** $\begin{array}{r} 888 \\ -\ 215 \\ \hline 673 \end{array}$

11. $\begin{array}{r} 6821 \\ -\ 610 \\ \hline 6211 \end{array}$ **12.** $\begin{array}{r} 4420 \\ -\ 310 \\ \hline 4110 \end{array}$ **13.** $\begin{array}{r} 5546 \\ -\ 2134 \\ \hline 3412 \end{array}$ **14.** $\begin{array}{r} 1875 \\ -\ 1362 \\ \hline 513 \end{array}$ **15.** $\begin{array}{r} 6259 \\ -\ 4148 \\ \hline 2111 \end{array}$

16. $\begin{array}{r} 8732 \\ -\ 1621 \\ \hline 7111 \end{array}$ **17.** $\begin{array}{r} 24{,}392 \\ -\ 11{,}232 \\ \hline 13{,}160 \end{array}$ **18.** $\begin{array}{r} 57{,}921 \\ -\ 34{,}801 \\ \hline 23{,}120 \end{array}$ **19.** $\begin{array}{r} 46{,}253 \\ -\ 5\ 143 \\ \hline 41{,}110 \end{array}$ **20.** $\begin{array}{r} 75{,}904 \\ -\ 3\ 702 \\ \hline 72{,}202 \end{array}$

Check the following subtractions. If an answer is not correct, give the correct answer.

Example:

$$\begin{array}{r} 725 \\ -\ 413 \\ \hline 212 \end{array}$$

Add. 413 + 212 = 625, which does not match 725. The answer does not check. Rework the subtraction to get the correct answer of **312.**

21. $\begin{array}{r} 47 \\ -\ 32 \\ \hline 15 \end{array}$ correct

22. $\begin{array}{r} 57 \\ -\ 26 \\ \hline 31 \end{array}$ correct

23. $\begin{array}{r} 89 \\ -\ 27 \\ \hline 63 \end{array}$ incorrect; should be 62

24. $\begin{array}{r} 47 \\ -\ 35 \\ \hline 13 \end{array}$ incorrect; should be 12

25. $\begin{array}{r} 382 \\ -\ 261 \\ \hline 131 \end{array}$ incorrect; should be 121

26. $\begin{array}{r} 515 \\ -\ 304 \\ \hline 211 \end{array}$ correct

27. $\begin{array}{r} 2984 \\ -\ 1321 \\ \hline 1663 \end{array}$ correct

28. $\begin{array}{r} 5217 \\ -\ 4105 \\ \hline 1132 \end{array}$ incorrect; should be 1112

29. $\begin{array}{r} 8643 \\ -\ 1421 \\ \hline 7212 \end{array}$ incorrect; should be 7222

30. $\begin{array}{r} 9428 \\ -\ 3124 \\ \hline 6324 \end{array}$ incorrect; should be 6304

Subtract by borrowing as necessary.

Example: 63 − 47 **Solution:** $\overset{5}{\not 6}\ \overset{13}{\not 3} - 4\ 7 = \mathbf{1\ 6}$

31. 25 − 17 = 8	**32.** 78 − 49 = 29	**33.** 61 − 32 = 29	**34.** 78 − 49 = 29	**35.** 45 − 29 = 16
36. 93 − 37 = 56	**37.** 719 − 658 = 61	**38.** 916 − 618 = 298	**39.** 771 − 252 = 519	**40.** 973 − 788 = 185
41. 9861 − 684 = 9177	**42.** 6171 − 1182 = 4989	**43.** 9988 − 2399 = 7589	**44.** 3576 − 1658 = 1918	**45.** 38,335 − 29,476 = 8859
46. 61,278 − 3 559 = 57,719	**47.** 40 − 37 = 3	**48.** 80 − 73 = 7	**49.** 60 − 37 = 23	**50.** 70 − 27 = 43
51. 300 − 299 = 1	**52.** 208 − 199 = 9	**53.** 4041 − 1208 = 2833	**54.** 4602 − 2063 = 2539	**55.** 9305 − 1530 = 7775
56. 7120 − 6033 = 1087	**57.** 1580 − 1077 = 503	**58.** 3068 − 2105 = 963	**59.** 2006 − 1850 = 156	**60.** 8203 − 5365 = 2838
61. 6020 − 4078 = 1942	**62.** 7050 − 6045 = 1005	**63.** 8503 − 2816 = 5687	**64.** 16,004 − 5 087 = 10,917	**65.** 80,705 − 61,667 = 19,038
66. 36,000 − 22,117 = 13,883	**67.** 33,000 − 17,222 = 15,778	**68.** 77,000 − 65,308 = 11,692	**69.** 20,080 − 13,496 = 6584	**70.** 80,056 − 23,869 = 56,187

Check the following subtractions. If an answer is incorrect, give the correct answer.

Example: 3084 − 1278 = 1806 **Solution:** 1278 + 1806 = **3084** (match) correct

71. 7582 − 1628 = 5954	**72.** 1671 − 1325 = 1346	**73.** 1439 − 1169 = 270	**74.** 5274 − 1130 = 4144
correct	incorrect; should be 346	correct	correct

75. 78,213 − 17,346 = 60,867

correct

76. 82,357 − 14,396 = 68,961

incorrect; should be 67,961

77. 27,689 − 22,306 = 5 383

correct

78. 34,821 − 17,735 = 17,735

incorrect; should be 17,086

79. An addition problem can be changed to a subtraction problem and a subtraction problem can be changed to an addition problem. Give an example to demonstrate this.

Answer varies.

80. A simple way to check an answer is to simply rework the problem. There are also other methods for checking. Using your own example, show how to check a subtraction problem using addition.

Answer varies.

Solve the following word problems.

81. A nursery had 98 rose bushes. It sold 40. How many rose bushes are left?

58 bushes

82. Lynn Couch has $553 in her checking account. She writes a check for $308 for school fees. How much is left in her account?

$245

83. An airplane is carrying 254 passengers. When it lands in Atlanta 133 passengers get off the plane. How many passengers are left on the plane?

121 passengers

84. A bottle contained 36 doses of a medicine. How many doses are left in the bottle after 12 doses have been given?

24 doses

Refer to the table to answer Exercises 85 and 86.

Annual Student Expense

Item	*Amount*
Rent	$3900
Food	$2220
Fees	$1670
Books	$ 615
Other	$1895
Total	

85. Find the total student expenses for the year.

$10,300

86. The amounts spent on fees and books were both school expenses. Find the total amount of school expenses.

$2285

87. Ruth Ng drove her truck 829 miles, while Jason Orr drove his car 517 miles. How many more miles did Ms. Ng drive?

312 miles

88. On Tuesday, 5822 people went to a soccer game, and on Friday, 7994 people went to a soccer game. How many more people went to the game on Friday?

2172 people

89. Last fall 12,625 students enrolled in classes. In the spring semester, 11,296 students enrolled. How many more students enrolled in classes in the fall semester than in the spring?

1329 students

90. One bid for painting a house was $1954. A second bid was $1742. How much would be saved using the second bid?

$212

91. Patriot Flag Company manufactured 14,608 flags and sold 5069. How many flags remain?

9539 flags

92. Eye exams have been given to 14,679 children in the school district. If there are 23,156 students in the school district, find the number of children who have not received eye exams.

8477 children

Solve each of the following word problems. Add or subtract as necessary.

▲ **93.** Fred Reyes had $1523 withheld from his paychecks last year for income tax, but he owes only $1379 in tax. What refund should he receive?

$144

▲ **94.** A car now goes 374 miles on a tank of gas. After a tune-up, the same car will go 401 miles on a tank of gas. How many additional miles will it go on a tank of gas after the tune-up?

27 miles

▲ **95.** The Jordanos now pay rent of $650 per month. If they buy a house their housing expense will increase by $263 per month. How much will they pay each month?

$913

▲ **96.** A retired couple used to receive a social security payment of $1479 per month. Recently, benefits were increased by $89 per month. How much money does the couple now receive?

$1568

▲ **97.** On Monday, 11,594 people visited Arcade Amusement Park, and 12,352 people visited the park on Tuesday. How many more people visited the park on Tuesday?

758 people

▲ **98.** Last month Alice Blake earned $2382 while this month she earned $2671. How much more did she earn this month than last month?

$289

1.4 MULTIPLICATION OF WHOLE NUMBERS

Adding the number 3 a total of 4 times gives 12.

$$3 + 3 + 3 + 3 = 12$$

This result can also be shown with a figure.

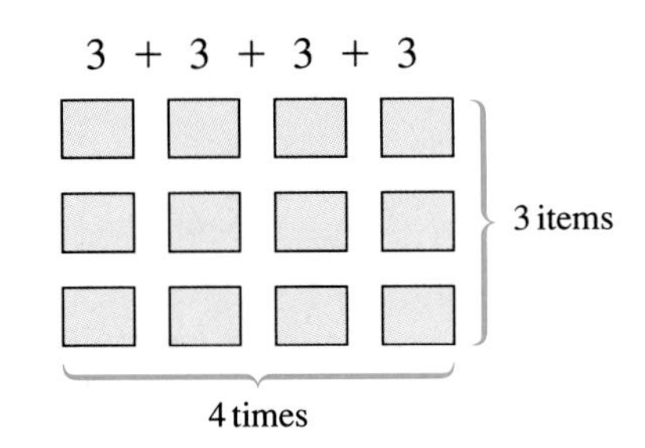

1 Multiplication is a shortcut for this repeated addition. The numbers being multiplied are called **factors.** The answer is called the **product.** For example, the product of 3 and 4 can be written with the symbol ×, a raised dot, or parentheses, as follows.

3	factor (also called *multiplicand*)
× 4	factor (also called *multiplier*)
12	product

$3 \times 4 = 12 \quad 3 \cdot 4 = 12 \quad (3)(4) = 12$

WORK PROBLEM 1 AT THE SIDE.

COMMUTATIVE PROPERTY OF MULTIPLICATION

By the **commutative property of multiplication,** the answer or product remains the same when the order of the factors is changed.

For example,

$$3 \times 5 = 15 \quad \text{and} \quad 5 \times 3 = 15$$

Note Recall that addition also has a commutative property. Remember that 4 + 2 is the same as 2 + 4.

EXAMPLE 1 *Multiplying Two Numbers*

Multiply. (Remember that a raised dot means to multiply.)

(a) $3 \times 4 = 12$

(b) $6 \cdot 0 = 0$ (The product of any number and 0 is 0; if you give no money to each of 6 relatives, you give no money.)

(c) $(4)(8) = 32$ ■

WORK PROBLEM 2 AT THE SIDE.

OBJECTIVES

1. Know the parts of a multiplication problem.
2. Do chain multiplications.
3. Multiply by single-digit numbers.
4. Multiply quickly by numbers ending in zeros.
5. Multiply by numbers having more than one digit.
6. Solve word problems with multiplication.

FOR EXTRA HELP

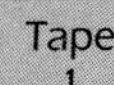
Tape 1

SSM pp. 7–10

MAC: A IBM: A

1. Identify the factors and the product in each multiplication problem.

(a) $3 \times 6 = 18$

(b) $5 \cdot 7 = 35$

(c) $(3)(9) = 27$

2. Multiply.

(a) 6×8

(b) 0×9

(c) $7 \cdot 4$

(d) $5 \cdot 5$

(e) $(3)(8)$

ANSWERS

1. (a) factors: 3, 6; product: 18
(b) factors: 5, 7; product: 35
(c) factors: 3, 9: product: 27
2. (a) 48 (b) 0 (c) 28 (d) 25 (e) 24

3. Multiply.

(a) $3 \times 3 \times 2$

(b) $6 \times 1 \times 4$

(c) $8 \times 3 \times 0$

ANSWERS
3. (a) 18 (b) 24 (c) 0

2 Some multiplications contain more than two factors.

ASSOCIATIVE PROPERTY OF MULTIPLICATION

By the associative property of multiplication, the grouping of numbers in any order gives the same product.

EXAMPLE 2 *Multiplying Three Numbers*
Multiply: $2 \times 3 \times 5$.

$$(2 \times 3) \times 5 \quad \text{Parentheses tell what to do first.}$$
$$6 \times 5 = 30$$

Also,

$$2 \times (3 \times 5)$$
$$2 \times 15 = 30$$

Either grouping results in the same answer. ■

A problem with more than two factors, such as the one in Example 2, is called a **chain multiplication.**

WORK PROBLEM 3 AT THE SIDE.

3 Carrying may be needed in multiplication problems with larger factors.

EXAMPLE 3 *Carrying with Multiplication*
Multiply.

(a) 53×4

Start by multiplying in the ones column.

$$\begin{array}{r} \scriptstyle 1 \\ 53 \\ \times \ \ 4 \\ \hline 2 \end{array} \quad 4 \times 3 = 12$$

Carry the 1 to the tens column.
Write 2 in the ones column.

Next, multiply 4 ones and 5 tens.

$$\begin{array}{r} \scriptstyle 1 \\ 53 \\ \times \ \ 4 \\ \hline 2 \end{array} \quad 4 \times 5 = \mathbf{20} \text{ tens}$$

Add the 1 that was carried to the tens column.

$$\begin{array}{r} \scriptstyle 1 \\ 53 \\ \times \ \ 4 \\ \hline 212 \end{array} \quad 20 + 1 = \mathbf{21} \text{ tens}$$

(b)
$$\begin{array}{r} 724 \\ \times \quad 5 \\ \hline \end{array}$$

Work as shown.

$$\begin{array}{r} \scriptstyle 12 \\ 724 \\ \times \quad 5 \\ \hline 3620 \end{array}$$

3620 ← 5 × 4 = **20** ones; write 0 ones and carry 2 tens.

5 × 2 = **10** tens; add the 2 tens to get 12 tens; write 2 tens and carry 1 hundred.

5 × 7 = **35** hundreds; add the 1 hundred to get 36 hundreds ■

WORK PROBLEM 4 AT THE SIDE.

4. Multiply.

(a) $\begin{array}{r} 63 \\ \times \;\; 4 \\ \hline \end{array}$

(b) $\begin{array}{r} 47 \\ \times \;\; 0 \\ \hline \end{array}$

(c) $\begin{array}{r} 862 \\ \times \;\; 9 \\ \hline \end{array}$

(d) $\begin{array}{r} 2831 \\ \times \quad 7 \\ \hline \end{array}$

(e) $\begin{array}{r} 5146 \\ \times \quad 8 \\ \hline \end{array}$

4 The product of two whole-number factors is also called a **multiple** of either factor. For example, since $4 \cdot 2 = 8$, the whole number 8 is a multiple of both 4 and 2. *Multiples of 10* are very useful when multiplying. A **multiple of 10** is a whole number that ends in zero, such as 10, 20 or 30; 100, 200, or 300; 1000, 2000, or 3000. There is a short way to multiply by these multiples of 10. For example,

$$8 \times 10 = 80, \quad 74 \times 100 = 7400,$$
and
$$953 \times 1000 = 953{,}000.$$

These examples suggest the following rule.

MULTIPLYING BY MULTIPLES OF 10

Multiply a whole number by 10, 100, or 1000, by attaching one, two, or three zeros to the right of the whole number.

■ **EXAMPLE 4** *Using Multiples of 10 to Multiply*

Multiply.

(a) $59 \times 10 + 590$ ← Attach 0.

(b) $74 \times 100 = 7400$ ← Attach 00.

(c) $803 \times 1000 = 803{,}000$ ← Attach 000 ■

WORK PROBLEM 5 AT THE SIDE.

5. Multiply.

(a) 36×10

(b) 102×100

(c) 571×1000

You can also find the product of other multiples of ten by attaching zeros.

ANSWERS

4. (a) 252 (b) 0 (c) 7758 (d) 19,817 (e) 41,168

5. (a) 360 (b) 10,200 (c) 571,000

6. Multiply.

(a) 12×80

(b) 53×300

(c) 180×30

(d) 6100×90

(e) 800×200

7. Complete each multiplication.

(a)
```
   54
 × 35
  270
 162
```

(b)
```
   76
 × 49
  684
 304
```

ANSWERS
6. (a) 960 (b) 15,900 (c) 5400 (d) 549,000 (e) 160,000
7. (a) 1890 (b) 3724

■ **EXAMPLE 5** *Multiplying by Using Other Multiples*
Multiply.

(a) 75×3000
Multiply 75 by 3 and attach 3 zeros.

```
  75     75 × 3000 = 225,000
×  3
 225                 ← Attach 000.
```

(b) 150×70
Multiply 15 by 7, and then attach 2 zeros.

```
  15     150 × 70 = 10,500 ← Attach 00. ■
×  7
 105
```

WORK PROBLEM 6 AT THE SIDE.

5 The next example shows multiplication when a factor has more than one digit.

■ **EXAMPLE 6** *Multiplying with More Than One Digit*
Multiply 46 and 23.

First multiply 46 by 3.

```
   1
  46
×  3
 138 ← 46 × 3 = 138
```

Now multiply 46 by 20.

```
   1
  46
× 20
 920 ← 46 × 20 = 920
```

Add the results.

```
    46
  × 23
   138 ← 46 × 3
 + 920 ← 46 × 20
  1058
  ↑ Add.
```

Both 138 and 920 are called **partial products.** To save time, the zero in 920 is usually not written.

```
   46
 × 23
  138
  92  ← 0 not written. Be very careful to
 1058      place the 2 in the tens column. ■
```

WORK PROBLEM 7 AT THE SIDE.

EXAMPLE 7 *Using Partial Products*
Multiply.

(a)
```
    233
  × 132
    466
   6 99    (tens lined up)
  23 3     (hundreds lined up)
 30,756    product
```

(b)
```
    538
  ×  46
```

First multiply by 6.

```
    24
    538
  ×  46     Carrying is
   3228       needed here.
```

Now multiply by 4, being careful to line up the tens.

```
    13
    24
    538
  ×  46
  3 228 ┐
  21 52 ┘ Finally, add the results.
  24,748
```

WORK PROBLEM 8 AT THE SIDE.

When zero appears in the multiplier, be sure to move the partial products to the left to account for the position held by the zero.

EXAMPLE 8 *Multiplication with Zeros*
Multiply.

(a)
```
    137
  × 306
    822
   0 00    (tens lined up)
  41 1     (hundreds lined up)
  41,922
```

(b)
```
      1406
    × 2001
      1406
     0000    ← (0 to line up tens)
    0000     ← (0 to line up hundreds)
   2 812
 2,813,406
```

```
     1 406
   × 2 001
     1 406
  2 812 00   ← Zeros are written
               so this partial
               product starts in
               the thousands
 2,813,406     column.
```

8. Multiply.

(a) 45 × 14

(b) 21 × 32

(c) 67 × 59

(d) 234 × 73

(e) 764 × 272

ANSWERS
8. (a) 630 (b) 672 (c) 3953 (d) 17,082 (e) 207,808

9. Multiply.

(a)
$$\begin{array}{r} 28 \\ \times\ 60 \\ \hline \end{array}$$

(b)
$$\begin{array}{r} 932 \\ \times\ 50 \\ \hline \end{array}$$

(c)
$$\begin{array}{r} 481 \\ \times\ 206 \\ \hline \end{array}$$

(d)
$$\begin{array}{r} 3526 \\ \times\ 6002 \\ \hline \end{array}$$

10. Find the total cost of the following items.

(a) 289 redwood planters at $12 per planter

(b) 274 electric shavers at $35 per shaver

(c) 15 forklifts at $8218 per forklift

ANSWERS

9. (a) 1680 (b) 46,600 (c) 99,086 (d) 21,163,052

10. (a) $3468 (b) $9590 (c) $123,270

Note In Example 8 (b) in the solution on the right, zeros were inserted so that thousands were placed in the thousands column.

WORK PROBLEM 9 AT THE SIDE.

6 The next example shows how multiplication can be used to solve a word problem.

■ **EXAMPLE 9** *Applying Multiplication Skills*

Find the total cost of 54 pieces of lumber that cost $24 each.

Approach To find the cost of all the lumber, multiply the number of pieces (54) by the cost of one piece ($24).

Solution Multiply 54 by 24.

$$\begin{array}{r} 54 \\ \times\ 24 \\ \hline 216 \\ 108 \\ \hline 1296 \end{array}$$

The total cost of the lumber is $1296. ■

WORK PROBLEM 10 AT THE SIDE.

1.4 EXERCISES

NAME DATE HOUR

Work each of the following chain multiplications.

Example: $3 \times 2 \times 9$

Solution: $3 \times 2 = 6 \rightarrow 6 \times 9 = \mathbf{54}$

1. $2 \times 5 \times 2$ 20

2. $3 \times 7 \times 3$ 63

3. $7 \times 1 \times 8$ 56

4. $2 \times 4 \times 5$ 40

5. $8 \cdot 9 \cdot 0$ 0

6. $7 \cdot 0 \cdot 9$ 0

7. $4 \cdot 1 \cdot 6$ 24

8. $1 \cdot 5 \cdot 7$ 35

9. $(2)(3)(6)$ 36

10. $(4)(1)(9)$ 36

11. $(3)(0)(7)$ 0

12. $(0)(9)(4)$ 0

13. Explain in your own words the commutative property of multiplication. How do the commutative properties of addition and multiplication compare to each other? **Answer varies.**

14. Explain in your own words the associative property of multiplication. How do the associative properties of addition and multiplication compare to each other? **Answer varies.**

Multiply.

Example: 24×2

Solution: $\mathbf{48}$ ← $2 \times 4 = 8$; $2 \times 2 = 4$

Example: 37×8 (carry 5)

Solution: $\mathbf{296}$ ← $8 \times 7 = 56$; write 6 and carry 5. $8 \times 3 = 24$, $24 + 5 = 29$

15. 45×6 = 270

16. 63×8 = 504

17. 32×7 = 224

18. 79×4 = 316

19. 512×4 = 2048

20. 472×4 = 1888

21. 624×3 = 1872

22. 852×7 = 5964

23. 2153×4 = 8612

24. 1137×3 = 3411

25. 2521×4 = 10,084

26. 2544×3 = 7632

27. 3182×6 = 19,092

28. 7326×5 = 36,630

29. $36,921 \times 7$ = 258,447

30. $28,116 \times 4$ = 112,464

Writing Conceptual Challenging ≈ Estimation

Multiply.

Example: $\begin{array}{r} 110 \\ \times\ 50 \\ \hline \end{array}$

Solution: First $\begin{array}{r} 11 \\ \times\ 5 \\ \hline 55 \end{array} \rightarrow \begin{array}{r} 110 \\ \times\ 50 \\ \hline \mathbf{5500} \end{array}$ Attach 00.

31. $\begin{array}{r} 30 \\ \times\ 3 \\ \hline 90 \end{array}$

32. $\begin{array}{r} 20 \\ \times\ 5 \\ \hline 100 \end{array}$

33. $\begin{array}{r} 40 \\ \times\ 8 \\ \hline 320 \end{array}$

34. $\begin{array}{r} 90 \\ \times\ 7 \\ \hline 630 \end{array}$

35. $\begin{array}{r} 740 \\ \times\ 3 \\ \hline 2220 \end{array}$

36. $\begin{array}{r} 200 \\ \times\ 9 \\ \hline 1800 \end{array}$

37. $\begin{array}{r} 500 \\ \times\ 4 \\ \hline 2000 \end{array}$

38. $\begin{array}{r} 86 \\ \times\ 7 \\ \hline 602 \end{array}$

39. $\begin{array}{r} 125 \\ \times\ 30 \\ \hline 3750 \end{array}$

40. $\begin{array}{r} 246 \\ \times\ 50 \\ \hline 12,300 \end{array}$

41. $\begin{array}{r} 1255 \\ \times\ 20 \\ \hline 25,100 \end{array}$

42. $\begin{array}{r} 8522 \\ \times\ 50 \\ \hline 426,100 \end{array}$

43. $\begin{array}{r} 900 \\ \times\ 300 \\ \hline 270,000 \end{array}$

44. $\begin{array}{r} 400 \\ \times\ 700 \\ \hline 280,000 \end{array}$

45. $\begin{array}{r} 43,000 \\ \times\ 2\ 000 \\ \hline 86,000,000 \end{array}$

46. $\begin{array}{r} 11,000 \\ \times\ 9\ 000 \\ \hline 99,000,000 \end{array}$

47. $970 \cdot 50$

48,500

48. $730 \cdot 40$

29,200

49. $500 \cdot 900$

450,000

50. $560 \cdot 800$

448,000

51. $9700 \cdot 200$

1,940,000

52. $10,050 \cdot 300$

3,015,000

Multiply.

Example: $\begin{array}{r} 63 \\ \times\ 28 \\ \hline \end{array}$

Solution: $\begin{array}{r} 63 \\ \times\ 28 \\ \hline 504 \\ 126 \\ \hline \mathbf{1764} \end{array}$

53. $\begin{array}{r} 38 \\ \times\ 17 \\ \hline 646 \end{array}$

54. $\begin{array}{r} 15 \\ \times\ 42 \\ \hline 630 \end{array}$

55. $\begin{array}{r} 72 \\ \times\ 33 \\ \hline 2376 \end{array}$

56. $\begin{array}{r} 79 \\ \times\ 49 \\ \hline 3871 \end{array}$

57. $\begin{array}{r} 83 \\ \times\ 45 \\ \hline 3735 \end{array}$

58. $(43)(27)$

1161

59. $(58)(41)$

2378

60. $(82)(67)$

5494

61. $(67)(92)$

6164

62. $(26)(33)$

858

63. (24)(758)
18,192

64. (71)(152)
10,792

65. (729)(45)
32,805

66. (681)(47)
32,007

67. (44)(331)
14,564

68. 332×772
256,304

69. 735×112
82,320

70. 621×415
257,715

71. 538×342
183,996

72. 3228×751
2,424,228

73. 7783×265
2,062,495

74. 528×106
55,968

75. 215×307
66,005

76. 218×106
23,108

77. 428×201
86,028

78. 2108×304
640,832

79. 6310×3078
19,422,180

80. 3533×5001
17,668,533

81. 2195×1038
2,278,410

82. 1502×2009
3,017,518

83. A classmate of yours is not clear on how to multiply a whole number by 10, by 100, or by 1000. Write a short note explaining how this can be done.

Answer varies.

84. Show two ways to multiply when a zero is in the factor that is multiplying. Use the problem 307 × 291 to show this.

Answer varies.

Solve the following word problems by using multiplication.

85. An encyclopedia has 30 volumes. Each volume has 800 pages. What is the total number of pages in the encyclopedia?

24,000 pages

86. A hospital has 20 bottles of thyroid medication, with each bottle containing 2500 tablets. How many of these tablets does the hospital have in all?

50,000 tablets

87. There are 12 tomato plants to a flat. If there are 18 flats, find the total number of tomato plants.

216 plants

88. Audia Employment Agency purchased 8 new computers at $4638 per unit. Find the total cost of the computers.

$37,104

89. A new Saturn automobile gets 38 miles per gallon on the highway. How many miles can it go on 11 gallons of gas?

418 miles

90. A small clinic has 8 patients. Each patient pays $144 per day. How much does the clinic collect from these patients in one day?

$1152

Find the total cost of each of the following.

Example: 54 hammers at $8 per hammer

Solution: Multiply.

$$\begin{array}{r} 54 \\ \times\ 8 \\ \hline 432 \end{array}$$

The total cost is **$432.**

91. 16 gallons of paint at $18 per gallon

$288

92. 27 cordless phones at $87 per phone

$2349

93. 84 tires at $49 per tire

$4116

94. 76 flats of flowers at $22 per flat

$1672

95. 108 sets of wrenches at $37 per set

$3996

96. 305 compact discs at $12 per disc

$3660

Multiply.

* **97.** $21 \cdot 43 \cdot 56$ 50,568

98. (600)(8)(75)(40) 14,400,000

Use addition, subtraction, or multiplication, as needed, to solve each of the following.

99. Debbie Real counted 53 joggers one day and 122 joggers the next day. How many joggers did she count altogether during the two days?

175 joggers

100. Find the cost of fifteen clocks at $38 per clock.

$570

101. A large meal contains 1406 calories, while a small meal contains 348 calories. How many more calories are in the large meal than the small one?

1058 calories

102. One company pays its shop workers $9 for every assembly. How much would be paid to a person who had 104 assemblies?

$936

▲ **103.** Michelle Dyas bought 2 bath towels at $7 each, 4 hand towels at $4 each and 6 guest towels at $3 each. Find the total cost of the towels.

$48

▲ **104.** The distance from Reno, Nevada, to the Atlantic Ocean is 2695 miles, while the distance from Reno to the Pacific Ocean is 255 miles. How much farther is it to the Atlantic Ocean than it is to the Pacific Ocean?

2440 miles

* Color exercise numbers are used to indicate exercises designed for calculator use.

1.5 DIVISION OF WHOLE NUMBERS

Suppose \$12 is to be divided into 3 equal parts. Each part would be \$4, as shown here.

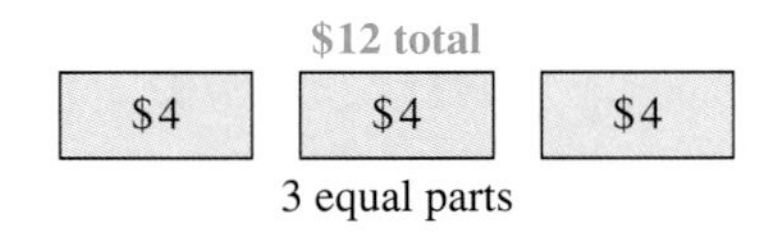

1 Just as $3 \cdot 4$, 3×4, and $(3)(4)$ are different ways of indicating the multiplication of 3 and 4, there are several ways to write 12 divided by 3.

$$12 \div 3 = 4 \qquad 3\overline{)12}^{\,4} \qquad \frac{12}{3} = 4$$

(12 is the number being divided; 3 is the number divided by.)

We will use all three division symbols, $\div$, $\overline{)}$, and —. In more advanced courses such as algebra, a slash symbol, /, or a bar, —, is most often used.

■ EXAMPLE 1 *Using Division Symbols*
Write each division by using two other symbols.

(a) $12 \div 4 = 3$
This division can also be written as

$$4\overline{)12}^{\,3} \quad \text{or} \quad \frac{12}{4} = 3.$$

(b) $\frac{15}{5} = 3$

$$15 \div 5 = 3 \quad \text{or} \quad 5\overline{)15}^{\,3}$$

(c) $5\overline{)20}^{\,4}$

$$20 \div 5 = 4 \quad \text{or} \quad \frac{20}{5} = 4 \quad ■$$

WORK PROBLEM 1 AT THE SIDE. ▶▶

2 In division, the number being divided is the **dividend,** the number divided by is the **divisor,** and the answer is the **quotient.**

$$\text{dividend} \div \text{divisor} = \text{quotient}$$

$$\text{divisor}\overline{)\text{dividend}}^{\,\text{quotient}} \qquad \frac{\text{dividend}}{\text{divisor}} = \text{quotient}$$

■ EXAMPLE 2 *Identifying the Parts in a Division Problem*
Identify the dividend, divisor, and quotient.

(a) $35 \div 7 = 5$

$$35 \div 7 = 5 \leftarrow \text{quotient}$$

(35 ↗ dividend; 7 ↖ divisor)

OBJECTIVES

1. Write division problems in three ways.
2. Identify the parts of a division problem.
3. Divide zero by a number.
4. Divide a number by itself.
5. Use short division.
6. Check the answer to a division problem.
7. Use tests for divisibility.

FOR EXTRA HELP

Tape 1

SSM pp. 10–14

MAC: A IBM: A

1. Write each division problem using two other symbols.

(a) $70 \div 7 = 10$

(b) $24 \div 6 = 4$

(c) $9\overline{)36}^{\,4}$

(d) $\frac{42}{6} = 7$

ANSWERS

1. (a) $7\overline{)70}^{\,10}$ and $\frac{70}{7} = 10$
(b) $6\overline{)24}^{\,4}$ and $\frac{24}{6} = 4$
(c) $36 \div 9 = 4$ and $\frac{36}{9} = 4$
(d) $6\overline{)42}^{\,7}$ and $42 \div 6 = 7$

2. Identify the dividend, divisor, and quotient.

(a) $8 \div 2 = 4$

(b) $30 \div 5 = 6$

(c) $\frac{28}{7} = 4$

(d) $2\overline{)36}$ with quotient 18

3. Divide.

(a) $0 \div 10$

(b) $\frac{0}{6}$

(c) $\frac{0}{36}$

(d) $57\overline{)0}$

ANSWERS

2. (a) dividend: 8; divisor: 2; quotient: 4
(b) dividend: 30; divisor: 5; quotient: 6
(c) dividend: 28; divisor: 7; quotient: 4
(d) dividend: 36; divisor: 2; quotient: 18
3. all 0

(b) $\frac{100}{20} = 5$

dividend ↓ $\frac{100}{20} = 5$ ← quotient; ↑ divisor

(c) $12\overline{)72}$ with quotient 6

6 ← quotient
$12\overline{)72}$ ← dividend
↑ divisor ■

◀◀ WORK PROBLEM 2 AT THE SIDE.

3 If no money, or \$0, is divided equally among five people; each person gets \$0. There is a general rule for dividing zero.

DIVIDING ZERO

> **Zero** divided by any nonzero number is **zero.**

■ **EXAMPLE 3** *Dividing Zero by a Number*
Divide.

(a) $0 \div 12 = 0$

(b) $0 \div 1728 = 0$

(c) $\frac{0}{375} = 0$

(d) $129\overline{)0}$ with quotient 0 ■

◀◀ WORK PROBLEM 3 AT THE SIDE.

Just as a subtraction such as $8 - 3 = 5$ can be written as the addition $8 = 3 + 5$, any division can be written as a multiplication. For example, $12 \div 3 = 4$ can be written as

$$3 \times 4 = 12 \quad \text{or} \quad 4 \times 3 = 12$$

■ **EXAMPLE 4** *Converting Division to Multiplication*
Convert each division to a multiplication.

(a) $\frac{20}{4} = 5$ becomes $4 \cdot 5 = 20$

(b) $8\overline{)48}$ with quotient 6 becomes $8 \cdot 6 = 48$

(c) $72 \div 9 = 8$ becomes $9 \cdot 8 = 72$ ■

WORK PROBLEM 4 AT THE SIDE.

Division by zero is not possible. To see why, try to find

$$9 \div 0.$$

As we have just seen, all division problems can be converted to a multiplication problem so that

$$\text{divisor} \cdot \text{quotient} = \text{dividend}.$$

If you convert the problem $9 \div 0 = ?$ to its multiplication counterpart, it reads

$$0 \cdot ? = 9$$

You already know that zero times any number must always equal zero. Try any number you like to replace the "?" and you'll always get 0 instead of 9. Therefore, the division problem

$$9 \div 0$$

is impossible to compute. No matter what number is used as the dividend, it is impossible to divide by zero.

DIVIDING BY ZERO

Division by **zero** is meaningless. It is impossible to compute an answer.

■ **EXAMPLE 5** *Dividing by Zero Is Meaningless*
All the following are meaningless.

(a) $\frac{6}{0}$ **(c)** $18 \div 0$

(b) $0\overline{)21}$ **(d)** $\frac{0}{0}$ ■

Division involving 0 is summarized below.

$$\frac{0}{\text{nonzero number}} = 0$$

$$\frac{\text{number}}{0} \text{ is meaningless}$$

WORK PROBLEM 5 AT THE SIDE.

Note When "0" is the divisor in a problem we say the answer is meaningless.

4 What happens when a number is divided by itself? For example, $4 \div 4$ or $97 \div 97$?

DIVIDING A NUMBER BY ITSELF

Any nonzero number divided by itself is **one.**

4. Write each division problem as a multiplication problem.

(a) $6\overline{)18}$ with quotient 3

(b) $\frac{28}{4} = 7$

(c) $48 \div 8 = 6$

5. Work the following problems whenever possible.

(a) $\frac{8}{0}$

(b) $\frac{0}{8}$

(c) $0\overline{)32}$

(d) $32\overline{)0}$

(e) $100 \div 0$

ANSWERS

4. (a) $6 \cdot 3 = 18$ (b) $4 \cdot 7 = 28$ (c) $8 \cdot 6 = 48$

5. (a) meaningless (b) 0 (c) meaningless (d) 0 (e) meaningless

6. Divide.

(a) $8 \div 8$

(b) $12\overline{)12}$

(c) $\frac{23}{23}$

7. Divide.

(a) $2\overline{)24}$

(b) $3\overline{)69}$

(c) $4\overline{)88}$

(d) $2\overline{)462}$

8. Divide.

(a) $2\overline{)225}$

(b) $3\overline{)275}$

(c) $4\overline{)538}$

(d) $\frac{819}{5}$

ANSWERS
6. all 1
7. (a) 12 (b) 23 (c) 22 (d) 231
8. (a) 112 R1 (b) 91 R2 (c) 134 R2 (d) 163 R4

■ **EXAMPLE 6** *Dividing a Nonzero Number by Itself*
Divide.

(a) $16 \div 16 = 1$

(b) $\begin{array}{r} 1 \\ 32\overline{)32} \end{array}$

(c) $\frac{57}{57} = 1$ ■

WORK PROBLEM 6 AT THE SIDE.

5 **Short division** is a method of dividing a number by a one-digit divisor.

■ **EXAMPLE 7** *Using Short Division*
Divide: $3\overline{)96}$.

First, divide 9 by 3.

$$\begin{array}{r} 3 \\ 3\overline{)96} \end{array} \leftarrow \frac{9}{3} = 3$$

Next, divide 6 by 3.

$$\begin{array}{r} 32 \\ 3\overline{)96} \end{array} \leftarrow \frac{6}{3} = 2$$ ■

WORK PROBLEM 7 AT THE SIDE.

When two numbers do not divide exactly, the leftover portion is called the **remainder.**

■ **EXAMPLE 8** *Using Short Division With a Remainder*
Divide 147 by 4.
Write the problem.

$$4\overline{)147}$$

Because 1 cannot be divided by 4, divide 14 by 4.

$$\begin{array}{r} 3 \\ 4\overline{)14^{2}7} \end{array} \quad \frac{14}{4} = 3 \text{ with 2 left over}$$

Next, divide 27 by 4. The final number left over is the remainder. Write the remainder to the side. "R" stands for remainder.

$$\begin{array}{r} 3\ 6\ \mathbf{R}3 \\ 4\overline{)14^{2}7} \end{array} \quad \frac{27}{4} = 6 \text{ with 3 left over}$$ ■

WORK PROBLEM 8 AT THE SIDE.

EXAMPLE 9 *Dividing and Having a Remainder*
Divide 1809 by 7.
Divide 7 into 18.

$$7\overline{)18^{4}09} \quad \text{quotient } 2 \qquad \frac{18}{7} = 2 \text{ with 4 left over}$$

Divide 7 into 40.

$$7\overline{)18^{4}0^{5}9} \quad \text{quotient } 2\ 5 \qquad \frac{40}{7} = 5 \text{ with 5 left over}$$

Divide 7 into 59.

$$7\overline{)18^{4}0^{5}9} \quad \text{quotient } 2\ 5\ 8\ \mathbf{R}3 \qquad \frac{59}{7} = 8 \text{ with 3 left over}$$

WORK PROBLEM 9 AT THE SIDE.

Note Short division takes practice but is useful in many situations.

6 **Check** the answer to a division problem as follows.

CHECKING DIVISION

divisor × quotient + remainder = dividend

EXAMPLE 10 *Checking Division by Using Multiplication*
Check each answer.

(a) $9\overline{)6512}$ = 723 **R**5

divisor × quotient + remainder = dividend

$$9 \times 723 + 5$$

$$6507 + 5 = 6512$$

matches original dividend

The division was done correctly.

(b) $6\overline{)1437}$ = 239 **R**4

divisor × quotient + remainder = dividend

$$6 \times 239 + 4$$

$$1434 + 4 = 1438$$

does not match original dividend

The answer does not check. Rework the original problem to get the correct answer, 239 **R**3.

9. Divide.

(a) $5\overline{)1874}$

(b) $\frac{675}{7}$

(c) $3\overline{)1885}$

(d) $8\overline{)1135}$

ANSWERS
9. (a) 374 R4 (b) 96 R3 (c) 628 R1 (d) 141 R7

10. Check each division. If an answer is incorrect, give the correct answer.

(a) $3\overline{)115}$ = 38 R1

(b) $8\overline{)739}$ = 92 R2

(c) $8\overline{)2624}$ = 328

(d) $5\overline{)2383}$ = 476 R3

11. Which are divisible by 2?

(a) 714

(b) 513

(c) 5128

(d) 21,000

ANSWERS
10. (a) correct
(b) incorrect, should be 92 R3
(c) correct (d) correct
11. all but (b)

Note A common error when checking division is forgetting to add the remainder. Be sure to add any remainder when checking a division problem.

WORK PROBLEM 10 AT THE SIDE.

7 It is often important to know whether a number is divisible by another number. You will find this useful in Chapter 2 when writing fractions in lowest terms.

DIVISIBILITY

One whole number is **divisible** by another if the remainder is zero.

Decide whether one number is exactly divisible by another by using the following tests for divisibility.

TESTS FOR DIVISIBILITY

A number is divisible by

2 if it ends in 0, 2, 4, 6, or 8.
3 if the sum of its digits is divisible by 3.
4 if the last two digits make a number that is divisible by 4.
5 if it ends in 0 or 5.
6 if it is divisible by both 2 and 3.
8 if the last three digits make a number that is divisible by 8.
9 if the sum of its digits is divisible by 9.
10 if it ends in 0.

The most commonly used tests are those for 2, 3, 5, and 10.

DIVISIBILITY BY 2

A number is divisible by **2** if the number ends in 0, 2, 4, 6, or 8.

EXAMPLE 11 *Testing for Divisibility by 2*
Are the following numbers divisible by 2?

(a) 986
↑ ends in 6

Because the number ends in 6, which is in the list above, the number 986 is divisible by 2.

(b) 3255 is not divisible by 2.
↑ ends in 5, and not in 0, 2, 4, 6, or 8 ■

WORK PROBLEM 11 AT THE SIDE.

DIVISIBILITY BY 3

A number is divisible by **3** if the sum of its digits is divisible by 3.

■ **EXAMPLE 12** *Testing for Divisibility by 3*
Are the following numbers divisible by 3?

(a) 4251
Add the digits.

4 2 5 1
$4 + 2 + 5 + 1 = 12$

Because 12 is divisible by 3, the number 4251 is divisible by 3.

(b) 29,806
Add the digits.

2 9 8 0 6
$2 + 9 + 8 + 0 + 6 = 25$

Because 25 is not divisible by 3, the number 29,806 is not divisible by 3. ■

Note Be careful when testing for divisibility by adding the digits. This method works only for the numbers 3 and 9.

WORK PROBLEM 12 AT THE SIDE.

DIVISIBILITY BY 5 AND BY 10

A number is divisible by **5** if it ends in 0 or 5.
A number is divisible by **10** if it ends in 0.

■ **EXAMPLE 13** *Determining Divisibility by 5*
Are the following numbers divisible by 5?

(a) 12,900 ends in 0 and is divisible by 5.

(b) 4325 ends in 5 and is divisible by 5.

(c) 392 ends in 2 and is not divisible by 5. ■

WORK PROBLEM 13 AT THE SIDE.

■ **EXAMPLE 14** *Determining Divisibility by 10*
Are the following numbers divisible by 10?

(a) 80, 700, and 9140 end in 0 and are divisible by 10.

(b) 29, 355, and 18,743 do not end in 0 and are not divisible by 10. ■

WORK PROBLEM 14 AT THE SIDE.

12. Which are divisible by 3?

(a) 2921

(b) 7545

(c) 242,913

(d) 508,396

13. Which are divisible by 5?

(a) 320

(b) 545

(c) 5551

(d) 32,705

14. Which are divisible by 10?

(a) 230

(b) 115

(c) 3030

(d) 13,780

ANSWERS
12. (b) and (c)
13. all but (c)
14. all but (b)

QUEST FOR NUMERACY

Math Lessons From the Past

In the 1800s students didn't have calculators or computers to help with their math homework. They used slates and chalk to complete assignments in class, and they had never seen a white board with erasable felt markers.

The following word problems are excerpts from *A Primary Arithmetic,* published in 1880 by Sheldon and Company. Notice that the language and statistics may seem outdated and sometimes humorous. Although these differences appear to be immense, the basic mathematics lessons and skills required, including many appearing in this book, have changed very little.

1. In 1870 there were 8 cities in the United States which had over 200,000 population each, viz., New York, 942,292; Philadelphia, 674,022; Brooklyn, 396,099; St. Louis, 310,864; Chicago, 298,977; Baltimore, 267,354; Boston, 250,526; Cincinnati, 216,239. What was the total population of these cities? What of New York and Brooklyn together?

3,356,373 population of cities
1,338,391 New York and Brooklyn

2. I bought a house for $3500, paid $1525 for improvements, then sold it so as to gain $860. How much did I sell it for?

$5885

3. This large basket contains 42 eggs. How many times can the little girl fill her small basket from it, if her small basket holds 6 eggs? How many times can she fill her basket if it holds 7 eggs? How many 6's in 42? How many times 6 make 42? 42 divided by 6 are how many? Why? How many 7's in 42? How many times 7 make 42? 42 divided by 7 make how many? Why?

7 times 6 equals 42
6 times 7 equals 42
The commutative property of multiplication is demonstrated here.

Courtesy Stan Salzman.

To extend this activity, ask students to update the population figures in problem 1 using library reference material. How much has each city's population increased? What is each city's current population rounded to the nearest ten thousand? To the nearest hundred thousand? What other large cities now exist that were not included in the 1870 list?

EXERCISES

Write each division problem by using two other symbols.

Example: $14 \div 2 = 7$

Solution: $\frac{14}{2} = 7 \qquad 2\overline{)14}$ (quotient 7)

1. $15 \div 5 = 3$

   $5\overline{)15}$ (quotient 3) $\quad \frac{15}{5} = 3$

2. $24 \div 8 = 3$

   $8\overline{)24}$ (quotient 3) $\quad \frac{24}{8} = 3$

3. $\frac{45}{9} = 5$

   $9\overline{)45}$ (quotient 5) $\quad 45 \div 9 = 5$

4. $\frac{56}{8} = 7$

   $56 \div 8 = 7$

   $8\overline{)56}$ (quotient 7)

5. $2\overline{)16}$ (quotient 8)

   $16 \div 2 = 8$

   $\frac{16}{2} = 8$

6. $8\overline{)48}$ (quotient 6)

   $48 \div 8 = 6$

   $\frac{48}{8} = 6$

Divide.

Examples:

$21 \div 3 \qquad \frac{0}{5} \qquad 18 \div 0$

Solutions:

$21 \div 3 = \mathbf{7} \qquad \frac{0}{5} = \mathbf{0} \qquad 18 \div 0$ is **meaningless**

7. $6 \div 6$ 1

8. $40 \div 5$ 8

9. $\frac{10}{2}$ 5

10. $\frac{8}{0}$ meaningless

11. $24 \div 0$ meaningless

12. $4 \div 4$ 1

13. $\frac{0}{4}$ 0

14. $\frac{24}{8}$ 3

15. $12\overline{)0}$ 0

16. $\frac{0}{7}$ 0

17. $0\overline{)15}$ meaningless

18. $\frac{2}{0}$ meaningless

19. $\frac{0}{3}$ 0

20. $\frac{0}{0}$ meaningless

21. $\frac{8}{1}$ 8

22. $\frac{0}{5}$ 0

Divide by using short division. Check each answer.

Examples:

(a) $8\overline{)376}$ (b) $6\overline{)1487}$

Solutions: $\begin{array}{r}\mathbf{4\ 7}\\ 8\overline{)37^{5}6}\end{array}$ $\begin{array}{r}\mathbf{2\ 4\ 7\ \ R5}\\ 6\overline{)14^{2}8^{4}7}\end{array}$

Check: $8 \times 47 = \mathbf{376}$ $6 \times 247 + 5 = 1482 + 5 = \mathbf{1487}$

23. $5\overline{)130}$ 26

24. $6\overline{)126}$ 21

25. $8\overline{)376}$ 47

26. $6\overline{)228}$ 38

27. $5\overline{)2005}$ 401

28. $4\overline{)3212}$ 803

29. $4\overline{)2509}$ 627 R1

30. $8\overline{)1335}$ 166 R7

31. $6\overline{)9137}$ 1522 R5

32. $9\overline{)8371}$ 930 R1

33. $6\overline{)1854}$ 309

34. $8\overline{)856}$ 107

35. $4024 \div 4$ 1006

36. $16{,}024 \div 8$ 2003

37. $15{,}018 \div 3$ 5006

38. $32{,}008 \div 8$ 4001

39. $4867 \div 6$ 811 R1

40. $5993 \div 7$ 856 R1

41. $12{,}947 \div 5$ 2589 R2

42. $33{,}285 \div 9$ 3698 R3

43. $29{,}298 \div 4$ 7324 R2

44. $17{,}937 \div 6$ 2989 R3

45. $12{,}630 \div 4$ 3157 R2

46. $46{,}560 \div 7$ 6651 R3

47. $\frac{52{,}569}{9}$ 5841

48. $\frac{26{,}348}{4}$ 6587

49. $\frac{74{,}751}{6}$ 12,458 R3

50. $\frac{72{,}543}{5}$ 14,508 R3

51. $\frac{71{,}776}{7}$ 10,253 R5

52. $\frac{77{,}621}{3}$ 25,873 R2

53. $\frac{128{,}645}{7}$ 18,377 R6

54. $\frac{172{,}255}{4}$ 43,063 R3

Check each answer. If an answer is incorrect, give the correct answer.

Example: $9\overline{)1609}$ with quotient **178 R7**

Solution: divisor × quotient + remainder = dividend

$$\underbrace{9 \times 178}_{1602} + 7 = \mathbf{1609}$$

The answer checks.

55. $5\overline{)397}$ = 79 **R**2

correct

56. $8\overline{)439}$ = 54 **R**7

correct

57. $3\overline{)5725}$ = 1908 **R**2

incorrect;
should be 1908 R1

58. $5\overline{)2158}$ = 432 **R**3

incorrect;
should be 431 R3

59. $7\overline{)4692}$ = 650 **R**2

incorrect;
should be 670 R2

60. $9\overline{)5974}$ = 663 **R**5

incorrect;
should be 663 R7

61. $6\overline{)21,409}$ = 3 568 **R**2

incorrect;
should be 3568 R1

62. $4\overline{)103,516}$ = 25,879

correct

63. $6\overline{)18,023}$ = 3 003 **R**5

correct

64. $8\overline{)33,664}$ = 4 208

correct

65. $6\overline{)69,140}$ = 11,523 **R**2

correct

66. $3\overline{)82,598}$ = 27,532 **R**1

incorrect;
should be 27,532 R2

67. $9\overline{)86,655}$ = 9 628 **R**7

incorrect;
should be 9628 R3

68. $7\overline{)50,809}$ = 7 258 **R**4

incorrect;
should be 7258 R3

69. $8\overline{)222,576}$ = 27,822

correct

70. $4\overline{)311,216}$ = 77,804

correct

71. Explain in your own words how to check a division problem using multiplication. Be sure to include what must be done if the quotient includes a remainder.

Answer varies.

72. Describe the three divisibility rules that you feel might be most useful to you and tell why.

Answer varies.

Solve each word problem.

73. A carton of antifreeze holds 4 one-gallon jugs. Find the number of cartons needed to package 624 one-gallon jugs.

156 cartons

74. Eight people invested a total of $244,224 to buy a condominium. Each person invested the same amount of money. How much did each person invest?

$30,528

75. If 8 identical metal shredders cost $115,528, find the cost of each shredder.

$14,441

76. One gallon of beverage will serve 9 people. How many gallons are needed for 3483 people?

387 gallons

77. A scholarship fund of $29,250 is divided evenly among 9 students. Find the amount received by each student.

$3250

78. How many 5-pound bags of rice can be filled from 8750 pounds of rice?

1750 bags

79. If 36 gallons of fertilizer are needed for each acre of land, find the number of acres that can be fertilized with 7380 gallons of fertilizer.

205 acres

80. A roofing contractor has purchased 2268 squares (10 feet by 10 feet) of roofing material. If each home needs 21 squares of material, find the number of homes that can be roofed.

108 homes

81. The Bel-Air Market has a profit of $105,676. If this profit is to be divided evenly among the 58 employees, find the amount received by each employee.

$1822

82. A box holds 24 bottles of aspirin. How many cartons are needed to hold 25,392 bottles of aspirin?

1058 boxes

Put a √ mark in the blank if the number at the left is divisible by the number at the top. Put an X in the blank if the number is not divisible by the number at the top.

Example: 40
The number 40 can be divided by 2, 5, and 10 but not by 3.

Solution:	**2**	**3**	**5**	**10**
40	√	X	√	√

	2	3	5	10
83. 30	√	√	√	√
84. 25	X	X	√	X
85. 184	√	X	X	X
86. 192	√	√	X	X
87. 445	X	X	√	X
88. 897	X	√	X	X
89. 903	X	√	X	X
90. 500	√	X	√	√
91. 5166	√	√	X	X
92. 8302	√	X	X	X
93. 21,763	X	X	X	X
94. 32,472	√	√	X	X

▲ **95.** Kaci Salmon, a supervisor at Albany Electric, earns $36,540 per year. Find the amount of her earnings each month.

$3045

▲ **96.** A worker assembles 168 light diffusers in an 8-hour shift. Find the number assembled in 3 hours.

63 light diffusers

1.6 LONG DIVISION

Long division is used to divide by a number with more than one digit.

1 In long division, estimate the various numbers by using a **trial divisor,** which is used to get a **trial quotient.**

■ **EXAMPLE 1** *Using a Trial Divisor and a Trial Quotient*
Divide: $42\overline{)3066}$.

Because 42 is closer to 40 than to 50, use the first digit of the divisor as a trial divisor.

42
↑—— trial divisor

Try to divide the first digit of the dividend by 4. Since 3 cannot be divided by 4, use the first *two* digits, 30.

$\frac{30}{4}$ = 7 with remainder 2

```
    7    ← trial quotient
42)3066
   ↑—7 goes over the 6, because 306/42 is about 7.
```

Multiply 7 and 42 to get 294; next, subtract 294 from 306.

```
    7
42)3066
   294   ← 7 × 42
    12   ← 306 − 294
```

Bring down the 6 at the right.

```
    7
42)3066
   294↓
    126  ← 6 brought down
```

Use the trial divisor, 4.

first two digits of 126 → $\frac{12}{4} = 3$

```
   73
42)3066
   294
    126
    126  ← 3 × 42 = 126
      0
```

Check the answer by multiplying 42 and 73. The product should be 3066. ■

Note The first digit (left hand) of the answer in long division must be placed in the proper position over the dividend.

WORK PROBLEM 1 AT THE SIDE. ▶▶

OBJECTIVES

1. Do long division.
2. Divide numbers ending in zero by numbers ending in zero.
3. Check answers.

FOR EXTRA HELP

Tape 2 | SSM pp. 14–17 | MAC: A IBM: A

1. Divide.

(a) $27\overline{)1728}$

(b) $52\overline{)4264}$

(c) $51\overline{)2295}$

(d) $\frac{6552}{84}$

ANSWERS
1. (a) 64 **(b)** 82 **(c)** 45 **(d)** 78

2. Divide.

(a) $28\overline{)1176}$

(b) $48\overline{)2691}$

(c) $65\overline{)5416}$

(d) $89\overline{)6649}$

ANSWERS
2. (a) 42 (b) 56 R3 (c) 83 R21 (d) 74 R63

■ **EXAMPLE 2** *Dividing to Find a Trial Quotient*
Divide: $58\overline{)2730}$.

Use 6 as a trial divisor, since 58 is closer to 60 than to 50.

first two digits of dividend → $\frac{27}{6}$ = 4 with 3 left over

```
    4     ← trial quotient
58)2730
   232    ← 4 × 58 = 232
    41    ← 273 − 232 = 41 (smaller than 58, the divisor)
```

Bring down the 0.

```
    4
58)2730
   232
    410   ← 0 brought down
```

first two digits of 410 → $\frac{41}{6}$ = 6 with 5 left over

```
    46    ← trial quotient
58)2730
   232
    410
    348   ← 6 × 58 = 348
     62   ← greater than 58
```

The remainder, 62, is greater than the divisor, 58, so 7 should be used instead of 6.

```
    47 R4
58)2730
   232
    410
    406   ← 7 × 58 = 406
      4   ← 410 − 406
```
■

WORK PROBLEM 2 AT THE SIDE.

Sometimes it is necessary to insert a zero in the quotient.

■ **EXAMPLE 3** *Inserting Zeros in the Quotient*
Divide: $42\overline{)8734}$.

Start as above.

```
    2
42)8734
   84    ← 2 × 42 = 84
    3    ← 87 − 84 = 3
```

Bring down the 3.

```
    2
42)8734
   84
    33   ← 3 brought down
```

Since 33 cannot be divided by 42, place a 0 in the quotient as a placeholder.

```
       20   ← 0 in quotient
   42)8734
      84
       33
```

Bring down the final digit, the 4.

```
       20
   42)8734
      84
       334  ← 4 brought down
```

Complete the problem.

```
       207 R40
   42)8734
      84
       334
       294
        40
```

The answer is 207 **R**40. ■

Note There must be a digit in the quotient (answer) above every digit in the dividend once the answer has begun. Notice that in Example 3 a zero was used to assure an answer digit above every digit in the dividend.

WORK PROBLEM 3 AT THE SIDE.

2 When the divisor and dividend both contain zeros at the far right, recall that these numbers are multiples of 10. There is a short way to divide these multiples of 10. For example,

$$90 \div 10 = 9, \quad 6400 \div 100 = 64, \quad 857{,}000 \div 1000 = 875.$$

These examples suggest the following rule

DIVIDING A WHOLE NUMBER BY 10, 100, or 1000

Divide a whole number by 10, 100, or 1000 by dropping the appropriate number of zeros from the whole number.

■ **EXAMPLE 4** *Dividing by Multiples of 10*

Divide.

(a) 60 ÷ 10 = 6 (1 zero in divisor; 0 dropped)

(b) 3500 ÷ 100 = 35 (2 zeros in divisor; 00 dropped)

(c) 915,000 ÷ 1000 = 915 (3 zeros in divisor; 000 dropped) ■

3. Divide.

(a) $24\overline{)3127}$

(b) $63\overline{)12{,}663}$

(c) $39\overline{)15{,}933}$

(d) $78\overline{)23{,}462}$

ANSWERS

3. (a) 130 R7 (b) 201 (c) 408 R21 (d) 300 R62

4. Divide.

(a) 30 ÷ 10

(b) 2100 ÷ 100

(c) 608,000 ÷ 1000

5. Divide.

(a) 70)11,200

(b) 150)180,750

(c) 2600)195,000

6. Decide whether the following divisions are correct.

(a)

```
     43
18)774
   72
    54
    54
     0
```

(b)

```
        45
426)19,170
    17 04
     2 130
     2 130
         0
```

(c)

```
      57 R18
514)29,316
    25 70
     3 616
     3 598
        18
```

ANSWERS
4. (a) 3 (b) 21 (c) 608
5. (a) 160 (b) 1205 (c) 75
6. (a) correct (b) correct (c) correct

◀ WORK PROBLEM 4 AT THE SIDE.

Find the quotient for other multiples of 10 by dropping zeros.

■ **EXAMPLE 5** *Dividing by Multiples of 10*
Divide.

(a) 40)11,000 Drop 1 zero from the divisor and the dividend.

```
    275
4)1100
  8
  30
  28
   20
   20
    0
```

(b) 3500)31,500 2 zeros dropped from the divisor and the dividend

```
     9
35)315
   315
     0
```
■

Note Dropping zeros when dividing by multiples of 10 does not change the answer (quotient).

◀ WORK PROBLEM 5 AT THE SIDE.

3 Answers in long division can be checked just as answers in short division were checked.

■ **EXAMPLE 6** *Checking Division*
Check each answer.

(a) 48)5324 with quotient 114 **R**43

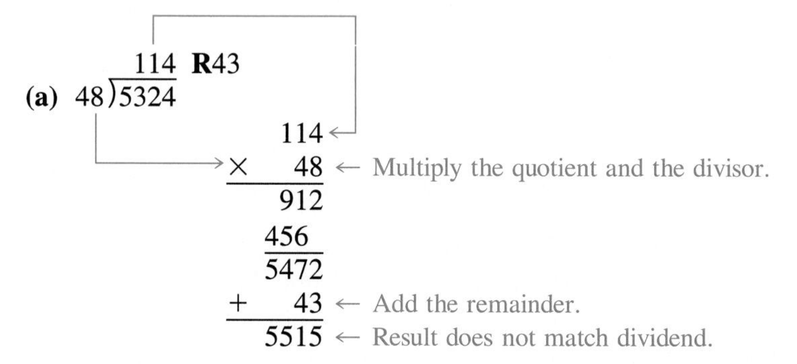

The answer does not check. Rework the original problem to get 110 **R**44.

(b)

```
       37            716
716)26,492         ×  37
    21 48          5 012
     5 012        21 48
     5 012        26,492   correct
         0
```
■

Note When checking a division problem, first multiply the quotient and the divisor. Then be sure to add any remainder before checking it against the original dividend.

◀ WORK PROBLEM 6 AT THE SIDE.

1.6 NAME DATE HOUR

EXERCISES

Divide by using long division. Check each answer.

Example: $41\overline{)2388}$

Solution:

$$\begin{array}{r} \mathbf{58\ R10} \\ 41\overline{)2388} \\ \underline{205} \\ 338 \\ \underline{328} \\ 10 \end{array}$$

Check:

$$\begin{array}{r} 58 \\ \times\ 41 \\ \hline 58 \\ \underline{232} \\ 2378 \\ +\ \ 10 \leftarrow \text{remainder (added)} \\ \hline \mathbf{2388} \leftarrow \text{matches dividend} \end{array}$$

1. $24\overline{)768}$ — 32
2. $34\overline{)1632}$ — 48
3. $53\overline{)2501}$ — 47 R10
4. $72\overline{)6308}$ — 87 R44
5. $59\overline{)2190}$ — 37 R7
6. $39\overline{)4095}$ — 105
7. $46\overline{)7045}$ — 153 R7
8. $42\overline{)8699}$ — 207 R5
9. $58\overline{)2204}$ — 38
10. $77\overline{)1650}$ — 21 R33
11. $47\overline{)11{,}121}$ — 236 R29
12. $83\overline{)39{,}692}$ — 478 R18
13. $26\overline{)62{,}583}$ — 2407 R1
14. $28\overline{)84{,}249}$ — 3008 R25
15. $63\overline{)78{,}072}$ — 1239 R15
16. $86\overline{)10{,}327}$ — 120 R7
17. $46\overline{)24{,}026}$ — 522 R14
18. $52\overline{)68{,}025}$ — 1308 R9
19. $12\overline{)116{,}953}$ — 9746 R1
20. $21\overline{)149{,}826}$ — 7134 R12
21. $32\overline{)247{,}892}$ — 7746 R20
22. $238\overline{)186{,}948}$ — 785 R118
23. $153\overline{)509{,}725}$ — 3331 R82
24. $402\overline{)29{,}346}$ — 73
25. $821\overline{)17{,}241}$ — 21
26. $523\overline{)638{,}075}$ — 1220 R15
27. $657\overline{)732{,}094}$ — 1114 R196
28. $420\overline{)357{,}000}$ — 850
29. $900\overline{)153{,}000}$ — 170
30. $230\overline{)253{,}230}$ — 1101

Check each answer. If an answer is incorrect, give the correct answer.

31. $56\overline{)5943}$ = 106 **R**17

incorrect; should be 106 R7

32. $87\overline{)3254}$ = 37 **R**37

incorrect; should be 37 R35

33. $28\overline{)18{,}424}$ = 658 **R**9

incorrect should be 658

34. $191\overline{)88{,}604}$ = 463 **R**171

correct

35. $614\overline{)38{,}068}$ = 62 **R**3

incorrect; should be 62

36. $557\overline{)97{,}286}$ = 174 **R**368

correct

37. Describe in your own words how you can divide a whole number by 10, by 100 or by 1000 by dropping zeros from the whole number. Write an example problem and solve it.

Answer varies.

38. Explain how to check a division problem using multiplication. Use an example division problem that has a remainder in your explanation.

Answer varies.

Solve each word problem by using addition, subtraction, multiplication, or division as needed.

39. A car travels 1350 miles at 54 miles per hour. How many hours did it travel?

25 hours

40. The total cost for 27 baseball uniforms is $2106. Find the cost of each uniform.

$78

41. A person borrows $3200 and pays interest of $706. Find the total amount that must be repaid.

$3906

42. Two separated parents each share some of the education costs of their child which amount to $3718. If one parent pays $1880, find the amount paid by the other parent.

$1838

43. Judy Martinez owes $3888 on a loan. Find her monthly payment if the loan is to be paid off in 36 months.

$108

44. A consultant charged $13,050 for studying a school's compliance with the Americans with Disabilities Act. If the consultant worked 225 hours, find the rate charged per hour.

$58

45. Clarence Hanks can assemble 42 circuits in 1 hour. How many circuits can he assemble in a 5-day workweek of 8 hours per day?

1680 circuits

46. There are two conveyer lines in a factory each of which packages 240 sacks of salt per hour. If the lines operate for 8 hours, find the total number of sacks of salt packaged by the two lines.

3840 sacks

47. A youth soccer association raised $7588 in fund-raising projects. There were expenses of $838 that had to be paid first, with the balance of the money divided evenly among the 18 teams. How much did each team receive?

$375

48. Feather Farms Egg Ranch collects 3545 eggs in the morning and 2575 eggs in the afternoon. If the eggs are packed in flats containing 30 eggs each, find the number of flats needed for packing.

204 flats

1.7 ROUNDING WHOLE NUMBERS

One way to get a rough check on an answer is to *round* the numbers in the problem. **Rounding** a number means finding a number that is close to the original number, but easier to work with.

For example, a superintendent of schools in a large city might be discussing the need to build new schools. In making her point, it probably would not be necessary to say that the school district has 152,807 students—it would probably be sufficient to say there are 153,000 students, or even 150,000 students.

OBJECTIVES

1. Locate the place to which a number is to be rounded.
2. Round numbers.
3. Round numbers to estimate an answer.
4. Use front end rounding to estimate an answer.

FOR EXTRA HELP

Tape 2	SSM pp. 17–22	MAC: A IBM: A

1 The first step in rounding a number is to locate the *place to which the number is to be rounded.*

■ **EXAMPLE 1** *Finding the Place to Which a Number Is to Be Rounded*

Locate and draw a line after the place to which each number is to be rounded.

(a) Round 83 to the nearest ten.

8|3 ← tens place

(b) Round 54,702 to the nearest thousand.

54|,702 ← thousands place

(c) Round 2,906,124 to the nearest hundred thousand.

2,9|06,124 ← hundred thousands place ■

WORK PROBLEM 1 AT THE SIDE. ▶▶

1. Locate and draw a line after the place to which the number is to be rounded.

(a) 746 (nearest ten)

(b) 2412 (nearest thousand)

(c) 89,512 (nearest hundred)

(d) 546,325 (nearest ten thousand)

2 Use the following rules for rounding whole numbers.

Step 1 Locate the **place** to which the number is to be rounded. Draw a line after that place to show that you are rounding off the rest of the digits.

Step 2A Look at the first digit to the right of the line. If the first digit is 5 or more, increase by one the digit in the place to which you are rounding.

Step 2B If the first digit to the right of the line is 4 or less, do not change the digit in the place to which you are rounding.

Step 3 **Change** all digits to the right of the line to zeroes.

■ **EXAMPLE 2** *Using Rounding Rules*

Round 349 to the nearest hundred.

Step 1 Locate the place to which the number is being rounded. Draw a line after that place to show that you are rounding off the rest of the digits.

3|49 ← place to which number is rounded

ANSWERS

1. (a) 74|6 **(b)** 2|412 **(c)** 89,5|12 **(d)** 54|6,325

2. Round to the nearest ten.

(a) 64

(b) 78

(c) 135

(d) 5729

3. Round to the nearest thousand.

(a) 2834

(b) 6511

(c) 78,099

(d) 83,408

ANSWERS
2. (a) 60 (b) 80 (c) 140 (d) 5730
3. (a) 3000 (b) 7000 (c) 78,000 (d) 83,000

Step 2 Because the first digit to the right of the line is 4, which is 4 or less, do not change the digit in the place to which the number is rounded.

3 remains 3

Step 3 Change all digits to the right of the 3 to zeros.

349 rounded to the nearest hundred is 300. ■

◀◀ WORK PROBLEM 2 AT THE SIDE.

■ **EXAMPLE 3** *Rounding Rules for 5 or Greater*
Round 36,833 to the nearest thousand.

Step 1 Find the place to which the number is to be rounded and draw a line after the thousands place.

36,|833
↑ place to which number is rounded

Step 2 Because the first digit to the right of the line is 8, which is 5 or more, 1 must be added to the thousands place.

Change 6 to 7. (6 + 1 = 7)

Step 3 All digits to the right of this 7 must be changed to zeros. Therefore,

36,833 rounded to the nearest thousand is 37,000. ■

◀◀ WORK PROBLEM 3 AT THE SIDE.

■ **EXAMPLE 4** *Using Rules for Rounding*
Round as shown.

(a) 2382 to the nearest ten

Step 1 238|2
↑ place to which number is rounded

Step 2 The first digit to the right is a 2.

238|2 4 or less

Step 3 238|2 Change to 0.
Leave as 8.

2382 rounded to the nearest ten is 2380.

(b) 13,961 to the nearest hundred

Step 1 13,9|61
↑ place to which number is rounded.

Step 2 The first digit to the right is a 6.

13,9|61 5 or more

Step 3 139|61 Change to 0's.
9 + 1 = 10 (write 0); 1 is carried.

14,000
Write 0.
3 + 1 = 4
carried

13,961 rounded to the nearest hundred is 14,000. ■

WORK PROBLEM 4 AT THE SIDE.

■ EXAMPLE 5 *Rounding Large Numbers*
Round as shown.

(a) 37,892 to the nearest ten thousand

Step 1 3|7,892
place to which number is rounded

Step 2 The first digit to the right is a 7.

3|7892 5 or more

Step 3 3|7,892 Change to 0's
3 + 1 = 4

37,892 rounded to the nearest ten thousand is 40,000.

(b) 528,498,675 to the nearest million

Step 1 528,|498,675
place to which number is rounded

Step 2 528,|498,675 4 or less

Step 3 528,|498,675 Change to 0's
Leave as 8.

528,498,675 rounded to the nearest million is 528,000,000. ■

WORK PROBLEM 5 AT THE SIDE.

Sometimes a number must be rounded to different places.

■ EXAMPLE 6 *Rounding to Different Places*
Round 648 (**a**) to the nearest ten and (**b**) to the nearest hundred.

(a) to the nearest ten

64|8 5 or more
tens position (4 + 1 = 5)

648 to the nearest ten is 650.

(b) to the nearest hundred

6|48 4 or less
hundreds position

648 to the nearest hundred is 600. ■

4. Round as shown.

(a) 6536 to the nearest ten

(b) 37,549 to the nearest hundred

(c) 73,077 to the nearest hundred

(d) 259,596 to the nearest thousand

5. Round each of the following as shown.

(a) 14,671 to the nearest ten thousand

(b) 724,518,715 to the nearest million

ANSWERS
4. (a) 6540 (b) 37,500 (c) 73,100 (d) 260,000
5. (a) 10,000 (b) 725,000,000

If 648 is rounded to the nearest ten at 650, and then 650 is rounded to the nearest hundred, the result is 700. If, however, 648 is rounded to the nearest hundred, the result is 600.

Note Always go back to the original number when rounding to a different place.

WORK PROBLEM 6 AT THE SIDE.

EXAMPLE 7 *Applying Rounding Rules*

Round each of the following to the nearest ten, nearest hundred, and nearest thousand.

(a) 4358 **(b)** 680,914

(a) First round 4358 to the nearest ten.

435|8 5 or more
tens position (5 + 1 = 6)

4358 rounded to the nearest ten is 4360.

Now go back to the original number to round to the nearest hundred.

43|58 5 or more
hundreds position (3 + 1 = 4)

4358 rounded to the nearest hundred is 4400.

Again, go back to the original number to round, this time to the nearest thousand.

4|358 4 or less
thousands position

4358 rounded to the nearest thousand is 4000.

(b) First, round to the nearest ten

680,91|4 4 or less
tens position

680,914 rounded to the nearest ten is 680,910.

Go back to the original number to round to the nearest hundred.

680,9|14 4 or less
hundreds position

680,914 rounded to the nearest hundred is 680,900.

Go back to the original number to round to the nearest thousand.

680,|914 5 or more
thousands position (0 + 1 = 1)

680,914 rounded to the nearest thousand is 681,000. ■

WORK PROBLEM 7 AT THE SIDE.

6. Round each of the following numbers to the nearest ten and then to the nearest hundred.

(a) 156

(b) 394

(c) 9809

7. Round each of the following numbers to the nearest ten, nearest hundred, and nearest thousand.

(a) 7065

(b) 37,754

(c) 178,419

ANSWERS

6. (a) 160; 200
(b) 390; 400
(c) 9810; 9800

7. (a) 7070; 7100; 7000
(b) 37,750; 37,800; 38,000
(c) 178,420; 178,400; 178,000

3 Numbers are rounded to estimate an answer. An estimated answer is one that is close to the exact answer and may be used as a check when the exact answer is found.

■ **EXAMPLE 8** *Using Rounding to Estimate an Answer*
Estimate the following answers by rounding to the nearest ten.

(a)

76	80	rounded to the nearest ten
53	50	
38	40	
+ 91	+ 90	
	260	estimated answer

(b)

27	30	rounded to the nearest ten
− 14	− 10	
	20	estimated answer

(c)

16	20	rounded to the nearest ten
× 21	× 20	
	400	estimated answer ■

WORK PROBLEM 8 AT THE SIDE. ▶▶

■ **EXAMPLE 9** *Using Rounding to Estimate an Answer*
Estimate the following answers by rounding to the nearest hundred.

(a)

152	200	rounded to the nearest hundred
749	700	
576	600	
+ 819	+ 800	
	2300	estimated answer

(b)

780	800	rounded to the nearest hundred
− 536	− 500	
	300	estimated answer

(c)

664	700	rounded to the nearest hundred
× 843	× 800	
	560,000	estimated answer ■

WORK PROBLEM 9 AT THE SIDE. ▶▶

4 **Front end rounding** is used to estimate an answer. With front end rounding, each number is rounded so that all the digits are changed to zero except the first digit, which is rounded. Only one non-zero digit remains.

■ **EXAMPLE 10** *Using Front End Rounding to Estimate an Answer*
Estimate the following answers by using front end rounding.

(a)

3825	4000	all digits changed to zero except first digit, which is rounded
72	70	
565	600	
+ 2389	+ 2000	
	6670	estimated answer

8. ≈ Estimate the following answers by rounding to the nearest ten.

(a) 23 + 57 + 81 + 36

(b) 44 − 18

(c) 64 × 76

9. ≈ Estimate the following answers by rounding to the nearest hundred.

(a) 175 + 618 + 739 + 865

(b) 891 − 542

(c) 723 × 478

ANSWERS
8. (a) 200 (b) 20 (c) 4800
9. (a) 2400 (b) 400 (c) 350,000

10. ≈ Use front end rounding to estimate each answer.

(a)
$$\begin{array}{r} 4782 \\ 86 \\ 372 \\ +\ 1438 \\ \hline \end{array}$$

(b)
$$\begin{array}{r} 2583 \\ -\ 765 \\ \hline \end{array}$$

(c)
$$\begin{array}{r} 639 \\ \times\ 55 \\ \hline \end{array}$$

ANSWERS
10.(a) 6490 (b) 2200 (c) 36,000

(b)
$$\begin{array}{r} 6712 \\ -\ 825 \\ \hline \end{array} \qquad \begin{array}{r} 7000 \\ -\ 800 \\ \hline 6200 \end{array}$$

7000 and 800: first digit rounded and all others changed to zero
6200: estimated answer

(c)
$$\begin{array}{r} 725 \\ \times\ 86 \\ \hline \end{array} \qquad \begin{array}{r} 700 \\ \times\ 90 \\ \hline 63{,}000 \end{array}$$

700 and 90: only one non-zero digit remains
63,000: estimated answer ■

Note When using front end rounding, only 1 non-zero digit (first digit) remains. All digits to the right are zeros.

◀◀ WORK PROBLEM 10 AT THE SIDE.

1.7 EXERCISES

NAME DATE HOUR

Round as shown.

Example: 4336 to the nearest ten **4340**

Solution: 433|6 5 or more, so add 1 to 3.
↑ tens place Change 6 to 0.

1. 514 to the nearest ten

510

2. 215 to the nearest ten

220

3. 1276 to the nearest ten

1280

4. 3928 to the nearest ten

3930

5. 7862 to the nearest hundred

7900

6. 6746 to the nearest hundred

6700

7. 86,813 to the nearest hundred

86,800

8. 17,211 to the nearest hundred

17,200

9. 42,495 to the nearest hundred

42,500

10. 18,273 to the nearest hundred

18,300

11. 7998 to the nearest hundred

8000

12. 9263 to the nearest hundred

9300

13. 15,758 to the nearest hundred

15,800

14. 28,065 to the nearest hundred

28,100

15. 78,499 to the nearest thousand

78,000

16. 14,314 to the nearest thousand

14,000

17. 5847 to the nearest thousand

6000

18. 49,706 to the nearest thousand

50,000

19. 53,182 to the nearest thousand

53,000

20. 13,124 to the nearest thousand

13,000

21. 595,008 to the nearest ten thousand

600,000

22. 725,182 to the nearest ten thousand

730,000

23. 8,906,422 to the nearest million

9,000,000

24. 13,713,409 to the nearest million

14,000,000

Round each of the following to the nearest ten, nearest hundred, and nearest thousand.

Example: 6375

Solution:

ten	*hundred*	*thousand*
5 or more	5 or more	4 or less
637|5 ← Change to 0.	63|75 Change to 0's.	6|375 Change to 0's.
↑ Add 1 to 7.	↑ Add 1 to 3.	↑ Leave original number.
6380	**6400**	**6000**

Remember to round from the original number.

	Ten	*Hundred*	*Thousand*
25. 2365	2370	2400	2000
26. 6482	6480	6500	6000
27. 5392	5390	5400	5000
28. 8624	8620	8600	9000
29. 5049	5050	5000	5000
30. 7065	7070	7100	7000
31. 3132	3130	3100	3000
32. 7456	7460	7500	7000
33. 19,539	19,540	19,500	20,000
34. 59,806	59,810	59,800	60,000
35. 26,292	26,290	26,300	26,000
36. 78,519	78,520	78,500	79,000
37. 23,502	23,500	23,500	24,000
38. 84,639	84,640	84,600	85,000

39. Write in your own words the three steps that you will use to round a number when the digit to the right of the place to which you are rounding is 5 or more.

Answer varies.

40. Write in your own words the three steps that you will use to round a number when the digit to the right of the place to which you are rounding is 4 or less.

Answer varies.

NAME ______________________________

* ≈ *Estimate the following answers by rounding to the nearest ten. Then find the exact answers.*

41. estimate — exact

estimate	rounds to	exact
40	←	37
60	←	62
90	←	93
+ 60	←	+ 58
250		250

42. estimate — exact

estimate	exact
50	49
10	13
80	76
+ 50	+ 52
190	190

43.

estimate	exact
100	97
− 30	− 26
70	71

44.

estimate	exact
60	57
− 20	− 24
40	33

45.

estimate	exact
80	76
× 20	× 22
1600	1672

46.

estimate	exact
50	53
× 80	× 75
4000	3975

≈ *Estimate the following answers by rounding to the nearest hundred. Then find the exact answers.*

47. estimate — exact

estimate	rounds to	exact
800	←	786
800	←	823
300	←	342
+ 700	←	+ 684
2600		2635

48. estimate — exact

estimate	exact
600	623
400	362
200	189
+ 700	+ 736
1900	1910

49.

estimate	exact
700	677
− 400	+ 361
300	316

50.

estimate	exact
600	614
− 300	− 276
300	338

51.

estimate	exact
300	279
× 500	× 518
150,000	144,522

52.

estimate	exact
700	739
× 500	× 487
350,000	359,893

≈ *Estimate the following answers by using front end rounding. Then find the exact answers.*

53. estimate — exact

estimate	rounds to	exact
8000	←	8215
60	←	56
700	←	729
+ 4000	←	+ 3605
12,760		12,605

54. estimate — exact

estimate	exact
3000	2685
70	73
600	592
+ 7000	+ 7183
10,670	10,533

*This symbol is used to indicate exercises for which you should estimate your answer.

55. estimate

```
  800
- 200
  600
```

exact

```
  783
- 238
  545
```

56. estimate

```
  900
- 300
  600
```

exact

```
  942
- 286
  656
```

57.

```
  600
×  50
30,000
```

```
  638
×  47
29,986
```

58.

```
  900
×  70
63,000
```

```
  864
×  74
63,936
```

59. The number 648 rounded to the nearest ten is 650, and 650 rounded to the nearest hundred is 700. But when 648 is rounded to the nearest hundred it becomes 600. Why is this true? Explain. **Answer varies.**

60. The use of rounding is helpful when estimating the answer to a problem. Why is this true? Give an example using either addition, subtraction, multiplication or division to show how this works. **Answer varies.**

61. Round 70,987,652 to the nearest ten, nearest ten thousand, and nearest ten million.

70,987,650; 70,990,000; 70,000,000

62. Round 621,999,652 to the nearest thousand, nearest ten thousand, and nearest hundred thousand.

622,000,000; 622,000,000; 622,000,000

63. The gross national product for the United States (sum of all goods and services sold) was $5,465,485,362,159. Round this amount to the nearest hundred thousand, nearest hundred million, and nearest billion.

$5,465,485,400,000; $5,465,500,000,000; $5,465,000,000,000

64. The total Aid to Families With Dependent Children (AFDC) in the United States last year was $18,630,604,733. Round this amount to the nearest hundred thousand, nearest hundred million, and nearest ten billion.

$18,630,600,000; $18,600,000,000; $20,000,000,000

1.8 ROOTS AND ORDER OF OPERATIONS

OBJECTIVES

1. Identify an exponent and a base.
2. Find the square root of a number.
3. Use the order of operations.

FOR EXTRA HELP

Tape 2 | SSM pp. 22–25 | MAC: A IBM: A

1 The product $3 \cdot 3$ can be written as 3^2 (read as "3 squared"). The small raised number 2, called an **exponent** says to use 2 factors of 3. The number 3 is called the **base.** Writing 3^2 as 9 is called *simplifying the expression.*

■ **EXAMPLE 1** *Simplifying an Expression*
Identify the exponent and the base. Simplify each expression.

(a) 4^3

base → 4^3 ← exponent $4^3 = 4 \times 4 \times 4 = 64$

(b) $2^5 = 2 \times 2 \times 2 \times 2 \times 2 = 32$

The base is 2 and the exponent is 5. ■

WORK PROBLEM 1 AT THE SIDE.

2 Because $3^2 = 9$, the number 3 is called the **square root** of 9. The square root of a number is one of two identical factors of that number. Square roots of numbers are written with the symbol $\sqrt{\ }$, so

$$\sqrt{9} = 3.$$

By definition,

SQUARE ROOT

$$\sqrt{\text{number} \cdot \text{number}} = \text{number}.$$

For example: $\sqrt{36} = \sqrt{6 \cdot 6} = 6$

To find the square root of 64 ask, "What number can be multiplied by itself (that is, *squared)* to give 64?" The answer is 8, so $\sqrt{\mathbf{64}} = \sqrt{8 \cdot 8} = 8$.

A **perfect square** is a number that is the square of a whole number. The first few perfect squares are listed here.

PERFECT SQUARES TABLE

0 $= 0^2$	**16** $= 4^2$	**64** $= 8^2$	**144** $= 12^2$
1 $= 1^2$	**25** $= 5^2$	**81** $= 9^2$	**169** $= 13^2$
4 $= 2^2$	**36** $= 6^2$	**100** $= 10^2$	**196** $= 14^2$
9 $= 3^2$	**49** $= 7^2$	**121** $= 11^2$	**225** $= 15^2$

■ **EXAMPLE 2** *Using Perfect Squares*
Find each square root.

(a) $\sqrt{16}$ Because $4^2 = 16$, $\sqrt{16} = 4$.   **(b)** $\sqrt{49} = 7$

(c) $\sqrt{0} = 0$ **(d)** $\sqrt{169} = 13$ ■

WORK PROBLEM 2 AT THE SIDE.

3 Frequently problems may have parentheses, exponents, and square roots, and may involve more than one operation. Work these problems with the following order of operations.

1. Identify the exponent and the base. Simplify each expression.

(a) 3^2

(b) 6^3

(c) 2^4

(d) 3^4

2. Find each square root.

(a) $\sqrt{4}$

(b) $\sqrt{25}$

(c) $\sqrt{81}$

(d) $\sqrt{196}$

(e) $\sqrt{1}$

ANSWERS
1. (a) 2; 3; 9 (b) 3; 6; 216 (c) 4; 2; 16 (d) 4; 3; 81
2. (a) 2 (b) 5 (c) 9 (d) 14 (e) 1

ORDER OF OPERATIONS

1 Do all operations inside **parentheses.**
2. Simplify any expressions with **exponents** and find any **square roots.**
3. **Multiply** or **divide,** proceeding from left to right.
4. **Add** or **subtract,** proceeding from left to right.

■ **EXAMPLE 3** *Understanding Order of Operations*
Work each problem.

(a) $8^2 + 5 + 2$

2 factors of 8

$$8^2 + 5 + 2 = 8 \cdot 8 + 5 + 2 \quad \text{Evaluate exponent first.}$$
$$= 64 + 5 + 2$$
$$= 69 + 2 \leftarrow \text{Add from the left.}$$
$$= 71$$

(b) $35 \div 5 \cdot 6 = 7 \cdot 6$ Divide first (start at left).
$= 42$ Multiply.

(c) $9 + (20 - 4) \cdot 3 = 9 + 16 \cdot 3$ Work inside parentheses first.
$= 9 + 48$ Then multiply
$= 57$ Add last

(d) $12 \cdot \sqrt{16} - 8 \cdot 4 = 12 \cdot 4 - 8 \cdot 4$ Find square root first.
$= 48 - 32$ Multiply from the left.
$= 16$ Subtract. ■

WORK PROBLEM 3 AT THE SIDE.

■ **EXAMPLE 4** *Using Order of Operations*
Work each problem.

(a) $15 - 4 + 2 = 11 + 2$ Subtract first (start at left).
$= 13$ Add.

(b) $8 + (7 - 3) \div 2 = 8 + 4 \div 2$ Work inside parentheses first.
$= 8 + 2$ Divide.
$= 10$ Add.

(c) $4^2 \cdot 2^2 + (7 + 3) \cdot 2 = 4^2 \cdot 2^2 + 10 \cdot 2$ Parentheses
$= 16 \cdot 4 + 10 \cdot 2$ Exponents
$= 64 + 20$ Multiply.
$= 84$ Add.

(d) $4 \cdot \sqrt{25} - 7 \cdot 2 = 4 \cdot 5 - 7 \cdot 2$ Find square root first.
$= 20 - 14$ Multiply from the left.
$= 6$ Subtract. ■

Note Getting a correct answer depends on using the order of operations.

WORK PROBLEM 4 AT THE SIDE.

3. Work each problem.

(a) $4 + 5 + 3^2$

(b) $3^3 + 2^2$

(c) $5 \cdot 8 \div 20 - 1$

(d) $40 \div 5 \div 2$

(e) $8 + (14 \div 2) \cdot 6$

4. Work each problem.

(a) $7 - 2 + 3^2$

(b) $2^2 + 3^2 - (5 \cdot 2)$

(c) $3 \cdot \sqrt{100} - 9 \cdot 1$

(d) $20 \div 2 + (7 - 5)$

(e) $15 \cdot \sqrt{9} - 8 \cdot \sqrt{4}$

ANSWERS
3. (a) 18 (b) 31 (c) 1 (d) 4 (e) 50
4. (a) 14 (b) 3 (c) 21 (d) 12 (e) 29

1.8 EXERCISES

NAME DATE HOUR

Use the Perfect Squares Table on page 65 to find each square root.

> **Example** $\sqrt{169}$
>
> **Solution:**
>
> From the table, $13^2 = 169$, so $\sqrt{169} = \mathbf{13}$.

1. $\sqrt{4}$ 2

2. $\sqrt{25}$ 5

3. $\sqrt{16}$ 4

4. $\sqrt{9}$ 3

5. $\sqrt{144}$ 12

6. $\sqrt{100}$ 10

7. $\sqrt{121}$ 11

8. $\sqrt{225}$ 15

Identify the exponent and the base. Simplify each expression.

> **Example:** 4^2
>
> **Solution:**
>
> 4^2 ← exponent
>
> ↑ base
>
> Simplify. $4^2 = 4 \times 4 = \mathbf{16}$

9. 3^2 2; 3; 9

10. 2^3 3; 2; 8

11. 6^2 2; 6; 36

12. 5^3 3; 5; 125

13. 12^2 2; 12; 144

14. 10^3 3; 10; 1000

15. 15^2 2; 15; 225

16. 11^3 3; 11; 1331

Complete each blank.

> **Example:** $23^2 =$ ____ so $\sqrt{} = 23$
>
> **Solution:** $23^2 = 23 \cdot 23 = \mathbf{529}$, so $\sqrt{\mathbf{529}} = 23$

17. $20^2 =$ 400 so $\sqrt{400} = 20$

18. $10^2 =$ 100 so $\sqrt{100} = 10$

19. $25^2 =$ 625 so $\sqrt{625} = 25$

20. $30^2 =$ 900 so $\sqrt{900} = 30$

21. $35^2 =$ 1225 so $\sqrt{1225} = 35$

22. $38^2 =$ 1444 so $\sqrt{1444} = 38$

23. $40^2 =$ 1600 so $\sqrt{1600} = 40$

24. $50^2 =$ 2500 so $\sqrt{2500} = 50$

25. $54^2 =$ 2916 so $\sqrt{2916} = 54$

26. $60^2 =$ 3600 so $\sqrt{3600} = 60$

27. Describe in your own words a perfect square. Of the two numbers 25 and 50, which is a perfect square and why.

Answer varies.

28. Use the following list of words and terms to write the four steps in the order of operations.

add	square root
exponents	subtract
multiply	divide
parentheses	

Answer varies.

Work each problem by using the order of operations.

Examples: $15 - 9 + 4 = 6 + 4$ $\quad$ $28 \div 7 + 3^2 = 28 \div 7 + 9$

Solutions: $= \mathbf{10}$ $\quad$ $= 4 + 9$ $\quad$ $= \mathbf{13}$

29. $8^2 + 4 - 3$ 65

30. $2^3 + 10 - 8$ 10

31. $4 \cdot 6 - 5$ 19

32. $2 \cdot 7 - 4$ 10

33. $15 \cdot 2 \div 6$ 5

34. $8 \cdot 9 \div 6$ 12

35. $25 \div 5(8 - 4)$ 20

36. $36 \div 18(7 - 3)$ 8

37. $6 \cdot 2^2 + \frac{0}{6}$ 24

38. $8 \cdot 3^2 - \frac{10}{2}$ 67

39. $4 \cdot 1 + 8(9 - 2) + 3$ 63

40. $3 \cdot 2 + 7(3 + 1) + 5$ 39

41. $3^3 \cdot 2^2 + (10 - 5) \cdot 2$ 118

42. $4^2 \cdot 5^2 + (20 - 9) \cdot 3$ 433

43. $6 \cdot \sqrt{144} - 8 \cdot 6$ 24

44. $9 \cdot \sqrt{100} - 3 \cdot 9$ 63

45. $5 \cdot 3 + 8 \cdot 3 - 4$ 35

46. $7 \cdot 2 + 3 \cdot 6 + 10$ 42

47. $2^3 \cdot 3^2 + (14 - 4) \cdot 3$ 102

48. $3^2 \cdot 4^2 + (15 - 6) \cdot 2$ 162

49. $8 + 10 \div 5 + \frac{0}{3}$ 10

50. $6 + 8 \div 2 + \frac{0}{8}$ 10

51. $3^2 + 6^2 + (30 - 21) \cdot 2$ 63

52. $4^2 + 5^2 + (25 - 9) \cdot 3$ 89

53. $7 \cdot \sqrt{81} - 5 \cdot 6$ 33

54. $5 \cdot \sqrt{144} - 5 \cdot 7$ 25

55. $7 \cdot 2 + 8(2 \cdot 3) - 4$ 58

56. $5 \cdot 2 + 3(5 + 3) - 6$ 28

57. $4 \cdot \sqrt{49} - 7(5 - 2)$ 7

58. $3 \cdot \sqrt{25} - 6(3 - 1)$ 3

59. $6 \cdot (5 - 1) + \sqrt{4}$ 26

60. $5 \cdot (4 - 3) + \sqrt{9}$ 8

61. $6^2 + 2^2 - 6 + 2$ 36

62. $3^2 - 2^2 + 3 - 2$ 6

63. $5^2 \cdot 2^2 + (8 - 4) \cdot 2$ 108

64. $5^2 \cdot 3^2 + (30 - 20) \cdot 2$ 245

65. $7 + 6 \div 3 + 5 \cdot 2$ 19

66. $8 + 3 \div 3 + 6 \cdot 3$ 27

67. $5 \cdot \sqrt{36} - 7(7 - 4)$ 9

68. $9 \cdot \sqrt{64} - 5(4 + 2)$ 42

69. $6^2 - 2^2 + 3 \cdot 4$ 44

70. $4^2 + 3^2 - 5 \cdot 3$ 10

71. $8 + 5 \div 5 + 7 + \frac{0}{3}$ 16

72. $2 + 12 \div 6 + 5 + \frac{0}{5}$ 9

73. $4 \cdot \sqrt{25} - 6 \cdot 2$ 8

74. $8 \cdot \sqrt{36} - 4 \cdot 6$ 24

75. $3 \cdot \sqrt{25} - 4 \cdot \sqrt{9}$ 3

76. $8 \cdot \sqrt{100} - 6 \cdot \sqrt{36}$ 44

77. $7 \div 1 \cdot 8 \cdot 2 \div (21 - 5)$ 7

78. $12 \div 4 \cdot 5 \cdot 4 \div (15 - 13)$ 30

79. $15 \div 3 \cdot 2 \cdot 6 \div (14 - 11)$ 20

80. $9 \div 1 \cdot 4 \cdot 2 \div (11 - 5)$ 12

81. $4 \cdot \sqrt{16} - 3 \cdot \sqrt{9}$ 7

82. $10 \cdot \sqrt{49} - 4 \cdot \sqrt{64}$ 38

83. $5 \div 1 \cdot 10 \cdot 4 \div (17 - 9)$ 25

84. $15 \div 3 \cdot 8 \cdot 9 \div (12 - 8)$ 90

▲ **85.** $8 \cdot 9 \div \sqrt{36} - 4 \div 2 + (14 - 8)$ 16

▲ **86.** $3 - 2 + 5 \cdot 4 \cdot \sqrt{144} \div \sqrt{36}$ 41

▲ **87.** $1 + 3 - 2 \cdot \sqrt{1} + 3 \cdot \sqrt{121} - 5 \cdot 3$ 20

▲ **88.** $6 - 4 + 2 \cdot 9 - 3 \cdot \sqrt{225} \div \sqrt{25}$ 11

▲ **89.** $6 \cdot \sqrt{25} \cdot \sqrt{100} \div 3 \cdot \sqrt{4} + 9$ 209

▲ **90.** $9 \cdot \sqrt{36} \cdot \sqrt{81} \div 2 + 6 - 3 - 5$ 241

1.9 SOLVING WORD PROBLEMS

OBJECTIVES

1. Find indicator words in word problems.
2. Solve word problems.
3. Estimate an answer.

FOR EXTRA HELP

Tape 2

SSM pp. 25–29

MAC: A
IBM: A

Most problems involving applications of mathematics are presented as word problems. You must read the words carefully to decide how to solve the problem.

1 You have to look for **indicator words** in the word problem—words that indicate the necessary operations—either addition, subtraction, multiplication, or division. Some of these word indicators appear below.

Addition	*Subtraction*
plus	less
more	subtract
more than	subtracted from
added to	difference
increased by	less than
sum	fewer
total	decreased by
sum of	loss of
increase of	minus
gain of	take away

Multiplication	*Division*
product	divided by
double	divided into
triple	quotient
times	goes into
of	divide
twice	divided equally
twice as much	per

Equals
is
the same as
equals
equal to
yields
results in
are

Note The word "and" does not indicate addition and does not appear as an indicator word above. Notice how the "and" shows the location of an operation.

The sum of 6 *and* 2 is $6 + 2$
The difference of 6 *and* 2 is $6 - 2$
The product of 6 *and* 2 is $6 \cdot 2$
The quotient of 6 *and* 2 is $6 \div 2$

2 Solve application problems by using the following steps.

1. ≈ Pick the most reasonable answer for each problem.

(a) an hourly wage \$2; \$6; \$60

(b) a score on a 100-point test 6; 20; 74; 109

(c) the cost of heart bypass surgery \$500; \$48,000; \$400,000

2. (a) On a recent geology field trip, 84 fossils were collected. If the fossils are divided equally among Gilbert, Sean, Noella, and Sue, how many fossils will each receive?

(b) A company advertising campaign generates 264 sales leads. If there are twelve sales people who will divide these leads equally, how many will each receive?

ANSWERS
1. (a) \$6 (b) 74 (c) \$48,000
2. (a) 21 fossils (b) 22 leads

STEPS FOR SOLVING APPLICATION PROBLEMS

Step 1 Read the problem carefully and be certain you *understand* what the problem is asking. It may be necessary to read the problem several times.

Step 2 Before doing any calculations, work out a *plan* and try to visualize the problem. Know which facts are given and which must be found. Use *indicators* to help decide on the *plan.*

Step 3 Estimate a *reasonable answer* by using rounding.

Step 4 *Solve* the problem by using the facts given and your plan. If the answer is reasonable, *check* your work. If the answer is not reasonable, begin again by rereading the problem.

3 These steps give a systematic approach for solving word problems. Each of the steps is important, but special emphasis should be placed on Step 3, estimating a *reasonable answer.* Many times an "answer" just does not fit the problem.

What is a reasonable answer? Read the problem and try to determine the approximate size of the answer. Should the answer be part of a dollar, a few dollars, hundreds, thousands, or even millions of dollars? For example, if a problem asks for the cost of a man's shirt, would an answer of \$20 be reasonable? \$1000? \$0.65? \$65?

Always make an estimate of a reasonable answer; then check the answer you get to see if it is close to your estimate.

◀◀ WORK PROBLEM 1 AT THE SIDE.

■ EXAMPLE 1 *Applying Division*

At a recent garage sale, the total sales were \$584. If the money was divided equally among Paul, Roietta, Maryann, and Jose, how much did each person get?

Approach To find the amount received by each person, divide the total amount of sales by the number of people.

Solution

Step 1 A reading of the problem shows that the four members in the group divided \$584 equally.

Step 2 The word indicators, ***divided equally,*** show that the amount each received can be found by dividing \$584 by 4.

Step 3 A reasonable answer would be a little less than \$150 each, since $\$600 \div 4 = \$150$ (\$584 rounded to \$600).

Step 4 Find the actual answer by dividing \$584 by 4.

$$4\overline{)584}\quad = 146$$

Each person should get \$146.

The answer \$146 is reasonable, as \$146 is close to the estimated answer of \$150.

Is the answer \$146 correct? Check the work.

$$\begin{array}{r} \$146 \\ \times\quad 4 \\ \hline \$584 \end{array}$$ ■

◀◀ WORK PROBLEM 2 AT THE SIDE.

EXAMPLE 2 *Applying Addition*
Matt earns \$46 on Monday, \$36 on Tuesday, \$48 on Wednesday, \$50 on Thursday, and \$32 on Friday. Find his total earnings for the week.

Approach To find the total for the week, add the earnings for each day.

Solution

Step 1 In this problem, the earnings for each day are given and the total earnings for the week must be found.
Step 2 Add the daily earnings to arrive at the weekly total.
Step 3 Because the earnings were about \$40 per day for a week of 5 days, a reasonable estimate would be around \$200 ($5 \times \$40 = \$200$).
Step 4 Find the actual answer by adding the earnings for the 5 days.

$$\begin{array}{rl} \$212 & \text{check by adding up} \\ \$\ 46 & \\ \$\ 36 & \\ \$\ 48 & \\ \$\ 50 & \\ +\ \$\ 32 & \\ \hline \$212 & \text{earnings for the week} \end{array}$$

This answer is reasonable and correct. ■

WORK PROBLEM 3 AT THE SIDE.

EXAMPLE 3 *Determining Whether Subtraction Is Necessary*
The number of students enrolled in Chabot College this year is 4084 fewer than the number enrolled last year. Enrollment last year was 21,382. Find the enrollment this year.

Approach To find the number of students enrolled this year, the enrollment decrease (fewer students) must be subtracted from last year.

Solution

Step 1 In this problem, the enrollment has decreased from last year to this year. The enrollment last year and the decrease in enrollment are given. This year's enrollment must be found.
Step 2 The word indicator, ***fewer,*** shows that subtraction must be used to find the number of students enrolled this year.
Step 3 Because the enrollment was about 21,000 students, and the decrease in enrollment is about 4000 students, a reasonable estimate would be 17,000 students ($21{,}000 - 4000 = 17{,}000$).
Step 4 Find the actual answer by subtracting 3684 from 21,382.

$$\begin{array}{r} 21{,}382 \\ -\ \ 4{,}084 \\ \hline 17{,}298 \end{array}$$

The enrollment this year is 17,298. The answer 17,298 is reasonable, as it is close to the estimate of 17,000. Check by adding.

$$\begin{array}{rl} 17{,}298 & \text{enrollment this year} \\ +\ \ 4{,}084 & \text{decrease in enrollment} \\ \hline 21{,}382 & \text{enrollment last year} \end{array}$$ ■

3. (a) During the semester, Cindy receives the following points on examinations and quizzes: 92, 81, 83, 98, 15, 14, 15, and 12. Find her total points for the semester.

(b) Stephanie Dixon works at the telephone order desk of a catalog sales company. One week she has the following number of customer contacts; Monday 78; Tuesday 64; Wednesday 118; Thursday 102; and Friday 196. Find her total number of customer contacts for the week.

ANSWERS
3. (a) 410 points
(b) 558 customer contacts

4. (a) One home occupies 1450 square feet, while an apartment occupies 980 square feet. Find the difference in the number of square feet of the two living units.

(b) A library had 25,622 books. After a loss of 1367 books, how many books were left?

5. (a) Gwen is paid $215 for each car that she sells. If she sells five cars and has $180 in sales expense, find the amount remaining after the expenses are deducted.

(b) During a 4-hour period, 125 cars enter a parking lot each hour. In the same time period, 271 cars leave the lot. Find the number of cars remaining in the lot.

ANSWERS
4. (a) 470 square feet (b) 24,255 books
5. (a) $895 (b) 229

WORK PROBLEM 4 AT THE SIDE.

EXAMPLE 4 *Solving a Two-Step Problem*

A landlord receives $680 from each of five tenants. After paying $1880 in expenses, how much rent money does the landlord have left?

Approach To find the amount remaining, first find the total rent received. Next, subtract the expenses paid to find the amount remaining.

Solution

Step 1 There are five tenants and each pays the same rent.

Step 2 The wording from *each* indicates that the five rents must be totaled. Since the rents are all the same, use multiplication to find the total rent received. Finally, subtract expenses.

Step 3 The amount of rent is about $700, making the total rent received about $3500 ($700 × 5). The expenses are about $1900. A reasonable estimate of the amount remaining is $1600 ($3500 − $1900).

Step 4 Find the exact amount by first multiplying $680 by 5 (the number of tenants).

$$\begin{array}{r} \$680 \\ \times \quad 5 \\ \hline \$3400 \end{array}$$

Finally, subtract the $1880 in expenses from $3400.

$$\$3400 - \$1880 = \$1520$$

The amount remaining is $1520.

The answer $1520 is reasonable, since it is close to the estimated answer of $1600. Check the amount by adding the expenses and then dividing by 5.

$$\$1520 + \$1880 = \$3400$$

$$5\overline{)3400} \quad \text{quotient } \$680$$ ■

WORK PROBLEM 5 AT THE SIDE.

≈ *Solve the following problems. First use front end rounding to estimate the answer. Then find the exact answer.*

1. The Top Side Cycling Club has 582 members in the amateur division, 208 members in the novice division, and 46 members in the professional division. Find the total number of members.

estimate: 600 + 200 + 50 = 850
exact: 836 members

2. Beth Anderson is applying for a home loan. The lender charges \$1950 in loan fees, \$175 for an appraisal, and \$360 in miscellaneous fees. Find the total lender charges.

estimate: 2000 + 200 + 400 = 2600
exact: \$2485

3. John and Will Kellogg invented the first cold flakes cereal in 1894. Of the total 200 different types of cold cereals produced today, a large supermarket decides to sell all but 62 types. How many types of cereal does the supermarket sell?

estimate: 200 − 60 = 140
exact: 138 sold

4. A truck weighs 9250 pounds when empty. After being loaded with firewood, it weighs 21,375 pounds. What is the weight of the firewood?

estimate: 20,000 − 9000 = 11,000
exact: 12,125 pounds

5. A packing machine can package 236 first-aid kits each hour. At this rate, find the number of first-aid kits packaged in 24 hours.

estimate: 200 × 20 = 4000
exact: 5664 kits

6. If 450 admission tickets to the Custom Car Show are sold each day, how many tickets are sold in a 14-day period?

estimate: 500 × 10 = 5000
exact: 6300 tickets

7. Ted Slauson, coordinator of Toys for Tots, has collected 2628 toys. If his group can give the same number of toys to each of 657 children, how many toys will each child receive?

estimate: 3000 ÷ 700 ≈ 4
exact: 4 toys

8. If profits of \$680,000 are divided evenly among a firm's 1000 employees, how much money will each employee receive?

estimate: 700,000 ÷ 1000 = 700
exact: \$680

9. The number of boaters and campers at the lake was 8392 on Friday. If this was 4218 more than the number of people at the lake on Wednesday, find the number of people at the lake on Wednesday.

estimate: 8000 − 4000 = 4000
exact: 4174 people

10. The community has raised \$52,882 for the homeless shelter. If the amount needed for the shelter is \$75,650, find the additional amount needed to be collected.

estimate: 80,000 − 50,000 = 30,000
exact: \$22,768

11. To qualify for a real estate loan at Uptown Bank, a borrower must have a monthly income of at least 4 times the monthly payment. For a monthly payment of $675, what must the borrower's minimum monthly income be?

estimate: 700 × 4 = 2800
exact: $2700

12. The cost of tuition and fees is $785 per quarter. If Gale Klein has five quarters of college remaining, find the total amount that she will need for tuition and fees.

estimate: 800 × 5 = 4000
exact: $3925

13. The total number of miles covered on a cross-country bicycle trip was 3150. If the trip took 18 days and the same number of miles was traveled each day, how many miles were traveled per day?

estimate: 3000 ÷ 20 = 150
exact: 175 miles

14. Erich Means completed a 2146-mile trip on his motorcycle and used 37 gallons of gasoline. How many miles did he travel on each gallon?

estimate: 2000 ÷ 40 = 50
exact: 58 miles

15. Dorene Cox decides to establish a budget. She will spend $450 for rent, $325 for food, $320 for child care, $182 for transportation, $150 for other expenses, and she will put the remainder in savings. If her monthly take-home pay is $1620, find her monthly savings.

estimate: 2000 − 500 − 300 − 300 − 200 − 200 = 500
exact: $193

16. Jared Ueda had $2874 in his checking account. He wrote checks for $308 for auto repairs, $580 for child support and $778 for an insurance payment. Find the amount remaining in his account.

estimate: 3000 − 300 − 600 − 800 = 1300
exact: $1208

17. There are 43,560 square feet in one acre. How many square feet are there in 138 acres?

estimate: 40,000 × 100 = 4,000,000
exact: 6,011,280 square feet

18. The number of gallons of water polluted each day in an industrial area is 209,670. How many gallons of water are polluted each year? (Use a 365-day year.)

estimate: 200,000 × 400 = 80,000,000
exact: 76,529,550 gallons

19. The Natural Chocolate Works melts 385 pounds of light chocolate, 100 pounds of dark chocolate, and 22 pounds of peanut butter. They add 18 pounds of almonds and 1 pound of confectioner's wax. What is the total weight of the candy made from these ingredients?

estimate: 400 + 100 + 20 + 20 + 1 = 541
exact: 526 pounds

20. Rent-a-Clunker Auto Rental owns 352 old and dented cars, 408 not so bad cars and 137 fairly acceptable cars. How many vehicles does it have in all?

estimate: 400 + 400 + 100 = 900
exact: 897 cars

21. The Enabling Supply House purchases 6 wheelchairs at $1256 each and 15 speech compression recorder-players at $895 each. Find the total cost.

estimate: $(1000 \times 6) + (900 \times 20) = 24{,}000$
exact: $20,961

22. Find the total cost if a college bookstore buys 17 computers at $506 each and 13 printers at $482 each.

estimate: $(500 \times 20) + (500 \times 10) = 15{,}000$
exact: $14,868

23. A group of college students on a field trip counted 18 endangered birds in one area, 137 in another and 263 in a third. How many endangered birds did the students count?

estimate: $20 + 100 + 300 = 420$
exact: 418 birds

24. Adrian is a waiter in a restaurant and earned tips of $32 on Monday, $47 on Wednesday and $106 on Saturday. Find the total tips earned.

estimate: $30 + 50 + 100 = 180$
exact: $185

25. Being able to identify indicator words is helpful in determining how to solve a word problem. Write three indicator words for each of these operations: add, subtract, multiply and divide. Write two indicator words that mean equals.

Answer varies.

26. Identify and explain the four steps used to solve a word problem. You may refer to the text if you need help.

Answer varies.

27. Write in your own words why it is important to estimate a reasonable answer. Give three examples of what might be a reasonable answer to a math problem in your daily activities.

Answer varies.

28. First estimate by rounding to thousands, then find the exact answer to the following problem.

$$7438 + 6493 + 2380$$

Do the two answers vary by more than 1000? Why? Will estimated answers always vary from exact answers?

Answer varies.

Solve the following problems.

29. A package of 3 undershirts costs $12, and a package of 6 pairs of socks costs $15. Find the total cost of 30 undershirts and 18 pairs of socks.

$165

30. In one week, Brian earned $8 per hour for 38 hours. Maria earned $9 per hour for working 39 hours. Find their total combined income.

$655

31. A car weighs 2425 pounds. If its 582-pound engine is removed and replaced with a 634-pound engine, find the weight of the car after the engine change.

2477 pounds

32. Barbara has $2324 in her preschool operating account. After spending $734 from this account the class parents raise $568 in a rummage sale. Find the balance in the account after depositing the money from the rummage sale.

$2158

33. Robert sold 80 orange trees for $10 each and 50 lemon trees for $20 each. Find his total sales.

$1800

34. Action Auto Supply purchased 70 cases of single grade oil for $9 per case and 60 cases of multi-grade oil for $12 per case. Find the total cost of the oil.

$1350

▲ **35.** Jim Peppa's vending machine company had 325 machines on hand at the beginning of the month. At different times during the month, machines were distributed to new locations; 35 machines were taken at one time, then 23 machines, and then 76 machines. During the same month additional machines were returned; 15 machines were returned at one time, then 38 machines, and then 108 machines. How many machines were on hand at the end of the month?

352 machines

▲ **36.** Mike Fitzgerald owns 70 acres of land that he leases to an alfalfa farmer for $150 per acre per year. If property taxes are $28 per acre per year, find the total amount of yearly lease income he has left after taxes are paid.

$8540

▲ **37.** A theater owner wants to provide enough seating for 1250 people. The main floor has 30 rows of 25 seats in each row. If the balcony has 25 rows, how many seats must be in each row to satisfy the owner's seating requirements?

20 seats

▲ **38.** Jennie makes 24 grapevine wreaths per week to sell to gift shops. She works 40 weeks a year and packages six wreaths per box. If she ships equal quantities to each of five shops, find the number of boxes each store will receive.

32 boxes

CHAPTER 1 SUMMARY

KEY TERMS

1.1	**whole numbers**	The whole numbers are 0, 1, 2, 3, 4, 5, 6, 7, 8, and so on.
1.2	**sum (total)**	The answer in an addition problem is called the sum (total).
	commutative property of addition	The commutative property of addition states that the order of numbers in an addition problem can be changed without changing the sum.
	carrying	The process of carrying is used in an addition problem when the sum of the digits in a column is greater than 9.
1.3	**difference**	The answer in a subtraction problem is called the difference.
	borrowing	The method of borrowing is used in subtraction if a digit is smaller than the one directly below.
1.4	**factors**	The numbers being multiplied are called factors. For example, in $3 \times 4 = 12$, both 3 and 4 are factors.
	product	The answer in a multiplication problem is called the product.
	commutative property of multiplication	The commutative property of multiplication states that the product in a multiplication problem remains the same when the order of the factors is changed.
	multiple	The product of two whole-number factors is a multiple of those numbers.
1.5	**quotient**	The answer in a division problem is called the quotient.
	remainder	The remainder is the number left over when two numbers do not divide exactly.
1.6	**long division**	The process of long division is used to divide by a number with more than one digit.
1.7	**rounding**	To find a number that is close to the original number, but easier to work with, we use rounding.
	front end rounding	Rounding a number so that only one nonzero digit remains. The front digit (left hand digit) is rounded and all other digits become zeros.
1.8	**square root**	The square root of a whole number is the number that can be multiplied by itself to produce the given (larger) number.
	perfect square	A number that is the square of a whole number is a perfect square.
1.9	**indicator words**	Words in a problem that indicate the necessary operations—either addition, subtraction, multiplication or division.

QUICK REVIEW

Concepts	Examples
1.1 Reading and Writing Whole Numbers Do not use the word "and" with a whole number. Commas help divide the periods for ones, thousands, millions, and billions. A comma is not needed with a number having four digits or fewer.	795 is written *seven hundred ninety-five.* 9,768,002 is written *nine million, seven hundred sixty-eight thousand, two.*

Concepts	Examples
1.2 Addition of Whole Numbers Add from top to bottom, starting with the ones column and working left. To check, add from bottom to top.	Problem (add up to check) $\begin{array}{r} 1140 \\ 687 \\ 26 \\ 9 \\ +\ 418 \\ \hline 1140 \end{array}$ 687, 26, 9, 418: addends; 1140: sum
1.2 Commutative Property of Addition The order of numbers in an addition problem can be changed without changing the sum.	$4 + 2 = 6$ $2 + 4 = 6$ By the commutative property, the sum is the same.
1.3 Subtraction of Whole Numbers Subtract the subtrahend from minuend to get the difference by borrowing when necessary. To check, add the difference to the subtrahend to get the minuend.	Problem $\begin{array}{r} \scriptstyle 6\ 12\,18 \\ 4738 \\ -\ 649 \\ \hline 4089 \end{array}$ 4738: minuend; 649: subtrahend; 4089: difference Check $\begin{array}{r} 4089 \\ +\ 649 \\ \hline 4738 \end{array}$
1.4 Multiplication of Whole Numbers The numbers being multiplied are called *factors.* The multiplicand is being multiplied by the multiplier, giving the product. When the multiplier has more than one digit, partial products must be used and added to find the product.	$\begin{array}{r} 78 \\ \times\ 24 \\ \hline 312 \\ 156\ \\ \hline 1872 \end{array}$ 78: multiplicand; 24: multiplier (78 and 24: factors) 312: partial product 156: partial product (one position left) 1872: product
1.4 Commutative Property of Multiplication The answer or product in multiplication remains the same when the order of the factors is changed.	$3 \times 4 = 12$ $4 \times 3 = 12$ By the commutative property, the product is the same.
1.5 Division of Whole Numbers ÷ and $\overline{)\ }$ mean divide. Also a —, as in $\frac{25}{5}$, means to divide the top number (dividend) by the bottom number (divisor).	$\begin{array}{r} 22 \\ 4\overline{)88} \\ 88 \\ \hline 0 \end{array}$ 4: divisor; 88: dividend; 22: quotient

Concepts	Examples
1.7 Rounding Whole Numbers	
Rules for rounding: **1.** Identify the position to be rounded, and draw a line after it. **2.** If the digit to the right is 5 or greater, increase by 1; if 4 or smaller, do not change. **3.** Change to zero all digits to the right of the place being rounded.	Round 72|6 to the nearest ten ↑ tens position; 5 or more, so add 1 to tens position. 726 rounds to 730. Round 1,|498, 586 to the nearest million. ↑ millions position; 4 or less, so do not change. 1,498,586 rounds to 1,000,000.
1.7 Front End Rounding	
Front end rounding leaves only the first digit as a nonzero digit. All other digits are changed to zero.	Round each of the following using front end rounding. 76 rounds to 80 348 rounds to 300 6512 rounds to 7000 23,751 rounds to 20,000 652,179 rounds to 700,000
1.8 Order of Operations	
Problems may have several operations. Work these problems with the following order of operations. **1.** Do all operations inside parentheses. **2.** Simplify any expressions with exponents and find any square roots. **3.** Multiply or divide from left to right. **4.** Add or subtract from left to right.	Solve, using the order of operations. $7 \cdot \sqrt{9} - 4 \cdot 5$ Find square root. $7 \cdot 3 - 4 \cdot 5$ Multiply from left to right. $21 - 20 = 1$ Subtract.

Concepts	Examples
1.9 Application Problems	
Follow these steps.	Manuel earns \$118 on Sunday, \$87 on Monday, and \$63 on Tuesday. Find total earnings for the three days. *Total* means to add.
1. Read the problem carefully, perhaps several times.	**1.** The earnings for each day are given, and the total for the 3 days must be found.
2. Work out a plan before starting.	**2.** Add the daily earnings to find the total.
3. Estimate a reasonable answer.	**3.** Since the earnings were about \$100 + \$90 + \$60 = \$250, a reasonable estimate would be approximately \$250.
4. Solve the problem. If the answer is reasonable, check; if not, start over.	**4.** $\begin{array}{r} \$268 \\ \hline \$118 \\ 87 \\ +\ 63 \\ \hline \$268 \end{array}$ check by adding up; total earnings Manuel's total earnings are \$268.

NAME DATE HOUR

CHAPTER 1 REVIEW EXERCISES

If you need help with any of these review exercises, look in the section indicated in brackets.

[1.1] *Fill in the digits for the given period in each of the following numbers.*

1. 2318
thousands 2
ones 318

2. 56,478
thousands 56
ones 478

3. 206,792
thousands 206
ones 792

4. 1,768,710,618
billions 1
millions 768
thousands 710
ones 618

Rewrite the following numbers in words.

5. 725 seven hundred twenty-five

6. 12,412 twelve thousand, four hundred twelve

7. 319,215 three hundred nineteen thousand, two hundred fifteen

8. 62,500,005 sixty-two million, five hundred thousand, five

Rewrite each of the following numbers in digits.

9. nine thousand, one hundred fifty 9150

10. two hundred million, four hundred fifty-five 200,000,455

[1.2] *Add.*

11. 63 + 19 = 82

12. 48 + 86 = 134

13. 914 + 3708 + 34 = 4656

14. 8215 + 9 + 7433 = 15,657

15. 1108 + 566 + 7201 + 304 = 9179

16. 187 + 5543 + 246 + 1003 = 6979

17. 5 732 + 11,069 + 37 + 1 595 + 22,169 = 40,602

18. 3 451 + 12,286 + 43 + 1 291 + 32,784 = 49,855

[1.3] *Subtract.*

19. $25 - 13$ 12

20. $64 - 32$ 32

21. $238 - 199$ 39

22. $573 - 389$ 184

23. $4380 - 577$ 3803

24. $5210 - 883$ 4327

25. $2210 - 1986$ 224

26. $99{,}704 - 73{,}838$ 25,866

[1.4] *Multiply.*

27. 6×6 36

28. 7×0 0

29. 3×5 15

30. 9×9 81

31. $(6)(6)$ 36

32. $(4)(9)$ 36

33. $7 \cdot 8$ 56

34. $9 \cdot 9$ 81

Work the following chain multiplications.

35. $2 \times 4 \times 6$ 48

36. $9 \times 1 \times 5$ 45

37. $4 \times 4 \times 3$ 48

38. $2 \times 2 \times 2$ 8

39. $(8)(0)(6)$ 0

40. $(8)(8)(1)$ 64

41. $6 \cdot 1 \cdot 8$ 48

42. $7 \cdot 7 \cdot 0$ 0

Multiply.

43. 34×3 102

44. 88×6 528

45. 36×5 180

46. 89×1 89

47. 639×6 3834

48. 781×7 5467

49. 1349×4 5396

50. 9163×5 45,815

51. 7259×2 14,518

52. 5440×6 32,640

53. $93{,}105 \times 5$ 465,525

54. $21{,}873 \times 8$ 174,984

55. 34×18
612

56. 52×36
1872

57. 98×12
1176

58. 68×75
5100

59. 655×21
13,755

60. 392×77
30,184

61. 4051×219
887,169

62. 1527×328
500,856

Find the total cost of each of the following.

63. 19 bicycle helmets at $42 per helmet
$798

64. 24 soccer balls at $13 per ball
$312

65. 278 batteries at $48 per battery
$13,344

66. 168 welders masks at $9 per mask
$1512

Multiply by using multiples of ten.

67. 250×70
17,500

68. 380×80
30,400

69. 324×600
194,400

70. 752×400
300,800

71. $16{,}000 \times 8\,000$
128,000,000

72. $43{,}000 \times 2\,100$
90,300,000

[1.5] *Divide whenever possible.*

73. $18 \div 3$
6

74. $35 \div 7$
5

75. $48 \div 8$
6

76. $36 \div 4$
9

77. $\frac{72}{8}$ 9

78. $\frac{36}{9}$ 4

79. $\frac{54}{6}$ 9

80. $\frac{0}{6}$ 0

81. $\frac{125}{0}$ meaningless

82. $\frac{0}{35}$ 0

83. $\frac{64}{8}$ 8

84. $\frac{81}{9}$ 9

[1.5–1.6] *Divide.*

85. $8\overline{)648}$ 81

86. $5\overline{)180}$ 36

87. $9\overline{)56,259}$ 6251

88. $76\overline{)26,752}$ 352

89. 2704 ÷ 18

150 R4

90. 15,525 ÷ 125

124 R25

[1.7] *Round as shown.*

91. 215 to the nearest ten

220

92. 18,602 to the nearest hundred

18,600

93. 19,721 to the nearest thousand

20,000

94. 67,485 to the nearest ten thousand

70,000

Round each of the following to the nearest ten, nearest hundred, and nearest thousand. Remember to round from the original number.

	ten	*hundred*	*thousand*
95. 1496	1500	1500	1000
96. 20,065	20,070	20,100	20,000
97. 98,201	98,200	98,200	98,000
98. 352,118	352,120	352,100	352,000

[1.8] *Find each square root by using the Perfect Squares Table on page 65.*

99. $\sqrt{25}$ 5

100. $\sqrt{144}$ 12

101. $\sqrt{81}$ 9

102. $\sqrt{196}$ 14

Identify the exponent and the base. Simplify each expression.

103. 2^3 3; 2; 8

104. 5^3 3; 5; 125

105. 2^5 5; 2; 32

106. 4^5 5; 4; 1024

Work each problem by using the order of operations.

107. $9^2 - 9$ 72

108. $3^2 - 5$ 4

109. $2 \cdot 3^2 \div 2$ 9

110. $9 \div 1 \cdot 2 \cdot 2 \div (11 - 2)$ 4

111. $\sqrt{9} + 2 \cdot 3$ 9

112. $6 \cdot \sqrt{16} - 6 \cdot \sqrt{9}$ 6

[1.9] ≈ *Solve each of the following word problems. First use front end rounding to estimate the answer. Then find the exact answer.*

113. Find the cost of 35 lawn chairs at $12 per chair.

estimate: $40 \times 10 = 400$
exact: $420

114. A pulley turns 1400 revolutions per minute. How many revolutions will the pulley turn in 60 minutes?

estimate: $1000 \times 60 = 60{,}000$
exact: 84,000 revolutions

115. A coffee pot makes 120 cups of coffee. Find the number of cups in 6 pots.

estimate: $100 \times 6 = 600$
exact: 720 cups

116. A keg contains 8000 nails. How many nails are in 40 kegs?

estimate: $8000 \times 40 = 320{,}000$
exact: 320,000 nails

117. It takes 2000 hours of work to build 1 home. How many hours of work are needed to build 12 homes?

estimate: $2000 \times 10 = 20{,}000$
exact: 24,000 hours

118. A train travels 80 miles in 1 hour. Find the number of miles traveled in 5 hours.

estimate: $80 \times 5 = 400$
exact: 400 miles

119. An amusement park charges $30 for each adult admission and $25 for each child. Find the total cost to admit a group of 20 adults and 38 children.

estimate: (30 × 20) + (30 × 40) = 1800
exact: $1550

120. A newspaper girl has 56 customers who take the paper daily and 23 customers who take the paper on weekends only. A daily customer pays $15 per month and a weekend-only customer pays $8 per month. Find the total monthly collections.

estimate: (60 × 20) + (20 × 10) = 1400
exact: $1024

121. Sam Lee had $1279 withheld from his paycheck last year for income tax. If he owed only $1080 in taxes, find the amount of refund he should receive.

estimate: 1000 − 1000 = 0 or 1300 − 1100 = 200
exact: $199

122. Susan Hessney has $382 in her checking account. She writes a check for $135. How much does she have left in her account?

estimate: 400 − 100 = 300
exact: $247

123. A food canner uses 1 pound of pork for every 175 cans of pork and beans. How many pounds of pork are needed for 8750 cans of pork and beans?

estimate: 9000 ÷ 200 = 45
exact: 50 pounds

124. A stamping machine produces 936 license plates each hour. How long will it take to produce 30,888 license plates?

estimate: 30,000 ÷ 900 ≈ 33
exact: 33 hours

125. If an acre needs 250 pounds of fertilizer, how many acres can be fertilized with 5750 pounds of fertilizer?

estimate: 6000 ÷ 300 = 20
exact: 23 acres

126. Each home in a subdivision requires 180 feet of fencing. Find the number of homes that can be fenced with 5760 feet of fencing material.

estimate: 6000 ÷ 200 = 30
exact: 32 homes

Mixed Review Exercises

Solve each of the following as indicated.

127. $\begin{array}{r} 64 \\ \times\ 5 \\ \hline 320 \end{array}$

128. $\begin{array}{r} 86 \\ \times\ 7 \\ \hline 602 \end{array}$

129. $\begin{array}{r} 179 \\ -\ 64 \\ \hline 115 \end{array}$

130. $\begin{array}{r} 286 \\ -\ 78 \\ \hline 208 \end{array}$

131. $\begin{array}{r} 662 \\ +\ 379 \\ \hline 1041 \end{array}$

132. $\begin{array}{r} 352 \\ +\ 678 \\ \hline 1030 \end{array}$

133. $\begin{array}{r} 38{,}140 \\ -\ 6\,078 \\ \hline 32{,}062 \end{array}$

134. $\begin{array}{r} 29{,}156 \\ -\ 4\,209 \\ \hline 24{,}947 \end{array}$

135. $21 \div 7$ 3

136. $\frac{42}{6}$ 7

137.
$$\begin{array}{r} 7\,218 \\ 3 \\ 18 \\ 1\,791 \\ 82{,}623 \\ +\ \ 1\,982 \\ \hline 93{,}635 \end{array}$$

138.
$$\begin{array}{r} 3\,812 \\ 5 \\ 22 \\ 1\,836 \\ 75{,}134 \\ +\ \ 2\,369 \\ \hline 83{,}178 \end{array}$$

139. $\frac{2}{0}$

meaningless

140. $\frac{5}{1}$ 5

141. $55{,}200 \div 4$

13,800

142. $49{,}509 \div 9$

5501

143.
$$\begin{array}{r} 8430 \\ \times\ \ 128 \\ \hline 1{,}079{,}040 \end{array}$$

144.
$$\begin{array}{r} 38{,}571 \\ \times\ \ \ \ 3 \\ \hline 115{,}713 \end{array}$$

145. $34\overline{)3672}$ 108

146. $68\overline{)14{,}076}$ 207

147. Rewrite 286,753 in words.

two hundred eighty-six thousand, seven hundred fifty-three

148. Rewrite 108,210 in words.

one hundred eight thousand, two hundred ten

149. Round 7245 to the nearest hundred.

7200

150. Round 500,196 to the nearest thousand.

500,000

Find each square root.

151. $\sqrt{36}$ 6

152. $\sqrt{100}$ 10

Find the total cost of each of the following.

153. 56 skateboards at $38 per board

$2128

154. 65 refrigerators at $520 per refrigerator.

$33,800

155. 185 team shirts at $12 per shirt

$2220

156. 607 boxes of avocados at $26 per box

$15,782

Solve each of the following word problems.

157. There are 52 cards in a deck. How many cards are there in 9 decks?

468 cards

158. Each member sells 15 magazine subscriptions. How many magazine subscriptions are sold by 600 members?

9000 subscriptions

159. John Miller travels 635 miles on a tank of Super Unleaded and 583 miles on a tank of Regular Unleaded. How many more miles does he travel on a tank of Super Unleaded?

52 miles

160. The Village School wants to raise $115,280 for a new library. If $87,340 has already been raised, how much more must be raised to reach the goal?

$27,940

161. Eight coworkers divided evenly the $182,000 won in a Super Lotto. How much did each receive?

$22,750

162. Building a lighted tennis court costs $26,950. If seven families divide the cost of a court equally, how much will each family pay?

$3850

163. If a family watches television 5 hours each day, how many hours of television will the family watch in 2 years of 365 days each?

3650 hours

164. An athlete trains 4 hours each day for 313 days of the year. Find the number of hours trained in a 2-year period.

2504 hours

NAME DATE HOUR

CHAPTER 1 TEST

Write the following numbers in words.

1. 5680

2. 52,008

3. Use digits to write "one hundred thirty-eight thousand, eight."

Add the following.

4.
$$\begin{array}{r} 834 \\ 72 \\ 5718 \\ +\ 9862 \\ \hline \end{array}$$

5.
$$\begin{array}{r} 17{,}063 \\ 7 \\ 12 \\ 1\ 505 \\ 93{,}710 \\ +\quad 333 \\ \hline \end{array}$$

Subtract.

6.
$$\begin{array}{r} 9001 \\ -\ 5936 \\ \hline \end{array}$$

7.
$$\begin{array}{r} 5062 \\ -\ 1978 \\ \hline \end{array}$$

Multiply.

8. $6 \times 5 \times 4$

9. $63 \cdot 2000$

10. $(85)(21)$

11.
$$\begin{array}{r} 7381 \\ \times\ \ 603 \\ \hline \end{array}$$

Divide whenever possible.

12. $16\overline{)123{,}952}$

13. $\dfrac{791}{0}$

14. $38{,}472 \div 84$

15. $280\overline{)44{,}800}$

Round as shown.

16. 3568 to the nearest ten

17. 98,587 to the nearest thousand

1. five thousand, six hundred eighty
2. fifty-two thousand, eight
3. 138,008
4. 16,486
5. 112,630
6. 3065
7. 3084
8. 120
9. 126,000
10. 1785
11. 4,450,743
12. 7747
13. meaningless
14. 458
15. 160
16. 3570
17. 99,000

Work each problem.

18. $6^2 + 8 + 7$

18. 51

19. $7 \cdot \sqrt{64} - 14 \cdot 2$

19. 28

Solve each of the following word problems. First use front end rounding to estimate the answer. Then find the exact answer.

20. The monthly rents collected from the four units in an apartment building are $485, $500, $515, and $425. After expenses of $785 are paid, find the amount that remains.

20. estimate: 500 + 500 + 500 + 400 − 800 = 1100
exact: $1140

21. A factory produces 288 automobiles each day. How many days would be needed to manufacture 35,424 autos?

21. estimate: $40{,}000 \div 300 \approx 133$
exact: 123 days

22. Kenée Shadbourne paid $690 for tuition, $185 on books, and $68 on supplies. If this money was withdrawn from her checking account, which had a balance of $1108, find her new balance.

22. estimate: 1000 − 700 − 200 − 70 = 30
exact: $165

23. An appliance manufacturer assembles 118 self-cleaning ovens each hour for 4 hours and 139 standard ovens each hour for the next 4 hours. Find the total number of ovens assembled in the 8-hour period.

23. estimate: $(100 \times 4) + (100 \times 4) = 800$
exact: 1028 ovens

24. Explain in your own words the rules for rounding numbers. Give an example of rounding a number to the nearest ten thousand.

24. Answer varies.

25. List and describe the four steps for solving word problems. Be sure to include estimating a reasonable answer and checking your work in your description.

25. Answer varies.

2 Multiplying and Dividing Fractions

Chapter 1 discussed whole numbers. Many times, however, we find that parts of whole numbers are considered. One way to write parts of a whole is with **fractions.** (Another way is with decimals, which is discussed in a later chapter.)

2.1 BASICS OF FRACTIONS

OBJECTIVES

1. Use a fraction to show which part of a whole is shaded.
2. Identify the numerator and denominator.
3. Identify proper and improper fractions.

FOR EXTRA HELP

Tape 2

SSM pp. 43–44

MAC: A
IBM: A

1 The number $\frac{1}{8}$ is a fraction that represents 1 of 8 equal parts. Read $\frac{1}{8}$ as "one eighth."

EXAMPLE 1 *Identifying Fractions*
Use a fraction to represent the shaded portions.

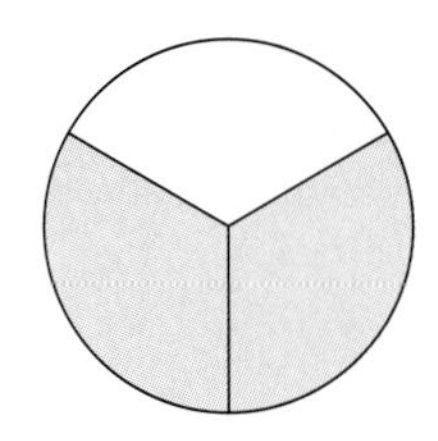

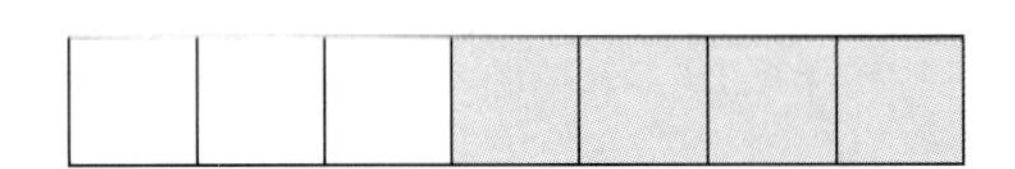

(a) The figure on the left has 3 equal parts. The 2 shaded parts are represented by the fraction $\frac{2}{3}$.

(b) The 4 shaded parts of the 7-part figure on the right are represented by the fraction $\frac{4}{7}$. ■

WORK PROBLEM 1 AT THE SIDE.

Fractions can be used to show more than one whole object.

1. Write fractions for the shaded portions.

(a)

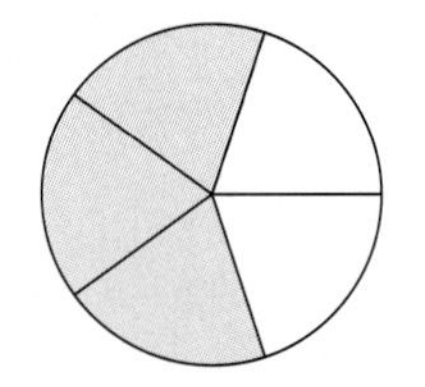

(b)

(c)

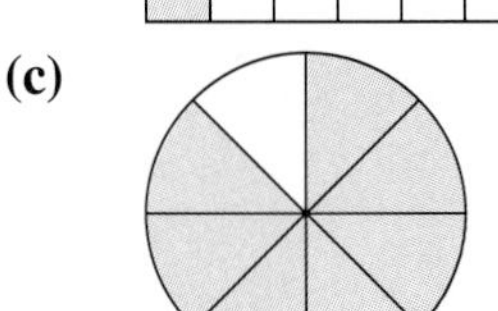

ANSWERS
1. (a) $\frac{3}{5}$ (b) $\frac{1}{6}$ (c) $\frac{7}{8}$

2. Write fractions for the shaded portions.

(a)

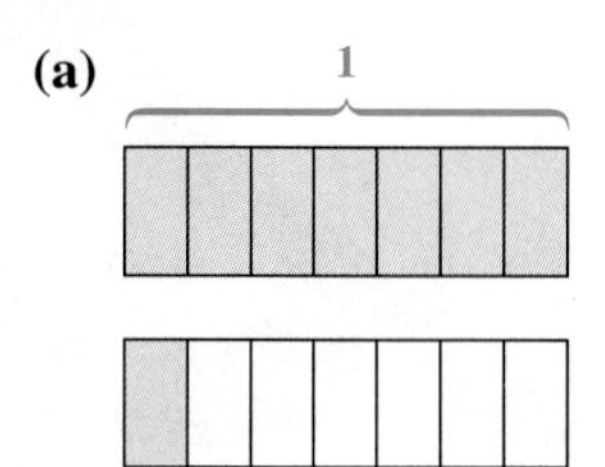

(b)

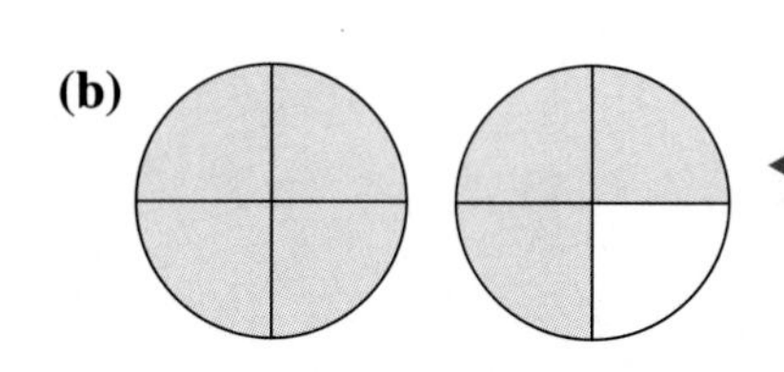

3. Identify the numerator and the denominator.

(a) $\frac{2}{3}$

(b) $\frac{1}{4}$

(c) $\frac{9}{7}$

(d) $\frac{106}{10}$

ANSWERS

2. (a) $\frac{8}{7}$ (b) $\frac{7}{4}$

3. (a) N: 2; D: 3
   (b) N: 1; D: 4
   (c) N: 9; D: 7
   (d) N: 106; D: 10

■ **EXAMPLE 2** *Representing Fractions Greater Than One*

Use a fraction to represent the shaded part.

(a)

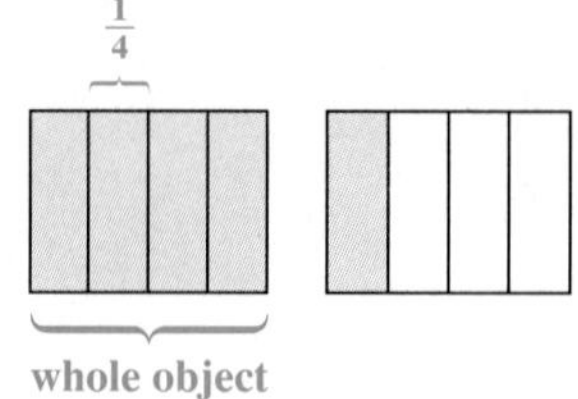

(b)

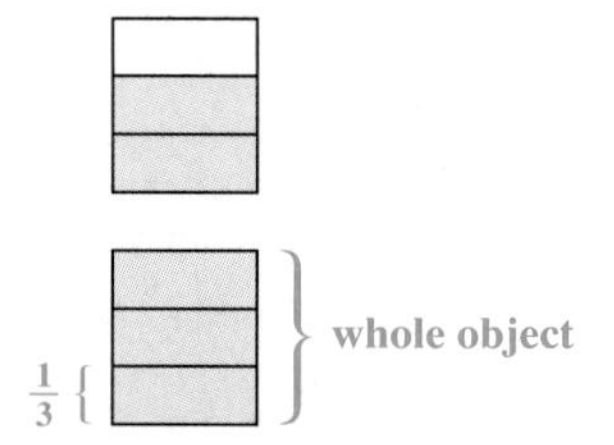

(a) An area equal to 5 of the $\frac{1}{4}$ parts is shaded. Write this as $\frac{5}{4}$.

(b) An area equal to 5 of the $\frac{1}{3}$ parts is shaded, so $\frac{5}{3}$ is shaded. ■

WORK PROBLEM 2 AT THE SIDE.

2 In the fraction $\frac{2}{3}$, the number 2 is the **numerator,** and 3 is the **denominator.** The bar between the numerator and the denominator is the *fraction bar.*

$$\text{fraction bar} \rightarrow \frac{2}{3} \begin{array}{l} \leftarrow \text{numerator} \\ \leftarrow \text{denominator} \end{array}$$

THE NUMERATOR AND DENOMINATOR

The denominator of a fraction shows the number of equivalent parts in the whole, and the numerator shows how many parts are being considered.

Note Recall that because division by 0 is meaningless, a fraction with a denominator of 0 is meaningless.

■ **EXAMPLE 3** *Identifying Numerator and Denominator*

Identify the numerator and denominator in each fraction.

(a) $\frac{5}{9}$ **(b)** $\frac{11}{7}$

(a) $\frac{5}{9} \begin{array}{l} \leftarrow \text{numerator} \\ \leftarrow \text{denominator} \end{array}$ **(b)** $\frac{11}{7} \begin{array}{l} \leftarrow \text{numerator} \\ \leftarrow \text{denominator} \end{array}$

WORK PROBLEM 3 AT THE SIDE.

3 Fractions are sometimes called *proper* or *improper* fractions.

PROPER AND IMPROPER FRACTIONS

If the numerator of a fraction is *smaller* than the denominator, the fraction is a **proper fraction.**
If the numerator is *greater than or equal to* the denominator, the fraction is an **improper fraction.**

proper fractions	*improper fractions*
$\frac{1}{2}, \frac{5}{11}, \frac{35}{36}$	$\frac{9}{7}, \frac{126}{125}, \frac{7}{7}$

■ **EXAMPLE 4** *Classifying Types of Fractions*

(a) Name all proper fractions in this list.

$$\frac{3}{4}, \frac{5}{9}, \frac{17}{5}, \frac{9}{7}, \frac{3}{3}, \frac{12}{25}, \frac{1}{9}, \frac{5}{3}$$

(b) Name all improper fractions in the list above.

(a) Proper fractions have a numerator that is smaller than the denominator. The proper fractions are:

$$\frac{3}{4} \leftarrow \text{3 is smaller than 4} \quad \frac{5}{9} \quad \frac{12}{25} \quad \text{and} \quad \frac{1}{9}$$

(b) Improper fractions have a numerator that is equal to or greater than the denominator. The improper fractions are:

$$\frac{17}{5} \leftarrow \text{17 is greater than 5} \quad \frac{9}{7} \quad \frac{3}{3} \quad \text{and} \quad \frac{5}{3}.$$ ■

WORK PROBLEM 4 AT THE SIDE.

4. From the following list of fractions:

$$\frac{3}{4}, \frac{8}{7}, \frac{5}{7}, \frac{6}{6}, \frac{1}{2}, \frac{2}{1}$$

(a) name all proper fractions

(b) name all improper fractions.

ANSWERS

4. (a) $\frac{3}{4}, \frac{5}{7}, \frac{1}{2}$

(b) $\frac{8}{7}, \frac{6}{6}, \frac{2}{1}$

QUEST FOR NUMERACY

Number Words

Use this chart to assign a value to each letter in the words you write.

A = 1	F = 6	K = 11	P = 16	U = 21	Z = 26
B = 2	G = 7	L = 12	Q = 17	V = 22	
C = 3	H = 8	M = 13	R = 18	W = 23	
D = 4	I = 9	N = 14	S = 19	X = 24	
E = 5	J = 10	O = 15	T = 20	Y = 25	

Example Write a three-letter word whose sum is greater than 45.

J O Y

↓ ↓ ↓

$10 + 15 + 25 = 50$

Do not use words that are proper names, slang, abbreviations, or from a foreign language.

1. Write a three-letter word whose sum is greater than 50.

Answer varies. Some possibilities are TOY, WIT, TWO, ZOO.

2. Write as many three-letter words as possible whose sums are less than 12.

Answer varies. Some possibilities are BAD, BAG, CAB.

3. Write a four-letter word whose sum is greater than 60.

Answer varies. Some possibilities are SPOT, STOP, TOOT, RUST.

4. What is the largest sum you can make with a four-letter word?

Answer varies. Try to top TOYS (79).

5. Write a five-letter word whose sum is between 65 and 75.

Answer varies. Some possibilities are PROUD (74), MONEY (72), PLAYS (73).

6. Write two four-letter words that have the same sum but do not use any of the same letters.

One possibility is FOUR and LIST. Both total 60.

7. Make up a problem for a friend to solve. Write it here.

Answer varies.

Extend this activity by having students exchange and solve the problems they make up. Or ask students to devise and use a different set of values for the letters of the alphabet, such as reversing it (Z = 1, A = 26) or using fractional values ($A = \frac{1}{8}$, $B = \frac{1}{4}$, $C = \frac{3}{8}$, etc.).

2.1 EXERCISES

NAME DATE HOUR

Write the fraction that represents the shaded area.

Example:

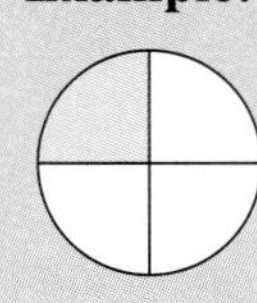

Solution:

$\frac{1}{4}$ (There are four parts, and one is shaded.)

1. 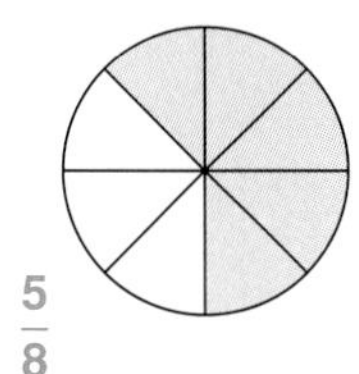

$\frac{5}{8}$

2. 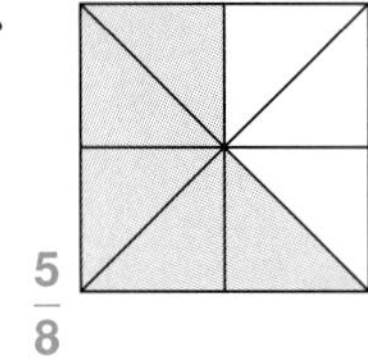

$\frac{5}{8}$

3. 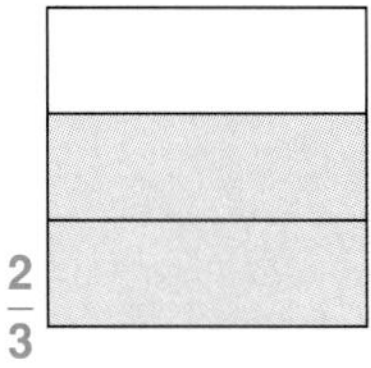

$\frac{2}{3}$

4. 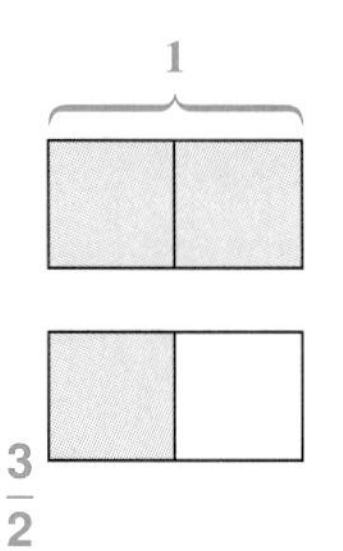

$\frac{3}{2}$

5. 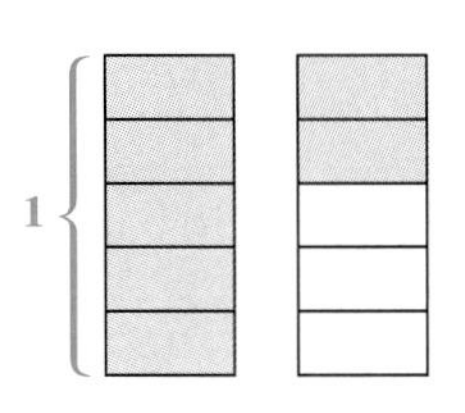

$\frac{7}{5}$

6. 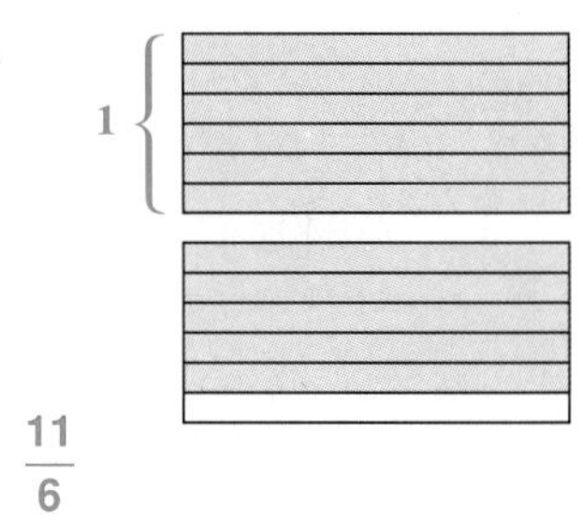

$\frac{11}{6}$

7. What fraction of these 11 coins are dimes?

$\frac{2}{11}$

8. What fraction of these 6 cups have handles?

$\frac{1}{6}$

9. In an American Sign Language (A.S.L.) class of 25 students, 8 are hearing impaired. What fraction of the students are hearing impaired?

$\frac{8}{25}$

10. Of 35 motor cycles in the parking lot, 17 are Harley Davidsons. What fraction of the motorcycles are *not* Harley Davidsons?

$\frac{18}{35}$

11. Of 71 cars making up a freight train, 58 are boxcars. What fraction of the cars are *not* boxcars?

$\frac{13}{71}$

12. On a basketball team of 10 players, 7 players are sophomores and the rest are freshmen. What fraction of the players are freshmen?

$\frac{3}{10}$

Identify the numerator and denominator.

Example: $\frac{9}{11}$ **Solution:** $\frac{\mathbf{9}}{\mathbf{11}}$ ← *numerator* (on top) ← *denominator* (on bottom)

	numerator	denominator
13. $\frac{3}{8}$	3	8
14. $\frac{9}{10}$	9	10
15. $\frac{12}{7}$	12	7
16. $\frac{8}{3}$	8	3

Name the proper and improper fractions in each of the following lists.

Example: $\frac{3}{8}, \frac{7}{4}, \frac{5}{6}, \frac{2}{3}, \frac{9}{4}$

Solution: *proper* $\frac{3}{8}, \frac{5}{6}, \frac{2}{3}$ *improper* $\frac{7}{4}, \frac{9}{4}$

	proper	improper
17. $\frac{9}{7}, \frac{1}{4}, \frac{3}{8}, \frac{7}{12}, \frac{11}{4}, \frac{8}{8}$	$\frac{1}{4}, \frac{3}{8}, \frac{7}{12}$	$\frac{9}{7}, \frac{11}{4}, \frac{8}{8}$
18. $\frac{1}{6}, \frac{5}{8}, \frac{15}{14}, \frac{11}{9}, \frac{7}{7}, \frac{3}{4}$	$\frac{1}{6}, \frac{5}{8}, \frac{3}{4}$	$\frac{15}{14}, \frac{11}{9}, \frac{7}{7}$
19. $\frac{3}{4}, \frac{3}{2}, \frac{5}{5}, \frac{9}{11}, \frac{7}{15}, \frac{19}{18}$	$\frac{3}{4}, \frac{9}{11}, \frac{7}{15}$	$\frac{3}{2}, \frac{5}{5}, \frac{19}{18}$
20. $\frac{12}{12}, \frac{15}{11}, \frac{13}{12}, \frac{11}{8}, \frac{17}{17}, \frac{19}{12}$	none	$\frac{12}{12}, \frac{15}{11}, \frac{13}{12}, \frac{11}{8}, \frac{17}{17}, \frac{19}{12}$

21. Write a fraction of your own choice. Label the parts of the fraction and write a sentence describing what each part represents. Answer varies.

22. Give one example of a proper fraction and one example of an improper fraction. What determines whether a fraction is proper or improper? Answer varies.

Complete the following sentences.

23. The fraction $\frac{9}{16}$ represents 9 of the 16 equal parts into which a whole is divided.

24. The fraction $\frac{23}{24}$ represents 23 of the 24 equal parts into which a whole is divided.

PREVIEW EXERCISES

Almost every exercise set in the rest of the book ends with a brief set of preview exercises. These exercises are designed to help you review ideas needed for the next few sections in the chapter. If you need help with these preview exercises, look in the chapter sections indicated.

Multiply each of the following. (For help, see **Section 1.4.***)*

25. $2 \times 3 \times 3$ 18 **26.** $3 \times 5 \times 2$ 30 **27.** $4 \cdot 3 \cdot 8$ 96 **28.** $2 \cdot 4 \cdot 24$ 192

Divide each of the following. (For help, see **Section 1.5.***)*

29. $21 \div 3$ 7 **30.** $56 \div 8$ 7 **31.** $209 \div 11$ 19 **32.** $115 \div 23$ 5

2.2 MIXED NUMBERS

1 When a fraction and a whole number are written together the result is a **mixed number.** For example, the mixed number

$$3\frac{1}{2} \quad \text{represents} \quad 3 + \frac{1}{2},$$

or 3 wholes and $\frac{1}{2}$ of a whole. Read $3\frac{1}{2}$ as "three and one half." As this figure shows, the mixed number $3\frac{1}{2}$ is equal to the improper fraction $\frac{7}{2}$.

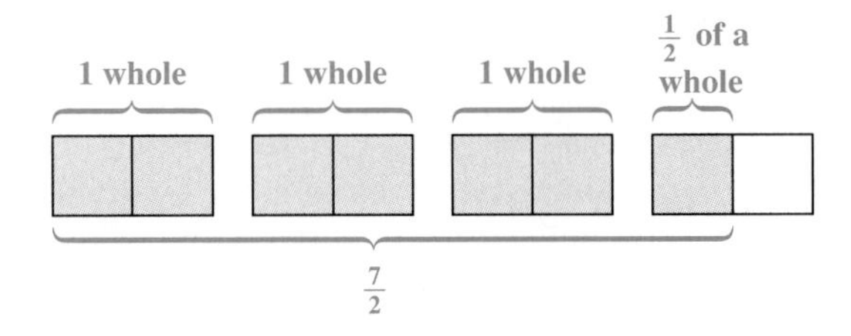

WORK PROBLEM 1 AT THE SIDE.

2 Use the following steps to write $3\frac{1}{2}$ as an improper fraction without drawing a figure.

Step 1 Multiply 3 and 2.

$$3\frac{1}{2} \quad 3 \cdot 2 = 6$$

Step 2 Add 1 to the product.

$$3\frac{1}{2} \quad 6 + 1 = 7$$

Step 3 Use 7, from Step 2, as the numerator and 2 as the denominator.

$$3\frac{1}{2} = \frac{7}{2}$$

same denominator

In summary, use the following steps to *write a mixed number as an improper fraction.*

WRITE A MIXED NUMBER AS AN IMPROPER FRACTION

Step 1 **Multiply** the denominator of the fraction and the whole number.
Step 2 **Add** to this product the numerator of the fraction.
Step 3 Write the result of Step 2 as the **numerator** and the original denominator as the **denominator.**

■ **EXAMPLE 1** *Changing Mixed Numbers to Improper Fractions*
Write $7\frac{2}{3}$ as an improper fraction (numerator greater than denominator).

Step 1 $7\frac{2}{3}$ $\quad 7 \cdot 3 = 21$ $\quad$ Multiply 7 and 3.

OBJECTIVES

1. Identify mixed numbers.
2. Write mixed numbers as improper fractions.
3. Write improper fractions as mixed numbers.

FOR EXTRA HELP

1. Use these diagrams to write $1\frac{2}{3}$ as an improper fraction.

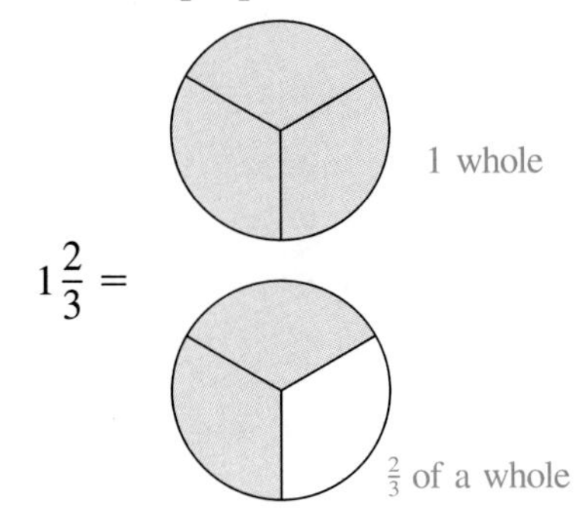

ANSWER

1. $\frac{5}{3}$

Step 2 $7\frac{2}{3}$ $21 + 2 = 23$ Add 2.

Step 3 $7\frac{2}{3} = \frac{23}{3}$ same denominator ■

WORK PROBLEM 2 AT THE SIDE.

3 Write an improper fraction as a mixed number as follows.

WRITE AN IMPROPER FRACTION AS A MIXED NUMBER

Write an **improper fraction** as a mixed number by dividing the numerator by the denominator. The quotient is the whole number (of the mixed number), the remainder is the numerator of the fraction part, and the denominator remains unchanged.

■ **EXAMPLE 2** *Changing Improper Fractions to Mixed Numbers*

Write as mixed numbers.

(a) $\frac{17}{5}$ **(b)** $\frac{24}{4}$

(a) Divide 17 by 5.

$$\begin{array}{r} 3 \\ 5\overline{)17} \\ \underline{15} \\ 2 \end{array} \leftarrow \text{remainder}$$

The quotient **3** is the whole number of the mixed number. The remainder **2** is the numerator of the fraction and the denominator remains as 5.

$$\frac{17}{5} = 3\frac{2}{5} \quad \begin{array}{l} \leftarrow \text{remainder} \\ \text{same denominator} \end{array}$$

$\frac{1}{5}$

$\frac{5}{5}$ = 1(whole) $\frac{5}{5}$ = 1(whole) $\frac{5}{5}$ = 1(whole) $\frac{2}{5}$

(b) Divide 24 by 4.

$$\begin{array}{r} 6 \\ 4\overline{)24} \\ \underline{24} \\ 0 \end{array} \quad \text{so} \quad \frac{24}{4} = 6$$

0 ← no remainder ■

Note A proper fraction has a value that is smaller than 1, while an improper fraction has a value that is 1 or greater.

WORK PROBLEM 3 AT THE SIDE.

2. Write as improper fractions.

(a) $2\frac{3}{4}$

(b) $3\frac{5}{8}$

(c) $5\frac{3}{4}$

(d) $8\frac{5}{6}$

3. Write as mixed numbers.

(a) $\frac{4}{3}$

(b) $\frac{18}{7}$

(c) $\frac{35}{5}$

(d) $\frac{75}{8}$

ANSWERS

2. (a) $\frac{11}{4}$ (b) $\frac{29}{8}$ (c) $\frac{23}{4}$ (d) $\frac{53}{6}$

3. (a) $1\frac{1}{3}$ (b) $2\frac{4}{7}$ (c) 7 (d) $9\frac{3}{8}$

2.2 NAME DATE HOUR

EXERCISES

Write each mixed number as an improper fraction.

Example: $6\frac{3}{7}$

Solution: $6 \cdot 7 = 42$ $\quad 42 + 3 = 45$ $\quad 6\frac{3}{7} = \mathbf{\frac{45}{7}}$

1. $1\frac{3}{4}$ $\frac{7}{4}$

2. $2\frac{1}{2}$ $\frac{5}{2}$

3. $3\frac{2}{3}$ $\frac{11}{3}$

4. $4\frac{1}{4}$ $\frac{17}{4}$

5. $3\frac{1}{4}$ $\frac{13}{4}$

6. $5\frac{1}{4}$ $\frac{21}{4}$

7. $6\frac{3}{4}$ $\frac{27}{4}$

8. $8\frac{1}{2}$ $\frac{17}{2}$

9. $1\frac{7}{11}$ $\frac{18}{11}$

10. $5\frac{4}{7}$ $\frac{39}{7}$

11. $6\frac{1}{3}$ $\frac{19}{3}$

12. $8\frac{2}{3}$ $\frac{26}{3}$

13. $11\frac{1}{3}$ $\frac{34}{3}$

14. $12\frac{2}{3}$ $\frac{38}{3}$

15. $10\frac{3}{4}$ $\frac{43}{4}$

16. $6\frac{1}{6}$ $\frac{37}{6}$

17. $3\frac{3}{8}$ $\frac{27}{8}$

18. $2\frac{8}{9}$ $\frac{26}{9}$

19. $8\frac{4}{5}$ $\frac{44}{5}$

20. $3\frac{4}{7}$ $\frac{25}{7}$

21. $4\frac{10}{11}$ $\frac{54}{11}$

22. $13\frac{5}{9}$ $\frac{122}{9}$

23. $22\frac{7}{8}$ $\frac{183}{8}$

24. $12\frac{9}{10}$ $\frac{129}{10}$

25. $17\frac{12}{13}$ $\frac{233}{13}$

26. $19\frac{8}{11}$ $\frac{217}{11}$

27. $17\frac{14}{15}$ $\frac{269}{15}$

28. $8\frac{17}{24}$ $\frac{209}{24}$

29. $6\frac{7}{18}$ $\frac{115}{18}$

30. $9\frac{7}{12}$ $\frac{115}{12}$

Write each improper fraction as a mixed number.

Example: $\frac{13}{5}$ **Solution:** $5\overline{)13}$ with quotient 2, 10, remainder 3; $\frac{13}{5} = 2\frac{3}{5}$

31. $\frac{9}{2}$ $4\frac{1}{2}$ **32.** $\frac{7}{5}$ $1\frac{2}{5}$ **33.** $\frac{8}{3}$ $2\frac{2}{3}$ **34.** $\frac{23}{10}$ $2\frac{3}{10}$ **35.** $\frac{77}{11}$ 7

36. $\frac{56}{7}$ 8 **37.** $\frac{27}{8}$ $3\frac{3}{8}$ **38.** $\frac{27}{7}$ $3\frac{6}{7}$ **39.** $\frac{19}{4}$ $4\frac{3}{4}$ **40.** $\frac{40}{9}$ $4\frac{4}{9}$

41. $\frac{27}{3}$ 9 **42.** $\frac{78}{6}$ 13 **43.** $\frac{58}{5}$ $11\frac{3}{5}$ **44.** $\frac{19}{5}$ $3\frac{4}{5}$ **45.** $\frac{47}{9}$ $5\frac{2}{9}$

46. $\frac{47}{9}$ $5\frac{2}{9}$ **47.** $\frac{50}{7}$ $7\frac{1}{7}$ **48.** $\frac{30}{7}$ $4\frac{2}{7}$ **49.** $\frac{84}{5}$ $16\frac{4}{5}$ **50.** $\frac{92}{3}$ $30\frac{2}{3}$

51. $\frac{123}{4}$ $30\frac{3}{4}$ **52.** $\frac{118}{5}$ $23\frac{3}{5}$ **53.** $\frac{183}{7}$ $26\frac{1}{7}$ **54.** $\frac{212}{11}$ $19\frac{3}{11}$

55. Your classmate asks you how to change a mixed number to an improper fraction. Write a couple of sentences and give an example showing how this is done.

Answer varies.

56. Explain in a sentence or two how to change an improper fraction to a mixed number. Give an example showing how this is done.

Answer varies.

Write each mixed number as an improper fraction.

▲ **57.** $255\frac{1}{8}$ $\frac{2041}{8}$ ▲ **58.** $218\frac{3}{5}$ $\frac{1093}{5}$ ▲ **59.** $333\frac{1}{3}$ $\frac{1000}{3}$

Write each improper fraction as a mixed number.

▲ **60.** $\frac{2017}{8}$ $252\frac{1}{8}$ * ▲ **61.** $\frac{2565}{15}$ 171 ▲ **62.** $\frac{2915}{16}$ $182\frac{3}{16}$

PREVIEW EXERCISES

Work each of the following problems. (For help, see **Section 1.8.***)*

63. $2^2 + 3^2$ 13 **64.** $2^2 \cdot 3^2$ 36 **65.** $15 \cdot 4 + 6$ 66

66. $5 \cdot 6 + 8$ 38 **67.** $6 \cdot 3^2 - 5$ 49 **68.** $5 \cdot 5^2 - 9$ 116

---

* Color exercise numbers are used to indicate exercises designed for calculator use.

2.3 FACTORS

OBJECTIVES

1. Find factors of a number.
2. Identify primes.
3. Find prime factorizations.

FOR EXTRA HELP

Tape 3 | SSM pp. 46–49 | MAC: A IBM: A

1 You will recall that numbers multiplied to give a product are called **factors.** Because $2 \cdot 5 = 10$, both 2 and 5 are factors of 10. The numbers 1 and 10 are also factors of 10, because

$$1 \cdot 10 = 10$$

The various tests for divisibility show 1, 2, 5, and 10 are the only whole-number factors of 10. The products $2 \cdot 5$ and $1 \cdot 10$ are called *factorizations* of 10.

> *Note* You might want to review the tests for divisibility in Section 1.5. The ones that you will want to remember are those for 2, 3, 5, and 10.

EXAMPLE 1 *Using Factors*
Find all possible two-number factorizations of each number.

(a) 12 **(b)** 60

(a) $1 \cdot 12$ $2 \cdot 6$ $3 \cdot 4$

The factors of 12 are 1, 2, 3, 4, 6, and 12.

(b) $1 \cdot 60$ $2 \cdot 30$
$3 \cdot 20$ $4 \cdot 15$
$5 \cdot 12$ $6 \cdot 10$

The factors of 60 are 1, 2, 3, 4, 5, 6, 10, 12, 15, 20, 30, and 60. ■

WORK PROBLEM 1 AT THE SIDE.

1. Find all the factors of the following numbers.

(a) 9

(b) 18

(c) 36

(d) 80

COMPOSITE NUMBERS

A number with a factor other than itself or 1 is called a **composite number.**

EXAMPLE 2 *Identifying Composite Numbers*
Which of the following numbers is (are) composite?

(a) 16
Because 16 has a factor of 8, a number other than 16 or 1, the number 16 is composite.

(b) 17
The number 17 has only two factors—17 and 1. It is not composite.

(c) 25
A factor of 25 is 5, so 25 is composite. ■

WORK PROBLEM 2 AT THE SIDE.

2. Which of these numbers is (are) composite?

2, 4, 5, 6, 8, 10, 11, 13, 19, 21, 27, 28, 33, 36, 42

2 Whole numbers that are not composite are called **prime numbers,** except 0 and 1, which are neither prime nor composite.

PRIME NUMBERS

A prime number is a whole number that has exactly two different factors, itself and 1.

ANSWERS
1. (a) 1, 3, 9 (b) 1, 2, 3, 6, 9, 18
(c) 1, 2, 3, 4, 6, 9, 12, 18, 36
(d) 1, 2, 4, 5, 8, 10, 16, 20, 40, 80
2. 4, 6, 8, 10, 21, 27, 28, 33, 36, 42

3. Which of the following are prime?

2, 3, 4, 7, 9, 13, 19, 29

4. Find the prime factorization of each number.

(a) 6

(b) 14

(c) 18

(d) 20

ANSWERS
3. 2, 3, 7, 13, 19, 29
4. (a) $2 \cdot 3$ (b) $2 \cdot 7$
(c) $2 \cdot 3 \cdot 3$ (d) $2 \cdot 2 \cdot 5$

The number 3 is a prime number, since it can be divided evenly only by itself and 1. The number 6 is not a prime number (it is composite), since 6 can be divided evenly by 2 and 3, as well as by itself and 1.

■ **EXAMPLE 3** *Finding Prime Numbers*
Which of the following numbers are prime?

2 5 8 11 15

The number 8 can be divided by 4 and 2, so it is not prime. Also, because 15 can be divided by 5 and 3, 15 is not prime. All the other numbers in the list are divisible by only themselves and 1, and are prime. ■

◀◀ **WORK PROBLEM 3 AT THE SIDE.**

3 For reference, here are the primes smaller than 100.

2, 3, 5, 7, 11,
13, 17, 19, 23, 29,
31, 37, 41, 43, 47,
53, 59, 61, 67, 71,
73, 79, 83, 89, 97,

Note All prime numbers are odd numbers except the number 2. Be careful though, because all odd numbers are not prime numbers. For example, 9, 15, and 21 are odd numbers but are *not* prime numbers.

The *prime factorization* of a number is especially useful.

PRIME FACTORIZATION

A **prime factorization** of a number is a factorization in which every factor is a prime number.

■ **EXAMPLE 4** *Determining the Prime Factorization*
Find the prime factorization of 12.

Try to divide 12 by the first prime, 2.

$$12 \div 2 = 6,$$
(↑ first prime)

so

$$12 = 2 \cdot 6.$$

Try to divide 6 by the prime, 2.

$$6 \div 2 = 3,$$

so

$$12 = 2 \cdot \underbrace{2 \cdot 3}_{\text{factorization of 6}}.$$

Because all factors are prime, the prime factorization of 12 is

$$2 \cdot 2 \cdot 3.$$ ■

◀◀ **WORK PROBLEM 4 AT THE SIDE.**

■ **EXAMPLE 5** *Factoring by Using the Division Method*
Find the prime factorization of 48.

all prime factors	$2\overline{)48}$	Divide 48 by 2 (first prime).
	$2\overline{)24}$	Divide 24 by 2.
	$2\overline{)12}$	Divide 12 by 2.
	$2\overline{)6}$	Divide 6 by 2.
	$3\overline{)3}$	Divide 3 by 3.
	1	Continue to divide until the quotient is 1.

Because all factors (divisors) are prime, the prime factorization of 48 is

$$2 \cdot 2 \cdot 2 \cdot 2 \cdot 3.$$

As shown in Chapter 1, $2 \cdot 2 = 2^2$, so the prime factorization of 48 can can be written, using exponents, as

$$2 \cdot 2 \cdot 2 \cdot 2 \cdot 3 = 2^4 \cdot 3.$$ ■

WORK PROBLEM 5 AT THE SIDE.

Note When using the division method of factoring, the last quotient found is 1. The "1" is never used as a prime factor because 1 is neither prime nor composite. Besides, 1 times any number is the number itself.

■ **EXAMPLE 6** *Using Exponents with Prime Factorization*
Find the prime factorization of 225.

all prime factors	$3\overline{)225}$	225 is not divisible by 2; use 3.
	$3\overline{)75}$	Divide 75 by 3.
	$5\overline{)25}$	25 is not divisible by 3; use 5.
	$5\overline{)5}$	Divide by 5.
	1	

Write the prime factorization,

$$3 \cdot 3 \cdot 5 \cdot 5$$

with exponents, as

$$3^2 \cdot 5^2$$ ■

WORK PROBLEM 6 AT THE SIDE.

Another method of factoring uses what is called a factor tree.

■ **EXAMPLE 7** *Factoring by Using a Factor Tree*
Find the prime factorization of each number.
(a) 30 **(b)** 24 **(c)** 45

(a) Try to divide by the first prime, 2. Write the factors under the 30. Circle the 2, since it is a prime.

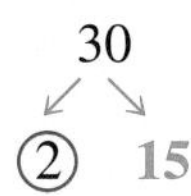

5. Find the prime factorization of each number. Write the factorizations with exponents.

(a) 24

(b) 36

(c) 60

(d) 126

6. Write the prime factorization of each number by using exponents.

(a) 50

(b) 88

(c) 90

(d) 150

(e) 280

ANSWERS
5. (a) $2^3 \cdot 3$ (b) $2^2 \cdot 3^2$ (c) $2^2 \cdot 3 \cdot 5$ (d) $2 \cdot 3^2 \cdot 7$
6. (a) $2 \cdot 5^2$ (b) $2^3 \cdot 11$ (c) $2 \cdot 3^2 \cdot 5$ (d) $2 \cdot 3 \cdot 5^2$ (e) $2^3 \cdot 5 \cdot 7$

7. Complete each factor tree and give the prime factorization.

(a)

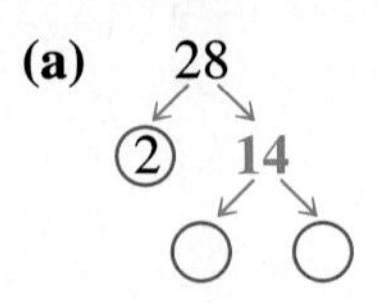

(b)

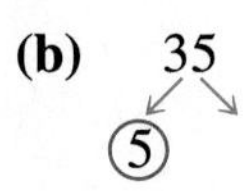

(c) 90

ANSWERS

7. (a) 28 → ② 14 → ② ⑦

$28 = 2 \cdot 2 \cdot 7$

(b) 35 → ⑤ ⑦

$35 = 5 \cdot 7$

(c) 90 → ② 45 → ③ 15 → ③ ⑤

$90 = 2 \cdot 3 \cdot 3 \cdot 5$

Since 15 cannot be divided by 2 (with remainder of 0), try the next prime, 3.

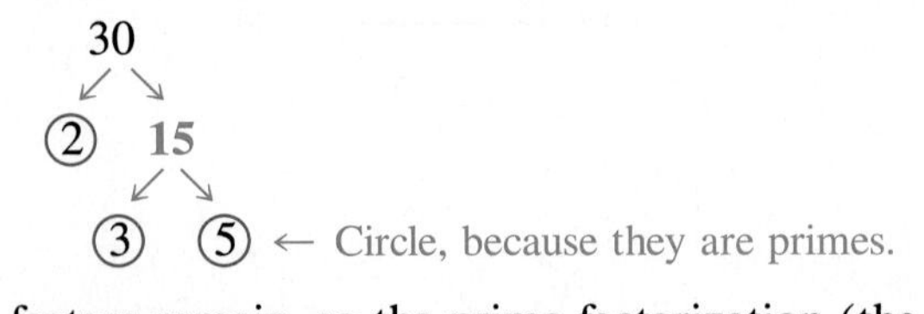

No uncircled factors remain, so the prime factorization (the circled factors) has been found.

$$2 \cdot 3 \cdot 5$$

(b) Divide by 2.

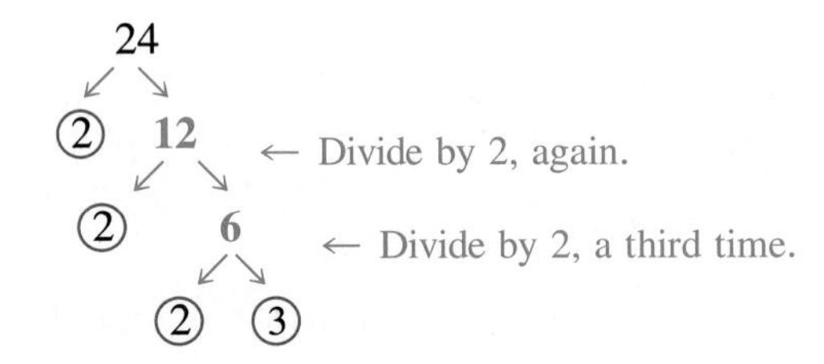

$2 \cdot 2 \cdot 2 \cdot 3$ with exponents is $2^3 \cdot 3$.

(c) Because 45 cannot be divided by 2, try 3.

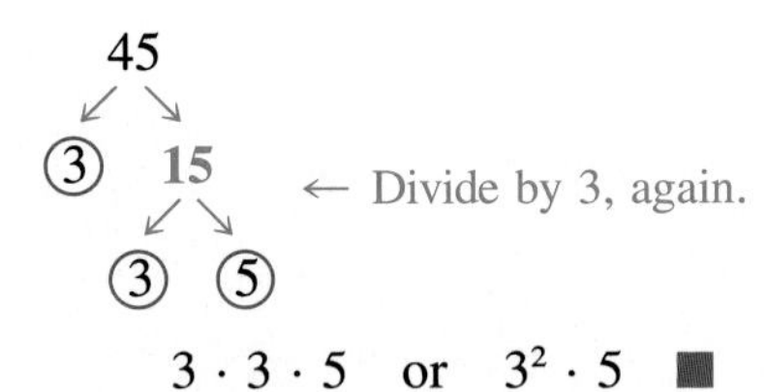

$$3 \cdot 3 \cdot 5 \quad \text{or} \quad 3^2 \cdot 5 \quad ■$$

Note The diagrams used in Example 7 look like trees, and that is why this method is referred to as factoring by using a factor tree.

◀ WORK PROBLEM 7 AT THE SIDE.

Find all the factors of each number.

Example: 18

Solution:

Write all factorizations of 18 that have two factors.

$1 \cdot 18 \quad 2 \cdot 9 \quad 3 \cdot 6$

The factors of 18 are **1, 2, 3, 6, 9,** and **18.**

1. 6

1, 2, 3, 6

2. 10

1, 2, 5, 10

3. 8

1, 2, 4, 8

4. 24

1, 2, 3, 4, 6, 8, 12, 24

5. 25

1, 5, 25

6. 30

1, 2, 3, 5, 6, 10, 15, 30

7. 18

1, 2, 3, 6, 9, 18

8. 20

1, 2, 4, 5, 10, 20

9. 40

1, 2, 4, 5, 8, 10, 20, 40

10. 60

1, 2, 3, 4, 5, 6, 10, 12, 15, 20, 30, 60

11. 64

1, 2, 4, 8, 16, 32, 64

12. 80

1, 2, 4, 5, 8, 10, 16, 20, 40, 80

Which numbers are prime and which are composite?

Example: 5, 12

Solution:

Because it can be divided by only itself and 1, *5 is prime.* Because it can be divided by 2 and 3, *12 is composite.*

13. 4

composite

14. 10

composite

15. 2

prime

16. 5

prime

17. 9

composite

18. 15

composite

19. 13

prime

20. 21

composite

21. 19

prime

22. 23

prime

23. 25

composite

24. 26

composite

25. 34

composite

26. 43

prime

27. 45

composite

28. 47

prime

Find the prime factorization of the following numbers. Write answers with exponents when repeated factors appear.

Example: 40

Solution:

Division method

$2\overline{)40}$

$2\overline{)20}$

$2\overline{)10}$

$5\overline{)5}$

1

$2 \cdot 2 \cdot 2 \cdot 5 = \mathbf{2^3 \cdot 5}$

Factor tree

40 ← Divide by 2; circle 2, since it is prime.

② 20 ← Divide by 2, again.

② 10 ← Divide by 2, a third time.

② ⑤

29. 4

2^2

30. 8

2^3

31. 15

$3 \cdot 5$

32. 30

$2 \cdot 3 \cdot 5$

33. 25

5^2

34. 18

$2 \cdot 3^2$

35. 32

2^5

36. 52

$2^2 \cdot 13$

37. 39

$3 \cdot 13$

38. 50

$2 \cdot 5^2$

39. 88

$2^3 \cdot 11$

40. 64

2^6

41. 75

$3 \cdot 5^2$

42. 80

$2^4 \cdot 5$

43. 100

$2^2 \cdot 5^2$

44. 120

$2^3 \cdot 3 \cdot 5$

45. 125

5^3

46. 180

$2^2 \cdot 3^2 \cdot 5$

47. 225

$3^2 \cdot 5^2$

48. 300

$2^2 \cdot 3 \cdot 5^2$

49. 320

$2^6 \cdot 5$

50. 340

$2^3 \cdot 5 \cdot 17$

51. 360

$2^3 \cdot 3^2 \cdot 5$

52. 400

$2^4 \cdot 5^2$

53. Give a definition in your own words of both a composite number and a prime number. Give three examples of each.

Answer varies.

54. With the exception of the number two, all prime numbers are odd numbers. Nevertheless, all odd numbers are not prime numbers. Explain why these statements are true.

Answer varies.

As a review, solve each of the following.

Examples: 2^4, 4^3, $2^2 \cdot 3^4$

$2^4 = 2 \cdot 2 \cdot 2 \cdot 2 \leftarrow$ 4 factors of 2
$= \mathbf{16}$

$4^3 = 4 \cdot 4 \cdot 4 \leftarrow$ 3 factors of 4
$= \mathbf{64}$

$2^2 \cdot 3^4 = 2 \cdot 2 \cdot 3 \cdot 3 \cdot 3 \cdot 3$
$= 4 \cdot 81$
$= \mathbf{324}$

55. 4^3 64

56. 5^3 125

57. 8^3 512

58. 6^3 216

59. 6^4 1296

60. 7^4 2401

61. $2^2 \cdot 3^3$ 108

62. $2^4 \cdot 3^2$ 144

63. $5^3 \cdot 3^2$ 1125

64. $2^4 \cdot 5^2$ 400

65. $3^5 \cdot 2^2$ 972

66. $5^2 \cdot 4^3$ 1600

Find the prime factorization of each number. Write answers by using exponents.

▲ **67.** 280 $2^3 \cdot 5 \cdot 7$

▲ **68.** 520 $2^3 \cdot 5 \cdot 13$

▲ **69.** 960 $2^6 \cdot 3 \cdot 5$

▲ **70.** 1125 $3^2 \cdot 5^3$

▲ **71.** 1600 $2^6 \cdot 5^2$

▲ **72.** 1575 $3^2 \cdot 5^2 \cdot 7$

Preview Exercises

Multiply each of the following. (For help, see ***Section 1.4.****)*

73. $5 \cdot 1 \cdot 2$ 10

74. $3 \cdot 2 \cdot 4$ 24

75. $5 \cdot 2 \cdot 70$ 700

76. $3 \cdot 5 \cdot 9$ 135

Divide each of the following. (For help, see ***Section 1.5.****)*

77. $24 \div 3$ 8

78. $27 \div 9$ 3

79. $135 \div 5$ 27

80. $45 \div 15$ 3

2.4 WRITING A FRACTION IN LOWEST TERMS

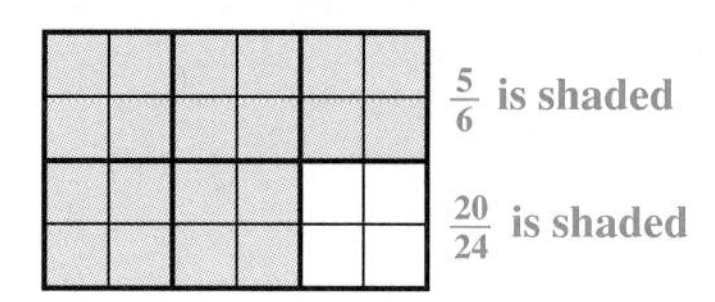

The figure shows areas that are $\frac{5}{6}$ shaded and $\frac{20}{24}$ shaded. Because the shaded areas are equivalent, the fractions $\frac{5}{6}$ and $\frac{20}{24}$ are **equivalent fractions.**

$$\frac{5}{6} = \frac{20}{24}$$

Because the numbers 20 and 24 both have 4 as a factor, 4 is called a **common factor** of the numbers. Other common factors of 20 and 24 are 1 and 2.

WORK PROBLEM 1 AT THE SIDE.

1 The fraction $\frac{5}{6}$ is in lowest terms because the numerator and denominator have no common factor other than 1; however, the fraction $\frac{20}{24}$ is *not* in lowest terms because its numerator and denominator have a common factor of 4.

WRITING A FRACTION IN LOWEST TERMS

A fraction is written in ***lowest terms*** when the numerator and the denominator have no common factor other than 1.

EXAMPLE 1 *Understanding Lowest Terms*

Are the following fractions in lowest terms?

(a) $\frac{3}{8}$

The numerator and denominator have no common factor other than 1, so the fraction is in lowest terms.

(b) $\frac{21}{36}$

The numerator and denominator have a common factor of 3, so the fraction is not in lowest terms. ■

WORK PROBLEM 2 AT THE SIDE.

2 There are two common methods for writing a fraction in lowest terms. These methods are shown in the next examples. The first method works best when the numerator and denominator are small numbers.

OBJECTIVES

1. Tell whether a fraction is written in lowest terms.
2. Write a fraction in lowest terms using common factors.
3. Write a fraction in lowest terms using prime factors.
4. Tell whether two fractions are equivalent.

FOR EXTRA HELP

Tape 3

SSM pp. 49–51

MAC: A IBM: A

1. Decide whether the given factor is a common factor of the given numbers.

(a) 14, 20; 2

(b) 32, 48; 16

(c) 24, 36; 8

(d) 56, 73; 1

2. Are the following fractions in lowest terms?

(a) $\frac{2}{3}$ **(b)** $\frac{3}{12}$

(c) $\frac{11}{15}$ **(d)** $\frac{15}{51}$

ANSWERS
1. (a) yes (b) yes (c) no (d) yes
2. (a) yes (b) no (c) yes (d) no

3. Write in lowest terms.

(a) $\frac{3}{6}$

(b) $\frac{6}{8}$

(c) $\frac{24}{30}$

(d) $\frac{15}{40}$

(e) $\frac{32}{80}$

ANSWERS

3. (a) $\frac{1}{2}$ (b) $\frac{3}{4}$ (c) $\frac{4}{5}$ (d) $\frac{3}{8}$ (e) $\frac{2}{5}$

■ **EXAMPLE 2** *Changing to Lowest Terms*

Write each fraction in lowest terms.

(a) $\frac{20}{24}$ **(b)** $\frac{30}{50}$ **(c)** $\frac{24}{42}$ **(d)** $\frac{60}{72}$

(a) The largest common factor of 20 and 24 is 4. Divide both numerator and denominator by **4.**

$$\frac{20}{24} = \frac{20 \div 4}{24 \div 4} = \frac{5}{6}$$

(b) The largest common factor of 30 and 50 is 10.

$$\frac{30}{50} = \frac{30 \div 10}{50 \div 10} \quad \text{Divide both numerator and denominator by 10.}$$

$$= \frac{3}{5}$$

(c) $\frac{24}{42} = \frac{24 \div 6}{42 \div 6} = \frac{4}{7}$

(d) Suppose we made an error and thought 4 was the largest common factor of 60 and 72. Dividing by 4 would give

$$\frac{60}{72} = \frac{60 \div 4}{72 \div 4} = \frac{15}{18}.$$

But $\frac{15}{18}$ is not in lowest terms, because 15 and 18 have a common factor of 3. Divide by 3.

$$\frac{15}{18} = \frac{15 \div 3}{18 \div 3} = \frac{5}{6}$$

The fraction $\frac{60}{72}$ could have been written in lowest terms in one step by dividing by 12, the largest common factor of 60 and 72.

$$\frac{60}{72} = \frac{60 \div 12}{72 \div 12} = \frac{5}{6} \quad ■$$

Note Dividing the numerator and denominator by the same number results in an equivalent fraction.

WORK PROBLEM 3 AT THE SIDE.

3 The method of writing a fraction in lowest terms by division works well for fractions with small numerators and denominators. For larger numbers, it is better to use the method of **prime factors,** which is shown in the next example.

■ **EXAMPLE 3** *Using Prime Factors*

Write each of the following in lowest terms.

(a) $\frac{24}{42}$ **(b)** $\frac{180}{54}$ **(c)** $\frac{54}{90}$

(a) Write the prime factorization of both numerator and denominator. Use one of the methods in **Section 2.3.**

$$\frac{24}{42} = \frac{2 \cdot 2 \cdot 2 \cdot 3}{2 \cdot 3 \cdot 7}$$

Just as with the other method, divide numerator and denominator by any common factors. Use a shortcut called **cancellation** to show this division. Place a **1** by each factor that is canceled.

$$\frac{24}{42} = \frac{\overset{1}{\cancel{2}} \cdot 2 \cdot 2 \cdot \overset{1}{\cancel{3}}}{\underset{1}{\cancel{2}} \cdot \underset{1}{\cancel{3}} \cdot 7}$$

Multiply the remaining factors in both numerator and denominator.

$$\frac{24}{42} = \frac{1 \cdot 2 \cdot 2 \cdot 1}{1 \cdot 1 \cdot 7} = \frac{4}{7}$$

Finally, $\frac{24}{42}$, written in lowest terms, is $\frac{4}{7}$.

(b) Write the prime factorization of both numerator and denominator.

$$\frac{180}{54} = \frac{2 \cdot 2 \cdot 3 \cdot 3 \cdot 5}{2 \cdot 3 \cdot 3 \cdot 3}$$

Now cancel the common factors. **Do not forget the 1's.**

$$\frac{180}{54} = \frac{\overset{1}{\cancel{2}} \cdot 2 \cdot \overset{1}{\cancel{3}} \cdot \overset{1}{\cancel{3}} \cdot 5}{\underset{1}{\cancel{2}} \cdot \underset{1}{\cancel{3}} \cdot \underset{1}{\cancel{3}} \cdot 3}$$

$$= \frac{1 \cdot 2 \cdot 1 \cdot 1 \cdot 5}{1 \cdot 1 \cdot 1 \cdot 3} = \frac{10}{3} \quad \text{or} \quad 3\frac{1}{3}$$

(c) $\frac{54}{90} = \frac{\overset{1}{\cancel{2}} \cdot \overset{1}{\cancel{3}} \cdot \overset{1}{\cancel{3}} \cdot 3}{\underset{1}{\cancel{2}} \cdot \underset{1}{\cancel{3}} \cdot \underset{1}{\cancel{3}} \cdot 5} = \frac{1 \cdot 1 \cdot 1 \cdot 3}{1 \cdot 1 \cdot 1 \cdot 5} = \frac{3}{5}$ ■

WORK PROBLEM 4 AT THE SIDE. ▶▶

This method of writing a fraction in lowest terms is summarized below.

THE METHOD OF PRIME FACTORS

Step 1 Write the **prime factorization** of both numerator and denominator.
Step 2 Use **cancellation** to divide numerator and denominator by any common factors.
Step 3 **Multiply** the remaining factors in numerator and denominator.

4 The next example shows how to use the *equivalency test* to tell whether two fractions are equivalent.

4. Use the method of prime factors to write each fraction in lowest terms.

(a) $\frac{12}{36}$

(b) $\frac{28}{60}$

(c) $\frac{74}{111}$

(d) $\frac{124}{340}$

ANSWERS
4. (a) $\frac{1}{3}$ (b) $\frac{7}{15}$ (c) $\frac{2}{3}$ (d) $\frac{31}{85}$

5. Are the following fractions equivalent?

(a) $\frac{2}{3}$ and $\frac{3}{4}$

(b) $\frac{10}{16}$ and $\frac{15}{24}$

(c) $\frac{6}{50}$ and $\frac{9}{75}$

(d) $\frac{12}{22}$ and $\frac{18}{32}$

ANSWERS

5. (a) not equivalent (b) equivalent (c) equivalent (d) not equivalent

EXAMPLE 4 *Using the Equivalency Test ("Cross Multiplication")*

Are the following fractions equivalent?

(a) $\frac{6}{15}$ and $\frac{8}{20}$ **(b)** $\frac{8}{12}$ and $\frac{21}{30}$

(a) Find each *cross product.*

$\frac{6}{15}$ $\frac{8}{20}$

$15 \cdot 8 = 120$

$6 \cdot 20 = 120$

equivalent $(120 = 120)$

Since the cross products are equal (=), the fractions are equivalent.

(b) Find the cross products.

$\frac{8}{12}$ $\frac{21}{30}$

$12 \cdot 21 = 252$

$8 \cdot 30 = 240$

not equivalent $(252 \neq 240)$

The cross products are *not equal* (≠), so the fractions are *not equivalent.* ■

Note Cross multiply to determine whether two fractions are *equivalent.* This is *not* used to multiply or divide fractions.

WORK PROBLEM 5 AT THE SIDE.

2.4 EXERCISES

NAME DATE HOUR

Write each fraction in lowest terms.

Example: $\frac{12}{15}$ **Solution:** $\frac{12}{15} = \frac{12 \div 3}{15 \div 3} = \mathbf{\frac{4}{5}}$

1. $\frac{10}{20}$ $\frac{1}{2}$ **2.** $\frac{9}{12}$ $\frac{3}{4}$ **3.** $\frac{15}{20}$ $\frac{3}{4}$ **4.** $\frac{12}{36}$ $\frac{1}{3}$

5. $\frac{30}{48}$ $\frac{5}{8}$ **6.** $\frac{16}{24}$ $\frac{2}{3}$ **7.** $\frac{36}{42}$ $\frac{6}{7}$ **8.** $\frac{22}{33}$ $\frac{2}{3}$

9. $\frac{63}{70}$ $\frac{9}{10}$ **10.** $\frac{27}{45}$ $\frac{3}{5}$ **11.** $\frac{180}{210}$ $\frac{6}{7}$ **12.** $\frac{72}{80}$ $\frac{9}{10}$

13. $\frac{36}{63}$ $\frac{4}{7}$ **14.** $\frac{73}{146}$ $\frac{1}{2}$ **15.** $\frac{12}{600}$ $\frac{1}{50}$ **16.** $\frac{54}{90}$ $\frac{3}{5}$

17. $\frac{96}{132}$ $\frac{8}{11}$ **18.** $\frac{165}{180}$ $\frac{11}{12}$ **19.** $\frac{60}{108}$ $\frac{5}{9}$ **20.** $\frac{112}{128}$ $\frac{7}{8}$

Write the numerator and denominator of each fraction as a product of prime factors. Then write in lowest terms.

Example: $\frac{24}{36}$ **Solution:** $\frac{24}{36} = \frac{\overset{1}{\cancel{2}} \cdot \overset{1}{\cancel{2}} \cdot 2 \cdot \overset{1}{\cancel{3}}}{\underset{1}{\cancel{2}} \cdot \underset{1}{\cancel{2}} \cdot \underset{1}{\cancel{3}} \cdot 3} = \frac{1 \cdot 1 \cdot 2 \cdot 1}{1 \cdot 1 \cdot 1 \cdot 3} = \mathbf{\frac{2}{3}}$

21. $\frac{10}{16}$ $\frac{2 \cdot 5}{2 \cdot 2 \cdot 2 \cdot 2} = \frac{5}{8}$ **22.** $\frac{18}{27}$ $\frac{2 \cdot 3 \cdot 3}{3 \cdot 3 \cdot 3} = \frac{2}{3}$ **23.** $\frac{35}{40}$ $\frac{5 \cdot 7}{2 \cdot 2 \cdot 2 \cdot 5} = \frac{7}{8}$

24. $\frac{36}{48}$ $\frac{2 \cdot 2 \cdot 3 \cdot 3}{2 \cdot 2 \cdot 2 \cdot 2 \cdot 3} = \frac{3}{4}$ **25.** $\frac{75}{150}$ $\frac{3 \cdot 5 \cdot 5}{2 \cdot 3 \cdot 5 \cdot 5} = \frac{1}{2}$ **26.** $\frac{16}{64}$ $\frac{2 \cdot 2 \cdot 2 \cdot 2}{2 \cdot 2 \cdot 2 \cdot 2 \cdot 2 \cdot 2} = \frac{1}{4}$

27. $\frac{36}{12}$ $\frac{2 \cdot 2 \cdot 3 \cdot 3}{2 \cdot 2 \cdot 3} = 3$ **28.** $\frac{192}{48}$ $\frac{2 \cdot 2 \cdot 2 \cdot 2 \cdot 2 \cdot 2 \cdot 3}{2 \cdot 2 \cdot 2 \cdot 2 \cdot 3} = 4$

29. $\frac{77}{264}$ $\frac{7 \cdot 11}{2 \cdot 2 \cdot 2 \cdot 3 \cdot 11} = \frac{7}{24}$ **30.** $\frac{65}{234}$ $\frac{5 \cdot 13}{2 \cdot 3 \cdot 3 \cdot 13} = \frac{5}{18}$

Decide whether the following pairs of fractions are equivalent or not equivalent.

Example:

$\frac{16}{20}$ and $\frac{36}{45}$

Solution: Find cross products.

$\frac{16}{20}$ $\frac{36}{45}$

$20 \cdot 36 = 720$

$16 \cdot 45 = 720$

equivalent

The fractions are equivalent.

31. $\frac{3}{4}$ and $\frac{18}{24}$

equivalent

32. $\frac{2}{3}$ and $\frac{26}{39}$

equivalent

33. $\frac{12}{15}$ and $\frac{35}{45}$

not equivalent

34. $\frac{11}{16}$ and $\frac{32}{48}$

not equivalent

35. $\frac{15}{24}$ and $\frac{35}{52}$

not equivalent

36. $\frac{7}{11}$ and $\frac{9}{12}$

not equivalent

37. $\frac{14}{16}$ and $\frac{35}{40}$

equivalent

38. $\frac{9}{30}$ and $\frac{12}{40}$

equivalent

39. $\frac{7}{52}$ and $\frac{9}{40}$

not equivalent

40. $\frac{21}{28}$ and $\frac{54}{72}$

equivalent

41. $\frac{25}{30}$ and $\frac{65}{78}$

equivalent

42. $\frac{24}{72}$ and $\frac{30}{90}$

equivalent

43. What does it mean when a fraction is expressed in lowest terms?

Answer varies.

44. Explain what equivalent fractions are and give an example of a pair of equivalent fractions. Show that they are equivalent.

Answer varies.

Write each fraction in lowest terms.

▲ **45.** $\frac{224}{256}$ $\frac{7}{8}$

▲ **46.** $\frac{363}{528}$ $\frac{11}{16}$

▲ **47.** $\frac{312}{975}$ $\frac{8}{25}$

▲ **48.** $\frac{492}{1025}$ $\frac{12}{25}$

PREVIEW EXERCISES

Find all the factors of each number. (For help, see ***Section 2.3.****)*

49. 12 1, 2, 3, 4, 6, 12

50. 18 1, 2, 3, 6, 9, 18

51. 64 1, 2, 4, 8, 16, 32, 64

52. 55 1, 5, 11, 55

2.5 MULTIPLICATION OF FRACTIONS

OBJECTIVES

1. Multiply fractions.
2. Use cancellation.
3. Multiply a fraction and a whole number.
4. Find the area of a rectangle.

FOR EXTRA HELP

1 Multiply the fractions $\frac{2}{3}$ and $\frac{1}{3}$.

$$\frac{2}{3} \cdot \frac{1}{3}$$

Find $\frac{2}{3}$ of $\frac{1}{3}$ by starting with a figure showing $\frac{1}{3}$.

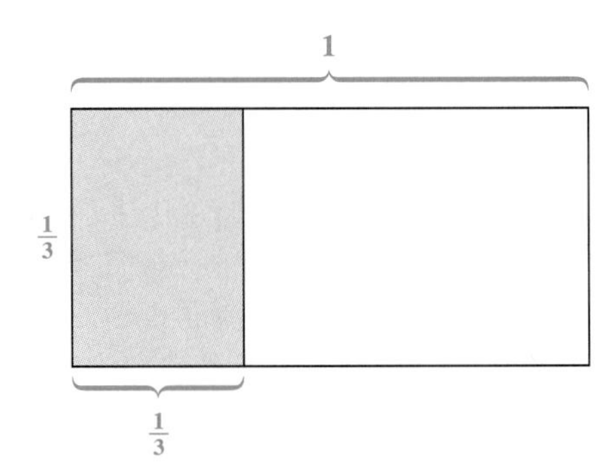

Next, take $\frac{2}{3}$ of the shaded area. (Here we are dividing $\frac{1}{3}$ into 3 equal parts and shading two of them.)

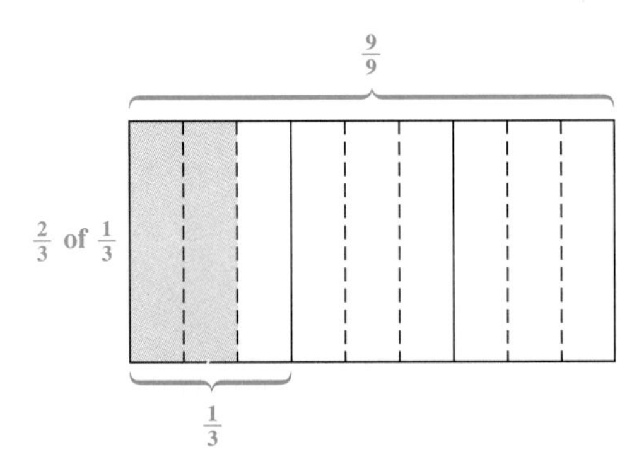

The shaded area in this second figure is equal to $\frac{2}{9}$ of the entire figure, so that

$$\frac{2}{3} \cdot \frac{1}{3} = \frac{2}{9}.$$

WORK PROBLEM 1 AT THE SIDE.

The rule for multiplying fractions follows.

MULTIPLYING FRACTIONS

Multiply two fractions by multiplying the numerators and multiplying the denominators.

Use this rule to find the product of $\frac{2}{3}$ and $\frac{1}{3}$.

$$\frac{2}{3} \cdot \frac{1}{3} = \frac{2 \cdot 1}{3 \cdot 3} \quad \begin{matrix} \text{Multiply numerators.} \\ \text{Multiply denominators.} \end{matrix}$$

$$= \frac{2}{9}$$

1. Use these figures to find the product of $\frac{1}{4}$ and $\frac{1}{2}$.

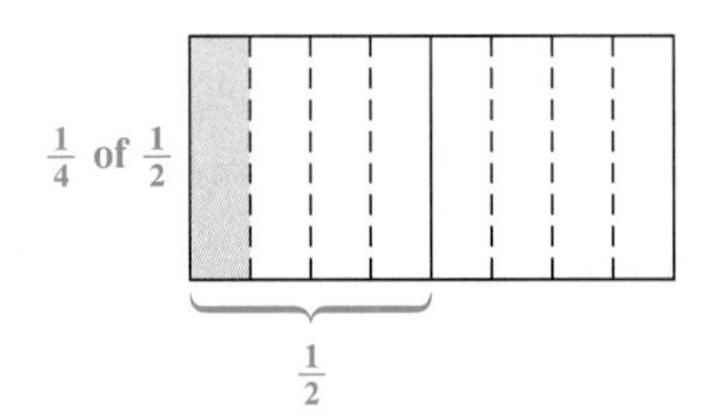

ANSWER

1. $\frac{1}{8}$

2. Multiply. Write answers in lowest terms.

(a) $\frac{2}{3} \cdot \frac{1}{8}$

(b) $\frac{5}{8} \cdot \frac{3}{4}$

(c) $\frac{1}{4} \cdot \frac{5}{9} \cdot \frac{1}{2}$

(d) $\frac{3}{4} \cdot \frac{1}{2} \cdot \frac{7}{8}$

ANSWERS

2. (a) $\frac{1}{12}$ (b) $\frac{15}{32}$ (c) $\frac{5}{72}$ (d) $\frac{21}{64}$

You will see that $\frac{2}{9}$ is in lowest terms.

$$\frac{2}{3} \cdot \frac{1}{3} = \frac{2 \cdot 1}{3 \cdot 3} = \frac{2}{9} \begin{array}{l} \leftarrow 2 \cdot 1 = 2 \\ \leftarrow 3 \cdot 3 = 9 \end{array}$$

■ **EXAMPLE 1** *Multiplying Fractions*

Multiply.

(a) $\frac{5}{8} \cdot \frac{3}{4}$

(b) $\frac{4}{7} \cdot \frac{2}{5}$

(c) $\frac{5}{8} \cdot \frac{3}{4} \cdot \frac{1}{2}$

(a) Multiply the numerators and multiply the denominators.

$$\frac{5}{8} \cdot \frac{3}{4} = \frac{5 \cdot 3}{8 \cdot 4} = \frac{15}{32}$$

Notice that there are no common factors other than 1 for 15 and 32, so the answer is in lowest terms.

(b) $\frac{4}{7} \cdot \frac{2}{5} = \frac{4 \cdot 2}{7 \cdot 5} = \frac{8}{35}$

(c) $\frac{5}{8} \cdot \frac{3}{4} \cdot \frac{1}{2} = \frac{5 \cdot 3 \cdot 1}{8 \cdot 4 \cdot 2} = \frac{15}{64}$ ■

◀ **WORK PROBLEM 2 AT THE SIDE.**

2 It is often easier to cancel before multiplying, as shown in Example 2.

■ **EXAMPLE 2** *Understanding Cancellation*

Multiply $\frac{5}{6}$ and $\frac{9}{10}$.

$$\frac{5}{6} \cdot \frac{9}{10} = \frac{5 \cdot 9}{6 \cdot 10} = \frac{45}{60} \quad \text{(not in lowest terms)}$$

The numerator and denominator have a common factor other than 1, so write the prime factorization of each number.

$$\frac{5}{6} \cdot \frac{9}{10} = \frac{5 \cdot 9}{6 \cdot 10} = \frac{5 \cdot 3 \cdot 3}{2 \cdot 3 \cdot 2 \cdot 5}$$

Next, cancel common factors.

$$\frac{5}{6} \cdot \frac{9}{10} = \frac{5 \cdot 9}{6 \cdot 10} = \frac{\overset{1}{\cancel{5}} \cdot \overset{1}{\cancel{3}} \cdot 3}{2 \cdot \underset{1}{\cancel{3}} \cdot 2 \cdot \underset{1}{\cancel{5}}}$$

Finally, multiply the remaining factors in the numerator and in the denominator.

$$\frac{5}{6} \cdot \frac{9}{10} = \frac{1 \cdot 1 \cdot 3}{2 \cdot 1 \cdot 2 \cdot 1} = \frac{3}{4} \quad \text{(lowest terms)}$$ ■

As a shortcut, instead of writing the prime factorization of each number, find the product of $\frac{5}{6}$ and $\frac{9}{10}$ as follows.

First, divide both 5 and 10 by 5.

$$\frac{\overset{1}{\cancel{5}}}{6} \cdot \frac{9}{\underset{2}{\cancel{10}}}$$

Next, divide both 6 and 9 by 3.

$$\frac{\overset{1}{\cancel{5}}}{\underset{2}{\cancel{6}}} \cdot \frac{\overset{3}{\cancel{9}}}{\underset{2}{\cancel{10}}}$$

Finally, multiply.

$$\frac{1 \cdot 3}{2 \cdot 2} = \frac{3}{4}$$

Note During **cancellation** you are dividing a numerator and a denominator. Be certain that you divide a numerator and a denominator by the same number. If you are able to do all possible cancellations your answer will be in lowest terms.

EXAMPLE 3 *Using Cancellation*

Use cancellation to multiply. Write in lowest terms or as mixed numbers.

(a) $\frac{6}{11} \cdot \frac{7}{8}$

(b) $\frac{7}{10} \cdot \frac{20}{21}$

(c) $\frac{35}{12} \cdot \frac{32}{25}$

(d) $\frac{2}{3} \cdot \frac{8}{15} \cdot \frac{3}{4}$

(a) Divide both 6 and 8 by 2. Next, multiply.

$$\frac{\overset{3}{\cancel{6}}}{11} \cdot \frac{7}{\underset{4}{\cancel{8}}} = \frac{3 \cdot 7}{11 \cdot 4} = \frac{21}{44} \quad \text{lowest terms}$$

(b) Divide 7 and 21 by 7, and divide 10 and 20 by 10.

$$\frac{\overset{1}{\cancel{7}}}{\underset{1}{\cancel{10}}} \cdot \frac{\overset{2}{\cancel{20}}}{\underset{3}{\cancel{21}}} = \frac{1 \cdot 2}{1 \cdot 3} = \frac{2}{3} \quad \text{lowest terms}$$

(c) $$\frac{\overset{7}{\cancel{35}}}{\underset{3}{\cancel{12}}} \cdot \frac{\overset{8}{\cancel{32}}}{\underset{5}{\cancel{25}}} = \frac{7 \cdot 8}{3 \cdot 5} = \frac{56}{15} \quad \text{or} \quad 3\frac{11}{15} \quad \text{mixed number}$$

(d) $$\frac{\overset{1}{\cancel{2}}}{\underset{1}{\cancel{3}}} \cdot \frac{\overset{4}{\cancel{8}}}{15} \cdot \frac{\overset{1}{\cancel{3}}}{\underset{\underset{1}{\cancel{2}}}{\cancel{4}}} = \frac{1 \cdot 4 \cdot 1}{1 \cdot 15 \cdot 1} = \frac{4}{15}$$ ■

Cancellation is especially helpful when the fractions involve large numbers.

3. Use cancellation to find each of the following.

(a) $\frac{3}{4} \cdot \frac{2}{3}$

(b) $\frac{6}{11} \cdot \frac{33}{21}$

(c) $\frac{20}{4} \cdot \frac{3}{40} \cdot \frac{1}{3}$

(d) $\frac{18}{17} \cdot \frac{1}{36} \cdot \frac{2}{3}$

4. Multiply. Write all answers in lowest terms.

(a) $8 \cdot \frac{1}{8}$

(b) $15 \cdot \frac{3}{5} \cdot \frac{4}{3}$

(c) $\frac{7}{10} \cdot 50$

(d) $\frac{5}{11} \cdot 99 \cdot \frac{3}{25}$

ANSWERS

3. (a) $\frac{\overset{1}{\cancel{3}}}{\underset{2}{\cancel{4}}} \cdot \frac{\overset{1}{\cancel{2}}}{\underset{1}{\cancel{3}}} = \frac{1}{2}$ (b) $\frac{\overset{2}{\cancel{6}}}{\underset{1}{\cancel{11}}} \cdot \frac{\overset{3}{\cancel{33}}}{\underset{7}{\cancel{21}}} = \frac{6}{7}$

(c) $\frac{\overset{1}{\cancel{20}}}{4} \cdot \frac{\overset{1}{\cancel{3}}}{\underset{2}{\cancel{40}}} \cdot \frac{1}{\underset{1}{\cancel{3}}} = \frac{1}{8}$

(d) $\frac{\overset{1}{\cancel{18}}}{17} \cdot \frac{1}{\underset{\underset{1}{\cancel{2}}}{\cancel{36}}} \cdot \frac{\overset{1}{\cancel{2}}}{3} = \frac{1}{51}$

4. (a) 1 (b) 12 (c) 35 (d) $\frac{27}{5}$ or $5\frac{2}{5}$

> *Note* There is no specific order that must be used in cancellation so long as both a numerator and a denominator are divided by the same number.

◀ WORK PROBLEM 3 AT THE SIDE.

3 The rule for multiplying a fraction and a whole number follows.

MULTIPLYING A WHOLE NUMBER AND A FRACTION

Multiply a whole number and a fraction by writing the whole number as a fraction with a denominator of 1.

For example, write the whole numbers 8, 10, and 25 as follows.

$$8 = \frac{8}{1}, \quad 10 = \frac{10}{1}, \quad \text{and} \quad 25 = \frac{25}{1}$$

■ **EXAMPLE 4** *Multiplying by a Whole Number*

Multiply. Write all answers in lowest terms or whole numbers where possible.

(a) $8 \cdot \frac{3}{4}$ **(b)** $12 \cdot \frac{5}{6}$

(a) Write 8 as $\frac{8}{1}$ and multiply.

$$8 \cdot \frac{3}{4} = \frac{\overset{2}{\cancel{8}}}{1} \cdot \frac{3}{\underset{1}{\cancel{4}}} = \frac{2 \cdot 3}{1 \cdot 1} = \frac{6}{1} = 6$$

(b) $12 \cdot \frac{5}{6} = \frac{\overset{2}{\cancel{12}}}{1} \cdot \frac{5}{\underset{1}{\cancel{6}}} = \frac{2 \cdot 5}{1 \cdot 1} = \frac{10}{1} = 10$ ■

◀ WORK PROBLEM 4 AT THE SIDE.

4 To find the area of a rectangle (the amount of space in the rectangle), use the following formula.

THE AREA OF A RECTANGLE

The area of a rectangle is equal to the length multiplied by the width.

area = length · width

For example, the rectangle shown here has an area of 12 square feet.

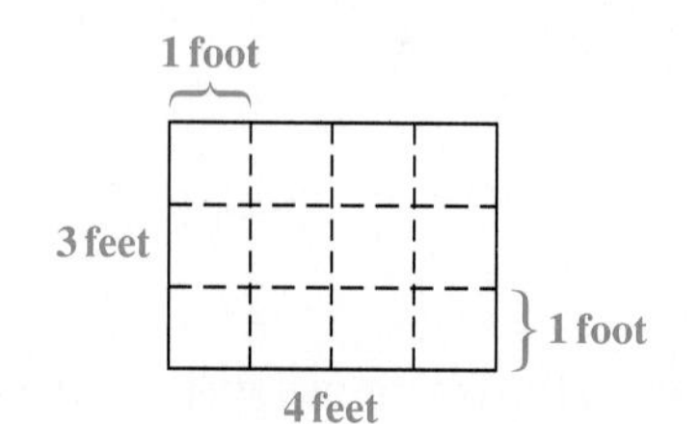

area = length · width
area = 4 · 3
area = 12 square feet

EXAMPLE 5 *Applying Fraction Skills*
Find the area of each rectangle.

(a) $\frac{3}{8}$ foot

$\frac{11}{12}$ foot

(b) a rectangle, $\frac{7}{9}$ inch by $\frac{3}{14}$ inch

(a) area = length · width

$$\text{area} = \frac{11}{12} \cdot \frac{3}{8}$$

$$= \frac{11}{\underset{4}{\cancel{12}}} \cdot \frac{\overset{1}{\cancel{3}}}{8} \quad \text{Cancel.}$$

$$= \frac{11}{32} \text{ square foot}$$

(b) Multiply the length and width.

$$\text{area} = \frac{7}{9} \cdot \frac{3}{14}$$

$$= \frac{\overset{1}{\cancel{7}}}{\underset{3}{\cancel{9}}} \cdot \frac{\overset{1}{\cancel{3}}}{\underset{2}{\cancel{14}}} \quad \text{Cancel.}$$

$$= \frac{1}{6} \text{ square inch}$$ ■

WORK PROBLEM 5 AT THE SIDE. ▶▶

5. Find the area of each rectangle.

(a) $\frac{1}{3}$ foot

$\frac{9}{11}$ foot

(b) $\frac{1}{10}$ inch

$\frac{7}{8}$ inch

(c) a rectangle, $\frac{7}{5}$ yard by $\frac{5}{8}$ yard

ANSWERS

5. (a) $\frac{3}{11}$ **square foot**

(b) $\frac{7}{80}$ **square inch**

(c) $\frac{7}{8}$ **square yard**

QUEST FOR NUMERACY

The Missing Keys

Use *only* the calculator keys listed to get the desired number. You must use each of the listed keys at least once.

Example Use only the 2, 3, +, and = keys to get 100.
One possible answer is 33 + 33 + 32 + 2 = 100. (There are many other possibilities.)

1. Use only the 1, 2, 3, +, and = keys to get 300. Find at least three different solutions.

   Some possible solutions are: 233 + 33 + 32 + 2 = 300
   222 + 33 + 33 + 2 = 300
   33 + 33 + 33 + 33 + 33 + 33 + 33 + 33 + 33 + 3 = 300

2. Use only the 4, 7, 9, +, and = keys to get 1000. Find at least three different solutions.

   Some possible solutions are: 979 + 7 + 7 + 7 = 1000
   977 + 9 + 7 + 7 = 1000
   797 + 97 + 97 + 9 = 1000

3. Use only the 1, 4, 9, −, and = keys to get 45. Find at least three different solutions.

   Some possible solutions are: 94 − 49 = 45
   99 − 44 − 9 − 1 = 45
   49 − 1 − 1 − 1 − 1 = 45

4. Use only the 2, 5, 7, −, and = keys to get 4000. Find at least three different solutions.

   Some possible solutions are: 7555 − 2555 − 775 − 225 = 4000
   7222 − 2222 − 755 − 75 − 55 − 55 − 55 − 5 = 4000
   5225 − 775 − 225 − 225 = 4000

Extend this activity by asking students to make up additional problems involving multiplication and division. Students can exchange and solve these problems, or you can collect them and create another activity page.

2.5 EXERCISES

NAME DATE HOUR

Multiply. Write all answers in lowest terms.

Example: $\frac{9}{16} \cdot \frac{8}{27} \cdot \frac{9}{10}$ **Solution:** $\frac{\overset{1}{\cancel{9}}}{\underset{2}{\cancel{16}}} \cdot \frac{\overset{1}{\cancel{8}}}{\underset{\underset{1}{\cancel{3}}}{\cancel{27}}} \cdot \frac{\overset{3}{\cancel{9}}}{10} = \frac{1 \cdot 1 \cdot 3}{2 \cdot 1 \cdot 10} = \mathbf{\frac{3}{20}}$

1. $\frac{1}{2} \cdot \frac{3}{4}$ $\frac{3}{8}$

2. $\frac{1}{3} \cdot \frac{2}{5}$ $\frac{2}{15}$

3. $\frac{1}{8} \cdot \frac{1}{4}$ $\frac{1}{32}$

4. $\frac{1}{3} \cdot \frac{1}{8}$ $\frac{1}{24}$

5. $\frac{3}{8} \cdot \frac{12}{5}$ $\frac{9}{10}$

6. $\frac{4}{9} \cdot \frac{12}{7}$ $\frac{16}{21}$

7. $\frac{5}{6} \cdot \frac{12}{25} \cdot \frac{3}{4}$ $\frac{3}{10}$

8. $\frac{7}{8} \cdot \frac{16}{21} \cdot \frac{1}{2}$ $\frac{1}{3}$

9. $\frac{3}{4} \cdot \frac{5}{6} \cdot \frac{2}{3}$ $\frac{5}{12}$

10. $\frac{2}{5} \cdot \frac{3}{8} \cdot \frac{2}{3}$ $\frac{1}{10}$

11. $\frac{9}{22} \cdot \frac{11}{16}$ $\frac{9}{32}$

12. $\frac{5}{12} \cdot \frac{7}{10}$ $\frac{7}{24}$

13. $\frac{21}{30} \cdot \frac{5}{7}$ $\frac{1}{2}$

14. $\frac{6}{11} \cdot \frac{22}{15}$ $\frac{4}{5}$

15. $\frac{14}{25} \cdot \frac{65}{48} \cdot \frac{15}{28}$ $\frac{13}{32}$

16. $\frac{32}{15} \cdot \frac{27}{64} \cdot \frac{35}{72}$ $\frac{7}{16}$

17. $\frac{16}{25} \cdot \frac{35}{32} \cdot \frac{15}{64}$ $\frac{21}{128}$

18. $\frac{39}{42} \cdot \frac{7}{13} \cdot \frac{7}{24}$ $\frac{7}{48}$

Multiply. Write all answers in lowest terms; change answers to whole or mixed numbers where possible.

Example: $27 \cdot \frac{5}{9}$

Solution: $27 \cdot \frac{5}{9} = \frac{\overset{3}{\cancel{27}}}{1} \cdot \frac{5}{\underset{1}{\cancel{9}}} = \frac{3 \cdot 5}{1 \cdot 1} = \frac{15}{1} = \mathbf{15}$

19. $8 \cdot \frac{3}{4}$ 6

20. $30 \cdot \frac{4}{5}$ 24

21. $72 \cdot \frac{5}{9}$ 40

22. $38 \cdot \frac{2}{19}$ 4

23. $40 \cdot \frac{5}{8}$ 25

24. $30 \cdot \frac{3}{10}$ 9

25. $42 \cdot \frac{7}{10} \cdot \frac{5}{7}$ 21

26. $35 \cdot \frac{3}{5} \cdot \frac{1}{2}$ $10\frac{1}{2}$

27. $100 \cdot \frac{21}{50} \cdot \frac{3}{4}$ $31\frac{1}{2}$

28. $300 \cdot \frac{5}{6}$ 250

29. $\frac{3}{5} \cdot 400$ 240

30. $\frac{5}{9} \cdot 360$ 200

31. $\frac{2}{3} \cdot 320$ $213\frac{1}{3}$

32. $\frac{12}{25} \cdot 430$ $206\frac{2}{5}$

33. $\frac{28}{21} \cdot 640 \cdot \frac{15}{32}$ 400

34. $\frac{21}{13} \cdot 520 \cdot \frac{7}{20}$ 294

35. $\frac{54}{38} \cdot 684 \cdot \frac{5}{6}$ 810

36. $\frac{76}{43} \cdot 473 \cdot \frac{5}{19}$ 220

Find the area of each rectangle.

Example:

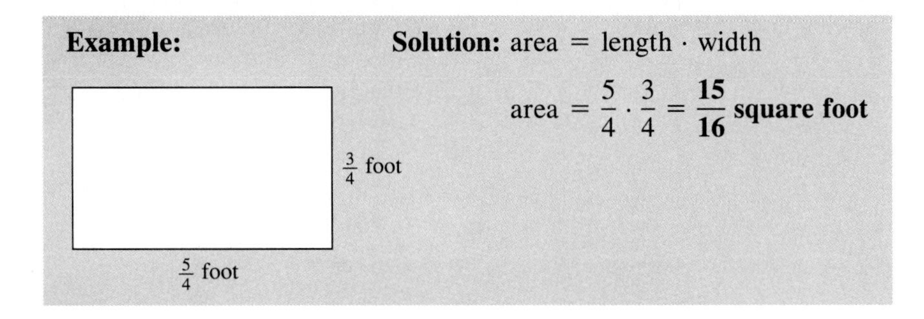

Solution: area = length · width

$$\text{area} = \frac{5}{4} \cdot \frac{3}{4} = \mathbf{\frac{15}{16}} \textbf{ square foot}$$

37. $\frac{5}{8}$ inch, $\frac{4}{5}$ inch — $\frac{1}{2}$ square inch

38. $\frac{8}{9}$ foot, $\frac{1}{4}$ foot — $\frac{2}{9}$ square foot

39. $\frac{2}{3}$ yard, 6 yards — 4 square yards

40. $\frac{3}{8}$ foot, 8 feet — 3 square feet

41. $\frac{3}{14}$ inch, $\frac{7}{5}$ inch — $\frac{3}{10}$ square inch

42. $\frac{9}{16}$ yard, $\frac{14}{15}$ yard — $\frac{21}{40}$ square yard

43. Write in your own words the rule for multiplying fractions. Make up an example problem of your own showing how this works.
Answer varies.

44. A useful shortcut when multiplying fractions involves dividing a numerator and a denominator before multiplying. This is often called cancellation. Describe how this works. Give an example to show how this works.
Answer varies.

Solve each of the following word problems. Write answers in lowest terms or as mixed numbers.

▲ 45. Find the area of a rectangle having a length of 16 inches and a width of $\frac{1}{4}$ inch.

4 square inches

▲ 46. Find the area of a rectangle having a length of 28 inches and a width of $\frac{5}{7}$ inch.

20 square inches

▲ 47. Find the floor area of a rabbit cage having a length of 2 yards and a width of $\frac{2}{3}$ yard.

$1\frac{1}{3}$ square yards

▲ 48. Find the floor area of a table having a length of 3 feet and a width of $\frac{11}{12}$ feet.

$2\frac{3}{4}$ square feet

▲ 49. A parcel of land measures $\frac{1}{2}$ mile by 2 miles. Find the total area of the parcel.

1 square mile

▲ 50. A motorcycle race course is $\frac{3}{4}$ mile wide by 6 miles long. Find the area of the race course.

$4\frac{1}{2}$ square miles

PREVIEW EXERCISES

Solve each of the following word problems. (For help, see **Section 1.9.***)*

51. If 480 free hot lunches are served each day, how many lunches are served in 7 days?

3360 lunches

52. There are 24 bandages per box. Find the number of bandages in 56 boxes.

1344 bandages

2.6 APPLICATIONS OF MULTIPLICATION

OBJECTIVE

1 Solve word problems using multiplication.

FOR EXTRA HELP

Tape 3 | SSM pp. 53–54 | MAC: A IBM: A

1 Many word problems are solved by multiplying fractions. Use the following indicator words for multiplication.

product
double
triple
times
of

Look for these indicator words in the following examples.

■ **EXAMPLE 1** *Applying Indicator Words*

Lois Stevens gives $\frac{1}{10}$ of her income to her church. One month she earned \$1980. How much did she give to the church that month?

Approach To find the amount given to the church, the fraction $\frac{1}{10}$ must be multiplied by the monthly earnings (\$1980).

Solution The indicator word is *of:* Stevens gave $\frac{1}{10}$ *of* her income. The word *of* indicates multiplication, so find the amount given to the church by multiplying $\frac{1}{10}$ and \$1980.

$$\text{amount} = \frac{1}{\cancel{10}_{1}} \cdot \frac{\overset{198}{\cancel{1980}}}{1} = \frac{198}{1} = 198$$

Stevens gave \$198 to the church that month. ■

WORK PROBLEM 1 AT THE SIDE. ▶▶

1. (a) At Frink Chevrolet $\frac{1}{3}$ of the new car buyers purchase the extended warranty. If the dealership sold 8397 new cars last year, find the number of extended warranties sold.

(b) Suppose $\frac{3}{8}$ of the value of a house is taxed. How much of the value of a \$68,000 house is taxed?

■ **EXAMPLE 2** *Solving a Fraction Word Problem*

Of the 42 students in a biology class, $\frac{2}{3}$ went on a field trip. How many went on the trip?

Approach Find the number of students who went on the field trip by multiplying the fraction $\frac{2}{3}$ by the number of students in the class (42).

Solution Reword the problem to read

$$\frac{2}{3} \text{ of the students went.}$$

↑ indicator word

Find the number who went by multiplying $\frac{2}{3}$ and 42.

$$\text{number who went} = \frac{2}{3} \cdot 42$$

$$= \frac{2}{\cancel{3}_{1}} \cdot \overset{14}{\cancel{42}} = \frac{28}{1} = 28$$

28 students went on the trip. ■

WORK PROBLEM 2 AT THE SIDE. ▶▶

2. At one pharmacy, $\frac{3}{16}$ of the prescriptions are paid by a third party (insurance company paid). If 2816 prescriptions are filled, find the number paid by a third party.

ANSWERS
1. (a) 2799 warranties (b) \$25,500
2. 528 prescriptions

EXAMPLE 3 *Finding a Fraction of a Fraction*
In her will, a woman divides her estate into 6 equal parts. 5 of the 6 parts are given to relatives. Of the sixth part, $\frac{1}{3}$ goes to the Salvation Army. What fraction of her total estate goes to the Salvation Army?

Approach To find the fraction of the estate going to the Salvation Army, the part not going to relatives ($\frac{1}{6}$) is multiplied by the fractional part going to the Salvation Army ($\frac{1}{3}$).

Solution The Salvation Army gets $\frac{1}{3}$ of $\frac{1}{6}$.
↑
indicator word

To find the fraction that the Salvation Army is to receive, multiply $\frac{1}{3}$ and $\frac{1}{6}$.

$$\text{fraction to Salvation Army} = \frac{1}{3} \cdot \frac{1}{6} = \frac{1}{18}$$

The Salvation Army gets $\frac{1}{18}$ of the total estate. ■

WORK PROBLEM 3 AT THE SIDE.

3. In a certain community $\frac{1}{3}$ of the residents speak a foreign language. Of those speaking a foreign language, $\frac{3}{4}$ speak Spanish. What fraction of the residents speak Spanish?

ANSWER
3. $\frac{1}{4}$ speak Spanish

2.6 EXERCISES

NAME DATE HOUR

Solve each of the following word problems. Look for indicator words.

Example: Of the 96 units in an apartment building, $\frac{5}{8}$ of the units have two bedrooms. How many units have two bedrooms?
Approach: To find the number of two-bedroom units, multiply the total number of units (96) by the fraction that are two-bedroom units ($\frac{5}{8}$). The indicator word for multiplication is *of* in $\frac{5}{8}$ *of* the units.

Solution: number of two-bedroom units $= \frac{5}{8} \cdot 96$

$$= \frac{5}{\underset{1}{8}} \cdot \overset{12}{96}$$

$$= \frac{60}{1} = 60$$

There are **60 two-bedroom units.**

1. A computer desk top is $\frac{2}{3}$ yard by $\frac{5}{4}$ yard. Find its area.

   $\frac{5}{6}$ square yard

2. A watering trough is $\frac{7}{8}$ yard by $\frac{10}{9}$ yards. Find its area.

   $\frac{35}{36}$ square yard

3. Here is a rectangle that is $\frac{4}{3}$ feet by $\frac{1}{2}$ foot. Find the area.

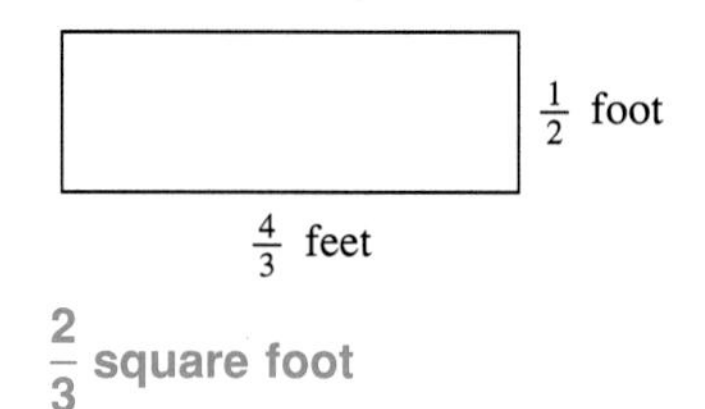

   $\frac{2}{3}$ square foot

4. Find the area of the top of the lamp table below.

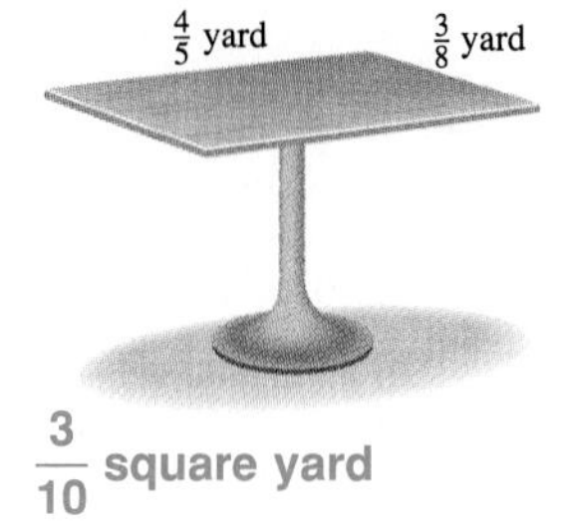

   $\frac{3}{10}$ square yard

5. At a convenience store, $\frac{1}{5}$ of the items sold are taxable. The store sells 1900 items. How many are taxable?

   380 items are taxable

6. A packaged goods store sells 2500 items, of which $\frac{3}{25}$ are classified as junk food. How many of the items are junk food?

   300 items

 Writing  Conceptual  ≈ Estimation

7. Danielle needs to earn \$3600 to go to school for one year. She earns $\frac{3}{4}$ of this amount during the summer. How much money does she earn during the summer?

\$2700

8. Adam paid \$165 for textbooks this term. Of this amount, the bookstore kept $\frac{1}{5}$ as profit. How much money did the bookstore keep?

\$33

9. A school gives scholarships to $\frac{5}{24}$ of its 1800 freshmen. How many freshman students received scholarships?

375 students

10. Jason estimates that it will cost him \$7800, including living expenses, to attend college full time for one year. If he must earn $\frac{5}{8}$ of the cost and borrow the balance, find the amount that he must earn.

\$4875

11. At the local hospital, $\frac{5}{11}$ of the volunteers are men. If there are 165 volunteers, how many are men?

75 men

12. A hotel has 408 rooms. Of these rooms, $\frac{9}{17}$ are for non-smokers. How many rooms are for non-smokers?

216 rooms

The following table shows the earnings for the Gomes family last year. Use this information to solve Exercises 13–18.

Month	*Earnings*	*Month*	*Earnings*
January	$2550	July	$2660
February	$2375	August	$1855
March	$2825	September	$2280
April	$2520	October	$3175
May	$2380	November	$2810
June	$2765	December	$3805

13. Find their total income for the year.

$32,000

14. They paid $\frac{1}{4}$ of their income in taxes. How much were their taxes?

$8000

15. They paid $\frac{1}{5}$ of their income for rent. How much was their rent?

$6400

16. They spent $\frac{5}{16}$ on food. How much money did they spend on food?

$10,000

17. The family saved $\frac{1}{16}$ of their income. How much did they save?

$2000

18. They spent $\frac{1}{8}$ of their income on clothing. How much did they spend on clothing?

$4000

19. Of the indicator words that mean multiplication, the word "of" seems to be the most common. List two other indicator words that indicate multiplication.

Answer varies.

20. When two whole numbers are multiplied, the product is always larger than the numbers being multiplied. When two common fractions are multiplied the answer is always smaller than the numbers being multiplied. Are these statements true? Why or why not?

Answer varies.

Solve each of the following word problems.

▲ **21.** Kerry earned \$120 in one 8-hour day. How much money did she earn in 5 hours?

\$75

▲ **22.** Jim ran 24 miles in 6 hours. How far did he run in 5 hours?

20 miles

▲ **23.** Elaine Silverstein is running for city council. She needs to get $\frac{5}{8}$ of her votes from the south side of town. Ms. Silverstein will need 2400 votes to win. How many votes does she need from areas other than the south side?

900 votes

▲ **24.** The start-up cost of a Subs and Sandwich Shop is \$32,000. If the bank will loan $\frac{9}{16}$ of the start up and the balance must be paid by the business owner, how much must be paid by the business owner?

\$14,000

▲ **25.** A will states that $\frac{7}{8}$ of the estate is to be divided among relatives. Of the remaining $\frac{1}{8}$, $\frac{1}{4}$ goes to the American Cancer Society. What fraction of the estate goes to the American Cancer Society?

$\frac{1}{32}$ of the estate

▲ **26.** A couple has invested $\frac{1}{5}$ of their total investment in stocks. Of the $\frac{1}{5}$ invested in stocks, $\frac{1}{8}$ is invested in General Motors. What fraction of the total investment is invested in General Motors?

$\frac{1}{40}$ of the total investment

PREVIEW EXERCISES

Solve each of the following word problems. (For help, see ***Section 1.9.****)*

27. There are 18 test kits per carton. Find the number of cartons needed to supply 1332 test kits.

74 cartons

28. Kathy Tracy drives 396 miles on 9 gallons of gas. Find the number of miles that she gets per gallon.

44 miles per gallon

2.7 DIVIDING FRACTIONS

As shown in Chapter 1, the division problem 12 ÷ 3 asks how many 3's are in 12. In the same way, the divison problem $\frac{2}{3} \div \frac{1}{6}$ asks how many $\frac{1}{6}$'s are in $\frac{2}{3}$. Look at the figure.

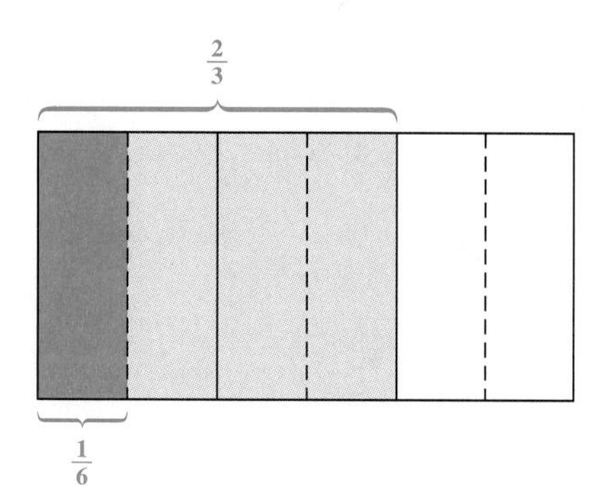

OBJECTIVES

1. Divide fractions.
2. Solve word problems in which fractions are divided.

FOR EXTRA HELP

Tape 4 — SSM pp. 54–56 — MAC: A IBM: A

The figure shows that there are 4 of the $\frac{1}{6}$'s in $\frac{2}{3}$, or

$$\frac{2}{3} \div \frac{1}{6} = 4.$$

1 Compare: $\frac{2}{3} \div \frac{1}{6} = 4$ and $\frac{2}{3} \cdot \frac{6}{1} = \frac{2}{\cancel{3}_1} \cdot \frac{\cancel{6}^2}{1} = 4$

Invert $\frac{1}{6}$ to get $\frac{6}{1}$.

DIVIDING FRACTIONS

Divide two fractions by inverting the second fraction (divisor) and multiplying.

■ **EXAMPLE 1** *Dividing One Fraction by Another*

Divide. Write answers in lowest terms.

(a) $\frac{7}{8} \div \frac{15}{16}$

(b) $\frac{\frac{4}{5}}{\frac{3}{10}}$

1. Divide. Write all answers in lowest terms.

(a) $\frac{1}{8} \div \frac{3}{8}$

(b) $\frac{2}{3} \div \frac{4}{5}$

(c) $\dfrac{\frac{9}{10}}{\frac{3}{5}}$

(d) $\dfrac{\frac{5}{6}}{\frac{25}{24}}$

ANSWERS

1. (a) $\frac{1}{3}$ (b) $\frac{5}{6}$ (c) $1\frac{1}{2}$ (d) $\frac{4}{5}$

Multiply.

(a) $\frac{7}{8} \div \frac{15}{16} = \frac{7}{8} \cdot \frac{16}{15}$ ← Invert $\frac{15}{16}$ to get $\frac{16}{15}$.

$= \frac{7}{\cancel{8}_1} \cdot \frac{\cancel{16}^2}{15}$ Cancel by dividing a numerator and a denominator by 8.

$= \frac{7 \cdot 2}{1 \cdot 15}$ Multiply.

$= \frac{14}{15}$

(b) $\dfrac{\frac{4}{5}}{\frac{3}{10}} = \frac{4}{5} \div \frac{3}{10}$ Rewrite by using the ÷ symbol for division.

$= \frac{4}{\cancel{5}_1} \cdot \frac{\cancel{10}^2}{3}$ Invert and cancel.

$= \frac{4 \cdot 2}{1 \cdot 3}$ Multiply.

$= \frac{8}{3} = 2\frac{2}{3}$ Mixed number ■

Note Be certain that the divisor fraction is inverted before doing any cancelling.

◀◀ WORK PROBLEM 1 AT THE SIDE.

■ **EXAMPLE 2** ***Dividing with a Whole Number***

Divide. Write all answers in lowest terms.

(a) $5 \div \frac{1}{4}$ **(b)** $\frac{2}{3} \div 6$

(a) Write 5 as $\frac{5}{1}$. Next, invert $\frac{1}{4}$ and multiply.

$5 \div \frac{1}{4} = \frac{5}{1} \cdot \frac{4}{1}$ Invert $\frac{1}{4}$ to $\frac{4}{1}$.

$= \frac{5 \cdot 4}{1 \cdot 1}$ Multiply.

$= \frac{20}{1} = 20$ Whole number

(b) Write 6 as $\frac{6}{1}$. Next, invert $\frac{6}{1}$ and multiply.

$$\frac{2}{3} \div \frac{6}{1} = \frac{2}{3} \cdot \frac{1}{6}$$

$$\frac{\overset{1}{\cancel{2}}}{3} \cdot \frac{1}{\underset{3}{\cancel{6}}} = \frac{1 \cdot 1}{3 \cdot 3} \quad \text{Cancel and multiply.}$$

$$= \frac{1}{9} \quad ■$$

WORK PROBLEM 2 AT THE SIDE. 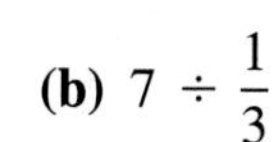

2 Many word problems require division of fractions. Recall typical indicator words for division such as *quotient, divide, divided by,* or *divided into.*

■ **EXAMPLE 3** *Applying Fraction Skills*
Mary must fill a 12-gallon barrel with a chemical. She has only a $\frac{2}{3}$-gallon container to use. How many times must she fill the $\frac{2}{3}$-gallon container and empty it into the 12-gallon barrel?

Approach To find the number of times Mary needs to fill the container, we need to divide the size of the barrel (12 gallons) by the size of the container ($\frac{2}{3}$ gallon).

Solution This problem can be solved by finding the number of times 12 can be divided by $\frac{2}{3}$.

$$12 \div \frac{2}{3} = \frac{12}{1} \cdot \frac{3}{2} \quad \text{Invert } \tfrac{2}{3} \text{ to } \tfrac{3}{2} \text{ and change “}\div\text{” to “}\cdot\text{”.}$$

$$= \frac{\overset{6}{\cancel{12}}}{1} \cdot \frac{3}{\underset{1}{\cancel{2}}} \quad \text{Cancel and multiply.}$$

$$= \frac{18}{1} = 18$$

The container must be filled 18 times. ■

WORK PROBLEM 3 AT THE SIDE. ▶▶

■ **EXAMPLE 4** *Applying Fraction Skills*
At the Happi-Time Day Care Center, $\frac{6}{7}$ of the total operating fund goes to classroom operation. If there are 18 classrooms, what fraction of the classroom operating amount does each one receive?

Approach To find the fraction of the classroom operating amount received by each classroom, we must divide the fraction of the total operating funds going to classroom operation ($\frac{6}{7}$) by the number of classrooms (18).

2. Divide. Write all answers in lowest terms.

(a) $3 \div \frac{3}{4}$

(b) $7 \div \frac{1}{3}$

(c) $\frac{7}{8} \div 3$

(d) $\frac{7}{10} \div 3$

3. (a) How many times must a $\frac{2}{3}$-quart spray bottle be filled in order to use up 18 quarts of cleaner?

(b) How many $\frac{3}{4}$-quart iced tea glasses may be filled from 24 quarts of iced tea?

ANSWERS
2. (a) 4 (b) 21 (c) $\frac{7}{24}$ (d) $\frac{7}{30}$
3. (a) 27 times (b) 32 glasses

Solution This problem can be solved by dividing $\frac{6}{7}$ by 18.

$$\frac{6}{7} \div 18 = \frac{6}{7} \div \frac{18}{1}$$

$$= \frac{\overset{1}{\cancel{6}}}{7} \cdot \frac{1}{\underset{3}{\cancel{18}}} \quad \text{Invert, change “÷” to “·”, and cancel.}$$

$$= \frac{1}{21} \quad \text{Multiply.}$$

Each classroom receives $\frac{1}{21}$ of the total operating funds. ■

◀◀ **WORK PROBLEM 4 AT THE SIDE.**

4. (a) The Sweepstakes Lottery pays out $\frac{7}{8}$ of the total revenue to 14 top winners. What fraction of the total revenue does each winner receive?

(b) The 6 super salespeople for Voyages Marketing will divide $\frac{1}{3}$ of the bonus money paid to employees. What fraction of the bonus money will each of the super salespeople receive?

ANSWERS

4. (a) $\frac{1}{16}$ **of total revenue**

(b) $\frac{1}{18}$ **of bonus money**

EXERCISES

Divide. Write all answers in lowest terms, as mixed numbers or whole numbers.

Example: $\frac{3}{4} \div \frac{1}{2}$ **Solution:** $\frac{3}{\underset{2}{\cancel{4}}} \cdot \frac{\overset{1}{\cancel{2}}}{1} = \frac{3}{2} = \mathbf{1\frac{1}{2}}$

1. $\frac{1}{4} \div \frac{1}{2}$ $\frac{1}{2}$

2. $\frac{1}{2} \div \frac{2}{3}$ $\frac{3}{4}$

3. $\frac{5}{8} \div \frac{4}{3}$ $\frac{15}{32}$

4. $\frac{3}{4} \div \frac{5}{3}$ $\frac{9}{20}$

5. $\frac{5}{8} \div \frac{3}{16}$ $3\frac{1}{3}$

6. $\frac{4}{5} \div \frac{9}{4}$ $\frac{16}{45}$

7. $\frac{7}{12} \div \frac{14}{15}$ $\frac{5}{8}$

8. $\frac{13}{20} \div \frac{4}{5}$ $\frac{13}{16}$

9. $\dfrac{\frac{7}{9}}{\frac{7}{36}}$ 4

10. $\dfrac{\frac{15}{32}}{\frac{5}{64}}$ 6

11. $\dfrac{\frac{36}{35}}{\frac{15}{14}}$ $\frac{24}{25}$

12. $\dfrac{\frac{28}{15}}{\frac{21}{5}}$ $\frac{4}{9}$

13. $6 \div \frac{2}{3}$ 9

14. $7 \div \frac{1}{4}$ 28

15. $\dfrac{15}{\frac{2}{3}}$ $22\frac{1}{2}$

16. $\dfrac{6}{\frac{5}{8}}$ $9\frac{3}{5}$

17. $\dfrac{\frac{4}{7}}{8}$ $\frac{1}{14}$

18. $\dfrac{\frac{11}{5}}{3}$ $\frac{11}{15}$

Solve each of the following word problems by using division.

19. Ms. Shaffer has a piece of property with an area that is $\frac{8}{9}$ acre. She wishes to divide it into 4 equal parts for her children. How many acres of land will each child get?

$\frac{2}{9}$ acre

20. Sophie wants to make stuffed geese to sell at a craft fair. Each goose requires $\frac{2}{3}$ yard of material. She has 22 yards of material. Find the number of geese she can make.

33 geese

21. It takes $\frac{4}{5}$ pound of salt to fill a large salt shaker. How many salt shakers can be filled with 28 pounds of salt?

35 shakers

22. Jason has 10 quarts of lubricating oil. If each lubricating reservoir holds $\frac{1}{3}$ quart of oil, how many reservoirs can be filled?

30 reservoirs

23. How many $\frac{1}{8}$-ounce eye drop dispensers can be filled with 11 ounces of eye drops?

88 dispensers

24. Each guest at a party will eat $\frac{5}{16}$ pound of peanuts. How many guests may be served with 10 pounds of peanuts?

32 guests

25. Pam had a small pickup truck that would carry $\frac{2}{3}$-cord of firewood. Find the number of trips needed to deliver 40 cords of wood.

60 trips

26. Matthew has a reel of steel cable 600 yards long. Find the number of cable sections $\frac{3}{4}$ yard in length that may be cut from the reel.

800 sections

27. A batch of double chocolate chip cookies requires $\frac{3}{4}$ pound of chocolate chips. If you have 9 pounds of chocolate chips, how many batches of cookies can be made?

12 batches

28. Find the number of $\frac{2}{3}$-quart cans of fruit that can be filled from a vat holding 82 quarts of fruit.

123 cans

29. Your classmate is confused on how to divide by a fraction. Write a short note telling how this should be done.

Answer varies.

30. If you multiply common fractions the answer is smaller than the fractions multiplied. When you divide by a common fraction is the answer smaller than the numbers in the problem? Prove your answer with examples.

Answer varies.

▲ **31.** Walt has driven $\frac{6}{7}$ of the distance to his freight pickup destination. He has driven 648 miles so far. How many *more* miles must he drive to get to his destination?

108 miles

▲ **32.** Sheila has been working on a job for 63 hours. The job is $\frac{7}{9}$ finished. How many *more* hours must she work to finish the job?

18 hours

▲ **33.** The Bridge Lighting Committee has raised $\frac{7}{8}$ of the funds necessary for their lighting project. If this amounts to $840,000, how much additional money must be raised?

$120,000

▲ **34.** A mountain guide has used pack animals for $\frac{14}{15}$ of a trip and must finish the trip on foot. The distance covered with pack animals is 98 miles. Find the number of miles to be completed on foot.

7 miles

PREVIEW EXERCISES

Write each mixed number as an improper fraction. (For help, see **Section 2.2.***)*

35. $3\frac{3}{4}$ $\frac{15}{4}$

36. $2\frac{7}{8}$ $\frac{23}{8}$

37. $15\frac{1}{3}$ $\frac{46}{3}$

38. $25\frac{3}{5}$ $\frac{128}{5}$

39. $112\frac{8}{9}$ $\frac{1016}{9}$

40. $221\frac{3}{7}$ $\frac{1550}{7}$

2.8 MULTIPLICATION AND DIVISION OF MIXED NUMBERS

OBJECTIVES

1. Multiply mixed numbers.
2. Divide mixed numbers.
3. Solve word problems with mixed numbers.

FOR EXTRA HELP

Tape 4

SSM pp. 56–58

MAC: A IBM: A

1 Multiply mixed numbers by using the following rule.

MULTIPLYING MIXED NUMBERS

Step 1 **Change** each mixed number to an improper fraction.
Step 2 **Multiply** as fractions.
Step 3 Write the answer in lowest terms and change to a mixed number or whole number where possible.

EXAMPLE 1 *Multiplying Mixed Numbers*

Multiply. Write all answers in lowest terms.

(a) $2\frac{1}{2} \cdot 3\frac{1}{5}$

(b) $3\frac{5}{8} \cdot 4\frac{4}{5}$

(c) $1\frac{3}{5} \cdot 3\frac{1}{3}$

(a) Change each mixed number to an improper fraction.

Step 1 $\quad 2\frac{1}{2} = \frac{5}{2}$ and $3\frac{1}{5} = \frac{16}{5}$

Next, multiply.

$$2\frac{1}{2} \cdot 3\frac{1}{5} = \frac{5}{2} \cdot \frac{16}{5} = \frac{\overset{1}{\cancel{5}}}{\underset{1}{\cancel{2}}} \cdot \frac{\overset{8}{\cancel{16}}}{\underset{1}{\cancel{5}}} = \underbrace{\frac{1 \cdot 8}{1 \cdot 1}}_{\text{Step 2}} = \underbrace{\frac{8}{1} = 8}_{\text{Step 3}}$$

(b) $$3\frac{5}{8} \cdot 4\frac{4}{5} = \underbrace{\frac{29}{8} \cdot \frac{24}{5}}_{\text{Step 1}} = \frac{29}{\underset{1}{\cancel{8}}} \cdot \frac{\overset{3}{\cancel{24}}}{5} = \underbrace{\frac{29 \cdot 3}{1 \cdot 5} = \frac{87}{5}}_{\text{Step 2}}$$

As a mixed number,

$$\underbrace{\frac{87}{5} = 17\frac{2}{5}}_{\text{Step 3}}.$$

(c) $$1\frac{3}{5} \cdot 3\frac{1}{3} = \frac{8}{\underset{1}{\cancel{5}}} \cdot \frac{\overset{2}{\cancel{10}}}{3} = \frac{8 \cdot 2}{1 \cdot 3} = \frac{16}{3} = 5\frac{1}{3}$$ ■

WORK PROBLEM 1 AT THE SIDE. ▶▶

1. Multiply. Write answers in lowest terms.

(a) $2\frac{1}{4} \cdot 7\frac{1}{3}$

(b) $4\frac{1}{2} \cdot 1\frac{2}{3}$

(c) $3\frac{3}{5} \cdot 4\frac{4}{9}$

(d) $3\frac{1}{5} \cdot 5\frac{3}{8}$

ANSWERS

1. (a) $16\frac{1}{2}$ (b) $7\frac{1}{2}$ (c) 16 (d) $17\frac{1}{5}$

2. Divide. Write answer in lowest terms.

(a) $1\frac{1}{4} \div 3\frac{1}{3}$

(b) $3\frac{3}{8} \div 2\frac{4}{7}$

(c) $8 \div 5\frac{1}{3}$

(d) $4\frac{1}{2} \div 6$

2 Divide mixed numbers by using the following rule.

DIVIDING MIXED NUMBERS

Step 1 **Change** each mixed number to an improper fraction.
Step 2 **Invert** the second fraction (divisor).
Step 3 **Multiply**.
Step 4 Write the answer in lowest terms and change to a mixed number or whole number where possible.

EXAMPLE 2 *Dividing Mixed Numbers*
Divide. Write answers in lowest terms.

(a) $2\frac{2}{5} \div 1\frac{1}{2}$ **(b)** $3\frac{1}{3} \div 2\frac{1}{2}$

(c) $8 \div 3\frac{3}{5}$ **(d)** $4\frac{3}{8} \div 5$

(a) First, change each mixed number to an improper fraction.

Step 1

$$2\frac{2}{5} \div 1\frac{1}{2} = \frac{12}{5} \div \frac{3}{2}$$

Next, invert the second fraction and multiply.

Step 2 Step 3 Step 4

$$\frac{12}{5} \div \frac{3}{2} = \frac{\overset{4}{\cancel{12}}}{5} \cdot \frac{2}{\underset{1}{\cancel{3}}} = \frac{4 \cdot 2}{5 \cdot 1} = \frac{8}{5} = 1\frac{3}{5}$$

inverted

Step 1 Step 2 Step 3 Step 4

(b) $3\frac{1}{3} \div 2\frac{1}{2} = \frac{10}{3} \div \frac{5}{2} = \frac{\overset{2}{\cancel{10}}}{3} \cdot \frac{2}{\underset{1}{\cancel{5}}} = \frac{2 \cdot 2}{3 \cdot 1} = \frac{4}{3} = 1\frac{1}{3}$

inverted

inverted

(c) $8 \div 3\frac{3}{5} = \frac{8}{1} \div \frac{18}{5} = \frac{\overset{4}{\cancel{8}}}{1} \cdot \frac{5}{\underset{9}{\cancel{18}}} = \frac{20}{9} = 2\frac{2}{9}$

Write 8 as $\frac{8}{1}$

inverted

(d) $4\frac{3}{8} \div 5 = \frac{35}{8} \div \frac{5}{1} = \frac{\overset{7}{\cancel{35}}}{8} \cdot \frac{1}{\underset{1}{\cancel{5}}} = \frac{7}{8}$ ■

Write 5 as $\frac{5}{1}$

◀ **WORK PROBLEM 2 AT THE SIDE.**

ANSWERS
2. (a) $\frac{3}{8}$ (b) $1\frac{5}{16}$ (c) $1\frac{1}{2}$ (d) $\frac{3}{4}$

3 The next two examples show how to solve word problems involving mixed numbers.

EXAMPLE 3 *Applying Multiplication Skills*
Suppose 11 building contractors each donate $3\frac{1}{4}$ days of labor to a community building project. How many days of labor will be donated in all?

Approach Because several contractors each donate the same amount of labor, multiply to get the total amount donated.

Solution Multiply the number of contractors and the amount of labor that each donates.

$$11 \cdot 3\frac{1}{4} = 11 \cdot \frac{13}{4}$$

$$= \frac{11}{1} \cdot \frac{13}{4} = \frac{143}{4} = 35\frac{3}{4}$$

The community building project will receive $35\frac{3}{4}$ days of donated labor. ■

WORK PROBLEM 3 AT THE SIDE.

EXAMPLE 4 *Applying Division Skills*
One tent requires $7\frac{1}{4}$ yards of nylon cloth. How many tents can be made from $65\frac{1}{4}$ yards of the cloth?

Approach Division must be used to find the number of times one number is in another number.

Solution Divide the number of yards of cloth by the number of yards needed for one tent.

$$65\frac{1}{4} \div 7\frac{1}{4} = \frac{261}{4} \div \frac{29}{4}$$

$$= \frac{\overset{9}{\cancel{261}}}{\underset{1}{\cancel{4}}} \cdot \frac{\overset{1}{\cancel{4}}}{\underset{1}{\cancel{29}}} = \frac{9}{1} = 9$$

9 tents can be made from $65\frac{1}{4}$ yards of cloth. ■

WORK PROBLEM 4 AT THE SIDE.

3. (a) Suppose a dress requires $2\frac{3}{4}$ yards of material. How much material would be needed for 7 dresses?

(b) Clare earns $\$9\frac{1}{4}$ per hour. How much would she earn in $6\frac{1}{2}$ hours? Write the answer as a mixed number.

4. (a) An airplane needs $2\frac{3}{8}$ pounds of a special metal. How many airplanes could be built from $28\frac{1}{2}$ pounds of the metal?

(b) Student help is paid $\$6\frac{1}{4}$ per hour. Find the number of hours of student help that can be paid for with $150.

ANSWERS
3. (a) $19\frac{1}{4}$ **yards (b)** $\$60\frac{1}{8}$
4. (a) 12 airplanes (b) 24 hours

QUEST FOR NUMERACY

Pattern Search

Work the first three or four problems in each table, using your calculator to help you. Look for a pattern in your answers and use the pattern to complete the table. Then describe the pattern.

$6 \times 1 = 6$
$14 \times 1 = 14$
$197 \times 1 = 197$
$999 \times 1 = 999$
$4003 \times 1 = 4003$
$50{,}927 \times 1 = 50{,}927$
The pattern is: Any number times 1 is the same number.

$8 \times 0 = 0$
$45 \times 0 = 0$
$444 \times 0 = 0$
$807 \times 0 = 0$
$9000 \times 0 = 0$
$630{,}725 \times 0 = 0$
The pattern is: Any number times 0 is 0.

$3 \div 3 = 1$
$75 \div 75 = 1$
$309 \div 309 = 1$
$500 \div 500 = 1$
$6636 \div 6636 = 1$
$85{,}333 \div 85{,}333 = 1$
The pattern is: Any number divided by itself is 1.

$2 \times 9 = 18$ and	$1 + 8 = 9$
$3 \times 9 = 27$ and	$2 + 7 = 9$
$4 \times 9 = 36$ and	$3 + 6 = 9$
$5 \times 9 = 45$ and	$4 + 5 = 9$
$6 \times 9 = 54$ and	$5 + 4 = 9$
$7 \times 9 = 63$ and	$6 + 3 = 9$
$8 \times 9 = 72$ and	$7 + 2 = 9$
$9 \times 9 = 81$ and	$8 + 1 = 9$
The pattern is: The sum of the digits is 9.	

$3 \times 37 = 111$
$6 \times 37 = 222$
$9 \times 37 = 333$
$12 \times 37 = 444$
$15 \times 37 = 555$
$18 \times 37 = 666$
$21 \times 37 = 777$
$24 \times 37 = 888$
$27 \times 37 = 999$
The pattern is: Digits are identical in each answer and increase by 1 in each successive answer.

$60 \times 37 = 2220$
$63 \times 37 = 2331$
$66 \times 37 = 2442$
$69 \times 37 = 2553$
$72 \times 37 = 2664$
$75 \times 37 = 2775$
The pattern is: First digit of the answer is always 2. The other digits each increase by 1 in successive answers.

$9 \times 9 = 81$
$99 \times 99 = 9801$
$999 \times 999 = 998{,}001$
$9999 \times 9999 = 99{,}980{,}001$
$99{,}999 \times 99{,}999 = 9{,}999{,}800{,}001$
$999{,}999 \times 999{,}999 = 999{,}998{,}000{,}001$
The pattern is: The second answer has one 9 before the 8 and one 0 after the 8. The next answer has two 9s before the 8 and two 0s after the 8. And so on.

Extend this activity by asking students to add more problems to each of the last four tables. Does the pattern in the answers continue? How long will it continue?

2.8 EXERCISES

NAME DATE HOUR

Multiply. Write answers as mixed numbers or whole numbers.

1. $2\frac{1}{4} \cdot 3\frac{1}{2}$ $7\frac{7}{8}$

2. $1\frac{1}{2} \cdot 3\frac{3}{4}$ $5\frac{5}{8}$

3. $1\frac{2}{3} \cdot 2\frac{7}{10}$ $4\frac{1}{2}$

4. $1\frac{1}{4} \cdot 2\frac{1}{2}$ $3\frac{1}{8}$

5. $3\frac{1}{9} \cdot 1\frac{2}{7}$ 4

6. $6\frac{1}{4} \cdot 3\frac{1}{5}$ 20

7. $10 \cdot 7\frac{1}{4}$ $72\frac{1}{2}$

8. $6 \cdot 2\frac{1}{3}$ 14

9. $4\frac{1}{2} \cdot 2\frac{1}{5} \cdot 5$ $49\frac{1}{2}$

10. $2\frac{2}{3} \cdot 4\frac{1}{2} \cdot 3\frac{1}{4}$ 39

11. $3 \cdot 1\frac{1}{2} \cdot 2\frac{2}{3}$ 12

12. $\frac{2}{3} \cdot 3\frac{2}{3} \cdot \frac{6}{11}$ $1\frac{1}{3}$

Divide. Write answers in lowest terms and change to mixed numbers or whole numbers where possible.

13. $3\frac{1}{4} \div 2\frac{5}{8}$ $1\frac{5}{21}$

14. $2\frac{1}{4} \div 1\frac{1}{8}$ 2

15. $2\frac{1}{2} \div 3$ $\frac{5}{6}$

16. $5\frac{1}{2} \div 4$ $1\frac{3}{8}$

17. $6 \div 1\frac{1}{4}$ $4\frac{4}{5}$

18. $5 \div 1\frac{7}{8}$ $2\frac{2}{3}$

19. $\frac{1}{2} \div 2\frac{1}{4}$ $\frac{2}{9}$

20. $\frac{1}{4} \div 2\frac{1}{2}$ $\frac{1}{10}$

21. $1\frac{7}{8} \div 6\frac{1}{4}$ $\frac{3}{10}$

22. $7\frac{1}{2} \div \frac{2}{3}$ $11\frac{1}{4}$

23. $5\frac{2}{3} \div 6$ $\frac{17}{18}$

24. $5\frac{3}{4} \div 2$ $2\frac{7}{8}$

Solve each of the following word problems by using multiplication or division.

25. Shirley wants to make 8 stuffed dolls to sell at the craft fair. Each doll needs $1\frac{5}{8}$ yards of material. How many yards does she need?

13 yards

26. Babbette worked $36\frac{1}{2}$ hours at \$9 per hour. How much money did she make?

$\$328\frac{1}{2}$

27. Each home of a certain design needs $109\frac{1}{2}$ yards of prefinished baseboard. How many homes can be fitted with baseboard if there are 1314 yards of baseboard available?

12 homes

28. For 1 acre of a crop, $7\frac{1}{2}$ gallons of fertilizer must be applied. How many acres can be fertilized with 1200 gallons of fertilizer?

160 acres

29. Insect spray is mixed $1\frac{3}{4}$ ounces of chemical per gallon of water. How many ounces of chemical are needed for $12\frac{1}{2}$ gallons of water?

$21\frac{7}{8}$ ounces

30. Each home requires $37\frac{3}{4}$ pounds of roofing nails. How many pounds of roofing nails are needed for 36 homes?

1359 pounds

31. A dictionary requires $2\frac{3}{8}$ pounds of paper. How many can be published with 11,875 pounds of paper?

5000 dictionaries

32. Each apartment unit requires $62\frac{1}{2}$ square yards of carpet. Find the number of apartment units that can be carpeted with 6750 square yards of carpet.

108 units

33. A manufacturer of bird feeders needs spacers that are to be cut from a tube that is $5\frac{1}{2}$ inches long. How many spacers can be cut from the tube if each spacer has to be $\frac{1}{2}$ inch thick?

11 spacers

34. A building contractor must move 12 tons of sand. If his truck can carry $\frac{3}{4}$ ton of sand, how many trips must be made to move the sand?

16 trips

▲ **35.** A photographer uses $12\frac{3}{4}$ rolls of film at a wedding and $7\frac{1}{8}$ rolls of film at a retirement party. Find the total number of rolls needed for 28 weddings and 16 retirement parties.

471 rolls

▲ **36.** One necklace can be completed in $6\frac{1}{2}$ minutes, while a bracelet takes $3\frac{1}{8}$ minutes. Find the total time that it takes to complete 36 necklaces and 22 bracelets.

$302\frac{3}{4}$ minutes

▲ **37.** A water tank contains 35 gallons when it is $\frac{5}{8}$ full. How much water will it hold when it is full?

56 gallons

▲ **38.** The fuel gauge on a piece of earth-moving equipment shows $\frac{3}{4}$ full when there are 156 gallons of fuel in the tank. Find the number of gallons in the tank when it is full.

208 gallons

PREVIEW EXERCISES

Write each fraction in lowest terms. (For help, see ***Section 2.4.****)*

39. $\frac{4}{8}$ $\frac{1}{2}$

40. $\frac{6}{8}$ $\frac{3}{4}$

41. $\frac{25}{40}$ $\frac{5}{8}$

42. $\frac{27}{45}$ $\frac{3}{5}$

43. $\frac{72}{80}$ $\frac{9}{10}$

44. $\frac{36}{63}$ $\frac{4}{7}$

CHAPTER 2 SUMMARY

KEY TERMS

	Term	Definition
2.1	**numerator**	The number above the division bar in a fraction is called the numerator. It shows how many of the equivalent parts are being considered.
	denominator	The number below the division bar in a fraction is called the denominator. It shows the number of equal parts in a whole.
	proper fraction	In a proper fraction, the numerator is smaller than the denominator. The fraction is less than one.
	improper fraction	In an improper fraction, the numerator is greater than or equal to the denominator. The fraction is equal to or greater than one.
2.2	**mixed number**	A mixed number includes a fraction and a whole number written together.
2.3	**factors**	Numbers that are multiplied to give a product are factors.
	composite number	A composite number has at least one factor other than itself and 1.
	prime number	A prime number is a whole number other than 0 and 1 that has exactly two factors, itself and 1.
	prime factorization	In a prime factorization every factor is a prime number.
2.4	**common factor**	A common factor is a number that can be divided into two or more whole numbers.
	lowest terms	A fraction is written in lowest terms when its numerator and denominator have no common factor other than 1.
2.5	**cancellation**	When multiplying or dividing fractions, the process of dividing a numerator and denominator by a common factor is called cancellation.

QUICK REVIEW

Concepts	Examples
2.1 Types of Fractions	
Proper Numerator smaller than denominator.	**Proper** $\frac{2}{3}, \frac{3}{4}, \frac{15}{16}, \frac{1}{8}$
Improper Numerator equal to or greater than denominator.	**Improper** $\frac{17}{8}, \frac{19}{12}, \frac{11}{2}, \frac{5}{3}, \frac{7}{7}$
2.2 Converting Fractions	
Mixed to improper Multiply denominator by whole number and add numerator.	**Mixed to improper** $7\frac{2}{3} = \frac{23}{3}$ $3 \times 7 + 2$ same denominator
Improper to mixed Divide numerator by denominator and place remainder over denominator.	**Improper to mixed** $\frac{17}{5} = 3\frac{2}{5}$ same denominator
2.3 Prime Numbers	
Determine whether a whole number is evenly divisible only by itself and 1. (By definition, 0 and 1 are not prime.)	The prime numbers less than 100 are 2, 3, 5, 7, 11, 13, 17, 19, 23, 29, 31, 37, 41, 43, 47, 53, 59, 61, 67, 71, 73, 79, 83, 89, and 97.

Concepts	Examples
2.3 Finding the Prime Factorization of a Number Divide each factor by a prime number by using a diagram that forms the shape of a tree.	Find the prime factorization of 30. Use a factor tree. 30 → (2), 15; 15 → (3), (5) Prime factors are circled. $30 = 2 \cdot 3 \cdot 5$
2.4 Writing Fractions in Lowest Terms Divide the numerator and denominator by the same number.	$\frac{30}{42} = \frac{30 \div 6}{42 \div 6} = \frac{5}{7}$
2.5 Multiplying Fractions **1.** Multiply numerators and denominators. **2.** Reduce answers to lowest terms if cancelling was not done.	$\frac{6}{11} \cdot \frac{7}{8} = \frac{\overset{3}{\cancel{6}}}{11} \cdot \frac{7}{\underset{4}{\cancel{8}}} = \frac{21}{44}$
2.7 Dividing Fractions Invert the divisor (second fraction) and multiply as fractions.	$\frac{25}{36} \div \frac{15}{18} = \frac{\overset{5}{\cancel{25}}}{\underset{2}{\cancel{36}}} \cdot \frac{\overset{1}{\cancel{18}}}{\underset{3}{\cancel{15}}} = \frac{5 \cdot 1}{2 \cdot 3} = \frac{5}{6}$
2.8 Multiplying Mixed Numbers **1. Change** each mixed number to an improper fraction. **2. Multiply.** **3.** Write the answer in lowest terms and change the answer to a mixed number if desired.	$1\frac{3}{5} \cdot 3\frac{1}{3} = \frac{8}{\underset{1}{\cancel{5}}} \cdot \frac{\overset{2}{\cancel{10}}}{3}$ $= \frac{8 \cdot 2}{1 \cdot 3}$ $= \frac{16}{3} = 5\frac{1}{3}$
2.8 Dividing Mixed Numbers **1. Change** each mixed number to an improper fraction. **2. Invert** the divisor (second fraction). **3. Multiply.** **4.** Write the answer in lowest terms and change the answer to a mixed number if desired.	$3\frac{5}{9} \div 2\frac{2}{5} = \frac{32}{9} \div \frac{12}{5}$ $= \frac{\overset{8}{\cancel{32}}}{9} \cdot \frac{5}{\underset{3}{\cancel{12}}} = \frac{40}{27} = 1\frac{13}{27}$

NAME DATE HOUR

CHAPTER 2 REVIEW EXERCISES

[2.1] *Write the fraction that represents the shaded area.*

1. 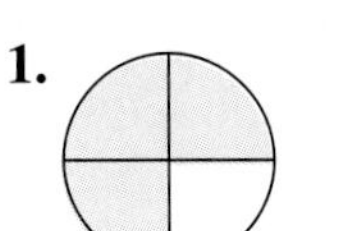 $\frac{3}{4}$

2. 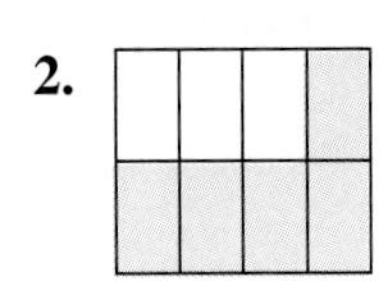 $\frac{5}{8}$

3. 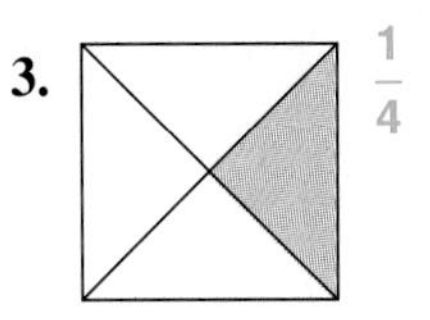 $\frac{1}{4}$

Name the proper and improper fractions in each list.

	proper	*improper*
4. $\frac{1}{2}, \frac{2}{1}, \frac{6}{5}, \frac{2}{3}, \frac{3}{8}$	$\frac{1}{2}, \frac{2}{3}, \frac{3}{8}$	$\frac{2}{1}, \frac{6}{5}$
5. $\frac{6}{5}, \frac{15}{16}, \frac{16}{13}, \frac{1}{8}, \frac{5}{3}$	$\frac{15}{16}, \frac{1}{8}$	$\frac{6}{5}, \frac{16}{13}, \frac{5}{3}$

[2.2] *Write each mixed number as an improper fraction. Write each improper fraction as a mixed number.*

6. $5\frac{3}{8}$ $\frac{43}{8}$

7. $11\frac{5}{16}$ $\frac{181}{16}$

8. $\frac{19}{2}$ $9\frac{1}{2}$

9. $\frac{63}{5}$ $12\frac{3}{5}$

[2.3] *Find all factors of each number.*

10. 6

1, 2, 3, 6

11. 12

1, 2, 3, 4, 6, 12

12. 55

1, 5, 11, 55

13. 90

1, 2, 3, 5, 6, 9, 10, 15, 18, 30, 45, 90

Write the prime factorization of each number by using exponents.

14. 25 5^2

15. 120 $2^3 \cdot 3 \cdot 5$

16. 225 $3^2 \cdot 5^2$

Write each of the following without exponents.

17. 6^2 36

18. $2^2 \cdot 3^2$ 36

19. $8^2 \cdot 3^3$ 1728

20. $4^3 \cdot 2^5$ 2048

[2.4] *Write each fraction in lowest terms.*

21. $\frac{18}{24}$ $\frac{3}{4}$

22. $\frac{35}{40}$ $\frac{7}{8}$

23. $\frac{175}{190}$ $\frac{35}{38}$

Write the numerator and denominator of each fraction as a product of prime factors. Next, write the fraction in lowest terms.

24. $\frac{25}{60}$ $\frac{5 \cdot 5}{2 \cdot 2 \cdot 3 \cdot 5}$; $\frac{5}{12}$

25. $\frac{384}{96}$ $\frac{2 \cdot 2 \cdot 2 \cdot 2 \cdot 2 \cdot 2 \cdot 2 \cdot 3}{2 \cdot 2 \cdot 2 \cdot 2 \cdot 2 \cdot 3}$; 4

Decide whether the following pairs of fractions are equivalent or not equivalent, using cross multiplication.

26. $\frac{2}{3}$ and $\frac{34}{51}$ equivalent

27. $\frac{3}{8}$ and $\frac{15}{40}$ equivalent

[2.5–2.8] *Multiply. Write all answers in lowest terms, and as mixed numbers or whole numbers where possible.*

28. $\frac{2}{3} \cdot \frac{3}{4}$ $\frac{1}{2}$

29. $\frac{4}{5} \cdot \frac{5}{12}$ $\frac{1}{3}$

30. $\frac{70}{175} \cdot \frac{5}{14}$ $\frac{1}{7}$

31. $\frac{44}{63} \cdot \frac{3}{11}$ $\frac{4}{21}$

32. $\frac{5}{16} \cdot 48$ 15

33. $\frac{5}{8} \cdot 1000$ 625

Divide. Write answers in lowest terms, and as mixed numbers or whole numbers where possible.

34. $\frac{3}{8} \div \frac{2}{3}$ $\frac{9}{16}$

35. $\frac{5}{6} \div \frac{1}{2}$ $\frac{5}{3} = 1\frac{2}{3}$

36. $\dfrac{\frac{15}{18}}{\frac{10}{30}}$ $\frac{5}{2} = 2\frac{1}{2}$

37. $\dfrac{\frac{3}{10}}{\frac{6}{40}}$ 2

38. $10 \div \frac{5}{8}$ 16

39. $9 \div \frac{3}{4}$ 12

40. $1 \div \frac{7}{8}$ $\frac{8}{7} = 1\frac{1}{7}$

41. $\frac{2}{3} \div 5$ $\frac{2}{15}$

42. $\dfrac{\frac{12}{13}}{3}$ $\frac{4}{13}$

Find the area of each rectangle.

43. $\frac{15}{32}$ square yard

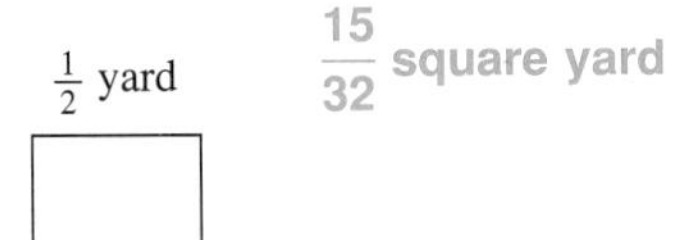

44. $\frac{7}{12}$ square inch

$\frac{2}{3}$ inch

$\frac{7}{8}$ inch

45. Find the area of a rectangle having a length of 15 feet and a width of $\frac{2}{3}$ foot.

10 square feet

46. Find the area of a rectangle having a length of 54 yards and a width of $\frac{2}{3}$ yard.

36 square yards

Multiply. Write answers in lowest terms, and as mixed numbers where possible.

47. $2\frac{3}{8} \times 1\frac{1}{2}$ $3\frac{9}{16}$

48. $2\frac{1}{4} \cdot 7\frac{1}{8} \cdot 1\frac{1}{3}$ $21\frac{3}{8}$

Divide. Write answers in lowest terms, and as mixed numbers where possible.

49. $15\frac{1}{2} \div 3$ $5\frac{1}{6}$

50. $3\frac{1}{8} \div 5\frac{5}{7}$ $\frac{35}{64}$

Solve each of the following word problems by using multiplication or division.

51. How many $\frac{3}{4}$-pound honey-roasted peanut cans can be filled with 15 pounds of honey roasted peanuts?

20 cans

52. An estate is divided so that each of 5 children receives equal shares of $\frac{2}{3}$ of the estate. What fraction of the total estate will each receive?

$\frac{2}{15}$ of the estate

53. Find the number of window blind pull cords that can be made from $157\frac{1}{2}$ yards of cord if $4\frac{3}{8}$ yards of cord are needed for each blind.

36 pull cords

54. Neta Fitzgerald worked 42 hours at $\$7\frac{1}{2}$ per hour. How much money did she earn?

$315

55. Joanna Wilson purchases 100 pounds of detergent. After selling $\frac{1}{2}$ of this to her neighbor, she gives $\frac{2}{5}$ of the remaining detergent to her parents. How many pounds of detergent does she have left?

30 pounds

56. Mary received a check for $1200. After paying $\frac{3}{8}$ of this amount for room and board, she paid $\frac{1}{2}$ of the remaining amount for school fees. How much money does she have left after paying the fees?

$375

57. The Springvale Parish will divide $\frac{5}{8}$ of its library budget evenly with the 4 largest parish libraries. What fraction of the total library budget will each of these libraries receive?

$\frac{5}{32}$ of the budget

58. Play It Now Sports Center has decided to divide $\frac{2}{3}$ of the company's profit sharing funds evenly among the top 8 store managers. What fraction of the total profit sharing amount will each receive?

$\frac{1}{12}$ of the total

MIXED REVIEW EXERCISES

Multiply or divide as indicated. Write answers in lowest terms, and as mixed numbers or whole numbers where possible.

59. $\frac{1}{3} \cdot \frac{3}{4}$ $\frac{1}{4}$

60. $\frac{2}{3} \cdot \frac{1}{4}$ $\frac{1}{6}$

61. $10\frac{1}{4} \cdot 2\frac{1}{2}$ $25\frac{5}{8}$

62. $12\frac{1}{2} \cdot 2\frac{1}{4}$ $28\frac{1}{8}$

63. $\frac{\frac{7}{8}}{6}$ $\frac{7}{48}$

64. $\frac{\frac{5}{8}}{4}$ $\frac{5}{32}$

65. $\frac{15}{31} \cdot 62$ 30

66. $3\frac{1}{4} \div 1\frac{1}{2}$ $2\frac{1}{6}$

Write each mixed number as an improper fraction. Write each improper fraction as a mixed number.

67. $\frac{7}{4}$ $1\frac{3}{4}$

68. $\frac{187}{4}$ $46\frac{3}{4}$

69. $5\frac{2}{3}$ $\frac{17}{3}$

70. $38\frac{3}{8}$ $\frac{307}{8}$

Write the numerator and denominator of each fraction as a product of prime factors, then write the fraction in lowest terms.

71. $\frac{8}{12}$ $\frac{2 \cdot 2 \cdot 2}{2 \cdot 2 \cdot 3} = \frac{2}{3}$

72. $\frac{108}{210}$ $\frac{2 \cdot 2 \cdot 3 \cdot 3 \cdot 3}{2 \cdot 3 \cdot 5 \cdot 7} = \frac{18}{35}$

Write each fraction in lowest terms.

73. $\frac{14}{42}$ $\frac{1}{3}$

74. $\frac{36}{96}$ $\frac{3}{8}$

75. $\frac{44}{110}$ $\frac{2}{5}$

76. $\frac{87}{261}$ $\frac{1}{3}$

Solve each of the following word problems.

77. The directions on a can of fabric glue say to apply $3\frac{1}{2}$ ounces of glue to each square yard. How many ounces are needed for $43\frac{5}{9}$ square yards?

$152\frac{4}{9}$ ounces

78. A gasoline additive says to use $6\frac{1}{2}$ quarts of additive with each in-ground tank of gasoline. How many quarts are need for $22\frac{1}{4}$ in-ground tanks?

$144\frac{5}{8}$ quarts

79. A postage stamp is $\frac{2}{3}$ inch by $\frac{3}{4}$ inch. Find its area.

$\frac{1}{2}$ square inch

80. A rectangle is $\frac{3}{8}$ meter by $\frac{4}{5}$ meter. Find its area.

$\frac{3}{10}$ square meter

NAME DATE HOUR

CHAPTER 2 TEST

Use a fraction to represent the shaded portions.

1.

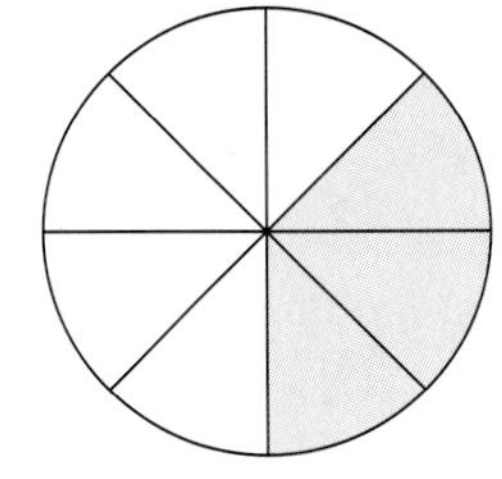

2.

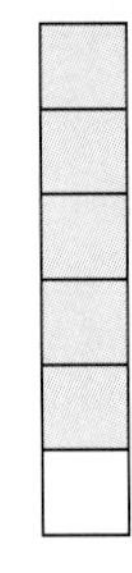

Name all the proper fractions.

3. $\frac{5}{8}, \frac{6}{6}, \frac{7}{16}, \frac{10}{7}, \frac{2}{3}, \frac{3}{14}, \frac{17}{5}$

4. Write $2\frac{7}{8}$ as an improper fraction.

5. Write $\frac{125}{6}$ as a mixed number.

6. Find all factors of 15.

Find the prime factorization of each number. Write the answers using exponents.

7. 45

8. 96

9. 500

Write each fraction in lowest terms.

10. $\frac{15}{18}$

11. $\frac{56}{84}$

12. The method of prime factors is used to write a fraction in lowest terms. Briefly explain how this is done. Use the fraction $\frac{56}{84}$ to show how this works.

1. $\frac{3}{8}$

2. $\frac{5}{6}$

3. $\frac{5}{8}, \frac{7}{16}, \frac{2}{3}, \frac{3}{14}$

4. $\frac{23}{8}$

5. $20\frac{5}{6}$

6. 1, 3, 5, 15

7. $3^2 \cdot 5$

8. $2^5 \cdot 3$

9. $2^2 \cdot 5^3$

10. $\frac{5}{6}$

11. $\frac{2}{3}$

12. Answer varies.

13. Answer varies.

14. $\frac{3}{8}$

15. 18

16. $\frac{1}{4}$ square inch

17. 483 students

18. $\frac{7}{10}$

19. $15\frac{3}{4}$

20. 100 vehicles

21. $17\frac{23}{32}$

22. $7\frac{17}{18}$

23. $4\frac{4}{15}$

24. $5\frac{1}{10}$

25. $30\frac{5}{8}$ grams

13. Explain how to multiply fractions. What additional step must be taken when dividing fractions?

Multiply or divide. Write answers in lowest terms, and as mixed numbers or whole numbers where possible.

14. $\frac{3}{4} \cdot \frac{1}{2}$

15. $24 \cdot \frac{3}{4}$

16. Find the area of a rectangle measuring $\frac{4}{5}$ inch by $\frac{5}{16}$ inch.

17. There are 690 students in the class registration line. If $\frac{7}{10}$ of them are registering for general education classes, how many students are registering for the general education classes?

18. $\frac{3}{5} \div \frac{6}{7}$

19. $\frac{7}{\frac{4}{9}}$

20. There are 60 tanks of hydraulic fluid in the airport supply depot. If each maintenance vehicle has a container that holds $\frac{3}{5}$ of a tank of fluid, how many maintenance vehicles can be filled?

21. $5\frac{1}{4} \cdot 3\frac{3}{8}$

22. $1\frac{5}{6} \cdot 4\frac{1}{3}$

23. $4\frac{4}{5} \div 1\frac{1}{8}$

24. $\frac{8\frac{1}{2}}{1\frac{2}{3}}$

25. A new vaccine is synthesized at the rate of $2\frac{1}{2}$ grams per day. How many grams can be synthesized in $12\frac{1}{4}$ days?

NAME DATE HOUR

CUMULATIVE REVIEW EXERCISES CHAPTERS 1–2

Name the digit that has the given place value in each of the following problems.

1. 638
hundreds 6
tens 3

2. 3,781,586
millions 3
ten thousands 8

Add, subtract, multiply, or divide as indicated.

3. $36 + 25 + 72 + 13$ 146

4. $82{,}121 + 5\,468 + 316 + 61{,}294$ 149,199

5. $2628 - 1056$ 1572

6. $4{,}819{,}604 - 1{,}597{,}783$ 3,221,821

7. 64×7 448

8. $5 \cdot 3 \cdot 7$ 105

9. 3784×573 2,168,232

10. 629×700 440,300

11. $\frac{72}{8}$ 9

12. $18\overline{)136{,}458}$ 7 581

13. $16{,}942 \div 4$ 4235 R2

14. $492\overline{)10{,}850}$ 22 R26

Round each of the following to the nearest ten, nearest hundred, and nearest thousand.

	ten	hundred	thousand
15. 8626	8630	8600	9000
16. 78,154	78,150	78,200	78,000

Simplify each problem by using the order of operations.

17. $5^2 - 9 \cdot 2$ 7

18. $\sqrt{36} - 2 \cdot 3 + 5$ 5

Solve each word problem.

19. The Beadworks purchased 5 cases of beads costing $25 per case and 8 cases of beads costing $40 per case. Find the total cost of the 13 cases of beads.

$445

20. Home blood-pressure monitors sell for as little as $20 and as much as $150. If Scott bought the cheapest model and Jenn the most expensive model, how much more did Jenn pay than Scott?

$130

21. A typical adult loses 100 hairs a day out of approximately 120,000 hairs. If the lost hairs were not replaced, find the number of hairs remaining after 2 years. (1 year = 365 days).

47,000 hairs

22. The cost of renting a group camp for one week is $3150. If the cost is to be split evenly among 18 families, find the cost for each family for the week.

$175

23. A lamp base is made of marble and is $\frac{3}{4}$ foot by $\frac{7}{12}$ foot. Find its area.

$\frac{7}{16}$ square foot

24. A large container holds 80 quarts of a thirst quencher. Find the number of $\frac{4}{5}$-quart sports bottles that can be filled from the container.

100 sports bottles

Write proper *or* improper *for each fraction.*

25. $\frac{5}{8}$ proper

26. $\frac{4}{3}$ improper

27. $\frac{1}{2}$ proper

Write each mixed number as an improper fraction. Write each improper fraction as a mixed number.

28. $1\frac{1}{2}$ $\frac{3}{2}$

29. $8\frac{3}{8}$ $\frac{67}{8}$

30. $\frac{11}{2}$ $5\frac{1}{2}$

31. $\frac{103}{8}$ $12\frac{7}{8}$

Find the prime factorization of each number. Write answers by using exponents.

32. 40 $2^3 \cdot 5$

33. 75 $3 \cdot 5^2$

34. 350 $2 \cdot 5^2 \cdot 7$

Simplify each of the following.

35. $2^2 \cdot 3^2$ 36

36. $3^3 \cdot 5^2$ 675

37. $2^3 \cdot 4^2 \cdot 5$ 640

Write each fraction in lowest terms.

38. $\frac{15}{40}$ $\frac{3}{8}$

39. $\frac{9}{21}$ $\frac{3}{7}$

40. $\frac{30}{54}$ $\frac{5}{9}$

Multiply or divide as indicated. Write all answers in lowest terms, and as mixed numbers or whole numbers where possible.

41. $\frac{2}{3} \cdot \frac{3}{4}$ $\frac{1}{2}$

42. $25 \cdot \frac{3}{5} \cdot \frac{2}{3}$ 10

43. $7\frac{1}{2} \cdot 3\frac{1}{3}$ 25

44. $\frac{3}{8} \div \frac{2}{3}$ $\frac{9}{16}$

45. $\frac{3}{8} \div 1\frac{1}{4}$ $\frac{3}{10}$

46. $3 \div 1\frac{1}{4}$ $2\frac{2}{5}$

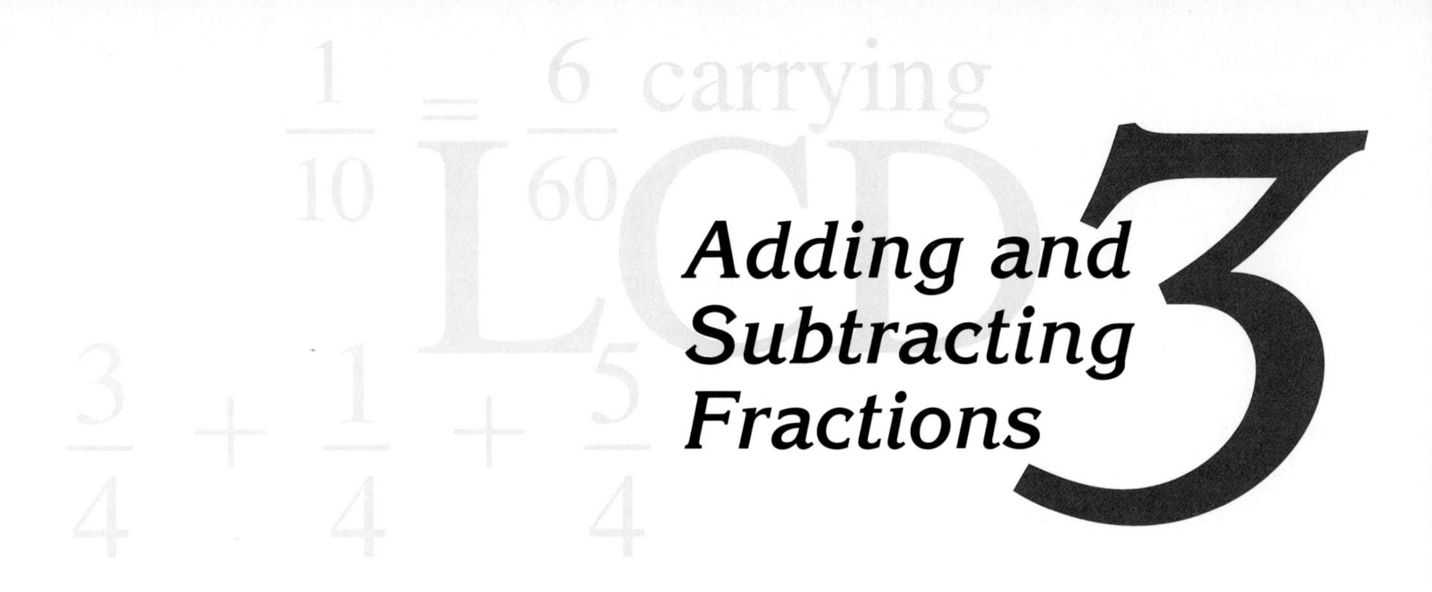

3 Adding and Subtracting Fractions

3.1 ADDING AND SUBTRACTING LIKE FRACTIONS

In Chapter 2 we looked at the basics of fractions and then practiced with multiplication and division of common fractions and mixed numbers. In this chapter we will work with addition and subtraction of common fractions and mixed numbers.

1 Fractions with the same denominators are **like fractions.** Fractions with different denominators are **unlike fractions.**

■ **EXAMPLE 1** *Identifying Like and Unlike Fractions*

(a) $\frac{3}{4}, \frac{1}{4}, \frac{5}{4}, \frac{6}{4}$, and $\frac{4}{4}$ are like fractions.

All denominators are the same.

(b) $\frac{7}{12}$ and $\frac{12}{7}$ are unlike fractions.

different denominators ■

WORK PROBLEM 1 AT THE SIDE. ▶▶

2 The following figures show you how to add the fractions $\frac{2}{7}$ and $\frac{4}{7}$.

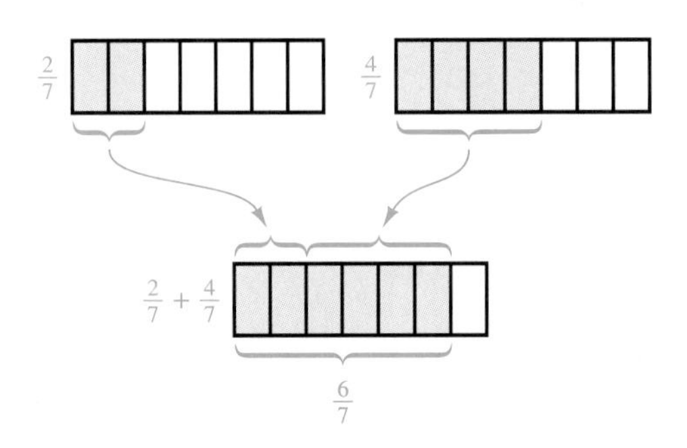

OBJECTIVES

1. Define like and unlike fractions.
2. Add like fractions.
3. Subtract like fractions.

FOR EXTRA HELP

Tape 4

SSM pp. 70–71

MAC: A IBM: A

1. Write *like* or *unlike* for each pair of fractions.

(a) $\frac{7}{8}$ $\frac{3}{8}$

(b) $\frac{7}{4}$ $\frac{6}{5}$

(c) $\frac{11}{12}$ $\frac{9}{12}$

(d) $\frac{8}{3}$ $\frac{7}{4}$

ANSWERS

1. (a) like (b) unlike (c) like (d) unlike

2. Add. Write answers in lowest terms.

(a) $\frac{1}{4} + \frac{2}{4}$

(b) $\frac{5}{9} + \frac{2}{9}$

(c) $\frac{1}{8} + \frac{3}{8}$

(d) $\frac{3}{9} + \frac{1}{9} + \frac{2}{9}$

ANSWERS

2. (a) $\frac{3}{4}$ (b) $\frac{7}{9}$ (c) $\frac{1}{2}$ (d) $\frac{2}{3}$

As the figures show,

$$\frac{2}{7} + \frac{4}{7} = \frac{6}{7}.$$

Add like fractions as follows.

ADDING LIKE FRACTIONS

Step 1 Find the numerator of the answer (the **sum**) by adding the numerators of the fractions.
Step 2 Write the denominator of the like fractions as the denominator of the sum.
Step 3 Write the answer in lowest terms.

EXAMPLE 2 *Adding Like Fractions*

Add. Write answers in lowest terms.

(a) $\frac{1}{5} + \frac{2}{5}$

(b) $\frac{1}{12} + \frac{7}{12} + \frac{1}{12}$

(a) $\frac{1}{5} + \frac{2}{5} = \frac{1 + 2}{5} = \frac{3}{5}$ ← Add numerators. ← same denominator

(b)

Step 1 $\frac{1 + 7 + 1}{12}$ ← Add numerators

Step 2 $= \frac{9}{12}$ ← Same denominator

Step 3 $= \frac{9}{12} = \frac{3}{4}$ (in lowest terms) ■

Note Fractions may be added only if they have like denominators.

◀ **WORK PROBLEM 2 AT THE SIDE.**

3 The figures show $\frac{7}{8}$ broken into $\frac{4}{8}$ and $\frac{3}{8}$.

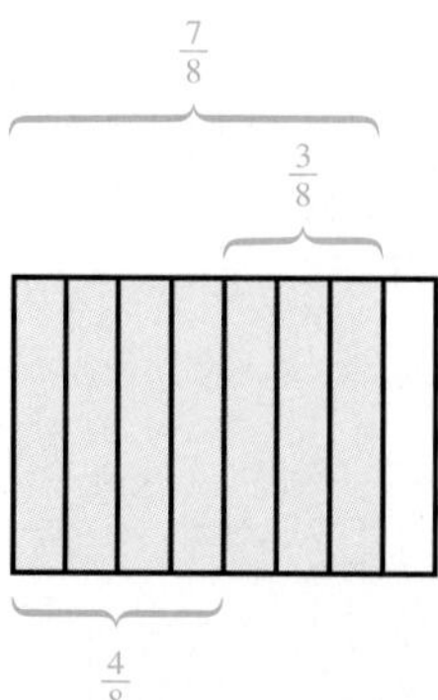

Subtracting $\frac{3}{8}$ from $\frac{7}{8}$ gives the answer $\frac{4}{8}$, or

$$\frac{7}{8} - \frac{3}{8} = \frac{4}{8}.$$

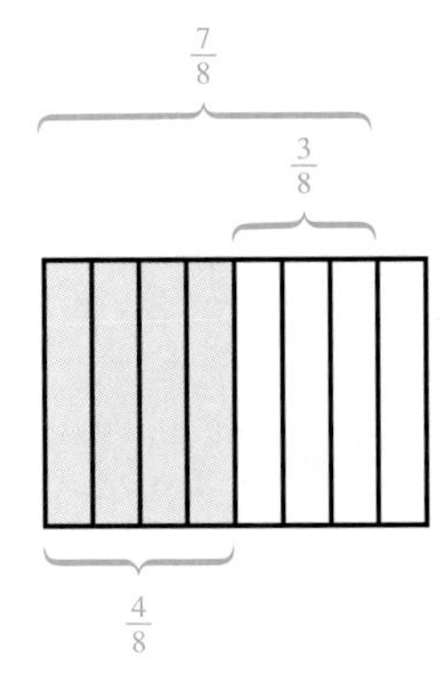

Write $\frac{4}{8}$ in lowest terms.

$$\frac{7}{8} - \frac{3}{8} = \frac{4}{8} = \frac{1}{2}$$

The steps for subtracting like fractions are very similar to those for adding like fractions.

SUBTRACTING LIKE FRACTIONS

Step 1 Find the numerator of the answer (the *difference*) by subtracting the numerators.
Step 2 Write the denominator of the like fractions as the denominator of the difference.
Step 3 Write the answer in lowest terms.

EXAMPLE 3 *Subtracting Like Fractions*
Subtract. Write answers in lowest terms or as a mixed number when possible.

(a) $\frac{11}{12} - \frac{7}{12}$

(b) $\frac{15}{3} - \frac{5}{3}$

(a)

Step 1 $\frac{11 - 7}{12}$ ← Subtract numerators.

Step 2 $= \frac{4}{12}$ ← denominator of like fractions

Write in lowest terms.

Step 3 $= \frac{4}{12} = \frac{1}{3}$

(b) $\frac{15}{3} - \frac{5}{3} = \frac{15 - 5}{3}$ ← Subtract numerators. ← denominator of like fractions

$= \frac{10}{3}$

Write as a mixed number.

$$\frac{10}{3} = 3\frac{1}{3}$$ ■

Note Fractions may be subtracted only if they have like denominators.

◀ **WORK PROBLEM 3 AT THE SIDE.**

3. Subtract. Write answers in lowest terms.

(a) $\frac{11}{15} - \frac{4}{15}$

(b) $\frac{8}{9} - \frac{5}{9}$

(c) $\frac{21}{5} - \frac{18}{5}$

(d) $\frac{101}{145} - \frac{17}{145}$

ANSWERS
3. (a) $\frac{7}{15}$ (b) $\frac{1}{3}$ (c) $\frac{3}{5}$ (d) $\frac{84}{145}$

3.1 EXERCISES

NAME DATE HOUR

Add. Write answers in lowest terms.

Example: $\frac{2}{9} + \frac{4}{9}$ **Solution:** Add numerators. $\frac{2}{9} + \frac{4}{9} = \frac{2+4}{9} = \frac{6}{9}$

Write in lowest terms. $\frac{6}{9} = \mathbf{\frac{2}{3}}$

1. $\frac{1}{3} + \frac{1}{3}$ $\frac{2}{3}$

2. $\frac{3}{5} + \frac{1}{5}$ $\frac{4}{5}$

3. $\frac{1}{10} + \frac{6}{10}$ $\frac{7}{10}$

4. $\frac{9}{11} + \frac{1}{11}$ $\frac{10}{11}$

5. $\frac{1}{4} + \frac{1}{4}$ $\frac{1}{2}$

6. $\frac{1}{8} + \frac{1}{8}$ $\frac{1}{4}$

7. $\frac{1}{9} + \frac{2}{9}$ $\frac{1}{3}$

8. $\frac{4}{10} + \frac{1}{10}$ $\frac{1}{2}$

9. $\frac{7}{12} + \frac{3}{12}$ $\frac{5}{6}$

10. $\frac{8}{15} + \frac{4}{15}$ $\frac{4}{5}$

11. $\frac{6}{20} + \frac{4}{20} + \frac{3}{20}$ $\frac{13}{20}$

12. $\frac{1}{7} + \frac{2}{7} + \frac{3}{7}$ $\frac{6}{7}$

13. $\frac{3}{17} + \frac{2}{17} + \frac{5}{17}$ $\frac{10}{17}$

14. $\frac{5}{11} + \frac{1}{11} + \frac{4}{11}$ $\frac{10}{11}$

15. $\frac{3}{8} + \frac{1}{8} + \frac{2}{8}$ $\frac{3}{4}$

16. $\frac{4}{9} + \frac{1}{9} + \frac{1}{9}$ $\frac{2}{3}$

17. $\frac{2}{54} + \frac{8}{54} + \frac{12}{54}$ $\frac{11}{27}$

18. $\frac{7}{120} + \frac{9}{120} + \frac{18}{120}$ $\frac{17}{60}$

Subtract. Write answers in lowest terms.

Example: $\frac{9}{12} - \frac{5}{12}$ **Solution:** Subtract numerators.

$\frac{9}{12} - \frac{5}{12} = \frac{9-5}{12} = \frac{4}{12}$

Write in lowest terms. $\frac{4}{12} = \mathbf{\frac{1}{3}}$

19. $\frac{2}{3} - \frac{1}{3}$ $\frac{1}{3}$

20. $\frac{4}{5} - \frac{1}{5}$ $\frac{3}{5}$

21. $\frac{16}{21} - \frac{8}{21}$ $\frac{8}{21}$

22. $\frac{28}{32} - \frac{19}{32}$ $\frac{9}{32}$

23. $\frac{9}{10} - \frac{3}{10}$ $\frac{3}{5}$

24. $\frac{7}{8} - \frac{5}{8}$ $\frac{1}{4}$

25. $\dfrac{14}{15} - \dfrac{4}{15}$ $\frac{2}{3}$

26. $\dfrac{8}{25} - \dfrac{3}{25}$ $\frac{1}{5}$

27. $\dfrac{27}{40} - \dfrac{19}{40}$ $\frac{1}{5}$

28. $\dfrac{38}{55} - \dfrac{16}{55}$ $\frac{2}{5}$

29. $\dfrac{43}{72} - \dfrac{25}{72}$ $\frac{1}{4}$

30. $\dfrac{71}{100} - \dfrac{31}{100}$ $\frac{2}{5}$

31. $\dfrac{87}{144} - \dfrac{71}{144}$ $\frac{1}{9}$

32. $\dfrac{356}{220} - \dfrac{235}{220}$ $\frac{11}{20}$

33. $\dfrac{746}{400} - \dfrac{506}{400}$ $\frac{3}{5}$

34. Describe in your own words the difference between like fractions and unlike fractions. Give three examples of each type.
Answer varies.

35. Write an explanation of either how to add or subtract like fractions. Consider using three steps in your explanation. Answer varies.

Solve each word problem. Write answers in lowest terms.

36. The Robinsons have saved $\frac{3}{16}$ of the amount needed for the down payment on a home. If they save another $\frac{5}{16}$ of the amount needed, find the total fraction of the amount needed that they have saved.

$\frac{1}{2}$

37. Maria jogged $\frac{5}{12}$ mile down a road and then $\frac{1}{12}$ mile along a creek. How far did she jog altogether?

$\frac{1}{2}$ mile

38. The Thompsons owe $\frac{7}{12}$ of a debt. If they pay $\frac{5}{12}$ of it this month, what fraction of the debt will they still owe?

$\frac{1}{6}$

39. Sam must inspect $\frac{11}{16}$ of a mile of high voltage line. He has already inspected $\frac{5}{16}$ of a mile. How much additional line must he inspect?

$\frac{3}{8}$ mile

40. Matt purchased $\frac{3}{8}$ acre of land one year and $\frac{7}{8}$ acre the next year. He then sold $\frac{5}{8}$ acre of land. How much land does he now have?

$\frac{5}{8}$ acre

41. A forester planted $\frac{5}{12}$ acre in seedlings in the morning and $\frac{11}{12}$ acre in the afternoon. If $\frac{7}{12}$ acre of seedlings were destroyed by frost, how many acres remained?

$\frac{3}{4}$ acre

PREVIEW EXERCISES

Find the prime factorization. Do not use exponents when writing the answer. (For help, see **Section 2.3.***)*

42. 6 $2 \cdot 3$

43. 10 $2 \cdot 5$

44. 30 $2 \cdot 3 \cdot 5$

45. 100 $2 \cdot 2 \cdot 5 \cdot 5$

46. 45 $3 \cdot 3 \cdot 5$

47. 75 $3 \cdot 5 \cdot 5$

48. 81 $3 \cdot 3 \cdot 3 \cdot 3$

49. 125 $5 \cdot 5 \cdot 5$

50. 200 $2 \cdot 2 \cdot 2 \cdot 5 \cdot 5$

3.2 LEAST COMMON MULTIPLES

OBJECTIVES

1. Find the least common multiple.
2. Find the least common multiple by using prime factorization and a table.
3. Find the least common multiple by using an alternative method.
4. Write a fraction with an indicated denominator.

FOR EXTRA HELP

Tape 4 | SSM pp. 71–74 | MAC: A IBM: A

Only *like* fractions can be added or subtracted. Because of this, *unlike* fractions must be rewritten as like fractions before adding or subtracting.

1 Unlike fractions can be written as like fractions by finding the *least common multiple* of the denominators.

LEAST COMMON MULTIPLE

The **least common multiple (LCM)** of two whole numbers is the smallest whole number divisible by both those numbers.

EXAMPLE 1 *Finding the Multiples of a Number*
The list shows you multiples of 6.

6, 12, 18, 24, 30, . . .

(The three dots show the list continues in the same pattern without stopping.) The next list shows multiples of 9.

9, 18, 27, 36, 45, . . .

The smallest number found in both lists is 18, so 18 is the **least common multiple** of 6 and 9; the number 18 is the smallest whole number divisible by both 6 and 9.

multiples of 6   6, 12, 18, 24, 30, . . .
multiples of 9   9, 18, 27, 36, 45, . . .

18 is the smallest number found in both lists. ■

WORK PROBLEM 1 AT THE SIDE.

1. (a) List the multiples of 8.
8, ____, ____, ____, ____, ____, . . .

(b) List the multiples of 10.
10, ____, ____, ____, ____, . . .

(c) Find the least common multiple of 8 and 10.

2 Example 1 shows how to find the least common multiple by making a list of the common multiples of each number. Although this method works, it is usually easier to find the least common multiple by using prime factorization, as shown in the next example.

EXAMPLE 2 *Applying Prime Factorization Knowledge*
Use prime factorization to find the least common multiple of 18 and 60.

We start by finding the prime factorization of each number.

$$18 = 2 \cdot 3 \cdot 3 \quad 60 = 2 \cdot 2 \cdot 3 \cdot 5$$

Next, place the factorizations in a table, as shown below.

Prime	2	3	5
18 =	2 ·	3 · 3	
60 =	2 · 2 ·	3 ·	5

ANSWERS
1. (a) 16, 24, 32, 40, 48, . . .
(b) 20, 30, 40, 50, . . .
(c) 40

2. Find the least common multiple of 36 and 54.

(a) Find the prime factorization of each number.

(b) Complete this table.

Prime	2	3
36 =		
54 =		

(c) Identify the largest product in each column.

Prime	2	3
36 =	$2 \cdot 2 \cdot$	$3 \cdot 3$
54 =	$2 \cdot$	$3 \cdot 3 \cdot 3$

(d) Find the least common multiple.

$2 \cdot 2 \cdot$ ____ $\cdot$ ____ $\cdot$ ____ $=$

3. Find the least common multiple of the denominators in these fractions.

(a) $\frac{2}{3}$ and $\frac{1}{6}$

(b) $\frac{3}{10}$ and $\frac{6}{5}$

(c) $\frac{5}{8}$ and $\frac{1}{6}$

(d) $\frac{2}{15}$ and $\frac{3}{10}$

ANSWERS

2. (a) $36 = 2 \cdot 2 \cdot 3 \cdot 3$
$54 = 2 \cdot 3 \cdot 3 \cdot 3$

(b)

	2	3
36 =	$2 \cdot 2 \cdot$	$3 \cdot 3$
54 =	$2 \cdot$	$3 \cdot 3 \cdot 3$

(c) $2 \cdot 2$; $3 \cdot 3 \cdot 3$ **(d)** 3, 3, 3; 108

3. (a) 6 **(b)** 10 **(c)** 24 **(d)** 30

Then circle the largest product in each column and write this product in the bottom row of the table.

Prime	2	3	5
18 =	$2 \cdot$	($3 \cdot 3$)	
60 =	($2 \cdot 2 \cdot$)	$3 \cdot$	(5)
LCM =	($2 \cdot 2$)	($3 \cdot 3$)	(5)

Now multiply the circled products to find the least common multiple.

$$\text{least common multiple (LCM)} = 2 \cdot 2 \cdot 3 \cdot 3 \cdot 5 = 180$$

The smallest whole number divisible by both 18 and 60 is 180. ■

◀ WORK PROBLEM 2 AT THE SIDE.

■ **EXAMPLE 3** *Using Prime Factorization*

Find the least common multiple of 12, 18, and 40.

Write each prime factorization.

$$12 = 2 \cdot 2 \cdot 3 \quad 18 = 2 \cdot 3 \cdot 3 \quad 40 = 2 \cdot 2 \cdot 2 \cdot 5$$

Prepare the following table.

Prime	2	3	5
12 =	$2 \cdot 2 \cdot$	3	
18 =	$2 \cdot$	$3 \cdot 3$	
40 =	$2 \cdot 2 \cdot 2 \cdot$		5

Circle the largest product in each column.

Prime	2	3	5
12 =	$2 \cdot 2 \cdot$	3	
18 =	$2 \cdot$	($3 \cdot 3$)	
40 =	($2 \cdot 2 \cdot 2 \cdot$)		(5)
LCM =	($2 \cdot 2 \cdot 2$)	($3 \cdot 3$)	(5)

Now multiply the circled products.

$$\text{least common multiple (LCM)} = 2 \cdot 2 \cdot 2 \cdot 3 \cdot 3 \cdot 5 = 360$$

The smallest whole number divisible by 12, 18, and 40 is 360. ■

Note If two of the products in a column are equal, circle either one but not both. Only one product in each column will be used.

◀ WORK PROBLEM 3 AT THE SIDE.

■ EXAMPLE 4 *Finding the Least Common Multiple*

Find the least common multiple for each set of numbers.

(a) 5 and 35 **(b)** 20, 24, 42

(a) Write each prime factorization.

$$5 = 5 \qquad 35 = 5 \cdot 7$$

Prepare the table.

Prime	5	7
5 =	5	
35 =	5 ·	7

Circle the largest product in each column, in this case, the only prime number in the column.

Prime	5	7
5 =	5	
35 =	(5) ·	(7)
LCM =	(5)	(7)

least common multiple (LCM) $= 5 \cdot 7 = 35$

(b) Write each prime factorization

$$20 = 2 \cdot 2 \cdot 5 \quad 24 = 2 \cdot 2 \cdot 2 \cdot 3 \quad 42 = 2 \cdot 3 \cdot 7$$

Prepare the table.

Prime	2	3	5	7
20 =	2 · 2 ·		5	
24 =	2 · 2 · 2 ·	3		
42 =	2 ·	3 ·		7

Circle the largest product in each column.

Prime	2	3	5	7
20 =	2 · 2 ·		(5)	
24 =	(2 · 2 · 2 ·)	3		
42 =	2 ·	(3) ·		(7)
LCM =	(2 · 2 · 2)	(3)	(5)	(7)

least common multiple (LCM) $= 2 \cdot 2 \cdot 2 \cdot 3 \cdot 5 \cdot 7 = 840$ ■

WORK PROBLEM 4 AT THE SIDE. ▶▶

4. Find the least common multiple for each set of numbers.

(a) 12, 15

(b) 8, 9, 12

(c) 18, 20, 30

(d) 14, 15, 18, 45

ANSWERS

4. (a) 60 (b) 72 (c) 180 (d) 630

3 Some people like the following *alternative method* for finding the least common multiple. Try both methods, and use the one you prefer. As a review, a list of the first few primes follows.

2, 3, 5, 7, 11, 13, 17

EXAMPLE 5 *Alternative Method for Finding the Least Common Multiple*

Find the least common multiple of each set of numbers.

(a) 14 and 21 **(b)** 6, 15, 18

(a) Start by trying to divide 14 and 21 by the prime numbers listed above. Use the following shortcut.

Divide by 2, the first prime.

$$\begin{array}{r|rr} 2 & 14 & \cancel{21} \\ \hline & 7 & 21 \end{array}$$

Because 21 cannot be divided evenly by 2, cross 21 out and bring it down.

Divide by 3, the second prime.

$$\begin{array}{r|rr} 2 & 14 & \cancel{21} \\ 3 & \cancel{7} & 21 \\ \hline & 7 & 7 \end{array}$$

Since 7 cannot be divided evenly by 5, the third prime, skip 5, and divide by the next prime, 7.

Divide by 7, the fourth prime.

$$\begin{array}{r|rr} 2 & 14 & \cancel{21} \\ 3 & \cancel{7} & 21 \\ 7 & 7 & 7 \\ \hline & 1 & 1 \end{array}$$

All quotients are 1.

When all quotients are 1, multiply the prime numbers on the left side.

$$\text{least common multiple} = 2 \cdot 3 \cdot 7 = 42$$

The least common multiple of 14 and 21 is 42.

(b) Divide by 2.

$$\begin{array}{r|rrr} 2 & 6 & \cancel{15} & 18 \\ \hline & 3 & 15 & 9 \end{array}$$

Cross out 15 and bring it down.

Divide by 3.

$$\begin{array}{r|rrr} 2 & 6 & \cancel{15} & 18 \\ 3 & 3 & 15 & 9 \\ \hline & 1 & 5 & 3 \end{array}$$

Divide by 3 again.

$$\begin{array}{r|rrr} 2 & 6 & 15 & 18 \\ \hline 3 & 3 & 15 & 9 \\ \hline 3 & 1 & 5 & 3 \\ \hline & 1 & 5 & 1 \end{array}$$

Finally, divide by 5.

$$\begin{array}{r|rrr} 2 & 6 & 15 & 18 \\ \hline 3 & 3 & 15 & 9 \\ \hline 3 & 1 & 5 & 3 \\ \hline 5 & 1 & 5 & 1 \\ \hline & 1 & 1 & 1 \end{array} \quad \text{All quotients are 1.}$$

Multiply the prime numbers on the side.

$$2 \cdot 3 \cdot 3 \cdot 5 = 90 \leftarrow \text{least common multiple}$$ ■

WORK PROBLEMS 5 AND 6 AT THE SIDE.

4 When adding and subtracting unlike fractions, the least common multiple is used as the denominator of the fractions.

■ **EXAMPLE 6** *Writing a Fraction with an Indicated Denominator*
Write the fraction $\frac{2}{3}$ by using a denominator of 15.

Find a numerator, so that

$$\frac{2}{3} = \frac{}{15}.$$

To find the new numerator, first divide 15 by 3.

$$\frac{2}{3} = \frac{}{15} \qquad 15 \div 3 = 5$$

Multiply both numerator and denominator of the fraction $\frac{2}{3}$ by 5.

$$\frac{2}{3} = \frac{2 \cdot 5}{3 \cdot 5} = \frac{10}{15}$$ ■

This process is just the opposite of writing a fraction in lowest terms. Check the answer by writing $\frac{10}{15}$ in lowest terms; you should get $\frac{2}{3}$.

■ **EXAMPLE 7** *Changing to a New Denominator*
Write each fraction with the indicated denominator.

(a) $\frac{3}{8} = \frac{}{48}$ **(b)** $\frac{5}{6} = \frac{}{42}$

5. In the problems below, the divisions have already been worked out. Multiply the prime numbers on the left to find the least common multiple.

(a)
$$\begin{array}{r|rr} 2 & 6 & 15 \\ \hline 3 & 3 & 15 \\ \hline 5 & 1 & 5 \\ \hline & 1 & 1 \end{array}$$

(b)
$$\begin{array}{r|rr} 2 & 20 & 36 \\ \hline 2 & 10 & 18 \\ \hline 3 & 5 & 9 \\ \hline 3 & 5 & 3 \\ \hline 5 & 5 & 1 \\ \hline & 1 & 1 \end{array}$$

6. Find the least common multiple of each set of numbers.

(a) 8 and 18

(b) 25 and 30

(c) 4, 8, and 12

(d) 25, 20, 35

ANSWERS
5. (a) 30 (b) 180
6. (a) 72 (b) 150 (c) 24 (d) 700

7. Write each fraction by using the indicated denominator.

(a) $\frac{1}{2} = \frac{\quad}{12}$

(b) $\frac{7}{9} = \frac{\quad}{27}$

(c) $\frac{4}{5} = \frac{\quad}{50}$

(d) $\frac{6}{11} = \frac{\quad}{55}$

ANSWERS

7. (a) $\frac{6}{12}$ (b) $\frac{21}{27}$ (c) $\frac{40}{50}$ (d) $\frac{30}{55}$

(a) Divide 48 by 8, getting 6. Now multiply both numerator and denominator of $\frac{3}{8}$ by 6.

$$\frac{3}{8} = \frac{3 \cdot 6}{8 \cdot 6} = \frac{18}{48}$$ Multiply numerator and denominator by 6.

That is, $\frac{3}{8} = \frac{18}{48}$. As a check, write $\frac{18}{48}$ in lowest terms.

(b) Divide 42 by 6, getting 7. Next, multiply both numerator and denominator of $\frac{5}{6}$ by 7.

$$\frac{5}{6} = \frac{5 \cdot 7}{6 \cdot 7} = \frac{35}{42}$$ Multiply numerator and denominator by 7. ■

Note In Example 6 the fraction $\frac{2}{3}$ was multiplied by $\frac{5}{5}$. In Example 7 the fraction $\frac{3}{8}$ was multiplied by $\frac{6}{6}$ and $\frac{5}{6}$ was multiplied by $\frac{7}{7}$. The fractions $\frac{5}{5}$, $\frac{6}{6}$ and $\frac{7}{7}$ are all equal to 1.

$$\frac{5}{5} = 1 \qquad \frac{6}{6} = 1 \qquad \frac{7}{7} = 1$$

Recall that any number multiplied by 1 is the number itself.

WORK PROBLEM 7 AT THE SIDE.

3.2 NAME DATE HOUR

EXERCISES

Find the least common multiple of each set of numbers.

Example: 18, 30

Solution:
Complete a table.

Prime	2	3	5
18 =	2 ·	3 · 3	
30 =	2 ·	3 ·	5

Identify the largest product in each column.

Prime	2	3	5
18 =	2 ·	(3 · 3)	
30 =	(2) ·	3 ·	(5)
LCM =	(2)	(3 · 3)	(5)

The least common multiple (LCM) = **2 · 3 · 3 · 5 = 90.**

Alternative Method:

Divide by 2	2	18	30
Divide by 3	3	9	15
Divide by 3	3	3	5
Divide by 5	5	1	5
		1	1

Multiply the prime numbers on the side.
2 · 3 · 3 · 5 = **90** = least common multiple

1. 6, 12 12

2. 5, 10 10

3. 7, 21 21

4. 6, 18 18

5. 18, 24 72

6. 6, 14 42

7. 25, 40 200

8. 15, 35 105

9. 36, 45 180

10. 21, 30 210

11. 6, 9, 10 90

12. 9, 15, 20 180

13. 15, 24, 30 120

14. 6, 10, 12 60

15. 18, 20, 24 360

16. 20, 24, 30 120

17. 6, 8, 10, 12 120

18. 8, 9, 12, 18 72

19. 10, 15, 20, 25 300

20. 12, 15, 18, 20 180

21. 6, 9, 27, 36 108

22. 15, 20, 30, 40 120

23. 9, 10, 25, 27 1350

24. 5, 18, 25, 30 450

Rewrite each of the following fractions, so that it has a denominator of 24.

Example: $\frac{7}{12} = \frac{}{24}$

Solution: Divide 24 by 12, getting 2. Next, multiply both numerator and denominator by 2.

$$\frac{7}{12} = \frac{7 \cdot 2}{12 \cdot 2} = \frac{\mathbf{14}}{24}$$

25. $\frac{1}{2} = \frac{12}{24}$

26. $\frac{5}{8} = \frac{15}{24}$

27. $\frac{3}{4} = \frac{18}{24}$

28. $\frac{5}{12} = \frac{10}{24}$

29. $\frac{3}{8} = \frac{9}{24}$

30. $\frac{1}{4} = \frac{6}{24}$

Rewrite each of the following fractions with the indicated denominators.

31. $\frac{3}{8} = \frac{6}{16}$

32. $\frac{5}{16} = \frac{10}{32}$

33. $\frac{9}{10} = \frac{36}{40}$

34. $\frac{7}{11} = \frac{28}{44}$

35. $\frac{7}{8} = \frac{28}{32}$

36. $\frac{5}{12} = \frac{20}{48}$

37. $\frac{5}{6} = \frac{55}{66}$

38. $\frac{7}{8} = \frac{84}{96}$

39. $\frac{9}{8} = \frac{45}{40}$

40. $\frac{6}{5} = \frac{48}{40}$

41. $\frac{9}{7} = \frac{72}{56}$

42. $\frac{3}{2} = \frac{96}{64}$

43. $\frac{8}{3} = \frac{136}{51}$

44. $\frac{9}{5} = \frac{216}{120}$

45. $\frac{8}{11} = \frac{96}{132}$

46. $\frac{7}{15} = \frac{98}{210}$

47. $\frac{3}{16} = \frac{27}{144}$

48. $\frac{9}{32} = \frac{63}{224}$

49. There are two methods shown to find the least common multiple (LCM). Of the two methods, the prime factorization method and the alternative method, which will you use? Why? Would you ever use the other method?
Answer varies.

50. Explain in your own words how to write a fraction with an indicated denominator. Give an example changing $\frac{3}{4}$ to a fraction having 12 as a denominator showing how this is done.
Answer varies.

Find the least common multiple of the denominators of the following fractions.

▲ **51.** $\frac{17}{800}, \frac{23}{3600}$ 7200

▲ **52.** $\frac{53}{288}, \frac{115}{1568}$ 14,112

▲ **53.** $\frac{109}{1512}, \frac{47}{392}$ 10,584

▲ **54.** $\frac{61}{810}, \frac{37}{1170}$ 10,530

PREVIEW EXERCISES

Write each improper fraction as a mixed number. (For help, see **Section 2.2.***)*

55. $\frac{5}{3}$ $1\frac{2}{3}$

56. $\frac{8}{5}$ $1\frac{3}{5}$

57. $\frac{9}{5}$ $1\frac{4}{5}$

58. $\frac{14}{9}$ $1\frac{5}{9}$

59. $\frac{27}{7}$ $3\frac{6}{7}$

60. $\frac{28}{9}$ $3\frac{1}{9}$

3.3 ADDING AND SUBTRACTING UNLIKE FRACTIONS

OBJECTIVES

1. Add unlike fractions.
2. Add fractions vertically.
3. Subtract unlike fractions.

FOR EXTRA HELP

Tape 4

SSM pp. 74–76

MAC: A
IBM: A

1 To add unlike fractions we must first change them to like fractions (fractions with the same denominator). For example, the diagrams show $\frac{3}{8}$ and $\frac{1}{4}$.

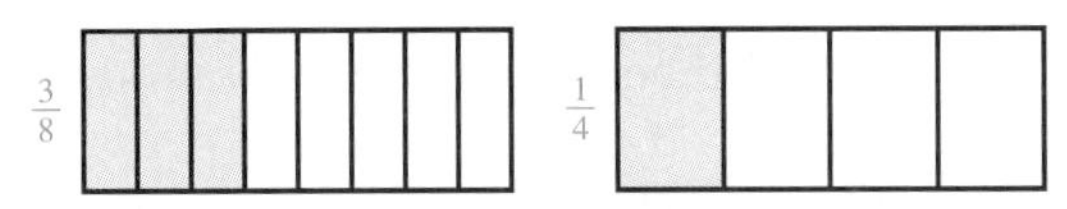

These fractions can be added by changing them to like fractions. Make like fractions by changing $\frac{1}{4}$ to the equivalent fraction $\frac{2}{8}$.

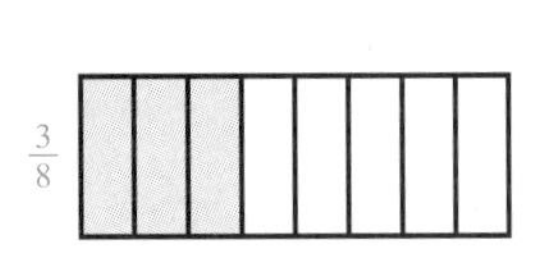

$\frac{1}{4} = \frac{2}{8}$

Next, add.

$$\frac{3}{8} + \frac{1}{4} = \frac{3}{8} + \frac{2}{8} = \frac{5}{8}$$

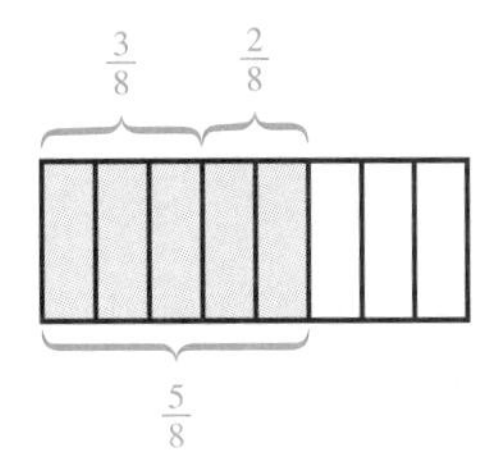

Unlike fractions may be added or subtracted using the following steps.

ADDING OR SUBTRACTING UNLIKE FRACTIONS

Step 1 Rewrite the **unlike fractions** as **like fractions** having the least common multiple as a denominator. This new denominator is called the **least common denominator (LCD).**

Step 2 Add or subtract as you did like fractions.

Step 3 Write the answer in lowest terms.

■ **EXAMPLE 1** *Adding Unlike Fractions*

Add $\frac{2}{3}$ and $\frac{1}{9}$.

The least common multiple of 3 and 9 is 9, so write the fractions as like fractions with a denominator of 9. This denominator is called the **least common denominator** of 3 and 9. First,

Step 1 $$\frac{2}{3} = \frac{\ }{9}.$$

Divide 9 by 3, getting 3. Next, multiply numerator and denominator by 3.

$$\frac{2}{3} = \frac{2 \cdot 3}{3 \cdot 3} = \mathbf{\frac{6}{9}}$$

Now, add the like fractions $\frac{6}{9}$ and $\frac{1}{9}$.

Step 2 $$\frac{2}{3}+\frac{1}{9}=\frac{6}{9}+\frac{1}{9}=\frac{6+1}{9}=\frac{7}{9}$$

Step 3 Step 3 is not needed because $\frac{7}{9}$ is already in lowest terms. ■

WORK PROBLEM 1 AT THE SIDE.

■ EXAMPLE 2 *Adding Fractions*

Add the following fractions using the three steps. Write all answers in lowest terms.

(a) $\frac{1}{4}+\frac{1}{6}$ **(b)** $\frac{6}{15}+\frac{3}{10}$

(a) The least common multiple of 4 and 6 is 12. Write both fractions as fractions with a least common denominator of 12.

$$\frac{1}{4}=\frac{3}{12} \quad \text{and} \quad \frac{1}{6}=\frac{2}{12}$$

Now add.

Add numerators.

$$\frac{1}{4}+\frac{1}{6}=\frac{3}{12}+\frac{2}{12}=\frac{3+2}{12}=\frac{5}{12}$$

(b) The least common multiple of 15 and 10 is 30, so write both fractions with a least common denominator of 30.

Rewrite as like fractions.

Step 1 $$\frac{6}{15}+\frac{3}{10}=\frac{12}{30}+\frac{9}{30}$$

Add numerators.

Step 2 $$\frac{12}{30}+\frac{9}{30}=\frac{21}{30}$$

Step 3 $$\frac{21}{30}=\frac{7}{10}$$ ← Lowest terms ■

WORK PROBLEM 2 AT THE SIDE.

2 Fractions can also be added vertically.

■ EXAMPLE 3 *Vertical Addition*

Add. Write answers in lowest terms.

(a)

$$\begin{array}{rcl} & \frac{3}{8} = & \frac{9}{24} \\ + & \frac{7}{12} = & \frac{14}{24} \\ \hline & & \frac{23}{24} \end{array}$$

← Add the numerators.

1. Add.

(a) $\frac{1}{4}+\frac{1}{2}$

(b) $\frac{3}{8}+\frac{1}{2}$

(c) $\frac{3}{10}+\frac{2}{5}$

(d) $\frac{1}{12}+\frac{5}{6}$

2. Add. Write answers in lowest terms.

(a) $\frac{3}{5}+\frac{1}{10}$

(b) $\frac{1}{3}+\frac{1}{12}$

(c) $\frac{1}{10}+\frac{1}{3}+\frac{1}{6}$

ANSWERS

1. (a) $\frac{3}{4}$ (b) $\frac{7}{8}$ (c) $\frac{7}{10}$ (d) $\frac{11}{12}$

2. (a) $\frac{7}{10}$ (b) $\frac{5}{12}$ (c) $\frac{3}{5}$

(b) $\dfrac{2}{9} = \dfrac{8}{36}$

$+\dfrac{1}{4} = \dfrac{9}{36}$

$\dfrac{17}{36}$ ■

WORK PROBLEM 3 AT THE SIDE. ▶▶

3 The next example shows subtraction of unlike fractions.

■ **EXAMPLE 4** *Subtracting Unlike Fractions*

Subtract the following fractions. Write answers in lowest terms. As with addition, rewrite unlike fractions with a least common denominator.

(a) $\dfrac{3}{4} - \dfrac{3}{8}$ **(b)** $\dfrac{3}{4} - \dfrac{5}{9}$

(a)

Step 1 $\dfrac{3}{4} - \dfrac{3}{8} = \dfrac{6}{8} - \dfrac{3}{8}$ Rewrite as like fractions.

Step 2 $\dfrac{6}{8} - \dfrac{3}{8} = \dfrac{3}{8}$ Subtract numerators.

Step 3 Not needed. ($\frac{3}{8}$ is in lowest terms.)

(b)

Step 1 $\dfrac{3}{4} - \dfrac{5}{9} = \dfrac{27}{36} - \dfrac{20}{36}$ Rewrite as like fractions.

Step 2 $\dfrac{27}{36} - \dfrac{20}{36} = \dfrac{27-20}{36} = \dfrac{7}{36}$ Subtract numerators.

Step 3 Not needed. ($\frac{7}{36}$ is in lowest terms.) ■

Note Step 3 was not needed in Example 4 because the answers $\frac{3}{8}$ and $\frac{7}{36}$ are both in lowest terms.

WORK PROBLEM 4 AT THE SIDE. ▶▶

3. Add.

(a) $\dfrac{1}{8} + \dfrac{3}{4}$

(b) $\dfrac{2}{3} + \dfrac{2}{9}$

4. Subtract. Write answers in lowest terms.

(a) $\dfrac{1}{2} - \dfrac{3}{8}$

(b) $\dfrac{5}{8} - \dfrac{1}{12}$

(c) $\dfrac{17}{18} - \dfrac{20}{27}$

ANSWERS

3. (a) $\dfrac{7}{8}$ (b) $\dfrac{8}{9}$

4. (a) $\dfrac{1}{8}$ (b) $\dfrac{13}{24}$ (c) $\dfrac{11}{54}$

QUEST FOR NUMERACY

Math and Music

The *time signature* at the beginning of a piece of music looks like a fraction. Commonly used time signatures are $\frac{2}{4}$, $\frac{3}{4}$, $\frac{4}{4}$, and $\frac{6}{8}$. Musicians use the time signature to tell how long to hold each note. Here is one way to think about notes when the time signature is $\frac{4}{4}$.

𝅝 $= 1$ 𝅗𝅥 $= \frac{1}{2}$ 𝅘𝅥 $= \frac{1}{4}$ 𝅘𝅥𝅮 $= \frac{1}{8}$ 𝅘𝅥𝅯 $= \frac{1}{16}$

Music is divided into measures. In $\frac{4}{4}$ time, each measure contains notes that add up to $\frac{4}{4}$. (Remember that $\frac{4}{4} = 1$.)

Directions: *Write one or more notes to make each measure add up to $\frac{4}{4}$.* **There are many solutions. Some possibilities are shown.**

Find *two different* ways to complete this measure.

Find *three different* solutions.

Find *three different* solutions.

Find *two different* solutions.

Find *three different* solutions.

Directions: *Divide each line of notes into measures. The first measure is done for you. Hint: There are eight measures on the first line of music and six measures on the second line of music.*

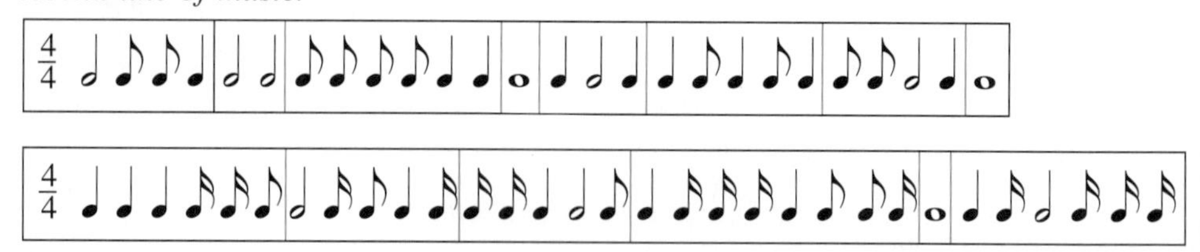

To extend this activity, ask students to make up similar problems in $\frac{4}{4}$ time to exchange with classmates. If one or more students are musicians, ask them to tap out the rhythm expressed by the note patterns, and to explain $\frac{3}{4}$, $\frac{2}{4}$, and $\frac{6}{8}$ time signatures.

3.3 EXERCISES

NAME DATE HOUR

Add the following fractions. Write answers in lowest terms.

Example: $\frac{2}{3} + \frac{1}{6}$ **Solution:** $\frac{2}{3} + \frac{1}{6} = \frac{4}{6} + \frac{1}{6} = \frac{4+1}{6} = \mathbf{\frac{5}{6}}$

Least common denominator is 6.

1. $\frac{2}{5} + \frac{1}{5}$ $\frac{3}{5}$

2. $\frac{3}{7} + \frac{2}{7}$ $\frac{5}{7}$

3. $\frac{5}{16} + \frac{3}{16}$ $\frac{1}{2}$

4. $\frac{5}{9} + \frac{1}{9}$ $\frac{2}{3}$

5. $\frac{9}{20} + \frac{3}{10}$ $\frac{3}{4}$

6. $\frac{5}{8} + \frac{1}{4}$ $\frac{7}{8}$

7. $\frac{9}{11} + \frac{1}{22}$ $\frac{19}{22}$

8. $\frac{5}{7} + \frac{3}{14}$ $\frac{13}{14}$

9. $\frac{2}{9} + \frac{5}{12}$ $\frac{23}{36}$

10. $\frac{5}{8} + \frac{1}{12}$ $\frac{17}{24}$

11. $\frac{1}{3} + \frac{3}{5}$ $\frac{14}{15}$

12. $\frac{2}{5} + \frac{3}{7}$ $\frac{29}{35}$

13. $\frac{1}{4} + \frac{2}{9} + \frac{1}{3}$ $\frac{29}{36}$

14. $\frac{3}{7} + \frac{2}{5} + \frac{1}{10}$ $\frac{13}{14}$

15. $\frac{3}{10} + \frac{2}{5} + \frac{3}{20}$ $\frac{17}{20}$

16. $\frac{1}{3} + \frac{3}{8} + \frac{1}{4}$ $\frac{23}{24}$

17. $\frac{4}{15} + \frac{1}{6} + \frac{1}{3}$ $\frac{23}{30}$

18. $\frac{5}{12} + \frac{2}{9} + \frac{1}{6}$ $\frac{29}{36}$

19. $\frac{1}{4} + \frac{2}{3}$ $\frac{11}{12}$

20. $\frac{5}{12} + \frac{3}{8}$ $\frac{19}{24}$

21. $\frac{8}{15} + \frac{3}{10}$ $\frac{5}{6}$

22. $\frac{1}{6} + \frac{5}{9}$ $\frac{13}{18}$

Subtract the following fractions. Write answers in lowest terms.

Example: $\frac{3}{5} - \frac{1}{2}$ **Solution:** $\frac{3}{5} - \frac{1}{2} = \frac{6}{10} - \frac{5}{10} = \frac{6-5}{10} = \mathbf{\frac{1}{10}}$

Least common denominator is 10.

23. $\frac{5}{6} - \frac{1}{6}$ $\frac{2}{3}$

24. $\frac{11}{12} - \frac{5}{12}$ $\frac{1}{2}$

25. $\frac{2}{3} - \frac{1}{6}$ $\frac{1}{2}$

26. $\frac{7}{8} - \frac{1}{2}$ $\frac{3}{8}$

27. $\frac{5}{12} - \frac{1}{4}$ $\frac{1}{6}$

28. $\frac{5}{6} - \frac{7}{9}$ $\frac{1}{18}$

29. $\frac{7}{8} - \frac{2}{3}$ $\frac{5}{24}$

30. $\frac{5}{7} - \frac{1}{3}$ $\frac{8}{21}$

31. $\frac{8}{9} - \frac{7}{15}$ $\frac{19}{45}$

32. $\frac{7}{8} - \frac{2}{3}$ = $\frac{5}{24}$

33. $\frac{4}{5} - \frac{2}{3}$ = $\frac{2}{15}$

34. $\frac{5}{8} - \frac{1}{3}$ = $\frac{7}{24}$

35. $\frac{4}{9} - \frac{5}{12}$ = $\frac{1}{36}$

36. $\frac{7}{10} - \frac{1}{4}$ = $\frac{9}{20}$

Solve each of the following word problems.

Example: The county painted a white line on $\frac{1}{6}$ of the road on Monday and on $\frac{1}{4}$ of the road on Tuesday. What fraction of the white line had been painted by the end of Tuesday?
Approach: To find the fraction painted during the two days, add the fraction painted on Monday to the fraction painted on Tuesday.
Solution: To find the total amount painted, add:

$$\frac{1}{6} \text{ on Monday} + \frac{1}{4} \text{ on Tuesday}$$

$$\frac{1}{6} + \frac{1}{4} = \frac{2}{12} + \frac{3}{12} = \frac{5}{12}$$

By the end of Tuesday, $\frac{5}{12}$ of the white line had been painted.

37. A tile mason purchased $\frac{1}{3}$ ton of white sand, $\frac{3}{8}$ ton of black sand and $\frac{1}{4}$ ton of pea gravel. How many tons of material were purchased?

$\frac{23}{24}$ ton

38. Fred Thompson paid $\frac{1}{8}$ of a debt in January, $\frac{1}{3}$ in February, $\frac{1}{4}$ in March, and $\frac{1}{12}$ in April. What fraction of the debt was paid in these 4 months?

$\frac{19}{24}$

39. The Weiner Works has $\frac{3}{4}$ acre of land. If $\frac{1}{6}$ acre must remain as a green belt and the remainder is buildable, find the amount of land that is buildable.

$\frac{7}{12}$ acre

40. Lori wants to open a day care center and has saved $\frac{2}{5}$ of the amount needed for start-up costs. If she saves another $\frac{1}{8}$ of the amount needed and then $\frac{1}{6}$ more, find the total portion of the start-up costs she has saved.

$\frac{83}{120}$ goal

41. A parcel of land is being developed into a park and sports field. Find the total distance around (perimeter of) this parcel of land.

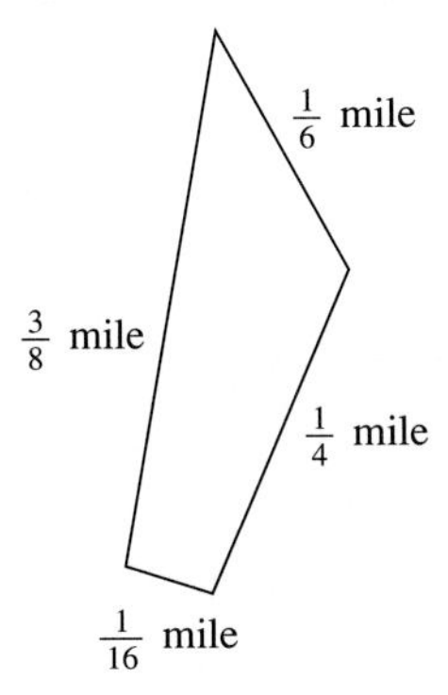

$\frac{41}{48}$ mile

42. Find the total length of this bolt.

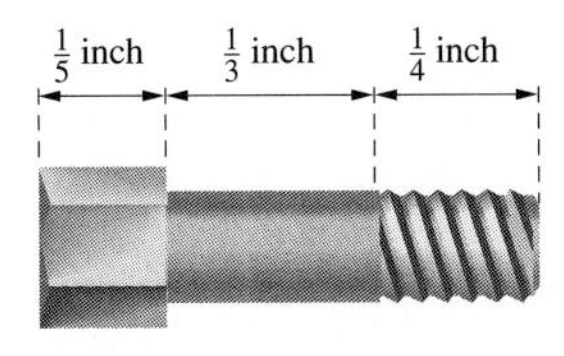

$\frac{47}{60}$ inch

43. A hydraulic jack contains $\frac{7}{8}$ gallon of hydraulic fluid. A cracked seal resulted in a loss of $\frac{1}{6}$ gallon of fluid in the morning and another $\frac{1}{3}$ gallon in the afternoon. Find the amount of fluid remaining.

$\frac{3}{8}$ gallon

44. Mike Odesa drives a gasoline truck. He leaves the refinery with his tanker filled to $\frac{3}{4}$ of capacity. If he delivers $\frac{1}{3}$ of the tank capacity at the first stop and $\frac{3}{8}$ of the tank capacity at the second stop, find the fraction of the tanker's capacity remaining.

$\frac{1}{24}$ capacity

45. Step 1 in adding or subtracting unlike fractions is to rewrite the fractions so they have the least common multiple as a denominator. Explain in your own words why this is necessary. **Answer varies.**

46. Briefly list the three steps used for addition and subtraction of unlike fractions.
Answer varies.

Refer to the circle graph to answer exercises 47–50.

The Day of the Student

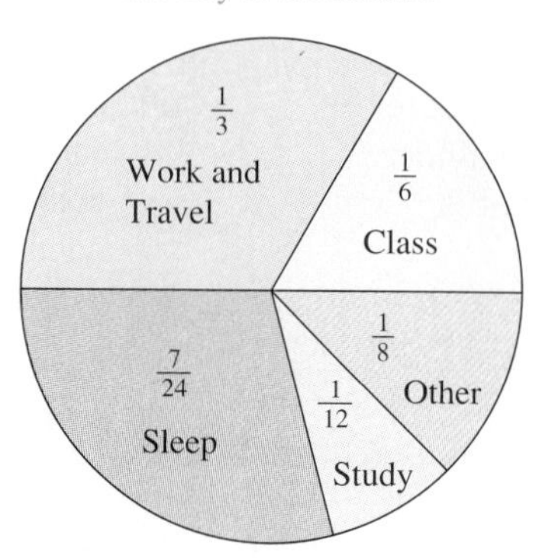

47. What fraction of the day was spent in class and study?

$\frac{1}{4}$

48. What fraction of the day was spent in work and travel and other?

$\frac{11}{24}$

49. In which activity was the greatest amount of time spent?

work and travel

50. In which activity was the least amount of time spent?

study

51. Sheri Minkner will swim $\frac{7}{8}$ of a mile in a five-day period. If she swims $\frac{1}{8}$ mile on both Monday and Wednesday, $\frac{1}{6}$ mile on both Tuesday and Thursday, and the balance on Friday, find the distance she must swim on Friday.

$\frac{7}{24}$ **mile**

52. Find the diameter of the hole in the bracket pictured below.

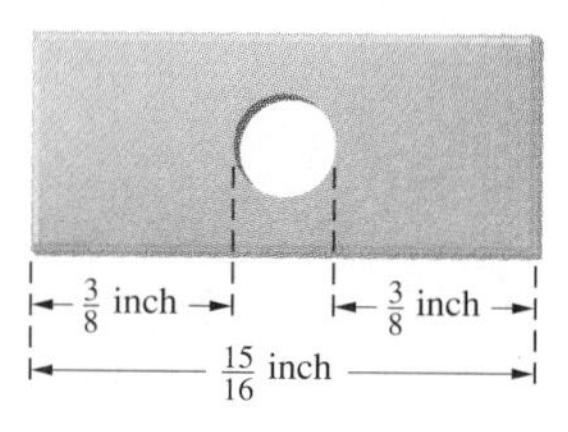

$\frac{3}{16}$ **inch**

Preview Exercises

Multiply or divide as indicated. Write answers as mixed numbers. (For help, see ***Section 2.8.****)*

53. $6\frac{1}{2} \cdot 3\frac{1}{4}$ $21\frac{1}{8}$

54. $5\frac{1}{4} \cdot 2\frac{1}{3}$ $12\frac{1}{4}$

55. $5 \cdot 2\frac{7}{10}$ $13\frac{1}{2}$

56. $3\frac{1}{4} \div 2\frac{5}{8}$ $1\frac{5}{21}$

57. $1\frac{1}{2} \div 3\frac{3}{4}$ $\frac{2}{5}$

58. $5\frac{3}{4} \div 2$ $2\frac{7}{8}$

3.4 ADDING AND SUBTRACTING MIXED NUMBERS

Recall that a mixed number is the sum of a whole number and a fraction. For example,

$$3\frac{2}{5} \quad \text{means} \quad 3 + \frac{2}{5}.$$

1 Add or subtract mixed numbers by adding or subtracting the fraction parts and then the whole number parts.

WORK PROBLEM 1 AT THE SIDE.

EXAMPLE 1 *Adding and Subtracting Mixed Numbers*

Add or subtract.

(a) $\begin{array}{r} 16\frac{1}{8} \\ +\ 5\frac{5}{8} \\ \hline \end{array}$ **(b)** $\begin{array}{r} 8\frac{5}{8} \\ -\ 3\frac{1}{12} \\ \hline \end{array}$

(a) $\begin{array}{r} 16\frac{1}{8} \\ +\ 5\frac{5}{8} \\ \hline 21\frac{6}{8} \end{array}$ ← sum of fractions; 21 ← sum of whole numbers

In lowest terms $\frac{6}{8}$ is $\frac{3}{4}$, so the final answer is $21\frac{3}{4}$.

(b) $\begin{array}{rcr} 8\frac{5}{8} & = & 8\frac{15}{24} \\ -\ 3\frac{1}{12} & = & 3\frac{2}{24} \\ \hline & & 5\frac{13}{24} \end{array}$

24 ← least common denominator; $\frac{13}{24}$ ← Subtract fractions.; 5 ← Subtract whole numbers.

Just as before, check by adding $5\frac{13}{24}$ and $3\frac{1}{12}$; the sum should be $8\frac{5}{8}$. ■

Note Mixed numbers usually are arranged vertically when adding or subtracting.

WORK PROBLEM 2 AT THE SIDE.

When you add the fraction parts of mixed numbers, the answer may be greater than 1. If this happens, **carrying** from the fraction column to the whole number is the best procedure. (You wouldn't want a whole number along with an improper fraction.)

OBJECTIVES

1 Add or subtract mixed numbers.

2 Subtract by using borrowing.

FOR EXTRA HELP

Tape 5

SSM pp. 77–80

MAC: A IBM: A

1. As a review of mixed numbers, convert mixed numbers to improper fractions and improper fractions to mixed numbers.

(a) $\frac{5}{2}$ **(b)** $\frac{14}{3}$

(c) $6\frac{3}{4}$ **(d)** $4\frac{5}{8}$

2. Add or subtract.

(a) $\begin{array}{r} 4\frac{5}{8} \\ +\ 3\frac{1}{4} \\ \hline \end{array}$

(b) $25\frac{3}{5} + 12\frac{3}{10}$

(c) $\begin{array}{r} 5\frac{4}{9} \\ -\ 3\frac{1}{3} \\ \hline \end{array}$

(d) $22\frac{1}{2} - 12\frac{1}{3}$

ANSWERS

1. (a) $2\frac{1}{2}$ (b) $4\frac{2}{3}$ (c) $\frac{27}{4}$ (d) $\frac{37}{8}$

2. (a) $7\frac{7}{8}$ (b) $37\frac{9}{10}$ (c) $2\frac{1}{9}$ (d) $10\frac{1}{6}$

3. Add.

(a) $\begin{array}{r} 9\frac{3}{4} \\ +\ 7\frac{1}{2} \\ \hline \end{array}$

(b) $\begin{array}{r} 18\frac{2}{3} \\ +\ 16\frac{3}{5} \\ \hline \end{array}$

(c) $\begin{array}{r} 45\frac{5}{8} \\ 36\frac{3}{4} \\ +\ 51\frac{1}{2} \\ \hline \end{array}$

ANSWERS

3. (a) $17\frac{1}{4}$ (b) $35\frac{4}{15}$ (c) $133\frac{7}{8}$

■ **EXAMPLE 2** *Carrying When Adding Mixed Numbers*

Add $9\frac{5}{8} + 13\frac{7}{8}$.

$$\begin{array}{r} 9\ \frac{5}{8} \\ +\ 13\ \frac{7}{8} \\ \hline 22\frac{12}{8} \end{array}$$

$\frac{12}{8}$ ← sum of fractions; 22 ← sum of whole numbers

The improper fraction $\frac{12}{8}$ can be written in lowest terms as $\frac{3}{2}$. Because $\frac{3}{2} = 1\frac{1}{2}$, the sum

$$22\frac{12}{8} = 22 + \frac{12}{8} = 22 + 1\frac{1}{2} = 23\frac{1}{2}. \quad ■$$

Note When adding mixed numbers, first add the fraction parts, then add the whole number parts. Then combine the two answers.

◀ **WORK PROBLEM 3 AT THE SIDE.**

2 Borrowing is sometimes necessary when subtracting mixed numbers.

■ **EXAMPLE 3** *Borrowing When Subtracting Mixed Numbers*

Subtract.

(a) $\begin{array}{r} 7 \\ -\ 2\frac{5}{6} \\ \hline \end{array}$ **(b)** $\begin{array}{r} 8\frac{1}{3} \\ -\ 4\frac{3}{5} \\ \hline \end{array}$

(a) $\begin{array}{r} 7 \\ -\ 2\frac{5}{6} \\ \hline \end{array}$ There is no fraction from which to subtract $\frac{5}{6}$.

It is not possible to subtract $\frac{5}{6}$ without borrowing from the whole number 7 first.

$$7 = 6 + 1 \quad \text{Borrow 1.}$$

$$= 6 + \frac{6}{6} \quad 1 = \frac{6}{6}$$

$$= 6\frac{6}{6}$$

Next, subtract.

$$\begin{array}{rcr} 7 & = & 6\frac{6}{6} \\ -\ 2\frac{5}{6} & = & 2\frac{5}{6} \\ \hline & & 4\frac{1}{6} \end{array}$$

(b)

$$\begin{array}{rcr} 8\frac{1}{3} & = & 8\frac{5}{15} \\ -\ 4\frac{3}{5} & = & 4\frac{9}{15} \\ \hline \end{array}$$

least common denominator

It is not possible to subtract $\frac{9}{15}$ from $\frac{5}{15}$, so borrow from the whole number **8.**

Borrow 1.

$$8\frac{5}{15} = 8 + \frac{5}{15} = 7 + 1 + \frac{5}{15}$$

1 is $\frac{15}{15}$

$$= 7 + \frac{15}{15} + \frac{5}{15}$$

$$= 7 + \frac{\mathbf{20}}{\mathbf{15}} \longleftarrow \frac{15}{15} + \frac{5}{15}$$

$$= 7\frac{20}{15}$$

Next, subtract.

$$\begin{array}{rcrcr} 8\frac{1}{3} & = & 8\frac{5}{15} & = & 7\frac{20}{15} \\ -\ 4\frac{3}{5} & = & 4\frac{9}{15} & = & 4\frac{9}{15} \\ \hline & & & & 3\frac{11}{15} \end{array}$$

The answer is $3\frac{11}{15}$ (lowest terms). ■

WORK PROBLEM 4 AT THE SIDE. ▶▶

4. Subtract.

(a) $\begin{array}{r} 4\frac{1}{3} \\ -\ 1\frac{5}{6} \\ \hline \end{array}$

(b) $\begin{array}{r} 2\frac{5}{8} \\ -\ 1\frac{15}{16} \\ \hline \end{array}$

(c) $\begin{array}{r} 25\frac{1}{6} \\ -\ 18\frac{11}{15} \\ \hline \end{array}$

ANSWERS

4. (a) $2\frac{1}{2}$ **(b)** $\frac{11}{16}$ **(c)** $6\frac{13}{30}$

QUEST FOR NUMERACY

Estimation and Fractions

It's always a good idea to *estimate* the answer to a problem. This will help you catch mistakes. When working with *mixed numbers*, one simple way to estimate is to round to the nearest whole number, then add, subtract, multiply, or divide the whole numbers.

Examples: $2\frac{1}{8}$ rounds to 2 $\quad$ $5\frac{7}{10}$ rounds to 6 $\quad$ $1\frac{1}{2}$ rounds to 2

HINT: If the numerator is *half* of the denominator *or more*, round up the whole number part. If the numerator is *less* than half the denominator, leave the whole number part as it is.

Try estimating the answers to the Example problems from the previous three pages.

Example 1

(a) *exact* → *estimate*

$16\frac{1}{8}$ rounds to $\longrightarrow$ 16

$+\ 5\frac{5}{8} \longrightarrow +\ 6$

$21\frac{3}{4}$ $\quad$ 22

(b) *exact* → *estimate*

$8\frac{5}{8}$ rounds to $\longrightarrow$ 9

$-\ 3\frac{1}{12} \longrightarrow -\ 3$

$5\frac{13}{24}$ $\quad$ 6

Example 2 *exact* → *estimate*

$9\frac{5}{8} \longrightarrow 10$

$+\ 13\frac{7}{8} \longrightarrow +\ 14$

$23\frac{1}{2}$ $\quad$ 24

Example 3

(a) *exact* → *estimate*

$7 \longrightarrow 7$

$-\ 2\frac{5}{6} \longrightarrow -\ 3$

$4\frac{1}{6}$ $\quad$ 4

(b) *exact* → *estimate*

$8\frac{1}{3} \longrightarrow 8$

$-\ 4\frac{3}{5} \longrightarrow -\ 5$

$3\frac{11}{15}$ $\quad$ 3

Now round to the nearest whole number and *estimate* each answer. On separate paper, work out the exact answers.

$5 \longrightarrow 5$; $-\ 2\frac{11}{16} \longrightarrow -\ 3$; estimate: 2

$1\frac{2}{9} \longrightarrow 1$; $+\ 6\frac{1}{3} \longrightarrow +\ 6$; estimate: 7

$4\frac{5}{8} \longrightarrow 5$; $-\ 3\frac{9}{10} \longrightarrow -\ 4$; estimate: 1

$6\frac{2}{5} \longrightarrow 6$; $+\ 6\frac{5}{9} \longrightarrow +\ 7$; estimate: 13

$4\frac{3}{4} \cdot 4\frac{2}{7}$ $\downarrow$ $5 \cdot 4 = 20$

$8\frac{1}{6} \div 1\frac{11}{12}$ $\downarrow$ $8 \div 2 = 4$

$2\frac{4}{5} \cdot 10\frac{5}{8}$ $\downarrow$ $3 \cdot 11 = 33$

LaRae needs $2\frac{3}{4}$ cups of flour for one recipe, $1\frac{2}{3}$ cup for a second recipe, and $3\frac{1}{4}$ cups for a third recipe. She has about $5\frac{1}{3}$ cups of flour in her kitchen. Estimate how much flour LaRae will need to borrow from her neighbor.

She needs 3 + 2 + 3 = 8 cups. $\quad$ **8 − 5 = 3 cups to borrow**

EXERCISES

Add. Write answers as mixed numbers.

Examples:

$$\begin{array}{r} 8\frac{3}{4} \\ +\ 2\frac{1}{8} \\ \hline \end{array}$$

Solution:

$$\begin{array}{rcl} 8\frac{3}{4} & = & 8\frac{6}{8} \\ +\ 2\frac{1}{8} & = & 2\frac{1}{8} \\ \hline & & \mathbf{10\frac{7}{8}} \end{array}$$

$\frac{7}{8}$ ← Add the fraction parts.

10 ← Add the whole number parts.

$$\begin{array}{r} 2\frac{3}{5} \\ +\ 9\frac{2}{3} \\ \hline \end{array}$$

Solution:

$$\begin{array}{rcl} 2\frac{3}{5} & = & 2\frac{9}{15} \\ +\ 9\frac{2}{3} & = & 9\frac{10}{15} \\ \hline & & 11\frac{19}{15} \end{array}$$

$\frac{19}{15} = 1\frac{4}{15}$, so

$11\frac{19}{15} = 11 + 1\frac{4}{15}$

$= \mathbf{12\frac{4}{15}}$

1. $8\frac{3}{8} + 9\frac{1}{8}$ — $17\frac{1}{2}$

2. $7\frac{1}{10} + 2\frac{3}{10}$ — $9\frac{2}{5}$

3. $1\frac{1}{4} + 9\frac{1}{2}$ — $10\frac{3}{4}$

4. $5\frac{1}{6} + 6\frac{2}{3}$ — $11\frac{5}{6}$

5. $\frac{3}{8} + 15\frac{1}{4}$ — $15\frac{5}{8}$

6. $26\frac{5}{8} + \frac{1}{12}$ — $26\frac{17}{24}$

7. $82\frac{3}{5} + 15\frac{4}{5}$ — $98\frac{2}{5}$

8. $24\frac{5}{6} + 18\frac{5}{6}$ — $43\frac{2}{3}$

9. $14\frac{6}{7} + 5\frac{1}{2}$ — $20\frac{5}{14}$

10. $3\frac{3}{5} + 8\frac{1}{2}$ — $12\frac{1}{10}$

11. $68\frac{3}{5} + 5\frac{3}{8}$ — $73\frac{39}{40}$

12. $2\frac{3}{4} + 15\frac{3}{7}$ — $18\frac{5}{28}$

13. $7\frac{1}{4} + 25\frac{3}{8} + 9\frac{1}{2}$ — $42\frac{1}{8}$

14. $18\frac{3}{5} + 47\frac{7}{10} + 25\frac{8}{15}$ — $91\frac{5}{6}$

15. $28\frac{1}{4} + 23\frac{3}{5} + 19\frac{9}{10}$ — $71\frac{3}{4}$

16. $32\frac{3}{4} + 6\frac{1}{3} + 14\frac{5}{8}$ — $53\frac{17}{24}$

17. $27\frac{7}{15} + 30\frac{1}{2} + 9\frac{3}{10}$ — $67\frac{4}{15}$

18. $16\frac{7}{10} + 26\frac{1}{5} + 8\frac{3}{8}$ — $51\frac{11}{40}$

Subtract. Write answers as mixed numbers.

Example: $5\frac{3}{5} - 2\frac{1}{10}$

Solution:

$$5\frac{3}{5} = 5\frac{6}{10}$$
$$-\,2\frac{1}{10} = 2\frac{1}{10}$$
$$3\frac{5}{10} = \mathbf{3\frac{1}{2}} \text{ (lowest terms)}$$

$6 - 4\frac{7}{8}$

Solution: $6 - 4\frac{7}{8}$

Borrow.
$$6 = 5 + 1 = 5 + \frac{8}{8} = 5\frac{8}{8}$$

Subtract.
$$6 = 5\frac{8}{8}$$
$$-\,4\frac{7}{8} = 4\frac{7}{8}$$
$$\mathbf{1\frac{1}{8}}$$

19. $6\frac{4}{5} - 3\frac{3}{5}$ — $3\frac{1}{5}$

20. $9\frac{3}{4} - 6\frac{1}{4}$ — $3\frac{1}{2}$

21. $6\frac{7}{12} - 2\frac{1}{3}$ — $4\frac{1}{4}$

22. $11\frac{9}{20} - 4\frac{3}{5}$ — $6\frac{17}{20}$

23. $28\frac{3}{10} - 6\frac{1}{15}$ — $22\frac{7}{30}$

24. $15\frac{7}{20} - 6\frac{1}{8}$ — $9\frac{9}{40}$

25. $19 - 8\frac{7}{8}$ — $10\frac{1}{8}$

26. $35 - 17\frac{3}{8}$ — $17\frac{5}{8}$

27. $68\frac{3}{8} - 6\frac{4}{5}$ — $61\frac{23}{40}$

28. $47\frac{3}{8} - 6\frac{7}{12}$ — $40\frac{19}{24}$

29. $25\frac{13}{24} - 18\frac{15}{16}$ — $6\frac{29}{48}$

30. $26\frac{5}{18} - 12\frac{11}{24}$ — $13\frac{59}{72}$

31. $157 - 86\frac{14}{15}$ — $70\frac{1}{15}$

32. $374 - 211\frac{5}{6}$ — $162\frac{1}{6}$

33. $429\frac{15}{16} - 57$ — $372\frac{15}{16}$

34. $625\frac{11}{12} - 319$ — $306\frac{11}{12}$

35. $15 - 8\frac{3}{7}$ — $6\frac{4}{7}$

36. $21 - 5\frac{7}{8}$ — $15\frac{1}{8}$

37. $415 - 198\frac{3}{4}$ — $216\frac{1}{4}$

38. $232 - 58\frac{20}{21}$ — $173\frac{1}{21}$

39. $410 - 203\frac{24}{25}$ — $206\frac{1}{25}$

40. Use three short sentences to explain how to add unlike fractions.

Answer varies.

41. When subtracting mixed numbers explain when you need to borrow. Explain how to borrow using an example.

Answer varies.

Solve each word problem by using addition or subtraction.

42. Adele studied $15\frac{1}{8}$ hours over the weekend. If she studied $6\frac{1}{2}$ hours on Saturday, find the number of hours she studied on Sunday.

$8\frac{5}{8}$ hours

43. The gas tank on a 4×4 pick-up truck has a capacity of $18\frac{2}{3}$ gallons. Starting with a full tank, Andre used $12\frac{3}{4}$ gallons of gasoline. Find the number of gallons that remain.

$5\frac{11}{12}$ gallons

44. A carpenter has two pieces of oak trim. One piece of trim is $12\frac{1}{2}$ feet long and the other is $8\frac{2}{3}$ feet in length. How many feet of oak trim does he have in all?

$21\frac{1}{6}$ ft

45. On Monday, $5\frac{3}{4}$ tons of cans were recycled, and $9\frac{3}{5}$ tons were recycled on Tuesday. How many tons were recycled in total on these two days?

$15\frac{7}{20}$ tons

46. Mike Kane worked $6\frac{3}{8}$ hours on Monday, $7\frac{1}{2}$ hours on Tuesday, $8\frac{3}{4}$ hours on Wednesday, $7\frac{3}{8}$ hours on Thursday, and 8 hours on Friday. How many hours did he work altogether?

38 hours

47. Hernando Ramirez drove for $5\frac{1}{2}$ hours on the first day of his vacation, $6\frac{1}{4}$ hours on the second day, $3\frac{3}{4}$ hours on the third day, and 7 hours on the fourth day. How many hours did he drive altogether?

$22\frac{1}{2}$ hours

48. A landscaper has $9\frac{5}{8}$ cubic yards of peat moss in a truck. If he unloads $1\frac{1}{2}$ cubic yards at the first stop, $2\frac{3}{4}$ cubic yards at the second stop, and 3 cubic yards at the third stop, how much peat moss remains in the truck?

$2\frac{3}{8}$ cubic yards

49. Marv Levenson bought 15 yards of material at a sale. He made two tops with $3\frac{3}{4}$ yards of the material, a suit for his wife with $4\frac{1}{8}$ yards, and a jacket with $3\frac{7}{8}$ yards. Find the number of yards of material remaining.

$3\frac{1}{4}$ yards

50. The exercise yard at the correction center has four sides and is enclosed with $527\frac{1}{24}$ feet of security fencing around it. If three sides of the yard measure $107\frac{2}{3}$ feet, $150\frac{3}{4}$ feet, and $138\frac{5}{8}$ feet, find the length of the fourth side.

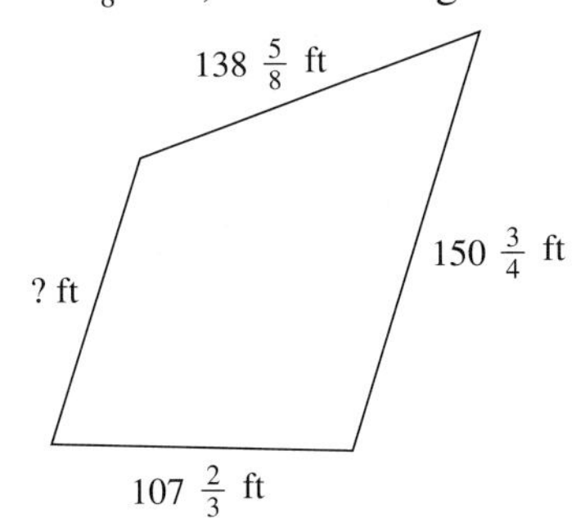

130 ft

51. Three sides of a parking lot are $108\frac{1}{4}$ feet, $162\frac{3}{8}$ feet, and $143\frac{1}{2}$ feet. If the distance around the lot is $518\frac{3}{4}$ feet, find the length of the fourth side.

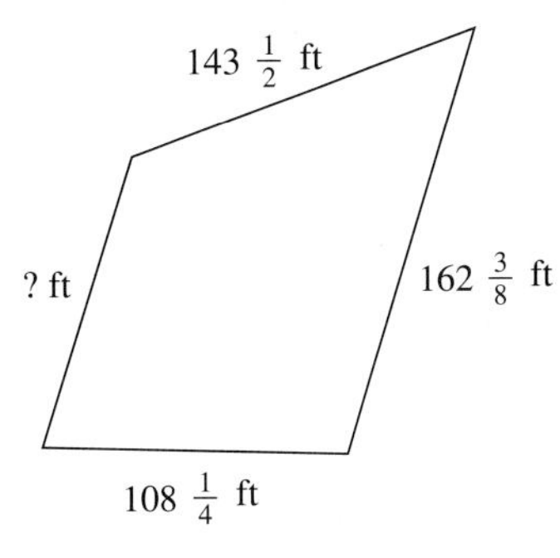

$104\frac{5}{8}$ ft

52. The Eastside Wholesale Vegetable Market sold $3\frac{1}{4}$ tons of broccoli, $2\frac{3}{8}$ tons of spinach, $7\frac{1}{2}$ tons of corn, and $1\frac{5}{6}$ tons of turnips last month. Find the total number of tons of these vegetables sold by the market last month.

$14\frac{23}{24}$ tons

53. Comet Auto Supply sold $16\frac{1}{2}$ cases of generic brand oil last week, $12\frac{1}{8}$ cases of Havoline Oil, $8\frac{3}{4}$ cases of Valvoline Oil, and $12\frac{5}{8}$ cases of Castrol Oil. Find the total number of cases of oil that Comet Auto Supply sold during the week.

50 cases

Find the length of the section represented by x *in the following figures.*

▲ **54.**

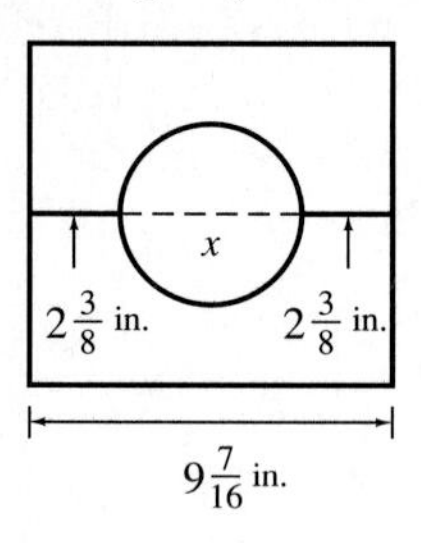

$4\frac{11}{16}$ in

▲ **55.**

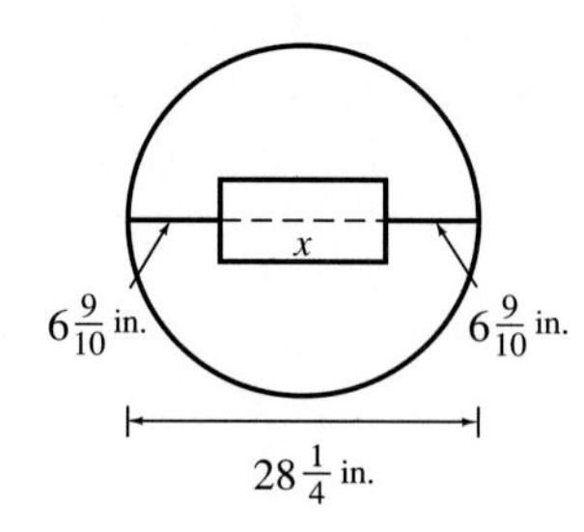

$14\frac{9}{20}$ in

▲ **56.**

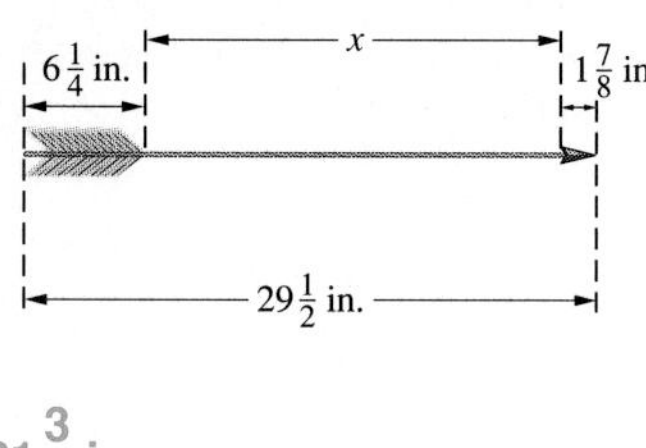

$21\frac{3}{8}$ in

▲ **57.**

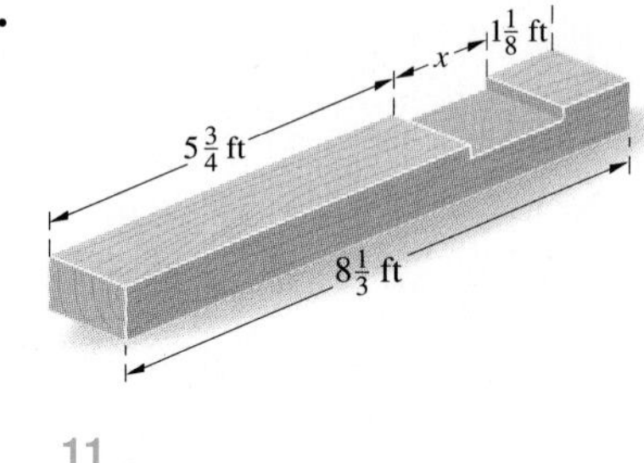

$1\frac{11}{24}$ ft

PREVIEW EXERCISES

Work each problem, using the order of operations. (For help, see **Section 1.8.***)*

58. $9^2 + 5 - 2$ 84

59. $2 \cdot 7 - 4$ 10

60. $4 \cdot 1 + 8 \cdot 7 + 3$ 63

61. $8 + 9 \div 3 + 6 \cdot 2$ 23

62. $3^2 \cdot (5 - 2)$ 27

63. $8 \cdot 4 - (15 - 8)$ 25

64. $(16 - 6) \cdot 2^3 - (3 \cdot 9)$ 53

3.5 ORDER RELATIONS AND THE ORDER OF OPERATIONS

Fractions, like whole numbers, can be located on a number line. For fractions, divide the space between whole numbers into equal parts.

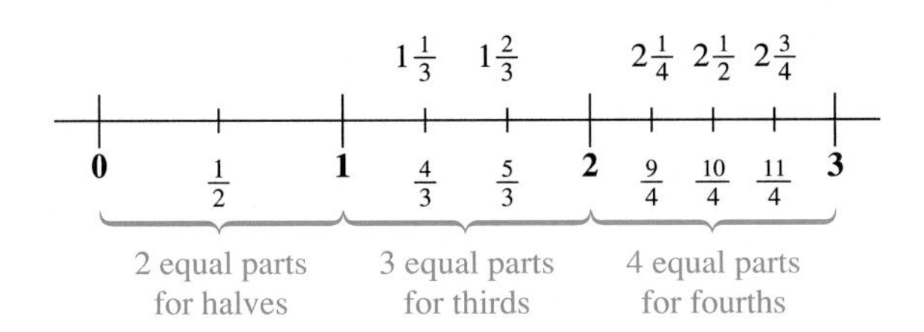

1 To compare the size of two numbers place the two numbers on a number line.

COMPARING THE SIZE OF TWO NUMBERS

The number farther to the left on the number line is always smaller and the number farther to the right on the number line is always larger.

For example, on the number line above, $\frac{1}{2}$ is to the left of $\frac{4}{3}$ ($1\frac{1}{3}$), so $\frac{1}{2}$ is smaller than $\frac{4}{3}$ ($1\frac{1}{3}$). Write these *order relations* by using the following symbols.

SYMBOLS FOR LESS THAN AND GREATER THAN

$<$	is less than
$>$	is greater than

■ EXAMPLE 1 *Using Less Than and Greater Than Symbols*

Write the following using $<$ and $>$ symbols.

(a) $\frac{1}{2}$ is less than $\frac{4}{3}$

(b) $\frac{9}{4}$ is greater than 1

(c) $\frac{5}{3}$ is less than $\frac{11}{4}$

(a) Write $\frac{1}{2}$ is less than $\frac{4}{3}$ as $\frac{1}{2} < \frac{4}{3}$

(b) Write $\frac{9}{4}$ is greater than 1 as $\frac{9}{4} > 1$

(c) Write $\frac{5}{3}$ is less than $\frac{11}{4}$ as $\frac{5}{3} < \frac{11}{4}$ ■

Note Using the number line above helps with Example 1. A number line is a very useful tool when working with order relations.

WORK PROBLEM 1 AT THE SIDE.

OBJECTIVES

1. Identify the greater of two fractions.
2. Use exponents with fractions.
3. Use the order of operations.

FOR EXTRA HELP

Tape 5	SSM pp. 80–83	MAC: A IBM: A

1. Use the number line in the text to help place $<$ or $>$ in each blank to make a true statement.

(a) 2 ____ $\frac{9}{4}$

(b) $\frac{11}{4}$ ____ $\frac{4}{3}$

(c) 0 ____ 2

(d) $\frac{11}{4}$ ____ $\frac{5}{3}$

ANSWERS
1. (a) $<$ (b) $>$ (c) $<$ (d) $>$

2. Place < or > in each blank to make a true statement.

(a) $\frac{3}{8}$ ____ $\frac{7}{12}$

(b) $\frac{11}{18}$ ____ $\frac{5}{9}$

(c) $\frac{17}{24}$ ____ $\frac{5}{6}$

(d) $\frac{13}{15}$ ____ $\frac{8}{9}$

The fraction $\frac{7}{8}$ represents 7 of 8 equivalent parts, while $\frac{3}{8}$ means 3 of 8 equivalent parts. Because $\frac{7}{8}$ represents more of the equivalent parts, $\frac{7}{8}$ is greater than $\frac{3}{8}$, or

$$\frac{7}{8} > \frac{3}{8}.$$

To identify the greater fraction use the following steps.

IDENTIFYING THE GREATER FRACTION

Step 1 Write the fractions as like fractions.
Step 2 Compare the numerators. The fraction with the greater numerator is the greater fraction.

EXAMPLE 2 *Identifying the Greater Fraction*
Decide which fraction in each pair is greater.

(a) $\frac{7}{8}, \frac{9}{10}$ **(b)** $\frac{8}{5}, \frac{23}{15}$

(a) First, write the fractions as like fractions. The least common multiple for 8 and 10 is 40, so

$$\frac{7}{8} = \frac{7 \cdot 5}{8 \cdot 5} = \frac{35}{40} \quad \text{and} \quad \frac{9}{10} = \frac{9 \cdot 4}{10 \cdot 4} = \frac{36}{40}.$$

Look at the numerators. Because 36 is greater than 35, $\frac{36}{40}$ is greater than $\frac{35}{40}$. Because $\frac{36}{40}$ is equivalent to $\frac{9}{10}$,

$$\frac{9}{10} > \frac{7}{8} \quad \text{or} \quad \frac{7}{8} < \frac{9}{10}.$$

The greater fraction is $\frac{9}{10}$.

(b) The least common multiple of 5 and 15 is 15.

$$\frac{8}{5} = \frac{8 \cdot 3}{5 \cdot 3} = \frac{24}{15} \quad \text{and} \quad \frac{23}{15} = \frac{23}{15}$$

This shows that $\frac{8}{5}$ is greater than $\frac{23}{15}$, or

$$\frac{8}{5} > \frac{23}{15}. \blacksquare$$

◀ WORK PROBLEM 2 AT THE SIDE.

2 Exponents were used in Chapter 1 to write repeated products. For example,

$$\underset{\text{two factors of 3}}{3^2 = \underbrace{3 \cdot 3}} = 9 \quad \text{and} \quad \underset{\text{three factors of 5}}{5^3 = \underbrace{5 \cdot 5 \cdot 5}} = 125.$$

(exponent labels point to the 2 in 3^2 and the 3 in 5^3)

The next example shows exponents used with fractions.

ANSWERS
2. (a) < (b) > (c) < (d) <

■ **EXAMPLE 3** *Using Exponents with Fractions*
Simplify each of the following.

(a) $\left(\frac{1}{2}\right)^3$ **(b)** $\left(\frac{5}{8}\right)^2$ **(c)** $\left(\frac{3}{4}\right)^2 \cdot \left(\frac{2}{3}\right)^3$

three factors of $\frac{1}{2}$

(a) $\left(\frac{1}{2}\right)^3 = \frac{1}{2} \cdot \frac{1}{2} \cdot \frac{1}{2} = \frac{1}{8}$

two factors of $\frac{5}{8}$

(b) $\left(\frac{5}{8}\right)^2 = \frac{5}{8} \cdot \frac{5}{8} = \frac{25}{64}$

(c) $\left(\frac{3}{4}\right)^2 \cdot \left(\frac{2}{3}\right)^3 = \left(\frac{3}{4} \cdot \frac{3}{4}\right) \cdot \left(\frac{2}{3} \cdot \frac{2}{3} \cdot \frac{2}{3}\right)$

$$= \frac{\overset{1}{\cancel{3}} \cdot \overset{1}{\cancel{3}} \cdot \overset{1}{\cancel{2}} \cdot \overset{1}{\cancel{2}} \cdot \overset{1}{\cancel{2}}}{\underset{2}{\cancel{4}} \cdot \underset{\underset{1}{\cancel{2}}}{\cancel{4}} \cdot \underset{1}{\cancel{3}} \cdot \underset{1}{\cancel{3}} \cdot 3}$$ Use cancellation.

$$= \frac{1}{6}$$ ■

WORK PROBLEM 3 AT THE SIDE.

3. Simplify each of the following.

(a) $\left(\frac{1}{2}\right)^2$

(b) $\left(\frac{7}{8}\right)^2$

(c) $\left(\frac{3}{4}\right)^3$

(d) $\left(\frac{1}{4}\right)^2 \cdot \left(\frac{8}{3}\right)^2$

ANSWERS
3. (a) $\frac{1}{4}$ (b) $\frac{49}{64}$ (c) $\frac{27}{64}$ (d) $\frac{4}{9}$

3 Recall the *order of operations* from Chapter 1.

ORDER OF OPERATIONS

1. Do all operations inside **parentheses.**
2. Simplify any expressions with **exponents** and find any **square roots.**
3. **Multiply** or **divide** from left to right.
4. **Add** or **subtract** from left to right.

The next example shows you how to apply the order of operations with fractions.

■ **EXAMPLE 4** *Using Order of Operations with Fractions*
Simplify by using the order of operations.

(a) $\frac{1}{3} + \frac{1}{2} \cdot \frac{4}{5}$ **(b)** $\frac{3}{8} \cdot \left(\frac{1}{2} + \frac{1}{3}\right)$ **(c)** $\left(\frac{2}{3}\right)^2 - \frac{4}{5} \cdot \frac{1}{2}$

(a) Multiply first.

$$\frac{1}{3} + \frac{1}{\underset{1}{\cancel{2}}} \cdot \frac{\overset{2}{\cancel{4}}}{5} = \frac{1}{3} + \frac{2}{5}$$

Next, add. The least common denominator of 3 and 5 is 15.

$$\frac{1}{3} + \frac{2}{5} = \frac{5}{15} + \frac{6}{15} = \frac{11}{15}$$

4. Simplify by using the order of operations.

(a) $\frac{3}{4} - \frac{7}{8} \cdot \frac{2}{3}$

(b) $\frac{1}{4} \cdot \left(\frac{2}{3} - \frac{1}{2}\right)$

(c) $\frac{3}{4} \cdot \frac{2}{3} - \left(\frac{1}{2}\right)^2$

(d) $\dfrac{\left(\frac{5}{6}\right)^2}{\frac{4}{3}}$

ANSWERS

4. (a) $\frac{1}{6}$ (b) $\frac{1}{24}$ (c) $\frac{1}{4}$ (d) $\frac{25}{48}$

(b) $\frac{3}{8} \cdot \left(\frac{1}{2} + \frac{1}{3}\right) = \frac{3}{8} \cdot \underbrace{\left(\frac{3}{6} + \frac{2}{6}\right)}$ Work in parentheses first.

$= \frac{3}{8} \cdot \frac{5}{6}$

$= \frac{\overset{1}{\cancel{3}} \cdot 5}{8 \cdot \underset{2}{\cancel{6}}}$ Use cancellation.

$= \frac{5}{16}$ Multiply.

(c) $\left(\frac{2}{3}\right)^2 - \frac{4}{5} \cdot \frac{1}{2} = \frac{4}{9} - \frac{4}{5} \cdot \frac{1}{2}$ Evaluate exponential expression first.

$= \frac{4}{9} - \frac{\overset{2}{\cancel{4}}}{5} \cdot \frac{1}{\underset{1}{\cancel{2}}}$ Next, multiply.

$= \frac{4}{9} - \frac{2}{5}$

$= \frac{20}{45} - \frac{18}{45}$ Subtract. (Least common denominator is 45.)

$= \frac{2}{45}$ ■

◀ WORK PROBLEM 4 AT THE SIDE.

3.5 EXERCISES

NAME DATE HOUR

Locate each fraction on the number line.

$\frac{1}{4}$ $\frac{3}{8}$ $\frac{7}{9}$ $\frac{7}{8}$ $\frac{5}{4}$ $1\frac{7}{8}$ $2\frac{1}{6}$ $\frac{7}{3}$ $\frac{11}{4}$ $3\frac{1}{4}$ $\frac{7}{2}$ $3\frac{4}{5}$

0 1 2 3 4

1. 2. 3. 10. 4. 12. 7. 5. 6. 11. 9. 8.

1. $\frac{1}{4}$ **2.** $\frac{3}{8}$ **3.** $\frac{7}{9}$ **4.** $\frac{5}{4}$

5. $\frac{7}{3}$ **6.** $\frac{11}{4}$ **7.** $2\frac{1}{6}$ **8.** $3\frac{4}{5}$

9. $\frac{7}{2}$ **10.** $\frac{7}{8}$ **11.** $3\frac{1}{4}$ **12.** $1\frac{7}{8}$

Write $<$ or $>$ to make a true statement.

Example: $\frac{7}{4}$ $\frac{13}{6}$

Solution:

The least common multiple of 4 and 6 is 12

$$\frac{7}{4} = \frac{21}{12} \qquad \frac{13}{6} = \frac{26}{12}$$

Because 21 is smaller than 26, $\frac{21}{12}$ or $\frac{7}{4}$ is smaller, so write $<$.

13. $\frac{1}{2} < \frac{3}{4}$ **14.** $\frac{2}{5} > \frac{1}{5}$ **15.** $\frac{5}{6} < \frac{11}{12}$ **16.** $\frac{5}{6} > \frac{2}{3}$

17. $\frac{3}{8} < \frac{5}{12}$ **18.** $\frac{7}{15} > \frac{9}{20}$ **19.** $\frac{7}{12} < \frac{11}{18}$ **20.** $\frac{19}{24} > \frac{17}{36}$

21. $\frac{19}{27} < \frac{13}{18}$

22. $\frac{21}{40} < \frac{17}{30}$

23. $\frac{37}{50} > \frac{13}{20}$

24. $\frac{7}{12} > \frac{11}{20}$

Evaluate each of the following.

Example: $\left(\frac{2}{3}\right)^4 = \frac{2}{3} \cdot \frac{2}{3} \cdot \frac{2}{3} \cdot \frac{2}{3} = \mathbf{\frac{16}{81}}$

25. $\left(\frac{2}{3}\right)^2$ $\frac{4}{9}$

26. $\left(\frac{3}{4}\right)^2$ $\frac{9}{16}$

27. $\left(\frac{7}{8}\right)^2$ $\frac{49}{64}$

28. $\left(\frac{9}{10}\right)^2$ $\frac{81}{100}$

29. $\left(\frac{2}{3}\right)^3$ $\frac{8}{27}$

30. $\left(\frac{3}{5}\right)^3$ $\frac{27}{125}$

31. $\left(\frac{5}{6}\right)^3$ $\frac{125}{216}$

32. $\left(\frac{4}{7}\right)^3$ $\frac{64}{343}$

33. $\left(\frac{3}{2}\right)^4$ $\frac{81}{16} = 5\frac{1}{16}$

34. $\left(\frac{4}{3}\right)^4$ $\frac{256}{81} = 3\frac{13}{81}$

35. $\left(\frac{1}{2}\right)^5$ $\frac{1}{32}$

36. $\left(\frac{2}{3}\right)^5$ $\frac{32}{243}$

37. Describe in your own words what a number line is. Be sure to include how it works and how it can be used.

Answer varies.

38. You have used the order of operations in two chapters now. List from memory the steps in the order of operations.

Answer varies.

Use the order of operations to simplify each of the following.

Example: $\left(\frac{2}{3}\right)^2 \cdot \left(\frac{1}{2} + \frac{1}{4}\right)$ **Solution:** $= \left(\frac{2}{3}\right)^2 \cdot \left(\frac{3}{4}\right)$ Work in parentheses first.

$= \frac{4}{9} \cdot \frac{3}{4}$ Evaluate exponential expression.

$= \mathbf{\frac{1}{3}}$ Multiply.

39. $3^2 + 2 - 3$ 8

40. $2 \cdot 4 + 3^2$ 17

41. $8 \cdot 3^2 - \frac{10}{2}$ 67

42. $3 \cdot 4^2 - \frac{6}{3}$ 46

43. $\left(\frac{1}{2}\right)^2 \cdot 4$ 1

44. $5 \cdot \left(\frac{1}{4}\right)^2$ $\frac{5}{16}$

45. $\left(\frac{3}{4}\right)^2 \cdot \left(\frac{1}{3}\right)$ $\frac{3}{16}$

46. $\left(\frac{2}{3}\right)^3 \cdot \left(\frac{1}{2}\right)$ $\frac{4}{27}$

47. $\left(\frac{3}{4}\right)^2 \cdot \left(\frac{2}{3}\right)^2$ $\frac{1}{4}$

48. $\left(\frac{5}{8}\right)^2 \cdot \left(\frac{4}{25}\right)^2$ $\frac{1}{100}$

49. $6 \cdot \left(\frac{2}{3}\right)^2 \cdot \left(\frac{1}{2}\right)^3$ $\frac{1}{3}$

50. $9 \cdot \left(\frac{1}{3}\right)^3 \cdot \left(\frac{4}{3}\right)^2$ $\frac{16}{27}$

51. $\frac{1}{4} \cdot \frac{3}{4} + \frac{3}{8} \cdot \frac{4}{3}$ $\frac{11}{16}$

52. $\frac{3}{4} \cdot \frac{2}{5} + \frac{1}{3} \cdot \frac{3}{5}$ $\frac{1}{2}$

53. $\frac{1}{2} + \left(\frac{1}{2}\right)^2 - \frac{3}{8}$ $\frac{3}{8}$

54. $\frac{2}{3} + \left(\frac{1}{3}\right)^2 - \frac{5}{9}$ $\frac{2}{9}$

55. $\left(\frac{1}{3} + \frac{1}{6}\right) \cdot \frac{1}{2}$ $\frac{1}{4}$

56. $\left(\frac{3}{5} - \frac{3}{20}\right) \cdot \frac{4}{3}$ $\frac{3}{5}$

57. $\frac{9}{8} \div \left(\frac{2}{3} + \frac{1}{12}\right)$ $1\frac{1}{2}$

58. $\frac{6}{5} \div \left(\frac{3}{5} - \frac{3}{10}\right)$ 4

59. $\left(\frac{5}{6} - \frac{1}{12}\right) \div \frac{3}{2}$

$\frac{1}{2}$

60. $\left(\frac{8}{5} - \frac{7}{10}\right) \div \frac{3}{5}$

$\frac{3}{2} = 1\frac{1}{2}$

61. $\frac{3}{8} \cdot \left(\frac{1}{4} + \frac{1}{2}\right) \cdot \frac{32}{3}$

3

62. $\frac{1}{6} \cdot \left(\frac{3}{5} - \frac{1}{10}\right) \cdot \frac{3}{2}$

$\frac{1}{8}$

63. $\left(\frac{3}{4}\right)^2 - \left(\frac{3}{4} - \frac{1}{8}\right) \div \frac{7}{4}$

$\frac{23}{112}$

64. $\left(\frac{2}{3}\right)^2 - \left(\frac{4}{5} - \frac{3}{10}\right) \div \frac{5}{4}$

$\frac{2}{45}$

65. $\left(\frac{7}{8} - \frac{1}{4}\right) - \left(\frac{3}{4}\right)^2 \cdot \frac{2}{3}$

$\frac{1}{4}$

66. $\left(\frac{5}{6} - \frac{7}{12}\right) - \left(\frac{1}{3}\right)^2 \cdot \frac{3}{4}$

$\frac{1}{6}$

▲ **67.** $\left(\frac{3}{5}\right)^2 \cdot \left(\frac{1}{3} + \frac{2}{9}\right) - \frac{1}{2} \cdot \frac{1}{5}$

$\frac{1}{10}$

▲ **68.** $\left(\frac{2}{3}\right)^2 \cdot \left(\frac{1}{2} - \frac{1}{8}\right) - \frac{2}{3} \cdot \frac{1}{8}$

$\frac{1}{12}$

PREVIEW EXERCISES

Rewrite the following numbers in words. (For help, see **Section 1.1.***)*

69. 8436 eight thousand, four hundred thirty-six

70. 625,115 six hundred twenty-five thousand, one hundred fifteen

71. 4,071,280 four million, seventy-one thousand, two hundred eighty

72. 220,518,315 two hundred twenty million, five hundred eighteen thousand, three hundred fifteen

CHAPTER 3 SUMMARY

KEY TERMS

3.1	**like fractions**	Fractions with the same denominator are called like fractions.
	unlike fractions	Fractions with different denominators are called unlike fractions.
3.2	**least common multiple**	Given two or more whole numbers, the least common multiple is the smallest whole number that is divisible by all the numbers.
3.3	**least common denominator**	When unlike fractions are rewritten as like fractions having the least common multiple as the denominator, the new denominator is the least common denominator.
3.4	**carrying**	The method used when the sum of the fractions of mixed numbers is greater than 1 is called carrying. Carry from the fraction to the whole number.

QUICK REVIEW

Concepts	Examples
3.1 Adding Like Fractions Add numerators and write in lowest terms.	$\frac{3}{4}+\frac{1}{4}+\frac{5}{4}=\frac{3+1+5}{4}=\frac{9}{4}=2\frac{1}{4}$
3.1 Subtracting Like Fractions Subtract numerators and write in lowest terms.	$\frac{7}{8}-\frac{5}{8}=\frac{7-5}{8}=\frac{2}{8}=\frac{1}{4}$
3.2 Finding the Least Common Multiple Method of prime numbers: Use prime numbers to find the least common multiple.	$\frac{1}{3}+\frac{1}{4}+\frac{1}{10}$ (see table below) least common multiple (LCM) = $2 \cdot 2 \cdot 3 \cdot 5 = 60$
3.3 Adding Unlike Fractions **1.** Find the least common multiple (LCM). **2.** Rewrite fractions with the least common multiple as the denominator. **3.** Add numerators, placing the answer over the least common denominator.	$\frac{1}{3}+\frac{1}{4}+\frac{1}{10}$ LCM = 60 $\frac{1}{3}=\frac{20}{60}$, $\frac{1}{4}=\frac{15}{60}$, $\frac{1}{10}=\frac{6}{60}$ $\frac{20}{60}+\frac{15}{60}+\frac{6}{60}=\frac{41}{60}$
3.3 Subtracting Unlike Fractions **1.** Find the least common multiple (LCM) **2.** Rewrite fractions with the least common multiple as the denominator. **3.** Subtract numerators, placing the difference over the common denominator.	$\frac{5}{8}-\frac{1}{3}=\frac{15}{24}-\frac{8}{24}=\frac{7}{24}$

Prime	2	3	5
3 =		(3)	
4 =	(2 · 2)		
10 =	2 ·		(5)
LCM =	(2 · 2)	(3)	(5)

Concepts	Examples
3.4 Adding Mixed Numbers 1. Add fractions. 2. Add whole numbers. 3. Combine the sums of whole numbers and fractions. Write the answer in lowest terms.	$9\frac{2}{3} = 9\frac{8}{12}$ $+\ 6\frac{3}{4} = 6\frac{9}{12}$ $15\frac{17}{12} = 16\frac{5}{12}$
3.4 Subtracting Mixed Numbers 1. Subtract fractions by using borrowing if necessary. 2. Subtract whole numbers. 3. Combine the differences of whole numbers and fractions. 4. Reduce fraction to lowest terms.	$8\frac{5}{8} = 8\frac{15}{24} = 7\frac{39}{24}$ $-\ 3\frac{11}{12} = 3\frac{22}{24} = 3\frac{22}{24}$ $4\frac{17}{24}$
3.5 Identifying the Larger of Two Fractions With unlike fractions, change to like fractions first. The fraction with the greater numerator is the greater fraction. $<$ is less than $>$ is greater than	Identify the larger fraction. $\frac{7}{8}, \frac{9}{10}$ $\frac{7}{8} = \frac{7 \cdot 5}{8 \cdot 5} = \frac{35}{40}$ $\frac{9}{10} = \frac{9 \cdot 4}{10 \cdot 4} = \frac{36}{40}$ $\frac{35}{40}$ is smaller than $\frac{36}{40}$, so $\frac{7}{8} < \frac{9}{10}$ or $\frac{9}{10} > \frac{7}{8}$. $\frac{9}{10}$ is greater.
3.5 Using the Order of Operations with Fractions Follow the order of operations. 1. Do all operations inside parentheses. 2. Simplify any expressions with exponents and find any square roots. 3. Multiply or divide from left to right. 4. Add or subtract from left to right.	Simplify by using the order of operations. $\frac{1}{2} \cdot \frac{2}{3} - \left(\frac{1}{4}\right)^2$ $= \frac{1}{2} \cdot \frac{2}{3} - \frac{1}{16}$ Simplify exponents. $= \frac{2}{6} - \frac{1}{16}$ Next, multiply. $= \frac{16}{48} - \frac{3}{48}$ Change to common denominator. $= \frac{13}{48}$ Subtract.

NAME DATE HOUR

CHAPTER 3 REVIEW EXERCISES

[3.1] *Add or subtract. Write answers in lowest terms.*

1. $\frac{1}{6} + \frac{2}{6}$ $\frac{1}{2}$

2. $\frac{5}{8} + \frac{2}{8}$ $\frac{7}{8}$

3. $\frac{1}{12} + \frac{2}{12} + \frac{1}{12}$ $\frac{1}{3}$

4. $\frac{8}{14} - \frac{3}{14}$ $\frac{5}{14}$

5. $\frac{3}{10} - \frac{1}{10}$ $\frac{1}{5}$

6. $\frac{5}{16} - \frac{1}{16}$ $\frac{1}{4}$

7. $\frac{36}{62} - \frac{10}{62}$ $\frac{13}{31}$

8. $\frac{79}{108} - \frac{47}{108}$ $\frac{8}{27}$

Solve each word problem. Write answers in lowest terms.

9. Tyrone milled $\frac{3}{16}$ of the lumber on the first day and $\frac{5}{16}$ of the lumber on the second day. What fraction of the lumber did he mill on these two days?

$\frac{1}{2}$ of the lumber

10. Diana did $\frac{7}{12}$ of her writing in the morning and $\frac{5}{12}$ of her writing in the afternoon. How much less writing did she do in the afternoon than in the morning?

$\frac{1}{6}$ less

[3.2] *Find the least common multiple of each set of numbers.*

11. 10, 8 40

12. 5, 12 60

13. 10, 12, 20 60

14. 9, 20, 15 180

15. 6, 8, 5, 15 120

16. 24, 5, 16 240

Rewrite each of the following fractions by using the indicated denominators.

17. $\frac{2}{3} = \frac{8}{12}$

18. $\frac{3}{4} = \frac{15}{20}$

19. $\frac{2}{5} = \frac{10}{25}$

20. $\frac{5}{9} = \frac{45}{81}$

21. $\frac{7}{16} = \frac{63}{144}$

22. $\frac{3}{22} = \frac{12}{88}$

[3.1–3.3] *Add or subtract. Write answers in lowest terms.*

23. $\frac{1}{3} + \frac{1}{2}$ $\frac{5}{6}$

24. $\frac{1}{5} + \frac{4}{15}$ $\frac{7}{15}$

25. $\frac{3}{8} + \frac{1}{5} + \frac{3}{10}$ $\frac{7}{8}$

26. $\frac{1}{2} + \frac{3}{8} + \frac{1}{16}$ $\frac{15}{16}$

27. $\frac{1}{4} + \frac{2}{3}$ $\frac{11}{12}$

28. $\frac{2}{5} + \frac{3}{7}$ $\frac{29}{35}$

29. $\frac{9}{16} + \frac{1}{12}$ $\frac{31}{48}$

30. $\frac{4}{9} - \frac{1}{3}$ $\frac{1}{9}$

31. $\frac{7}{8} - \frac{7}{16}$ $\frac{7}{16}$

32. $\frac{5}{8} - \frac{3}{16}$ $\frac{7}{16}$

33. $\frac{3}{4} - \frac{1}{3}$ $\frac{5}{12}$

34. $\frac{11}{12} - \frac{4}{9}$ $\frac{17}{36}$

Solve each of the following word problems.

35. A dump truck contains $\frac{1}{4}$ cubic yard of fine gravel, $\frac{1}{3}$ cubic yard of pea gravel, and $\frac{3}{8}$ cubic yard of coarse gravel. How many cubic yards of gravel are on the truck?

$\frac{23}{24}$ cubic yard

36. The Madison High School Drill Team is raising money to attend the state competition. They have raised $\frac{1}{3}$ of the amount needed through donations, $\frac{3}{8}$ of the amount needed has been earned through car washes and another $\frac{1}{5}$ has been raised through raffle ticket sales. Find the portion of the total that has been raised.

$\frac{109}{120}$ of the amount needed

[3.4] *Add or subtract. Write answers as mixed numbers.*

37. $6\frac{2}{3} + 5\frac{1}{6}$ $11\frac{5}{6}$

38. $25\frac{3}{4} + 16\frac{3}{8}$ $42\frac{1}{8}$

39. $78\frac{3}{7} + 17\frac{6}{7}$ $96\frac{2}{7}$

40. $12\frac{3}{5} + 8\frac{5}{8} + 10\frac{5}{16}$ $31\frac{43}{80}$

41. $6\frac{2}{3} - 1\frac{1}{2}$ $5\frac{1}{6}$

42. $18\frac{1}{3} - 12\frac{3}{4}$ $5\frac{7}{12}$

43. $73\frac{1}{2} - 55\frac{2}{3}$ $17\frac{5}{6}$

44. $215\frac{7}{16} - 136$ $79\frac{7}{16}$

Solve each word problem.

45. The lab had $14\frac{1}{3}$ gallons of distilled water. If $5\frac{1}{2}$ gallons were used in the morning and $6\frac{3}{4}$ gallons were used in the afternoon, find the number of gallons remaining.

$2\frac{1}{12}$ gallons

46. The Scouts collected $6\frac{4}{5}$ tons of newspaper on Saturday and $9\frac{2}{3}$ tons on Sunday. Find the total amount of newspaper collected.

$16\frac{7}{15}$ tons

47. At birth, the Bolton triplets weigh $5\frac{3}{4}$ pounds, $4\frac{7}{8}$ pounds, and $5\frac{1}{3}$ pounds. Find their total weight.

$15\frac{23}{24}$ pounds

48. A developer wants to build a shopping center. She bought two parcels of land, one, $1\frac{11}{16}$ acres, and the other, $2\frac{3}{4}$ acres. If she needs a total of $8\frac{1}{2}$ acres for the center, how much additional land does she need to buy?

$4\frac{1}{16}$ acres

[3.5] *Locate each fraction on the number line.*

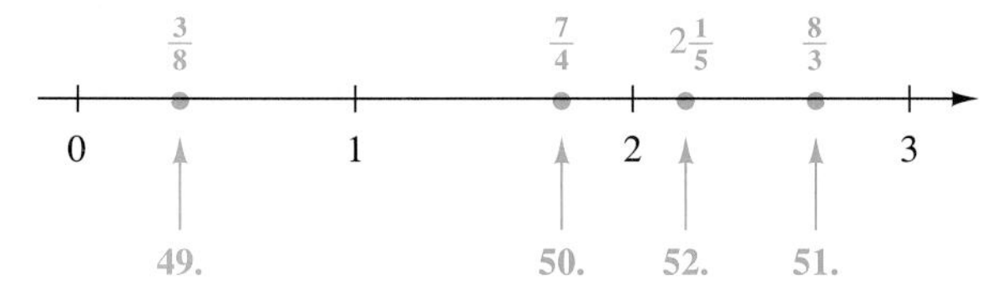

49. $\frac{3}{8}$ **50.** $\frac{7}{4}$ **51.** $\frac{8}{3}$ **52.** $2\frac{1}{5}$

Write < or > to make a true statement.

53. $\frac{1}{3} < \frac{2}{5}$ **54.** $\frac{2}{3} < \frac{5}{6}$ **55.** $\frac{3}{4} < \frac{7}{9}$ **56.** $\frac{7}{10} > \frac{8}{15}$

57. $\frac{5}{12} < \frac{8}{18}$ **58.** $\frac{7}{20} > \frac{8}{25}$ **59.** $\frac{19}{36} < \frac{29}{54}$ **60.** $\frac{19}{132} > \frac{7}{55}$

Simplify each of the following.

61. $\left(\frac{1}{2}\right)^2$ $\frac{1}{4}$ **62.** $\left(\frac{3}{4}\right)^2$ $\frac{9}{16}$ **63.** $\left(\frac{3}{5}\right)^3$ $\frac{27}{125}$ **64.** $\left(\frac{3}{8}\right)^4$ $\frac{81}{4096}$

Simplify by using the order of operations.

65. $\left(\frac{1}{3}\right)^2 \cdot 6$ $\frac{2}{3}$ **66.** $\left(\frac{2}{3}\right)^2 \cdot 12$ $5\frac{1}{3}$ **67.** $\left(\frac{3}{4}\right)^2 \cdot \left(\frac{8}{9}\right)^2$ $\frac{4}{9}$

68. $\frac{3}{5} \div \left(\frac{1}{10} + \frac{1}{5}\right)$ 2 **69.** $\left(\frac{1}{2}\right)^2 \cdot \left(\frac{1}{4} + \frac{1}{2}\right)$ $\frac{3}{16}$ **70.** $\left(\frac{1}{4}\right)^3 + \left(\frac{5}{8} + \frac{3}{4}\right)$ $1\frac{25}{64}$

Mixed Review Exercises

Solve by using the order of operations as necessary. Write answers in lowest terms or as mixed numbers.

71. $\frac{9}{15} + \frac{4}{15}$ $\frac{13}{15}$

72. $\frac{3}{4} - \frac{1}{8}$ $\frac{5}{8}$

73. $\frac{75}{86} - \frac{4}{43}$ $\frac{67}{86}$

74. $\frac{1}{4} + \frac{1}{8} + \frac{5}{16}$ $\frac{11}{16}$

75. $8\frac{1}{2} - 3\frac{2}{3}$ $4\frac{5}{6}$

76. $8\frac{3}{4} + 15\frac{1}{2}$ $24\frac{1}{4}$

77. $7 - 1\frac{5}{8}$ $5\frac{3}{8}$

78. $2\frac{3}{5} + 8\frac{5}{8} + \frac{5}{16}$ $11\frac{43}{80}$

79. $92\frac{5}{16} - 27$ $65\frac{5}{16}$

80. $\frac{7}{22} + \frac{3}{22} + \frac{3}{11}$ $\frac{8}{11}$

81. $\left(\frac{1}{4}\right)^2 \cdot \left(\frac{2}{5}\right)^3$ $\frac{1}{250}$

82. $\frac{1}{4} \div \left(\frac{1}{3} + \frac{1}{6}\right)$ $\frac{1}{2}$

83. $\left(\frac{2}{3}\right)^2 \cdot \left(\frac{1}{3} + \frac{1}{6}\right)$ $\frac{2}{9}$

84. $\left(\frac{2}{3}\right)^3 + \left(\frac{2}{3} - \frac{5}{9}\right)$ $\frac{11}{27}$

Write < or > to make a true statement.

85. $\frac{11}{9} < \frac{11}{6}$

86. $\frac{10}{11} < \frac{32}{33}$

87. $\frac{19}{40} < \frac{29}{60}$

88. $\frac{5}{8} > \frac{17}{30}$

Find the least common multiple of each set of numbers.

89. 12, 22 132

90. 3, 5, 7, 12 420

91. 2, 16, 36, 42 1008

Rewrite each of the following fractions by using the indicated denominators.

92. $\frac{3}{5} = \frac{36}{60}$

93. $\frac{9}{12} = \frac{108}{144}$

94. $\frac{3}{7} = \frac{180}{420}$

Solve each word problem.

95. A carpet layer needs $13\frac{1}{2}$ feet of carpet for a bedroom and $22\frac{3}{8}$ feet of carpet for a living room. If the roll from which the carpet layer is cutting is $92\frac{3}{4}$ feet long, find the number of feet remaining after the two rooms have been carpeted.

$56\frac{7}{8}$ feet

96. The business department has $1\frac{5}{8}$ positions for student help and the science department has $4\frac{5}{6}$ positions. If the college wishes to fill 10 positions, find the number of positions that remain.

$3\frac{13}{24}$ positions

NAME DATE HOUR

CHAPTER 3 TEST

Add. Write answers in lowest terms.

1. $\frac{3}{4} - \frac{1}{4}$

2. $\frac{3}{10} + \frac{5}{10}$

Subtract. Write answers in lowest terms.

3. $\frac{7}{8} - \frac{3}{8}$

4. $\frac{9}{15} - \frac{6}{15}$

Find the least common multiple of each set of numbers.

5. 4, 8, 2, 16

6. 7, 15, 3, 5

7. 3, 4, 5, 7

Add. Write answers in lowest terms.

8. $\frac{1}{3} + \frac{5}{16}$

9. $\frac{2}{7} + \frac{3}{5}$

Subtract. Write answers in lowest terms.

10. $\frac{5}{9} - \frac{1}{6}$

11. $\frac{3}{8} - \frac{1}{5}$

1. $\frac{1}{2}$

2. $\frac{4}{5}$

3. $\frac{1}{2}$

4. $\frac{1}{5}$

5. 16

6. 105

7. 420

8. $\frac{31}{48}$

9. $\frac{31}{35}$

10. $\frac{7}{18}$

11. $\frac{7}{40}$

12. $8\frac{5}{8}$
13. $2\frac{3}{4}$
14. $40\frac{29}{60}$
15. $\frac{5}{8}$
16. Answer varies.
17. Answer varies.
18. $18\frac{1}{4}$ hours
19. $35\frac{7}{8}$ gallons
20. $<$
21. $<$
22. $\frac{1}{3}$
23. $\frac{13}{48}$
24. $1\frac{1}{4}$
25. $1\frac{1}{2}$

Add or subtract. Write answers as mixed numbers.

12. $5\frac{7}{8} + 2\frac{3}{4}$

13. $7\frac{2}{3} - 4\frac{11}{12}$

14. $18\frac{3}{4} + 9\frac{2}{5} + 12\frac{1}{3}$

15. $6\frac{1}{4} - 5\frac{5}{8}$

16. Most of my students say that "addition and subtraction of fractions is more difficult than multiplication and division of fractions." Why do you think they say this? Do you agree with these students?

17. Devise and explain a method of estimating an answer to addition and subtraction problems involving mixed numbers. Might your estimated answer vary from the exact answer? If it did, what would the estimation accomplish?

Solve the following word problems.

18. Howard studied $3\frac{1}{4}$ hours on Monday, $4\frac{1}{6}$ hours on Tuesday, $2\frac{1}{3}$ hours on Wednesday, $3\frac{5}{6}$ hours on Thursday, and $4\frac{2}{3}$ hours on Friday. Find the total number of hours that he studied.

19. A commercial painting contractor arrived at a 6-unit apartment complex with $147\frac{1}{2}$ gallons of exterior paint. If his crew sprayed $68\frac{1}{2}$ gallons on the wood siding, rolled $37\frac{3}{8}$ gallons on the masonry exterior and brushed $5\frac{3}{4}$ gallons on the trim, find the number of gallons of paint remaining.

Write $<$ or $>$ to make a true statement.

20. $\frac{3}{5} \quad \frac{13}{20}$

21. $\frac{11}{18} \quad \frac{17}{24}$

Simplify each of the following. Use the order of operations as needed.

22. $\left(\frac{1}{3}\right)^3 \cdot 9$

23. $\left(\frac{3}{4}\right)^2 - \left(\frac{7}{8} \cdot \frac{1}{3}\right)$

24. $\left(\frac{5}{6} - \frac{5}{12}\right) \cdot 3$

25. $\frac{2}{3} + \frac{5}{8} \cdot \frac{4}{3}$

NAME DATE HOUR

CUMULATIVE REVIEW EXERCISES CHAPTERS 1-3

Name the digit that has the given place value in each of the following problems.

1. 371
hundreds 3
ones 1

2. 3,528,630
millions 3
thousands 8

Round each of the following to the nearest ten, nearest hundred, and nearest thousand.

	ten	hundred	thousand
3. 2847	2850	2800	3000
4. 59,803	59,800	59,800	60,000

*≈ *Round the numbers in each of these problems so that there is only one non-zero digit in each number. Then add, subtract, multiply, or divide as indicated to first estimate the answer. Finally, solve for the exact answer.*

5.

estimate		exact
10,000	← rounds to	9 834
300	←	279
50,000	←	51,506
+ 50,000	←	+ 51,702
110,300		113,321

6.

estimate	exact
20,000	24,276
− 10,000	− 9 887
10,000	14,389

7.

estimate	exact
1000	1258
× 400	× 420
400,000	528,360

8. estimate: $40\overline{)100{,}000}$ = 2 500   exact: $35\overline{)112{,}385}$ = 3 211

Add, subtract, multiply or divide as indicated.

9. 8 + 5 + 6 + 7 = 26

10. 375,899 + 521,742 + 357,968 = 1,255,609

11. 2857 − 1936 = 921

12. 3,896,502 − 1,094,807 = 2,801,695

13. 4 × 8 × 6 192

14. 5 × 3 × 9 135

15. 9 × 4 × 6 216

* This symbol is used to indicate exercises for which you should estimate your answer.

16. $\begin{array}{r} 79 \\ \times\ 8 \\ \hline \end{array}$

632

17. $\begin{array}{r} 802 \\ \times\ 261 \\ \hline \end{array}$

209,322

18. $\begin{array}{r} 370 \\ \times\ 40 \\ \hline \end{array}$

14,800

19. $9\overline{)1422}$

158

20. $13{,}467 \div 5$

2693 R2

21. $506\overline{)16{,}358}$

32 R166

Solve each word problem.

22. The Americans With Disabilities Act gives the single parking space design below. Find the perimeter (distance around) this parking space including the accessible aisle.

64 feet

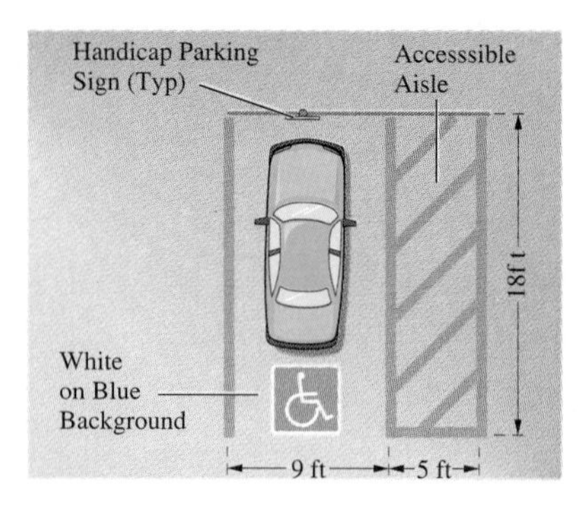

23. The single parking space design in Exercise 22 measures 18 feet by 14 feet. Find its area.

252 square feet

24. How many 16-ounce cans of beverage can be filled from a vat holding 9280 ounces of the beverage?

580 cans

25. A fan blade makes 1800 revolutions in one minute. How many revolutions would it make in 50 minutes?

90,000 revolutions

26. A rectangle is $\frac{3}{8}$ inch by $\frac{4}{5}$ inch. Find its area.

$\frac{3}{10}$ square inch

27. A woman has an estate of \$10,000. She leaves $\frac{2}{5}$ to a charity. Of the remainder, $\frac{2}{3}$ goes to her son. How much money does her son get?

\$4000

28. Larry Foxworthy cuts, splits, and delivers firewood. If his truck, when fully loaded, holds $5\frac{1}{4}$ cords of firewood, find the number of cords he could deliver with $3\frac{1}{2}$ loads.

$18\frac{3}{8}$ cords

29. The Sears Tower in Chicago is 110 stories tall. If the total height of the building is $1536\frac{7}{8}$ feet including a flagpole at the top of the building which is $82\frac{1}{2}$ feet tall, find the height of the building itself.

$1454\frac{3}{8}$ feet

Find the prime factorization of each number. Write answers by using exponents.

30. 30 $2 \cdot 3 \cdot 5$

31. 144 $2^4 \cdot 3^2$

32. 250 $2 \cdot 5^3$

Solve each of the following.

33. $5^2 \cdot 2^3$ 200

34. $2^4 \cdot 3^2$ 144

35. $4^2 \cdot 3^3$ 432

Find each square root.

36. $\sqrt{9}$ 3

37. $\sqrt{64}$ 8

38. $\sqrt{225}$ 15

Simplify each of the following by using the order of operations.

39. $5^2 - 6 \cdot 3$ 7

40. $\sqrt{25} + 5 \cdot 9 - 6$ 44

41. $\left(\frac{3}{8} - \frac{1}{3}\right) \cdot \frac{1}{2}$ $\frac{1}{48}$

42. $\frac{3}{4} \div \left(\frac{1}{3} + \frac{1}{2}\right)$ $\frac{9}{10}$

43. $\frac{2}{3} + \left(\frac{7}{8}\right)^2 - \frac{1}{4}$ $1\frac{35}{192}$

Write proper or improper for each fraction.

44. $\frac{3}{4}$ proper

45. $\frac{5}{8}$ proper

46. $\frac{7}{4}$ improper

Write each fraction in lowest terms.

47. $\frac{25}{40}$ $\frac{5}{8}$

48. $\frac{38}{50}$ $\frac{19}{25}$

49. $\frac{105}{300}$ $\frac{7}{20}$

Add, subtract, multiply, or divide as indicated.

50. $\frac{3}{4} \times \frac{2}{3}$ $\frac{1}{2}$

51. $\frac{9}{11} \cdot \frac{5}{18}$ $\frac{5}{22}$

52. $34 \times \frac{5}{8}$ $21\frac{1}{4}$

53. $\frac{3}{8} \div \frac{1}{3}$ $1\frac{1}{8}$

54. $\frac{25}{40} \div \frac{10}{35}$ $2\frac{3}{16}$

55. $9 \div \frac{2}{3}$ $13\frac{1}{2}$

56. $\frac{2}{3} + \frac{1}{9}$ $\frac{7}{9}$

57. $\frac{5}{16} + \frac{1}{4} + \frac{3}{8}$ $\frac{15}{16}$

58. $\frac{11}{18} - \frac{5}{12}$ $\frac{7}{36}$

59. $2\frac{1}{4} + 3\frac{5}{8}$ $5\frac{7}{8}$

60. $21\frac{7}{8} + 4\frac{5}{12}$ $26\frac{7}{24}$

61. $5 - 2\frac{3}{8}$ $2\frac{5}{8}$

Find the least common multiple of each set of numbers.

62. 25, 30 150

63. 15, 20, 50 300

64. 12, 16, 18 144

Write each fraction by using the indicated denominator.

65. $\frac{5}{9} = \frac{40}{72}$

66. $\frac{7}{12} = \frac{77}{132}$

67. $\frac{9}{56} = \frac{27}{168}$

68. $\frac{5}{7} = \frac{60}{84}$

Locate each fraction on the number line.

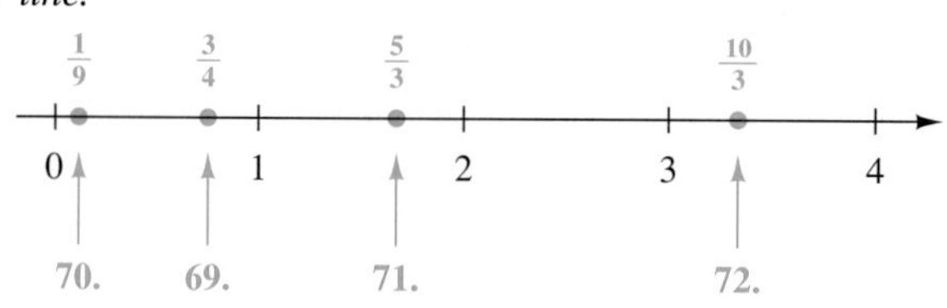

69. $\frac{3}{4}$

70. $\frac{1}{9}$

71. $\frac{5}{3}$

72. $\frac{10}{3}$

Write < or > to make a true statement.

73. $\frac{7}{10} < \frac{37}{50}$

74. $\frac{19}{25} < \frac{23}{30}$

75. $\frac{7}{12} < \frac{11}{18}$

estimating

$0.45 = \frac{9}{20}$

place value

$52.8 \div 0.75$

Decimals 4

Fractions are used to represent parts of a whole. In this chapter, decimals are used as another way to show parts of a whole. For example, our money system is based on decimals. One dollar is divided into 100 equivalent parts. One cent is one of the parts, and a dime is 10 of the parts.

4.1 READING AND WRITING DECIMALS

OBJECTIVES

1. Write parts of a whole as decimals.
2. Find the place value of a digit.
3. Read decimals.
4. Write decimals as fractions.

FOR EXTRA HELP

Tape 5

SSM pp. 99–101

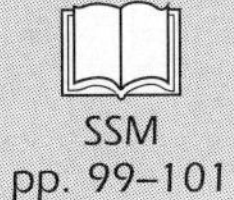

MAC: A IBM: A

1 Decimals are used when a whole is divided into 10 equivalent parts or into 100 or 1000 or 10,000 equivalent parts. In other words, decimals are fractions with denominators that are a power of 10. For example, the square below is cut into 10 equivalent parts. Written as a fraction, each part is $\frac{1}{10}$ of the whole. Written as a decimal, each part is 0.1. Both are read as "*one tenth.*"

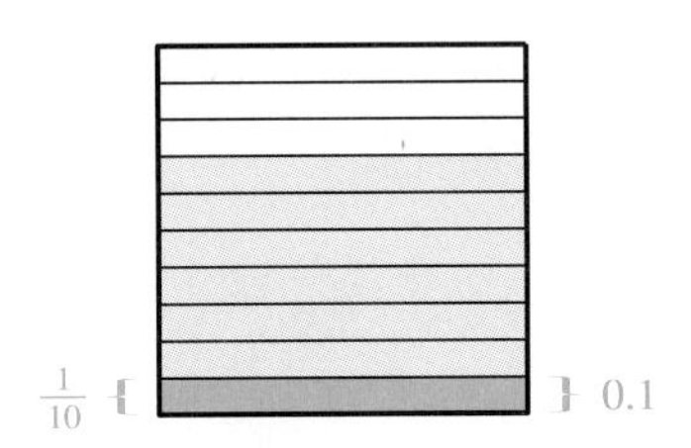

The dot in 0.1 is called the **decimal point.**

0.1
↑ decimal point

The square above has 7 of its 10 parts shaded.

Written as a fraction, $\frac{7}{10}$ of the square is shaded.

Written as a decimal, 0.7 of the square is shaded.

Both are read as "*seven tenths.*"

1. There are 10 dimes in one dollar. Each dime is $\frac{1}{10}$ of a dollar. Write a fraction and a decimal that represent the shaded portion of each dollar.

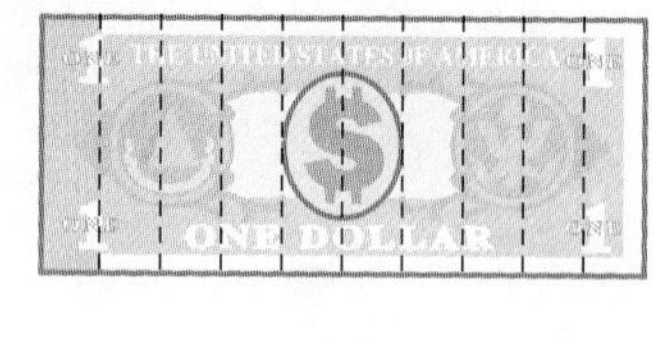

(a)

(b)

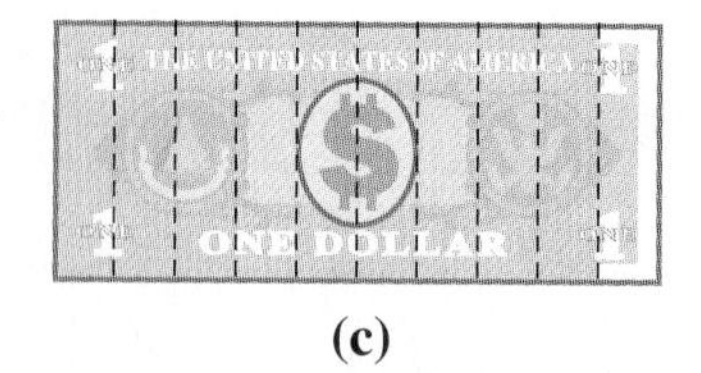

(c)

2. Write the portion of each square that is shaded as a fraction and as a decimal.

(a)

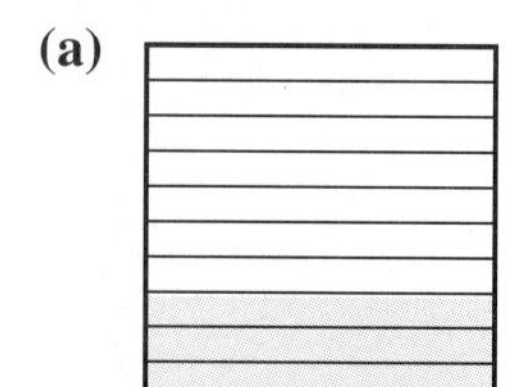

(b)

ANSWERS

1. (a) $\frac{1}{10}$; 0.1 (b) $\frac{3}{10}$; 0.3 (c) $\frac{9}{10}$; 0.9

2. (a) $\frac{3}{10}$; 0.3 (b) $\frac{41}{100}$; 0.41

WORK PROBLEM 1 AT THE SIDE.

The square below is cut into 100 equivalent parts. Written as a fraction, each part is $\frac{1}{100}$ of the whole.

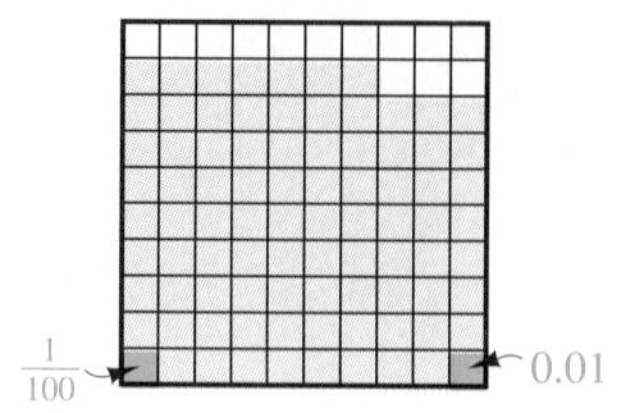

Written as a decimal, each part is

0.01 of the whole.
↑
(Read "one hundredth.")

The square has 87 parts shaded.

Written as a fraction, $\frac{87}{100}$ of the total area is shaded.

Written as a decimal, **0.87** of the total area is shaded.

Both are read as "*eighty-seven hundredths.*"

WORK PROBLEM 2 AT THE SIDE.

The example below shows several numbers written as both fractions and decimals.

EXAMPLE 1 *Using the Decimal Forms of Fractions*

	fraction	*decimal*	*read as*
(a)	$\frac{3}{10}$	0.3	three tenths
(b)	$\frac{9}{100}$	0.09	nine hundredths
(c)	$\frac{71}{100}$	0.71	seventy-one hundredths
(d)	$\frac{832}{1000}$	0.832	eight hundred thirty-two thousandths ■

WORK PROBLEM 3 AT THE SIDE.

2 The decimal point separates the *whole-number part* from the *fractional part* in a decimal number. In the chart below, you see that the **place value names** for fractional parts are similar to those on the whole number side but end in "*ths.*"

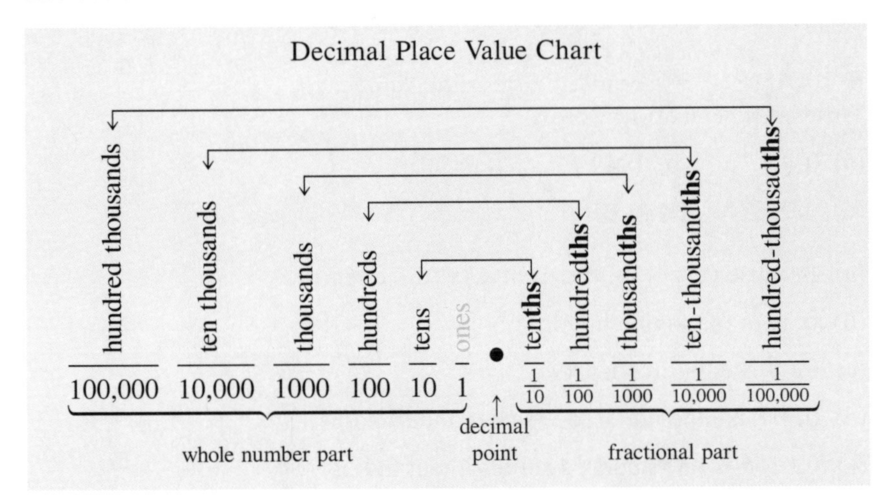

Notice that the ones place is at the center. (There is no "oneths" place.) Also notice that each place is 10 times the value of the place to its right.

> *Note* In this chapter, if a number does *not* have a decimal point, it is a *whole number.* (There is no fractional part.) If you want to show the decimal point in a whole number, it is just to the right of the digit in the ones place. For example:
>
> $8 \rightarrow 8. \qquad 306 \rightarrow 306.$

EXAMPLE 2 *Identifying the Place Value of a Digit*

Give the place values of the digits in each decimal.

(a) 78.36 **(b)** 0.0093

(a)

tens	ones		tenths	hundredths
7	8	.	3	6

(b)

ones		tenths	hundredths	thousandths	ten-thousandths
0	.	0	0	9	3

■

WORK PROBLEM 4 AT THE SIDE.

3. Write each decimal as a fraction.

(a) 0.7

(b) 0.9

(c) 0.03

(d) 0.69

(e) 0.119

4. Identify the place value of each digit in these decimals.

(a) 971.54

(b) 5.602

(c) 0.4

(d) 0.083

ANSWERS

3. (a) $\frac{7}{10}$ (b) $\frac{9}{10}$ (c) $\frac{3}{100}$ (d) $\frac{69}{100}$ (e) $\frac{119}{1000}$

4. (a)

hundreds	tens	ones		tenths	hundredths
9	7	1	.	5	4

(b)

ones		tenths	hundredths	thousandths
5	.	6	0	2

(c)

ones		tenths
0	.	4

(d)

ones		tenths	hundredths	thousandths
0	.	0	8	3

3 A decimal is read according to its form as a fraction. We read 0.9 as "nine tenths" because 0.9 is the same as $\frac{9}{10}$. Notice that 0.9 ends in the tenths place.

ones tenths
0. 9

We read 0.02 as "two hundredths" because 0.02 is the same as $\frac{2}{100}$. Notice that 0.02 ends in the hundredths place.

ones tenths hundredths
0. 0 2

EXAMPLE 3 *Reading a Decimal Number*

Write each decimal in words.

(a) 0.3 **(b)** 0.49 **(c)** 0.08

(d) 0.918 **(e)** 0.0106

(a) Because $0.3 = \frac{3}{10}$, the decimal is three tenths.

(b) 0.49 is forty-nine hundredths.

(c) 0.08 is eight hundredths.

(d) 0.918 is nine hundred eighteen thousandths.

(e) 0.0106 is one hundred six ten-thousandths. ■

WORK PROBLEM 5 AT THE SIDE.

READING A DECIMAL NUMBER

Step 1 Read any whole number part to the *left* of the decimal point as you normally would.

Step 2 Read the decimal point as "*and*."

Step 3 Read the part of the number to the *right* of the decimal point as if it was an ordinary whole number.

Step 4 Finish with the place value name of the right-most digit.

Note If there is *no whole number part,* you will use only Steps 3 and 4.

EXAMPLE 4 *Reading a Decimal*

Read each decimal.

(a)

16.9 (9 is in tenths place)

sixteen **and** nine **tenths**

16.9 is read "sixteen and nine tenths."

(b)

482.35 (5 is in hundredths place)

four hundred eighty-two **and** thirty-five **hundredths**

482.35 is read "four hundred eighty-two and thirty-five hundredths."

5. Write each decimal in words.

(a) 0.3

(b) 0.46

(c) 0.09

(d) 0.409

(e) 0.0003

(f) 0.0703

(g) 0.088

ANSWERS

5. (a) three tenths
(b) forty-six hundredths
(c) nine hundredths
(d) four hundred nine thousandths
(e) three ten-thousandths
(f) seven hundred three ten-thousandths
(g) eighty-eight thousandths

→3 is in thousandths place

(c) 0.063 is sixty-three **thousandths.** (No whole number part.)

(d) 11.1085 is eleven **and** one thousand eighty-five **ten-thousandths.** ■

WORK PROBLEM 6 AT THE SIDE. ▶▶

4 Knowing how to read decimals will help you when writing decimals as fractions.

WRITING DECIMALS AS FRACTIONS OR MIXED NUMBERS

Step 1 The digits to the right of the decimal point are the numerator of the fraction.

Step 2 The denominator is 10 for tenths, 100 for hundredths, 1000 for thousandths, 10,000 for ten-thousandths, and so on.

Step 3 If the decimal has a whole-number part, the fraction will be a mixed number with the same whole-number part.

■ **EXAMPLE 5** *Writing a Decimal as a Fraction or Mixed Number*

Write each decimal as a fraction or mixed number.

(a) 0.19 **(b)** 0.863 **(c)** 4.0099

(a) The digits to the right of the decimal point, 19, are the numerator of the fraction. The denominator is 100 for hundredths because the right-most digit is in the hundredths place.

$$0.19 = \frac{19}{100}$$ ← 100 for hundredths.

↑ hundredths place

(b) $$0.863 = \frac{863}{1000}$$ ← 1000 for thousandths.

↑ thousandths place

The whole number part stays the same.

(c) $$4.0099 = 4\frac{99}{10{,}000}$$ ← 10,000 for ten-thousandths. ■

↑ ten-thousandths place

WORK PROBLEM 7 AT THE SIDE. ▶▶

Note After you write a decimal as a fraction or a mixed number, check to see if the fraction is in lowest terms.

6. Write each decimal in words.

(a) 3.8

(b) 15.1

(c) 0.72

(d) 64.309

7. Write each decimal as a fraction or mixed number.

(a) 0.7

(b) 9.89

(c) 0.101

(d) 0.007

(e) 1.3717

ANSWERS

6. (a) three and eight tenths **(b)** fifteen and one tenth **(c)** seventy-two hundredths **(d)** sixty-four and three hundred nine thousandths

7. (a) $\frac{7}{10}$ **(b)** $9\frac{89}{100}$ **(c)** $\frac{101}{1000}$ **(d)** $\frac{7}{1000}$ **(e)** $1\frac{3717}{10{,}000}$

8. Write each decimal as a fraction or mixed number in lowest terms.

(a) 0.2

(b) 12.6

(c) 0.85

(d) 3.05

(e) 0.225

(f) 420.0802

ANSWERS

8. (a) $\frac{1}{5}$ (b) $12\frac{3}{5}$ (c) $\frac{17}{20}$ (d) $3\frac{1}{20}$ (e) $\frac{9}{40}$ (f) $420\frac{401}{5000}$

■ **EXAMPLE 6** *Writing a Decimal as a Fraction or Mixed Number*

Write each decimal as a fraction or mixed number in lowest terms.

(a) $0.4 = \frac{4}{10}$ ← 10 for tenths.

Write $\frac{4}{10}$ in lowest terms. $\frac{4}{10} = \frac{4 \div 2}{10 \div 2} = \frac{2}{5}$

(b) $0.75 = \frac{75}{100} = \frac{3}{4}$ (lowest terms)

(c) $18.105 = 18\frac{105}{1000} = 18\frac{21}{200}$ (lowest terms)

(d) $42.8085 = 42\frac{8085}{10{,}000} = 42\frac{1617}{2000}$ (lowest terms) ■

◀ WORK PROBLEM 8 AT THE SIDE.

NAME DATE HOUR

4.1 EXERCISES

Name the digit that has the given place value.

Example: 75.6382
tenths **6**
thousandths **8**

1. 37.602
ones 7
tenths 6
tens 3

2. 135.296
ones 5
tenths 2
tens 3

3. 0.417
tenths 4
hundredths 1
thousandths 7

4. 0.7694
tenths 7
hundredths 6
thousandths 9

5. 93.01472
thousandths 4
ten-thousandths 7
tenths 0

6. 0.51968
tenths 5
ten-thousandths 6
hundredths 1

7. 314.658
tens 1
tenths 6
hundreds 3

8. 51.325
tens 5
tenths 3
hundredths 2

9. 149.0832
hundreds 1
hundredths 8
ones 9

10. 3458.712
hundreds 4
hundredths 1
tenths 7

11. 6285.7125
thousands 6
thousandths 2
hundredths 1

12. 5417.6832
thousands 5
thousandths 3
ones 7

Give the place value of each digit in the following decimals.

Example: 5.309 5 **ones** 3 **tenths** 0 **hundredths** 9 **thousandths**

13. 0.93 0 ones 9 tenths 3 hundredths

14. 0.81 0 ones 8 tenths 1 hundredths

15. 8.965 8 ones 9 tenths 6 hundredths 5 thousandths

16. 5.173 5 ones 1 tenths 7 hundredths 3 thousandths

17. 60.372 6 tens 0 ones
3 tenths 7 hundredths 2 thousandths

18. 89.325 8 tens 9 ones
3 tenths 2 hundredths 5 thousandths

Write each decimal as a fraction or mixed number in lowest terms.

Example: 0.68

Solution: $0.68 = \frac{68}{100} = \frac{17}{25}$ (lowest terms)

Example: 4.005

Solution: $4.005 = 4\frac{5}{1000}$
$= \mathbf{4\frac{1}{200}}$ (lowest terms)

19. 0.7 $\frac{7}{10}$

20. 0.1 $\frac{1}{10}$

21. 13.4 $13\frac{2}{5}$

22. 9.8 $9\frac{4}{5}$

23. 0.35 $\frac{7}{20}$

24. 0.85 $\frac{17}{20}$

25. 0.66 $\frac{33}{50}$

26. 0.33 $\frac{33}{100}$

27. 10.17 $10\frac{17}{100}$

28. 31.99 $31\frac{99}{100}$

29. 0.06 $\frac{3}{50}$

30. 0.08 $\frac{2}{25}$

31. 0.205 $\frac{41}{200}$

32. 0.805 $\frac{161}{200}$

33. 5.002 $5\frac{1}{500}$

34. 4.008 $4\frac{1}{125}$

35. 0.686 $\frac{343}{500}$

36. 0.492 $\frac{123}{250}$

Write each decimal in words.

Example: 16.028

Solution:

16 . 028 — 8 is in thousandths place.

sixteen **and** twenty-eight thousandths

37. 0.5

five tenths

38. 0.9

nine tenths

39. 0.78

seventy-eight hundredths

40. 0.55

fifty-five hundredths

41. 0.105

one hundred five thousandths

42. 0.609

six hundred nine thousandths

43. 12.04

twelve and four hundredths

44. 86.09

eighty-six and nine hundredths

45. 1.075

one and seventy-five thousandths

46. 4.025

four and twenty-five thousandths

Write each decimal in numbers.

47. six and seven tenths

6.7

48. eight and twelve hundredths

8.12

49. thirty-two hundredths

0.32

50. one hundred eleven thousandths

0.111

51. four hundred twenty and eight thousandths

420.008

52. two hundred and twenty-four thousandths

200.024

53. seven hundred three ten-thousandths

0.0703

54. eight hundred and six hundredths

800.06

55. seventy-five and thirty thousandths

75.030

56. sixty and fifty hundredths

60.50

57. Anne read the number 4302 as "four thousand three hundred and two." Explain what is wrong with the way Anne read the number.

Answer varies.

58. Jerry read the number 9.0106 as "nine and one hundred and six thousandths." Explain the error he made.

Answer varies.

Suppose your job is to take phone orders for precision parts. Use the table below to complete Exercises 59–62. Write the correct part number that matches what you hear the customer say over the phone.

Part number	*Size in centimeters*
3-A	0.06
3-B	0.26
3-C	0.6
3-D	0.86
4-A	1.006
4-B	1.026
4-C	1.06
4-D	1.6

59. "Please send the six-tenths centimeter bolt."

3-C

60. "The part missing from our order was the one and six hundredths size."

4-C

61. "The size we need is one and six thousandths centimeters."

4-A

62. "Do you still stock the twenty-six hundredths centimeter bolt?"

3-B

▲ **63.** Write 8006.5001 in words.

eight thousand six and five thousand one ten-thousandths

▲ **64.** Write 20,060.00303 in words.

twenty thousand sixty and three hundred three hundred-thousandths

▲ **65.** Write 625.4284 as a mixed number in lowest terms.

$625\frac{1071}{2500}$

▲ **66.** Write 714.1372 as a mixed number in lowest terms.

$714\frac{343}{2500}$

PREVIEW EXERCISES

Round each of the following to the nearest ten, nearest hundred, and nearest thousand. (For help, see ***Section 1.7.****)*

	ten	*hundred*	*thousand*
67. 8235	8240	8200	8000
68. 3565	3570	3600	4000
69. 19,705	19,710	19,700	20,000
70. 89,604	89,600	89,600	90,000

4.2 ROUNDING DECIMALS

Section 1.7 showed how to round whole numbers. For example, 89 rounded to the nearest ten is 90, and 8512 rounded to the nearest hundred is 8500.

1 It is also important to be able to **round** decimals. For example, a store is selling 2 candy mints for $0.25 but you want only one mint. The price of each mint is $0.25 ÷ 2, which is $0.125, but you cannot pay part of a cent. So the store rounds the price to $0.13 for the mint.

ROUNDING DECIMALS

Step 1 Find the place to which the rounding is being done. Draw a line **after** that place to show that you are cutting off the rest of the digits.

Step 2 Look **only** at the **first** digit you are cutting off.

Step 3A If this digit is **less than 5,** the part of the number you are keeping **stays the same.**

Step 3B If this digit is **5 or more,** you must **round up** the part of the number you are keeping.

Note Do **not** move the decimal point when rounding.

2 These examples show you how to round decimals.

EXAMPLE 1 *Rounding a Decimal Number*

Round 14.39656 to the nearest thousandth.

Step 1 Draw a line after the thousandths place.

14.396|52 You will be cutting off the 5 and 2.

thousandths ↑

Step 2 Look *only* at the *first* digit you are cutting off. Ignore the other digits you are cutting off.

14.396|52 ← Look only at the 5. Ignore the 2.

Step 3 If the first digit you are cutting off is 5 or more, round up the part of the number you are keeping.

14.396|52 ← First digit cut is 5 or more, so round up by adding 1 thousandth to the part you are keeping.

$$\begin{array}{r} 14.396 \\ +\ 0.001 \\ \hline 14.397 \end{array}$$ ← To round up, add 1 thousandth.

So, 14.39652 rounded to the nearest thousandth is 14.397. ■

WORK PROBLEM 1 AT THE SIDE.

OBJECTIVES

1. Learn the rules for rounding decimals.
2. Round decimals to any given place.
3. Round money amounts to the nearest cent or nearest dollar.

FOR EXTRA HELP

Tape 5 | SSM pp. 101–103 | MAC: A IBM: A

1. Round to the nearest thousandth.

(a) 0.33492

(b) 8.00851

(c) 265.42068

(d) 10.70180

ANSWERS

1. (a) 0.335 (b) 8.009 (c) 265.421 (d) 10.702

EXAMPLE 2 *Rounding Decimals to Different Places*

Round to the place indicated.

(a) 5.3496 to the nearest tenth

(b) 0.69738 to the nearest hundredth

(c) 0.01806 to the nearest thousandth

(d) 57.976 to the nearest tenth

(a) *Step 1* Draw a line after the tenths place.

5.3|496 You will be cutting off the 4, 9, and 6.
tenths ↑

Step 2 5.3|496 ← Look only at the 4.
Ignore these digits.

Step 3 5.3|496 ← First digit cut is less than 5 so the part you are keeping stays the same.
5.3 ← stays the same

5.3496 rounded to the nearest tenth is 5.3.

(b) *Step 1* 0.69|738 Draw a line after the hundredths place.
hundredths ↑

Step 2 0.69|738 ← Look only at the 7.

Step 3 0.69|738 ← First digit cut is 5 or more, so round up by adding 1 hundredth to the part you are keeping.

$$\begin{array}{r} 0.69 \\ +\ 0.01 \\ \hline 0.70 \end{array}$$

← To round up, add 1 hundredth.
← 9 + 1 is 10; write 0 and carry 1 to the 6 in the tenths place.

0.69738 rounded to the nearest hundredth is 0.70. (You *must* write the zero in the hundredths place.)

(c) 0.018|06 ← First digit cut is less than 5 so the part you are keeping stays the same.
0.018

0.01806 rounded to the nearest thousandth is 0.018.

(d) 57.9|76 ← First digit cut is 5 or more so round up by adding 1 tenth to the part you are keeping.

$$\begin{array}{r} 57.\mathbf{9} \\ +\ 0.\mathbf{1} \\ \hline 58.\mathbf{0} \end{array}$$

← 9 + 1 is 10; write the 0 and carry 1 to the 7 in the ones place.

57.976 rounded to the nearest tenth is 58.0.
You must write the zero in the tenths place to show that the number was rounded to the nearest tenth. ■

Note Check that your rounded answer shows **exactly** the number of decimal places called for, even if a zero is in that place.

WORK PROBLEM 2 AT THE SIDE.

3 When you are shopping in a store, money amounts are usually rounded to the nearest cent. There are 100 cents in a dollar.

$$\text{Each cent is } \frac{1}{100} \text{ of a dollar.}$$

Another way to write $\frac{1}{100}$ is 0.01. So rounding to the *nearest cent* is the same as rounding to the *nearest hundredth of a dollar*.

■ **EXAMPLE 3** *Rounding to the Nearest Cent*

Round each of these money amounts to the nearest cent.

(a) $2.4238 **(b)** $0.695

(a) $2.42|38 ← Less than 5

$2.42 ← You pay

(b) $0.69|5 ← 5 or more; round up

$$\begin{array}{r} \$0.69 \\ +\ \$0.01 \\ \hline \$0.70 \end{array}$$

← To round up, add 1 hundredth
← You pay ■

WORK PROBLEM 3 AT THE SIDE.

It is also common to round money amounts to the nearest dollar. You can do that on your federal and state income tax, for example, to make the calculations easier.

2. Round to the place indicated.

(a) 0.8988 to the nearest hundredth

(b) 5.8903 to the nearest hundredth

(c) 11.0299 to the nearest thousandth

(d) 0.545 to the nearest tenth

3. Round each of the following money amounts to the nearest cent.

(a) $14.595

(b) $578.0663

(c) $0.849

(d) $0.0548

ANSWERS
2. (a) 0.90 (b) 5.89 (c) 11.030 (d) 0.5
3. (a) $14.60 (b) $578.07 (c) $0.85 (d) $0.05

4. Round to the nearest dollar.

(a) \$29.10

(b) \$136.49

(c) \$990.91

(d) \$5949.88

(e) \$49.60

(f) \$0.55

(g) \$1.08

ANSWERS
4. (a) \$29 (b) \$136 (c) \$991 (d) \$5950 (e) \$50 (f) \$1 (g) \$1

■ **EXAMPLE 4** *Rounding to the Nearest Dollar*
Round to the nearest dollar.

(a) \$48.69 **(b)** \$594.36 **(c)** \$349.88

(d) \$2689.50 **(e)** \$0.61

(a) \$48.|69 — First digit cut is 5 or more so round up by adding \$1.

\$48
+ 1
\$49

\$48.69 rounded to the nearest dollar is \$49. Write the answer as \$49 to show that the rounding is to the *nearest dollar*. Writing \$49.00 would show rounding to the nearest *cent*.

(b) \$594.|36 — Less than 5 so the part you keep stays the same.

\$594

\$594.36 rounded to the nearest dollar is \$594.

(c) \$349.|88 — 5 or more, so round up by adding \$1.

\$349
+ 1
\$350

\$349.88 rounded to the nearest dollar is \$350.

(d) \$2689.50 rounded to the nearest dollar is \$2690.

(e) \$0.61 rounded to the nearest dollar is \$1. ■

◀◀ WORK PROBLEM 4 AT THE SIDE.

Round each of the following to the place indicated.

1. 16.8974 to the nearest tenth

16.9

2. 193.845 to the nearest hundredth

193.85

3. 0.95647 to the nearest thousandth

0.956

4. 96.81584 to the nearest ten-thousandth

96.8158

5. 0.799 to the nearest hundredth

0.80

6. 0.952 to the nearest tenth

1.0

7. 3.66062 to the nearest thousandth

3.661

8. 1.5074 to the nearest hundredth

1.51

9. 793.988 to the nearest tenth

794.0

10. 476.1196 to the nearest thousandth

476.120

11. 0.09804 to the nearest ten-thousandth

0.0980

12. 176.004 to the nearest tenth

176.0

13. 48.512 to the nearest one

49

14. 3.385 to the nearest one

3

15. 9.0906 to the nearest hundredth

9.09

16. 30.1290 to the nearest thousandth

30.129

17. 82.000151 to the nearest ten-thousandth

82.0002

18. 0.400594 to the nearest ten-thousandth

0.4006

Nardos is grocery shopping. The store will round the amount she pays for each item to the nearest cent. Write the rounded amounts.

19. Soup is 3 cans for $1.25, so one can is $0.41666. Nardos pays __$0.42__

20. Orange juice is 2 cartons for $2.69, so one carton is $1.345. Nardos pays __$1.35__

21. Facial tissue is 4 boxes for $4.89, so one box is $1.2225. Nardos pays __$1.22__

22. Muffin mix is 3 packages for $1.75, so one package is $0.58333. Nardos pays __$0.58__

23. Candy bars are 6 for $1.79, so one bar is $0.2983. Nardos pays __$0.30__

24. Boxes of spaghetti are 4 for $1.99, so one box is $0.4975. Nardos pays __$0.50__

As she gets ready to do her income tax return. Ms. Chen rounds each amount to the nearest dollar. Write the rounded amounts.

25. Income from job, $17,249.70

$17,250

26. Income from interest on bank account, $69.58

$70

27. Union dues, $310.08

$310

28. Federal withholding, $2150.49

$2150

29. Donations to charity, $378.82

$379

30. Medical expenses, $609.38

$609

31. Explain what happens when you round $0.499 to the nearest dollar.

Answer varies.

32. Explain what happens when you round $0.0015 to the nearest cent.

Answer varies.

Round each of these money amounts.

33. $499.98 to the nearest dollar.

$500

34. $9899.59 to the nearest dollar.

$9900

35. $0.996 to the nearest cent.

$1.00

36. $0.09929 to the nearest cent.

$0.10

37. $999.73 to the nearest dollar.

$1000

38. $9999.80 to the nearest dollar.

$10,000

PREVIEW EXERCISES

≈ *Round each number so there is only one non-zero digit and* ***estimate*** *the total. Then add to get the* ***exact*** *answer. (For help, see* ***Sections 1.2 and 1.7.****)*

39.

estimate		*exact*
8000	← rounds to	7929
6000	←	6076
+ 8000	←	+ 8218
22,000		22,223

40.

estimate		*exact*
2000	←	2078
200	←	183
200	←	231
+ 7000	←	+ 7209
9400		9701

41. 80,000 + 100 + 800 = 80,900 *estimate*

81,976 + 98 + 785 = 82,859 *exact*

42. 2000 + 20,000 + 900 = 22,900 *estimate*

1750 + 18,763 + 918 = 21,431 *exact*

4.3 ADDING DECIMALS

OBJECTIVES

1. Add decimals.
2. Estimate the answer.

FOR EXTRA HELP

Tape 5 | SSM pp. 103–105 | MAC: A IBM: A

1 When adding *whole* numbers (**Section 1.2**), you lined up the numbers in columns so that you were adding ones to ones, tens to tens, and so on. A similar idea applies to adding *decimal* numbers. With decimals you line up the decimal points to make sure you are adding tenths to tenths, hundredths to hundredths, and so on.

ADDING DECIMALS

Step 1 Write the numbers in columns with the decimal points lined up.

Step 2 Add the numbers as if they were whole numbers.

Step 3 Line up the decimal point in the answer directly below the decimal points in the problem.

EXAMPLE 1 *Adding Decimal Numbers*

Add.

(a) 16.92 and 48.34 **(b)** 5.897 + 4.632 + 12.174

(a) Write the numbers in columns with the decimal points lined up.

tens	ones		tenths	hundredths
1	6	.	9	2
+ 4	8	.	3	4

Decimal points are lined up.

Add as if these were whole numbers. Then line up the decimal point in the answer under the decimal points in the problem.

$$\begin{array}{r} \scriptstyle 11 \\ 16.92 \\ +\ 48.34 \\ \hline 65.26 \end{array}$$

Decimal point in answer is lined up under decimal points in problem.

(b) Write the numbers vertically with decimal points lined up. Next, add.

$$\begin{array}{r} \scriptstyle 11\ \ 21 \\ 5.897 \\ 4.632 \\ +\ 12.174 \\ \hline 22.703 \end{array}$$

Decimal points are lined up. ■

WORK PROBLEM 1 AT THE SIDE.

1. Find each sum.

(a) 2.86 + 7.09

(b) 13.761 + 8.325

(c) 0.319 + 56.007 + 8.252

(d) 39.4 + 0.4 + 177.2

In Example 1(a), both numbers had *two* **decimal places.** Decimal places are the number of digits to the right of the decimal point. In Example 1(b), all the numbers had *three decimal places* (three digits to the right of the decimal point). That made it easy to add tenths to tenths, hundredths to hundredths, and so on.

ANSWERS
1. (a) 9.95 (b) 22.086 (c) 64.578 (d) 217.0

2. Find each of the following sums.

(a) 6.54 + 9.8

(b) 0.831 + 222.2 + 10

(c) 8.64 + 39.115 + 3.0076

(d) 5 + 429.823 + 0.76

3. ≈ First, round each number so there is only one non-zero digit and estimate the answer. Then add to find the exact answer.

(a) 2.83 + 5.009 + 76.1

(b) 398.81 + 47.658 + 4158.7

(c) 3217.6 + 5.4 + 37.288

ANSWERS
2. (a) 16.34 (b) 233.031 (c) 50.7626 (d) 435.583
3. (a) 3 + 5 + 80 = 88; 83.939
(b) 400 + 50 + 4000 = 4450; 4605.168
(c) 3000 + 5 + 40 = 3045; 3260.288

If the number of decimal places does *not* match, you can write in zeros as placeholders to make them match.

EXAMPLE 2 *Writing Zeros as Placeholders Before Adding*
Add.

(a) 7.3 + 0.85 **(b)** 6.42 + 9 + 2.576

In Example 2(a), there are two decimal places in 0.85 (tenths and hundredths), so write a zero in the hundredths place in 7.3 so that it has two decimal places also. In Example 2(b), make all the addends have three decimal places.

(a)
```
  7.30   ← One 0 is written in.
+ 0.85
  8.15
```

(b)
```
  6.4 2 0 ← One 0 is written in.
  9.0 0 0 ← 9 is a whole number;
            decimal point and
            three 0's are written in.
+ 2.5 7 6 ← No 0's are needed.
 17.9 9 6
```
■

In Example 2(a), 7.30 is equivalent to 7.3 because

$$7\frac{30}{100} \text{ in lowest terms is } 7\frac{3}{10}.$$

Writing zeros to the right of a decimal number does *not* change the value of the number.

In Example 2(b), notice how the whole number 9 is written with the decimal point at the *far right* side. (If you put the decimal point on the *left* side of the 9, you would turn it into the decimal fraction 0.9.)

WORK PROBLEM 2 AT THE SIDE.

2 A common error in working decimal problems by hand is to misplace the decimal point in the answer. Or, when using a calculator, you may accidentally press the wrong key. **Estimating** the answer will help you avoid these mistakes. Start by rounding each number so there is only one non-zero digit (as you did in **Section 1.7**). Here are several examples. Notice that in the rounded numbers only the left-most digit is something other than zero.

3.25 rounds to 3   6.812 rounds to 7
532.6 rounds to 500   26.397 rounds to 30

EXAMPLE 3 *Estimating a Decimal Answer*
Round each number so there is only one non-zero digit. Then add the rounded numbers to get an estimated answer. Finally, find the exact answer. Add 194.2 and 6.825

estimate		*exact*
200	← rounds to	194.200
+ 7	← rounds to	+ 6.825
207		201.025

The estimate goes out to the hundreds place (three places to the left of the decimal point), and so does the exact answer. Therefore, the decimal point is probably in the right place in the exact answer. ■

WORK PROBLEM 3 AT THE SIDE.

4.3 EXERCISES

NAME DATE HOUR

Find each sum.

Example:

826.28
0.6
+ 38.152

Solution:

Line up decimal points.

826.280
0.600 — Use zeros as placeholders.
+ 38.152
865.032

1. 5.69
0.24
+ 11.79
17.72

2. 372.1
33.7
+ 42.3
448.1

3. 224.008
0.325
+ 16.409
240.742

4. 0.77
306.26
+ 9.88
316.91

5. 8.763
0.5
+ 339.25
348.513

6. 76.5
89.39
+ 0.506
166.396

7. 0.38
7
+ 4.6
11.98

8. 3.7
0.812
+ 55
59.512

9. 14.23 + 8 + 74.63 + 18.715 + 0.286 115.861

10. 197.4 + 0.72 + 17.43 + 25 + 1.4 241.95

11. 27.65 + 18.714 + 9.749 + 3.21 59.323

12. 58.546 + 19.2 + 8.735 + 14.58 101.061

13. 39.76005 + 182 + 4.799 + 98.31 + 5.9999 330.86895

14. 489.76 + 0.9993 + 38 + 8.55087 + 80.697 618.00717

15. Explain and correct the error that a student made when he added 0.72 + 6 + 39.5 this way:

0.72
0.60
+ 39.50
40.82

Answer varies.

16. Explain and correct the error that a student made when she added 7.21 + 65 + 13.15 this way:

7.21
65
+ 13.15
21.01

Answer varies.

≈ *Round each number so there is only one non-zero digit and estimate the sum. Then add to find the exact answer.*

Example: 56.9 + 0.82 + 12.06

Solution:

estimate		*exact*
60	← rounds to	56.90
1	←	0.82
+ 10	←	+ 12.06
71		**69.78**

17.

estimate	*exact*
40	37.25
20	18.9
+ 8	+ 7.5
68	63.65

18.

estimate	*exact*
20	24.83
20	19.7
+ 50	+ 46.19
90	90.72

19.

estimate	*exact*
400	392.7
1	0.865
+ 20	+ 21.08
421	414.645

20.

estimate	*exact*
40	38.55
8	7.716
+ 1	+ 0.6
49	46.866

21.

estimate	*exact*
60	62.8173
500	539.99
+ 6	+ 5.629
566	608.4363

22.

estimate	*exact*
300	332.607
10	12.5
+ 800	+ 823.3949
1110	1168.5019

23.

estimate	*exact*
400	382.504
600	591.089
+ 600	+ 612.715
1600	1586.308

24.

estimate	*exact*
8000	8159.76
9000	9382.54
+ 7000	+ 7179.18
24,000	24,721.48

≈ *Round each number so there is only one non-zero digit and estimate the answer. Then find the exact answer for each word problem.*

25. Joann put two checks in the deposit envelope at the automated teller machine. There was a $310.14 paycheck and a $0.95 refund check. How much did she deposit in her account?

estimate: $300 + $1 = $301
exact: $311.09

26. Rodney Green's paycheck stub showed wages of $274.19 at the regular rate of pay and $72.94 at the overtime rate. What were his total wages?

estimate: $300 + $70 = $370
exact: $347.13

27. Chris Howard worked at Blockblaster Video 4.5 days one week, 6.25 days another week, and 3.74 days a third week. How many days did he work altogether?

estimate: 5 + 6 + 4 = 15 days
exact: 14.49 days

28. Mrs. Little Owl has three pieces of wire for an experiment in her physics class. The pieces measure 1.32 meters, 0.8 meter, and 1.5 meters. What is the total length of wire used in the experiment?

estimate: 1 + 1 + 2 = 4 meters
exact: 3.62 meters

29. At a bakery, Sue Chee bought $7.42 worth of muffins and $10.09 worth of croissants for a staff party and a $0.69 cookie for herself. How much money did she spend altogether?

estimate: $7 + $10 + $1 = $18
exact: $18.20

30. Jeff McGee wrote checks for $172.15, $0.75, $9.06, and $122.24. Find the total of the checks.

estimate: $200 + $1 + $9 + $100 = $310
exact: $304.20

31. At the beginning of a trip to Visalia, a car odometer read 7542.1 miles. It is 186.4 miles to Visalia. What should the odometer read after driving to Visalia *and back?*

estimate: 8000 + 200 + 200 = 8400 miles
exact: 7914.9 miles

32. Gonzalo runs his own package delivery service. On one trip he started from Atlanta and drove 226.6 miles to Charlotte, then 153.8 miles to Roanoke, and finally, 341.3 miles back to Atlanta. Find the total length of his trip.

estimate: 200 + 200 + 300 = 700 miles
exact: 721.7 miles

Yiangos works part-time at a factory. His time card for last week is shown below. Use the time card to solve Exercises 33–34.

Day	*Date*	*Hours*
Mon	6/1	4.5
Tue	6/2	0
Wed	6/3	0
Thr	6/4	6.2
Fri	6/5	5
Sat	6/6	9.5
Sun	6/7	4.8

33. Yiangos is paid a higher hourly wage for working on weekends. How many weekend hours did he work?

estimate: 10 + 5 = 15 hours
exact: 14.3 hours

34. How many hours did Yiangos work on weekdays?

estimate: 5 + 6 + 5 = 16 hours
exact: 15.7 hours

The accountant at Top Notch Lumber had the list of expenses shown below for one week. Use the list to solve Exercises 35–36.

Expenses	
payroll	$979.80
utilities	$108.11
radio ads	$366.03
newspaper ads	$253.79
TV ad	$740
mill payments	$6985.46

35. Find the amount spent on advertising during the week.

estimate: $400 + $300 + $700 = $1400
exact: $1359.82

36. Find the total expenses for the week.

estimate: $1000 + $100 + $400 + $300 + $700 + $7000 = $9500
exact: $9433.19

Solve each problem. There may be extra information in the problem, or you may need to do several steps to solve it.

37. Tameka keeps track of her business mileage so her company will pay for her travel. She is not paid for trips to lunch or for travel to and from home. Today she drove 12.6 miles to work, 35.4 miles to visit a client, 14.9 miles to visit another client, 8 miles to lunch, 40 miles to attend a business meeting, and 12.6 miles home. How many miles will her company pay for?

90.3 miles

38. Tony wrote a lot of checks today. His tuition at the community college was \$476.44 and textbooks were \$80.06. He also paid \$17.99 for an oil change on his car, \$20.75 at the grocery store, and \$31.62 for brushes and paint for an art class he is taking at the college. What were his total school expenses?

\$588.12

39. Ms. Hattori was trying to decide whether to buy a red sweater that cost \$21.89 or a blue one that cost \$4 less. She decided to buy both sweaters and give the blue one to her sister. How much did she pay for the two sweaters?

\$39.78

40. James jogged 3.25 kilometers this morning. His friend Anthony jogged with him and kept going another 1.4 kilometers after James stopped. What was the total distance run by the two men?

7.9 kilometers

Find the perimeter of (distance around) these figures by adding the lengths of the sides.

41.

19.75 inches
6.3 inches
6.3 inches
19.75 inches

52.1 inches

42.

2 meters
1 meter
0.9 meter
1.7 meters
1.18 meters
0.86 meter
2.095 meters

9.735 meters

PREVIEW EXERCISES

≈ *Round the numbers so there is only one non-zero digit and estimate each answer. Then find the exact answer. (For help, see* **Sections 1.3 and 1.7.**)

43. *estimate* / *exact*

estimate		exact
300	← rounds to	301
− 100	←	− 104
200		**197**

44. *estimate* / *exact*

estimate	exact
600	553
− **400**	− 386
200	**167**

45. **7000** − **100** = **6900** *estimate*

6708 − 139 = **6569** *exact*

46. **70,000** − **900** = **69,100** *estimate*

71,000 − 856 = **70,144** *exact*

4.4 SUBTRACTING DECIMALS

OBJECTIVES

1. Subtract decimals.
2. Estimate the answer.

FOR EXTRA HELP

Tape 6 | SSM pp. 105–107 | MAC: A IBM: A

1 Subtraction of decimals is done in much the same way as addition of decimals. Use the following steps.

SUBTRACTING DECIMALS

Step 1 Write the numbers in columns with the decimal points lined up.

Step 2 If necessary, write in zeros so both numbers have the same number of decimal places. Then subtract as if they were whole numbers.

Step 3 Line up the decimal point in the answer directly below the decimal points in the problem.

■ **EXAMPLE 1** *Subtracting Decimal Numbers*

Subtract each of the following. Check your answer using addition.

(a) 15.82 from 28.93 **(b)** 146.35 less 58.98

(a) *Step 1*

$$\begin{array}{r} 28.93 \\ -\ 15.82 \\ \hline \end{array}$$

Line up decimal points. Then you will be subtracting hundredths from hundredths and tenths from tenths.

Step 2

$$\begin{array}{r} 28.93 \\ -\ 15.82 \\ \hline 13\ 11 \end{array}$$

Both numbers have two decimal places; no need to write in zeros.

Subtract as if they were whole numbers.

Step 3

$$\begin{array}{r} 28.93 \\ -\ 15.82 \\ \hline 13.11 \end{array}$$

Decimal point in answer lined up.

Check the answer by adding 13.11 and 15.82. If the subtraction is done correctly, the sum will be 28.93.

(b) Borrowing is needed here.

$$\begin{array}{r} {}^{0\ 13\ 15\quad 12\ 15} \\ \not{1}\ \not{4}\ \not{6}\ .\ \not{3}\ \not{5} \\ -\quad 5\ 8\ .\ 9\ 8 \\ \hline 8\ 7\ .\ 3\ 7 \end{array}$$

Check the answer by adding 87.37 and 58.98. If you did the subtraction correctly, the sum will be 146.35. (If it *isn't*, you need to rework the problem.) ■

WORK PROBLEM 1 AT THE SIDE.

1. Subtract. Check your answers by addition.

(a) 22.7 from 72.9

(b) 6.425 from 11.813

(c) 20.15 − 19.67

■ **EXAMPLE 2** *Writing Zeros as Placeholders Before Subtracting*

Subtract each of the following.

(a) 16.5 from 28.362 **(b)** 59.7 − 38.914 **(c)** 12 less 5.83

(a) Use the same steps as above, remembering to write in zeros so both numbers have three decimal places.

$$\begin{array}{r} 28.362 \\ -\ 16.500 \\ \hline 11.862 \end{array}$$

← Write two 0's.

← Next, subtract as usual.

ANSWERS

1. (a) 50.2; 50.2 + 22.7 = 72.9
   (b) 5.388; 5.388 + 6.425 = 11.813
   (c) 0.48; 0.48 + 19.67 = 20.15

2. Subtract. Check your answers by addition.

(a) 18.651 from 25.3

(b) 5.816 − 4.98

(c) 40 less 3.66

(d) 1 − 0.325

3. ≈ Round the numbers so there is only one non-zero digit and estimate the answer. Then subtract to find an exact answer.

(a) 11.365 from 38

(b) 214.603 − 53.4

(c) $19.28 less $1.53

(d) Find the difference between 12.837 meters and 46.091 meters

ANSWERS

2. (a) 6.649; 6.649 + 18.651 = 25.3
(b) 0.836; 0.836 + 4.98 = 5.816
(c) 36.34; 36.34 + 3.66 = 40
(d) 0.675; 0.675 + 0.325 = 1

3. (a) 40 − 10 = 30; 26.635
(b) 200 − 50 = 150; 161.203
(c) $20 − 2 = $18; $17.75
(d) 50 − 10 = 40; 33.254 meters

(b)

```
  59.700  ← Write two 0's
− 38.914
  20.786  ← Subtract as usual.
```

(c)

```
  12.00  ← Write a decimal point and two 0's
−  5.83
   6.17  ← Subtract as usual. ■
```

WORK PROBLEM 2 AT THE SIDE.

2 *Estimating* the answer will help you check that the problem is set up properly and the decimal point is correctly placed in the answer.

■ EXAMPLE 3 *Estimating a Decimal Answer*

Estimate the answer. Then subtract to find the exact answer.

(a) $69.42 − $13.78 **(b)** 1.8614 from 7.3

(c) Find the difference between 0.92 feet and 8 feet.

Estimate the answer by rounding each number so there is only one non-zero digit (as you did in Section 4.3)

(a)

```
estimate      rounds to      exact
  $70      ←────────────     $69.42
−  10      ←────────────   −  13.78
  $60                        $55.64
```

Answers are close, so the problem is probably set up correctly.

(b)

```
estimate                     exact
   7       ←────────────     7.3000  ← Write three 0's.
 − 2       ←────────────   − 1.8614
   5                         5.4386
```

Answers are close, so the problem is probably set up correctly.

(c) Use subtraction to find the difference between two numbers. The larger number, 8, is written on top.

```
estimate                     exact
   8       ←────────────     8.00  ←Write a decimal point and two 0's.
 − 1       ←────────────   − 0.92
   7                         7.08 feet
```

Answers are close. ■

WORK PROBLEM 3 AT THE SIDE.

EXERCISES

Subtract. Check your answer by addition.

Example: 71 − 0.352

Solution:

$$\begin{array}{r} 71.000 \\ -\ 0.352 \\ \hline \mathbf{70.648} \end{array}$$

71.000 ← Write decimal point and three 0's

Add these numbers. If the sum is 71.000, the problem is done correctly.

1. $\begin{array}{r} 73.5 \\ -\ 19.2 \\ \hline \end{array}$ 54.3

2. $\begin{array}{r} 47.8 \\ -\ 36.5 \\ \hline \end{array}$ 11.3

3. $\begin{array}{r} 58.413 \\ -\ 25.847 \\ \hline \end{array}$ 32.566

4. $\begin{array}{r} 27.905 \\ -\ 18.176 \\ \hline \end{array}$ 9.729

5. $\begin{array}{r} 58.254 \\ -\ 19.7 \\ \hline \end{array}$ 38.554

6. $\begin{array}{r} 47.658 \\ -\ 20.9 \\ \hline \end{array}$ 26.758

7. $\begin{array}{r} 21 \\ -\ 0.896 \\ \hline \end{array}$ 20.104

8. $\begin{array}{r} 9 \\ -\ 1.183 \\ \hline \end{array}$ 7.817

9. $\begin{array}{r} 15.7 \\ -\ 2.852 \\ \hline \end{array}$ 12.848

10. $\begin{array}{r} 36.9 \\ -\ 14.582 \\ \hline \end{array}$ 22.318

11. 90.5 − 0.8 — 89.7

12. 303.72 − 0.68 — 303.04

13. 0.4 − 0.291 — 0.109

14. 0.35 − 0.088 — 0.262

15. 6 − 5.09 — 0.91

16. 80 − 16.3 — 63.7

17. 15 − 8.339 — 6.661

18. 44 − 0.08 — 43.92

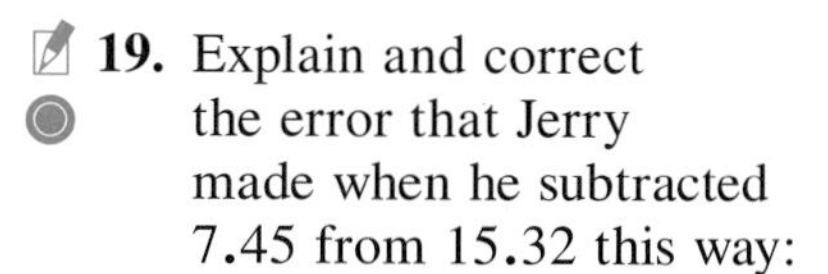

19. Explain and correct the error that Jerry made when he subtracted 7.45 from 15.32 this way:

$$\begin{array}{r} 7.45 \\ -\ 15.32 \\ \hline 12.13 \end{array}$$

Answer varies.

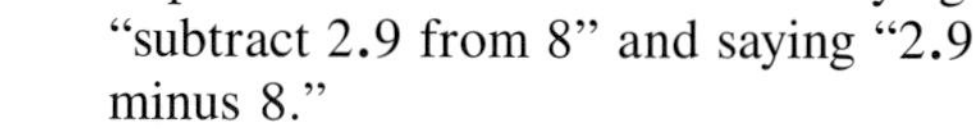

20. Explain the difference between saying "subtract 2.9 from 8" and saying "2.9 minus 8."

Answer varies.

≈ *Round the numbers and estimate the answer, rounding so there is only one non-zero digit. Then subtract to find the exact answer.*

Example: Subtract 4.962 from 7.3

Solution: ***estimate*** (rounds to) ***exact***

$$\begin{array}{r} 7 \\ -\ 5 \\ \hline \mathbf{2} \end{array} \quad \longleftarrow \quad \begin{array}{r} 7.300 \\ -\ 4.962 \\ \hline \mathbf{2.338} \end{array}$$

21. *estimate* / *exact*

$$\begin{array}{r} \$20 \\ -\ 7 \\ \hline \$13 \end{array} \qquad \begin{array}{r} \$19.74 \\ -\ 6.58 \\ \hline \$13.16 \end{array}$$

22. *estimate* / *exact*

$$\begin{array}{r} \$30 \\ -\ 8 \\ \hline \$22 \end{array} \qquad \begin{array}{r} \$27.96 \\ -\ 8.39 \\ \hline \$19.57 \end{array}$$

Writing · Conceptual · ▲ Challenging · ≈ Estimation

23. What is 8.6 less 3.751?

estimate 9 − 4 = 5 *exact* 4.849

24. What is 31.7 less 4.271?

estimate 30 − 4 = 26 *exact* 27.429

25. Find the difference between 1.981 inches and 2 inches.

estimate 2 − 2 = 0 *exact* 0.019 inch

26. Find the difference between 13.582 meters and 28 meters.

estimate 30 − 10 = 20 *exact* 14.418 meters

27. What is 9.006 liters from 384.2 liters?

estimate 400 − 9 = 391 *exact* 375.194 liters

28. What is 23.607 kilograms from 786.1 kilograms?

estimate 800 − 20 = 780 *exact* 762.493 kilograms

Subtract.

29. 12 − 11.7251

0.2749

30. 20 − 1.37009

18.62991

31. 6004.003 − 52.7172

5951.2858

32. 803.25 − 0.69815

802.55185

≈ *Round the numbers so there is only one non-zero digit and estimate the answer. Then find the exact answer.*

33. Tom has agreed to work 42.5 hours a week as a car wash attendant. So far this week he has worked 16.35 hours. How many more hours must he work?

estimate: 40 − 20 = 20 hours
exact: 26.15 hours

34. Cathy Eastes bought $187.12 worth of clothes. She returned a dress worth $37.95. Find the value of the clothes she kept.

estimate: $200 − $40 = $160
exact: $149.17

35. Steven One Feather gave the cashier a $20 bill to pay for $9.12 worth of groceries. How much change did he get?

estimate: $20 − $9 = $11
exact: $10.88

36. The cost of Julie's tennis racket, with tax, is $41.09. She gave the clerk two $20 bills and a $10 bill. What amount of change did Julie receive?

estimate: $50 − $40 = $10
exact: $8.91

37. Namiko is comparing two boxes of chicken nuggets. One box weighs 9.85 ounces and the other weighs 10.5 ounces. What is the difference in the weight of the two boxes?

estimate: 11 − 10 = 1 ounce
exact: 0.65 ounce

38. Sammy works in a veterinarian's office. He weighed two newborn kittens. One was 3.9 ounces and the other was 4.05 ounces. What was the difference in the weight of the two kittens?

estimate: 4 − 4 = 0 ounce
exact: 0.15 ounce

39. At the beginning of January, the odometer of Maria DeRisi's company car read 29,086.1 miles. At the end of December, it read 51,561.9 miles. How many miles did Ms. DeRisi drive during the year?

estimate: 50,000 − 30,000 = 20,000 miles
exact: 22,475.8 miles

40. Refer to Exercise 39. During the year, Ms. DeRisi drove the car 11,237.4 miles on personal business. How many miles was the car driven on company business?

estimate: 20,000 − 10,000 = 10,000 miles
exact: 11,238.4 miles

Subtract.

▲ **41.** 386.021 − 179.68231

206.33869

▲ **42.** 2.5006 − 0.005318

2.495282

▲ **43.** 221.04 − 218.528683

2.511317

▲ **44.** 128.3506 − 97.009398

31.341202

Work each problem.

45. Mitch Albers had a checking account balance of $129.86 on September 1. During the month, he deposited an additional $1749.82 to the account, and wrote checks totaling $1802.15. The bank charged him a $2 service charge. Find the amount in the account at the end of the month.

$75.53

46. On February 1, Lynn Fiorentino had $1009.24 in her checking account. During the month she deposited a tax refund check of $704.42 and her paycheck of $1258.94. She wrote checks totaling $1389.54 and had $200 transferred to her savings account. Find her checking account balance at the end of the month.

$1383.06

≈ *Solve each problem. First round the numbers so there is only one non-zero digit and estimate the answer. Then find the exact answer.*

▲ **47.** The manual for Jason's car says the gas tank holds 16.6 gallons. Jason knows that the tank actually holds an extra 1.4 gallons. The gas station pump showed that Jason bought 8.628 gallons of gas to fill the tank. How much gas was in the tank before he filled it?

estimate: 20 + 1 = 21; 21 − 9 = 12 gallons
exact: 9.372 gallons

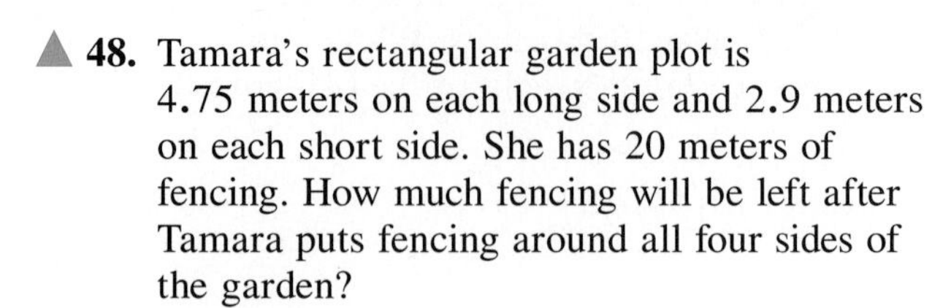

▲ **48.** Tamara's rectangular garden plot is 4.75 meters on each long side and 2.9 meters on each short side. She has 20 meters of fencing. How much fencing will be left after Tamara puts fencing around all four sides of the garden?

estimate: 5 + 5 + 3 + 3 = 16; 20 − 16 = 4 meters
exact: 4.7 meters

Find the missing measurement in each rectangle or circle.

49.

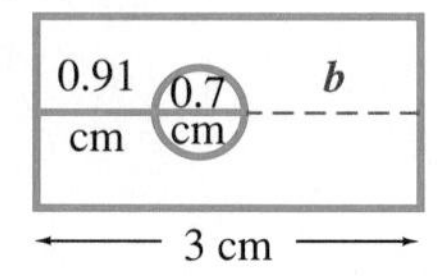

b = 1.39 cm

50.

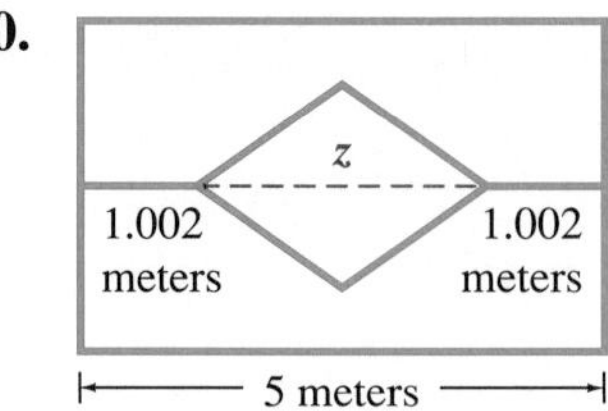

z = 2.996 meters

51.

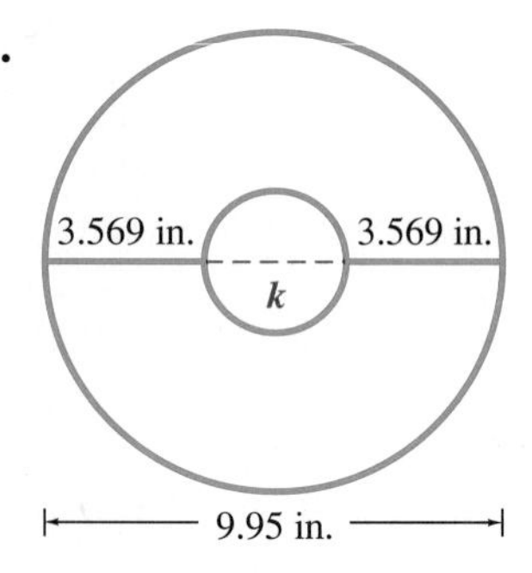

k = 2.812 inches

52.

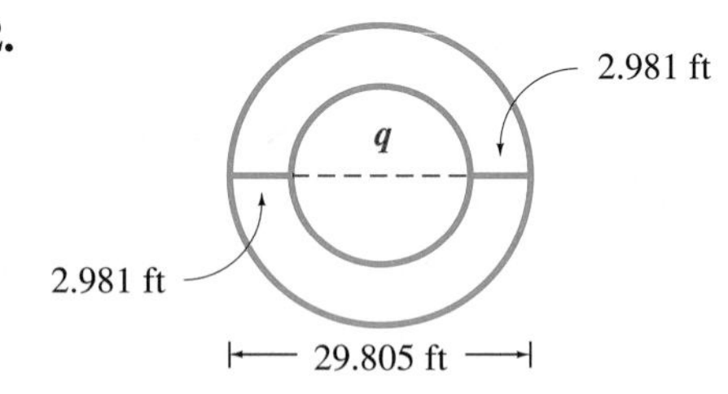

q = 23.843 feet

PREVIEW EXERCISES

First round the numbers and estimate each answer. Round so there is only one non-zero digit. Then find the exact answer. (For help, see **Sections 1.4 and 1.7.**)

53.

estimate		*exact*
80	← rounds to	83
× 30	←	× 28
2400		2324

54.

estimate	*exact*
70	67
× 70	× 72
4900	4824

55. 4000 × 200 = 800,000 *estimate*
3789 × 205 = 776,745 *exact*

56. 6000 × 700 = 4,200,000 *estimate*
6381 × 709 = 4,524,129 *exact*

4.5 MULTIPLYING DECIMALS

1 The decimals 0.3 and 0.07 can be multiplied by writing them as fractions.

$$0.\underline{3} \times 0.\underline{07} = \frac{3}{10} \times \frac{7}{100} = \frac{21}{1000} = 0.\underline{021}$$

1 decimal place + 2 decimal places → 3 decimal places

Can you see a way to multiply decimals without writing them as fractions? Try these steps. Remember that each number in a multiplication problem is called a *factor*.

MULTIPLYING DECIMALS

Step 1 Multiply the numbers (the factors) as if they were whole numbers.

Step 2 Find the *total* number of decimal places in *both* factors.

Step 3 Write the decimal point in the answer (the product) so it has the same number of decimal places as the total from Step 2. You may need to write in extra zeros on the left side of the product in order to get the correct number of decimal places.

Note When multiplying decimals, you do **not** need to line up decimal points. (You **do** need to line up decimal points when adding or subtracting decimals.)

■ EXAMPLE 1 *Multiplying Decimal Numbers*

Multiply 8.34 and 4.2.

Step 1 Multiply the numbers as if they were whole numbers.

$$\begin{array}{r} 8.34 \\ \times \quad 4.2 \\ \hline 1668 \\ 3336 \\ \hline 35028 \end{array}$$

Step 2 Count the total number of decimal places in both factors.

$$\begin{array}{rl} 8.34 & \leftarrow \text{2 decimal places} \\ \times \quad 4.2 & \leftarrow \underline{\text{1 decimal place}} \\ \hline 1668 & \quad \text{3 total decimal places} \\ 3336 & \\ \hline 35028 & \end{array}$$

Step 3 Count over 3 places and write the decimal point in the answer. Count from *right to left*.

$$\begin{array}{rl} 8.34 & \leftarrow \text{2 decimal places} \\ \times \quad 4.2 & \leftarrow \underline{\text{1 decimal place}} \\ \hline 1\,668 & \quad \text{3 total decimal places} \\ 33\,36 & \\ \hline 35.028 & \leftarrow \text{3 decimal places in answer} \end{array}$$

Count over 3 places from right to left to position the decimal point. ■

WORK PROBLEM 1 AT THE SIDE.

OBJECTIVES

1 Multiply decimals.

2 Estimate the answer.

FOR EXTRA HELP

Tape 6 | SSM pp. 107–110 | MAC: A IBM: A

1. Multiply.

(a) $\begin{array}{r} 2.6 \\ \times\ 0.4 \\ \hline \end{array}$

(b) $\begin{array}{r} 45.2 \\ \times\ 0.25 \\ \hline \end{array}$

(c) $\begin{array}{rl} 0.104 & \leftarrow \text{3 decimal places} \\ \times \quad 7 & \leftarrow \underline{\text{0 decimal places}} \\ \hline & \leftarrow \text{3 decimal places in the answer} \end{array}$

(d) $\begin{array}{r} 3.18 \\ \times\ 2.23 \\ \hline \end{array}$

(e) $\begin{array}{r} 611 \\ \times\ 3.7 \\ \hline \end{array}$

ANSWERS

1. (a) 1.04 (b) 11.300 (c) 0.728 (d) 7.0914 (e) 2260.7

2. Multiply.

(a) $0.04 \cdot 0.09$

(b) $0.2 \cdot 0.008$

(c) $0.063 \cdot 0.04$

(d) $0.0081 \cdot 0.003$

(e) $(0.11)(0.0005)$

3. ≈ First round the numbers and estimate the answer. Then multiply to find the exact answer.

(a) $11.62 \cdot 4.01$

(b) $5.986 \cdot 33$

(c) $8.31 \cdot 4.2$

(d) $58.6 \cdot 17.4$

ANSWERS

2. (a) 0.0036 (b) 0.0016 (c) 0.00252 (d) 0.0000243 (e) 0.000055

3. (a) 10 · 4 = 40; 46.5962 (b) 6 · 30 = 180; 197.538 (c) 8 · 4 = 32; 34.902 (d) 60 · 20 = 1200; 1019.64

■ **EXAMPLE 2** *Writing Zeros as Placeholders in the Answer*

Multiply 0.042 by 0.03.

Start by multiplying and counting decimal places.

$$\begin{array}{r l} 0.042 & \leftarrow 3 \text{ decimal places} \\ \times \ \ 0.03 & \leftarrow 2 \text{ decimal places} \\ \hline 126 & \leftarrow 5 \text{ decimal places needed in answer} \end{array}$$

The answer has only three decimal places, but five are needed. So write two zeros on the *left* side of the answer.

$$\begin{array}{r} 0.042 \\ \times \ \ 0.03 \\ \hline 00126 \end{array} \qquad \begin{array}{r l} 0.042 & \leftarrow 3 \text{ decimal places} \\ \times \ \ 0.03 & \leftarrow 2 \text{ decimal places} \\ \hline .00126 & \leftarrow 5 \text{ decimal places} \end{array}$$

Write two 0's on *left* side of answer.

Now count over 5 places and write in the decimal point.

The final answer is 0.00126, which has five decimal places. ■

◀ WORK PROBLEM 2 AT THE SIDE.

2 If you are doing multiplication problems by hand, estimating the answer helps you check that the decimal point is in the right place. When you are using a calculator, estimating helps you catch an error like pressing the ÷ key instead of the × key.

■ **EXAMPLE 3** *Estimating before Multiplying*

First estimate $76.34 \cdot 12.5$. Round each number so there is only one non-zero digit. Then multiply to find the exact answer.

estimate (rounds to) *exact*

$$\begin{array}{r} 80 \\ \times \ 10 \\ \hline 800 \end{array} \quad \longleftarrow \quad \begin{array}{r l} 76.34 & \leftarrow 2 \text{ decimal places} \\ \times \ 12.5 & \leftarrow 1 \text{ decimal place} \\ \hline 38\ 170 & \ \ 3 \text{ decimal places are in answer.} \\ 152\ 68\ \ & \\ 763\ 4\ \ \ \ & \\ \hline 954.250 & \end{array}$$

Both the estimate and the exact answer go out to the hundreds, so the decimal point in 954.250 is probably in the correct place.

◀ WORK PROBLEM 3 AT THE SIDE.

4.5 NAME DATE HOUR

EXERCISES

Multiply.

Example:
0.093
× 0.6

Solution:
0.093 ← 3 decimal places
× 0.6 ← 1 decimal place
0.0558 ← 4 decimal places in answer
Write a zero in order to get 4 decimal places.

1. 0.042 × 3.2
0.1344

2. 0.571 × 2.9
1.6559

3. 21.5 × 7.4
159.10

4. 85.4 × 3.5
298.90

5. 234 × 0.666
155.844

6. 896 × 0.799
715.904

7. 51.81 × 0.407
21.08667

8. 36.78 × 0.108
3.97224

Use the fact that 72 × 6 = 432 to help you solve Exercises 9–16.

9. 72 × 0.6
43.2

10. 7.2 × 6
43.2

11. (7.2)(0.06)
0.432

12. (0.72)(0.6)
0.432

13. 0.72 × 0.06
0.0432

14. 72 × 0.0006
0.0432

15. 0.0072 × 0.6
0.00432

16. 0.072 × 0.006
0.000432

17. (0.006)(0.0052)
0.0000312

18. (0.0052)(0.009)
0.0000468

19. 0.003 • 0.002
0.000006

20. 0.0079 • 0.006
0.0000474

21. Do these multiplications:
(5.96)(10) (3.2)(10)
(0.476)(10) (80.35)(10)
(722.6)(10) (0.9)(10)
What pattern do you see? Write a "rule" for multiplying by 10. What do you think the rule is for multiplying by 100? By 1000? Write the rules and try them out on the numbers above.

Answer varies.

22. Do these multiplications:
(59.6)(0.1) (3.2)(0.1)
(0.476)(0.1) (80.35)(0.1)
(65)(0.1) (523)(0.1)
What pattern do you see? Write a "rule" for multiplying by 0.1. What do you think the rule is for multiplying by 0.01? By 0.001? Write the rules and try them out on the numbers above.

Answer varies.

≈ *First round the numbers so there is only one non-zero digit and estimate the answer. Then multiply to find the exact answer.*

	estimate		*exact*
23.	40	← rounds to	39.6
	× 5	←	× 4.8
	200		190.08

	estimate	*exact*
24.	20	18.7
	× 2	× 2.3
	40	43.01

25. 40 × 40 = 1600 | 37.1 × 42 = 1558.2

26. 5 × 70 = 350 | 5.08 × 71 = 360.68

27. 7 × 5 = 35 | 6.53 × 4.6 = 30.038

28. 8 × 8 = 64 | 7.51 × 8.2 = 61.582

29. 3 × 7 = 21 | 2.809 × 6.85 = 19.24165

30. 70 × 20 = 1400 | 73.52 × 22.34 = 1642.4368

Solve. If the problem involves money, round to the nearest cent.

31. LaTasha worked 50.5 hours over the last two weeks. She earns $11.73 per hour. How much did she make?

$592.37

32. Michael's time card shows 42.2 hours at $10.03 per hour. What are his gross earnings?

$423.27

33. Sid needs 0.6 meter of canvas material to make a carry-all bag that fits on his wheelchair. If canvas is \$4.09 per meter, how much will Sid spend? (Note: \$4.09 *per* meter means \$4.09 for *one* meter.)

\$2.45

34. How much will Mrs. Nguyen pay for 3.5 yards of lace trim that costs \$0.87 per yard?

\$3.05

35. Michelle pumped 18.65 gallons of gas into her pickup truck. The price was \$1.15 per gallon. How much did she pay for gas?

\$21.45

36. Spicy chicken wings were on sale for \$0.98 per pound. Juma bought 1.7 pounds of wings. How much did the chicken wings cost?

\$1.67

37. Ms. Rolack is a real estate broker who helps people sell their homes. Her fee is 0.07 times the price of the home. What was her fee for selling a \$92,300 home?

\$6461.00

38. The Darby's state income tax is found by multiplying the family's \$22,900 income by 0.054. Find the amount of their tax.

\$1236.60

39. Judy Lewis pays \$28.96 per month for cable TV. How much will she pay for cable over one year?

\$347.52

40. Chuck's car payment is \$220.27 per month for three years. How much will he pay altogether?

\$7929.72

41. Paper for the copy machine at the library costs \$0.015 per sheet. How much will the library pay for 5100 sheets?

\$76.50

42. A student group collected 2200 pounds of plastic as a fund raiser. How much will they make if the recycling center pays \$0.142 per pound?

\$312.40

Use the list of prices below from the Look Smart mail order catalog to solve Exercises 43 and 44.

Knit Shirt Ordering Information		
43–2A	short sleeve, solid colors	$14.75 each
43–2B	short sleeve, stripes	$16.75 each
43–3A	long sleeve, solid colors	$18.95 each
43–3B	long sleeve, stripes	$21.95 each
Extra-large size, add $2 per shirt.		

43. Find the total cost of four long-sleeve, solid-color shirts and two short-sleeve, striped shirts, all in the extra-large size.

$121.30

44. What is the total cost of eight long-sleeve shirts, five in solid colors and three striped?

$160.60

45. Jack Burgess used 3.5 gallons of fertilizer on each of his 158.2 acres of corn. After he finished, how much fertilizer was left in a storage tank that originally contained 600 gallons?

46.3 gallons

46. Stan Johnson bought 7.8 yards of a Hawaiian print fabric at $5.62 per yard. He paid for it with three $20 bills. Find the amount of his change. (Ignore sales tax.)

$16.16

47. Ms. Sanchez paid $29.95 a day to rent a car, plus $0.29 per mile. Find the cost of her rental for a four-day trip of 926 miles.

$388.34

48. The Bell family rented a motor home for $375 per week plus $0.35 per mile. What was the rental cost for their three-week vacation trip of 2650 miles?

$2052.50

49. Barry bought 16.5 meters of rope at $0.47 per meter and three meters of wire at $1.05 per meter. How much change did he get from three $5 bills?

$4.09

50. Susan bought a VCR that cost $229.88. She paid $45 down and $37.98 per month for six months. How much could she have saved by paying cash?

$43.00

PREVIEW EXERCISES

≈ *First round the numbers and estimate each answer. Round so there is only one non-zero digit. Then find the exact answer using long division and an* ***R*** *to express a remainder. (For help, see* ***Sections 1.5, 1.6,*** *and* ***1.7.****)*

	estimate	*exact*
51.	5)1000 = 200	5)954 = 190 R4
52.	4)2000 = 500	4)2223 = 555 R3
53.	20)20,000 = 1000	21)19,020 = 905 R15
54.	30)90,000 = 3000	28)93,621 = 3343 R17

4.6 DIVIDING DECIMALS

There are two kinds of decimal division problems; those in which a decimal is divided by a whole number, and those in which a decimal is divided by a decimal. First recall the parts of a division problem from **Section 1.5.**

```
               8 ← quotient
divisor → 4)33 ← dividend
            32
             1 ← remainder
```

1 When the divisor is a whole number, use these steps.

DIVIDING DECIMALS BY WHOLE NUMBERS

Step 1 Write the decimal point in the quotient (answer) directly above the decimal point in the dividend.
Step 2 Divide as if both numbers were whole numbers.

■ **EXAMPLE 1** *Dividing a Decimal by a Whole Number*
Divide.

(a) 21.93 by 3 (21.93: dividend; 3: divisor)

(b) $9\overline{)470.7}$ (9: divisor; 470.7: dividend)

(a) Rewrite the division problem. $3\overline{)21.93}$

Write the decimal point in the answer directly above the decimal point in the dividend. $3\overline{)21.93}$ (decimal points lined up)

Divide as if the numbers were whole numbers.

$$\begin{array}{r} 7.31 \\ 3\overline{)21.93} \end{array}$$

Check by multiplying the divisor 3 and the quotient 7.31. The product should be equal to the dividend 21.93.

(b) Write the decimal point in the answer above the decimal point in the dividend. Then divide as if the numbers were whole numbers.

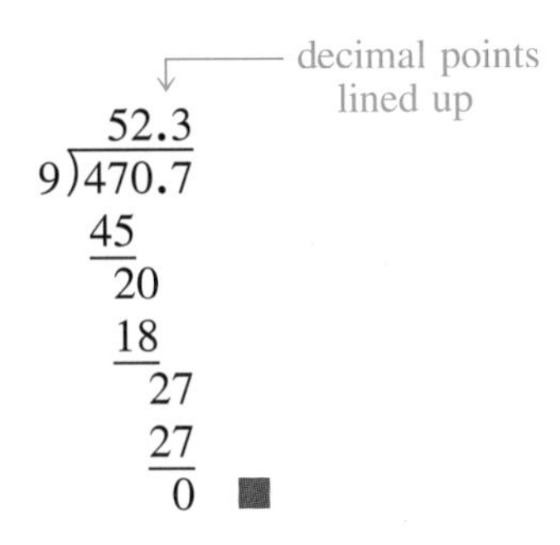

■

WORK PROBLEM 1 AT THE SIDE. ▶▶

OBJECTIVES

1. Divide a decimal by a whole number.
2. Divide a decimal by a decimal.
3. Estimate the answer.
4. Use the order of operations with decimals.

FOR EXTRA HELP

Tape 6 | SSM pp. 110–115 | MAC: A IBM: A

1. Divide. Check your answers by multiplying.

(a) $4\overline{)93.6}$

(b) $6\overline{)6.804}$

(c) $11\overline{)278.3}$

(d) $0.51835 \div 5$

(e) $213.45 \div 15$

ANSWERS
1. (a) 23.4; $23.4 \cdot 4 = 93.6$
(b) 1.134; $1.134 \cdot 6 = 6.804$
(c) 25.3; $25.3 \cdot 11 = 278.3$
(d) 0.10367; $0.10367 \cdot 5 = 0.51835$
(e) 14.23; $14.23 \cdot 15 = 213.45$

2. Divide. Check your answers by multiplying.

(a) $5\overline{)6.4}$

(b) $30.87 \div 14$

(c) $\dfrac{259.5}{30}$

(d) $0.3 \div 8$

ANSWERS

2. (a) 1.28; $1.28 \cdot 5 = 6.40$ or 6.4
(b) 2.205; $2.205 \cdot 14 = 30.870$ or 30.87
(c) 8.65; $8.65 \cdot 30 = 259.50$ or 259.5
(d) 0.0375; $0.0375 \cdot 8 = 0.3000$ or 0.3

■ **EXAMPLE 2** *Writing Extra Zeros to Complete a Division*

Divide 1.5 by 8.

Keep dividing until the remainder is zero, or until the digits in the answer begin to repeat in a pattern. In Example 1(b), you ended up with a remainder of 0. But sometimes you run out of digits in the dividend before that happens. If so, write extra zeros on the right side of the dividend so you can continue dividing.

$$\begin{array}{r} 0.1 \\ 8\overline{)1.5} \\ \underline{8} \\ 7 \end{array}$$

← All digits have been used.
← Remainder is not yet 0.

Write a zero after the 5 so you can continue dividing. Keep writing more zeros in the dividend if needed. Recall that writing zeros to the right of a decimal number does **not** change its value.

$$\begin{array}{r} 0.1875 \\ 8\overline{)1.5000} \\ \underline{8} \\ 70 \\ \underline{64} \\ 60 \\ \underline{56} \\ 40 \\ \underline{40} \\ 0 \end{array}$$

← Three 0's needed to complete the division.
← Remainder is finally 0. ■

◀ WORK PROBLEM 2 AT THE SIDE.

The next example shows a quotient (answer) that must be rounded because you will never get a remainder of zero.

■ **EXAMPLE 3** *Rounding a Decimal Quotient*

Divide 4.7 by 3. Round to the nearest thousandth.
Write extra zeros in the dividend so you can continue dividing.

$$\begin{array}{r} 1.5666 \\ 3\overline{)4.7000} \\ \underline{3} \\ 1\,7 \\ \underline{1\,5} \\ 20 \\ \underline{18} \\ 20 \\ \underline{18} \\ 20 \\ \underline{18} \\ 2 \end{array}$$

← Three 0's added so far
← Remainder is still not 0.

Notice that the digit 6 is repeating. It will continue to do so. The remainder will never be zero. There are two ways to show that the answer is a **repeating decimal** that goes on forever. Write three dots after the answer, or, write a bar above the digits that repeat (in this case, the 6).

$$1.5666\ldots \quad \text{or} \quad 1.5\overline{6}$$

three dots ← bar above repeating digit

When repeating decimals occur, round the answer according to the directions in the problem. In this example, to round to thousandths, divide out one *more* place, to ten-thousandths.

$$4.7 \div 3 = 1.5666\ldots \text{ rounds to } 1.567$$

Check the answer by multiplying 1.567 by 3. Because 1.567 is a rounded answer, the check will not give exactly 4.7, but it should be very close.

$$1.567 \cdot 3 = 4.701$$ (does not exactly equal 4.7 because 1.567 was rounded) ■

Note When checking answers that you've rounded, the check will not match the dividend exactly, but it should be very close.

WORK PROBLEM 3 AT THE SIDE.

2 To divide by a *decimal* divisor, first change the divisor to a whole number. Then divide as before. To see how this is done, write the problem in fraction form. For example:

$$1.2\overline{)6.36} \quad \text{can be written} \quad \frac{6.36}{1.2}$$

In **Section 3.2** you learned that multiplying the numerator and denominator by the same number gives an equivalent fraction. We want the divisor (1.2) to be a whole number. Multiplying by 10 will accomplish that.

$$\frac{6.36}{1.2} = \frac{6.36 \cdot 10}{1.2 \cdot 10} = \frac{63.6}{12}$$

In short, move the decimal point one place to the right in both the divisor and the dividend.

$$1.2\overline{)6.36} \quad \text{is equivalent to} \quad 12\overline{)63.6}$$

Moving the decimal points the **same** number of places in **both** the divisor and dividend will **not** change the answer.

DIVIDING BY DECIMALS

Step 1 Count the number of decimal places in the divisor and move the decimal point that many places to the *right*. (This changes the divisor to a whole number.)

Step 2 Move the decimal point in the dividend the *same* number of places to the *right*. (Write in extra zeros if needed.)

Step 3 Write the decimal point in the answer directly above the decimal point in the dividend. Then divide as usual.

3. Divide. Round answers to the nearest thousandth. If it is a repeating decimal, also write the answer using a bar. Check your answers by multiplying.

(a) $13\overline{)267.01}$

(b) $6\overline{)20.5}$

(c) $10.22 \div 9$

(d) $16.15 \div 3$

(e) $116.3 \div 7$

ANSWERS

3. (a) 20.539; not a repeating decimal
(b) 3.417; $3.41\overline{6}$
(c) 1.136; $1.13\overline{5}$
(d) 5.383; $5.38\overline{3}$
(e) 16.614; not a repeating decimal

4. Divide. If the quotient does not come out even, round to the nearest hundredth.

(a) $0.2\overline{)1.04}$

(b) $0.06\overline{)1.8072}$

(c) $0.005\overline{)32}$

(d) $8.1 \div 0.025$

(e) $7 \div 1.3$

(f) $5.3091 \div 6.2$

ANSWERS
4. (a) 5.2 (b) 30.12 (c) 6400 (d) 324 (e) 5.38 (rounded) (f) 0.86 (rounded)

EXAMPLE 4 *Dividing by a Decimal*

(a) $0.003\overline{)27.69}$ **(b)** Divide 5 by 4.2. Round to the nearest hundredth.

(a) Move the decimal point in the divisor *three* places to the *right* so 0.003 becomes the whole number 3. In order to move the decimal point in the dividend the same number of places, write in an extra zero.

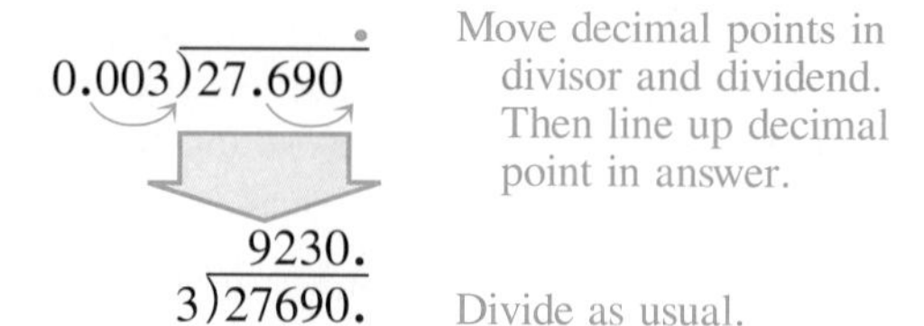

$$\begin{array}{r} 9230. \\ 3\overline{)27690.} \end{array}$$

Divide as usual.

(b) Move the decimal point in the divisor one place to the right so 4.2 becomes the whole number 42. The decimal point in the dividend starts on the right side of 5 and is also moved one place to the right.

$$\begin{array}{r} 1.190 \\ 4.2\overline{)5.0\ 000} \\ 4\ 2 \\ \hline 8\ 0 \\ 4\ 2 \\ \hline 3\ 80 \\ 3\ 78 \\ \hline 20 \end{array}$$

← In order to round to hundredths, divide out one *more* place, to thousandths.

Round the quotient to 1.19 (nearest hundredth). ■

WORK PROBLEM 4 AT THE SIDE.

3 Estimating the answer to a division problem helps you catch errors. Compare the estimate to your exact answer. If they are very different, recheck your work.

EXAMPLE 5 *Estimating before Dividing*

First round the numbers so there is only one non-zero digit and estimate the answer. Then divide to find the exact answer.

$$580.44 \div 2.8$$

Here is how one student solved this problem. She rounded 580.44 to 600 and 2.8 to 3 to estimate the answer.

estimate

$$\begin{array}{r} 200 \\ 3\overline{)600} \end{array}$$

exact

$$\begin{array}{r} 2\ 7.3 \\ 2.8\overline{)580.4\ 4} \\ 56 \\ \hline 20\ 4 \\ 19\ 6 \\ \hline 8\ 4 \\ 8\ 4 \\ \hline 0 \end{array}$$

Notice that the estimate, which is in the hundreds, is very different from the exact answer, which is only in the tens. This tells the student that she needs to rework the problem. Can you find the error? (The exact answer should be 207.3, which fits with the estimate of 200.) ■

WORK PROBLEM 5 AT THE SIDE.

4 Use the order of operations when a decimal problem involves more than one operation.

ORDER OF OPERATIONS

1. Do all operations inside parentheses.
2. Simplify any expressions with exponents and find any square roots.
3. Multiply or divide from left to right.
4. Add or subtract from left to right.

EXAMPLE 6 *Using the Order of Operations*

Simplify by using the order of operations.

(a) $2.5 + 6.3^2 + 9.62$ — Do exponents first.

$2.5 + 39.69 + 9.62$ — Add from left to right.

$42.19 + 9.62 = 51.81$

(b) $1.82 + (6.7 - 5.2) \cdot 5.8$ — Work inside parentheses.

$1.82 + 1.5 \cdot 5.8$ — Multiply next.

$1.82 + 8.7 = 10.52$ — Add last.

(c) $3.7^2 - 1.8 \times 5.1 \div 1.5$ — Do exponents first.

$13.69 - 1.8 \times 5.1 \div 1.5$ — Multiply and divide from left to right.

$13.69 - 9.18 \div 1.5$

$13.69 - 6.12 = 7.57$ — Subtract last. ■

WORK PROBLEM 6 AT THE SIDE.

5. ≈ Decide if each answer is reasonable by rounding the numbers and estimating the answer. If the exact answer is *not* reasonable, find and correct the error.

(a) $42.75 \div 3.8 = 1.125$
estimate:

(b) $807.1 \div 1.76 = 458.580$ to nearest thousandth
estimate:

(c) $48.63 \div 52 = 93.519$ to nearest thousandth
estimate:

(d) $9.0584 \div 2.68 = 0.338$
estimate:

6. Simplify by using the order of operations.

(a) $4.6 - 0.79 + 1.5^2$

(b) $3.64 \div 1.3 \cdot 3.6$

(c) $0.08 + 0.6 \cdot (3 - 2.99)$

(d) $10.85 - 2.3 \cdot 5.2 \div 3.2$

ANSWERS

5. (a) Estimate is $40 \div 4 = 10$; exact answer not reasonable, should be 11.25
(b) Estimate is $800 \div 2 = 400$; exact answer is reasonable.
(c) Estimate is $50 \div 50 = 1$; exact answer is not reasonable, should be 0.935.
(d) Estimate is $9 \div 3 = 3$; exact answer is not reasonable, should be 3.38.

6. (a) 6.06 **(b)** 10.08 **(c)** 0.086 **(d)** 7.1125

QUEST FOR NUMERACY

Slightly Funny Math Problems

Keep an open mind. Not all of these problems are quite what they seem to be.

7:21

The sum of the digits on the clock is $7 + 2 + 1 = 10$. What time will give the smallest sum? **1:00; $1 + 0 + 0 = 1$**

What time will give the largest sum? **9:59; $9 + 5 + 9 = 23$**

How many minutes will it take until the sum is 20? **in 28 minutes at 7:49**

10:01

The time reads the same forwards and backwards. How many minutes will it take until the time again reads the same forwards and backwards? **in 70 minutes at 11:11**

A $1 bill weighs 1 gram. What does a $20 bill weigh? **A $20 bill also weighs 1 gram.**

Two women play six games of chess and each one wins six games. How can that happen? **They were not playing against each other.**

There are 8 slices of pizza. If you eat all but $2\frac{1}{2}$ slices, how many slices are left? **$2\frac{1}{2}$ slices are left**

You bought two textbooks. When you add the prices you get $60. When you subtract the prices you get $10. What were the prices? **$35 and $25**

You bought two notebooks. When you add the prices you get $7. When you subtract the prices you get $1. What were the prices? **$4 and $3**

You bought two calculators. When you add the prices you get $21. When you subtract the prices you get $1.50. What were the prices? **$11.25 and $9.75**

Write a rule for solving this type of problem.

Answers vary. One possibility is to divide the sum by 2 as the first step. Also divide the difference by two, then add this amount to, and subtract this amount from, the result of the first step.

NAME DATE HOUR

4.6 EXERCISES

Divide. Round to the nearest thousandth if necessary.

Example:

$3 \div 0.08$

Solution:

$$\begin{array}{r} 37.5 \\ 0.08\overline{)3.000} \\ \underline{24} \\ 60 \\ \underline{56} \\ 40 \\ \underline{40} \\ 0 \end{array}$$

1. $7\overline{)27.3}$ 3.9

2. $8\overline{)50.4}$ 6.3

3. $4.23 \div 9$ 0.47

4. $1.62 \div 6$ 0.27

5. $0.05\overline{)20.01}$ 400.2

6. $0.08\overline{)16.04}$ 200.5

7. $1.5\overline{)54}$ 36

8. $2.4\overline{)132}$ 55

In Exercises 9–16, round your answers to the nearest hundredth, if necessary.

9. $163.57 \div 15$ 10.90

10. $75.813 \div 54$ 1.40

11. $9.3 \div 28$ 0.33

12. $5.53 \div 35$ 0.16

13. $4.6\overline{)116.38}$ 25.3

14. $2.6\overline{)4.992}$ 1.92

15. $3.1 \div 0.006$ 516.67

16. $1.7 \div 0.09$ 18.89

In Exercises 17–20, round your answers to the nearest thousandth.

17. $240 \div 9.88$ 24.291

18. $7643 \div 5.36$ 1425.933

19. $0.034\overline{)342.81}$ 10,082.647

20. $0.043\overline{)1748.4}$ 40,660.465

Writing Conceptual Challenging Estimation

21. Do these division problems:

$3.77 \div 10$ $9.1 \div 10$
$0.886 \div 10$ $30.19 \div 10$
$406.5 \div 10$ $6625.7 \div 10$

What pattern do you see? Write a "rule" for dividing by 10. What do you think the rule is for dividing by 100? By 1000? Write the rules and try them out on the numbers above.

Answer varies.

22. Do these division problems:

$40.2 \div 0.1$ $7.1 \div 0.1$
$0.339 \div 0.1$ $15.77 \div 0.1$
$46 \div 0.1$ $873 \div 0.1$

What pattern do you see? Write a "rule" for dividing by 0.1. What do you think the rule is for dividing by 0.01? By 0.001? Write the rules and try them out on the numbers above.

Answer varies.

≈ *Decide if each answer is reasonable or unreasonable by rounding the numbers and estimating the answer. If the exact answer is not reasonable, find the correct answer.*

23. $37.8 \div 8 = 47.25$ unreasonable

estimate: $40 \div 8 = 5$ $8\overline{)37.8}$ = 4.725

24. $345.6 \div 3 = 11.52$ unreasonable

estimate: $300 \div 3 = 100$ $3\overline{)345.6}$ = 115.2

25. $54.6 \div 48.1 = 1.135$ reasonable

estimate: $50 \div 50 = 1$

26. $2428.8 \div 4.8 = 50.6$ unreasonable

estimate: $2000 \div 5 = 400$ $4.8\overline{)2428.8}$ = 506

27. $307.02 \div 5.1 = 6.2$ unreasonable

estimate: $300 \div 5 = 60$ $5.1\overline{)307.02}$ = 60.2

28. $395.415 \div 5.05 = 78.3$ reasonable

estimate: $400 \div 5 = 80$

29. $9.3 \div 1.25 = 0.744$ unreasonable

estimate: $9 \div 1 = 9$ $1.25\overline{)9.3}$ = 7.44

30. $78 \div 14.2 = 0.182$ unreasonable

estimate: $80 \div 10 = 8$ $14.2\overline{)78}$ = 5.493

Solve each problem. Round money answers to the nearest cent.

31. Children's tights are on sale at six pairs for \$5.98 but Larisha wants only one pair for her daughter. How much will she pay for one pair?

\$1.00

32. The bookstore is selling four notepads for \$1.69. How much did Randall pay for one notepad?

\$0.42

33. It will take 21 months for Aimee to pay off her charge account balance of $408.66. How much is she paying each month?

$19.46

34. Marcella Anderson bought 2.6 meters of suede fabric for $18.19. How much did she pay per meter?

$7.00 per meter

35. Adrian Webb bought 619 bricks to build a barbecue pit, paying $185.70. Find the cost per brick. (Hint: Cost *per* brick means the cost for *one* brick.)

$0.30

36. Lupe Wilson is a newspaper distributor. Last week she paid the newspaper $130.51 for 842 copies. Find the cost per copy.

$0.16

37. Darren Jackson earned $235.60 for 40 hours of work. Find his earnings per hour.

$5.89 per hour

38. At a record manufacturing company, 400 records cost $289. Find the cost per record.

$0.72

39. It took 16.35 gallons of gas to fill Kim's car gas tank. She had driven 346.2 miles. How many miles per gallon did she get? Round to the nearest tenth.

21.2 miles per gallon

40. Mr. Rodriquez pays $53.19 each month to Household Finance. How many months will it take him to pay off $1436.13?

27 months

Use the table of Olympic long jump records to solve Exercises 41–42. To find an average, add up the values you are interested in and then divide the sum by the number of values. Round your answer to the nearest tenth.

Country	*Length*
U.S.	29.2 ft
U.S.	27.04 ft
U.S.	27.375 ft
Germany	28.02 ft
U.S.	28.02 ft
U.S.	28.6 ft
G.Britain	26.48 ft

41. Find the average length of the long jumps made by U.S. athletes. (Hint: Add the lengths of U.S. jumps, then divide by 5.)

28.0 ft

42. Find the average length of all the long jumps listed in the table. (Hint: First add, then divide by 7.)

27.8 ft

Simplify by using the order of operations.

Example: $\underbrace{5.2^2} + 7.9 \cdot 6.3$ Do exponents first.

$27.04 + \underbrace{7.9 \cdot 6.3}$ Multiply next.

$27.04 + 49.77 = \mathbf{76.81}$ Add last.

43. $7.2 - 5.2 + 3.5^2$

14.25

44. $6.2 + 4.3^2 - 9.72$

14.97

45. $38.6 + 11.6 \cdot (13.4 - 10.4)$

73.4

46. $2.25 - 1.06 \cdot (4.85 - 3.95)$

1.296

47. $8.68 - 4.6 \cdot 10.4 \div 6.4$

1.205

48. $25.1 + 11.4 \div 7.5 \cdot 3.75$

30.8

49. $33 - 3.2 \cdot (0.68 + 9) - 1.3^2$

0.334

50. $0.6 + (1.89 + 0.11) \div 0.004 \cdot 0.5$

250.6

Solve.

▲ **51.** Soup is on sale at six cans for \$3.25, or you can purchase individual cans for \$0.57. How much will you save per can if you buy six cans? Round to the nearest cent.

\$0.03

▲ **52.** Nadia's diet says she can eat 3.5 ounces of chicken nuggets. The package weighs 10.5 ounces and contains 15 nuggets. How many nuggets can Nadia eat?

5 nuggets

▲ **53.** The annual premium for Jenny's auto insurance policy is \$938. She can pay it in four quarterly installments, if she adds a \$2.75 service fee to each payment. Find the amount of each quarterly payment.

\$237.25

▲ **54.** Lock and Store charges rent of \$936 per year for 200 square feet of storage space. To pay the rent monthly, \$1.25 must be added to each payment. Find the amount of each monthly payment.

\$79.25

PREVIEW EXERCISES

Write < or > to make a true statement. (For help, see ***Section 3.5.****)*

55. $\frac{7}{12} < \frac{3}{4}$

56. $\frac{5}{8} < \frac{11}{16}$

57. $\frac{5}{6} > \frac{7}{9}$

58. $\frac{7}{8} < \frac{11}{12}$

59. $\frac{13}{24} < \frac{23}{36}$

60. $\frac{9}{20} > \frac{11}{30}$

4.7 WRITING FRACTIONS AS DECIMALS

Writing a fraction as an equivalent decimal can help you do calculations more easily or compare the size of two numbers.

1 Recall that a fraction is one way to show division (see **Section 1.5**). For example, $\frac{3}{4}$ means $3 \div 4$. If you do the division, you will get the decimal equivalent of $\frac{3}{4}$.

WRITING FRACTIONS AS DECIMALS

Step 1 Divide the numerator of the fraction by the denominator.
Step 2 If necessary, round the answer to the place indicated.

■ **EXAMPLE 1** *Writing a Fraction or Mixed Number as a Decimal*

(a) Write $\frac{1}{8}$ as a decimal. **(b)** Write $1\frac{3}{4}$ as a decimal.

(a) $\frac{1}{8}$ means $1 \div 8$. Write it as $8\overline{)1}$. The decimal point in the dividend is on the right side of the 1. Write extra zeros in the dividend so you can continue dividing until the remainder is 0.

Decimal points lined up.

$$\begin{array}{r} 0.125 \\ 8\overline{)1.000} \\ \underline{8} \\ 20 \\ \underline{16} \\ 40 \\ \underline{40} \\ 0 \end{array}$$

← Three extra 0's needed.

← Remainder is 0.

Therefore, $\frac{1}{8} = 0.125$.

(b) Write $1\frac{3}{4}$ as an improper fraction.

$$1\frac{3}{4} = \frac{7}{4} \quad \leftarrow \quad \frac{7}{4} \text{ means } 7 \div 4$$

$$\begin{array}{r} 1.75 \\ 4\overline{)7.00} \\ \underline{4} \\ 3\,0 \\ \underline{2\,8} \\ 20 \\ \underline{20} \\ 0 \end{array}$$

← Two extra 0's needed.

Whole number parts match.

So, $\mathbf{1\frac{3}{4} = 1.75}$

$\frac{3}{4}$ is equivalent to $\frac{75}{100}$ or 0.75.

You can also solve this problem by dividing 3 by 4 to get 0.75. Then add $1 + 0.75$ for a final answer of 1.75. ■

WORK PROBLEM 1 AT THE SIDE. ▶▶

OBJECTIVES

1. Change a fraction to a decimal.
2. Compare the size of fractions and decimals.

FOR EXTRA HELP

Tape 6 | SSM pp. 115–119 | MAC: A IBM: A

1. Write each fraction or mixed number as a decimal.

(a) $\frac{1}{4}$

(b) $2\frac{1}{2}$

(c) $\frac{5}{8}$

(d) $4\frac{3}{5}$

(e) $\frac{7}{8}$

ANSWERS
1. (a) 0.25 (b) 2.5 (c) 0.625 (d) 4.6 (e) 0.875

2. Write as decimals. Round to the nearest thousandth.

(a) $\frac{1}{3}$

(b) $2\frac{7}{9}$

(c) $\frac{10}{11}$

(d) $\frac{3}{7}$

(e) $3\frac{5}{6}$

3. Use the number lines in the text to help you decide whether to write <, >, or = in each blank.

(a) 0.4375 ____ 0.5

(b) 0.75 ____ 0.6875

(c) 0.625 ____ 0.0625

(d) $\frac{1}{4}$ ____ 0.375

(e) $0.8\overline{3}$ ____ $\frac{5}{6}$

(f) $\frac{1}{2}$ ____ $0.\overline{5}$

(g) $0.\overline{1}$ ____ $0.1\overline{6}$

(h) $\frac{8}{9}$ ____ $0.\overline{8}$

ANSWERS

2. (a) 0.333 **(b)** 2.778 **(c)** 0.909 **(d)** 0.429 **(e)** 3.833

3. (a) < **(b)** > **(c)** > **(d)** < **(e)** = **(f)** < **(g)** < **(h)** =

EXAMPLE 2 *Changing to a Decimal and Rounding*

Write $\frac{2}{3}$ as a decimal and round to the nearest thousandth.

$\frac{2}{3}$ means $2 \div 3$. To round to thousandths, divide out one *more* place, to ten-thousandths.

$$\begin{array}{r} 0.6666 \\ 3\overline{)2.0000} \\ \underline{1\ 8} \\ 20 \\ \underline{18} \\ 20 \\ \underline{18} \\ 20 \\ \underline{18} \\ 2 \end{array}$$

← Four 0's needed for ten-thousandths.

Written as a repeating decimal, $\frac{2}{3} = 0.\overline{6}$. Rounded to the nearest thousandth, $\frac{2}{3} = 0.667$. ■

WORK PROBLEM 2 AT THE SIDE.

2 You can use a number line to compare fractions and decimals. For example, the number line below shows the space between 0 and 1. The locations of some commonly used fractions are marked, along with their decimal equivalents.

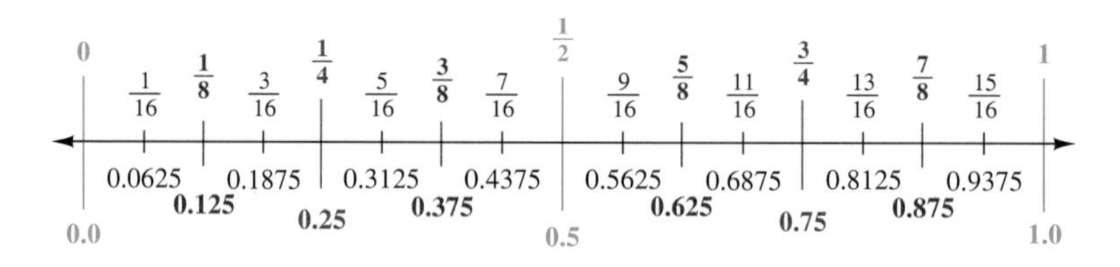

The next number line also shows the locations of some commonly used fractions. The decimal equivalents use a bar above repeating digits.

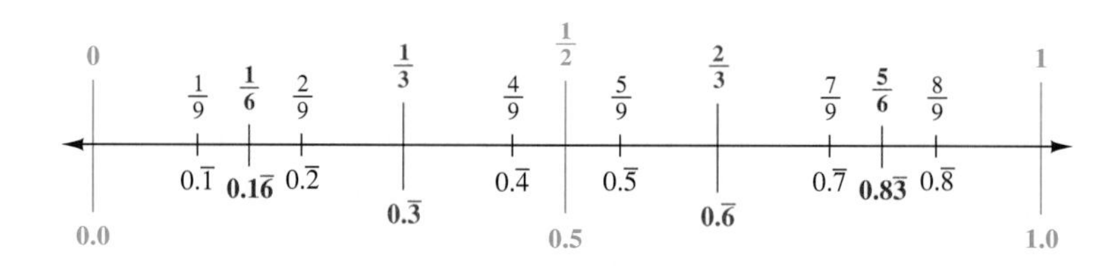

Recall that you can compare two numbers by locating them on the number line. The number farther to the left is smaller.

WORK PROBLEM 3 AT THE SIDE.

Fractions can also be compared by first writing each one as a decimal. The decimals can then be compared by writing each one with the same number of decimal places.

EXAMPLE 3 *Arranging Numbers in Order*

Write the following numbers in order, from smallest to greatest.

(a) 0.49   0.487   0.4903

(b) $2\frac{5}{8}$ 2.63   2.6

(a) It is easier to compare decimals if they are all tenths, or all hundredths, and so on. Because 0.4903 has four decimal places (ten-thousandths), write zeros to the right of 0.49 and 0.487 so they also have four decimal places. Writing zeros to the right of a decimal number does *not* change its value (see **Section 4.3**). Now find the smallest and largest number of ten-thousandths.

0.4900 = **4900** ten-thousandths
0.4870 = **4870** ten-thousandths ← 4870 is smallest.
0.4903 = **4903** ten-thousandths ← 4903 is largest.

From smallest to largest, the correct order is:

0.487   0.49   0.4903

(b) Write $2\frac{5}{8}$ as a decimal by dividing $8\overline{)21}$ to get 2.625. Then, because 2.625 has three decimal places, write in zeros so all the numbers have three decimal places.

2.625 = 2 and **625** thousandths
2.630 = 2 and **630** thousandths ← 630 is largest.
2.600 = 2 and **600** thousandths ← 600 is smallest.

From smallest to largest, the correct order is:

2.6 $2\frac{5}{8}$ 2.63 ■

WORK PROBLEM 4 AT THE SIDE.

4. Arrange in order, from smallest to largest.

(a) 0.7, 0.703, 0.7029

(b) 6.39, 6.309, 6.4, 6.401

(c) 1.085, $1\frac{3}{4}$, 0.9

(d) $\frac{1}{4}, \frac{2}{5}, \frac{3}{7}$, 0.428

ANSWERS
4. (a) 0.7, 0.7029, 0.703
(b) 6.309, 6.39, 6.4, 6.401
(c) 0.9, 1.085, $1\frac{3}{4}$
(d) $\frac{1}{4}, \frac{2}{5}$, 0.428, $\frac{3}{7}$

QUEST FOR NUMERACY

Is the Answer Reasonable?

Even with most of the problem missing, you can tell that these answers are *not* reasonable. (They are actual student answers to test questions.) Write a sentence explaining why each answer does not make sense. Then give a reasonable range for the correct answer.

What is the teacher's monthly take-home pay?

Answer: $19.57

Unreasonable because:

Answer should be between

_______ and _______.

How many hours did you work each day?

Answer: $24\frac{3}{4}$ hours

Unreasonable because:

Answer should be between

_______ and _______.

Joe had $2\frac{2}{3}$ cans of paint. How much paint did he use on each window?

Answer: $4\frac{1}{2}$ cans

Unreasonable because:

Answer should be between

_______ and _______.

How far can the car travel on one gallon of gas?

Answer: 3703.86 miles

Unreasonable because:

Answer should be between

_______ and _______.

A $36\frac{1}{4}$ mile hike is planned. How many miles are left to travel?

Answer: $56\frac{3}{8}$ miles

Unreasonable because:

Answer should be between

_______ and _______.

How much do you pay for rent?

Answer: $2.70

Unreasonable because:

Answer should be between

_______ and _______.

Extend this activity by taking actual student answers from the tests you give and asking students to evaluate their reasonableness. Include some that *are* reasonable and be sure to keep the source anonymous.

EXERCISES

Write each fraction or mixed number as a decimal. Round to the nearest thousandth if necessary.

Example: $\frac{9}{16}$

Solution: $16\overline{)9.0000}$ = 0.5625 rounds to **0.563** ← Four 0's needed to complete the division.

$$\begin{array}{r} 0.5625 \\ 16\overline{)9.0000} \\ \underline{8\,0} \\ 1\,00 \\ \underline{96} \\ 40 \\ \underline{32} \\ 80 \\ \underline{80} \\ 0 \end{array}$$

1. $\frac{1}{2}$ 0.5
2. $\frac{1}{4}$ 0.25
3. $\frac{7}{10}$ 0.7
4. $\frac{3}{5}$ 0.6
5. $\frac{1}{6}$ 0.167
6. $\frac{1}{3}$ 0.333
7. $\frac{7}{8}$ 0.875
8. $\frac{5}{8}$ 0.625
9. $\frac{2}{3}$ 0.667
10. $\frac{5}{6}$ 0.833
11. $1\frac{2}{5}$ 1.4
12. $2\frac{3}{4}$ 2.75
13. $3\frac{4}{7}$ 3.571
14. $1\frac{8}{9}$ 1.889
15. $\frac{22}{25}$ 0.88
16. $\frac{19}{20}$ 0.95
17. $\frac{11}{16}$ 0.688
18. $\frac{13}{16}$ 0.813
19. $12\frac{4}{5}$ 12.8
20. $14\frac{3}{10}$ 14.3
21. $\frac{1}{99}$ 0.010
22. $\frac{148}{149}$ 0.993
23. $78\frac{9}{25}$ 78.36
24. $69\frac{49}{50}$ 69.98

25. Explain and correct the error that Keith made when changing a fraction to an equivalent decimal.

$$\frac{5}{9} = \begin{array}{r} 1.8 \\ 5\overline{)9.0} \\ \underline{5} \\ 4\,0 \\ \underline{4\,0} \\ 0 \end{array} \quad \text{so} \quad \frac{5}{9} = 1.8$$

Answer varies.

26. Explain and correct the error Sandra made when writing $2\frac{7}{20}$ as a decimal.

$$2\frac{7}{20} = \begin{array}{r} 0.35 \\ 20\overline{)7.00} \\ \underline{6\,0} \\ 1\,00 \\ \underline{1\,00} \\ 0 \end{array} \quad \text{so} \quad 2\frac{7}{20} = 2.035$$

Answer varies.

27. Ving knows that $\frac{3}{8} = 0.375$. How can he write $1\frac{3}{8}$ as a decimal *without* having to do a division? How can he write $3\frac{3}{8}$ as a decimal? $295\frac{3}{8}$? Explain your answer.

Answer varies.

28. Iris has found a shortcut for writing mixed numbers as decimals:

$$2\frac{\mathbf{7}}{10} = 2.7 \qquad 1\frac{\mathbf{13}}{100} = 1.13$$

Does her shortcut work for all mixed numbers? Explain.

Answer varies.

Find the decimal or fraction equivalent for each of the following. Round decimals to the nearest thousandth and write fractions in lowest terms.

Example:

fraction	*decimal*
______	0.375

Solution: $0.375 = \frac{375}{1000} = \frac{375 \div 125}{1000 \div 125} = \frac{\mathbf{3}}{\mathbf{8}}$ in lowest terms

thousandths place so write 1000 in denominator

	fraction	*decimal*		*fraction*	*decimal*
29.	$\frac{2}{5}$	0.4	**30.**	$\frac{3}{4}$	0.75
31.	$\frac{3}{8}$	0.375	**32.**	$\frac{111}{1000}$	0.111
33.	$\frac{7}{20}$	0.35	**34.**	$\frac{9}{10}$	0.9
35.	$\frac{7}{20}$	0.35	**36.**	$\frac{1}{40}$	0.025
37.	$\frac{1}{25}$	0.04	**38.**	$\frac{13}{25}$	0.52
39.	$\frac{3}{20}$	0.15	**40.**	$\frac{17}{20}$	0.85
41.	$\frac{1}{5}$	0.2	**42.**	$\frac{1}{8}$	0.125
43.	$\frac{9}{100}$	0.09	**44.**	$\frac{1}{50}$	0.02

Solve each problem.

45. The label on the bottle of vitamins says that each capsule contains 0.5 gram of calcium. When checked, each capsule had 0.505 gram of calcium. Was there too much or too little calcium?

Too much.

46. The patient in room 406 is supposed to get 8.3 milligrams of medicine. She was actually given 8.03 milligrams. Did she get too much or too little medicine?

Too little.

47. The average length of a newborn baby is 20.8 inches. Charlene's baby is 20.08 inches long. Is her baby longer or shorter than the average?

Shorter.

48. A bolt for a space shuttle is supposed to be 3.045 inches long. The bolt was checked and measured 3.05 inches long. Was the bolt too long or too short?

Too long.

49. Precision Medical Parts makes an artificial heart valve that must measure between 0.998 centimeter and 1.002 centimeters. Circle the lengths that are acceptable: 1.01 cm, (0.9991 cm), (1.0007 cm), 0.99 cm.

50. The mice in a medical experiment must start out weighing between 2.95 ounces and 3.05 ounces. Circle the weights that can be used: (3.0 ounces), (2.995 ounces), 3.055 ounces, (3.005 ounces).

51. Ginny Brown hoped her crops would get $3\frac{3}{4}$ inches of rain this month. The newspaper said the area received 3.8 inches of rain. Was that more or less than Ginny hoped for?

More.

52. The mice in the experiment gained $\frac{3}{8}$ ounce. They were expected to gain 0.3 ounce. Was their actual gain more or less than expected?

More.

Arrange in order from smallest to largest.

Example: 0.8075, 0.875, 0.88, 0.808

Solution: 0.8075 = 8075 ten-thousandths
0.8750 = 8750 ten-thousandths
0.8800 = 8800 ten-thousandths
0.8080 = 8080 ten-thousandths

} Write zeros so all the numbers have four decimal places.

(smallest) **0.8075** **0.808** **0.875** **0.88** (largest)

53. 0.54, 0.5455, 0.5399

0.5399, 0.54, 0.5455

54. 0.76, 0.7, 0.7006

0.7, 0.7006, 0.76

55. 5.8, 5.79, 5.0079, 5.804

5.0079, 5.79, 5.8, 5.804

56. 12.99, 12.5, 13.0001, 12.77

12.5, 12.77, 12.99, 13.0001

57. 0.628, 0.62812, 0.609, 0.6009

0.6009, 0.609, 0.628, 0.62812

58. 0.27, 0.281, 0.296, 0.3

0.27, 0.281, 0.296, 0.3

59. 5.8751, 4.876, 28.902, 3.88

2.8902, 3.88, 4.876, 5.8751

60. 0.98, 0.89, 0.904, 0.9

0.89, 0.9, 0.904, 0.98

61. 0.043, 0.051, 0.006, $\frac{1}{20}$

0.006, 0.043, $\frac{1}{20}$, 0.051

62. 0.629, $\frac{5}{8}$, 0.65, $\frac{7}{10}$

$\frac{5}{8}$, 0.629, 0.65, $\frac{7}{10}$

63. $\frac{3}{8}$, $\frac{2}{5}$, 0.37, 0.4001

0.37, $\frac{3}{8}$, $\frac{2}{5}$, 0.4001

64. 0.1501, 0.25, $\frac{1}{10}$, $\frac{1}{5}$

$\frac{1}{10}$, 0.1501, $\frac{1}{5}$, 0.25

Some rulers show each inch divided into tenths. Change the measurements on this scale drawing to decimals and round them to the nearest tenth of an inch.

65. length (a) is 1.4 in

66. length (b) is 1.1 in

67. length (c) is 0.3 in

68. length (d) is 0.5 in

69. length (e) is 0.4 in

70. length (f) is 0.7 in

(a) $1\frac{7}{16}$ in.

(b) $1\frac{1}{8}$ in.

(c) $\frac{1}{4}$ in.

(d) $\frac{1}{2}$ in.

(e) $\frac{3}{8}$ in.

(f) $\frac{11}{16}$ in.

Arrange in order from smallest to largest.

▲ **71.** $\frac{6}{11}$, $\frac{5}{9}$, $\frac{4}{7}$, 0.571

$\frac{6}{11}$, $\frac{5}{9}$, 0.571, $\frac{4}{7}$

▲ **72.** $\frac{8}{13}$, $\frac{10}{17}$, 0.615, $\frac{11}{19}$

$\frac{11}{19}$, $\frac{10}{17}$, 0.615, $\frac{8}{13}$

▲ **73.** $\frac{3}{11}$, $\frac{4}{15}$, 0.25, $\frac{1}{3}$

0.25, $\frac{4}{15}$, $\frac{3}{11}$, $\frac{1}{3}$

▲ **74.** 0.223, $\frac{2}{11}$, $\frac{2}{9}$, $\frac{1}{4}$

$\frac{2}{11}$, $\frac{2}{9}$, 0.223, $\frac{1}{4}$

▲ **75.** $\frac{3}{16}$, $\frac{1}{6}$, $\frac{1}{5}$, 0.188

$\frac{1}{6}$, $\frac{3}{16}$, 0.188, $\frac{1}{5}$

▲ **76.** $\frac{7}{20}$, $\frac{1}{3}$, $\frac{3}{8}$, 0.375

$\frac{1}{3}$, $\frac{7}{20}$, 0.375, $\frac{3}{8}$

or $\frac{1}{3}$, $\frac{7}{20}$, $\frac{3}{8}$, 0.375

PREVIEW EXERCISES

Write each fraction in lowest terms. (For help, see ***Section 2.4.****)*

77. $\frac{9}{12}$ $\frac{3}{4}$

78. $\frac{30}{60}$ $\frac{1}{2}$

79. $\frac{60}{80}$ $\frac{3}{4}$

80. $\frac{40}{75}$ $\frac{8}{15}$

81. $\frac{96}{132}$ $\frac{8}{11}$

82. $\frac{26}{98}$ $\frac{13}{49}$

QUEST FOR NUMERACY

Number Patterns

The first two examples use one consistent pattern. What do you need to do to get from one number to the next one? Add? Subtract? Multiply? Divide? Find the next two numbers that fit the pattern.

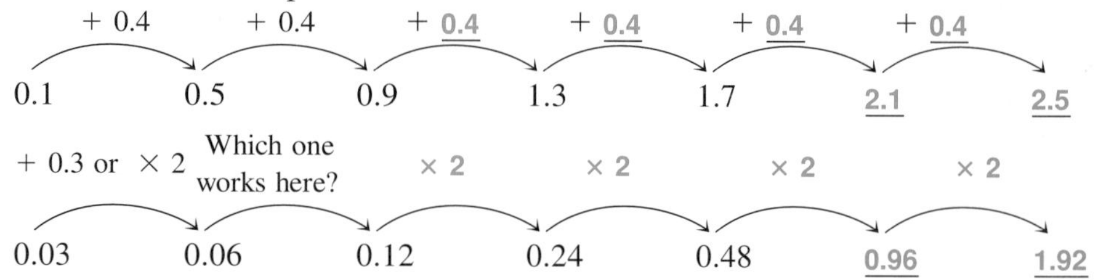

In this example, look at every other number. There are two separate patterns.

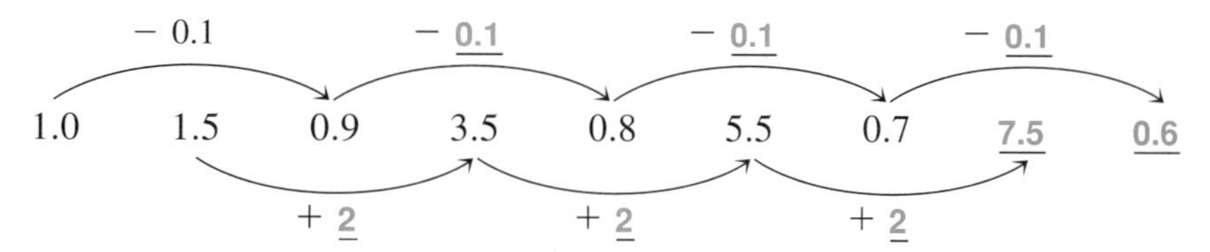

Now try these patterns. **A possible solution is shown, but there is more than one way to look at these patterns.**

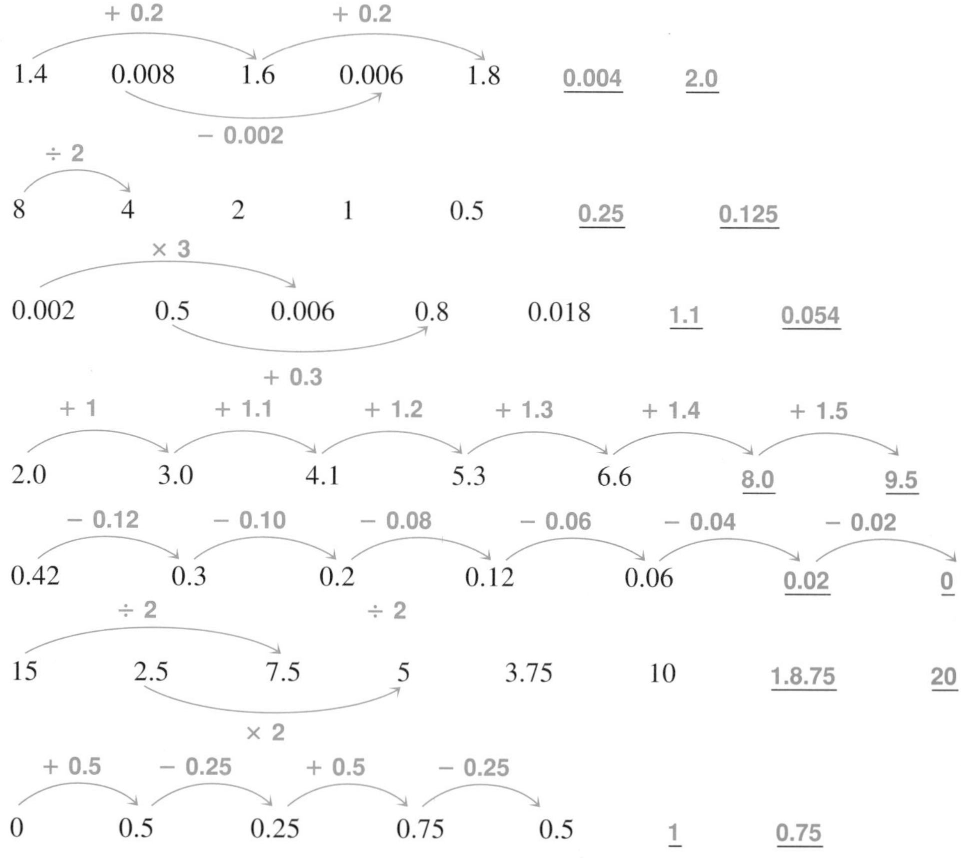

To extend this activity, ask students to create their own patterns using whole numbers, decimals, or fractions. Have students turn in their most interesting patterns to be made into an extra credit worksheet.

CHAPTER 4 SUMMARY

KEY TERMS

4.1	**decimals**	Decimals, like fractions, are used to show parts of a whole.
	decimal point	The dot that is used to separate the whole-number part from the fractional part of a decimal number.
	place value	The value assigned to each place to the right or left of the decimal point. Whole numbers, such as ones and tens, are to the *left* of the decimal point. Fractional parts, such as tenths and hundredths, are to the *right* of the decimal point.
4.2	**rounding**	"Cutting off" a number after a certain place, such as rounding to the nearest hundredth. The rounded number is less accurate than the original number.
4.3	**decimal places**	The number of digits to the *right* of the decimal point; for example, 6.37 has two decimal places.
	estimating	The process of rounding the numbers in a problem and getting an approximate answer. This helps you check that the decimal point is in the correct place in the exact answer.
4.6	**repeating decimal**	A decimal with one or more digits that repeat forever, such as the 6 in 0.1666 . . . . Use three dots to indicate that it is a repeating decimal; it never terminates (ends). You can also write it with a bar above the repeating digits, as in $0.1\overline{6}$.

QUICK REVIEW

Concepts	Example
4.1 Reading and Writing Decimals 5 thousands 8 hundreds 4 tens 6 ones . decimal point ("and") 0 ten**ths** 7 hundred**ths** 3 thousand**ths** 2 ten-thousand**ths**	Write each decimal in words. 15.38 (8 is in hundredths place) fifteen **and** thirty-eight hundredths 0.0103 (3 is in ten-thousandths place) one hundred three ten-thousandths
4.1 Writing Decimals as Fractions The digits to the right of the decimal point are the numerator. The place value of the right-most digit determines the denominator. Reduce to lowest terms.	Write 0.45 as a fraction. The numerator is 45. The right-most digit, 5, is in the hundredths place, so the denominator is 100. Then reduce. $\frac{45}{100} = \frac{45 \div 5}{100 \div 5} = \frac{9}{20}$ lowest terms

Concepts	Example
4.2 Rounding Decimals	
Find the place to which you are rounding. Draw a line to the right of that place; the rest of the digits will be cut off. Look only at the first digit being cut. If it is *less than 5*, the part you are keeping stays the same. If it is *5 or more*, the part you are keeping rounds up. Do not move the decimal point when rounding.	Round 0.17952 to the nearest thousandth. 0.179 | 52 — First digit cut is 5 or more so round up by adding 1 thousandth to the part you are keeping. 0.179 + 0.001 ← To round up, add 1 thousandth. 0.180 0.17952 rounds to 0.180.
4.3 and 4.4 Adding and Subtracting Decimals	
Estimate the answer by rounding each number so there is only one non-zero digit. To find the exact answer, line up the decimal points. If needed, write in zeros as placeholders. Add or subtract as if they were whole numbers. Line up the decimal point in the answer.	Add 5.68 + 785.3 + 12 + 2.007 *estimate* / *exact* 6 ← 5.680 Use zeros as placeholders so all numbers have three decimal places. 800 ← 785.300 10 ← 12.000 + 2 ← + 2.007 818 / 804.987 ↑ line up decimal points The estimate and exact answer are both in hundreds, so the decimal point is probably in the correct place.
4.5 Multiplying Decimals	
1. Multiply as you would for whole numbers. **2.** Count the total number of decimal places in both factors. **3.** Write the decimal point in the answer so it has the same number of decimal places as the total from Step 2. You may need to write extra zeros on the left side of the product in order to get enough decimal places in the answer.	Multiply 0.169 × 0.21 0.169 ← 3 decimal places × 0.21 ← 2 decimal places 169 — 5 total decimal places 338 .03549 ← 5 decimal places in answer Write in a zero so you can count over 5 decimal places. Final answer is 0.03549.
4.6 Dividing by a Decimal	
1. Change the divisor to a whole number by moving the decimal point to the right. **2.** Move the decimal point in the dividend the same number of places to the right. **3.** Write the decimal point in the answer directly above the decimal point in the dividend. **4.** Divide as with whole numbers.	Divide 52.8 by 0.75. 70.4 0.75)52.80 0 52 5 300 300 0 Move decimal point two places to the right in divisor and dividend. Write zeros in the dividend so you can move the decimal point and continue dividing until the remainder is zero.

Concepts	Example
4.7 Writing Fractions as Decimals	
Divide the numerator by the denominator. If necessary, round to the place indicated.	Write $\frac{1}{8}$ as a decimal. $\frac{1}{8}$ means $1 \div 8$. Write it as $8\overline{)1}$. The decimal point is on the right side of 1. $\begin{array}{r} 0.125 \\ 8\overline{)1.000} \\ \underline{8} \\ 20 \\ \underline{16} \\ 40 \\ \underline{40} \\ 0 \end{array}$ ← Decimal point and three zeros written in so you can continue dividing. Therefore, $\frac{1}{8}$ is equivalent to 0.125.

NAME DATE HOUR

CHAPTER 4 REVIEW EXERCISES

[4.1] *Name the digit that has the given place value.*

1. 36.587
tenths 5
hundredths 8

2. 0.4729
ones 0
thousandths 2

3. 435.621
hundreds 4
hundredths 2

4. 896.503
tenths 5
tens 9

5. 20.73861
tenths 7
ten-thousandths 6

Write each decimal as a fraction or mixed number in lowest terms.

6. 2.3 $2\frac{3}{10}$

7. 0.75 $\frac{3}{4}$

8. 4.05 $4\frac{1}{20}$

9. 0.875 $\frac{7}{8}$

10. 0.027 $\frac{27}{1000}$

11. 1.0016 $1\frac{1}{625}$

Write each decimal in words

12. 0.8 eight tenths

13. 400.29 four hundred and twenty-nine hundredths

14. 12.007 twelve and seven thousandths

15. 0.0306 three hundred six ten-thousandths

Write each decimal in numbers.

16. eight and three tenths 8.3

17. two hundred five thousandths 0.205

18. seventy and sixty-six ten-thousandths
70.0066

19. thirty hundredths 0.30

[4.2] *Round to the place indicated.*

20. 275.635 to the nearest tenth 275.6

21. 72.789 to the nearest hundredth 72.79

22. 0.1604 to the nearest thousandth 0.160

23. 0.0905 to the nearest thousandth 0.091

24. 0.98 to the nearest tenth 1.0

Round to the nearest cent.

25. $15.8333 $15.83

26. $0.698 $0.70

27. $17,625.7906 $17,625.79

Round each income or expense item to the nearest dollar.

28. Income from washers and dryers is $170.45. $170

29. Lawn care costs $39.28. $39

30. Water and garbage bill is $24.50 $25

31. Trash pickup costs $9.72. $10

[4.3] ≈ *First round the numbers so there is only one non-zero digit and estimate the answer. Then add to find the exact answer.*

32. *estimate* *exact*

$$\begin{array}{r} 6 \\ 400 \\ +\ 20 \\ \hline 426 \end{array} \qquad \begin{array}{r} 5.81 \\ 423.96 \\ +\ 15.09 \\ \hline 444.86 \end{array}$$

33. *estimate* *exact*

$$\begin{array}{r} 80 \\ 1 \\ 100 \\ 1 \\ +\ 30 \\ \hline 212 \end{array} \qquad \begin{array}{r} 75.6 \\ 1.29 \\ 122.045 \\ 0.88 \\ +\ 33.7 \\ \hline 233.515 \end{array}$$

[4.4] ≈ *First round the numbers so there is only one non-zero digit and estimate the answer. Then subtract to find the exact answer.*

34. *estimate* *exact*

$$\begin{array}{r} 300 \\ -\ 20 \\ \hline 280 \end{array} \qquad \begin{array}{r} 308.5 \\ -\ 17.8 \\ \hline 290.7 \end{array}$$

35. *estimate* *exact*

$$\begin{array}{r} 9 \\ -\ 8 \\ \hline 1 \end{array} \qquad \begin{array}{r} 9.2 \\ -\ 7.9316 \\ \hline 1.2684 \end{array}$$

[4.3–4.4] ≈ *First round the numbers so there is only one non-zero digit and estimate the answer. Then find the exact answer.*

36. Tim agreed to donate 12.5 hours of work at his children's school. He has already worked 9.75 hours. How many more hours will he work?

estimate: 13 − 10 = 3 hours
exact: 2.75 hours

37. Today Jasmin wrote a check to the daycare center for $215.53 and a check for $44.47 at the grocery store. What was the total of the two checks?

estimate: $200 + $40 = $240
exact: $260.00

38. Joey spent $1.59 for toothpaste, $5.33 for a gift, and $18.94 for a toaster. He gave the clerk three $10 bills. How much change did he get?

estimate: $2 + $5 + $20 = $27;
$30 − $27 = $3
exact: $4.14

39. Roseanne is training for a wheelchair race. She raced 2.3 kilometers on Monday, 4 kilometers on Wednesday, and 5.25 kilometers on Friday. How far did she race altogether?

estimate: 2 + 4 + 5 = 11 kilometers
exact: 11.55 kilometers

[4.5] ≈ *First round the numbers so there is only one non-zero digit and estimate an answer. Then multiply to find the exact answer.*

40. *estimate* *exact*

$$\begin{array}{r} 6 \\ \times\ 4 \\ \hline 24 \end{array} \qquad \begin{array}{r} 6.138 \\ \times\ 3.7 \\ \hline 22.7106 \end{array}$$

41. *estimate* *exact*

$$\begin{array}{r} 40 \\ \times\ 3 \\ \hline 120 \end{array} \qquad \begin{array}{r} 42.9 \\ \times\ 3.3 \\ \hline 141.57 \end{array}$$

Multiply.

42. (5.6)(0.002) 0.0112

43. (0.071)(0.005) 0.000355

[4.6] ≈ *Decide if each answer is reasonable by rounding the numbers and estimating the answer. If the exact answer is not reasonable, find the correct answer.*

44. $89.44 \div 13 = 6.88$ reasonable
estimate:

$90 \div 10 = 9$

45. $26.6 \div 2.8 = 0.95$ unreasonable
estimate:

$30 \div 3 = 10$ $\quad 2.8\overline{)26.6}$ = 9.5

Divide. Round to the nearest thousandth.

46. $3\overline{)43.4}$ 14.467

47. $0.05\overline{)775}$ 15,500

48. $0.00048 \div 0.0012$

0.4

[4.5–4.6] *Solve these word problems.*

49. Adrienne worked 36.5 hours this week. Her hourly wage is $6.57. Find her total earnings to the nearest dollar.

$240

50. A book of 12 tickets costs $16.50 at the amusement park. What is the cost per ticket, to the nearest cent?

$1.38

51. Stock in MathTronic sells for $3.75 per share. Kenneth is thinking of investing $500. How many whole shares could he buy?

133 shares

52. Hamburger meat is on sale at $0.89 per pound. How much will Ms. Lee pay for 3.5 pounds of hamburger, to the nearest cent?

$3.12

Simplify by using the order of operations.

53. $3.5^2 + 8.7 \cdot 1.95$

29.215

54. $11 - 3.06 \div (3.95 - 0.35)$

10.15

[4.7] *Write each fraction as a decimal. Round to the nearest thousandth.*

55. $3\frac{4}{5}$

3.8

56. $\frac{16}{25}$

0.64

57. $1\frac{7}{8}$

1.875

58. $\frac{1}{9}$

0.111

Arrange in order from smallest to largest.

59. 3.68, 3.806, 3.6008

3.6008, 3.68, 3.806

60. 0.215, 0.22, 0.209, 0.2102

0.209, 0.2102, 0.215, 0.22

61. $0.17, \frac{3}{20}, \frac{1}{8}, 0.159$

$\frac{1}{8}, \frac{3}{20}$, 0.159, 0.17

MIXED REVIEW EXERCISES

Solve each problem.

62. $89.19 + 0.075 + 310.6 + 5$

404.865

63. 72.8×3.5

254.8

64. $1648.3 \div 0.46$ Round to thousandths.

3583.261

65. $30 - 0.9102$

29.0898

66. $(4.38)(0.007)$

0.03066

67. $0.005\overline{)0.047}$

9.4

68. $72.105 + 8.2 + 95.37$

175.675

69. $81.36 \div 9$

9.04

70. $(5.6 - 1.22) + 4.8 \cdot 3.15$

19.50

71. 0.455×18

8.19

72. $1.6 \cdot 0.58$

0.928

73. $0.218\overline{)7.63}$

35

74. $21.059 - 20.8$

0.259

75. $18.3 - 3^2 \div 0.5$

0.3

Solve each word problem.

76. On her vacation, Suzanne paid $154.36 for 4 nights at a motel. What was the cost per night?

$38.59

77. Ray spent $5.49 at the grocery store. How much change should he get from two $5 bills?

$4.51

78. Last week, 8.25 inches of snow fell. This week we received another 14.9 inches of snow. How much snow has fallen during the two weeks?

23.15 inches

79. The state record for the weight of a sunfish is 1.08 kilograms. Chue Fan caught a sunfish weighing 1.009 kilograms. Did she beat the record?

No.

80. Gas went down to $1.09 per gallon this week. George pumped 14.5 gallons into his car. How much did he pay, to the nearest cent?

$15.81

81. Jorge wrapped eight boxes of the same size. He used 14.4 meters of paper in all. How much paper did he use per box?

1.8 meters

82. At birth, Maya's baby weighed 8.45 pounds. A month later he weighed 12.3 pounds. How much weight had he gained?

3.85 pounds

83. A legal secretary earns $12.25 per hour working part-time. How much did she earn by working 26.5 hours last week, to the nearest dollar?

$325

NAME DATE HOUR

CHAPTER 4 TEST

Write each decimal as a fraction or mixed number in lowest terms.

1. 3.16

2. 0.025

Write each decimal in words.

3. 5.0003

4. 0.705

≈ *Round to the place indicated.*

5. 725.6089 to the nearest tenth

6. 0.62951 to the nearest thousandth

7. $1.4945 to the nearest cent

8. $7859.51 to the nearest dollar

≈ *Round the numbers so there is only one non-zero digit and estimate each answer. Then find the exact answer.*

9. 7.6 + 82.0128 + 39.59

10. 79.1 − 3.602

11. 5.79 · 1.2

12. 20.04 ÷ 4.8

Solve.

13. 53.1 + 4.631 + 782 + 0.031

14. 670 − 0.996

15. (0.0069)(0.007)

16. 0.15)72

1. $3\frac{4}{25}$

2. $\frac{1}{40}$

3. five and three ten-thousandths

4. seven hundred five thousandths

5. 725.6

6. 0.630

7. $1.49

8. $7860

9. estimate: 8 + 80 + 40 = 128
exact: 129.2028

10. estimate: 80 − 4 = 76
exact: 75.498

11. estimate: 6 · 1 = 6
exact: 6.948

12. estimate: 20 ÷ 5 = 4
exact: 4.175

13. 839.762

14. 669.004

15. 0.0000483

16. 480

Arrange in order from smallest to largest.

17. 3.5, 3.508, 3.51, 3.5108

17. 3.508, 3.51, 3.5108, 3.5

18. 0.44, $\frac{9}{20}$, 0.4506, 0.451

18. 0.44, 0.451, $\frac{9}{20}$, 0.4506

Use the order of operations to simplify.

19. 35.49

19. $6.3^2 - 5.9 + 3.4 \cdot 0.5$

Solve these word problems.

20. $462.87

20. Jennifer had $271.15 in her checking account. Yesterday her account earned $0.95 interest for the month, and she deposited a paycheck for $190.77. What is the new balance in her account?

21. Davida

21. Davida ran a race in 3.059 minutes. Angela ran the race in 3.5 minutes. Who won?

22. $5.35

22. Mr. Yamamoto bought 1.85 pounds of cheese at $2.89 per pound. How much did he pay for the cheese, to the nearest cent?

23. 2.8 degrees

23. Loren's baby had a temperature of 102.7 degrees. Later in the day it was 99.9 degrees. How much had the temperature dropped?

24. $4.55 per meter

24. Pat bought 3.4 meters of fabric. She paid $15.47. What was the cost per meter?

25. Answer varies.

25. Write your own word problem using decimals. Make it different from problems 20–24. Then show how to solve your problem.

NAME DATE HOUR

CUMULATIVE REVIEW EXERCISES CHAPTERS 1-4

Name the digit that has the given place value.

1. 19,076,542
hundreds 5
millions 9
ones 2

2. 83.0754
tenths 0
thousandths 5
tens 8

Round each number as indicated.

3. 279,506 to the nearest thousand 280,000

4. 0.3908 to the nearest thousandth 0.391

5. $339.70 to the nearest dollar $340

6. $0.8522 to the nearest cent $0.85

≈ *Round the numbers in each problem so there is only one non-zero digit. Then add, subtract, multiply, or divide the rounded numbers, as indicated, to estimate the answer. Finally, solve for the exact answer.*

7. *estimate*
$$\begin{array}{r} 4000 \\ 600 \\ +\ 9000 \\ \hline 13{,}600 \end{array}$$
exact
$$\begin{array}{r} 3672 \\ 589 \\ +\ 9078 \\ \hline 13{,}339 \end{array}$$

8. *estimate*
$$\begin{array}{r} 4 \\ 16 \\ +\ 1 \\ \hline 21 \end{array}$$
exact
$$\begin{array}{r} 4.06 \\ 15.7 \\ +\ 0.923 \\ \hline 20.683 \end{array}$$

9. *estimate*
$$\begin{array}{r} 5000 \\ -\ 2000 \\ \hline 3000 \end{array}$$
exact
$$\begin{array}{r} 5018 \\ -\ 1809 \\ \hline 3209 \end{array}$$

10. *estimate*
$$\begin{array}{r} 50 \\ -\ 7 \\ \hline 43 \end{array}$$
exact
$$\begin{array}{r} 51.6 \\ -\ 7.094 \\ \hline 44.506 \end{array}$$

11. *estimate*
$$\begin{array}{r} 3000 \\ \times\ 200 \\ \hline 600{,}000 \end{array}$$
exact
$$\begin{array}{r} 3317 \\ \times\ 166 \\ \hline 550{,}622 \end{array}$$

12. *estimate*
$$\begin{array}{r} 7 \\ \times\ 7 \\ \hline 49 \end{array}$$
exact
$$\begin{array}{r} 6.82 \\ \times\ 7.3 \\ \hline 49.786 \end{array}$$

13. *estimate*
$$50\overline{)100{,}000} \quad 2000$$
exact
$$46\overline{)123{,}740} \quad 2690$$

14. *estimate*
$$8\overline{)40} \quad 5$$
exact
$$8.4\overline{)37.8} \quad 4.5$$

Add, subtract, multiply, or divide as indicated.

15. $10 - 0.329$

9.671

16. $2\frac{3}{5} \times \frac{5}{9}$

$1\frac{4}{9}$

17. $9 + 72{,}417 + 799$

73,225

18. $11\frac{1}{5} \div 8$

$1\frac{2}{5}$

19. $5006 - 92$

4914

20. $0.7 + 85 + 7.903$

93.603

21. $7\overline{)2831}$

404 R3

22. $\frac{5}{6} + \frac{7}{8}$

$1\frac{17}{24}$

23. 332×704

233,728

24. $(0.006)(5.44)$

0.03264

25. 3.2×2.5

8

26. $25.2 \div 0.56$

45

27. $\frac{2}{3} \div 5\frac{1}{6}$

$\frac{4}{31}$

28. $5\frac{1}{4} - 4\frac{7}{12}$

$\frac{2}{3}$

29. $4.7 \div 9.3$

Round to nearest hundredth.

0.51

Simplify by using the order of operations.

30. $24 - 16 \div (3 + 5)$

22

31. $\sqrt{36} + 3 \cdot 8 - 4^2$

14

32. $\frac{2}{3} \cdot \left(\frac{7}{8} - \frac{1}{2}\right)$

$\frac{1}{4}$

33. $0.9^2 + 10.6 \div 0.53$

20.81

34. Solve $4^3 \cdot 3^2$ 576

35. Find $\sqrt{196}$ 14

36. Find the prime factorization of 200. Write your answer using exponents.

$2^3 \cdot 5^2$

37. Write 40.035 in words.

forty and thirty-five thousandths

38. Write three hundred six ten-thousandths in numbers.

0.0306

Write each decimal as a fraction or mixed number in lowest terms.

39. 0.125 $\frac{1}{8}$

40. 3.08 $3\frac{2}{25}$

Write each fraction or mixed number as a decimal. Round to the nearest thousandth, if necessary.

41. $2\frac{3}{5}$ 2.6

42. $\frac{7}{11}$ 0.636

43. Write $<$ or $>$ to make a true statement: $\frac{5}{8} > \frac{4}{9}$

Arrange in order from smallest to largest.

44. 7.005, 7.5005, 7.5, 7.505

7.005, 7.5, 7.5005, 7.505

45. $\frac{7}{8}$, 0.8, $\frac{21}{25}$, 0.8015

0.8, 0.8015, $\frac{21}{25}$, $\frac{7}{8}$

Solve each word problem.

46. Lameck had two $10 bills. He spent $7.96 on gasoline and $0.87 for a candy bar at the convenience store. How much money does he have left?

$11.17

47. Manuela's daughter is 50 inches tall. Last year she was $46\frac{5}{8}$ inches tall. How much has she grown?

$3\frac{3}{8}$ inches

48. Sharon records textbooks on tape for students who are blind. Her hourly wage is $8.73. How much did she earn working 16.5 hours last week, to the nearest cent?

$144.05

49. The Farnsworth Elementary School has eight classrooms with 22 students in each one and 12 classrooms with 26 students in each one. How many students attend the school?

488 students

50. Toshihiro bought $2\frac{1}{3}$ yards of cotton fabric and $3\frac{7}{8}$ yards of wool fabric. How many yards did he buy in all?

$6\frac{5}{24}$ yards

51. About $\frac{7}{8}$ of all children are right-handed. How many of the 96 children in the daycare center would be expected to be right-handed?

84 children

52. Paulette bought 2.7 pounds of grapes for $2.56. What was the cost per pound, to the nearest cent?

$0.95

53. Kimberly had $29.44 in her checking account. She wrote a check for $40 and deposited a $220.06 paycheck into her account, but not in time to prevent a $15 overdraft charge. What is the new balance in her account?

$194.50

54. Carter Community College received a $75,000 grant from a local computer company to help students pay tuition for computer classes. How much money could be given to each of 135 students? Round to the nearest dollar.

$556

55. Carlos checks the lengths of precision parts for aircraft. A bolt that is supposed to measure 2.05 centimeters long is actually 2.1 centimeters. Is the bolt too long or too short?

Too long.

5 Ratio and Proportion

A **ratio** compares two quantities. You can compare two numbers, such as 8 and 4, or two measurements, such as 3 days and 12 days.

5.1 RATIOS

OBJECTIVES

1. Write ratios as fractions.
2. Solve ratio problems involving decimals or mixed numbers.
3. Solve ratio problems after converting units.

FOR EXTRA HELP

Tape 6 | SSM pp. 134–136 | MAC: A IBM: A

1 A ratio can be written in three ways.

WRITING A RATIO

The ratio of \$7 **to** \$3 can be written:

7 **to** 3 or 7:3 or $\frac{7}{3}$ ← fraction bar indicates **to**

":" indicates **to**

Writing a ratio as a fraction is the most common method, and the one we will use here. All three ways are read, "the ratio of 7 to 3." The word to separates the quantities being compared.

WRITING A RATIO AS A FRACTION

Order is important when writing a ratio. The quantity mentioned **first** is the **numerator**. The quantity mentioned **second** is the **denominator**. For example:

5 to **12** is written $\frac{\mathbf{5}}{\mathbf{12}}$

EXAMPLE 1 *Writing a Ratio*

Lucinda's living room is 20 feet long, 17 feet wide, and 9 feet high.

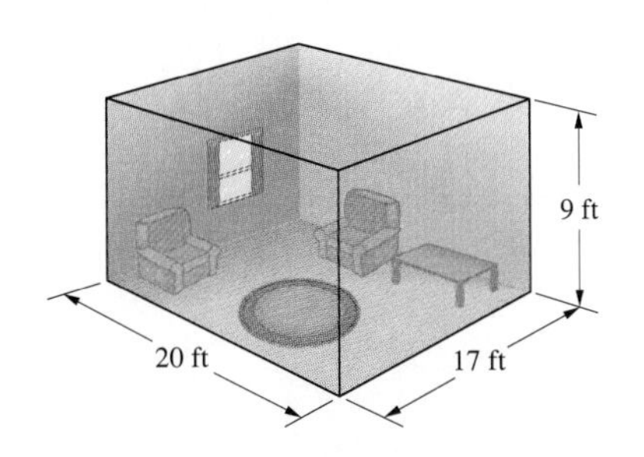

Write these ratios using the living room measurements:

(a) ratio of length to width

(b) ratio of height to length

(a) The ratio of **length** to **width** is $\frac{\textbf{20 feet}}{\textbf{17 feet}} = \frac{20}{17}$.

numerator (mentioned first) — denominator (mentioned second)

You can cancel common *units* just like you canceled common *factors* when writing fractions in lowest terms. (See **Section 2.4.**)

(b) The ratio of height to length is $\frac{9 \text{ feet}}{20 \text{ feet}} = \frac{9}{20}$. ■

Note Remember, the *order* of the numbers is important in a ratio. Look for the words "ratio of a to b." Write the ratio as $\frac{a}{b}$, **not** $\frac{b}{a}$. The quantity mentioned first is the numerator.

WORK PROBLEM 1 AT THE SIDE.

Any ratio can be written as a fraction. Therefore, you can write a ratio in lowest terms, just as you do with any fraction.

■ **EXAMPLE 2** *Writing a Ratio in Lowest Terms*

Write each ratio in lowest terms.

(a) 60 days to 20 days

(b) 50 ounces of medicine to 120 ounces of medicine.

(c) 18 people in a large van to 8 people in a small van.

(a) The ratio is $\frac{60}{20}$. Write this ratio in lowest terms by dividing numerator and denominator by 20.

$$\frac{60}{20} = \frac{60 \div 20}{20 \div 20} = \frac{3}{1}$$

(b) The ratio is $\frac{50}{120}$. Divide numerator and denominator by 10.

$$\frac{50}{120} = \frac{50 \div 10}{120 \div 10} = \frac{5}{12}$$

(c) The ratio is $\frac{18}{8} = \frac{18 \div 2}{8 \div 2} = \frac{9}{4}$. ■

Note Although $\frac{9}{4} = 2\frac{1}{4}$, ratios are *not* written as mixed numbers. Nevertheless, the ratio $\frac{9}{4}$ does mean the large van holds $2\frac{1}{4}$ times as many people as the small van.

WORK PROBLEM 2 AT THE SIDE.

2 Sometimes a ratio compares two decimal numbers or two fractions. It is easier to understand if we rewrite the ratio as a ratio of two whole numbers.

1. Shane spent \$14 on meat, \$5 on milk, and \$7 on fresh fruit. Write these ratios as fractions.

(a) The ratio of amount spent on fruit to amount spent on milk.

(b) The ratio of amount spent on milk to amount spent on meat.

(c) The ratio of amount spent on meat to amount spent on milk.

2. Write each ratio as a fraction in lowest terms.

(a) 6 liters to 9 liters

(b) 16 tons to 8 tons

(c) Write the ratio of width to length for this rectangle.

length 48 ft

width 24 ft

ANSWERS

1. (a) $\frac{7}{5}$ (b) $\frac{5}{14}$ (c) $\frac{14}{5}$

2. (a) $\frac{2}{3}$ (b) $\frac{2}{1}$ (c) $\frac{1}{2}$

EXAMPLE 3 *Using Decimal Numbers in a Ratio*
The price of a Sunday newspaper increased from \$1.20 to \$1.50. Find the ratio of the increase in price to the original price.

Approach The increase in price is the quantity mentioned first so it is the numerator. How much did the price go up? Use subtraction.

$$\text{new price} - \text{original price} = \text{increase}$$
$$\$1.50 - \$1.20 = \$0.30$$

Use the increase in price and the original price to form the ratio.

Solution The ratio of increase in price to original price is

$$\frac{0.30 \leftarrow \text{increase}}{1.20 \leftarrow \text{original price}}$$

Now we rewrite the ratio as a ratio of whole numbers. Recall that if you multiply both the numerator and denominator of a fraction by the same number, you get an equivalent fraction. The decimals in this example are hundredths, so multiply by 100 to get whole numbers. (If the decimals are tenths, multiply by 10. If thousandths, multiply by 1000.) Then write the ratio in lowest terms.

$$\frac{0.30}{1.20} = \frac{0.30 \times 100}{1.20 \times 100} = \frac{30}{120} \quad \text{ratio is now two whole numbers}$$

$$\frac{30}{120} = \frac{30 \div 30}{120 \div 30} = \frac{1}{4} \quad \text{ratio in lowest terms}$$ ■

WORK PROBLEM 3 AT THE SIDE.

3. Write each ratio as a ratio of whole numbers in lowest terms.

(a) The price of Tamar's favorite brand of lipstick increased from \$3.75 to \$4.25. Find the ratio of the increase in price to the original price.

(b) Last week Sean walked 0.9 kilometer each day. This week he increased it to 1.2 kilometers per day. Find the ratio of the increase in distance to the original distance.

ANSWERS
3. (a) $\frac{2}{15}$ (b) $\frac{1}{3}$

EXAMPLE 4 *Using a Mixed Number in a Ratio*
Write each ratio as a comparison of whole numbers in lowest terms.

(a) 2 days to $2\frac{1}{4}$ days

(b) $3\frac{1}{4}$ to $1\frac{1}{2}$

(a) Write the ratio as follows. Cancel the common units.

$$\frac{2 \text{ days}}{2\frac{1}{4} \text{ days}} = \frac{2}{2\frac{1}{4}}$$

Next, write 2 as $\frac{2}{1}$ and $2\frac{1}{4}$ as the improper fraction $\frac{9}{4}$.

$$\frac{2}{2\frac{1}{4}} = \frac{\frac{2}{1}}{\frac{9}{4}}$$

4. Write each ratio as a ratio of whole numbers in lowest terms.

(a) $3\frac{1}{2}$ to 4

(b) $5\frac{5}{8}$ pounds to $3\frac{3}{4}$ pounds

(c) $3\frac{1}{2}$ inches to $\frac{7}{8}$ inch

ANSWERS

4. (a) $\frac{7}{8}$ (b) $\frac{3}{2}$ (c) $\frac{4}{1}$

Rewrite the problem using the "÷" symbol for division. Then invert and multiply, as you did in **Section 2.7.**

$$\frac{\frac{2}{1}}{\frac{9}{4}} = \frac{2}{1} \div \frac{9}{4} = \frac{2}{1} \cdot \frac{4}{9} = \frac{8}{9}$$

invert

The ratio, in lowest terms, is $\frac{8}{9}$.

(b) Write the ratio as $\frac{3\frac{1}{4}}{1\frac{1}{2}}$. Then write $3\frac{1}{4}$ and $1\frac{1}{2}$ as improper fractions.

$$3\frac{1}{4} = \frac{13}{4} \quad \text{and} \quad 1\frac{1}{2} = \frac{3}{2}$$

The ratio is

$$\frac{3\frac{1}{4}}{1\frac{1}{2}} = \frac{\frac{13}{4}}{\frac{3}{2}}.$$

Write as a division problem using the "÷" symbol. Invert and multiply.

$$\frac{13}{4} \div \frac{3}{2} = \frac{13}{\cancel{4}_2} \cdot \frac{\cancel{2}^1}{3} = \frac{13}{6}$$ ■

◀ WORK PROBLEM 4 AT THE SIDE.

3 When a ratio compares measurements, both measurements must be in the *same* units. For example, *feet* must be compared to *feet, hours* to *hours, pints* to *pints,* and *inches* to *inches.*

■ **EXAMPLE 5** *Applications Using Measurement*

(a) Write the ratio of the length of the board on the left to the board on the right. Compare in inches.

(b) Write the ratio of 28 days to 3 weeks. Compare in days.

(a) First, express 2 feet in inches. Because 1 foot has 12 inches, 2 feet is

$$2 \cdot 12 \text{ inches} = 24 \text{ inches}.$$

The length of the board on the left is 24 inches, so the ratio of the lengths is

$$\frac{24 \text{ inches}}{30 \text{ inches}} = \frac{24}{30}.$$

Write the ratio in lowest terms.

$$\frac{24}{30} = \frac{24 \div 6}{30 \div 6} = \frac{4}{5}$$

The shorter board is $\frac{4}{5}$ the length of the longer board.

(b) First express 3 weeks in days. Because 1 week has 7 days, 3 weeks is

$$3 \cdot 7 \text{ days} = 21 \text{ days}$$

So the ratio in days is

$$\frac{28 \text{ days}}{21 \text{ days}} = \frac{28}{21} = \frac{4}{3} \leftarrow \text{lowest terms}$$ ■

The following table will help you set up ratios that compare measurements. You will work with these measurements again in Chapter 7.

Length	Capacity (Volume)
1 foot = 12 inches	1 pint = 2 cups
1 yard = 3 feet	1 quart = 2 pints
1 mile = 5280 feet	1 gallon = 4 quarts
Weight	**Time**
1 pound = 16 ounces	1 week = 7 days
1 ton = 2000 pounds	1 day = 24 hours
	1 hour = 60 minutes
	1 minute = 60 seconds

WORK PROBLEM 5 AT THE SIDE.

5. Write each ratio as a fraction in lowest terms.

(a) 9 inches to 6 feet
Compare in inches.

(b) 2 days to 8 hours
Compare in hours.

(c) 7 yards to 14 feet
Compare in feet.

(d) 8 pints to 9 quarts
Compare in pints.

(e) 25 minutes to 2 hours
Compare in minutes.

(f) 4 pounds to 12 ounces
Compare in ounces.

ANSWERS

5. (a) $\frac{1}{8}$ (b) $\frac{6}{1}$ (c) $\frac{3}{2}$ (d) $\frac{4}{9}$ (e) $\frac{5}{24}$ (f) $\frac{16}{3}$

QUEST FOR NUMERACY

The Roots of Math Vocabulary

Many words are built from Latin or Greek root words and prefixes. Knowing the meaning of the more common ones can help you figure out the meaning of terms in many subject areas, including math. Listed below are the prefixes related to number names. In parentheses is the meaning of the prefix and then several related words which use the prefix. Try to think of more math related words you've seen that use these prefixes.

mono (1)-*mono*mial, *mono*gram
uni (1)-*uni*cycle, *uni*t
bi (2)-*bi*nomial, *bi*ennium
di (2)-*di*vide, *di*ssect
tri (3)-*tri*angle, *tri*pod
quar/quad (4)-*quar*ter, *quad*rilateral
tetra (4)-*tetra*hedron
pent (5)-*pent*agon, *pent*house (originally the 5th floor luxury apartment)
quint (5)-*quint*illion, *quint*uplets

sex (6)-*sex*tet, *sex*tuplets
hex (6)-*hex*agon, *hex*ahedron
hept (7)-*hept*agon, *hept*ameter
sept (7)-*Sept*ember, *sept*illion
oct (8)-*Oct*ober, *oct*agon
non/nov (9)-*Nov*ember, *non*agon
dec (10)-*Dec*ember, *dec*ade
cent (100)-100 cents in a dollar, *cent*imeter
poly (many)-*poly*gon
multi (many)-*multi*faceted

The months on the old calendar matched their numbers (there were 10 months). July and August were added later.

Now use your knowledge of roots and prefixes to answer these questions.

1. A test question asks: How many *dec*ades are in two *cent*uries? What numbers will you use? decade = 10, century = 100
so 200 years ÷ 10 = 20 decades

2. To write the number 1 *tri*llion, start with 1000 and add *three* groups of zeros after it (because the *tri* in *tri*llion means three).

1 trillion = 1,000,000,000,000 (start with 1000; three groups of zeros)

Use this technique to write the following numbers.

1 quadrillion = 1,000,000,000,000,000
1 octillion = 1,000,000,000,000,000,000,000,000,000
1 billion = 1,000,000,000
1 quintillion = 1,000,000,000,000,000,000

3. The *Pent*agon is a building in Washington, D.C., that has *five* sides. It is the headquarters of the U.S. military forces. How many sides would the building have if it was called:

The Hexagon 6 sides   The Heptagon 7 sides
The Nonagon 9 sides   The Polygon many sides

Extend this activity by asking students to think of additional words that use these prefixes, then to use a dictionary to check the derivations. Additional prefixes are found on the Quest for Numeracy page in Section 8.7.

EXERCISES

Write each ratio as a fraction in lowest terms.

1. 8 to 9 $\frac{8}{9}$

2. 11 to 15 $\frac{11}{15}$

3. \$100 to \$50 $\frac{2}{1}$

4. 35¢ to 7¢ $\frac{5}{1}$

5. 30 minutes to 90 minutes $\frac{1}{3}$

6. 9 pounds to 36 pounds $\frac{1}{4}$

7. 80 miles to 50 miles $\frac{8}{5}$

8. 300 people to 450 people $\frac{2}{3}$

9. 6 hours to 16 hours $\frac{3}{8}$

10. 45 books to 35 books $\frac{9}{7}$

Write each ratio as a ratio of whole numbers in lowest terms.

11. \$1.50 to \$2.50 $\frac{3}{5}$

12. \$0.08 to \$0.06 $\frac{4}{3}$

13. 3 to $2\frac{1}{2}$ $\frac{6}{5}$

14. 5 to $1\frac{1}{4}$ $\frac{4}{1}$

15. $1\frac{1}{4}$ to $1\frac{1}{2}$ $\frac{5}{6}$

16. $2\frac{1}{3}$ to $2\frac{2}{3}$ $\frac{7}{8}$

Write each ratio as a fraction in lowest terms. For help, use the table of measurement relationships in Example 5.

17. 4 feet to 30 inches
Compare in inches. $\frac{8}{5}$

18. 6 yards to 12 feet
Compare in feet. $\frac{3}{2}$

19. 5 minutes to 1 hour
Compare in minutes. $\frac{1}{12}$

20. 8 quarts to 5 pints
Compare in pints. $\frac{16}{5}$

21. 15 hours to 2 days
Compare in hours. $\frac{5}{16}$

22. 4 ounces to 2 pounds
Compare in ounces. $\frac{1}{8}$

23. 5 gallons to 5 quarts
Compare in quarts. $\frac{4}{1}$

24. 3 cups to 3 pints
Compare in cups. $\frac{1}{2}$

Solve each word problem. Write each ratio as a fraction in lowest terms.

25. Mr. Wilkins is 35 years old, and his daughter is 10 years old. Find the ratio of his age to hers.

$\frac{7}{2}$

26. The Empire State Building is 1100 feet tall, and the Sears Tower in Chicago is 1300 feet tall. Write the ratio of the height of the Empire State Building to the height of the Sears Tower.

$\frac{11}{13}$

27. Our math class has 16 women and 20 men. What is the ratio of men to women?

$\frac{5}{4}$

28. Cherise sells souvenirs at baseball games. She sold 30 red hats and 40 blue hats. What is the ratio of blue hats to red hats?

$\frac{4}{3}$

29. The Sanchez Company made 400 washing machines. Four of them had defects. What is the ratio of defective washers to the total number of washers?

$\frac{1}{100}$

30. Andrew spends $500 per month on rent and $120 per month on utilities. Find the ratio of the amount spent on utilities to the amount spent on rent.

$\frac{6}{25}$

31. Would you prefer that the ratio of your income to your friend's income be 1 to 3 or 3 to 1? Explain your answer.

Answer varies.

32. Amelia said that the ratio of her age to her mother's age is 5 to 3. Is this possible? Explain your answer.

It is not possible. Amelia would have to be older than her mother to have a ratio of 5 to 3.

Use the circle graph below of one family's monthly budget to complete Exercises 33–36. Write each ratio as a fraction in lowest terms.

33. Find the ratio of taxes to transportation. $\frac{2}{1}$

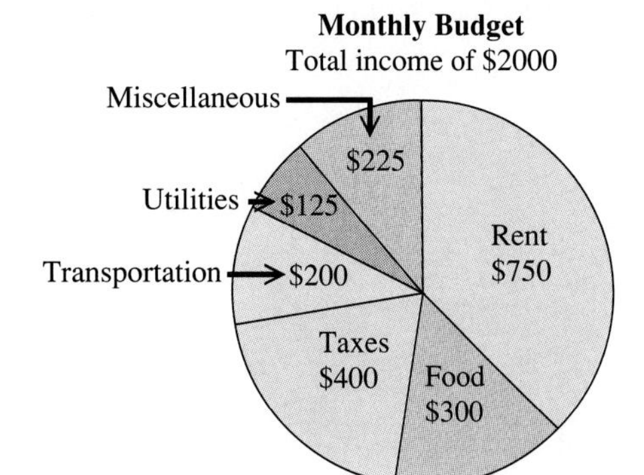

34. Find the ratio of rent to food. $\frac{5}{2}$

35. Find the ratio of rent to total income. $\frac{3}{8}$

36. Find the ratio of utilities to total income. $\frac{1}{16}$

For each figure, find the ratio of the length of the longest side to the length of the shortest side. Write each ratio as a fraction in lowest terms.

37.

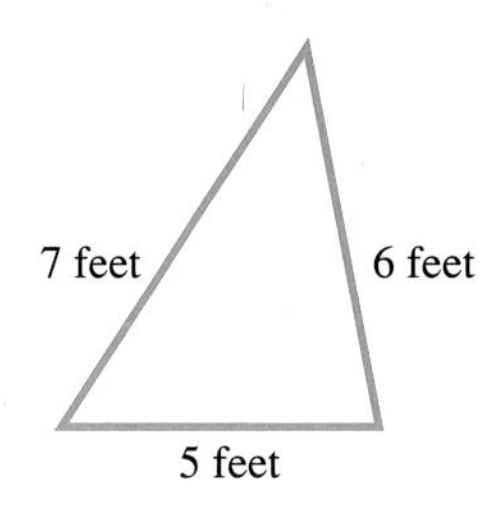

$\frac{7}{5}$

38.

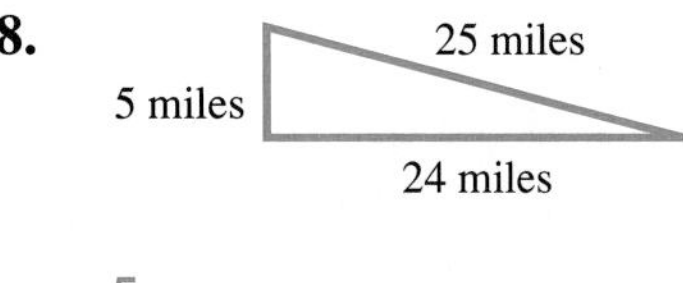

$\frac{5}{1}$

39.

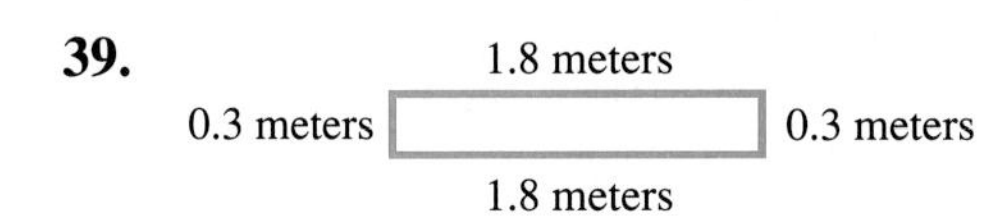

$\frac{6}{1}$

40.

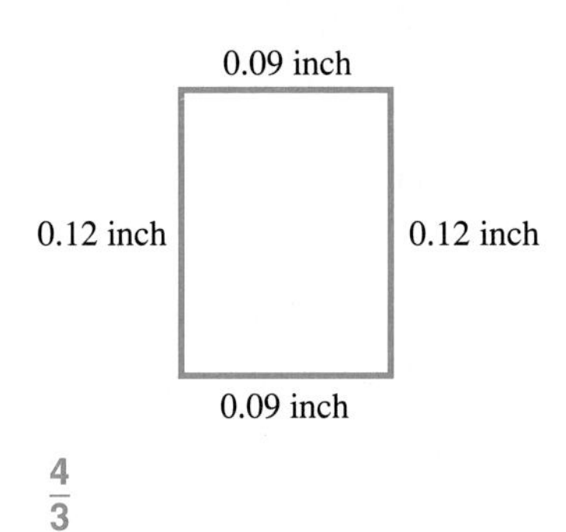

$\frac{4}{3}$

41.

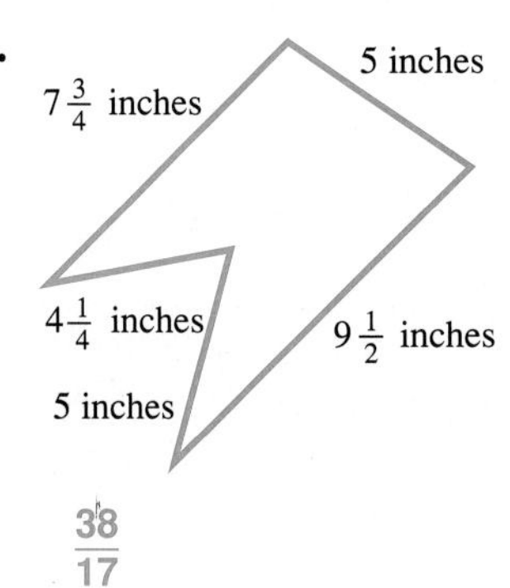

$\frac{38}{17}$

42.

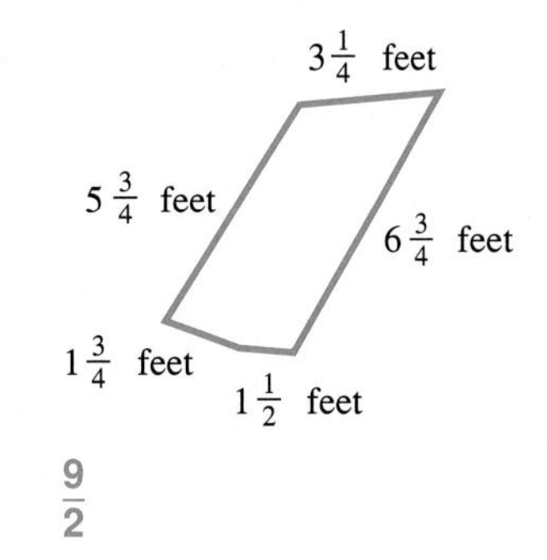

$\frac{9}{2}$

▲ **43.** The price of oil recently went from $6.60 to $9.90 per case of 12 quarts. Find the ratio of the increase in price to the original price.

$\frac{1}{2}$

▲ **44.** The price of an antibiotic decreased from $8.80 to $5.60 for a bottle of 100 tablets. Find the ratio of the decrease in price to the original price.

$\frac{4}{11}$

▲ **45.** What is the ratio of $59\frac{1}{2}$ days to $8\frac{3}{4}$ weeks? Compare in weeks.

$\frac{34}{35}$

▲ **46.** Find the ratio of $18\frac{1}{2}$ inches to $4\frac{1}{2}$ feet. Compare in inches.

$\frac{37}{108}$

47. The ratio of John's age to his sister's age is 4 to 5. One possibility is that John is 4 years old and his sister is 5 years old. Find three other possibilities that fit the 4 to 5 ratio.

Answer varies. Some possibilities are 8 and 10, 12 and 15, 16 and 20.

48. In this figure, what is the ratio of the length of the longest side to the length of the shortest side?

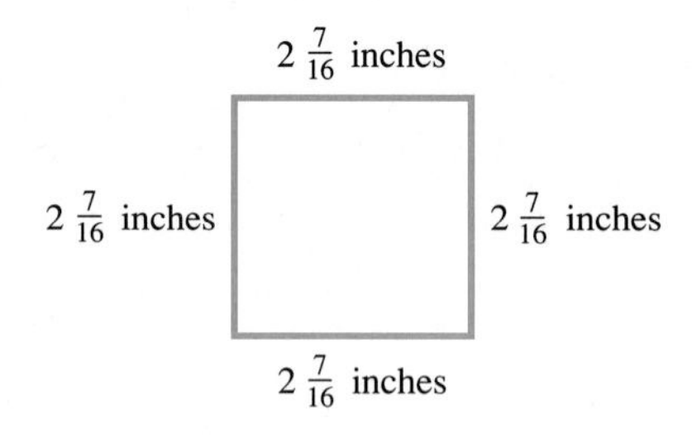

$\frac{1}{1}$

PREVIEW EXERCISES

Divide. Round to the nearest thousandth if necessary. (For help, see ***Section 4.6.****)*

49. $7\overline{)0.65}$ 0.093

50. $3\overline{)7.33}$ 2.443

51. $4\overline{)4.1}$ 1.025

52. $0.95\overline{)41.8}$ 44

53. $0.71\overline{)6.72}$ 9.465

54. $4.6\overline{)116.38}$ 25.3

5.2 RATES

A ratio compares two measurements with the same type of units, such as 9 feet to 12 feet (both length measurements). But many of the comparisons we make use measurements with different types of units, such as 40 dollars for 8 hours or 450 miles on 18 gallons of gas. This type of comparison is called a **rate.**

OBJECTIVES

1. Write rates as fractions.
2. Find unit rates.

FOR EXTRA HELP

Tape 7 | SSM pp. 136–138 | MAC: A IBM: A

1 For example, suppose you hiked 18 miles in 4 hours. The **rate** at which you hiked can be written as a fraction in lowest terms.

$$\frac{18 \text{ miles}}{4 \text{ hours}} = \frac{18 \text{ miles} \div 2}{4 \text{ hours} \div 2} = \frac{9 \text{ miles}}{2 \text{ hours}} \Big\} \text{ lowest terms}$$

In a rate, you often find these words separating the quantities you are comparing:

in for on per from

Notice that the units in a rate are different so they do *not* cancel.

Note When writing a rate, always include the units.

EXAMPLE 1 *Write a Rate in Lowest Terms*

Write each rate as a fraction in lowest terms.

(a) 5 gallons of chemical for $60.

(b) $1500 wages in 10 weeks

(c) 2225 miles on 75 gallons of gas

(a) $\frac{5 \text{ gallons}}{60 \text{ dollars}} = \frac{1 \text{ gallon}}{12 \text{ dollars}}$

(b) $\frac{1500 \text{ dollars}}{10 \text{ weeks}} = \frac{150 \text{ dollars}}{1 \text{ week}}$

(c) $\frac{2225 \text{ miles}}{75 \text{ gallons}} = \frac{89 \text{ miles}}{3 \text{ gallons}}$ ■

WORK PROBLEM 1 AT THE SIDE. ▶▶

1. Write each rate as a fraction in lowest terms.

(a) $6 for 30 packages

(b) 500 miles in 10 hours

(c) 4 teachers for 90 students

(d) 1270 bushels on 30 acres

2 When the *denominator* of a rate is 1, it is called a **unit rate.** We use unit rates frequently. For example, you earn $8.75 for *1 hour* of work. This unit rate is written:

$8.75 per hour or $8.75/hour

Use per or a / mark when writing unit rates. You drive 28 miles on *1 gallon* of gas. This unit rate is written 28 miles per gallon, or 28 miles/gallon.

ANSWERS

1. (a) $\frac{1 \text{ dollar}}{5 \text{ packages}}$ **(b)** $\frac{50 \text{ miles}}{1 \text{ hour}}$ **(c)** $\frac{2 \text{ teachers}}{45 \text{ students}}$ **(d)** $\frac{127 \text{ bushels}}{3 \text{ acres}}$

2. Find each unit rate.

(a) \$4.35 for 3 pounds of cheese

(b) 304 miles on 9.5 gallons of gas

(c) \$850 in 5 days

(d) 24-pound turkey for 15 people

ANSWERS
2. (a) \$1.45/pound (b) 32 miles/gallon (c) \$170/day (d) 1.6 pounds/person

■ **EXAMPLE 2** *Finding a Unit Rate*

Find each unit rate.

(a) 337.5 miles on 13.5 gallons of gas

(b) 549 miles in 18 hours

(c) \$810 in 6 days

(a) Write the rate as a fraction.

$$\frac{337.5 \text{ miles}}{13.5 \text{ gallons}} \leftarrow \text{fraction bar indicates division}$$

Divide by 13.5 to find the unit rate.

$$13.5\overline{)337.5} = 25.$$

$$\frac{337.5 \text{ miles} \div 13.5}{13.5 \text{ gallons} \div 13.5} = \frac{25 \text{ miles}}{1 \text{ gallon}}$$

The unit rate is 25 miles/gallon.

(b) $\frac{549 \text{ miles}}{18 \text{ hours}}$ Divide $18\overline{)549.0}$ = 30.5

The unit rate is 30.5 miles/hour.

(c) $\frac{810 \text{ dollars}}{6 \text{ days}}$ Divide $6\overline{)810}$ = 135

The unit rate is \$135/day. ■

◀◀ **WORK PROBLEM 2 AT THE SIDE.**

COST PER UNIT

Cost per unit is a rate that tells how much you pay for *one* item or *one* unit. Examples are \$1.25 per gallon, \$47 per shirt, and \$2.98 per pound. When shopping, you can save money by finding the lowest cost per unit.

■ **EXAMPLE 3** *Determining the Best Buy*

The local store charges the following prices for pancake syrup. Find the best buy.

Size	*Price*
12 ounces	\$1.28
24 ounces	\$1.79
36 ounces	\$2.73

Approach The best buy is the container with the lowest cost per unit. All the containers are measured in *ounces,* so you first need to find the *cost per ounce* for each one. Divide the price of the container by the number of ounces in it.

Solution

Size	*Cost Per Unit*
12 ounces	$\frac{\$1.28}{12 \text{ ounces}} = \$0.107$ per ounce (rounded)
24 ounces	$\frac{\$1.79}{24 \text{ ounces}} = \$0.075$ per ounce (rounded)
36 ounces	$\frac{\$2.73}{36 \text{ ounces}} = \$0.076$ per ounce (rounded)

Notice that we rounded to thousandths instead of hundredths, in order to see the difference between the 24-ounce and 36-ounce containers. The lowest cost per ounce is $0.075, so the 24-ounce container is the best buy. ■

WORK PROBLEM 3 AT THE SIDE. ▶▶

3. Find the best buy (lowest cost per unit) for each purchase.

(a) 2 quarts for $3.25
3 quarts for $4.95
4 quarts for $6.48

(b) 6 cans of cola for $1.39
12 cans of cola for $2.49
24 cans of cola for $5.05

ANSWERS
3. (a) 4 quarts, at $1.62 per quart
(b) 12 cans, at $0.208 per can

QUEST FOR NUMERACY

The Real "Best Buy"

Finding the "best buy" is sometimes a complicated process. Things that affect the cost per unit can include "cents off" coupons and differences in how much use you'll get out of each unit. Try these examples. You may need to round cost per unit to thousandths, or even ten-thousandths, in order to see the difference between different items.

Three brands of cornflakes are available. Brand G is priced at \$2.39 for 10 ounces. Brand K is \$3.99 for 20.3 ounces and Brand P is \$3.39 for 16.5 ounces. You have a coupon for 50¢ off Brand P and a coupon for 60¢ off Brand G. Which cereal is the best buy based on cost per unit?

Brand P with the 50¢ coupon is the best buy.
\$2.89 ÷ 16.5 ounces
= \$0.175/ounce
Brand G with the 60¢ coupon is close.
\$1.79 ÷ 10 ounces
= \$0.179/ounce

Two brands of facial tissue are available. Brand K is on special at three boxes of 175 tissues each for \$5. Brand S is priced at \$1.29 per box of 125 tissues. You have a coupon for 20¢ off one box of Brand K and a coupon for 45¢ off two boxes of Brand S. How can you get the best buy on tissue?

1 box Brand K with 20¢ coupon
\$1.47 ÷ 175 tissues
= \$0.0084/tissue
If you cannot buy just one box of Brand K at the special price, then two boxes of Brand S with 45¢ coupon is best buy.
(\$1.27 × 2) − 0.45 = \$2.13
\$2.13 ÷ 250 tissues
= \$0.0085/tissue

There are many brands of liquid laundry detergent. If you feel they all do a good job of cleaning your clothes, you can base your purchase on cost per unit. But be careful, because some brands are now "concentrated" so you use less detergent for each load of clothes. What is the best buy among these choices? What additional information do you need before calculating unit price on the last choice?

50 fluid ounces for \$3.99 Does same number of washloads as the old 64-ounce bottle.	150 fluid ounces for \$9.65 (not concentrated)	One gallon for \$8.89 Does twice the washloads of the old gallon bottle.
This bottle washes as much as 64 fl oz so unit cost is \$0.062/fl oz.	**\$0.064/fl oz**	**one gallon = 128 fl oz. This bottle washes as much as 256 fl oz so unit cost is \$0.035/fl oz. Best Buy**

Some batteries claim to last longer than others. If you believe these claims, which of the following brands is the best buy?

Four-pack of AA size batteries for \$2.79.	One AA size battery for \$1.19 Lasts twice as long.	Eight-pack of AA size batteries for \$7.49 Lasts 50% longer (half again as long).
\$2.79 ÷ 4 = \$0.698/battery	**Like getting 2 batteries. \$1.19 ÷ 2 = \$0.595/battery Best Buy**	**Like getting 12 batteries. \$7.49 ÷ 12 = 0.624/battery**

To extend this activity, ask students to bring in additional "best buy" problems that they encounter when shopping. You might also discuss ways to evaluate advertisers' claims.

EXERCISES

Write each rate in lowest terms.

1. 10 cups for 6 people $\frac{\text{5 cups}}{\text{3 people}}$

2. $12 for 30 pens $\frac{\text{\$2}}{\text{5 pens}}$

3. 15 feet in 35 seconds $\frac{\text{3 feet}}{\text{7 seconds}}$

4. 100 miles in 30 hours $\frac{\text{10 miles}}{\text{3 hours}}$

5. 14 people for 28 dresses $\frac{\text{1 person}}{\text{2 dresses}}$

6. 12 wagons for 48 horses $\frac{\text{1 wagon}}{\text{4 horses}}$

7. 25 letters in 5 minutes $\frac{\text{5 letters}}{\text{1 minute}}$

8. 68 pills for 17 people $\frac{\text{4 pills}}{\text{1 person}}$

9. $63 for 6 visits $\frac{\text{\$21}}{\text{2 visits}}$

10. 25 doctors for 310 patients $\frac{\text{5 doctors}}{\text{62 patients}}$

11. 72 miles on 4 gallons $\frac{\text{18 miles}}{\text{1 gallon}}$

12. 132 miles on 8 gallons $\frac{\text{33 miles}}{\text{2 gallons}}$

Find each unit rate.

13. $60 in 5 hours

$12 per hour or $12/hour

14. $2500 in 20 days

$125 per day or $125/day

15. 50 eggs from 10 chickens

5 eggs per chicken or 5 eggs/chicken

16. 36 children from 12 families

3 children per family or 3 children/family

17. 7.5 pounds for 6 people

1.25 pounds/person

18. 44 bushels from 8 trees

5.5 bushels/tree

19. $413.20 for 4 days

$103.30/day

20. $74.25 for 9 hours

$8.25/hour

Earl kept the record shown below of the gas he bought for his car. For each entry, find the number of miles he traveled and the unit rate. Round your answers to the nearest tenth.

	Date	Odometer at start	Odometer at end	Miles traveled	Gallons purchased	Miles per gallon
21.	2/4	27,432.3	27,758.2	325.9	15.5	21.0
22.	2/9	27,758.2	28,058.1	299.9	13.4	22.4
23.	2/16	28,058.1	28,396.7	338.6	16.2	20.9
24.	2/20	28,396.7	28,704.5	307.8	13.3	23.1

Find the best buy (based on the cost per unit) for each of the following.

25. black pepper
4 ounces for $0.89
8 ounces for $2.13

4 oz for $0.89

26. shampoo
8 ounces for $0.99
12 ounces for $1.47

12 ounces for $1.47

27. cereal
10 ounces for $1.60
15 ounces for $2.10
20 ounces for $3.03

15 ounces for $2.10

28. soup
2 cans for $0.75
3 cans for $1.17
5 cans for $1.79

5 cans for $1.79

29. chunky peanut butter
12 ounces for $1.09
18 ounces for $1.41
28 ounces for $2.29
40 ounces for $3.19

18 ounces for $1.41

30. pork and beans
8 ounces for $0.37
16 ounces for $0.77
21 ounces for $0.99
31 ounces for $1.50

8 ounces for $0.37

31. Suppose you are choosing between two brands of chicken noodle soup. Brand A is \$0.38 per can and Brand B is \$0.48 per can. But Brand B has more chunks of chicken in it. Which soup is the better buy? Explain your choice.

Answer varies.

32. A small bag of potatoes costs \$0.19 per pound. A large bag costs \$0.15 per pound. But there are only two people in your family, so half the large bag would probably rot before you use it up. Which bag is the better buy? Explain.

Answer varies.

Solve each word problem.

33. Makesha lost 10.5 pounds in six weeks. What was her rate of loss in pounds per week?

1.75 pounds/week

34. Enrique's taco recipe uses four pounds of meat to feed 10 people. Give the rate in pounds per person.

0.4 pounds/person

35. Russ works 7 hours to earn \$85.82. What is his rate per hour?

\$12.26/hour

36. Find the cost of 1 gallon of gas if 18 gallons cost \$20.88.

\$1.16/gallon

37. Ms. Johnson bought 150 shares of stock for \$1725. Find the cost of one share.

\$11.50/share

38. A company pays \$6450 in dividends for the 2500 shares of its stock. Find the dividend per share.

\$2.58/share

39. John can pack six crates of berries in 24 minutes. Give his rate in crates per minute and in minutes per crate.

0.25 crate/min or $\frac{1}{4}$ crate/min

4 min/crate

40. Sofia can clean and adjust five hearing aids in four hours. Give her rate in hearing aids per hour and in hours per hearing aid.

1.25 hearing aids/hour or $1\frac{1}{4}$ aids/hour

0.8 hour/hearing aid or $\frac{4}{5}$ hour/aid

41. The 4.6 yards of fabric needed for a dress coat costs \$51.75. Find the cost of 1 yard of fabric.

\$11.25/yard

42. The cost to lay 42.4 square yards of carpet is \$691.12. Find the cost of 1 square yard of carpet.

\$16.30/square yard

PREVIEW EXERCISES

Multiply. Write your answers as whole or mixed numbers. (For help, see ***Section 2.8.****)*

43. $4 \cdot 2\frac{3}{4}$ 11

44. $12 \cdot 5\frac{2}{3}$ 68

45. $5\frac{2}{5} \cdot 20$ 108

46. $3\frac{5}{8} \cdot 6$ $21\frac{3}{4}$

47. $1\frac{1}{6} \cdot 3$ $3\frac{1}{2}$

48. $12 \cdot 2\frac{5}{8}$ $31\frac{1}{2}$

5.3 PROPORTIONS

1 A **proportion** states that two ratios (or rates) are equivalent. For example,

$$\frac{\$20}{4 \text{ hours}} = \frac{\$40}{8 \text{ hours}}$$

is a proportion that says the rate $\frac{\$20}{4 \text{ hours}}$ is equivalent to the rate $\frac{\$40}{8 \text{ hours}}$. This proportion is read

20 dollars is to 4 hours as 40 dollars is to 8 hours.

EXAMPLE 1 *Writing a Proportion*
Write each of the following proportions.

(a) 6 feet is to 11 feet as 18 feet is to 33 feet

(b) \$9 is to 6 liters as \$3 is to 2 liters

(a) $\frac{6}{11} = \frac{18}{33}$ The common units (feet) cancel and are not written.

(b) $\frac{\$9}{6 \text{ liters}} = \frac{\$3}{2 \text{ liters}}$ Units must be written. ■

WORK PROBLEM 1 AT THE SIDE.

2 There are two ways to see whether a proportion is true. One way is to *write both of the ratios in lowest terms.*

EXAMPLE 2 *Writing Both Ratios in Lowest Terms*
Are the following proportions true?

(a) $\frac{5}{9} = \frac{18}{27}$

(b) $\frac{16}{12} = \frac{28}{21}$

(a) Write each fraction in lowest terms. $\frac{5}{9}$ is already in lowest terms, and

$$\frac{18}{27} = \frac{2}{3}. \quad \text{lowest terms}$$

Because $\frac{2}{3}$ is *not* equivalent to $\frac{5}{9}$, the proportion is *false.*

(b) Write both fractions in lowest terms.

$$\frac{16}{12} = \frac{4}{3} \quad \text{and} \quad \frac{28}{21} = \frac{4}{3}$$

Both fractions are equivalent to $\frac{4}{3}$, so the proportion is *true.* ■

WORK PROBLEM 2 AT THE SIDE.

3 The ratios in a proportion are written as fractions, so another way to test whether they are equivalent is to use cross multiplication, as you did in **Section 2.4.**

OBJECTIVES

1. Write proportions.
2. Decide whether proportions are true or false.
3. Find cross products.

FOR EXTRA HELP

Tape 7 | SSM pp. 138–140 | MAC: A IBM: A

1. Write each proportion.

(a) \$7 is to 3 cans as \$28 is to 12 cans

(b) 9 meters is to 16 meters as 18 meters is to 32 meters

(c) 5 is to 7 as 35 is to 49

(d) 10 is to 30 as 60 is to 180

2. Are these proportions true or false?

(a) $\frac{6}{12} = \frac{15}{30}$

(b) $\frac{20}{24} = \frac{3}{4}$

(c) $\frac{25}{40} = \frac{30}{48}$

(d) $\frac{35}{45} = \frac{12}{18}$

(e) $\frac{21}{45} = \frac{56}{120}$

ANSWERS

1. (a) $\frac{\$7}{3 \text{ cans}} = \frac{\$28}{12 \text{ cans}}$ **(b)** $\frac{9}{16} = \frac{18}{32}$ **(c)** $\frac{5}{7} = \frac{35}{49}$ **(d)** $\frac{10}{30} = \frac{60}{180}$

2. (a) true **(b)** false **(c)** true **(d)** false **(e)** true

3. Cross multiply to see whether the following proportions are true or false.

(a) $\frac{5}{9} = \frac{10}{18}$

(b) $\frac{32}{15} = \frac{16}{8}$

(c) $\frac{10}{17} = \frac{20}{34}$

(d) $\frac{2.4}{6} = \frac{5}{12}$ $\quad 6 \cdot 5 =$ $\quad 2.4 \cdot 12 =$

(e) $\frac{3}{4.25} = \frac{24}{34}$

(f) $\frac{1\frac{1}{6}}{2\frac{1}{3}} = \frac{4}{8}$

ANSWERS
3. **(a)** true **(b)** false **(c)** true **(d)** false **(e)** true **(f)** true

DECIDING WHETHER A PROPORTION IS TRUE OR FALSE

To see whether a proportion is true, first multiply along one diagonal, then multiply along the other diagonal, as shown here.

$$\frac{2}{5} = \frac{4}{10} \qquad \begin{matrix} 5 \cdot 4 = \mathbf{20} \\ 2 \cdot 10 = \mathbf{20} \end{matrix} \quad \text{cross products}$$

In this case the **cross products** are both 20. When cross products are *equal,* the proportion is *true.* If the cross products are *unequal,* the proportion is *false.*

■ EXAMPLE 3 *Using Cross Products*

Use cross multiplication to see whether the following proportions are true or false.

(a) $\frac{3}{5} = \frac{12}{20}$

(b) $\frac{2\frac{1}{3}}{3\frac{1}{3}} = \frac{9}{16}$

(a) Cross multiply one way and then the other way.

$$\frac{3}{5} = \frac{12}{20} \qquad \begin{matrix} 5 \cdot 12 = \mathbf{60} \\ 3 \cdot 20 = \mathbf{60} \end{matrix} \quad \text{equal}$$

The cross products are equal, so the proportion is *true.*

(b) Cross multiply.

changed to improper fractions

$$\frac{2\frac{1}{3}}{3\frac{1}{3}} = \frac{9}{16} \qquad \begin{matrix} 3\frac{1}{3} \cdot 9 = \frac{10}{\cancel{3}_1} \cdot \frac{\cancel{9}^3}{1} = \frac{30}{1} = \mathbf{30} \\ 2\frac{1}{3} \cdot 16 = \frac{7}{3} \cdot \frac{16}{1} = \frac{112}{3} = \mathbf{37\frac{1}{3}} \end{matrix} \quad \text{unequal}$$

The cross products are unequal, so the proportion is *false.* ■

Note The numbers in a proportion do not have to be whole numbers.

WORK PROBLEM 3 AT THE SIDE.

NAME DATE HOUR

5.3 EXERCISES

Write each proportion.

1. \$9 is to 12 cans as \$18 is to 24 cans

$\frac{\$9}{12 \text{ cans}} = \frac{\$18}{24 \text{ cans}}$

2. 28 people is to 7 cars as 16 people is to 4 cars

$\frac{28 \text{ people}}{7 \text{ cars}} = \frac{16 \text{ people}}{4 \text{ cars}}$

3. 20 dogs is to 45 cats as 4 dogs is to 9 cats

$\frac{20 \text{ dogs}}{45 \text{ cats}} = \frac{4 \text{ dogs}}{9 \text{ cats}}$

4. 20 hits is to 70 at bats as 6 hits is to 21 at bats

$\frac{20 \text{ hits}}{70 \text{ at bats}} = \frac{6 \text{ hits}}{21 \text{ at bats}}$

5. 120 feet is to 150 feet as 8 feet is to 10 feet

$\frac{120}{150} = \frac{8}{10}$

6. \$6 is to \$9 as \$10 is to \$15

$\frac{6}{9} = \frac{10}{15}$

7. 2.2 hours is to 3.3 hours as 3.2 hours is to 4.8 hours

$\frac{2.2}{3.3} = \frac{3.2}{4.8}$

8. 4 meters is to 4.75 meters as 6 meters is to 7.125 meters

$\frac{4}{4.75} = \frac{6}{7.125}$

9. $1\frac{1}{2}$ is to 8 as 6 is to 32 $\frac{1\frac{1}{2}}{8} = \frac{6}{32}$

10. 6 is to $3\frac{1}{4}$ as 24 is to 13 $\frac{6}{3\frac{1}{4}} = \frac{24}{13}$

Write each ratio in lowest terms in order to decide whether the following proportions are true or false.

11. $\frac{6}{10} = \frac{3}{5}$ true

12. $\frac{1}{4} = \frac{9}{36}$ true

13. $\frac{5}{8} = \frac{25}{40}$ true

14. $\frac{2}{3} = \frac{20}{27}$ false

15. $\frac{8}{20} = \frac{20}{30}$ false

16. $\frac{15}{18} = \frac{21}{24}$ false

17. $\frac{42}{15} = \frac{28}{10}$ true

18. $\frac{18}{16} = \frac{36}{32}$ true

19. $\frac{32}{18} = \frac{48}{27}$ true

20. $\frac{15}{48} = \frac{10}{24}$ false

21. $\frac{7}{6} = \frac{54}{48}$ false

22. $\frac{28}{21} = \frac{44}{33}$ true

Use cross multiplication to decide whether the following proportions are true or false. Circle the correct answer.

Example:

$$\frac{10.2}{15.3} = \frac{4}{6}$$

True (circled) False

Solution:

$15.3 \cdot 4 = \mathbf{61.2}$

equal

$10.2 \cdot 6 = \mathbf{61.2}$

Cross products are equal, so proportion is true.

23. $\frac{2}{9} = \frac{6}{27}$

True (circled) False

24. $\frac{20}{25} = \frac{4}{5}$

True (circled) False

25. $\frac{20}{28} = \frac{12}{16}$

True False (circled)

26. $\frac{16}{40} = \frac{22}{55}$

True (circled) False

27. $\frac{110}{18} = \frac{160}{27}$

True False (circled)

28. $\frac{600}{420} = \frac{20}{14}$

True (circled) False

29. $\frac{3.5}{4} = \frac{7}{8}$

(True) False

30. $\frac{36}{23} = \frac{9}{5.75}$

(True) False

31. $\frac{18}{16} = \frac{2.8}{2.5}$

True (False)

32. $\frac{0.26}{0.39} = \frac{1.3}{1.9}$

True (False)

33. $\frac{6}{3\frac{2}{3}} = \frac{18}{11}$

(True) False

34. $\frac{16}{13} = \frac{2}{1\frac{5}{8}}$

(True) False

35. $\frac{2\frac{5}{8}}{3\frac{1}{4}} = \frac{21}{26}$

(True) False

36. $\frac{28}{17} = \frac{9\frac{1}{3}}{5\frac{2}{3}}$

(True) False

37. In 180 times at bat, Carol had 63 hits. Bathsheba had 49 hits in 140 times at bat. The coach says they hit equally well. Show how you could use a proportion and cross products to see if the coach is correct.

Answer varies.

38. Jay worked 3.5 hours and packed 91 cartons. Craig packed 126 cartons in 5.25 hours. To see if the men worked equally fast, Barry set up this proportion:

$$\frac{3.5}{91} = \frac{126}{5.25}$$

Explain what is wrong with Barry's proportion and write a correct one. Is the correct proportion true or false?

Answer varies.

Decide whether each proportion is true or false. Circle the correct answer

39. $\dfrac{\frac{2}{3}}{2} = \dfrac{2.7}{8}$

True (False)

40. $\dfrac{3.75}{1\frac{1}{4}} = \dfrac{7.5}{2\frac{1}{2}}$

(True) False

PREVIEW EXERCISES

Write each fraction in lowest terms. Write your answers as mixed numbers when possible. (For help, see ***Sections 2.2 and 2.4.****)*

41. $\frac{16}{24}$ $\frac{2}{3}$

42. $\frac{24}{40}$ $\frac{3}{5}$

43. $\frac{60}{48}$ $1\frac{1}{4}$

44. $\frac{20}{15}$ $1\frac{1}{3}$

45. $\frac{36}{63}$ $\frac{4}{7}$

46. $\frac{30}{48}$ $\frac{5}{8}$

47. $\frac{65}{10}$ $6\frac{1}{2}$

48. $\frac{38}{8}$ $4\frac{3}{4}$

5.4 SOLVING PROPORTIONS

OBJECTIVES

1. Find the missing number in a proportion.
2. Find the missing number in a proportion with mixed numbers or decimals.

FOR EXTRA HELP

Tape 7 | SSM pp. 140–142 | MAC: A IBM: A

1 Four numbers are used in a proportion. If any three of these numbers are known, the fourth can be found. For example, find the missing number that will make this proportion true.

$$\frac{3}{5} = \frac{x}{40}$$

The x represents the unknown number. Start by finding the cross products.

$$\frac{3}{5} = \frac{x}{40} \qquad 5 \cdot x \qquad 3 \cdot 40$$

The cross products in a true proportion are equal, so

$$5 \cdot x = 3 \cdot 40$$

$$5 \cdot x = 120$$

The equal sign says that $5 \cdot x$ and 120 are equivalent. If both $5 \cdot x$ and 120 are divided by 5, the results will still be equivalent.

$$\frac{5 \cdot x}{5} = \frac{120}{5} \qquad \text{Divide both sides by 5.}$$

Cancel 5 in numerator and denominator.

$$\frac{\overset{1}{\cancel{5}} \cdot x}{\underset{1}{\cancel{5}}} = 24 \qquad \text{Divide 120 by 5.}$$

Multiplying by 1 does not change a number, so $1 \cdot x$ is the same as x.

$$\frac{x}{1} = 24$$

Dividing by 1 does not change a number, so $\frac{x}{1}$ is the same as x.

$$x = 24$$

The missing number in the proportion is 24. The complete proportion is shown below.

$$\frac{3}{5} = \frac{24}{40} \leftarrow x \text{ is } 24.$$

Check by finding the cross products. If they are equal, you solved the problem correctly. If they are unequal, rework the problem.

$$\frac{3}{5} = \frac{24}{40} \qquad 5 \cdot 24 = \mathbf{120} \qquad 3 \cdot 40 = \mathbf{120}$$

equal; proportion is true

Solve a proportion for a missing number with the following steps.

FINDING A MISSING NUMBER IN A PROPORTION

Step 1 Find the cross products.
Step 2 Show that the cross products are equivalent.
Step 3 Divide both products by the number multiplied by x (the number next to x).

EXAMPLE 1 *Solving for a Missing Number*
Find the missing number in each proportion. Round to hundredths, if necessary.

(a) $\frac{16}{x} = \frac{32}{20}$

(b) $\frac{7}{12} = \frac{15}{x}$

(a) Recall that ratios can be rewritten in lowest terms. If desired, you can do that before finding the cross products. In this example, write $\frac{32}{20}$ in lowest terms ($\frac{8}{5}$) to get $\frac{16}{x} = \frac{8}{5}$.

Step 1 $\frac{16}{x} = \frac{8}{5}$ — $x \cdot 8$; $16 \cdot 5 = 80$ — Find cross products.

Step 2 $x \cdot 8 = 80$ ← Show that cross products are equivalent.

Step 3 $\frac{x \cdot \overset{1}{\cancel{8}}}{\underset{1}{\cancel{8}}} = \frac{80}{8}$ ← Divide both sides by 8.

$x = 10$ ← Find x. (No rounding necessary.)

Write the complete proportion and check by finding the cross products.

$\frac{16}{10} = \frac{8}{5}$ — $10 \cdot 8 = \mathbf{80}$; $16 \cdot 5 = \mathbf{80}$ — equal; proportion is true

Note It is not necessary to write the ratios in lowest terms before solving. However, if you do, you will have smaller numbers to work with.

(b) Find cross products.

$\frac{7}{12} = \frac{15}{x}$ — $12 \cdot 15 = 180$; $7 \cdot x$

Show that cross products are equivalent.

$$7 \cdot x = 180$$

Divide both sides by 7.

$$\frac{\overset{1}{\cancel{7}} \cdot x}{\underset{1}{\cancel{7}}} = \frac{180}{7}$$

$$x = 25.71 \text{ (rounded to nearest hundredth)}$$

When the division does not come out even, check for directions on how to round your answer. Divide out one more place, then round.

$$7\overline{)180.000} = 25.714$$ ← Divide out to thousandths. Round to hundredths.

Write the complete proportion and check by finding cross products.

$$\frac{7}{12} = \frac{15}{25.71}$$

$12 \cdot 15 = \mathbf{180}$

$7 \cdot 25.71 = \mathbf{179.97}$

very close but not equal

The cross products are slightly different because you rounded the value of x. However, they are close enough to see that the problem was done correctly. ■

WORK PROBLEM 1 AT THE SIDE.

1. Find the missing number. Round to hundredths, if necessary. Check your answer by finding cross products.

(a) $\frac{1}{2} = \frac{x}{12}$

(b) $\frac{6}{10} = \frac{15}{x}$

(c) $\frac{28}{x} = \frac{21}{9}$

(d) $\frac{x}{8} = \frac{3}{5}$

(e) $\frac{14}{11} = \frac{x}{3}$

2 The following examples show how the numbers in a proportion can be mixed numbers or decimals.

■ **EXAMPLE 2** *Using Mixed Numbers and Decimals*

Find the missing number in each proportion.

(a) $\frac{2\frac{1}{5}}{6} = \frac{x}{10}$

(b) $\frac{1.5}{0.6} = \frac{2}{x}$

(a) Cross multiply.

$$\frac{2\frac{1}{5}}{6} = \frac{x}{10}$$

$6 \cdot x$

$2\frac{1}{5} \cdot 10$

Find $2\frac{1}{5} \cdot 10$.

$$2\frac{1}{5} \cdot 10 = \frac{11}{\underset{1}{\cancel{5}}} \cdot \frac{\overset{2}{\cancel{10}}}{1} = \frac{22}{1} = 22$$

changed to improper fractions

ANSWERS

1. (a) $x = 6$ **(b)** $x = 25$ **(c)** $x = 12$ **(d)** $x = 4.8$ **(e)** $x = 3.82$ **(rounded to nearest hundredth)**

Show that the cross products are equivalent.

$$6 \cdot x = 22$$

Divide both sides by 6.

$$\frac{\overset{1}{\cancel{6}} \cdot x}{\underset{1}{\cancel{6}}} = \frac{22}{6}$$

Write answer as a mixed number in lowest terms.

$$x = \frac{22}{6} = \frac{11}{3} = 3\frac{2}{3}$$

(b) Show that cross products are equivalent.

$$1.5 \cdot x = 0.6 \cdot 2$$

$$1.5 \cdot x = 1.2$$

Divide both sides by 1.5.

$$\frac{\overset{1}{\cancel{1.5}} \cdot x}{\underset{1}{\cancel{1.5}}} = \frac{1.2}{1.5}$$

$$x = \frac{1.2}{1.5}$$

Complete the division. $1.5\overline{)1.20}$ = .8

So the missing number is 0.8. ■

◀ **WORK PROBLEM 2 AT THE SIDE.**

2. Find the missing numbers. Round to hundredths on the decimal problems, if necessary.

(a) $\dfrac{3\frac{1}{4}}{2} = \dfrac{x}{8}$

(b) $\dfrac{x}{3} = \dfrac{1\frac{2}{3}}{5}$

(c) $\dfrac{0.06}{x} = \dfrac{0.3}{0.4}$

(d) $\dfrac{2.2}{5} = \dfrac{13}{x}$

(e) $\dfrac{x}{6} = \dfrac{0.5}{1.2}$

(f) $\dfrac{0}{2} = \dfrac{x}{7.092}$

ANSWERS

2. (a) $x = 13$ (b) $x = 1$ (c) $x = 0.08$ (d) $x = 29.55$ (rounded to nearest hundredth) (e) $x = 2.5$ (f) $x = 0$

5.4 EXERCISES

Find the missing number in each proportion. Round your answers to hundredths, if necessary. Check your answers by finding cross products.

Example: $\frac{x}{5} = \frac{6}{15}$

Solution:

$x \cdot 15 = 5 \cdot 6$ Show cross products are equivalent.

$x \cdot 15 = 30$

$\frac{x \cdot \cancel{15}^{1}}{\cancel{15}_{1}} = \frac{30}{15}$ Divide both sides by 15.

$x = 2$

Check:

$\frac{2}{5} = \frac{6}{15}$

$5 \cdot 6 = \mathbf{30}$
equal
$2 \cdot 15 = \mathbf{30}$

Cross products are equal, so the proportion is true.

1. $\frac{1}{3} = \frac{x}{12}$ 4

2. $\frac{x}{6} = \frac{15}{18}$ 5

3. $\frac{15}{10} = \frac{3}{x}$ 2

4. $\frac{5}{x} = \frac{20}{8}$ 2

5. $\frac{x}{11} = \frac{32}{4}$ 88

6. $\frac{12}{9} = \frac{8}{x}$ 6

7. $\frac{10}{5} = \frac{x}{20}$ 40

8. $\frac{49}{x} = \frac{14}{18}$ 63

9. $\frac{x}{25} = \frac{4}{20}$ 5

10. $\frac{6}{x} = \frac{4}{8}$ 12

11. $\frac{8}{x} = \frac{24}{30}$ 10

12. $\frac{32}{5} = \frac{x}{10}$ 64

13. $\frac{99}{55} = \frac{44}{x}$ 24.44

14. $\frac{x}{12} = \frac{101}{147}$ 8.24

15. $\frac{0.7}{9.8} = \frac{3.6}{x}$ 50.4

16. $\frac{x}{3.6} = \frac{4.5}{6}$ 2.7

17. $\frac{250}{24.8} = \frac{x}{1.75}$ 17.64

18. $\frac{4.75}{17} = \frac{43}{x}$ 153.89

Writing Conceptual Challenging Estimation

These proportions are not *true. Change* any one *of the numbers in each proportion to make them true.*

19. $\frac{10}{4} = \frac{5}{3}$

$\frac{6.67}{4} = \frac{5}{3}$ or $\frac{10}{6} = \frac{5}{3}$ or $\frac{10}{4} = \frac{7.5}{3}$ or $\frac{10}{4} = \frac{5}{2}$

20. $\frac{6}{8} = \frac{24}{30}$

$\frac{6.4}{8} = \frac{24}{30}$ or $\frac{6}{7.5} = \frac{24}{30}$ or $\frac{6}{8} = \frac{22.5}{30}$ or $\frac{6}{8} = \frac{24}{32}$

Find the missing number in each proportion. Write your answers as mixed numbers when possible.

21. $\frac{15}{1\frac{2}{3}} = \frac{9}{x}$ 1

22. $\frac{x}{\frac{3}{10}} = \frac{2\frac{2}{9}}{1}$ $\frac{2}{3}$

23. $\frac{2\frac{1}{3}}{1\frac{1}{2}} = \frac{x}{2\frac{1}{4}}$ $3\frac{1}{2}$

24. $\frac{1\frac{5}{6}}{x} = \frac{\frac{3}{14}}{\frac{6}{7}}$ $7\frac{1}{3}$

PREVIEW EXERCISES

Write each set of rates as a proportion and use cross multiplication to decide whether it is true or false. Circle the correct answer. (For help, see ***Sections 5.2 and 5.3.****)*

25. 25 feet in 18 seconds
15 feet in 10 seconds

True (False)

$\frac{25 \text{ feet}}{18 \text{ sec}} = \frac{15 \text{ feet}}{10 \text{ sec}}$

26. 50 children to 70 adults
15 children to 21 adults

(True) False

$\frac{50 \text{ children}}{70 \text{ adults}} = \frac{15 \text{ children}}{21 \text{ adults}}$

27. 170 miles on 6.8 gallons
330 miles on 13.2 gallons

(True) False

$\frac{170 \text{ miles}}{6.8 \text{ gallons}} = \frac{330 \text{ miles}}{13.2 \text{ gallons}}$

28. $14.75 for 2 hours
$33.25 for 4.5 hours

True (False)

$\frac{\$14.75}{2 \text{ hours}} = \frac{\$33.25}{4.5 \text{ hours}}$

5.5 APPLICATIONS OF PROPORTIONS

1 Proportions can be used to solve a wide variety of problems. Watch for problems in which you are given a ratio or rate and then asked to find part of a corresponding ratio or rate. Remember that a ratio or rate compares two quantities and often includes one of these indicator words:

in for on per from to

When setting up the proportion, use a letter to represent the unknown number. We have used the letter x, but you may use any letter you like.

EXAMPLE 1 *Using a Proportion*

Mike's car can go 163 miles **on** 6.4 **gallons** of gas. How far can it go on a full tank of 14 **gallons** of gas? Round to the nearest whole mile.

Approach Decide what is being compared. This example compares miles to **gallons**. Write the two rates described in the example. Be sure that *both* rates compare miles to gallons in the same order. In other words, miles is in both numerators and gallons is in both denominators. Use a letter to represent the missing number.

compares miles to **gallons** $\left\{ \frac{163 \text{ miles}}{6.4 \text{ gallons}} \qquad \frac{x \text{ miles}}{14 \text{ gallons}} \right\}$ compares miles to **gallons**

Solution Both rates compare miles to **gallons**, so you can set them up as a proportion.

> *Note* Do **not** mix up the units in the rates.
>
> compares miles to **gallons** $\left\{ \frac{163 \text{ miles}}{6.4 \text{ gallons}} \right.$ $\left. \frac{14 \text{ gallons}}{x \text{ miles}} \right\}$ compares **gallons** to miles
>
> These rates do **not** compare things in the same order and **cannot** be set up as a proportion.

With the proportion set up correctly, solve for the missing number.

$$\frac{163 \text{ miles}}{6.4 \text{ gallons}} = \frac{x \text{ miles}}{14 \text{ gallons}}$$

(matching units — numerators; matching units — denominators)

Ignore the units while finding the cross products and dividing both sides by 6.4.

$6.4 \cdot x = 163 \cdot 14$ Show that cross products are equivalent.

$6.4 \cdot x = 2282$

$\frac{\overset{1}{\cancel{6.4}} \cdot x}{\underset{1}{\cancel{6.4}}} = \frac{2282}{6.4}$ Divide both sides by 6.4.

$x = 356.5625$

Rounded to the nearest mile, the car can go 357 *miles* on a full tank of gas. Be sure to *include the units* in your answer. ■

WORK PROBLEM 1 AT THE SIDE.

OBJECTIVE

1 Use proportions to solve word problems.

FOR EXTRA HELP

Tape 7

SSM pp. 142–145

MAC: A IBM: A

1. Set up and solve a proportion for each problem.

(a) If 2 pounds of fertilizer will cover 50 square feet of garden, how many pounds are needed for 225 square feet?

(b) A U.S. map has a scale of 1 inch to 75 miles. Lake Superior is 4.75 inches long on the map. What is the lake's actual length in miles?

(c) Cough syrup is to be given at the rate of 30 milliliters for each 100 pounds of body weight. How much should be given to a 34-pound child? Round to the nearest whole milliliter.

ANSWERS

1. (a) $\frac{2 \text{ pounds}}{50 \text{ sq feet}} = \frac{x \text{ pounds}}{225 \text{ sq feet}}$
$x = 9$ pounds

(b) $\frac{1 \text{ inch}}{75 \text{ miles}} = \frac{4.75 \text{ inches}}{x \text{ miles}}$
$x = 356.25$ miles

(c) $\frac{30 \text{ milliliters}}{100 \text{ pounds}} = \frac{x \text{ milliliters}}{34 \text{ pounds}}$
$x = 10$ milliliters (rounded)

2. (a) A survey showed that 2 out of 3 people would like to lose weight. At this rate, how many people in a group of 150 want to lose weight?

(b) In one state, 3 out of 5 college students receive financial aid. At this rate, how many of the 4500 students at Central Community College receive financial aid?

(c) An advertisement says that 9 out of 10 dentists recommend sugarless gum. If the ad is true, how many of the 60 dentists in our city would recommend sugarless gum?

ANSWERS
2. (a) **100 people**
(b) **2700 students**
(c) **54 dentists**

■ **EXAMPLE 2** *More Proportion Applications*

A newspaper report says that 7 out of 10 people surveyed watch the news on TV. At that rate, how many of the 3200 people in town would you expect to watch the news?

Approach You are comparing people who watch the news to people surveyed. Write the two rates descibed in the example. Be sure that both rates make the same comparison. "People who watch the news" is mentioned first, so it should be in the numerator of *both* ratios.

Solution Set up the two rates as a proportion and solve for the missing number.

$$\frac{7 \leftarrow \text{people who watch news}}{10 \leftarrow \text{total group (people surveyed)}} = \frac{x \leftarrow \text{people who watch news}}{3200 \leftarrow \text{total group (people in town)}}$$

$$10 \cdot x = 7 \cdot 3200 \quad \text{Show that cross products are equivalent.}$$

$$10 \cdot x = 22{,}400$$

$$\frac{\overset{1}{\cancel{10}} \cdot x}{\underset{1}{\cancel{10}}} = \frac{22{,}400}{10} \quad \text{Divide both sides by 10.}$$

$$x = 2240$$

You would expect 2240 people in town to watch the news on TV. ■

◀ **WORK PROBLEM 2 AT THE SIDE.**

EXERCISES

Set up and solve a proportion for each problem.

Example:
8 pounds of vegetables cost \$5. Find the cost of 20 pounds.

Solution:
You are comparing pounds to dollars. Set up a proportion.

$$\frac{8 \text{ pounds}}{5 \text{ dollars}} = \frac{20 \text{ pounds}}{x \text{ dollars}}$$

matching units

$$\frac{8}{5} = \frac{20}{x}$$ Ignore the units while solving for x.

$$8 \cdot x = 5 \cdot 20$$ Show that cross products are equivalent.

$$8 \cdot x = 100$$

$$\frac{\overset{1}{8} \cdot x}{\underset{1}{8}} = \frac{100}{8}$$ Divide both sides by 8.

$$x = 12.5$$

The cost is \$12.50.

1. It took five hours to load four trucks. How long will it take to load 18 trucks?

22.5 hours

2. Ms. Smith can put up 18 meters of fence in 12 hours. How long will it take her to put up 99 meters of fence?

66 hours

3. 6 magazines cost \$15. Find the cost of 14 magazines.

\$35

4. 22 ties cost \$176. Find the cost of 12 ties.

\$96

5. Five pounds of grass seed cover 3500 square feet of ground. How many pounds are needed for 4900 square feet?

7 pounds

6. Anna earns \$1242.08 in 14 days. How much does she earn in 260 days?

\$23,067.20

7. Tom makes $255.75 in 5 days. How much does he make in 3 days?

$153.45

8. If 5 ounces of a medicine must be mixed with 11 ounces of water, how many ounces of medicine would be mixed with 99 ounces of water?

45 ounces

Use the floor plan shown below to complete Exercises 9–10. On the plan, one inch represents four feet.

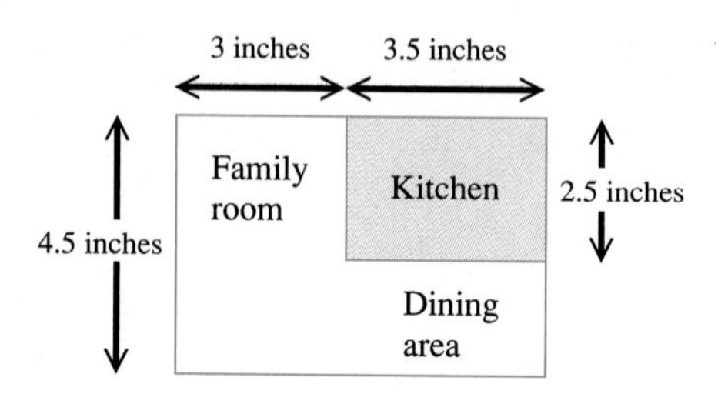

9. What is the actual length and width of the kitchen?

14 feet, 10 feet

10. What is the actual length and width of the family room?

18 feet, 12 feet

11. The Cardinals' pitcher gave up 78 runs in 234 innings. At that rate, how many runs will he give up in a 9-inning game?

3 runs

12. A quarterback completed 198 out of 318 passes last season. If he tries 30 passes in today's game, how many would you expect him to complete? Round to the nearest whole number of passes.

19 passes

13. Nearly 4 out of 5 people choose vanilla as their favorite ice cream flavor. If 238 people attend an ice cream social, how many would you expect to choose vanilla? Round to the nearest whole person.

190 people

14. In a test of 200 sewing machines, only one had a defect. At that rate, how many of the 5600 machines shipped from the factory have defects?

28 sewing machines

15. The tax on a $20 item is $1. Find the tax on a $110 item.

$5.50

16. A carpenter charges $150.50 to install a deck railing 10 feet long. How much would he charge to install a deck railing 18 feet long?

$270.90

17. The stock market report says that 5 stocks went up for every 6 stocks that went down. If 750 stocks went down yesterday, how many went up?

625 stocks

18. Raoul paid $15 for 14 cans of oil. How much would 8 cans cost? Round to the nearest cent.

$8.57

19. Terry's boat traveled 65 miles in 3 hours. At that rate, how long will it take her to travel 100 miles? Round to the nearest tenth.

4.6 hours

20. The human body contains 90 pounds of water for every 100 pounds of body weight. How many pounds of water are in a child who weighs 80 pounds?

72 pounds

21. The ratio of the length of an airplane wing to its width is 8 to 1. If the length of a wing is 32.5 meters, how wide must it be? Round to the nearest hundredth.

4.06 meters

22. The Rosebud School District wants a student-to-teacher ratio of 19 to 1. How many teachers are needed for 1850 students? Round to the nearest whole number.

97 teachers

23. At 3 P.M., Coretta's shadow is 1.05 meters long. Her height is 1.68 meters. At the same time, a tree's shadow is 6.58 meters long. How tall is the tree? Round to the nearest hundredth.

10.53 meters

24. Refer to Exercise 23. Later in the day, the same woman had a shadow that was 2.95 meters long. How long a shadow did the tree have at that time? Round to the nearest hundredth.

18.49 meters

25. Can you set up a proportion to solve this problem? Explain why or why not. Jim is 25 years old and weighs 180 pounds. How much will he weigh when he is 50 years old?

Answer varies.

26. Write your own word problem that can be solved by setting up a proportion. Also show the proportion and the steps needed to solve your problem.

Answer varies.

A box of instant mashed potatoes has the list of ingredients shown below. Use this information to find the amount of each ingredient you would need to make 15 servings.

Ingredient	*For 12 Servings*
Water	$3\frac{1}{2}$ cups
Margarine	6 tablespoons
Milk	$1\frac{1}{2}$ cups
Potato flakes	4 cups

27. Amount of water for 15 servings. $4\frac{3}{8}$ cups

28. Amount of milk for 15 servings. $1\frac{7}{8}$ cups

29. Amount of margarine for 15 servings. $7\frac{1}{2}$ tbsp

30. Amount of potato flakes for 15 servings. 5 cups

31. A survey of college students shows that 4 out of 5 drink coffee. Of the students who drink coffee, 1 out of 8 adds cream to it. How many of the 47,000 students at the University of Minnesota would be expected to use cream in their coffee?

4700 students

32. Nearly 9 out of 10 adults think it's a good idea to exercise regularly. But of the ones who think it is a good idea, only 1 in 6 actually exercise at least three times a week. At this rate, how many of the 300 employees in our company exercise regularly?

45 employees

Preview Exercises

Multiply or divide as indicated. (For help, see ***Sections 4.5 and 4.6.****)*

33. 0.06×100 6

34. 6.1×100 610

35. 2.87×1000 2870

36. $25.8 \div 100$ 0.258

37. $1.93 \div 100$ 0.0193

38. $5 \div 1000$ 0.005

CHAPTER 5 SUMMARY

KEY TERMS

5.1	**ratio**	A ratio compares two quantities. For example, the ratio of 6 apples to 11 apples is written as $\frac{6}{11}$.
5.2	**rate**	A rate compares two measurements with different types of units. Examples are 96 dollars for 8 hours or 450 miles on 18 gallons.
	unit rate	A unit rate has 1 in the denominator.
	cost per unit	Cost per unit is a rate that tells how much you pay for one item or one unit. The lowest cost per unit is the best buy.
5.3	**proportion**	A proportion states that two ratios or rates are equivalent.
	cross products	Cross multiply to get the cross products of a proportion. If the cross products are equal, the proportion is true.

QUICK REVIEW

Concepts | **Examples**

5.1 Ratios

Writing a Ratio
A ratio compares two quantities. Ratios are usually written as a fraction with the number that is mentioned first in the numerator. The units cancel.

Write this ratio as a fraction in lowest terms.

60 ounces of medicine to 160 ounces of medicine

$$\frac{60 \text{ ounces}}{160 \text{ ounces}} = \frac{60 \div 20}{160 \div 20} = \frac{3}{8}$$

↑ common units cancel

Using a Mixed Number in a Ratio
If a ratio has mixed numbers, change the mixed numbers to improper fractions. Rewrite the problem using the " ÷ " symbol for division. Finally, invert the divisor and multiply.

Write as a ratio of whole numbers in lowest terms.

$$2\frac{1}{2} \text{ to } 3\frac{3}{4}$$

$$\frac{2\frac{1}{2}}{3\frac{3}{4}} \quad \text{ratio in mixed numbers}$$

$$= \frac{\frac{5}{2}}{\frac{15}{4}} \quad \text{ratio in improper fractions}$$

$$= \frac{5}{2} \div \frac{15}{4} = \frac{5}{2} \cdot \frac{4}{15} \quad \text{Invert and multiply.}$$

$$= \frac{\overset{1}{\cancel{5}}}{\underset{1}{\cancel{2}}} \cdot \frac{\overset{2}{\cancel{4}}}{\underset{3}{\cancel{15}}} = \frac{2}{3} \quad \text{ratio in lowest terms}$$

Concepts	Examples
5.1 Ratios cont'd.	
Comparing Measurements When a ratio compares measurements, both measurements must be in the *same* units.	Write as a ratio in lowest terms. Compare in inches. 8 inches to 6 feet Because 1 foot has 12 inches, 6 feet is $6 \cdot 12 \text{ inches} = 72 \text{ inches}.$ The ratio is $\frac{8 \text{ inches}}{72 \text{ inches}} = \frac{8 \div 8}{72 \div 8} = \frac{1}{9}.$ ↑ common units cancel
5.2 Rates	
Writing a Rate A rate compares two measurements with different types of units. The units do not cancel, so you must write them as part of the rate.	Write the rate as a fraction in lowest terms. 475 miles in 10 hours $\frac{475 \text{ miles} \div 5}{10 \text{ hours} \div 5} = \frac{95 \text{ miles}}{2 \text{ hours}}$
Finding a Unit Rate A unit rate has 1 in the denominator. To find the unit rate, divide the numerator by the denominator. Write unit rates using the word "per" or a / mark.	Write as a unit rate. \$1278 in 9 days $\frac{\$1278}{9 \text{ days}}$ ← fraction bar indicates division $9\overline{)1278}$ = 142 so $\frac{\$1278}{9 \text{ days}} = \frac{\$142}{1 \text{ day}}$ Write answer as \$142 per day or \$142/day.
Finding the Best Buy The best buy is the item with the lowest cost per unit. Divide the price by the number of units. Round to thousandths. Then compare to find the lowest cost per unit.	Find the best buy on cheese. 2 pounds for \$2.25 3 pounds for \$3.40 Find cost per unit (pound). $\frac{\$2.25}{2} = \$1.125$ per pound $\frac{\$3.40}{3} = \$1.133$ per pound The lower cost per pound is \$1.125, so 2 pounds for \$2.25 is the better buy.

Concepts	Examples
5.3 Proportions	
A proportion states that two ratios or rates are equivalent. This proportion, $\frac{5}{6}=\frac{25}{30}$, is read as "5 is to 6 as 25 is to 30."	Write as a proportion. 8 is to 40 as 32 is to 160 $\frac{8}{40}=\frac{32}{160}$
To see whether a proportion is true or false, cross multiply one way, then cross multiply the other way. If the two products are equal, the proportion is true. If the two products are unequal, the proportion is false.	Cross multiply to see whether the following proportion is true or false. $\frac{6}{8\frac{1}{2}}=\frac{24}{34}$ Cross multiply. $8\frac{1}{2}\cdot 24=\frac{17}{\cancel{2}_1}\cdot\frac{\overset{12}{\cancel{24}}}{1}=\mathbf{204}$; $6\cdot 34=\mathbf{204}$ (equal). Cross products are equal, so the proportion is true.
5.4 Solving a Proportion	
Solve for a missing number by using these steps.	Find the value of the missing number. $\frac{12}{x}=\frac{6}{8}$; $\frac{12}{x}=\frac{3}{4}$ (lowest terms)
Step 1 Find the cross products. (If desired, you can rewrite the ratios in lowest terms before finding the cross products.)	*Step 1* $\frac{12}{x}=\frac{3}{4}$; $x\cdot 3$; $12\cdot 4=48$
Step 2 Show that the cross products are equivalent.	*Step 2* $x\cdot 3=48$ Show that cross products are equivalent.
Step 3 Divide both products by the number multiplied by x (the number next to x).	*Step 3* $\frac{x\cdot\overset{1}{\cancel{3}}}{\underset{1}{\cancel{3}}}=\frac{48}{3}$ Divide both sides by 3. $x=16$

Concepts	Examples
5.4 Solving a Proportion cont'd. Check your answer by writing the complete proportion and finding the cross products.	*Check* $\frac{12}{16} = \frac{6}{8}$ $\quad 16 \cdot 6 = \mathbf{96}$, $12 \cdot 8 = \mathbf{96}$ (equal) Cross products are equal, so the proportion is true.
5.5 Applications of Proportions Decide what is being compared, for example, pounds to square feet. Write the two rates described in the problem. Be sure that *both* rates compare things in the *same order*. Use a letter, like x, to represent the missing number. Set up a proportion. Check that the numerators have matching units and the denominators have matching units. Solve for the missing number.	If 3 pounds of grass seed cover 450 square feet of lawn, how much seed is needed for 1500 square feet of lawn? matching units $\frac{3 \text{ pounds}}{450 \text{ square feet}} = \frac{x \text{ pounds}}{1500 \text{ square feet}}$ matching units Both sides compare pounds to square feet. Ignore the units while finding cross products. $450 \cdot x = 3 \cdot 1500$ Show that cross products are equivalent. $450 \cdot x = 4500$ $\frac{\overset{1}{\cancel{450}} \cdot x}{\underset{1}{\cancel{450}}} = \frac{4500}{450}$ Divide both sides by 450. $x = 10$ 10 pounds of seed are needed.

NAME DATE HOUR

CHAPTER 5 REVIEW EXERCISES

[5.1] *Write each ratio as a fraction in lowest terms. Change to the same units when necessary, using the list of relationships in* ***Section 5.1.***

1. 3 oranges to 11 oranges $\frac{3}{11}$

2. 19 miles to 7 miles $\frac{19}{7}$

3. 9 doughnuts to 6 doughnuts $\frac{3}{2}$

4. 90 feet to 50 feet $\frac{9}{5}$

5. \$2.50 to \$1.25 $\frac{2}{1}$

6. \$0.30 to \$0.45 $\frac{2}{3}$

7. $2\frac{1}{2}$ yards to 10 yards $\frac{1}{4}$

8. $3\frac{1}{2}$ hours to $1\frac{1}{6}$ hours $\frac{3}{1}$

9. 5 hours to 100 minutes
Compare in minutes. $\frac{3}{1}$

10. 9 inches to 2 feet
Compare in inches. $\frac{3}{8}$

11. 20 hours to 3 days
Compare in hours. $\frac{5}{18}$

12. 3 pounds to 10 ounces
Compare in ounces. $\frac{24}{5}$

13. Jake sold \$350 worth of jewelry. Marcie sold \$500 worth of jewelry. What is the ratio of her sales to his?

$\frac{10}{7}$

14. Ms. Wei's new car gets 35 miles per gallon. Her old car got 25 miles per gallon. Find the ratio of the new car's mileage to the old car's mileage.

$\frac{7}{5}$

15. This fall, 60 students are taking math and 72 students are taking English. Find the ratio of math students to English students.

$\frac{5}{6}$

16. There are 9 players on a baseball team and 5 players on a basketball team. What is the ratio of basketball players to baseball players?

$\frac{5}{9}$

[5.2] *Write each rate as a fraction in lowest terms.*

17. \$100 for 20 hours $\frac{\$5}{\text{1 hour}}$

18. 290 miles in 6 hours $\frac{\text{145 miles}}{\text{3 hours}}$

19. Explain the difference between a ratio and a rate.

Answer varies.

20. Explain the term "unit rate." Give three examples of unit rates.

Answer varies.

Estimation

21. Patrick can type 4 pages in 20 minutes. Give his rate in pages per minute and minutes per page.

0.2 page/minute or $\frac{1}{5}$ page/minute

5 minutes/page

22. Elena made \$24 in 3 hours. Give her earnings in dollars per hour and hours per dollar.

\$8/hour

0.125 hour/dollar or $\frac{1}{8}$ hour/dollar

Find the best buy.

23. spice
16 ounces for \$2.80
8 ounces for \$1.45
3 ounces for \$1.15

16 oz for \$2.80

24. dog food
50 pounds for \$19.95
25 pounds for \$9.40
8 pounds for \$3.40

25 pounds for \$9.40

[5.3] *Write each proportion.*

25. 5 is to 10 as 20 is to 40.

$\frac{5}{10} = \frac{20}{40}$

26. 7 is to 2 as 35 is to 10.

$\frac{7}{2} = \frac{35}{10}$

27. $1\frac{1}{4}$ is to 5 as 3 is to 12.

$\frac{1\frac{1}{4}}{5} = \frac{3}{12}$

Use the method of writing in lowest terms or cross multiplication to decide whether the following proportions are true or false.

28. $\frac{6}{10} = \frac{9}{15}$

true

29. $\frac{16}{48} = \frac{9}{36}$

false

30. $\frac{47}{10} = \frac{98}{20}$

false

31. $\frac{64}{36} = \frac{96}{54}$

true

32. $\frac{1.5}{2.4} = \frac{2}{3.2}$

true

33. $\frac{3\frac{1}{2}}{2\frac{1}{3}} = \frac{6}{4}$

true

[5.4] *Find the missing number in each proportion. Round to hundredths, if necessary.*

34. $\frac{4}{42} = \frac{150}{x}$ 1575

35. $\frac{16}{x} = \frac{12}{15}$ 20

36. $\frac{100}{14} = \frac{x}{56}$ 400

37. $\frac{5}{8} = \frac{x}{20}$ 12.5

38. $\frac{x}{24} = \frac{11}{18}$ 14.67

39. $\frac{7}{x} = \frac{18}{21}$ 8.17

40. $\frac{x}{3.6} = \frac{9.8}{0.7}$ 50.4

41. $\frac{13.5}{1.7} = \frac{4.5}{x}$ 0.57

42. $\frac{0.82}{1.89} = \frac{x}{5.7}$ 2.47

[5.5] *Set up and solve a proportion for each problem.*

43. The ratio of cats to dogs at the animal shelter is 3 to 5. If there are 45 dogs, how many cats are there?

27 cats

44. Danielle had 8 hits in 28 times at bat during last week's games. If she continues to hit at the same rate, how many hits will she get in 161 times at bat?

46 hits

45. If 3.5 pounds of steak cost $13.79, what will 5.6 pounds cost? Round to the nearest cent.

$22.06

46. About 4 out of 10 students are expected to vote in campus elections. There are 8247 students. How many are expected to vote? Round to the nearest whole number.

3299 students

47. The scale on Brian's model railroad is 1 inch to 16 feet. One of the scale model boxcars is 4.25 inches long. What is the length of a real boxcar in feet?

68 feet

48. In the hospital pharmacy, Michiko sees that a certain medicine is to be given at the rate of 3.5 milligrams for every 50 pounds of body weight. How much medicine should be given to a patient who weighs 210 pounds?

14.7 milligrams

49. Damien earns $77 for 14 hours of part-time work at the convenience store. How long must he work to earn $121?

22 hours

50. Marvette makes ornaments to sell at a local gift shop. She made 3 dozen ornaments in $4\frac{1}{2}$ hours. How long will it take her to make 14 dozen ornaments?

21 hours

MIXED REVIEW EXERCISES

Find the missing number in each proportion. Round to hundredths, if necessary.

51. $\frac{x}{45} = \frac{70}{30}$ 105

52. $\frac{x}{52} = \frac{0}{20}$ 0

53. $\frac{64}{10} = \frac{x}{20}$ 128

54. $\frac{15}{x} = \frac{65}{100}$ 23.08

55. $\frac{7.8}{3.9} = \frac{13}{x}$ 6.5

56. $\frac{34.1}{x} = \frac{0.77}{2.65}$ 117.36

Use cross multiplication to decide whether the following proportions are true or false. Circle the correct answer.

57. $\frac{55}{18} = \frac{80}{27}$

True (False)

58. $\frac{5.6}{0.6} = \frac{18}{1.94}$

True (False)

59. $\frac{\frac{1}{5}}{2} = \frac{1\frac{1}{6}}{11\frac{2}{3}}$

(True) False

Write each ratio as a fraction in lowest terms. Change to the same units when necessary.

60. 4 dollars to 10 quarters
Compare in quarters.

$\frac{8}{5}$

61. $4\frac{1}{8}$ inches to 10 inches

$\frac{33}{80}$

62. 10 yards to 8 feet
Compare in feet.

$\frac{15}{4}$

63. \$3.60 to \$0.90 $\frac{4}{1}$

64. 12 eggs to 15 eggs $\frac{4}{5}$

65. 37 meters to 7 meters $\frac{37}{7}$

66. 3 pints to 4 quarts
Compare in pints.

$\frac{3}{8}$

67. 15 minutes to 3 hours
Compare in minutes.

$\frac{1}{12}$

68. $4\frac{1}{2}$ miles to $1\frac{3}{10}$ miles

$\frac{45}{13}$

69. Nearly 7 out of 8 fans buy something to drink at the ballpark. How many of the 28,500 fans at today's game would be expected to buy a beverage? Round to the nearest hundred fans.

24,900

70. Emily spent \$150 on car repairs and \$400 on car insurance. What is the ratio of amount spent on insurance to amount spent on repairs?

$\frac{8}{3}$

71. Antonio is choosing among three packages of plastic wrap. Is the best buy 25 feet for \$0.78; 75 feet for \$1.95; or 100 feet for \$2.79?

75 feet for \$1.95

72. On a scale model of a small town, 1 inch represents 25 feet. If a park on the model is 30.5 inches long, what is the actual length of the park in feet?

762.5 feet

73. An antibiotic is to be given at the rate of $1\frac{1}{2}$ teaspoons for every 24 pounds of body weight. How much should be given to an infant who weighs 8 pounds?

$\frac{1}{2}$ teaspoon or 0.5 teaspoon

74. Charles made 251 points during 169 minutes of playing time last year. If he plays 14 minutes in tonight's game, how many points would you expect him to make? Round to the nearest whole number.

21 points

75. Refer to Exercise 73. Explain each step you took in solving the problem. Be sure to tell how you decided which way to set up the proportion and how you checked your answer.

Answer varies.

76. A lawn mower uses 0.8 gallon of gas every 3 hours. The gas tank holds 2 gallons. How long can the mower run on a full tank?

7.5 hours or $7\frac{1}{2}$ hours

NAME DATE HOUR

CHAPTER 5 TEST

Write each rate or ratio as a fraction in lowest terms. Change to the same units when necessary.

1. 12 shirts to 18 shirts

1. $\frac{2}{3}$

2. 300 miles on 15 gallons

2. $\frac{20 \text{ miles}}{1 \text{ gallon}}$

3. $15 for 75 minutes

3. $\frac{\$1}{5 \text{ minutes}}$

4. The little theater has 320 seats. The auditorium has 1200 seats. Find the ratio of auditorium seats to theater seats.

4. $\frac{15}{4}$

5. 3 quarts to 60 gallons
Compare in quarts.

5. $\frac{1}{80}$

6. 3 hours to 40 minutes
Compare in minutes.

6. $\frac{9}{2}$

7. Find the best buy on cornflakes.
28 ounces for $3.15
18 ounces for $2.17
10 ounces for $1.49

7. 28 ounces for $3.15

8. Suppose the ratio of your income last year to your income this year is 3 to 2. Explain what this means. Give an example of the dollars earned last year and this year that fits the 3 to 2 ratio.

8. Answer varies.

Decide whether the following proportions are true or false.

9. $\frac{6}{14} = \frac{18}{45}$

10. $\frac{8.4}{2.8} = \frac{2.1}{0.7}$

9. False

10. True

11. 25

12. 2.67

13. 325

14. $10\frac{1}{2}$

15. 576 words

16. 6.4 hours

17. 87 students

18. Answer varies.

19. 23.8 grams

20. 60 feet

Find the missing number in each proportion. Round to hunredths, if necessary.

11. $\frac{5}{9} = \frac{x}{45}$

12. $\frac{3}{1} = \frac{8}{x}$

13. $\frac{x}{20} = \frac{6.5}{0.4}$

14. $\frac{2\frac{1}{3}}{x} = \frac{\frac{8}{9}}{4}$

Set up and solve a proportion for each problem.

15. Pedro types 240 words in 5 minutes. How many words can he type in 12 minutes?

16. A boat travels 75 miles in 4 hours. At that rate, how long will it take to travel 120 miles?

17. About 2 out of every 15 people are left-handed. How many of the 650 students in our school would you expect to be left-handed? Round to the nearest whole number.

18. A student set up the proportion for Exercise 17 this way and arrived at an answer of 4875.

$$\frac{2}{15} = \frac{650}{x} \qquad \text{Check: } \frac{2}{15} = \frac{650}{4875} \qquad \begin{array}{l} 15 \cdot 650 = 9750 \\ 2 \cdot 4875 = 9750 \end{array}$$

Because the cross products are equal, the student said the answer is correct. Is the student right? Explain why or why not.

19. A medication is given at the rate of 8.2 grams for every 50 pounds of body weight. How much should be given to a 145-pound person? Round to the nearest tenth.

20. On a scale model, 1 inch represents 8 feet. If a building in the model is 7.5 inches tall, what is the actual height in feet?

NAME DATE HOUR

CUMULATIVE REVIEW EXERCISES CHAPTERS 1-5

Name the digit that has the given place value.

1. 216,475,038
thousands 5
tens 3
millions 6

2. 340.6915
hundredths 9
ones 0
ten-thousandths 5

Round each number as indicated.

3. 16,952 to the nearest hundred 17,000

4. 0.0508 to the nearest hundredth 0.05

5. $79.45 to the nearest dollar $79

6. $2.5555 to the nearest cent $2.56

$\approx$ *Round the numbers in each problem so there is only one non-zero digit. Then add, subtract, multiply, or divide the rounded numbers, as indicated, to estimate the answer. Finally, solve for the exact answer.*

7. *estimate*
$$\begin{array}{r} 30 \\ 5000 \\ +\ 400 \\ \hline 5430 \end{array}$$
exact
$$\begin{array}{r} 28 \\ 5206 \\ +\ 351 \\ \hline 5585 \end{array}$$

8. *estimate*
$$\begin{array}{r} 60 \\ -\ 6 \\ \hline 54 \end{array}$$
exact
$$\begin{array}{r} 63.1 \\ -\ 5.692 \\ \hline 57.408 \end{array}$$

9. *estimate*
$$\begin{array}{r} 5000 \\ \times\ 800 \\ \hline 4{,}000{,}000 \end{array}$$
exact
$$\begin{array}{r} 4716 \\ \times\ 804 \\ \hline 3{,}791{,}664 \end{array}$$

10. *estimate*
$$\begin{array}{r} 1 \\ \times\ 18 \\ \hline 18 \end{array}$$
exact
$$\begin{array}{r} 0.982 \\ \times\ 17.8 \\ \hline 17.4796 \end{array}$$

11. *estimate* $50\overline{)50{,}000}$ = 1000

exact $53\overline{)48{,}071}$ = 907

12. *estimate* $5\overline{)2000}$ = 400

exact $4.5\overline{)1638}$ = 364

Add, subtract, multiply, or divide as indicated.

13. $1\frac{5}{6} \times 3\frac{3}{5}$

$6\frac{3}{5}$

14. $988 + 373{,}422 + 6$

374,416

15. $30 - 0.66$

29.34

16. $33\overline{)20{,}157}$

610 R 27

17. $\frac{7}{8} \div 5\frac{1}{4}$

$\frac{1}{6}$

18. $0.401 + 62.98 + 5$

68.381

19. $6\frac{3}{5} - \frac{2}{3}$

$5\frac{14}{15}$

20. (6392)(5609)

35,852,728

21. $1.39 \div 0.025$

55.6

22. $3020 - 708$

2312

23. $2\frac{9}{10} + 10\frac{1}{2}$

$13\frac{2}{5}$

24. (1.9)(0.004)

0.0076

Simplify by using the order of operations.

25. $36 + 18 \div 6$

39

26. $8 \div 4 + (10 - 3^2) \cdot 4^2$

18

27. $88 \div \sqrt{121} \cdot 2^3$

64

28. $(16.2 - 5.85) - 2.35 \cdot 4$

0.95

29. Write 0.0105 in words.

one hundred five ten-thousandths

30. Write sixty and seventy-one thousandths in numbers.

60.071

Write each fraction or mixed number as a decimal. Round to the nearest thousandth, if necessary.

31. $\frac{5}{16}$ 0.313

32. $4\frac{7}{9}$ 4.778

Arrange in order from smallest to largest.

33. 0.0711, 0.7, 0.707, 0.07

0.07, 0.0711, 0.7, 0.707

34. $\frac{3}{8}, \frac{7}{20}, 0.305, \frac{1}{3}$

$0.305, \frac{1}{3}, \frac{7}{20}, \frac{3}{8}$

Write each rate or ratio as a fraction in lowest terms. Change to the same units when necessary.

35. 20 cars to 5 cars $\frac{4}{1}$

36. $39 for 6 hours $\frac{\$13}{2 \text{ hours}}$

37. 20 minutes to 4 hours
Compare in minutes. $\frac{1}{12}$

38. 8 inches to 2 feet
Compare in inches. $\frac{1}{3}$

39. Ray is 25 years old. His father is 55 years old. Find the ratio of the father's age to Ray's age.

$\frac{11}{5}$

40. Find the best buy on instant mashed potatoes.
a box that makes 20 servings for $1.89
a box that makes 36 servings for $3.24
a box that makes 48 servings for $4.99

36 servings for $3.24

Find the missing number in each proportion. Round your answers to the nearest hundredth, if necessary.

41. $\frac{9}{12} = \frac{x}{28}$ 21

42. $\frac{7}{12} = \frac{10}{x}$ 17.14

43. $\frac{x}{\frac{3}{4}} = \frac{2\frac{1}{2}}{\frac{1}{6}}$ $11\frac{1}{4}$

44. $\frac{6.7}{x} = \frac{62.8}{9.15}$ 0.98

Solve each word problem.

45. The honor society has a goal of collecting 1500 pounds of food to fill Thanksgiving baskets. So far they've collected $\frac{5}{6}$ of their goal. How many more pounds do they need?

250 pounds

46. Tara has a photo that is 10 centimeters wide by 15 centimeters long. If the photo is enlarged to a length of 40 centimeters, find the new width, to the nearest tenth.

26.7 centimeters

Use the circle graph below of one college's enrollment to complete exercises 47–50.

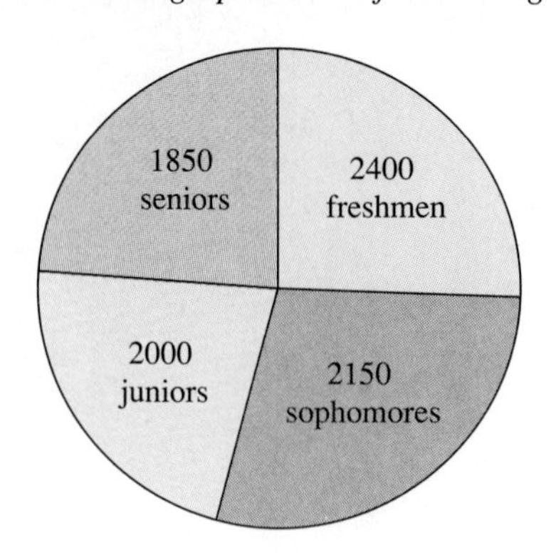

47. Find the total enrollment at the college.

8400 students

48. The college has budgeted \$186,400 for freshman orientation. How much is spent on each freshman, to the nearest dollar?

\$78

49. Write the ratio of freshmen to the total enrollment as a fraction in lowest terms.

$\frac{2}{7}$

50. The college collects a \$3.75 technology fee from each student to support the computer lab. What total amount is collected?

\$31,500

51. The distance around the running track is $\frac{3}{10}$ mile. Norma ran around the track $4\frac{2}{3}$ times in the morning and $3\frac{1}{2}$ times in the afternoon. How far did she run in all?

$2\frac{9}{20}$ miles

52. Rodney bought 49.8 gallons of gas for his truck while driving 896.5 miles on a vacation. How many miles per gallon did he get, rounded to the nearest tenth?

18.0 miles per gallon

53. In a survey, 5 out of 6 apartment residents said they are sometimes bothered by noise from their neighbors. How many of the 240 residents at Harris Towers would you expect to be bothered by noise?

200 residents

54. The directions on a can of plant food call for $\frac{1}{2}$ teaspoon in two quarts of water. How much plant food is needed for five quarts?

$1\frac{1}{4}$ teaspoons

Percent 6

$13 is 25% of $52

46% = 0.46

ratio

term 0.75 = 75%

6.1 BASICS OF PERCENT

Notice that the figure below has one hundred squares of equal size. Eleven of the squares are shaded. The shaded portion is $\frac{11}{100}$, or 0.11, of the total figure.

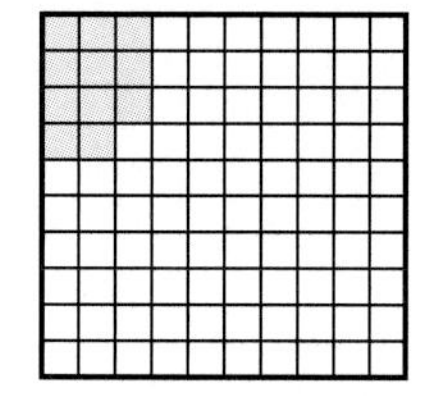

The shaded portion is also 11% of the total, or "eleven parts out of 100 parts." Read **11%** as "eleven percent."

1 As shown above, a percent is a ratio with a denominator of 100.

THE MEANING OF PERCENT

Percent means *per one hundred.* The "%" sign is used to show the number of parts per one hundred parts.

■ **EXAMPLE 1** *Understanding Percent*

(a) If 43 out of 100 students are men, then 43 per (out of) 100 or $\frac{43}{100}$ or **43%** of the students are men.

(b) If a person pays a tax of $7 per $100, then the tax rate is $7 per $100. The ratio is $\frac{7}{100}$ and the percent is 7%. ■

WORK PROBLEM 1 AT THE SIDE. ▶▶

OBJECTIVES

1. Learn the meaning of percent.
2. Write percents as decimals.
3. Write decimals as percents.

FOR EXTRA HELP

Tape	SSM	MAC: A
7	pp. 160–162	IBM: A

1. Write as percents.

(a) In a group of 100 people, 63 are unmarried.

(b) The tax is $14 per $100.

(c) Out of 100 students, 36 are attending school full time.

ANSWERS
1. (a) 63% (b) 14% (c) 36%

2. Write as a decimal.

(a) 53%

(b) 27%

(c) 79%

ANSWERS
2. (a) 0.53 (b) 0.27 (c) 0.79

2 If 8% means 8 parts out of 100 parts or $\frac{8}{100}$, then $p\%$ means p parts out of 100 parts or $\frac{p}{100}$. Because $\frac{p}{100}$ is another way to write the division $p \div 100$, we have

$$p\% = \frac{p}{100} = p \div 100$$

WRITE A PERCENT AS A DECIMAL

$$p\% = \underbrace{\frac{p}{100}}_{\text{as a fraction}} \quad or \quad p\% = p \div 100$$

■ **EXAMPLE 2** *Writing a Percent as a Decimal*
Write each percent as a decimal.

(a) 47% **(b)** 76%

Because $p\% = p \div 100$,

(a) $47\% = 47 \div 100 = 0.47$ decimal form

(b) $76\% = 76 \div 100 = 0.76$ decimal form ■

WORK PROBLEM 2 AT THE SIDE.

The resulting answers in Example 2 suggest the following rule for writing a percent as a decimal.

WRITING A PERCENT AS A DECIMAL

Step 1 Drop the percent sign.
Step 2 Divide by 100.

Note A quick way to divide a number by 100 is to move the decimal point two places to the left.

■ **EXAMPLE 3** *Changing to a Decimal by Moving the Decimal Point*
Write each percent as a decimal by moving the decimal point two places to the left.

(a) 17% **(b)** 83.4% **(c)** 4.9% **(d)** 0.6%

(a) 17.% Decimal point starts at far right side.
0.17 ← Percent sign is dropped. (Step 1)
Decimal point is moved two places to the left. (Step 2)

17% = 0.17

(b) 83.4% = 0.834

(c) 0.049 0 is attached so the decimal point can be moved two places to the left.

4.9% = 0.049

(d) 0.6% = 0.006 ■

> *Note* Look at Example 3(d) where 0.6% is less than 1%. Because 1% is equivalent to 0.01 or $\frac{1}{100}$, any fraction of a percent smaller than 1% is less than 0.01.

WORK PROBLEM 3 AT THE SIDE. ▶▶

3 You can write a decimal as a percent. For example, the decimal 0.78 is the same as the fraction

$$\frac{78}{100}.$$

This fraction means 78 of 100 parts, or 78%. The following steps give the same result.

WRITING A DECIMAL AS A PERCENT

Step 1 Multiply by 100.
Step 2 Attach a percent sign.

> *Note* A quick way to multiply a number by 100 is to move the decimal point two places to the right. Also notice that the rule for writing a decimal as a percent involves multiplying by 100, which is the opposite of writing a percent as a decimal (dividing by 100).

■ **EXAMPLE 4** *Changing to Percent by Moving the Decimal Point*
Write each decimal as a percent.

(a) 0.21 **(b)** 0.529 **(c)** 1.92 **(d)** 2.5 **(e)** 3

(a) 0.21% ← Percent sign is attached. (Step 1)
Decimal point is moved two places to the right. (Step 2)

0.21 = 21%
Decimal point is not written with whole number percents.

(b) 0.529 = 52.9%

(c) 1.92 = 192%

(d) 2.50% 0 is attached so the decimal point can be moved two places to the right.

2.5 = 250%

(e) 3. = 3.00% so 3 = 300% ■

> *Note* Look at Examples 4(c), 4(d) and 4(e) where 1.92, 2.5, and 3 are greater than 100%. Because the number 1 is equivalent to 100%, all whole numbers will be 100% or larger.

WORK PROBLEM 4 AT THE SIDE. ▶▶

3. Write each percent as a decimal.

(a) 88%

(b) 4%

(c) 21.6%

(d) 0.8%

4. Write as a percent

(a) 0.95

(b) 0.18

(c) 0.09

(d) 0.617

(e) 0.834

(f) 5.34

(g) 2.8

(h) 4

ANSWERS
3. (a) 0.88 (b) 0.04 (c) 0.216 (d) 0.008
4. (a) 95% (b) 18% (c) 9% (d) 61.7% (e) 83.4% (f) 534% (g) 280% (h) 400%

QUEST FOR NUMERACY

Percent Shortcuts

100%

100% means 100 parts out of 100 parts. That's *all* of the parts. If you pay 100% of a $45 dentist bill, you pay $45 (*all* of it). Try these problems.

100% of $34 is $34

100% of $3.95 is $3.95

100% of 4 cats is 4 cats

100% of 3000 students is 3000 students

100% of 95 people is 95 people

100% of $8\frac{1}{2}$ hours is $8\frac{1}{2}$ hours

There are 24 children in the daycare center. 100% of the children received flu shots. How many children got flu shots? 24 children

What is the "shortcut" way to find 100% of some number? Use the entire number.

50%

50% means 50 out of 100 parts, which is *half* of the parts. 50% of $10 is $5 (*half* of the money). Find a quick way to solve these problems.

50% of $20 is $10

50% of 36 cookies is 18 cookies

50% of 280 miles is 140 miles

50% of 6000 women is 3000 women

John owes $285 for tuition. Financial aid will pay 50% of the cost. How much will financial aid pay? $142.50

The Animal Humane Society took in 20,000 homeless animals last year. About 50% of them were dogs. How many dogs did they take in? 10,000 dogs

What is a "shortcut" way to find 50% of some number? Divide by 2.

25%

25% means 25 parts out of 100 parts. That's $\frac{1}{4}$ of the parts. If you give your friend 25% of $8, you give $2 ($\frac{1}{4}$ of the money). Find a quick way to solve these problems.

25% of $40 is $10

25% of 200 pounds is 50 pounds

25% of 4 cats is 1 cat

25% of 2400 books is 600 books

25% of 108 credits is 27 credits

25% of $6 is $1.50

LaRae owes $372 for tuition and books. She paid 25% of the cost herself. How much money did LaRae pay? $93

What is the "shortcut" way to find 25% of some number? Divide by 4.

Use your shortcuts to solve these problems.

50% of 800 people is 400 people

100% of $5 is $5

100% of $4436.75 is $4436.75

50% of $5 is $2.50

50% of 3200 college students is 1600 students

25% of $5 is $1.25

25% of 3200 college students is 800 students

50% of 150 pigs is 75 pigs

EXERCISES

Write each percent as a decimal.

Examples: 42% 1.4%

Solutions: 0.42 ← Percent sign is dropped. 0.01.4
Decimal point is moved two places to the left. — two places left

42% = **0.42** 1.4% = **0.014**

1. 50% 0.50 or 0.5

2. 70% 0.70 or 0.7

3. 18% 0.18

4. 57% 0.57

5. 45% 0.45

6. 83% 0.83

7. 140% 1.40 or 1.4

8. 250% 2.50 or 2.5

9. 7.8% 0.078

10. 6.7% 0.067

11. 100% 1.00 or 1

12. 600% 6.00 or 6

13. 0.5% 0.005

14. 0.2% 0.002

15. 0.35% 0.0035

16. 0.076% 0.00076

Write each decimal as a percent.

Examples: 0.23 | 8.6

Solutions: 0.23 % ← Percent sign is attached. | 8.60%

Decimal point is moved two places to the right.

0.23 = **23%** | 8.6 = **860%**

17. 0.6 60%

18. 0.7 70%

19. 0.91 91%

20. 0.31 31%

21. 0.07 7%

22. 0.06 6%

23. 0.125 12.5%

24. 0.875 87.5%

25. 0.629 62.9%

26. 0.494 49.4%

27. 2 200%

28. 5 500%

29. 2.6 260%

30. 1.8 180%

31. 0.0312 3.12%

32. 0.0625 6.25%

33. 4.162 416.2%

34. 8.715 871.5%

35. 0.0075 0.75%

36. 0.0013 0.13%

37. Fractions, decimals, and percents are all used to describe a part of something. The use of percent is much more common than fractions and decimals. Why do you suppose this is true?

Answer varies.

38. List five uses of percent that are or will be part of your life. Consider the activities of working, shopping, saving, and planning for the future.

Answer varies.

In each of the following, write percents as decimals and decimals as percents.

39. In Roseville, the property tax rate is 7%. 0.07

40. At Hinds Valley College, 23% of the students use child day care. 0.23

41. In one company, 65% of the salespeople are women. 0.65

42. Only 38.6% of those registered actually voted. 0.386

43. The sales tax rate in one parish (county) is 0.086. 8.6%

44. A church building fund has 0.49 of the money needed. 49%

45. The success rate in CPR training this session is 2 times that of the last session. 200%

46. Attendance at the picnic this year is 3 times last year's attendance. 300%

47. Only 0.005 of the total population has this genetic defect. 0.5%

48. Defects in cellular phone production remain at 0.0073 of total production. 0.73%

49. The patient's blood pressure was 153.6% of normal. 1.536

50. Success with the diet was 248.7% greater than anticipated. 2.487

Write a percent for both the shaded and unshaded part of each figure.

51.

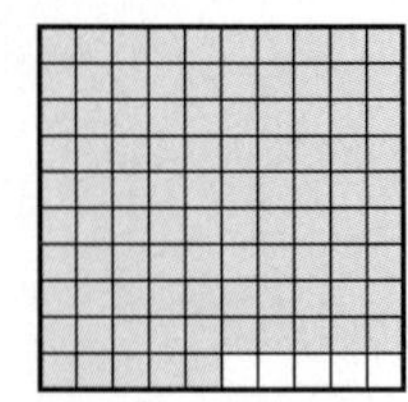

95%; 5%

52.

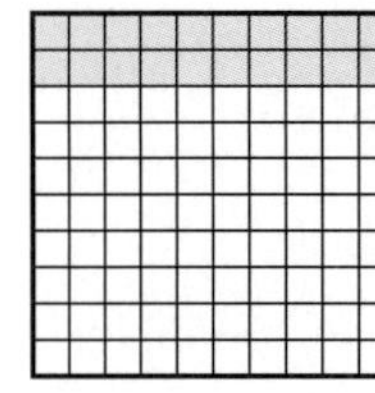

20%; 80%

53.

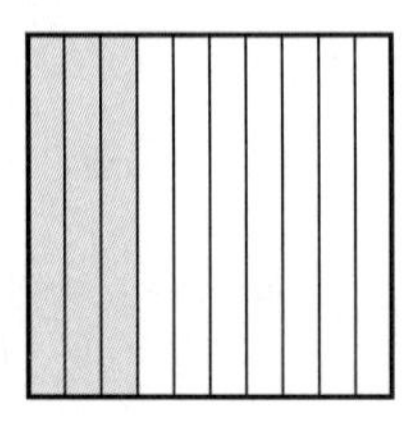

30%; 70%

54.

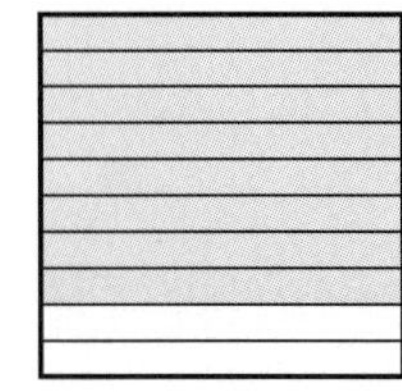

80%; 20%

55.

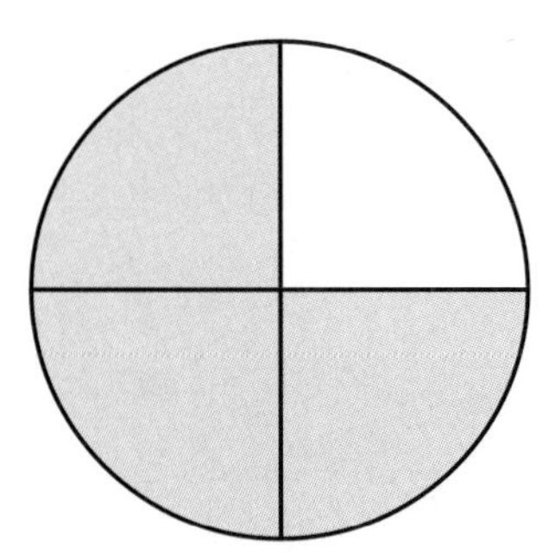

75%; 25%

56.

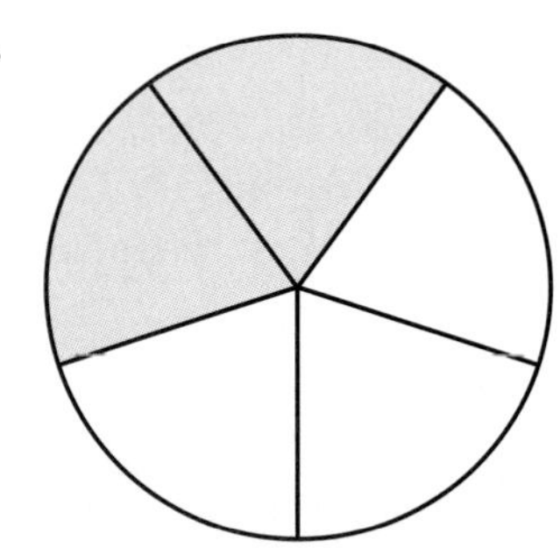

40%; 60%

57.

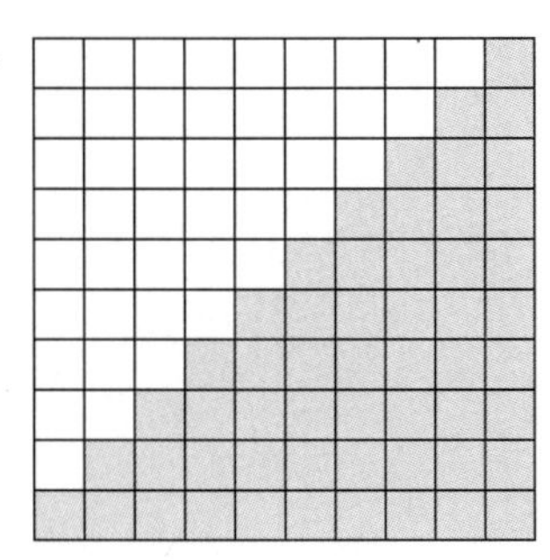

55%; 45%

58.

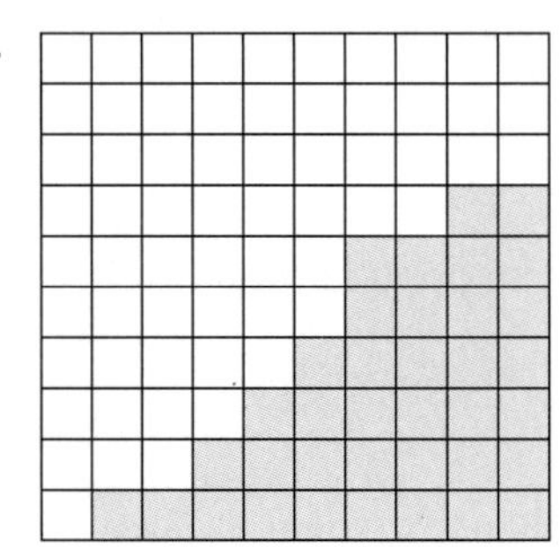

37%; 63%

PREVIEW EXERCISES

Change each of the following fractions to decimals. (For help, see **Section 4.7.***)*

59. $\frac{1}{2}$ 0.5

60. $\frac{1}{5}$ 0.2

61. $\frac{3}{4}$ 0.75

62. $\frac{1}{4}$ 0.25

63. $\frac{7}{8}$ 0.875

64. $\frac{5}{8}$ 0.625

65. $\frac{4}{5}$ 0.8

66. $\frac{7}{10}$ 0.7

6.2 PERCENTS AND FRACTIONS

1 Percents can be written as fractions by using what we learned in the previous section.

WRITE A PERCENT AS A FRACTION

$$p\% = \frac{p}{100}, \quad \text{as a fraction.}$$

■ **EXAMPLE 1** *Writing a Percent as a Fraction*
Write each percent as a fraction or mixed number in lowest terms.

(a) 45%

(b) 76%

(c) 150%

(a) As we saw in the last section, 45% can be written as a decimal.

$$45\% = 45 \div 100 = 0.45 \quad \text{(percent sign dropped)}$$

Because 0.45 means 45 hundredths,

$$0.45 = \frac{45}{100} = \frac{9}{20}. \quad \text{(lowest terms)}$$

It is not necessary, however, to write 45% as a decimal first. Just write

$$45\% = \frac{45}{100} \quad \text{(45 per 100)}$$

$$= \frac{9}{20} \quad \text{(lowest terms)}$$

(b) Write 76% as $\frac{76}{100}$.

The percent becomes the numerator.

The *denominator* is always 100 because percent means parts per 100.

Write $\frac{76}{100}$ in lowest terms.

$$\frac{76}{100} = \frac{19}{25} \quad \text{(lowest terms)}$$

(c) $150\% = \frac{150}{100} = \frac{3}{2} = 1\frac{1}{2}$ (mixed number) ■

Note Remember that percent means **per 100.**

WORK PROBLEM 1 AT THE SIDE.

OBJECTIVES

1. Write percents as fractions.
2. Write fractions as percents.
3. Use the table of percent equivalents.

FOR EXTRA HELP

Tape 8

SSM pp. 162–165

MAC: A IBM: A

1. Write each percent as a fraction or mixed number in lowest terms.

(a) 17%

(b) 22%

(c) 68%

(d) 25%

(e) 120%

(f) 210%

ANSWERS

1. (a) $\frac{17}{100}$ (b) $\frac{11}{50}$ (c) $\frac{17}{25}$ (d) $\frac{1}{4}$ (e) $1\frac{1}{5}$ (f) $2\frac{1}{10}$

2. Write as fractions in lowest terms.

(a) 18.5%

(b) 87.5%

(c) 8.5%

(d) $66\frac{2}{3}\%$

(e) $12\frac{1}{3}\%$

(f) $62\frac{1}{2}\%$

ANSWERS

2. (a) $\frac{37}{200}$ (b) $\frac{7}{8}$ (c) $\frac{17}{200}$ (d) $\frac{2}{3}$ (e) $\frac{37}{300}$ (f) $\frac{5}{8}$

The next example shows how to write decimal and fraction percents as fractions.

EXAMPLE 2 *Writing a Decimal or Fraction Percent as a Fraction*
Write each percent as a fraction in lowest terms.

(a) 15.5% **(b)** $33\frac{1}{3}\%$

(a) Place 15.5 over 100.

$$15.5\% = \frac{15.5}{100}$$

Multiply numerator and denominator by 10. This will result in a whole number in the numerator. (Multiplying by $\frac{10}{10}$ is the same as multiplying by 1.)

$$\frac{15.5}{100} = \frac{15.5 \cdot 10}{100 \cdot 10} = \frac{155}{1000}$$

Write in lowest terms.

$$\frac{155}{1000} = \frac{31}{200}$$

(b) Place $33\frac{1}{3}$ over 100.

$$33\frac{1}{3}\% = \frac{33\frac{1}{3}}{100}$$

Write $33\frac{1}{3}$ as the improper fraction $\frac{100}{3}$. Now,

$$\frac{33\frac{1}{3}}{100} = \frac{\frac{100}{3}}{100}.$$

Rewrite the division problem in a horizontal form. Then invert the divisor and multiply.

$$\frac{\frac{100}{3}}{100} = \frac{100}{3} \div 100 = \frac{100}{3} \div \frac{100}{1} = \frac{\overset{1}{\cancel{100}}}{3} \cdot \frac{1}{\underset{1}{\cancel{100}}} = \frac{1}{3}$$

(invert and multiply) ■

Note In Example 2 (a) we could have changed 15.5% to $15\frac{1}{2}\%$ and then written it as the improper fraction $\frac{31}{2}$ over 100 as was done in part (b). But it is usually best to leave decimal percents as they are.

WORK PROBLEM 2 AT THE SIDE.

2 We may use the result given at the beginning of this section to write fractions as percents.

$$p\% = \frac{p}{100}$$

EXAMPLE 3 *Writing a Fraction as a Percent*
Write each fraction as a percent. Round to the nearest tenth if necessary.

(a) $\frac{3}{5}$ **(b)** $\frac{7}{8}$ **(c)** $\frac{5}{6}$

(a) Write this fraction as a percent by solving for p in the proportion

$$\frac{3}{5} = \frac{p}{100}.$$

Find cross products.

$$5 \cdot p = 3 \cdot 100$$
$$5 \cdot p = 300$$

Divide both sides by 5.

$$\frac{\overset{1}{\cancel{5}} \cdot p}{\underset{1}{\cancel{5}}} = \frac{300}{5}$$
$$p = 60$$

This result means that $\frac{3}{5} = \frac{60}{100}$ or 60%.

(b) Write a proportion.

$$\frac{7}{8} = \frac{p}{100}$$

$$8 \cdot p = 7 \cdot 100 \quad \text{Cross multiply.}$$
$$8 \cdot p = 700$$
$$\frac{\overset{1}{\cancel{8}} \cdot p}{\underset{1}{\cancel{8}}} = \frac{700}{8} \quad \text{Divide both sides by 8.}$$
$$p = 87.5$$

Finally, $\frac{7}{8} = 87.5\%$.

Note If you think of $\frac{700}{8}$ as an improper fraction, changing it to a mixed number gives an answer of $87\frac{1}{2}$. So $\frac{7}{8} = 87.5\%$ or $87\frac{1}{2}\%$.

3. Write as percents. Round to the nearest tenth if necessary.

(a) $\frac{2}{5}$

(b) $\frac{7}{25}$

(c) $\frac{7}{10}$

(d) $\frac{3}{8}$

(e) $\frac{1}{6}$

(f) $\frac{2}{9}$

ANSWERS
3. (a) 40% (b) 28% (c) 70% (d) 37.5% (e) 16.7% (f) 22.2%

(c) Start with a proportion.

$$\frac{5}{6} = \frac{p}{100}$$

$$6 \cdot p = 5 \cdot 100 \quad \text{Cross multiply.}$$

$$6 \cdot p = 500$$

$$\frac{\overset{1}{\cancel{6}} \cdot p}{\underset{1}{\cancel{6}}} = \frac{500}{6} \quad \text{Divide.}$$

$$p = 83.3 \quad \text{rounded to the nearest tenth}$$

Solving this proportion shows

$$\frac{5}{6} = 83.3\% \quad \text{rounded.} \ ■$$

WORK PROBLEM 3 AT THE SIDE.

3 The table on the next two pages shows common fractions and mixed numbers and their decimal and percent equivalents.

■ **EXAMPLE 4** *Using a Conversion Table*

Read the following from the chart.

(a) $\frac{1}{12}$ as a percent

(b) 0.375 as a fraction

(c) $\frac{13}{16}$ as a percent

(a) Find $\frac{1}{12}$ in the "fraction" column. The percent is 8.33% or $8\frac{1}{3}\%$.

(b) Look in the "decimal" column for 0.375. The fraction is $\frac{3}{8}$.

(c) Find $\frac{13}{16}$ in the "fraction" column. The percent is 81.25% or $81\frac{1}{4}\%$. ■

Note Remember that the bar over a digit shows that the digit is repeating. In $0.08\overline{3}$, the **3** is repeating; in $0.16\overline{6}$, the **6** is repeating. When a repeating decimal is written as a percent, round the percent to the nearest tenth or hundredth.

WORK PROBLEM 4 AT THE SIDE.

PERCENT, DECIMAL, AND FRACTION EQUIVALENTS

Percent (rounded to hundredths when necessary)	*Decimal*	*Fraction*
1%	0.01	$\frac{1}{100}$
5%	0.05	$\frac{1}{20}$
6.25% or $6\frac{1}{4}\%$	0.0625	$\frac{1}{16}$
8.33% or $8\frac{1}{3}\%$	$0.08\overline{3}$	$\frac{1}{12}$
10%	0.1	$\frac{1}{10}$
12.5% or $12\frac{1}{2}\%$	0.125	$\frac{1}{8}$
16.67% or $16\frac{2}{3}\%$	$0.16\overline{6}$	$\frac{1}{6}$
18.75% or $18\frac{3}{4}\%$	0.1875	$\frac{3}{16}$
20%	0.2	$\frac{1}{5}$
25%	0.25	$\frac{1}{4}$
30%	0.3	$\frac{3}{10}$
31.25% or $31\frac{1}{4}\%$	0.3125	$\frac{5}{16}$
33.33% or $33\frac{1}{3}\%$	$0.33\overline{3}$	$\frac{1}{3}$
37.5% or $37\frac{1}{2}\%$	0.375	$\frac{3}{8}$
40%	0.4	$\frac{2}{5}$
43.75% or $43\frac{3}{4}\%$	0.4375	$\frac{7}{16}$
50%	0.5	$\frac{1}{2}$
56.25% or $56\frac{1}{4}\%$	0.5625	$\frac{9}{16}$
60%	0.6	$\frac{3}{5}$

4. Read the following common fractions, decimals, and percents from the table.

(a) $\frac{1}{2}$ as a percent

(b) $12\frac{1}{2}\%$ as a fraction

(c) $0.66\overline{6}$ as a fraction

(d) 20% as a fraction

(e) $\frac{7}{8}$ as a percent

(f) $\frac{1}{4}$ as a percent

(g) $33\frac{1}{3}\%$ as a fraction

(h) $\frac{4}{5}$ as a percent

ANSWERS

4. (a) 50% (b) $\frac{1}{8}$ (c) $\frac{2}{3}$ (d) $\frac{1}{5}$ (e) 87.5% (f) 25% (g) $\frac{1}{3}$ (h) 80%

Percent (rounded to hundredths when necessary)	*Decimal*	*Fraction*
62.5% or $62\frac{1}{2}\%$	0.625	$\frac{5}{8}$
66.67% or $66\frac{2}{3}\%$	$0.66\overline{6}$	$\frac{2}{3}$
68.75% or $68\frac{3}{4}\%$	0.6875	$\frac{11}{16}$
70%	0.7	$\frac{7}{10}$
75%	0.75	$\frac{3}{4}$
80%	0.8	$\frac{4}{5}$
81.25% or $81\frac{1}{4}\%$	0.8125	$\frac{13}{16}$
83.33% or $83\frac{1}{3}\%$	$0.83\overline{3}$	$\frac{5}{6}$
87.5% or $87\frac{1}{2}\%$	0.875	$\frac{7}{8}$
90%	0.9	$\frac{9}{10}$
93.75% or $93\frac{3}{4}\%$	0.9375	$\frac{15}{16}$
100%	1.0	1
110%	1.1	$1\frac{1}{10}$
125%	1.25	$1\frac{1}{4}$
133.33% or $133\frac{1}{3}\%$	$1.33\overline{3}$	$1\frac{1}{3}$
150%	1.5	$1\frac{1}{2}$
166.67% or $166\frac{2}{3}\%$	$1.66\overline{6}$	$1\frac{2}{3}$
175%	1.75	$1\frac{3}{4}$
200%	2.0	2

6.2 NAME DATE HOUR

EXERCISES

Write each percent as a fraction or mixed number in lowest terms.

Example: 20%

Solution: $20\% = \frac{20}{100} = \mathbf{\frac{1}{5}}$ ← in lowest terms

Denominator is always 100.

1. 25% $\frac{1}{4}$

2. 50% $\frac{1}{2}$

3. 10% $\frac{1}{10}$

4. 75% $\frac{3}{4}$

5. 85% $\frac{17}{20}$

6. 15% $\frac{3}{20}$

7. 37.5% $\frac{3}{8}$

8. 87.5% $\frac{7}{8}$

9. 6.25% $\frac{1}{16}$

10. 43.75% $\frac{7}{16}$

11. $16\frac{2}{3}\%$ $\frac{1}{6}$

12. $83\frac{1}{3}\%$ $\frac{5}{6}$

13. $6\frac{2}{3}\%$ $\frac{1}{15}$

14. $46\frac{2}{3}\%$ $\frac{7}{15}$

15. 0.4% $\frac{1}{250}$

16. 0.9% $\frac{9}{1000}$

17. 130% $1\frac{3}{10}$

18. 175% $1\frac{3}{4}$

19. 250% $2\frac{1}{2}$

20. 325% $3\frac{1}{4}$

Write each fraction as a percent. Round percents to the nearest tenth if necessary.

Examples: $\frac{1}{4}$ $\qquad$ $\frac{3}{7}$

Solutions:

$$\frac{1}{4} = \frac{p}{100} \qquad \frac{3}{7} = \frac{p}{100}$$

$$4 \cdot p = 1 \cdot 100 \qquad 7 \cdot p = 3 \cdot 100$$

$$4 \cdot p = 100 \qquad 7 \cdot p = 300$$

$$\frac{\overset{1}{\cancel{4}} \cdot p}{\underset{1}{\cancel{4}}} = \frac{100}{4} \qquad \frac{\overset{1}{\cancel{7}} \cdot p}{\underset{1}{\cancel{7}}} = \frac{300}{7}$$

$$p = 25 \qquad p = 42.9 \quad \text{(rounded)}$$

$$\frac{1}{4} = \mathbf{25\%} \qquad \frac{3}{7} = \mathbf{42.9\%} \quad \text{(rounded)}$$

21. $\frac{1}{4}$ 25%

22. $\frac{1}{2}$ 50%

23. $\frac{3}{10}$ 30%

24. $\frac{9}{10}$ 90%

25. $\frac{2}{5}$ 40%

26. $\frac{3}{5}$ 60%

27. $\frac{37}{100}$ 37%

28. $\frac{63}{100}$ 63%

29. $\frac{5}{8}$ 62.5%

30. $\frac{1}{8}$ 12.5%

31. $\frac{7}{8}$ 87.5%

32. $\frac{3}{8}$ 37.5%

33. $\frac{13}{25}$ 52%

34. $\frac{7}{25}$ 28%

35. $\frac{29}{50}$ 58%

36. $\frac{17}{20}$ 85%

37. $\frac{1}{20}$ 5%

38. $\frac{1}{50}$ 2%

39. $\frac{5}{6}$ 83.3%

40. $\frac{1}{6}$ 16.7%

41. $\frac{5}{9}$ 55.6%

42. $\frac{7}{9}$ 77.8%

43. $\frac{1}{7}$ 14.3%

44. $\frac{5}{7}$ 71.4%

Complete this chart. Round decimals to the nearest thousandth and percents to the nearest tenth of a percent.

Example:

fraction	*decimal*	*percent*
$\frac{3}{50}$	**0.06**	**6%**

	fraction	*decimal*	*percent*
45.	$\frac{1}{10}$	0.1	10%
46.	$\frac{3}{10}$	0.3	30%
47.	$\frac{1}{8}$	0.125	12.5%
48.	$\frac{3}{8}$	0.375	37.5%
49.	$\frac{1}{5}$	0.2	20%
50.	$\frac{3}{5}$	0.6	60%
51.	$\frac{1}{6}$	0.167	16.7%
52.	$\frac{1}{3}$	0.333	33.3%
53.	$\frac{1}{4}$	0.25	25%
54.	$\frac{1}{2}$	0.5	50%
55.	$\frac{7}{8}$	0.875	87.5%
56.	$\frac{5}{8}$	0.625	62.5%

	fraction	*decimal*	*percent*
57.	$\frac{2}{3}$	0.667	66.7%
58.	$\frac{5}{6}$	0.833	83.3%
59.	$\frac{3}{4}$	0.75	75%
60.	$\frac{4}{5}$	0.8	80%
61.	$\frac{1}{100}$	0.01	1%
62.	1	1.0 or 1	100%
63.	$\frac{1}{200}$	0.005	0.5%
64.	$\frac{7}{500}$	0.014	1.4%
65.	$2\frac{1}{2}$	2.5	250%
66.	$1\frac{7}{10}$	1.7	170%
67.	$3\frac{1}{4}$	3.25	325%
68.	$2\frac{4}{5}$	2.8	280%

69. Select a decimal percent and write it as a fraction. Select a fraction and write it as a percent. Write an explanation of each step of your work.

Answer varies.

70. Prepare a table showing percent, decimal, and fraction equivalents for eight common fractions and mixed numbers of your choice.

Answer varies.

In the following word problems, write the answer as a fraction, as a decimal, and as percent.

71. The license applicant answered 76 of 100 questions correctly. What portion was correct?

$\frac{19}{25}$; 0.76; 76%

72. Of 50 foods at a buffet, 14 are fat free. What portion are fat free?

$\frac{7}{25}$; 0.28; 28%

73. The price of a television set was $500. It was reduced $100. By what portion was the price reduced?

$\frac{1}{5}$; 0.2; 20%

74. To pass a French class, Carolyn must study 8 hours a week. So far this week, she has studied 5 hours. What portion of the necessary time must she still study?

$\frac{3}{8}$; 0.375; 37.5%

75. Of the fifteen people at the office, nine are single parents. What portion are single parents?

$\frac{3}{5}$; 0.6; 60%

76. A pizza shop has 25 employees. Of these, 14 are students. What portion are students?

$\frac{14}{25}$; 0.56; 56%

77. A real estate office has 40 salespeople. If 22 of these salespeople drive Cadillacs and the rest drive Lincolns, what portion drive Lincolns?

$\frac{9}{20}$; 0.45; 45%

78. A zoo has 125 animals, including 25 that are members of endangered species. What portion is not endangered?

$\frac{4}{5}$; 0.8; 80%

79. An antibiotic is used to treat 380 people. If 342 people do not have a side reaction to the antibiotic, find the portion that do have a side reaction.

$\frac{1}{10}$; 0.1; 10%

80. An apple grower's cooperative has 250 members. If 100 of the growers use a certain insecticide, find the portion that do not use the insecticide.

$\frac{3}{5}$; 0.6; 60%

The graph below shows the type of transportation used by 4200 students at Metro-Community College. Use this graph to answer the questions in Exercises 81–84, giving the answer as a fraction, as a decimal, and as a percent.

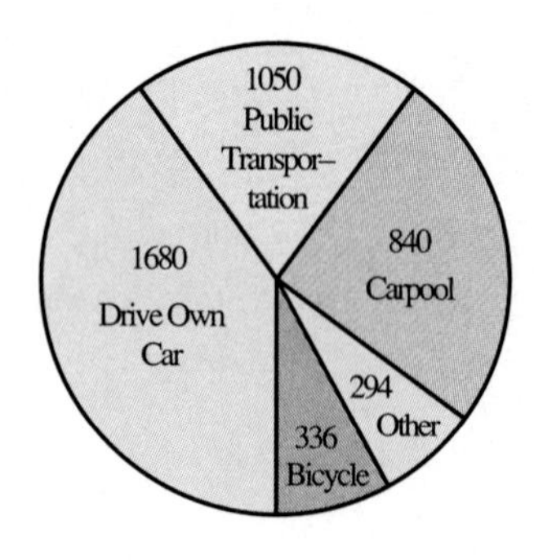

81. What portion of the students use public transportation?

$\frac{1}{4}$; 0.25; 25%

82. What portion of the students use a bicycle?

$\frac{2}{25}$; 0.08; 8%

83. Find the portion that drive their own cars.

$\frac{2}{5}$; 0.4; 40%

84. Find the portion that carpool.

$\frac{1}{5}$; 0.2; 20%

PREVIEW EXERCISES

Find the missing number in each proportion. (For help, see **Section 5.4.***)*

85. $\frac{10}{5} = \frac{x}{20}$ 40

86. $\frac{n}{50} = \frac{8}{20}$ 20

87. $\frac{4}{y} = \frac{12}{15}$ 5

88. $\frac{6}{x} = \frac{4}{8}$ 12

89. $\frac{42}{30} = \frac{14}{b}$ 10

90. $\frac{6}{22} = \frac{a}{220}$ 60

6.3 THE PERCENT PROPORTION

There are two ways to solve percent problems. One method uses ratios and is discussed in this section, while the percent equation method is explained in Section 6.6.

1 We have seen that a statement of two equivalent ratios is called a proportion.

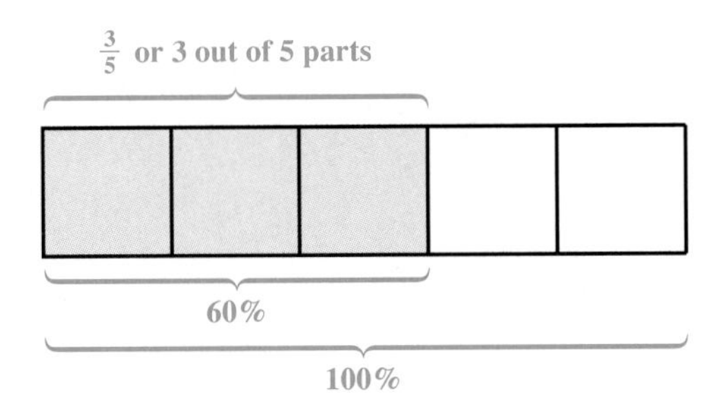

For example, the fraction $\frac{3}{5}$ is the same as the ratio 3 to 5, and 60% is the ratio 60 to 100. As the figure above shows, these two ratios are equivalent and make a proportion.

WORK PROBLEM 1 AT THE SIDE.

The **percent proportion** can be used to solve percent problems.

THE PERCENT PROPORTION

Amount is to *base* as percent is to 100, or

$$\frac{\text{amount}}{\text{base}} = \frac{\text{percent}}{100}. \leftarrow \text{always 100 because percent means per 100}$$

With letters, the proportion is

$$\frac{a}{b} = \frac{p}{100}. \quad \textit{percent proportion}$$

The final statement in the box is the **percent proportion.** In the figure at the top of the page, the **base** is 5 (the entire quantity), the **amount** is 3 (the part of the total), and the **percent** is 60. Write the percent proportion as follows.

$$\begin{matrix} a \rightarrow \\ b \rightarrow \end{matrix} \frac{3}{5} = \frac{60}{100} \begin{matrix} \leftarrow p \\ \leftarrow 100 \end{matrix}$$

2 As shown in **Section 5.4,** if any two of the three values in the percent proportion are known, the third can be found by solving the proportion.

EXAMPLE 1 *Using the Percent Proportion*

Use the percent proportion and solve for the missing number.

(a) $a = 12$, $p = 25$, find b **(b)** $a = 30$, $b = 50$, find p

(c) $b = 150$, $p = 18$, find a

OBJECTIVES

1. Learn the percent proportion.
2. Solve for a missing letter in a proportion.

FOR EXTRA HELP

Tape 8	SSM pp. 165–167	MAC: A IBM: A

1. As a review of proportions, use the method of cross products to decide whether these proportions are *true* or *false*.

(a) $\frac{3}{4} = \frac{75}{100}$

(b) $\frac{3}{5} = \frac{51}{85}$

(c) $\frac{17}{83} = \frac{68}{312}$

(d) $\frac{29}{83} = \frac{145}{415}$

(e) $\frac{104}{37} = \frac{515}{185}$

ANSWERS

1. (a) true (b) true (c) false (d) true (e) false

2. Find the value of the missing letter.

(a) $a = 20, p = 10$

(b) $a = 26, b = 104$

(c) $b = 350, p = 12$

(d) $b = 5000, p = 27$

(e) $a = 74, b = 185$

(a) Replace a with 12 and p with 25.

$$\frac{a}{b} = \frac{p}{100} \quad \text{percent proportion}$$

$$\frac{12}{b} = \frac{25}{100} \quad \text{or} \quad \frac{12}{b} = \frac{1}{4} \quad \text{(lowest terms)}$$

Find the cross products to solve this proportion.

$$\frac{12}{b} = \frac{1}{4} \qquad b \cdot 1 \qquad 12 \cdot 4$$

Show the products are equivalent.

$$b \cdot 1 = 12 \cdot 4$$
$$b = 48$$

The base is 48.

(b) Use the percent proportion.

$$\frac{30}{50} = \frac{p}{100} \quad \text{percent proportion}$$

$$\frac{3}{5} = \frac{p}{100} \quad \text{(lowest terms)}$$

$$5 \cdot p = 3 \cdot 100 \quad \text{cross products}$$

$$5 \cdot p = 300$$

$$\frac{\overset{1}{\cancel{5}} \cdot p}{\underset{1}{\cancel{5}}} = \frac{300}{5} \quad \text{Divide by 5.}$$

$$p = 60$$

The percent is 60, written as 60%.

(c) $\frac{a}{150} = \frac{18}{100}$ or $\frac{a}{150} = \frac{9}{50}$ (lowest terms)

$$a \cdot 50 = 150 \cdot 9 \quad \text{cross products}$$

$$a \cdot 50 = 1350$$

$$\frac{a \cdot \overset{1}{\cancel{50}}}{\underset{1}{\cancel{50}}} = \frac{1350}{50} \quad \text{Divide by 50.}$$

$$a = 27$$

The amount is 27. ■

◀◀ **WORK PROBLEM 2 AT THE SIDE.**

ANSWERS

2. (a) $b = 200$ (b) $p = 25$ (so, the percent is 25%) (c) $a = 42$ (d) $a = 1350$ (e) $p = 40$ (so, the percent is 40%)

NAME DATE HOUR

6.3 EXERCISES

Find the value of the missing letter in the percent proportion $\frac{a}{b} = \frac{p}{100}$. *Round to the nearest tenth if necessary.*

Examples:

$a = 10, p = 50$, find b

$a = 80, b = 120$, find p

$b = 90, p = 75$, find a

Solutions:

$\frac{10}{b} = \frac{50}{100}$

$\frac{10}{b} = \frac{1}{2}$ ← lowest terms

$b \cdot 1 = 10 \cdot 2$

$b = 20$

$b = \mathbf{20}$

$\frac{80}{120} = \frac{p}{100}$

$\frac{2}{3} = \frac{p}{100}$ ← lowest terms

$3 \cdot p = 2 \cdot 100$

$3 \cdot p = 200$

$\frac{\overset{1}{\cancel{3}} \cdot p}{\underset{1}{\cancel{3}}} = \frac{200}{3}$

$p = \mathbf{66.7}$ (rounded)

The percent is 66.7%.

$\frac{a}{90} = \frac{75}{100}$

$\frac{a}{90} = \frac{3}{4}$ ← lowest terms

$a \cdot 4 = 90 \cdot 3$

$a \cdot 4 = 270$

$\frac{a \cdot \overset{1}{\cancel{4}}}{\underset{1}{\cancel{4}}} = \frac{270}{4}$

$a = \mathbf{67.5}$

1. $a = 20, p = 25$

80

2. $a = 45, p = 50$

90

3. $a = 90, p = 20$

450

4. $a = 144, p = 75$

192

5. $a = 16, p = 40$

40

6. $a = 22, p = 10$

220

7. $a = 25, p = 6$

416.7

8. $a = 61, p = 12$

508.3

9. $a = 55, b = 110$

50

10. $a = 15, b = 60$

25

11. $a = 105, b = 35$

300

12. $a = 36, b = 24$

150

13. $a = 1.5, b = 4.5$

33.3

14. $a = 9.25, b = 27.75$

33.3

15. $b = 52, p = 50$

26

16. $b = 80, p = 25$

20

17. $b = 144, p = 15$

21.6

18. $b = 112, p = 38$

42.6

19. $b = 47.2, p = 28$

13.2

20. $b = 79.6, p = 13$

10.3

21. Give two examples of your own choosing—one showing a true proportion and why it is true and the other showing a false proportion and why it is false.

Answer varies.

22. Make up a problem that uses the percent proportion and that has an answer of $p = 15$. Hint: many different values may be used for a and b to get the result $p = 15$.

Answer varies.

Solve each of the following problems. Round answers to the nearest tenth.

23. Find b if a is 89 and p is 25.

356

24. p is 45 and b is 160. Find a.

72

25. b is 5000 and a is 20. Find p.

0.4

26. Suppose a is 15 and b is 2500. Find p.

0.6

▲ **27.** Find p if b is 1850 and a is 157.25.

8.5

▲ **28.** What is a, if p is $25\frac{1}{2}$ and b is 2800?

714

▲ **29.** b is 8116 and a is 994.21. Find p.

12.25

▲ **30.** Suppose a is 550 and p is $6\frac{1}{4}$. Find b.

8800

PREVIEW EXERCISES

Write each fraction as a percent. Round the percent to the nearest tenth if necessary. (For help, see ***Section 6.2.****)*

31. $\frac{3}{4}$ 75%

32. $\frac{4}{5}$ 80%

33. $\frac{57}{100}$ 57%

34. $\frac{1}{5}$ 20%

35. $\frac{7}{8}$ 87.5%

36. $\frac{1}{6}$ 16.7%

37. $\frac{2}{3}$ 66.7%

38. $\frac{17}{25}$ 68%

6.4 IDENTIFYING THE PARTS IN A PERCENT PROBLEM

OBJECTIVES

Identify the

1. percent;
2. base;
3. amount.

FOR EXTRA HELP

Tape 8

SSM pp. 167–168

MAC: A IBM: A

In this section you will learn how to solve percent problems. As a help in solving these problems it is good to remember what is involved in percent problems.

PERCENT PROBLEMS

All percent problems involve a comparison between a part of something and the whole.

Solving these problems requires identifying the three parts of a percent proportion: amount (a), base (b), and percent (p).

1 Look for p, percent, first. It is the easiest to identify.

PERCENT

The **percent** is the ratio of a part to a whole, with 100 as the denominator. In a problem, the percent **p** appears with the word "percent" or with the symbol "%" after it.

■ **EXAMPLE 1** *Finding Percent in a Percent Problem*

Find p in each of the following.

(a) 32% of the 900 men were retired.

32% → p

p is 32. The number 32 appears with the symbol %.

(b) $150 is 25 percent of what number?

25 → p

p is 25 because 25 appears with the word "percent."

(c) What percent of the 350 women will go?

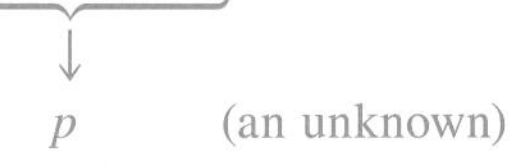

What percent → p (an unknown)

The word "percent" has no number with it, so the percent is the unknown part of the problem. ■

WORK PROBLEM 1 AT THE SIDE. ▶▶

2 Next, look for b, the base.

BASE

The **base** is the entire quantity, or the total. In a problem, the base often appears after the word **of.**

1. Identify p.

(a) Of the $1000, 23% will be spent on new tires.

(b) Of the 642 employees, 97% will be rehired.

(c) Find the sales tax by multiplying $590 and $6\frac{1}{2}$ percent.

(d) 600 is 55% of what number?

(e) What percent of the 110 rental cars will be rented today?

ANSWERS

1. (a) 23 **(b)** 97 **(c)** $6\frac{1}{2}$ **(d)** 55 **(e)** p is unknown.

■ **EXAMPLE 2** *Finding Base in a Percent Problem*

Find b in each of the following.

(a) 32% of the 900 men where too large for the imported car.

b is 900. The number 900 appears after the word *of*.

(b) $150 is 25 percent of what number?

The base is the unknown part of the proportion.

(c) 85% of 7000 is what number?

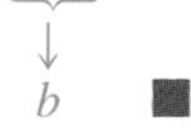

■

◀◀ **WORK PROBLEM 2 AT THE SIDE.**

3 Finally, look for a, the amount.

AMOUNT

The **amount** is the part being compared with the whole.

Note If you have trouble identifying amount, find base and percent first. The remaining number is amount.

■ **EXAMPLE 3** *Finding Amount in a Percent Problem*

Find a in each of the following.

(a) 54% of 700 students is 378 students.

Find p and b.

54% of 700 students is 378 students.

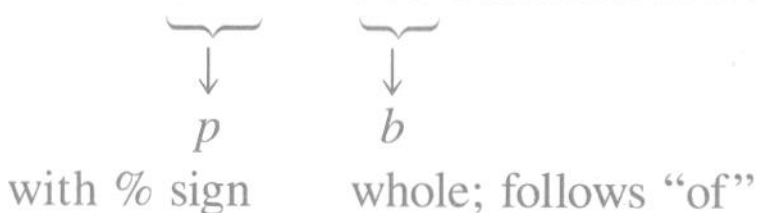

with % sign whole; follows "of"

54% of 700 students is 378 students.

p b a part

The amount, a, is 378.

(b) $150 is 25% of what number?

(the unknown part of the proportion)

150 is the remaining number, so $a = 150$.

(c) 85% of $7000 is what number?

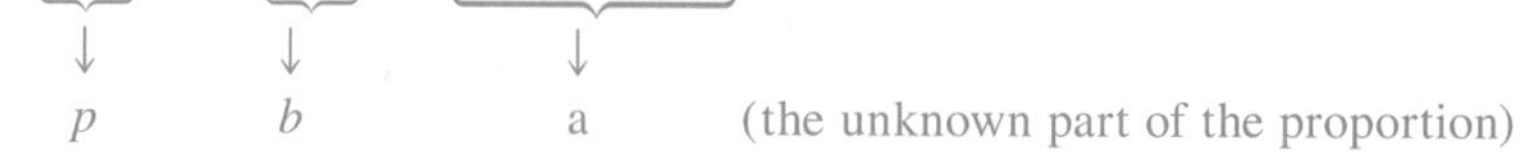

(the unknown part of the proportion)

◀◀ **WORK PROBLEM 3 AT THE SIDE.**

2. Find b.

(a) Of the $1000, 23% will be spent on new tires.

(b) Of the 642 employees, 97% will be rehired.

(c) Find the sales tax by multiplying $590 and $6\frac{1}{2}$ percent.

(d) $600 is 55% of what number?

(e) What percent of the 110 rental cars will be rented today?

3. Find a.

(a) 12.5% of $1500 is $187.50.

(b) 59 people is 25% of 236 people.

(c) $500 is what percent of $4250?

(d) The 5% tax on an item costing $650 is $32.50.

ANSWERS

2. (a) 1000 **(b)** 642 **(c)** 590 **(d)** what number (an unknown) **(e)** 110

3. (a) 187.50 **(b)** 59 **(c)** 500 **(d)** 32.50

6.4 EXERCISES

NAME DATE HOUR

Identify p, b, and a in each of the following. Do not try to solve for any unknowns.

Example: 60 is **75%** of **what number**?
Solution: 60 → a, 75% → p, what number → b is unknown

Example: Of the **592** tomato plants, **75%,** or **444,** are ready to be sold.
Solution: 592 → b, 75% → p, 444 → a

	p	b	a
1. 25% of how much money is \$200?	25	unknown	200
2. 35% of how many chairs is 700 chairs?	35	unknown	700
3. 81% of what number is 748?	81	unknown	748
4. 93% of what number is 11.5?	93	unknown	11.5
5. What is 15% of \$75?	15	75	unknown
6. What is 61% of 830 homes?	61	830	unknown
7. 18 is 72% of what number?	72	unknown	18
8. 46 is 13% of what number?	13	unknown	46

	p	b	a
9. 26 lamps is what percent of 52 lamps?	unknown	52	26
10. 182 plants is what percent of 546 plants?	unknown	546	182
11. What percent of \$148 is \$88.80?	unknown	148	88.8
12. What percent of \$95 is \$42.50?	unknown	95	42.5
13. 27.17 is 6.5% of what number?	6.5	unknown	27.17
14. 16.74 is 11.9% of what number?	11.9	unknown	16.74
15. 0.68% of 487 is what number?	0.68	487	unknown
16. What number is 12.42% of 1408.7?	12.42	1408.7	unknown

17. Identify the three parts in a percent problem. In your own words, write one sentence telling how you will identify each of these three parts.

Answer varies.

18. Write one short sentence or statement using numbers and words. The statement should include a percent, a base, and an amount. Identify each of these three parts.

Answer varies.

Find p, b, and a in the following word problems. Do not try to solve for any unknowns.

19. In a tree-planting project, 640 of the 810 trees planted were still living one year later. What percent of the trees planted were still living?

p is unknown; 810; 640

20. A popular bar of soap is $99\frac{44}{100}$% pure. If the bar of soap weighs 9 ounces, how many ounces are pure?

$99\frac{44}{100}$; 9 a is unknown

21. Sales tax of $0.90 is charged on a compact disc costing $15. What percent sales tax is charged?

p is unknown; 15; 0.90

22. From Armond's check of $340, 17% is withheld. How much is withheld?

17; 340; a is unknown

23. Of the baby salmon shipped from a hatchery, 7% do not arrive healthy. If 1500 salmon are shipped, how many do not arrive healthy?

7; 1500; a is unknown

24. There are 590 quarts of grape juice in a vat holding a total of 1700 quarts of fruit juice. What percent is grape juice?

p is unknown; 1700; 590

25. Of the total candy bars contained in a vending machine, 240 bars have been sold. If 25% of the bars have been sold, find the total number of candy bars that were in the machine.

25; b is unknown; 240

26. There have been 36 cups of coffee served from a banquet-sized coffee pot. If this is 30% of the capacity of the pot, find the capacity of the pot.

30; b is unknown; 36

27. A part-time employee earns $110 per week and has 18% of this amount withheld for taxes, Social Security, and Medicare. Find the amount withheld.

18; 110; a is unknown

28. A student needs 64 credits to graduate. If 48 of the credits have already been completed, what percent of the credits have already been completed?

p is unknown; 64; 48

Find p, b, and a in the following word problems. Do not try to solve for any unknowns.

29. At a recent health fair 32% of the people tested were found to have high blood cholesterol levels. If 272 people were tested, find the number having high blood cholesterol.

32; 272; *a* is unknown

30. The sales tax on a new car is $820. If the sales tax rate is 5%, find the price of the car before the sales tax is added.

5; *b* is unknown; 820

31. A medical clinic found that 16.8% of the patients were late for their appointments. The number of patients who were late was 504. Find the total number of patients.

16.8; *b* is unknown; 504

32. 1848 automobiles are tested for exhaust emissions and 231 do not pass the test. Find the percent that do not pass.

***p* is unknown; 1848; 231**

PREVIEW EXERCISES

Write a proportion for each of the following and then find the missing number. (For help, see **Section 5.4.***)*

	proportion	*missing number*
33. 7 is to x as 5 is to 10.	$\frac{7}{x} = \frac{5}{10}$	14
34. 50 is to 40 as y is to 80.	$\frac{50}{40} = \frac{y}{80}$	100
35. x is to 36 as $\frac{4}{3}$ is to 12.	$\frac{x}{36} = \frac{\frac{4}{3}}{12}$	4
36. 1.2 is to 10 as 3.6 is to a.	$\frac{1.2}{10} = \frac{3.6}{a}$	30

6.5 USING PROPORTIONS TO SOLVE PERCENT PROBLEMS

In the percent proportion,

$$\frac{\text{amount}}{\text{base}} = \frac{\text{percent}}{100}$$

or

$$\frac{a}{b} = \frac{p}{100},$$

three letters, a, b, and p are used. As discussed, if any two of these letters are known, the third can be found.

> *Note* Remember that in the percent proportion, b (base) is the entire quantity, a (amount) is part of the total, and p is the percent.

1 The first example shows the percent proportion used to find a, the amount. (Remember: the amount is a part of the whole.)

■ **EXAMPLE 1** *Finding Amount with the Percent Proportion*
Find 15% of \$160.

Here p (percent) is 15 and b (base) is 160. (Recall that the base often comes after the word *of*.) Now find a (amount).

$$\frac{a}{b} = \frac{p}{100} \quad \text{so} \quad \frac{a}{160} = \frac{15}{100} \quad \text{or} \quad \frac{a}{160} = \frac{3}{20} \quad \text{(lowest terms)}$$

Find the cross products in the proportion.

$$a \cdot 20 = 160 \cdot 3 \quad \text{Cross products}$$

$$a \cdot 20 = 480$$

$$\frac{a \cdot \overset{1}{\cancel{20}}}{\underset{1}{\cancel{20}}} = \frac{480}{20} \quad \text{Divide both sides by 20.}$$

$$a = 24 \quad \text{Amount}$$

15% of \$160 is \$24. ■

WORK PROBLEM 1 AT THE SIDE. ▶▶

Just as with the word problems given earlier, the word *of* is an indicator word meaning *multiply*. For example,

15% of 160

means

$15\% \cdot 160.$

Because of this, there is a more direct way to find the amount, a.

OBJECTIVES

1. Use the percent proportion to solve for a (amount).
2. Solve for b (base) using the percent proportion.
3. Find p (percent) using the percent proportion.

FOR EXTRA HELP

Tape 8 | SSM pp. 169–174 | MAC: A IBM: A

1. Use the percent proportion.

(a) Find 25% of 840 cars.

(b) Find 35% of \$1200.

(c) Find 9% of 3250 miles.

(d) Find 78% of 610 meters.

ANSWERS
1. (a) 210 cars (b) \$420 (c) 292.5 miles (d) 475.8 meters

2. Use multiplication to find a (amount).

(a) Find 45% of 6000 hogs.

(b) Find 18% of 80 feet.

(c) Find 125% of 78 acres.

(d) Find 0.6% of \$120.

ANSWERS

2. (a) 2700 hogs (b) 14.4 feet (c) 97.5 acres (d) \$0.72

FINDING AMOUNT

To find amount (a):

Step 1 Find p. Write the percent as a decimal.
Step 2 Multiply this decimal and b.

■ **EXAMPLE 2** *Finding Amount by Using Multiplication*
Use the shortcut to find a.

(a) Find 42% of 830 yards.

(b) Find 140% of 60 miles.

(c) Find 0.4% of 50 kilometers.

(a) *Step 1* Here p is 42. Write 42% as the decimal 0.42.
Step 2 Multiply 0.42 and b, which is 830.

$$a = 0.42 \cdot 830$$
$$a = 348.6 \text{ yards}$$

It is a good idea to estimate the answer, to make sure no mistakes were made with decimal points. Estimate 42% as 40% or 0.4, and estimate 830 as 800. Next, 40% of 800 is

$$0.4 \cdot 800 = 320,$$

So that 348.6 is a reasonable answer.

(b) In this problem, p is 140. Write 140% as the decimal 1.40. Next, multiply 1.40 and 60.

$$a = 1.40 \cdot 60 = 84 \text{ miles} \quad \text{Multiply.}$$

We can estimate because 140% is close to 150% (which is $1\frac{1}{2}$) and $1\frac{1}{2}$ times 60 is 90. So, 84 miles is a reasonable answer.

(c) $a = 0.004 \cdot 50 = 0.2$ kilometers   Multiply.
(0.004: Write 0.4% percent as a decimal.)

An estimate would not be very useful here. ■

WORK PROBLEM 2 AT THE SIDE.

■ **EXAMPLE 3** *Solving for Amount by Using Multiplication*
Video Production has 850 employees. Of these employees, 28% are students. How many of the employees are students?

Approach Look for the word *of* as an indicator word for multiplication.

Solution

28% of the employees are students
↑ indicator word

The total number of employees is 850, so $b = 850$. The percent is $28 (p = 28)$. Find a to find the number of students.

$$a = 0.28 \cdot 850 = 238 \quad \text{Multiply.}$$
↑ Write 28% as a decimal.

Video Productions has 238 student employees. ■

WORK PROBLEM 3 AT THE SIDE. ▶▶

2 The next example shows how to use the percent proportion to find b, the base.

Note Remember, the base is the entire quantity, or the total.

■ **EXAMPLE 4** *Finding b (base) with the Percent Proportion*

(a) 8 tables is 4% of what number of tables?

(b) 135 tourists is 15% of what number of tourists?

(a) Here $p = 4$, b is unknown, and $a = 8$. Use the percent proportion to find b.

$$\frac{a}{b} = \frac{p}{100} \quad \text{so} \quad \frac{8}{b} = \frac{4}{100} \quad \text{or} \quad \frac{8}{b} = \frac{1}{25} \quad \text{(lowest terms)}$$

Find cross products.

$$b \cdot 1 = 8 \cdot 25$$
$$b = 200$$

8 tables is 4% of 200 tables.

3. Use the shortcut way to find a.

(a) In a town of 1560 people, 25% are members of the Grange. How many people in the town belong to the Grange?

(b) There are 9250 students enrolled at the campus. If 28% of the students use tobacco products, find the number of tobacco users.

ANSWERS
3. (a) 390 people (b) 2590 students

(b) $p = 15$ and $a = 135$, so

$$\frac{135}{b} = \frac{15}{100}$$

$$\frac{135}{b} = \frac{3}{20} \quad \text{(lowest terms)}$$

$$b \cdot 3 = 135 \cdot 20 \quad \text{cross products}$$

$$b \cdot 3 = 2700$$

$$\frac{b \cdot \overset{1}{\cancel{3}}}{\underset{1}{\cancel{3}}} = \frac{2700}{3} \quad \text{Divide both sides by 3.}$$

$$b = 900$$

135 tourists is 15% of 900 tourists. ■

WORK PROBLEM 4 AT THE SIDE.

■ **EXAMPLE 5** *Applying the Percent Proportion*

At Newark Salt Works, 78 employees are absent because of illness. If this is 5% of the total number of employees, how many employees does the company have?

Approach From the information in the problem, the percent is 5 ($p = 5$) and the amount, or *part* of the total number of employees is 78 ($a = 78$). The total number of employees or entire quantity, which is the base (b), must be found.

Solution We can use the percent proportion to find b (the total number of employees).

$$\frac{78}{b} = \frac{5}{100}$$

$$\frac{78}{b} = \frac{1}{20} \quad \text{(lowest terms)}$$

Find cross products.

$$b \cdot 1 = 78 \cdot 20$$

$$b = 1560$$

Estimate the answer.

78 is approximately 80, and 5% is equivalent to the fraction $\frac{1}{20}$. Because 80 is $\frac{1}{20}$ of 1600, or

$$80 \cdot 20 = 1600,$$

1560 is a reasonable answer.

The company has 1560 employees. ■

> *Note* To estimate the answer to Example 5 the 5% was changed to its fraction equivalent $\frac{1}{20}$. Because 80 (rounded) is $\frac{1}{20}$ of the total employees, 80 was multiplied by 20 to get 1600, the estimated answer.

WORK PROBLEM 5 AT THE SIDE.

4. Use the percent proportion to find the missing part.

(a) 50 guests is 25% of what number of guests?

(b) 48 books is 15% of what number of books?

(c) 774 employees is 72% of what number of employees?

(d) 97.5 miles is 12.5% of what number of miles?

5. (a) A freeze resulted in a loss of 52% of an avocado crop. If the loss was 182 tons, find the total number of tons in the crop.

(b) A metal alloy contains 8% zinc. The alloy contains 450 pounds of zinc. Find the total weight of the alloy.

ANSWERS

4. (a) 200 guests (b) 320 books (c) 1075 employees (d) 780 miles

5. (a) 350 tons (b) 5625 pounds

3 Finally, if a and b are known, the percent proportion can be used to find p.

■ EXAMPLE 6 *Using the Percent Proportion to Find p*

(a) 13 roofs is what percent of 52 roofs?

(b) What percent of \$500 is \$100?

(a) The base is $b = 52$ (52 follows *of*) and $a = 13$. Next, find p.

$$\frac{a}{b} = \frac{p}{100}$$

$$\frac{13}{52} = \frac{p}{100}$$

$$\frac{1}{4} = \frac{p}{100} \quad \text{(lowest terms)}$$

Find cross products.

$$4 \cdot p = 1 \cdot 100$$

$$\frac{\overset{1}{\cancel{4}} \cdot p}{\underset{1}{\cancel{4}}} = \frac{100}{4} \quad \text{Divide both sides by 4.}$$

$$p = 25$$

13 roofs is 25% of 52 roofs.

(b) $b = 500$, $a = 100$

$$\frac{100}{500} = \frac{p}{100}$$

$$\frac{1}{5} = \frac{p}{100} \quad \text{(lowest terms)}$$

$$5 \cdot p = 1 \cdot 100 \quad \text{cross products}$$

$$5 \cdot p = 100$$

$$\frac{\overset{1}{\cancel{5}} \cdot p}{\underset{1}{\cancel{5}}} = \frac{100}{5} \quad \text{Divide both sides by 5.}$$

$$p = 20$$

20% of \$500 is \$100. ■

Note When finding p (percent), be sure to label your answer with the percent symbol (%).

WORK PROBLEM 6 AT THE SIDE. ▶▶

6. (a) \$14 is what percent of \$56?

(b) What percent of 300 teachers is 60 teachers?

(c) What percent of 1140 cases is 513 cases?

(d) 36 trophies is what percent of 9 trophies?

ANSWERS
6. (a) 25% (b) 20% (c) 45% (d) 400%

7. (a) The Cruisers Club has raised \$578 of the \$850 needed for an annual picnic. What percent of the total has been raised?

(b) A late-model domestic car gets 38 miles per gallon on the highway and 32.3 miles per gallon around town. What percent of the highway mileage does the car get around town?

8. (a) The number of students who usually take this class is 300. If 450 students are taking this class now, find the percent of the usual number who are now taking the class.

(b) Ann Shaffer set a sales goal of \$96,000. If her sales amounted to \$115,200, find the percent of her sales goal achieved.

ANSWERS
7. (a) 68% (b) 85%
8. (a) 150% (b) 120%

EXAMPLE 7 *Applying the Percent Proportion*

A roof is expected to last 20 years before needing replacement. If the roof is now 15 years old, what percent of the roof's life has been used?

Approach The expected life of the roof is the entire quantity or base ($b = 20$). The part of the roof that is already used is the amount ($a = 15$). You need to find the percent (p) of the roof's life that is already used.

Solution Use the percent proportion to find p, the percent of roof life used.

$$\frac{15}{20} = \frac{p}{100} \quad \text{or} \quad \frac{3}{4} = \frac{p}{100} \qquad \text{(lowest terms)}$$

Find cross products.

$$4 \cdot p = 3 \cdot 100$$
$$4 \cdot p = 300$$
$$\frac{\overset{1}{\cancel{4}} \cdot p}{\underset{1}{\cancel{4}}} = \frac{300}{4} \qquad \text{Divide both sides by 4.}$$
$$p = 75$$

75% of the roof's life has been used. ■

◀ WORK PROBLEM 7 AT THE SIDE.

EXAMPLE 8 *Applying the Percent Proportion*

Rainfall this year was 33 inches while normal rainfall is only 30 inches. What percent of normal rainfall is this year's rainfall?

Approach The normal rainfall is the base ($b = 30$). This year's rainfall is all of last year's rainfall and more, or 33 ($a = 33$). You need to find the percent (p) that this year's rainfall is of last year's rainfall.

Solution

$$\frac{33}{30} = \frac{p}{100} \quad \text{or} \quad \frac{11}{10} = \frac{p}{100} \qquad \text{(lowest terms)}$$

Find cross products.

$$10 \cdot p = 11 \cdot 100$$
$$10 \cdot p = 1100$$
$$\frac{\overset{1}{\cancel{10}} \cdot p}{\underset{1}{\cancel{10}}} = \frac{1100}{10}$$
$$p = 110$$

This year's rainfall is 110% of last year's rainfall. ■

◀ WORK PROBLEM 8 AT THE SIDE

6.5 EXERCISES

NAME DATE HOUR

Find the amount using the multiplication shortcut.

Example: 47% of 5000 bicycles

Solution: $0.47 \cdot 5000 = 2350$

$a = \mathbf{2350}$ **bicycles**

1. 10% of 520 adults

52 adults

2. 20% of 1500 tires

300 tires

3. 14% of 780 meters

109.2 meters

4. 12% of 350 miles

42 miles

5. 4% of 120 feet

4.8 feet

6. 9% of $150

$13.50

7. 150% of 38 lamps

57 lamps

8. 130% of 60 trees

78 trees

9. 22.5% of 1100 boxes

247.5 boxes

10. 38.2% of 4250 loads

1623.5 loads

11. 2% of $164

$3.28

12. 6% of $434

$26.04

13. 250% of 740 sales

1850 sales

14. 125% of 920 students

1150 students

15. 15.5% of 275 pounds

42.625 pounds

16. 46.1% of 843 kilograms

388.623 kilograms

17. 0.9% of $2400

$21.60

18. 0.3% of $1400

$4.20

Find the base using the percent proportion.

Example: 64% of what number is 1600?

Solution: $p = 64$ and $a = 1600$, so

$$\frac{1600}{b} = \frac{64}{100}$$

$$\frac{1600}{b} = \frac{16}{25}. \quad \text{(lowest terms)}$$

Solve, to get $b =$ **2500.**

19. 20 trucks is 10% of what number of trucks?

200 trucks

20. 5 pianos is 5% of what number of pianos?

100 pianos

21. 25% of what number is 76?

304

22. 65% of what number is 182?

280

23. 210 envelopes is 50% of what number of envelopes?

420 envelopes

24. 84 letters is 28% of what number of letters?

300 letters

25. 748 books is 110% of what number of books?

680 books

26. 77 hats is 140% of what number of hats?

55 hats

27. $12\frac{1}{2}\%$ of what number is 350?
(Hint: $12\frac{1}{2}\% = 12.5\%$)

2800

28. $5\frac{1}{2}\%$ of what number is 176?

3200

Find the percent using the percent proportion. Round your answers to the nearest tenth if necessary.

Example: What percent of 80 is 15?

Solution: $b = 80$ and $a = 15$, so

$$\frac{15}{80} = \frac{p}{100}$$

$$80 \cdot p = 1500$$

$$\frac{\overset{1}{\cancel{80}} \cdot p}{\underset{1}{\cancel{80}}} = \frac{1500}{80}$$

$$p = 18.75$$

$$p = 18.8 \quad \text{(rounded)}$$

18.8% of 80 is 15.

29. 32 jeeps is what percent of 64 jeeps?

50%

30. 25 containers is what percent of 125 containers?

20%

31. 26 desks is what percent of 50 desks?

52%

32. 325 bottles is what percent of 500 bottles?

65%

33. 8 doors is what percent of 400 doors?

2%

34. 7 bridges is what percent of 350 bridges?

2%

35. 9 rolls is what percent of 600 rolls?

1.5%

36. 30 tubes is what percent of 1200 tubes?

2.5%

37. What percent of \$172 is \$32?

18.6%

38. What percent of \$398 is \$14?

3.5%

39. What percent of 500 wheels is 46 wheels?

9.2%

40. What percent of 105 employees is 54 employees?

51.4%

41. Write a percent problem about something you have recently purchased being sure to include only two of the following: the price, the amount of sales tax, the sales tax rate. Identify each part of your percent problem and use the percent proportion to solve it.

Answer varies.

42. Write a percent problem on any topic you choose. Be sure to include only two of the three parts so that you can solve for the third part. Identify each part of the problem and then solve it.

Answer varies.

Solve each word problem. Round percent answers to the nearest tenth if necessary.

43. A library has 270 visitors on Saturday, 20% of whom are children. How many are children?

54 children

44. The Yale University marching band has 250 members. If 18.4% of the band members are senior class students, find the number of students who are seniors.

46 seniors

45. A survey at an intersection found that of 2200 drivers, 38% were wearing seat belts. How many drivers in the survey were wearing seat belts?

836 drivers

46. A home valued at $95,000 will gain 6% in value this year. Find the gain in value this year.

$5700

47. An Atlantic City casino advertises that it gives a 97.4% payback on slot machines, and the balance is retained by the casino. The amount retained by the casino is $4823. Find the total amount played on the slot machines.

$185,500

48. A resort hotel states that 35% of its rooms are for nonsmokers. The resort allows smoking in 468 rooms. Find the total number of rooms.

720 rooms

49. This year, there are 550 scholarship applicants. If 40% of the applicants will receive a scholarship, find the number of students who will receive a scholarship.

220 students

50. A U.S. Food and Drug Administration (FDA) biologist found that canned tuna is "relatively clean." Extraneous matter was found in 5% of the 1600 cans of tuna tested. How many cans of tuna contained extraneous matter?

80 cans

51. Meadow Vista Bottled Water estimates that $117,000 will be spent this year on delivery costs alone. If total sales are estimated at $755,000, what percent of total sales will be spent on delivery?

15.5%

52. There are 55,000-plus words in Webster's Dictionary, but most educated people can identify only 20,000 of these words. What percent of the words in the dictionary can these people identify?

36.4%

53. Rocky Mountain Water estimates 11,700 gallons of their water will be used in steam irons. If 755,000 gallons are sold, what percent will be used in steam irons?

1.5%

54. Barbara's Antiquery says that of its 3800 items in stock, 3344 are just plain junk, and the rest are antiques. What percent of the number of items in stock is antiques?

12%

55. This month's sales goal for Easy Writer Pen Company is 2,380,000 ball-point pens. If sales of 2,618,000 pens have been made, what percent of the goal has been reached?

110%

56. The number of apartment unit vacancies was predicted to be 2112 while the actual number of vacancies was 2640. The actual number was what percent of the predicted number of vacancies?

125%

57. Americans who are 65 years of age or older make up 12.7% of the total population. If there are 31.5 million Americans in this group, find the total U.S. population. (Round to the nearest tenth of a million.)

248.0 million

58. Woolworth Corporation will close, sell, or change the format of 900 stores, or about 9.7% of its total stores. How many stores does the Woolworth Corporation have? (Round to the nearest whole number.)

9278 stores

The graph below shows the market share of U.S. ready-to-eat cereal based on sales. The total annual U.S. sales of ready-to-eat cereal is $7,500,000,000. Use this information to do Exercises 59–62.

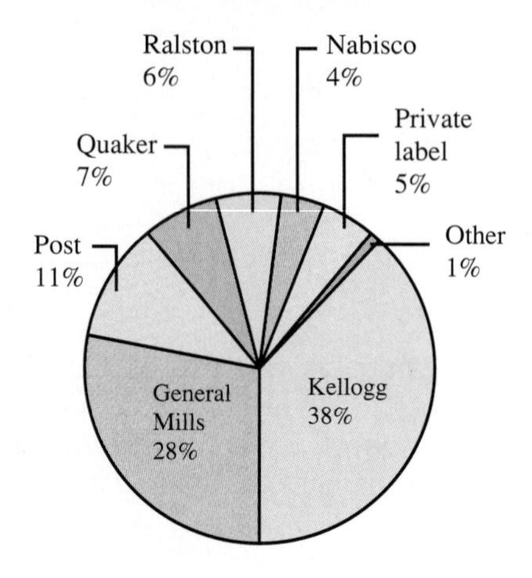

From "The Battle Over Breakfast." Reprinted by permission of *The Wall Street Journal*.

59. Find the annual U.S. sales for the Kellogg Company.

$2,850,000,000

60. Find the annual U.S. sales for the Nabisco Company

$300,000,000

61. What are the annual U.S. sales of Post cereals?

$825,000,000

62. What are the annual U.S. sales of Private label (the store's own brand) cereals?

$375,000,000

▲ **63.** A collection agency, specializing in collecting past-due child support, charges $25 as an application fee plus 20% of the amount collected. What is the total charge for collecting $3100 past-due child support?

$645

▲ **64.** Raw steel production by the nation's steel mills decreased by 2.5% from last week. The decrease amounted to 50,475 tons. Find the steel production last week.

2,019,000 tons

▲ **65.** Marketing Intelligence Service says that there were 15,401 new products introduced last year. If 86% of the products introduced last year failed to reach their business objectives, find the number of products that did reach their objectives. (Round to the nearest whole number.)

2156 products

▲ **66.** A family of four with a monthly income of $2900 spends 90% of its earnings and saves the balance. Find (*a*) the monthly savings and (*b*) the annual savings of this family.

$290; $3480

PREVIEW EXERCISES

Multiply or divide as indicated. (For help, see **Sections 4.5 and 4.6.***)*

67. 48.05×0.85 = 40.8425

68. 0.655×0.012 = 0.00786

69. 325.6×0.031 = 10.0936

70. 51.81×0.021 = 1.08801

71. $306 \div 0.085$ 3600

72. $24 \div 0.01$ 2400

73. $172 \div 688$ 0.25

74. $120 \div 24$ 5

6.6 THE PERCENT EQUATION

In the last section you were shown how to use a proportion to solve percent problems. This section shows another way to solve these problems by using the ***percent equation.*** The percent equation is just a rearrangement of the percent proportion.

OBJECTIVES

1. Use the percent equation to solve for amount (a).
2. Solve for base (b) using the percent equation.
3. Find percent (p) using the percent equation.

FOR EXTRA HELP

Tape 8

SSM pp. 174–177

MAC: A IBM: A

PERCENT EQUATION

$$\text{amount} = \text{percent} \cdot \text{base}$$

Be sure to write the percent as a decimal before using the equation.

In the percent proportion we did not have to write the percent as a decimal because of the 100 in the denominator. However, because there is no 100 in the percent *equation,* it is necessary for us first to write the percent as a decimal.

Some of the examples solved earlier will be reworked by using the percent equation. If you want to, you can look back at Section 6.5 to see how some of these same problems were solved using proportions. This will give you a comparison of the two methods.

1 The first example shows how to find the amount, a.

EXAMPLE 1 *Solving for the Amount (a)*

(a) Find 15% of 160 workers.

(b) Find 140% of \$60.

(c) Find 0.4% of 250 patients.

(a) Write 15% as the decimal 0.15. The base (the whole, which often comes after the word *of*) is 160. Next, use the percent equation.

$$\text{amount} = \text{percent} \cdot \text{base}$$
$$\text{amount} = 0.15 \cdot 160$$

Multiply 0.15 and 160 to get

$$\text{amount} = 24.$$

24 workers is 15% of 160 workers.

(b) Write 140% as the decimal 1.40. The base is \$60.

$$\text{amount} = \text{percent} \cdot \text{base}$$
$$a = 1.40 \cdot 60$$
$$a = \$84$$

\$84 is 140% of \$60.

(c) Write 0.4% as the decimal 0.004. The base is 250.

$$\text{amount} = \text{percent} \cdot \text{base}$$
$$a = 0.004 \cdot 250$$
$$a = 1$$

1 patient is 0.4% of 250 patients. ■

Note When using the percent equation, the percent must always be changed to a decimal before multiplying.

1. Find each of the following.

(a) 14% of 650 pictures

(b) 38% of 720 households

(c) 125% of $84

(d) 115% of $310

(e) 0.5% of 600 samples

(f) 0.25% of 160 pounds

ANSWERS

1. (a) 91 pictures (b) 273.6 households (c) $105 (d) $356.50 (e) 3 samples (f) 0.4 pound

WORK PROBLEM 1 AT THE SIDE.

2 The next example shows how to use the percent equation to find the base, *b*.

Note Remember that the word *of* is an indicator word for *multiply*.

EXAMPLE 2 *Solving for the Base (b)*

(a) 8 feet is 4% of what number of feet?

(b) 135 miles is 15% of what number of miles?

(c) $8\frac{1}{2}\%$ of what number is 102?

(a) The amount is 8 and the percent is 4% or the decimal 0.04. The base is unknown.

8 is 4% of what number?

↑ indicator word

Next, use the percent equation.

$$\text{amount} = \text{percent} \cdot \text{base}$$

$$8 = 0.04 \cdot b \quad b \text{ is the base.}$$

$$\frac{8}{0.04} = \frac{\overset{1}{\cancel{0.04}} \cdot b}{\underset{1}{\cancel{0.04}}} \quad \text{Divide both sides by 0.04.}$$

$$200 = b \quad \text{base}$$

8 feet is 4% of 200 feet.

(b) Write 15% as 0.15. The amount is 135. Next, use the percent equation to find the base.

$$\text{amount} = \text{percent} \cdot \text{base}$$

$$135 = 0.15 \cdot b$$

Divide both sides by 0.15.

$$\frac{135}{0.15} = \frac{\overset{1}{\cancel{0.15}} \cdot b}{\underset{1}{\cancel{0.15}}}$$

$$900 = b \quad \text{base}$$

135 miles is 15% of 900 miles.

(c) Write $8\frac{1}{2}\%$ as 8.5%, or the decimal 0.085. The amount is 102. Use the percent equation.

$$\text{amount} = \text{percent} \cdot \text{base.}$$

$$102 = 0.085 \cdot b$$

Divide both sides by 0.085.

$$\frac{102}{0.085} = \frac{\overset{1}{\cancel{0.085}} \cdot b}{\underset{1}{\cancel{0.085}}}$$

$$1200 = b \quad \text{base}$$

102 is $8\frac{1}{2}\%$ of 1200. ■

Note In Example 2(c) the $8\frac{1}{2}\%$ was changed to 8.5%, which is the decimal form of $8\frac{1}{2}\%$ ($8\frac{1}{2} = 8.5$). The percent sign still remained in 8.5%. Then 8.5% was changed to the decimal 0.085 before dividing.

WORK PROBLEM 2 AT THE SIDE.

3 The final example shows how to use the percent equation to find the percent, p.

EXAMPLE 3 *Solving For Percent (p)*

(a) \$13 is what percent of \$52?

(b) What percent of 500 feet is 100 feet?

(c) What percent of \$300 is \$390?

(d) 6 ladders is what percent of 1200 ladders?

(a) Because \$52 follows *of*, the base is 52. The amount is 13, and the percent is unknown. Use the percent equation,

$$\text{amount} = \text{percent} \cdot \text{base}$$

If p is the unknown percent, then

$$13 = p \cdot 52.$$

Divide both sides by 52.

$$\frac{13}{52} = \frac{p \cdot \overset{1}{\cancel{52}}}{\underset{1}{\cancel{52}}}$$

$0.25 = p$

0.25 is 25% Write the decimal as percent.

13 is 25% of 52.

This problem could also be solved by using *of* as an indicator word for multiplication, and *is* as an indicator word for "is equal to."

13 is what percent of 52?
↓ ↓ ↓ ↓ ↓
$13 = \quad p \quad \cdot 52$
$13 = p \cdot 52$

(b) The base is 500 and the amount is 100. Let p be the unknown percent.

$$\text{amount} = \text{percent} \cdot \text{base}$$
$$100 = p \cdot 500$$

Divide both sides by 500.

$$\frac{100}{500} = \frac{p \cdot \overset{1}{\cancel{500}}}{\underset{1}{\cancel{500}}}$$

$0.20 = p$

$0.20 = 20\%$ Write decimal as percent.

20% of 500 feet is 100 feet.

2. Find the base.

(a) 48 cows is 20% of what number of cows?

(b) 22.5 boxes is 18% of what number of boxes?

(c) 286 miles is 55% of what number of miles?

(d) $5\frac{1}{2}\%$ of what number of units is 33 units?

ANSWERS
2. (a) 240 cows (b) 125 boxes (c) 520 miles (d) 600 units

3. (a) What percent of 90 pallets is 18 pallets?

(b) 68 cartons is what percent of 170 cartons?

(c) What percent of 230 crates is 322 crates?

(d) 3 sacks is what percent of 375 sacks?

(c) The base is 300 and the amount is 390. Let p be the unknown percent.

$$\text{amount} = \text{percent} \cdot \text{base}$$
$$390 = p \cdot 300$$

Divide both sides by 300.

$$\frac{390}{300} = \frac{p \cdot \overset{1}{\cancel{300}}}{\underset{1}{\cancel{300}}}$$
$$1.3 = p$$
$$1.3 = 130\% \quad \text{Write decimal as percent.}$$

130% of \$300 is \$390.

(d) Since 1200 follows *of*, the base is 1200. The amount is 6. Let p be the unknown percent.

$$\text{amount} = \text{percent} \cdot \text{base}$$
$$6 = p \cdot 1200$$

Divide both sides by 1200.

$$\frac{6}{1200} = \frac{p \cdot \overset{1}{\cancel{1200}}}{\underset{1}{\cancel{1200}}}$$
$$0.005 = p$$
$$0.005 = 0.5\% \quad \text{Write decimal as percent.}$$

6 is 0.5% of 1200. ■

Note When you use the percent equation to solve for an unknown percent the answer will always be in decimal form. Notice that in Example 3(a), (b), (c), and (d) the decimal answer had to be changed to a percent by multiplying by 100. The answers became: (a) 0.25 = 25%; (b) 0.20 = 20%; (c) 1.3 = 130%; and (d) 0.005 = 0.5%.

◀◀ **WORK PROBLEM 3 AT THE SIDE.**

ANSWERS
3. (a) 20% **(b)** 40% **(c)** 140% **(d)** 0.8%

NAME DATE HOUR

6.6 EXERCISES

Find the amount using the percent equation.

Example: 14% of $750

Solution:

$$\text{amount} = \text{percent} \cdot \text{base}$$

$$\text{amount} = 0.14 \cdot 750 \quad \text{Write 14\% as the decimal 0.14.}$$

$$\text{amount} = 105$$

14% of $750 is **$105.**

1. 25% of $440

$110

2. 75% of 320 flowers

240 flowers

3. 85% of 900 fish

765 fish

4. 65% of 420 yards

273 yards

5. 32% of 260 quarts

83.2 quarts

6. 44% of 430 liters

189.2 liters

7. 175% of 360 sales

630 sales

8. 150% of 90 units

135 units

9. 12.4% of 8300 meters

1029.2 meters

10. 13.2% of 9400 acres

1240.8 acres

11. 0.8% of $520

$4.16

12. 0.3% of $480

$1.44

Find the base using the percent equation.

Example: 48% of what number of shirts is 2496 shirts?

Solution:

$$\text{amount} = \text{percent} \cdot \text{base}$$

$$2496 = 0.48 \cdot b \quad \text{Write 48\% as the decimal 0.48.}$$

$$\frac{2496}{0.48} = \frac{\overset{1}{\cancel{0.48}} \cdot b}{\underset{1}{\cancel{0.48}}} \quad \text{Divide by 0.48.}$$

$$5200 = b \quad \text{base}$$

48% of **5200 shirts** is 2496 shirts.

13. 30 apartments is 15% of what number of apartments?

200 apartments

14. 16 books is 20% of what number of books?

80 books

15. 50% of what number of cookies is 63 cookies?

126 cookies

16. 75% of what number of wrenches is 675 wrenches?

900 wrenches

17. 238 wires is 35% of what number of wires?

680 wires

18. 621 tons is 45% of what number of tons?

1380 tons

19. $12\frac{1}{2}\%$ of what number of people is 135 people?

1080 people

20. $6\frac{1}{2}\%$ of what number of bottles is 130 bottles?

2000 bottles

21. $1\frac{1}{4}\%$ of what number of gallons is 3.75 gallons?

300 gallons

22. $2\frac{1}{4}\%$ of what number of files is 9 files?

400 files

Find the percent using the percent equation.

Example: What percent of 80 inches is 20 inches?

Solution: amount = percent · base

$20 = p \cdot 80$ p is the unknown percent

$\frac{20}{80} = \frac{p \cdot \cancel{80}^{1}}{\cancel{80}_{1}}$ Divide by 80.

$0.25 = p$

$0.25 = 25\%$ Write the decimal as percent.

25% of 80 inches is 20 inches.

23. 35 cities is what percent of 70 cities?

50%

24. 60 boats is what percent of 150 boats?

40%

25. 19 videos is what percent of 25 videos?

76%

26. 75 offices is what percent of 125 offices?

60%

27. What percent of $264 is $330?

125%

28. What percent of $480 is $696?

145%

29. What percent of 80 is 1.2?

1.5%

30. What percent of 600 is 7.5?

1.25%

31. 999 is what percent of 740?

135%

32. 1224 is what percent of 850?

144%

NAME ____________________

33. When using the percent equation the percent must always be changed to a decimal before doing any calculations. Show and explain how to change a fraction percent to a decimal. Use $2\frac{1}{2}\%$ in your explanation.

Answer varies.

34. Suppose a problem on your homework assignment was, "Find $\frac{1}{2}\%$ of \$1300." Your classmates got answers of \$0.65, \$6.50, \$65, and \$650. Which answer is correct? How and why are they getting all of these answers? Explain.

Answer varies.

Solve each word problem.

35. The product known as WD-40 is used in 79% of U.S. homes. If there are 104.2 million homes in the U.S., find the number of homes that have WD-40 in them. Round to the nearest tenth of a million.

82.3 million homes

36. Most shampoos contain 75% to 90% water. If a 16-ounce bottle of shampoo contains 78% water, find the number of ounces of water in the 16-ounce bottle. Round to the nearest tenth of an ounce.

12.5 ounces

37. A gardener has 52 clients, 25% of whom are residential. Find the number of residential clients.

13 clients

38. For a tour of the eastern United States, a travel agent promised a trip of 3300 miles. Exactly 35% of the trip was by air. How many miles would be traveled by air?

1155 miles

39. In a test by *Consumer Reports*, six of the 123 cans of tuna that it analyzed contained more than the 30 microgram intake limit of mercury. What percent of the cans contained this level of mercury? Round to the nearest tenth of a percent.

4.9%

40. According to industry figures there are 44,500 hotels and motels in America. Economy hotels and motels account for 16,910 of this total. What percent of the total are economy hotels and motels?

38%

41. When a tank of an industrial chemical is 25% full it contains 175 gallons. How many gallons will it contain when it is completely full?

700 gallons

42. Sterling has completed 75% of the units needed for a degree. If he has completed 90 units, how many are needed for a degree?

120 units

43. The Chevy Camaro was introduced in 1967. Sales that year were 220,917 which was 46.2% of the number of Ford Mustangs sold in the same year. Find the number of Mustangs sold in the same year. Round to the nearest whole number.

478,175 Mustangs

44. Unemployment reported by the Bureau of Labor Statistics states that the number of unemployed people is 8.9 million or 7.1% of all workers. What is the size of the total workforce? Round to the nearest tenth of a million.

125.4 million

▲ **45.** J & K Mustang has increased the sale of auto parts by $32\frac{1}{2}$% over last year. If the sale of parts last year amounted to $385,200, find the volume of parts sales this year.

$510,390

▲ **46.** An ad for steel-belted radial tires promises 15% better mileage. If mileage has been 25.6 miles per gallon in the past, what mileage could be expected after new tires are installed? (Round to the nearest tenth of a mile.)

29.4 miles per gallon

▲ **47.** Electrical components originally priced at $14,000 are discounted 15%. Find the price after the discount.

$11,900

▲ **48.** A PC-Compatible Deskjet 500 printer priced at $599 is marked down 17%. Find the price of the printer after the markdown.

$497.17

PREVIEW EXERCISES

Identify p, b, and a in each of the following. Do not try to solve for any unknowns. (For help, see ***Section 6.4.****)*

	p	b	a
49. 15% of 375 drums is 56.25 drums.	15	375	56.25
50. What is 4% of 235 students?	4	235	unknown
51. 18 players is 36% of what number of players?	36	unknown	18
52. 3% of $2448 is $73.44.	3	2448	73.44
53. What percent of $830 is $128.65?	unknown	830	128.65
54. 364 stools is what percent of 1092 stools?	unknown	1092	364

6.7 APPLICATIONS OF PERCENT

Percent has many applications in our daily lives. This section discusses percent as it applies to sales tax, commissions, discounts, and the percent of change (increase and decrease).

OBJECTIVES

1. Find sales tax.
2. Find commissions.
3. Find the discount and sale price.
4. Find the percent of change.

FOR EXTRA HELP

Tape 9 | SSM pp. 177–181 | MAC: A IBM: A

1 States, counties, and cities often collect taxes on sales to customers. The **sales tax** is a percent of the total sale. You may use the following formula to find sales tax.

SALES TAX FORMULA

amount of sales tax = cost of item · rate of tax

■ **EXAMPLE 1** *Solving for Sales Tax*

Fit and Fine Cyclery sold a mountain bicycle for \$374. If the sales tax rate is 5%, how much tax was paid? What was the total cost of the bicycle?

Approach Use the formula above with the cost of the bicycle (\$374) as cost of the item and the tax rate (5%) to find the amount of sales tax.

Solution

$$\begin{aligned}\text{amount of sales tax} &= \$374 \cdot 5\% \\ &= \$374 \cdot 0.05 \\ &= \$18.70\end{aligned}$$

The tax paid on the bicycle is \$18.70. The customer buying the bicycle would pay a total cost of \$374 + \$18.70 = \$392.70. ■

WORK PROBLEM 1 AT THE SIDE.

1. Suppose the sales tax in your state is 6%. Find the amount of the tax for each item.

(a) \$100 watch

(b) \$542 television

(c) \$1287 sofa and chair

(d) \$14,520 automobile

■ **EXAMPLE 2** *Finding the Sales Tax Rate*

The sales tax on a \$10,800 pickup truck is \$702. Find the rate of the sales tax.

Approach Use the sales tax formula:

sales tax = cost of item · rate of tax

The rate of tax is the percent.

Solution Solve for the rate of tax. The cost of the pickup truck is \$10,800, and the amount of sales tax is \$702. Use r for the rate of tax (the percent).

$$\text{sales tax} = \text{cost of item} \cdot \text{rate of tax}$$

$$\$702 = 10{,}800 \cdot r$$

$$\frac{702}{10{,}800} = \frac{\overset{1}{\cancel{10{,}800}} \cdot r}{\underset{1}{\cancel{10{,}800}}} \quad \text{Divide both sides by 10,800.}$$

$$0.065 = r$$

$$0.065 = 6.5\% \quad \text{or} \quad 6\frac{1}{2}\% \quad \text{Write the decimal as percent.}$$

The sales tax rate is $6\frac{1}{2}\%$. ■

ANSWERS

1. (a) \$6 (b) \$32.52 (c) \$77.22 (d) \$871.20

2. Find the rate of sales tax.

(a) The tax on a $420 television set is $16.80.

(b) The tax on a $12 pair of running shorts is $0.78.

(c) The tax on an $18,800 car is $1316.

3. Find the commission.

(a) Eboni Perkins works on a commission rate of 8% and has sales for the month of $36,800.

(b) Last month the sales for Angie Gragg were $62,500 with a commission rate of 3%.

4. Find the rate of commission.

(a) A commission of $450 is earned on one sale of computer products worth $22,500.

(b) Jamal Story earns $2898 for selling office furniture worth $32,200.

ANSWERS

2. (a) 4% (b) 6.5% or $6\frac{1}{2}$% (c) 7%

3. (a) $2944 (b) $1875

4. (a) 2% (b) 9%

Note You can use the sales tax formula to find the amount of sales tax, the cost of an item, or the rate of sales tax (the percent).

WORK PROBLEM 2 AT THE SIDE.

2 Many salespeople are paid by **commission** rather than an hourly wage. In this method you are paid a certain percent of the total sales dollars. Use the following formula to find the commission.

COMMISSION FORMULA

amount of commission = rate or percent of commission · amount of sales.

EXAMPLE 3 *Determining the Amount of Commission*

Jill Beauteo sold dental tools worth $19,500. If her commission rate is 11%, find the amount of her commission.

Approach Use the commission formula: the rate of commission (11%) is multiplied by the amount of sales ($19,500).

Solution

$$\begin{aligned}\text{amount of commission} &= 11\% \cdot 19{,}500\\ &= 0.11 \cdot 19{,}500\\ &= 2145\end{aligned}$$

She earned a commission of $2145 for selling the dental tools. ■

WORK PROBLEM 3 AT THE SIDE.

EXAMPLE 4 *Finding the Rate of Commission*

A salesperson earned $510 for selling $17,000 worth of paper products. Find the rate of commission.

Approach You could use the commission formula. Or another approach is to use the percent proportion with $b = 17{,}000$, $a = 510$, and p unknown. (The rate of commission is the percent.)

Solution

$$\frac{510}{17{,}000} = \frac{p}{100}$$

Find cross products and solve this proportion to see that $p = 3$, so the rate of commission is 3%. ■

WORK PROBLEM 4 AT THE SIDE.

3 Most of us prefer buying things when they are on sale. A store will reduce prices, or **discount,** to attract additional customers. Use the following formula to find the discount and the sale price.

DISCOUNT FORMULA AND SALE PRICE FORMULA

amount of discount = original price · rate of discount
sale price = original price − amount of discount

EXAMPLE 5 *Application of a Sales Discount*
A furniture store has a sofa with an original price of $470 on sale at 15% off. Find the sale price of the sofa.

Approach This problem is solved in two steps. First, find the amount of the discount, that is, the amount that will be "taken off" (subtracted) by multiplying the original price ($470) by the rate of the discount (15%). The second step is to subtract the amount of discount from the original price. This gives you the sale price, what you will actually pay for the sofa.

Solution First find the amount of the discount.

amount of discount = 470 · 0.15   Write 15% as a decimal.
= 70.50

or $70.50. Find the sale price of the sofa by subtracting the amount of the discount from the original price.

sale price = $470 − $70.50
= $399.50

During the sale, you can buy the sofa for $399.50. ■

WORK PROBLEM 5 AT THE SIDE.

4 We are often interested in looking at increases or decreases in sales, production, population, and many other areas. This type of problem involves finding the percent of change. Use the following steps to find the **percent of increase.**

FINDING PERCENT OF INCREASE

Step 1 Use subtraction to find the amount of increase.
Step 2 Use the percent proportion to find the percent of increase.

$$\frac{\textbf{amount of increase}}{\textbf{original}} = \frac{\textbf{p}}{\textbf{100}}$$

5. Find the amount of the discount and the sale price.

(a) The Natural Leather Store offers leather jackets originally priced at $190 at a discount of 30%.

(b) Eastside Department Store has women's swimsuits on sale at 40% off. One swimsuit was originally priced at $34.

ANSWERS
5. (a) $57; $133 (b) $13.60; $20.40

6. Find the percent of increase.

(a) Production of aluminum boats increased from 9400 units last year to 12,690 this year.

(b) The number of flu cases rose from 496 cases last week to 620 this week.

7. Find the percent of decrease.

(a) The number of employees absent this week fell to 285 from 380 last week.

(b) The number of patients admitted to the hospital fell from 760 last month to 646 this month.

ANSWERS
6. (a) 35% (b) 25%
7. (a) 25% (b) 15%

■ **EXAMPLE 6** *Finding the Percent of Increase*
Attendance at county parks climbed from 18,300 last month to 56,730 this month. Find the percent of increase.

Approach Subtract the attendance last month (18,300) from the attendance this month (56,730) to find the amount of increase in attendance. Next, use the percent proportion, with $b = 18{,}300$ (last month's original attendance), $a = 38{,}430$ (amount of increase in attendance) and p as the unknown percent.

Solution

Step 1 $56{,}730 - 18{,}300 = 38{,}430$ amount of increase in attendance

Step 2 $\frac{38{,}430}{18{,}300} = \frac{p}{100}$ percent proportion

Solve this proportion to find that $p = 210$, so the percent of increase is 210%. ■

◀ WORK PROBLEM 6 AT THE SIDE.

Use the following steps to find the **percent of decrease.**

FINDING PERCENT OF DECREASE

Step 1 Use subtraction to find the amount of decrease.
Step 2 Use the percent proportion to find the percent of decrease

$$\frac{\textbf{amount of decrease}}{\textbf{original}} = \frac{\boldsymbol{p}}{\textbf{100}}$$

■ **EXAMPLE 7** *Finding the Percent of Decrease*
The number of production employees this week fell to 1406 people from 1480 people last week. Find the percent of decrease.

Approach Subtract the number of employees this week (1406) from the number of employees last week (1480) to find the amount of decrease. Next, use the percent proportion, with $b = 1480$ (last week's original number of employees), $a = 74$ (decrease in employees), and p as the unknown percent.

Solution

Step 1 $1480 - 1407 = 74$ decrease in number of employees

Step 2 $\frac{74}{1480} = \frac{p}{100}$ percent proportion

Solve this proportion to find that $p = 5$, so the percent of decrease is 5%. ■

Note When solving for percent of increase or decrease, the base is always the previous value or value before the change happened. The amount is the change in values, that is, how much something went up or went down.

◀ WORK PROBLEM 7 AT THE SIDE.

6.7 EXERCISES

NAME DATE HOUR

Find the amount of the sales tax or tax rate and the total cost (amount of sale + amount of tax = total cost). Round money answers to the nearest cent.

Example: The cost of a table is $235 and the sales tax rate is 5%. Find the amount of sales tax and the total cost.

Solution:

$$\begin{aligned}\text{sales tax} &= \text{cost of item} \cdot \text{rate of tax} \\ &= \$235 \cdot 5\% \\ &= \$235 \cdot 0.05 \\ &= \$11.75\end{aligned}$$

Sales tax is **$11.75** and the total cost is **$246.75** ($235 + $11.75).

	amount of sale	*tax rate*	*amount of tax*	*total cost*
1.	$100	5%	$5	$105
2.	$200	6%	$12	$212
3.	$75	4%	$3	$78
4.	$210	7%	$14.70	$224.70
5.	$365	6%	$21.90	$386.90
6.	$15	5%	$0.75	$15.75
7.	$220	$5\frac{1}{2}\%$	$12.10	$232.10
8.	$780	$7\frac{1}{2}\%$	$58.50	$838.50

Find the commission earned, or the rate of commission. Round money answers to the nearest cent.

Example: The sales are \$9850 and the commission rate is 3%. Find the amount of commission.

Solution: commission = rate of commission · sales

= 3% · \$9850

= 0.03 · \$9850

= \$295.50

The amount of commission is **\$295.50.**

	sales	*rate of commission*	*commission*
9.	\$200	8%	\$16
10.	\$450	6%	\$27
11.	\$1200	22%	\$264
12.	\$2800	11%	\$308
13.	\$5783	3%	\$173.49
14.	\$2275	7%	\$159.25
15.	\$45,250	10%	\$4525
16.	\$65,300	5%	\$3265

Find the amount or rate of discount and the amount paid after the discount. Round money answers to the nearest cent.

Example: An AM-FM, cassette car radio is normally priced at $278. If it is discounted 20%, find the amount of discount and the sale price.

Solution: discount = original price · rate of discount
= $278 · 20%
= $278 · 0.2
= $55.60

The amount of discount is **$55.60** and the sale price is **$222.40** ($278 − $55.60).

	original price	*rate of discount*	*amount of discount*	*sale price*
17.	$100	25%	$25	$75
18.	$200	10%	$20	$180
19.	$252	25%	$63	$189
20.	$19	30%	$5.70	$13.30
21.	$17.50	25%	$4.38	$13.12
22.	$76	60%	$45.60	$30.40
23.	$37.50	10%	$3.75	$33.75
24.	$49.90	40%	$19.96	$29.94

25. If you were trying to decide between Company A paying a 10% commission and Company B paying an 8% commission, for which company would you prefer to work? Are there considerations other than commission rate that would be important to you? What would they be?

Answer varies.

26. Give four examples of where you might use the percent of increase or the percent of decrease in your own personal activities. Think in terms of work, school, home, hobbies, and sports. Write an increase or a decrease problem about one of these four examples, then show how to solve it.

Answer varies.

Solve the following word problems. Round money answers to the nearest cent and rates to the nearest tenth of a percent.

27. The sales tax rate is 6% and the sales are \$725. Find the amount of sales tax.

\$43.50

28. An exercise machine sells for \$480 plus 5% sales tax. Find the amount of sales tax.

\$24

29. Alan's Shoes sells shoes at 30% off the regular price. Find the sale price of a shoe normally priced at \$45.

\$31.50

30. Stephen Louis can purchase a new car at 8% below sticker price. Find his cost on a car with a window sticker price of \$17,650.

\$16,238

31. A gold bracelet costs \$1450 with a sales tax of \$58. Find the rate of sales tax.

4%

32. Textbooks for three classes cost \$135 plus sales tax of \$8.10. Find the sales tax rate.

6%

33. Enrollment in computer science courses increased from 2480 students last semester to 3286 students this semester. Find the percent of increase.

32.5%

34. Americans are eating more fish. This year the average American will eat $15\frac{1}{2}$ pounds compared to only $12\frac{1}{2}$ pounds per year a decade ago. Find the percent of increase. (Hint: $15\frac{1}{2} = 15.5$; $12\frac{1}{2} = 12.5$.)

24%

35. The number of industrial accidents this month fell to 989 accidents from 1276 accidents last month. Find the percent of decrease.

22.5%

36. The average number of hours worked in manufacturing jobs last week fell from 41.1 to 40.9. Find the percent of decrease.

0.5%

37. A "super 40% off sale" begins today. What is the price of a hair dryer normally priced at $20?

$12

38. What is the sale price of a $995 bedroom set with a discount of 45%?

$547.25

39. Tara Chatard is a sales representative for a cosmetic company. If she was paid a commission of $459 on sales of $7650, find her rate of commission.

6%

40. Easthills Ski Center has just been sold for $1,692,804. The real estate agent selling the Center earned a commission of $47,400. Find the rate of commission, rounding to the nearest tenth of a percent.

2.8%

41. An 8-millimeter camcorder normally priced at $1180 is on sale for 35% off. Find the discount and the sales price.

$413; $767

42. This week minivans are offered at 15% off manufacturers' suggested price. Find the discount and the sale price of a minivan originally priced at $16,500.

$2475; $14,025

43. The price per share of Pacific Enterprise common stock fell from $\$38\frac{7}{8}$ per share to $\$26\frac{1}{4}$ per share in one year. Find the percent of decrease. (Hint: $\$38\frac{7}{8} = \$38.875$; $\$26\frac{1}{4} = \$26.25$)

32.5%

44. In the past five years, the cost of generating electricity from the sun has been brought down from 24 cents per kilowatt hour to 8 cents (less than the newest nuclear power plants). Find the percent of decrease.

66.7%

45. College students are offered a 6% discount on a dictionary that sells for $18.50. If the sales tax is 6%, find the cost of the dictionary including the sales tax.

$18.43

46. A FAX machine priced at $398 is marked down 7% to promote the new model. If the sales tax is also 7%, find the cost of the FAX machine including sales tax.

$396.05

47. A real estate agent sells a house for $129,605. A sales commission of 6% is charged. The agent gets 55% of this commission. How much money does the agent get?

$4276.97

48. The local real estate agents' association collects a fee of 2% on all money received by its members. The members charge 6% of the selling price of a property as their fee. How much does the association get, if its members sell property worth a total of $8,680,000?

$10,416

49. What is the total price of a boat with an original price of $13,905, if it is sold at an 18% discount? Sales tax is 4.75%.

$11,943,70

50. Annual business defaults climbed from $60 million one year to $840 million the next year. Find the percent of increase.

1300%

PREVIEW EXERCISES

Use the percent equation (amount = percent · base) to find amount, base, or percent. (For help, see ***Section 6.6.****)*

51. 15% of 280 screwdrivers

42 screwdrivers

52. 12.4% of $4150

$514.60

53. 0.3% of $960

$2.88

54. 0.5% of 700 barrels

3.5 barrels

55. $6\frac{1}{4}$% of what number is 50?

800

56. $5\frac{1}{2}$% of what number is 66?

1200

57. 147.2 meters is what percent of 460 meters? 32%

58. 125.8 yards is what percent of 740 yards? 17%

6.8 SIMPLE INTEREST

When we open a savings account we are actually lending money to the financial institution. It will in turn lend this money to individuals and businesses that become borrowers. The financial institution pays a fee to the savings account holders and charges a higher fee to its borrowers. This fee is called interest.

Interest is a fee paid or a charge made for lending or borrowing money. The amount of money borrowed is called the **principal.** The charge for interest is often given as a percent, called the interest rate or **rate of interest.** The rate of interest is assumed to be *per year,* unless stated otherwise.

OBJECTIVES

1. Find the simple interest on a loan.
2. Find the total amount due on a loan.

FOR EXTRA HELP

Tape 9 | SSM pp. 181–183 | MAC: A IBM: A

1 In most cases interest is computed on the original principal and is called **simple interest.** We use the following **interest formula** to find simple interest.

FORMULA FOR SIMPLE INTEREST

$$\text{interest} = \text{principal} \cdot \text{rate} \cdot \text{time}$$

The formula is usually written in letters.

$$I = p \cdot r \cdot t$$

Note Simple interest is used for most short-term business loans, most real estate loans, and many automobile and consumer loans.

EXAMPLE 1 *Finding Interest for a Year*

Find the interest on $2000 at 6% for 1 year.

The amount borrowed, or principal (p), is $2000. The interest rate (r) is 6%, which is 0.06 as a decimal, and the time of the loan (t) is 1 year. Using the formula

$$I = p \cdot r \cdot t,$$

gives

$$I = 2000 \cdot (0.06) \cdot 1$$

or

$$I = \$120.$$

The interest is $120. ■

WORK PROBLEM 1 AT THE SIDE.

EXAMPLE 2 *Finding Interest for More Than a Year*

Find the interest on $4200 at 8% for three and a half years.

The principal (p) is $4200. The rate ($r$) is 8% or 0.08 as a decimal, and the time (t) is $3\frac{1}{2}$ or 3.5 years. Use the formula

$$I = p \cdot r \cdot t.$$
$$I = 4200 \cdot (0.08) \cdot (3.5)$$
$$I = \$1176$$

The interest charge is $1176. ■

WORK PROBLEM 2 AT THE SIDE.

1. Find the interest.

(a) $600 at 3% for 1 year

(b) $1430 at 5% for 1 year

2. Find the interest.

(a) $85 at 4% for $4\frac{1}{2}$ years

(b) $2450 at 8% for $3\frac{1}{4}$ years

(c) $14,200 at 6% for $2\frac{3}{4}$ years

ANSWERS
1. (a) $18 (b) $71.50
2. (a) $15.30 (b) $637 (c) $2343

3. Find the interest.

(a) 1500 at 7% for 4 months

(b) $35,000 at 11% for 2 months

4. Find the total amount due on a loan of

(a) $480 at 8% for 8 months

(b) $10,800 at 6% for 4 years

(c) $4300 at 10% for $2\frac{1}{2}$ years

ANSWERS
3. (a) $35 (b) $641.67 (rounded)
4. (a) $505.60 (b) $13,392 (c) $5375

Interest rates are given *per year*. For loan periods of less than one year, be careful to express time as a fraction of a year.

If time is given in months, for example, use a denominator of 12, because there are 12 months in a year. A loan of 9 months would be for $\frac{9}{12}$ of a year.

EXAMPLE 3 *Finding Interest for Less Than 1 Year*
Find the interest on $840 at 8% for 9 months.

The principal is $840. The rate is 8% or 0.08, and the time is $\frac{9}{12}$ of a year. Use the formula $I = p \cdot r \cdot t$.

$$I = \underbrace{840 \cdot (0.08)}_{} \cdot \frac{9}{12} \quad \left(9 \text{ months} = \tfrac{9}{12} \text{ of a year}\right)$$

$$= 67.2 \cdot \frac{3}{4} \quad \left(\tfrac{9}{12} \text{ in lowest terms is } \tfrac{3}{4}\right)$$

$$= \frac{(67.2) \cdot 3}{4}$$

$$= \frac{201.6}{4} = 50.40$$

The interest is $50.40. ■

WORK PROBLEM 3 AT THE SIDE.

2 When a loan is repaid, the interest is added to the original principal to find the total amount due.

FORMULA FOR AMOUNT DUE

amount due = principal + interest

EXAMPLE 4 *Calculating the Total Amount Due*
A loan of $1080 has been made at 8% for three months. Find the total amount due.

First find the interest, then add the principal and the interest to find the total amount due.

$$I = 1080 \cdot (0.08) \cdot \frac{3}{12}$$

$$I = 21.60$$

The interest is $21.60.

amount due = principal + interest
= $1080 + $21.60 = $1101.60

The total amount due is $1101.60. ■

WORK PROBLEM 4 AT THE SIDE.

6.8 EXERCISES

NAME DATE HOUR

Find the interest.

Example: Find the interest on \$750 at 5% for 2 years.

Solution:

$$I = p \cdot r \cdot t$$
$$= 750 \cdot (0.05) \cdot 2$$
$$= \$75$$

The interest is **\$75.**

	principal	*rate*	*time in years*	*interest*
1.	\$100	5%	1	\$5
2.	\$300	5%	3	\$45
3.	\$600	6%	4	\$144
4.	\$800	7%	2	\$112
5.	\$60	4%	1	\$2.40
6.	\$190	3%	2	\$11.40
7.	\$1500	9%	6	\$810
8.	\$5280	7%	5	\$1848
9.	\$820	4%	$2\frac{1}{2}$	\$82
10.	\$430	3%	$1\frac{1}{2}$	\$19.35
11.	\$960	6%	$1\frac{1}{4}$	\$72
12.	\$1000	8%	$3\frac{1}{4}$	\$260

Find the interest. Round to the nearest cent if necessary.

Example: Find the interest on \$980 at 7% for 6 months.

Solution:

$$I = p \cdot r \cdot t$$

$$= \underbrace{980 \cdot (0.07)} \cdot \frac{6}{12} \qquad \text{(6 months} = \tfrac{6}{12} \text{ of a year)}$$

$$= \quad 68.6 \quad \cdot \frac{1}{2}$$

$$= \frac{68.6}{2} = 34.3$$

The interest is **\$34.30.**

	principal	*rate*	*time in months*	*interest*
13.	\$300	8%	6	\$12
14.	\$400	9%	9	\$27
15.	\$750	5%	12	\$37.50
16.	\$920	6%	18	\$82.80
17.	\$820	8%	24	\$131.20
18.	\$92	4%	5	\$1.53
19.	\$750	3%	10	\$18.75
20.	\$522	8%	6	\$20.88
21.	\$1160	6%	3	\$17.40
22.	\$2310	7%	3	\$40.43
23.	\$14,500	4%	7	\$338.33
24.	\$10,800	3%	5	\$135

Find the total amount due on the following loans.

Example: A loan of \$550 was made at 6% for 8 months. Find the total amount due.

Solution:

$$I = p \cdot r \cdot t$$
$$= 550 \cdot 0.06 \cdot \frac{8}{12}$$
$$= 33 \cdot \frac{2}{3}$$
$$= \frac{33 \cdot 2}{3} = \frac{66}{3} = 22$$

The interest is \$22.

amount due = principal + interest
= \$550 + \$22 = \$572

The total amount due is **\$572.**

	principal	*rate*	*time*	*total amount due*
25.	\$400	5%	1 year	\$420
26.	\$700	4%	6 months	\$714
27.	\$680	3%	3 months	\$685.10
28.	\$1020	8%	2 years	\$1183.20
29.	\$1500	10%	18 months	\$1725
30.	\$3000	5%	5 months	\$3062.50
31.	\$2450	7%	6 months	\$2535.75
32.	\$5400	4%	1 year	\$5616
33.	\$1780	10%	6 months	\$1869
34.	\$15,400	7%	5 years	\$20,790
35.	\$18,200	8%	8 months	\$19,170.67
36.	\$22,400	4%	9 months	\$23,072

37. Some lenders refer to the interest that they charge as the rent that must be paid for using their money. Name and describe the factors that are used in determining the amount of rent charged by the lender.

Answer varies.

38. Interest rates are usually given as a rate per year (annual rate). Explain what must be done when time is given in months. Write your own problem where time is given in months and then show how to solve it.

Answer varies.

Solve the following word problems. Round to the nearest cent.

39. Bill Monroe deposits $2850 at 7% for 1 year. How much interest will he earn?

$199.50

40. The Jidobu family invests $18,000 at 9% for 6 months. What amount of interest will the family earn?

$810

41. Joann Selzy lends $6500 to her daughter for the down payment on a home. The loan is for 18 months at 8%. How much interest will she earn?

$780

42. A retiree deposits $40,000 at 6% for 2 years. How much interest will be earned?

$4800

43. A student borrows $1000 at 10% for 3 months to pay tuition. Find the total amount due.

$1025

44. A loan of $1350 will be paid back with 12% interest at the end of 9 months. Find the total amount due.

$1471.50

45. Chadonna deposits $3850 in a credit union account for 11 months. If the credit union pays 5% interest, find the amount of interest she will earn.

$176.46

46. Silvo Di Loreto, owner of Sunset Realtors, borrows $27,000 to update his office computer system. If the loan is for 24 months at 7%, find the amount of interest he will owe.

$3780

47. An investment fund pays $7\frac{1}{4}\%$ interest. If Beverly Habecker deposits \$8800 in her account for $\frac{1}{4}$ year, find the amount of interest she will earn.

\$159.50

48. Ms. Henderson owes \$1900 in taxes. She is charged a penalty of $12\frac{1}{4}\%$ annual interest and pays the taxes and penalty after 6 months. Find the total amount she must pay.

\$2016.38

49. A gift shop owner has additional profits of \$11,500 that are invested at $8\frac{3}{4}\%$ interest for 9 months. Find the total amount in the account at the end of 9 months.

\$12,254.69

50. A pawn shop owner lends \$35,400 to another business for 6 months at an interest rate of 14.9%. How much interest will be earned on the loan?

\$2637.30

PREVIEW EXERCISES

Rewrite each of the following fractions with the indicated denominator. (For help, see ***Section 3.2.****)*

51. $\frac{3}{4} = \frac{6}{8}$

52. $\frac{2}{3} = \frac{6}{9}$

53. $\frac{4}{9} = \frac{16}{36}$

54. $\frac{3}{4} = \frac{36}{48}$

55. $\frac{5}{12} = \frac{25}{60}$

56. $\frac{4}{5} = \frac{80}{100}$

57. $\frac{15}{19} = \frac{60}{76}$

58. $\frac{7}{15} = \frac{98}{210}$

QUEST FOR NUMERACY

Figuring the Tip

A general guideline in sit-down restaurants is to leave a tip equal to 15% of your total bill. Suppose your total bill is $12.38.

Using your calculator: Enter the amount of the bill and use the percent key.

12.38 × 15 % (Do not press = key.)
Calculator shows 1.857.
Round to $1.90 or even $2.00.

Using a shortcut: Find 10% of $12.38 by moving the decimal point one place to the left. 12.38 = 1.238

Round 1.238 to the nearest dime. ⟶ $1.20
Add on half of $1.20 to get the other 5%.
Half of $1.20 is $0.60. ⟶ + 0.60
$1.80

Round up to $ 2.00 if you wish.

WHY IT WORKS

10% is $\frac{10}{100}$, which is equivalent to $\frac{1}{10}$. The fraction $\frac{1}{10}$ means something is divided into 10 parts. The shortcut for dividing by 10 is moving the decimal point one place to the left.

Try the shortcut on these restaurant bills. Then use your calculator to check your answer.

Total bill $24.77	
10% is	$2.47
Round to	$2.50
Add half	+ 1.25
Tip is	$3.75
Check by Calculator	$3.72

Total bill $8.05	
10% is	$0.805
Round to	$0.80
Add half	+ 0.40
Tip is	$1.20
Check by Calculator	$1.21

Total bill $142.16	
10% is	$14.216
Round to	$14.20
Add half	+ 7.10
Tip is	$21.30
Check by Calculator	$21.32

Extend this activity by asking students during one week to jot down their actual restaurant bills and the steps they took to find the tip. Discuss with students how to quickly find 20% tips for the time when they can afford to go to a posh restaurant.

6.9 COMPOUND INTEREST

OBJECTIVES

1. Understand compound interest.
2. Use a compound interest table.
3. Find the compound amount and the amount of compound interest.
4. Find interest compounded more often than annually.

FOR EXTRA HELP

Tape 9

SSM pp. 184–186

MAC: A IBM: A

The interest we studied in Section 6.8 was *simple interest* (interest only on the original principal). A common type of interest used with savings accounts and most investments is **compound interest,** or interest paid on past interest as well as on the principal.

1 Suppose Faith Lucio makes a single deposit of \$1000 in a savings account that earns 5% per year. What will happen to her savings over three years? At the end of the first year, one year's interest on the original deposit is found.

Interest = principal · rate · time

Year 1 \$1000 · 0.05 · 1 = \$50
Add the interest to the \$1000 to find the amount in her account at the end of the first year. \$1000 + \$50 = \$1050
The interest for the second year is found on \$1050; that is, the interest is **compounded.**

Year 2 \$1050 · 0.05 · 1 = \$52.50
Add this interest to the \$1050 to find the amount in her account at the end of the second year. \$1050 + \$52.50 = \$1102.50
The interest for the third year is found on \$1102.50.

Year 3 \$1102.50 · 0.05 · 1 = \$55.13
Add this interest to the \$1102.50. (\$1102.50 + \$55.13 = \$1157.63)

At the end of three years, Faith Lucio will have \$1157.63 in her savings account.

If she had earned only *simple* interest, then, for 3 years,

$$I = 1000 \cdot 0.05 \cdot 3$$
$$= 150$$

and she would have \$1000 + \$150 = \$1150 in her account. Compounding the interest increased her earnings by \$7.63.

With *compound* interest, the amount earned during the second year is greater than that earned during the first year, and the interest earned during the third year is greater than that earned during the second year.

This happens because the interest earned each year is added to the principal, and the new total is used to find the amount of interest in the next year.

COMPOUND INTEREST

Interest paid on principal plus past interest is **compound interest.**

2 Compound interest tables are used to calculate compound interest. The table on page 401 gives the **compound amount** (principal and interest) on a \$1 deposit for given lengths of time and interest rates.

1. Find the compound amount using the table on page 401.

(a) $1 at 4% for 12 years

(b) $1 at 3% for 24 years

(c) $1 at $5\frac{1}{2}$% for 8 years

2. Find the compound amount and the interest.

(a) $5000 at 6% for 14 years

(b) $14,100 at $3\frac{1}{2}$% for 10 years.

(c) $25,600 at 8% for 16 years.

ANSWERS
1. (a) $1.60 (b) $2.03 (c) $1.53
2. (a) $11,304.50; $6304.50
(b) $19,889.46; $5789.46
(c) $87,703.04; $62,103.04

EXAMPLE 1 *Using a Compound Interest Table*
Find the compound amount.

(a) $1 is deposited at a 5% interest rate for 10 years.
Look down the column headed 5%, and across to row 10 (10 years = 10 time periods). At the intersection of the column and row, read the compound amount, **1.6289,** which can be rounded to $1.63.

(b) $1 is deposited at $5\frac{1}{2}$% for 21 years.
The intersection of the $5\frac{1}{2}$% (5.50%) column and row 21 shows **3.0782** as the compound amount. Round this to $3.08. ■

WORK PROBLEM 1 AT THE SIDE.

3 Find the compound amount and interest as follows.

FINDING THE COMPOUND AMOUNT AND THE INTEREST

Compound Amount

Find the compound amount for any amount of principal by multiplying the principal by the compound amount for $1.

Interest

Find the amount of interest earned on a deposit by subtracting the amount originally deposited from the compound amount.

EXAMPLE 2 *Finding Compound Interest*
Find the compound amount and the interest.

(a) $1000 at $5\frac{1}{2}$% interest for 12 years

(b) $6400 at 8% for 7 years

(a) Look in the table for $5\frac{1}{2}$% (5.50%) and 12 periods; find the number 1.9012. Multiply this number and the principal of $1000.

$$\$1000 \cdot 1.9012 = \$1901.20$$

The account will contain $1901.20 after 12 years.

Find the amount of interest by subtracting the original deposit from the compound amount.

compound amount → original amount → amount of interest

$$\$1901.20 - \$1000 = \$901.20$$

(b) Look in the table for 8% and 7 periods, finding 1.7138. Multiply.

$$\$6400 \cdot 1.7138 = \$10{,}968.32 \quad \text{compound amount}$$

Subtract the original deposit from the compound amount.

$$\$10{,}968.32 - \$6400 = \$4568.32 \quad \text{interest}$$

A total of $4568.32 in interest was earned. ■

WORK PROBLEM 2 AT THE SIDE.

Compound Interest

time periods	*2.00%*	*2.50%*	*3.00%*	*3.50%*	*4.00%*	*4.50%*	*5.00%*	*5.50%*	*6.00%*	*8.00%*	*10.00%*	*12.00%*
1	1.0200	1.0250	1.0300	1.0350	1.0400	1.0450	1.0500	1.0550	1.0600	1.0800	1.1000	1.1200
2	1.0404	1.0506	1.0609	1.0712	1.0816	1.0920	1.1025	1.1130	1.1236	1.1664	1.2100	1.2544
3	1.0612	1.0769	1.0927	1.1087	1.1249	1.1412	1.1576	1.1742	1.1910	1.2597	1.3310	1.4049
4	1.0824	1.1038	1.1255	1.1475	1.1699	1.1925	1.2155	1.2388	1.2625	1.3605	1.4641	1.5735
5	1.1041	1.1314	1.1593	1.1877	1.2167	1.2462	1.2763	1.3070	1.3382	1.4693	1.6105	1.7623
6	1.1262	1.1597	1.1941	1.2293	1.2653	1.3023	1.3401	1.3788	1.4185	1.5869	1.7716	1.9738
7	1.1487	1.1887	1.2299	1.2723	1.3159	1.3609	1.4071	1.4547	1.5036	1.7138	1.9487	2.2107
8	1.1717	1.2184	1.2668	1.3168	1.3686	1.4221	1.4775	1.5347	1.5938	1.8509	2.1436	2.4760
9	1.1951	1.2489	1.3048	1.3629	1.4233	1.4861	1.5513	1.6191	1.6895	1.9990	2.3579	2.7731
10	1.2190	1.2801	1.3439	1.4106	1.4802	1.5530	1.6289	1.7081	1.7908	2.1589	2.5937	3.1058
11	1.2434	1.3121	1.3842	1.4600	1.5395	1.6229	1.7103	1.8021	1.8983	2.3316	2.8531	3.4785
12	1.2682	1.3449	1.4258	1.5111	1.6010	1.6959	1.7959	1.9012	2.0122	2.5182	3.1384	3.8960
13	1.2936	1.3785	1.4685	1.5640	1.6651	1.7722	1.8856	2.0058	2.1329	2.7196	3.4523	4.3635
14	1.3195	1.4130	1.5126	1.6187	1.7317	1.8519	1.9799	2.1161	2.2609	2.9372	3.7975	4.8871
15	1.3459	1.4483	1.5580	1.6753	1.8009	1.9353	2.0789	2.2325	2.3966	3.1722	4.1772	5.4736
16	1.3728	1.4845	1.6047	1.7340	1.8730	2.0224	2.1829	2.3553	2.5404	3.4259	4.5950	6.1304
17	1.4002	1.5216	1.6528	1.7947	1.9479	2.1134	2.2920	2.4848	2.6928	3.7000	5.0545	6.8660
18	1.4282	1.5597	1.7024	1.8575	2.0258	2.2085	2.4066	2.6215	2.8543	3.9960	5.5599	7.6900
19	1.4568	1.5987	1.7535	1.9225	2.1068	2.3079	2.5270	2.7656	3.0256	4.3157	6.1159	8.6128
20	1.4859	1.6386	1.8061	1.9898	2.1911	2.4117	2.6533	2.9178	3.2071	4.6610	6.7275	9.6463
21	1.5157	1.6796	1.8603	2.0594	2.2788	2.5202	2.7860	3.0782	3.3996	5.0338	7.4002	10.8038
22	1.5460	1.7216	1.9161	2.1315	2.3699	2.6337	2.9253	3.2475	3.6035	5.4365	8.1403	12.1003
23	1.5769	1.7646	1.9736	2.2061	2.4647	2.7522	3.0715	3.4262	3.8197	5.8715	8.9543	13.5523
24	1.6084	1.8087	2.0328	2.2833	2.5633	2.8760	3.2251	3.6146	4.0489	6.3412	9.8497	15.1786
25	1.6406	1.8539	2.0938	2.3632	2.6658	3.0054	3.3864	3.8134	4.2919	6.8485	10.8347	17.0001

4 In the previous examples, interest was calculated at the end of each year, or **compounded annually.** It is common for banks and other financial institutions to compound interest more often, such as every six months, or every quarter, or even every day.

EXAMPLE 3 *Compounding Semiannually*

Find the compound amount and the amount of interest earned on a deposit of $850 at 6% compounded semiannually for 8 years.

Approach "Semiannually" means twice a year. At 6% per year, the interest earned each six months is $6\% \div 2 = 3\%$. In 8 years, there are $8 \cdot 2 = 16$ semiannual periods. Thus, we look in the table for 3% and 16 periods, finding 1.6047.

Solution

$$\$850 \cdot 1.6047 = \mathbf{\$1364} \quad \text{compound amount (rounded)}$$

Subtract to find the amount of interest earned.

$$\mathbf{\$1364} - \$850 = \mathbf{\$514} \quad \text{interest}$$

WORK PROBLEM 3 AT THE SIDE.

EXAMPLE 4 *Compounding Quarterly*

Find the compound amount and the amount of interest earned on a deposit of $1500 at 12% compounded quarterly for 5 years.

Approach Because there are four quarters in one year, the interest earned per quarter is $12\% \div 4 = 3\%$ per quarter. In five years, there are $5 \cdot 4 = 20$ quarterly periods.

Solution Look for 3% and 20 periods in the table, finding 1.8061. The compound amount is

$$\$1500 \cdot 1.8061 = \mathbf{\$2709.15}$$

and the amount of interest earned is

$$\mathbf{\$2709.15} - \$1500 = \mathbf{\$1209.15}.$$

Note You can still use the table when the interest is compounded more often than annually. However, to find the correct column you must divide the annual interest rate by the number of times that the interest is compounded per year. To find the correct row you must multiply the number of years by the number of times the interest is compounded each year.

WORK PROBLEM 4 AT THE SIDE.

3. Find the compound amount and the amount of interest.

(a) $4000 at 8% compounded semiannually for 7 years

(b) $36,500 at 6% compounded semiannually for 9 years

4. Find the compound amount and the amount of interest.

(a) $1600 at 8% compounded quarterly for 6 years

(b) $14,500 at 10% for 3 years compounded quarterly

(c) $42,750 at 12% compounded quarterly for 6 years

ANSWERS

3. (a) $6926.80; $2926.80
(b) $62,137.60; $25,637.60

4. (a) $2573.44; $973.44
(b) $19,501.05; $5001.05
(c) $86,902.20; $44,152.20

6.9 EXERCISES

NAME DATE HOUR

Use the table on page 401 to find the compound amount. Interest is compounded annually.

Example: \$500 at 4% for 8 years

Solution: 4% column, row 8 of the table gives 1.3686.
Multiply. $\$500 \cdot 1.3686 =$ **\$684.30**

1. \$1000 at 6% for 4 years

   \$1262.50

2. \$10,000 at 3% for 15 years

   \$15,580

3. \$4000 at 5% for 9 years

   \$6205.20

4. \$5700 at 4% for 10 years

   \$8437.14

5. \$8428.17 at $4\frac{1}{2}$% for 6 years

   \$10,976.01

6. \$10,472.88 at $5\frac{1}{2}$% for 20 years

   \$30,557.77

Find the compound amount.

Example: \$600 at 6% compounded semiannually for 7 years

Solution: Look up 3% (6% ÷ 2) and $2 \cdot 7 = 14$ time periods.
Find 1.5126. Then multiply $\$600 \cdot 1.5126 =$ **\$907.56**

7. \$1000 at 8% compounded semiannually for 9 years

   \$2025.80

8. \$1000 at 10% compounded semiannually for 5 years

   \$1628.90

9. \$1400 at 5% compounded semiannually for 9 years.

   \$2183.58

10. \$8000 at 6% compounded semiannually for 11 years

    \$15,328.80

11. \$2800 at 8% compounded quarterly for 5 years

    \$4160.52

12. \$10,000 at 10% compounded quarterly for 6 years

    \$18,087

13. \$19,700 at 8% compounded quarterly for 3 years

    \$24,983.54

14. \$35,670 at 10% compounded quarterly for 6 years

    \$64,516.33

Find the missing numbers.

Example: $1250 at 8% compounded quarterly for 3 years

Solution: 8% ÷ 4 is 2% and 3 · 4 = 12 time periods. Look up 2%, 12 periods to find 1.2682.

Multiply. $1250 · 1.2682 = **$1585.25** compound amount
$1585.25 − $1250 = **$335.25** compound interest

	principal	*rate*	*compounded*	*time in years*	*compound amount*	*compound interest*
15.	$1000	8%	quarterly	4	$1372.80	$372.80
16.	$1000	4%	semiannually	6	$1268.20	$268.20
17.	$1270	6%	semiannually	10	$2293.75	$1023.75
18.	$1150	12%	quarterly	5	$2077.02	$927.02
19.	$1480	8%	quarterly	4	$2031.74	$551.74
20.	$1820	10%	quarterly	5	$2982.25	$1162.25
21.	$7700	7%	semiannually	7	$12,463.99	$4763.99
22.	$10,900	10%	semiannually	12	$35,153.59	$24,253.59

*Find the compound amount when interest is compounded **(a)** annually, **(b)** semiannually, and **(c)** quarterly.*

Example: $2000 8% 2 years

Solution: **(a)** annually 8%, 2 periods from the table is 1.1664
$2000 · 1.1664 = **$2332.80**

(b) semiannually 4%, 4 periods from the table is 1.1699
$2000 · 1.1699 = **$2339.80**

(c) quarterly 2%, 8 periods from the table is 1.1717
$2000 · 1.1717 = **$2343.40**

	principal	*rate*	*time*	**(a)** *annually*	**(b)** *semiannually*	**(c)** *quarterly*
23.	$2000	8%	3 years	$2519.40	$2530.60	$2536.40
24.	$4000	12%	6 years	$7895.20	$8048.80	$8131.20
25.	$17,600	10%	5 years	$28,344.80	$28,668.64	$28,839.36
26.	$22,400	8%	4 years	$30,475.20	$30,656.64	$30,750.72

27. Which period of compounding gives the highest return: annually, semiannually, or quarterly?

quarterly

28. Which period of compounding gives the lowest return?

annually

29. Write a definition for compound interest. Describe in your own words what compound interest means to you.

Answer varies.

30. What is the difference between the compound amount and compound interest?

Answer varies.

31. Explain in your own words how to use the compound interest table when interest is compounded quarterly. Make up a problem where interest is compounded quarterly and show how to solve it.

Answer varies.

32. How can you use the table to determine the time it will take a certain amount of principal to double at a particular interest rate? How long will it take for a principal amount to double given an annual compound rate of 5%, 8%, 12%?

Answer varies.

Solve each word problem. Round answers to the nearest cent if necessary.

33. Find the interest earned on \$10,000 for 4 years at 6% compounded (a) yearly and (b) semiannually.

(a) \$2625 (b) \$2668

34. Suppose \$32,000 is deposited for 2 years at 5% interest. Find the interest earned on the deposit if the interest is compounded (a) yearly and (b) semiannually.

(a) \$3280 (b) \$3321.60

35. Glenda Wong deposits \$5280 in an account that pays 8% interest, compounded quarterly. Find the amount she will have at the end of 5 years.

\$7845.55

36. Josh Mahoney loans \$9200 to the owner of a sporting goods store. He will be repaid at the end of 4 years, with interest at 10% compounded semiannually. Find the amount he will be repaid.

\$13,593

37. Al Granard lends $7500 to the owner of Rick's Limousine Service. He will be repaid at the end of 6 years at 8% interest compounded annually. Find (a) the total amount that he should be repaid and (b) the amount of interest earned.

(a) $11,901.75 (b) $4401.75

38. Sadie Simms has $28,500 in an Individual Retirement Account (IRA) that pays 6% interest compounded semiannually. Find (a) the total amount she will have at the end of 5 years and (b) the amount of interest earned.

(a) $38,301.15 (b) $9801.15

▲ **39.** There are two banks in Citrus Heights. One pays 8% interest compounded annually, and the other pays 8% compounded quarterly.

(a) If Bobbi deposits $10,000 in each bank, how much will she have in each bank at the end of 6 years?

(a) $15,869; $16,084

(b) How much more will she have in the bank that pays more interest?

(b) $215

▲ **40.** Which yields more interest for Barker Aluminum: $5000 deposited for 7 years at 10% simple interest or $4000 deposited for 7 years at 10% interest compounded semiannually? What is the difference in interest?

$4000 for 7 years at 10% semiannually; $419.60

▲ **41.** Jennifer Del Campo deposits $10,000 at 8% compounded quarterly. Two years after she makes the first deposit, she adds another $20,000, also at 8% compounded quarterly.

(a) What total amount will she have five years after her first deposit?

(b) What amount of interest will she have earned?

(a) $40,223.50 (b) $10,223.50

▲ **42.** Scott Striver invests $9000 at 10% compounded quarterly. Three years after he makes the first deposit, he adds another $15,000, also at 10% compounded quarterly.

(a) What total amount will he have five years after his first deposit?

(b) What amount of interest will he have earned?

(a) $33,023.64 (b) $9023.64

PREVIEW EXERCISES

Simplify by using the order of operations. (For help, see ***Section 4.6.****)*

43. $6.3 \div 4.2 \cdot 3.1$

4.65

44. $18.304 \div 8.32 \cdot 3$

6.6

45. $19.3 + (6.7 - 5.2) \cdot 58$

106.3

46. $2.12 + (9.7 - 7.9) \cdot 4.5$

10.22

47. $5.34 - 2.6 \cdot 5.2 \div 2.6$

0.14

48. $61.5 - 22.8 \cdot 15 \div 5.7$

1.5

KEY TERMS

6.1	**percent**	Percent means per one hundred. A percent is a ratio with a denominator of 100.
6.3–6.4	**percent proportion**	The proportion $\frac{\text{amount}}{\text{base}} = \frac{\text{percent}}{100}$ or $\frac{a}{b} = \frac{p}{100}$ is used to solve percent problems.
	base	The base in a percent problem is the entire quantity or the total.
	amount	The amount in a percent problem is the part being compared with the whole.
6.6	**percent equation**	The percent equation is amount = percent · base. It is another way to solve percent problems.
6.7	**sales tax**	Sales tax is a percent of the total sales charged as a tax.
	commission	Commission is a percent of the dollar value of total sales paid to a salesperson.
	discount	Discount is often expressed as a percent of the original price; it is then deducted from the original price, resulting in the sale price.
	percent of increase or decrease	Percent of increase or decrease is the amount of increase or decrease expressed as a percent of the original amount.
6.8	**interest**	Interest is a fee paid or a charge for lending or borrowing money.
	interest formula	The interest formula is used to calculate interest. It is interest = principal · rate · time or $I = p \cdot r \cdot t$.
	principal	Principal is the amount of money on which interest is earned.
	rate of interest	Often referred to as "rate," it is the charge for interest and is given as a percent.
6.9	**compound interest**	Compound interest is interest paid on past interest as well as on principal.
	compounding	Interest that is **compounded** once each year is compounded **annually;** interest that is compounded twice each year is compounded **semiannually;** and interest that is compounded four times each year is compounded **quarterly.**

QUICK REVIEW

Concepts	Examples
6.1 Basics of Percent	
Writing a Percent as a Decimal To write a percent as a decimal, move the decimal point two places to the left and drop the % sign.	50% (.50%) = 0.50 or just 0.5 3% (.03%) = 0.03
Writing a Decimal as a Percent To write a decimal as a percent, move the decimal point two places to the right and attach a % sign.	0.75 (0.75) = 75% 3.6 (3.60) = 360%

Concepts	Examples
6.2 Writing a Fraction as a Percent Use a proportion and solve for p to change a fraction to percent.	$\frac{2}{5} = \frac{p}{100}$ proportion $5 \cdot p = 2 \cdot 100$ cross products $5 \cdot p = 200$ $\frac{\overset{1}{\cancel{5}} \cdot p}{\underset{1}{\cancel{5}}} = \frac{200}{5}$ Divide by 5. $p = 40$ $\frac{2}{5} = 40\%$ Attach % sign.
6.3 Learning the Percent Proportion **Amount** is to **base** as percent is to 100, or $\frac{a}{b} = \frac{p}{100}.$	Use the percent proportion to solve for the missing number. $a = 30$, $b = 50$, find p $\frac{30}{50} = \frac{p}{100}$ percent proportion $\frac{3}{5} = \frac{p}{100}$ lowest terms $5 \cdot p = 3 \cdot 100$ percent proportion $5 \cdot p = 300$ $\frac{\overset{1}{\cancel{5}} \cdot p}{\underset{1}{\cancel{5}}} = \frac{300}{5}$ Divide by 5. $p = 60$ The percent is 60%.
6.4 Identifying Percent (p), Base (b), and Amount (a) in a Percent Problem The percent (p) appears with the word **percent** or with the symbol %. The base (b) often appears after the word **of.** Base is the entire quantity. The amount (a) is the part of the total. If $\boldsymbol{p}$ and $\boldsymbol{b}$ are found first, the remaining number is $\boldsymbol{a}$.	Find p, b, and a in each of the following. $\underbrace{\text{10\%}}_{p}$ **of** the $\underbrace{\text{500}}_{b}$ pies is $\underbrace{\text{how many pies?}}_{a \text{ (unknown)}}$ $\underbrace{\text{20 cats}}_{a}$ is $\underbrace{\text{5\%}}_{p}$ **of** $\underbrace{\text{what number of cats?}}_{b \text{ (unknown)}}$ $\underbrace{\text{What percent}}_{p \text{ (unknown)}}$ **of** $\underbrace{\text{\$220}}_{b}$ is $\underbrace{\text{\$33?}}_{a}$

Concepts	Examples
6.5 Applying the Percent Proportion	
Read the problem and identify p, b, and a. Use the percent proportion.	A tank contains 35% distilled water. 28 gallons of distilled water are in the tank when it is full. Find the volume of the tank. $p = 35$ and $a = 28$ Use the percent proportion to find b. $\frac{a}{b} = \frac{p}{100}$ $\frac{28}{b} = \frac{35}{100}$ $\frac{28}{b} = \frac{7}{20}$ lowest terms $b \cdot 7 = 560$ cross products $\frac{b \cdot \overset{1}{\cancel{7}}}{\underset{1}{\cancel{7}}} = \frac{560}{7}$ Divide by 7. $b = 80$ The volume of the tank is 80 gallons.
6.6 Using the Percent Equation	
The percent equation is amount = percent · base. Identify p, b, and a and solve for the missing quantity. Always write the percent as a decimal before using the equation.	Solve each of the following. **(a)** Find 15% of 160 drivers amount = percent · base $a = 0.15 \cdot 160$ $a = 24$ 15% of 160 drivers is 24 drivers. **(b)** 8 balls is 4% of what number of balls? amount = percent · base $8 = 0.04 \cdot b$ $\frac{8}{0.04} = \frac{\overset{1}{\cancel{0.04}} \cdot b}{\underset{1}{\cancel{0.04}}}$ $b = 200$ 8 balls is 4% of 200 balls. **(c)** 13 is what percent of 52 amount = percent · base $13 = p \cdot 52$ $\frac{13}{52} = \frac{p \cdot \overset{1}{\cancel{52}}}{\underset{1}{\cancel{52}}}$ $p = 0.25 = 25\%$ \$13 is 25% of \$52.

Concepts	Examples
6.7 Applications of Percent	
To solve for **sales tax,** use the formula amount of sales tax = cost of item · rate of tax.	The cost of an item is \$450, and the sales tax is 6%. Find the sales tax. amount of sales tax = \$450 · 6% = \$450 · 0.06 = \$27
To find **commissions,** use the formula amount of commission = rate or percent of commission · amount of sales.	The sales are \$92,000 with a commission rate of 3%. Find the commission. amount of commission = \$92,000 · 3% = \$92,000 · 0.03 = \$2760
To find the **discount** and the **sale price,** use the formulas amount of discount = original price · rate of discount; **sale price** = original price − amount of discount	A dishwasher originally priced at \$480 is offered at a 25% discount. Find the amount of the discount and the sale price. discount = \$480 · 0.25 = **\$120**; **sale price** = \$480 − **\$120** = **\$360**
To find the **percent of change,** subtract to find the amount of change (increase or decrease), which is the amount (a). Base (b) is the previous value or value before the change.	Enrollment rose from 3820 students to 5157 students. Find the percent of increase. $5157 - 3820 = 1337$ increase; $\frac{1337}{3820} = \frac{p}{100}$. Solve the proportion to find that $p = 35$ so the percent of increase is 35%.
6.8 Finding Simple Interest	
Use the formula $I = p \cdot r \cdot t$ interest = principal · rate · time. Time (t) is in years. When the time is given in months, use a fraction with 12 in the denominator because there are 12 months in a year.	\$2800 is deposited at 8% for 3 months. Find the amount of interest. $I = p \cdot r \cdot t$ $= 2800 \cdot (0.08) \cdot \frac{3}{12}$ $= 224 \cdot \frac{1}{4} = \frac{224 \cdot 1}{4} = \$56$
6.9 Finding Compound Amount and Compound Interest	
Find the number of compounding periods and the interest rate per period. Use the table to find the interest on \$1. Multiply the table value by the principal to obtain the compound amount. Subtract the principal from the compound amount to obtain interest.	What is the compound amount and interest, if \$1500 is deposited at 12% interest compounded quarterly for 6 years? interest of 12% per year $= \frac{12\%}{4} = 3\%$ per period. Interest compounded quarterly means there are $6 \cdot 4 = 24$ total periods. Locate 3% across the top of the table and 24 periods at left. Table value is 2.0328. compound amount = \$1500 · 2.0328 = **\$3049.20**; interest = **\$3049.20** − \$1500 = **\$1549.20**

NAME DATE HOUR

CHAPTER 6 REVIEW EXERCISES

[6.1] *Write each of the following percents as decimals and decimals as percents.*

1. 50% 0.5

2. 250% 2.5

3. 13.7% 0.137

4. 0.085% 0.00085

5. 3.75 375%

6. 0.02 2%

7. 0.375 37.5%

8. 0.002 0.2%

[6.2] *Write each percent as a fraction or mixed number in lowest terms and each fraction as percent.*

9. 24% $\frac{6}{25}$

10. 6.25% $\frac{1}{16}$

11. 325% $3\frac{1}{4}$

12. 0.025% $\frac{1}{4000}$

13. $\frac{4}{5}$ 80%

14. $\frac{5}{8}$ 62.5%

15. $1\frac{1}{2}$ 150%

16. $\frac{1}{400}$ 0.25%

Complete this chart.

fraction	*decimal*	*percent*
$\frac{1}{8}$	**17.** 0.125	**18.** 12.5%
19. $\frac{3}{20}$	0.15	**20.** 15%
21. $2\frac{2}{5}$	**22.** 2.4	240%

[6.3] *Find the value of the missing letter in the percent proportion* $\frac{a}{b} = \frac{p}{100}$.

23. $a = 50$, $p = 5$

1000

24. $b = 960$, $p = 10$

96

[6.4] *Identify percent, base, and amount in each of the following. Do not try to solve.*

25. 40% of 150 bulbs is 60 bulbs.

40; 150; 60

26. 73 brooms is what percent of 90 brooms?

unknown; 90; 73

27. Find 46% of 1040 folders.

46; 1040; unknown

28. 418 handles is 30% of what number of handles?

30; unknown; 418

29. A golfer lost 3 of his 8 balls. What percent were lost?

unknown; 8; 3

30. Only 88% of the door keys cut will operate properly. If there are 1280 keys cut, find the number of keys that will operate properly.

88; 1280; unknown

[6.5] *Find the amount using the percent proportion or the multiplication shortcut.*

31. 15% of 900 refrigerators

135 refrigerators

32. 78% of 2250 telephones

1755 telephones

33. 0.9% of 4800 miles

43.2 miles

34. 0.2% of 1400 kilograms

2.8 kilograms

Find the base using the percent proportion.

35. 25 tablets is 5% of what number of tablets?

500 tablets

36. 348 vials is 15% of what number of vials?

2320 vials

37. 338.8 meters is 140% of what number of meters?

242 meters

38. 2.5% of what number of cases is 425 cases?

17,000 cases

Find the percent using the percent proportion. Round percent answers to the nearest tenth if necessary.

39. 115 signs is what percent of 230 signs?

50%

40. What percent of 1850 reams is 75 reams?

4.1%

41. What percent of 190 dozen is 18 dozen?

9.5%

42. What percent of 650 cans is 200 cans?

30.8%

[6.1–6.5] *Solve each word problem. Round to the nearest tenth if necessary.*

43. Patricia Przynosch pays 32% of her total earnings in state and federal income tax. Her earnings are $26,500 per year. Find her tax.

$8480

44. A bank certificate of deposit amounting to $4800 paid $211.20 as interest. Find the percent of return.

4.4%

[6.6] *Use the percent equation to find each of the following.*

45. 22% of $118

$25.96

46. 125% of 64 dumpsters

80 dumpsters

47. 0.128 ounces is what percent of 32 ounces?

0.4%

48. 101.5 liters is what percent of 58 liters?

175%

49. 33.6 miles is 28% of what number of miles?

120 miles

50. $23 is 16% of what number?

$143.75

[6.7] *Find the amount of sales tax and the total cost. Round to the nearest cent.*

	amount of sale	*tax rate*	*amount of tax*	*total cost*
51.	$210	4%	$8.40	$218.40
52.	$110	$5\frac{1}{2}\%$	$6.05	$116.05

Find the commission earned.

	sales	*rate of commission*	*commission*
53.	$520	15%	$78
54.	$69,500	6%	$4170

Find the amount of discount and the sale price. Round to the nearest cent.

	original price	rate of discount	amount of discount	sale price
55.	\$100	10%	\$10	\$90
56.	\$732.50	20%	\$146.50	\$586

[6.8] *Find the simple interest due on each loan.*

	principal	rate	time	interest
57.	\$200	8%	1 year	\$16
58.	\$1250	5%	$4\frac{1}{2}$ years	\$281.25

Find the simple interest paid on each investment.

	principal	rate	time in months	interest
59.	\$100	7%	6 months	\$3.50
60.	\$1640	4%	18 months	\$98.40

Find the total amount due on the following simple interest loans.

	principal	rate	time	total amount due
61.	\$350	$4\frac{1}{2}\%$	3 years	\$397.25
62.	\$1180	6%	4 months	\$1203.60

[6.9] *Find the missing numbers in the following compound interest problems. Use the table on page 401.*

	principal	rate	compounded	time in years	compound amount	compound interest
63.	\$3550	7%	semiannually	10	\$7063.79	\$3513.79
64.	\$1320	4%	semiannually	5	\$1609.08	\$289.08
65.	\$4200	8%	quarterly	3	\$5326.44	\$1126.44
66.	\$10,500	10%	quarterly	6	\$18,991.35	\$8491.35

MIXED REVIEW EXERCISES

Find the value of the missing letter in the percent proportion $\frac{a}{b} = \frac{p}{100}$.

67. $b = 80$, $p = 60$

48

68. $a = 574$, $p = 35$

1640

Use the percent proportion or equation to find each of the following.

69. 24% of 97 meters

23.28 meters

70. 327 cars is what percent of 218 cars?

150%

71. 0.6% of $85

$0.51

72. 99 teachers is 60% of what number of teachers?

165 teachers

73. 76 chickens is what percent of 190 chickens?

40%

74. 107.242 liters is 43% of what number of liters?

249.4 liters

Write the percents as decimals and the decimals as percents.

75. 25% 0.25

76. 100% 1

77. 4 400%

78. 7.15 715%

79. 8.5% 0.085

80. 0.621 62.1%

81. 0.375% 0.00375

82. 0.0006 0.06%

Write each percent as a fraction in lowest terms and each fraction as a percent. Round to the nearest tenth of a percent if necessary.

83. $\frac{1}{2}$ 50%

84. 38% $\frac{19}{50}$

85. 87.5% $\frac{7}{8}$

86. $\frac{3}{8}$ 37.5%

87. $32\frac{1}{2}\%$ $\frac{13}{40}$

88. $\frac{1}{5}$ 20%

89. 0.5% $\frac{1}{200}$

90. $1\frac{3}{4}$ 175%

Solve each of the following word problems. Round percent answers to the nearest tenth if necessary.

91. The owner of Fair Oaks Hardware deposits \$8520 at $5\frac{1}{2}\%$ for 9 months. Find the amount of interest earned.

\$351.45

92. Clarence Hanks borrows \$1620 at 14% for 18 months to buy a toy train collection. Find the total amount due.

\$1960.20

▲ **93.** After spending \$38 on a textbook, a student has \$342 remaining. What percent of the original amount of money still remains?

90%

94. Tommy Downs invests the money he inherited from his aunt at 6% compounded semiannually for 4 years. If the amount of money invested is \$12,500, find
(a) the compound amount at the end of 4 years and
(b) the amount of interest that he earned.

(a) \$15, 835 (b) \$3335

95. Linda Freitas, a real estate agent, sold two properties, one for \$105,000 and the other for \$145,000. After all of her expenses she receives a commission of $1\frac{1}{2}\%$ of total sales. Find the commission that she earned.

\$3750

96. Our mail carrier, Norm, saw his route expand from 481 residential stops to 520 residential stops. Find the percent of increase.

8.1%

97. The mileage on a car dropped from 32.8 miles per gallon to 28.5 miles per gallon. Find the percent of decrease.

13.1%

98. The sales tax rate is 6% and the amount of tax collected is \$478.20. Find the total sales.

\$7970

99. A young couple established a budget allowing 25% for rent, 30% for food, 8% for clothing, 20% for travel and recreation, and the remainder for savings. The man takes home \$1450 per month, and the woman takes home \$24,500 per year. How much money will the couple save in a year?

\$7123

100. Raymelle Revel invests \$9800 at 8% compounded quarterly for 6 years. Find
(a) the total amount that she will have at the end of 6 years and
(b) the amount of interest earned.

(a) \$15,762.32 (b) \$5962.32

NAME DATE HOUR

CHAPTER 6 TEST

Write each percent as a decimal and each decimal as a percent.

1. 35% — **1.** 0.35

2. 0.05% — **2.** 0.0005

3. 200% — **3.** 2.00 or 2

4. 0.625 — **4.** 62.5%

5. 1.7 — **5.** 170%

6. 0.8 — **6.** 80%

Write as fractions in lowest terms.

7. 37.5% — **7.** $\frac{3}{8}$

8. 0.25% — **8.** $\frac{1}{400}$

Write each fraction or mixed number as a percent.

9. $\frac{1}{4}$ — **9.** 25%

10. $\frac{5}{8}$ — **10.** 62.5%

11. $2\frac{1}{2}$ — **11.** 250%

Solve each of the following.

12. 32 pens is 5% of what number of pens? — **12.** 640 pens

13. $125 is what percent of $625? — **13.** 20%

14. Erica has saved 72% of the amount needed for a down payment on a home. She has saved $12,096. Find the total down payment needed. — **14.** $16,800

15. $2854.20

15. The price of a copy machine is $2680 plus sales tax of $6\frac{1}{2}$ %. Find the total cost of the copy machine including sales tax.

16. $246.40

16. An encyclopedia company pays its salespeople a commission of 28% on all sales. Find the commission earned for selling a set of encyclopedias for $880.

17. 35%

17. Enrollment in mathematics courses increased from 1440 students last semester to 1944 students this semester. Find the percent of increase.

18. Answer varies.

18. Write a word problem about finding the percent of increase or decrease. Give the step-by-step solution to the problem. Be certain to include the first step of subtracting to find the amount of increase or decrease.

19. Answer varies.

19. Write the formula used to find interest. Explain the difference in what to do if the time is expressed in months or in years. Write a problem that involves finding interest for 9 months and another problem that involves finding interest for $2\frac{1}{2}$ years. Use your own numbers for the principal and the rate. Show how to solve your problems.

Find the amount of discount and the sale price. Round answers to the nearest cent.

	original price	***rate of discount***
20.	$48	8%

20. $3.84; $44.16

	original price	***rate of discount***
21.	$182	37.5%

21. $68.25; $113.75

Find the interest on each of the following.

	principal	***rate***	***time***
22.	$2110	7%	$3\frac{1}{2}$ years
23.	$6750	6%	3 months

22. $516.95

23. $101.25

24. $2996

24. A parent borrows $2800 to help her child finish college. The loan is for 6 months at 14% interest. Find the total amount due on the loan.

25. $3200 for 2 years at 8% quarterly; $110.19

25. Which yields more interest for the P.T.A emergency fund: $3500 deposited for 2 years at 6% compounded semiannually or $3200 deposited for 2 years at 8% compounded quarterly? What is the difference in interest?

NAME DATE HOUR

CUMULATIVE REVIEW EXERCISES CHAPTERS 1–6

≈ *Round the numbers in each problem so there is only one non-zero digit. Then add, subtract, multiply, or divide the rounded numbers, as indicated, to estimate the answer. Finally, solve for the exact answer.*

1. *estimate* $\begin{array}{r} 10{,}000 \\ 70 \\ +\ 600 \\ \hline 10{,}670 \end{array}$ *exact* $\begin{array}{r} 9804 \\ 72 \\ +\ 648 \\ \hline 10{,}524 \end{array}$

2. *estimate* $\begin{array}{r} 1 \\ 30 \\ +\ 4 \\ \hline 35 \end{array}$ *exact* $\begin{array}{r} 0.79 \\ 29.548 \\ +\ 4.4 \\ \hline 34.738 \end{array}$

3. *estimate* $\begin{array}{r} 60{,}000 \\ -\ 50{,}000 \\ \hline 10{,}000 \end{array}$ *exact* $\begin{array}{r} 61{,}033 \\ -\ 51{,}040 \\ \hline 9\,993 \end{array}$

4. *estimate* $\begin{array}{r} 6 \\ -\ 3 \\ \hline 3 \end{array}$ *exact* $\begin{array}{r} 6.2 \\ -\ 2.7055 \\ \hline 3.4945 \end{array}$

5. *estimate* $\begin{array}{r} 8000 \\ \times\ 600 \\ \hline 4{,}800{,}000 \end{array}$ *exact* $\begin{array}{r} 7749 \\ \times\ 603 \\ \hline 4{,}672{,}647 \end{array}$

6. *estimate* $\begin{array}{r} 30 \\ \times\ 9 \\ \hline 270 \end{array}$ *exact* $\begin{array}{r} 32.7 \\ \times\ 8.5 \\ \hline 277.95 \end{array}$

7. *estimate* $40\overline{)40{,}000}$ = 1000 *exact* $43\overline{)38{,}786}$ = 902

8. *estimate* $8\overline{)2000}$ = 250 *exact* $7.6\overline{)2432}$ = 320

9. *estimate* $1\overline{)7}$ = 7 *exact* $0.8\overline{)6.76}$ = 8.45

Simplify each problem by using the order of operations.

10. $6^2 - 7 \cdot 5$ 1

11. $\sqrt{49} + 3 \cdot 2 - 5$ 8

12. $9 + 6 \div 3 + 7 \cdot 4$ 39

Round each number to the place shown.

13. 7899 to the nearest ten 7900

14. 5,678,159 to the nearest hundred thousand 5,700,000

15. \$375.499 to the nearest dollar \$375

16. \$451.825 to the nearest cent \$451.83

Add, subtract, multiply, or divide as indicated. Write answers in lowest terms and as whole or mixed numbers when possible.

17. $\frac{3}{4} + \frac{5}{8}$ $1\frac{3}{8}$

18. $\frac{1}{2} + \frac{2}{3}$ $1\frac{1}{6}$

19. $5\frac{2}{3} + 7\frac{2}{5}$ $13\frac{1}{15}$

20. $\frac{7}{8} - \frac{1}{2}$ $\frac{3}{8}$

21. $3\frac{1}{2} - 1\frac{2}{3}$ $1\frac{5}{6}$

22. $26\frac{1}{3} - 17\frac{4}{5}$ $8\frac{8}{15}$

23. $\frac{5}{3} \cdot \frac{3}{8}$ $\frac{5}{8}$

24. $7\frac{3}{4} \cdot 3\frac{3}{8}$ $26\frac{5}{32}$

25. $36 \cdot \frac{4}{5}$ $28\frac{4}{5}$

26. $\frac{5}{9} \div \frac{5}{8}$ $\frac{8}{9}$

27. $10 \div \frac{2}{5}$ 25

28. $2\frac{3}{4} \div 7\frac{1}{2}$ $\frac{11}{30}$

Write < or > to make a true statement.

29. $\frac{3}{4} < \frac{7}{8}$

30. $\frac{5}{12} < \frac{7}{15}$

31. $\frac{7}{15} > \frac{9}{20}$

Simplify each of the following. Use the order of operations as needed.

32. $\left(\frac{5}{8} - \frac{1}{2}\right) \cdot \frac{1}{3}$ $\frac{1}{24}$

33. $\frac{3}{4} \div \left(\frac{2}{5} + \frac{1}{5}\right)$ $1\frac{1}{4}$

34. $\left(\frac{5}{6} - \frac{5}{12}\right) - \left(\frac{1}{2}\right)^2 \cdot \frac{2}{3}$ $\frac{1}{4}$

Write each fraction as a decimal. Round to the nearest thousandth if necessary.

35. $\frac{2}{5}$ 0.4

36. $\frac{7}{8}$ 0.875

37. $\frac{17}{20}$ 0.85

38. $\frac{12}{14}$ 0.857

Write each of the following ratios in lowest terms. Be sure to make all necessary conversions.

39. 2 hours to 40 minutes

Compare in minutes $\frac{3}{1}$

40. There are 8 boys and 12 girls in a class. What is the ratio of girls to boys? $\frac{3}{2}$

41. $1\frac{5}{8}$ to 13 $\frac{1}{8}$

Use cross multiplication to decide whether the following proportions are true or false. Circle the correct answer.

42. $\frac{8}{20} = \frac{40}{100}$

(True) False

43. $\frac{64}{144} = \frac{48}{108}$

(True) False

Find the missing numbers in each proportion. Write answers as mixed numbers whenever possible.

44. $\frac{1}{5} = \frac{x}{15}$ 3

45. $\frac{315}{45} = \frac{21}{x}$ 3

46. $\frac{9}{x} = \frac{57}{114}$ 18

47. $\frac{x}{120} = \frac{7.5}{30}$ 30

Write each of the following percents as decimals. Write each of the following decimals as percents.

48. 25% 0.25

49. 139.7% 1.397

50. 300% 3.00 or 3

51. 2.62% 0.0262

52. 0.68 68%

53. 2.7 270%

54. 0.023 2.3%

Write each percent as a fraction or mixed number in lowest terms. Write each fraction as a percent.

55. 4% $\frac{1}{25}$

56. $37\frac{1}{2}\%$ $\frac{3}{8}$

57. 150% $1\frac{1}{2}$

58. $\frac{7}{8}$ 87.5% or $87\frac{1}{2}\%$

59. $\frac{1}{20}$ 5%

60. $2\frac{3}{4}$ 275%

Solve these percent problems.

61. 35% of 800 folders

280 folders

62. 10.8% of $3600 is how much?

$388.80

63. 36 cans is 20% of what number of cans?

180 cans

64. $4\frac{1}{2}\%$ of what number of miles is 76.5 miles?

1700 miles

65. What percent of 220 workers is 110 workers?

50%

66. 72 hours is what percent of 180 hours?

40%

Find the amount of sales tax and the total cost. Round to the nearest cent.

	amount of sale	tax rate	amount of tax	total cost
67.	$53.99	4%	$2.16	$56.15
68.	$392	7%	$27.44	$419.44

Find the commission earned.

	sales	rate of commission	commission
69.	$14,622	5%	$731.10
70.	$358,560	3.4%	$12,191.04

Find the amount of discount and the sale price. Round to the nearest cent if necessary.

	original price	rate of discount	amount of discount	sale price
71.	$152	35%	$53.20	$98.80
72.	$238.50	22.5%	$53.66	$184.84

Find the total amount due on the following loans. Round to the nearest cent, if necessary.

	principal	rate	time	total amount to be repaid
73.	$714	9%	2 years	$842.52
74.	$18,350	11%	9 months	$19,863.88

Set up and solve a proportion for each problem.

75. 7 watches can be cleaned in 3 hours. Find the number of watches that can be cleaned in 12 hours.

28 watches

76. If 3.5 ounces of weed killer is needed to make 6 gallons of spray, how much weed killer is needed for 102 gallons of spray?

59.5 ounces

Solve the following word problems.

77. The number of entrants in this year's triathlon was 624. If 468 of these athletes entered the triathlon last year, find the percent of returning athletes.

75%

78. The number of electronic defects has increased from 660 defects last month to 891 defects this month. Find the percent of increase.

35%

79. After receiving a legal settlement, Joan Ong invests $13,440 in the ownership of a construction company. If this is 28% of the legal settlement, find the total amount that she received.

$48,000

80. Stephanie Hirata deposits $5750 in a savings account that pays 5% interest compounded semiannually. Find
(a) the amount that she will have in the account at the end of 4 years and
(b) the amount of interest earned.

(a) $7005.80 (b) $1255.80

7 Measurement

mL mg km metric 98.6°F

We measure things all the time: the distance traveled on vacation, the floor area we want to cover with carpet, the amount of milk in a recipe, the weight of the bananas we buy at the store, the number of hours we work, and many more.

In the United States we still use the **English system** of measurement for many everyday activities. Examples of English units are inches, feet, quarts, ounces, and pounds. However, the fields of science, medicine, sports, and manufacturing increasingly use the **metric system** (meters, liters, and grams). And, because the rest of the world uses only the metric system, U.S. businesses are beginning to change to the metric system in order to compete internationally.

7.1 THE ENGLISH SYSTEM

OBJECTIVES

1. Know the basic units in the English system.
2. Convert among units.
3. Use unit fractions to convert among units.

FOR EXTRA HELP

Tape 9

SSM pp. 207–210

MAC: A IBM: A

1 Until the switch to the metric system is complete, we still need to know how to use the English system of measurement. The table below lists the relationships you should memorize. The time relationships are used in both the English and metric systems.

LENGTH	WEIGHT
1 foot = 12 inches (in.)	1 pound (lb) = 16 ounces (oz)
1 yard (yd) = 3 feet (ft)	1 ton (T) = 2000 pounds (lb)
1 mile (mi) = 5280 feet (ft)	
CAPACITY	**TIME**
1 cup (c) = 8 fluid ounces	1 week (wk) = 7 days
1 pint (pt) = 2 cups	1 day = 24 hours (hr)
1 quart (qt) = 2 pints (pt)	1 hour (hr) = 60 minutes (min)
1 gallon (gal) = 4 quarts (qt)	1 minute (min) = 60 seconds (sec)

As you can see, there is no simple or "natural" way to convert among these various measures. The units evolved over hundreds of years and were based on a variety of "standards." For example, one yard was the distance from the tip of a king's nose to his thumb when his arm was outstretched. An inch was three dried barleycorns laid end to end.

1. After memorizing the measurement conversions, answer these questions.

(a) 24 hours = _______ day

(b) 1 gallon = _______ quarts

(c) 60 seconds = _______ minute

(d) 1 yard = _______ feet

(e) 1 foot = _______ inches

(f) 1 pound = _______ ounces

(g) 2000 pounds = _______ ton

(h) 1 hour = _______ minutes

(i) 2 cups = _______ pint

(j) 1 day = _______ hours

(k) 1 minute = _______ seconds

(l) 1 quart = _______ pints

(m) 1 mile = _______ feet

ANSWERS

1. (a) 1 (b) 4 (c) 1 (d) 3 (e) 12 (f) 16 (g) 1 (h) 60 (i) 1 (j) 24 (k) 60 (l) 2 (m) 5280

■ **EXAMPLE 1** *Knowing English Measurement Units*

Memorize the English measurement conversions. Then answer these questions.

(a) 7 days = _______ week Answer: 1 week

(b) 1 yard = _______ feet Answer: 3 feet ■

◀ WORK PROBLEM 1 AT THE SIDE.

2 You often need to convert from one unit of measure to another. Two methods of converting measurements are shown here. Study each way and use the method you prefer. Some conversions can be done by deciding whether to multiply or divide.

CONVERTING AMONG MEASUREMENT UNITS

1. *Multiply* when converting from a larger unit to a smaller unit.
2. *Divide* when converting from a smaller unit to a larger unit.

■ **EXAMPLE 2** *Converting from One Unit of Measure to Another*

Convert each measurement.

(a) 7 feet to inches

(b) 4 pounds to ounces

(c) 20 quarts to gallons

(a) You are converting from a larger unit to a smaller unit (feet to inches), so multiply.
Because *1 foot = 12 inches,* multiply by 12.

$$7 \text{ feet} = 7 \cdot 12 = 84 \text{ inches}$$

(b) You are converting from a larger unit to a smaller unit, pounds to ounces, so multiply.
Because *1 pound = 16 ounces*, multiply by 16.

$$4 \text{ pounds} = 4 \cdot 16 = 64 \text{ ounces}$$

(c) You are converting from a smaller unit to a larger unit (quarts to gallons), so divide.
Because *4 quarts = 1 gallon,* divide by 4.

$$20 \text{ quarts} = \frac{20}{4} = 5 \text{ gallons}$$ ■

WORK PROBLEM 2 AT THE SIDE.

3 If you have trouble deciding whether to multiply or divide when converting units, using **unit fractions** will solve the problem. You'll also find this method useful in science classes. A unit fraction is equivalent to 1. For example:

$$\frac{12 \text{ inches}}{12 \text{ inches}} = 1$$

You know that 12 inches is the same as 1 foot. So you can substitute 1 foot for 12 inches in the numerator, or you can substitute 1 foot for 12 inches in the denominator. This makes two useful unit fractions.

$$\frac{1 \text{ foot}}{12 \text{ inches}} = 1 \quad \text{or} \quad \frac{12 \text{ inches}}{1 \text{ foot}} = 1$$

To convert from one measurement unit to another, just multiply by the appropriate unit fraction. Remember, a unit fraction is equivalent to 1. Multiplying something by 1 does *not* change its value.

Use these guidelines to choose the correct unit fraction.

CHOOSING A UNIT FRACTION

The *numerator* should use the measurement unit you want in the *answer*.
The *denominator* should use the measurement unit you want to *change*.

EXAMPLE 3 *Using Unit Fractions with Length Measurement*

(a) Convert 60 inches to feet. **(b)** Convert 9 feet to inches.

(a) Use a unit fraction with feet (the unit for your answer) in the numerator, and inches (the unit being changed) in the denominator. Because *12 inches = 1 foot,* the necessary unit fraction is

$$\frac{1 \text{ foot}}{12 \text{ inches}} \begin{matrix} \leftarrow \text{unit for your answer} \\ \leftarrow \text{unit being changed} \end{matrix}$$

Next, multiply 60 inches times this unit fraction. Write 60 inches as the fraction $\frac{60 \text{ inches}}{1}$. Then cancel units and numbers wherever possible.

$$60 \text{ inches} \cdot \frac{1 \text{ foot}}{12 \text{ inches}} = \frac{\overset{5}{\cancel{60}} \, \cancel{\text{inches}}}{1} \cdot \frac{1 \text{ foot}}{\underset{1}{\cancel{12}} \, \cancel{\text{inches}}} = 5 \text{ feet}$$

These units should match.

cancel inches
cancel numbers

2. Convert each measurement.

(a) 9 feet to inches

(b) 64 ounces to pounds

(c) 6 yards to feet

(d) 2 tons to pounds

(e) 36 pints to quarts

(f) 180 minutes to hours

(g) 4 weeks to days

ANSWERS
2. (a) 108 inches (b) 4 pounds (c) 18 feet (d) 4000 pounds (e) 18 quarts (f) 3 hours (g) 28 days

3. First write the unit fraction needed to make each conversion. Then complete the conversion.

(a) 36 inches to feet

unit fraction $\left.\right\} \dfrac{1 \text{ foot}}{12 \text{ inches}}$

(b) 14 feet to inches

unit fraction $\left.\right\} \dfrac{\text{inches}}{\text{foot}}$

(c) 60 inches to feet

unit fraction } ______

(d) 4 yards to feet

unit fraction } ______

(e) 39 feet to yards

unit fraction } ______

(f) 2 miles to feet

unit fraction } ______

ANSWERS

3. (a) 3 feet (b) $\dfrac{12 \text{ inches}}{1 \text{ foot}}$; 168 inches

(c) $\dfrac{1 \text{ foot}}{12 \text{ inches}}$; 5 feet

(d) $\dfrac{3 \text{ feet}}{1 \text{ yard}}$; 12 feet

(e) $\dfrac{1 \text{ yard}}{3 \text{ feet}}$; 13 yards

(f) $\dfrac{5280 \text{ feet}}{1 \text{ mile}}$; 10,560 feet

(b) Select the correct unit fraction to change 9 feet to inches.

$\dfrac{12 \text{ inches}}{1 \text{ foot}}$ ← unit for your answer / ← unit being changed

Multiply 9 feet times the unit fraction.

$$9 \text{ feet} \cdot \frac{12 \text{ inches}}{1 \text{ foot}} = \frac{9 \cancel{\text{feet}}}{1} \cdot \frac{12 \text{ inches}}{1 \cancel{\text{foot}}} = 108 \text{ inches}$$

These units should match. cancel feet ■

Note If no units will cancel, you made a mistake in choosing the unit fraction.

◀ WORK PROBLEM 3 AT THE SIDE.

■ **EXAMPLE 4** *Using Unit Fractions with Capacity and Weight Measurement*

(a) Convert 9 pints to quarts.

(b) Convert $7\frac{1}{2}$ gallons to quarts.

(c) Convert 36 ounces to pounds.

(a) Select the correct unit fraction.

$\dfrac{1 \text{ quart}}{2 \text{ pints}}$ ← unit for answer / ← unit being changed

Next multiply.

$$\frac{9 \cancel{\text{pints}}}{1} \cdot \frac{1 \text{ quart}}{2 \cancel{\text{pints}}} = \frac{9}{2} \text{ quarts} = 4\frac{1}{2} \text{ quarts}$$

↑ cancel pints

write as improper fraction

(b) $$\frac{7\frac{1}{2} \cancel{\text{gallons}}}{1} \cdot \frac{4 \text{ quarts}}{1 \cancel{\text{gallon}}} = \frac{15}{2} \cdot \frac{4}{1} \text{ quarts}$$

$$= \frac{15}{\cancel{2}_1} \cdot \frac{\cancel{4}^2}{1} \text{ quarts}$$

$$= 30 \text{ quarts}$$

(c) $$\frac{\cancel{36}^9 \cancel{\text{ounces}}}{1} \cdot \frac{1 \text{ pound}}{\cancel{16}_4 \cancel{\text{ounces}}} = \frac{9}{4} \text{ pounds} = 2\frac{1}{4} \text{ pounds}$$

In example 2(c), if you do $9 \div 4$ on your calculator, you get 2.25 pounds. English measurements usually use fractions or mixed numbers, like $2\frac{1}{4}$ pounds, or in example 2(a), $4\frac{1}{2}$ quarts. However, 2.25 pounds is also correct and is the way grocery stores often show weights of fruit and meat. ■

WORK PROBLEM 4 AT THE SIDE.

■ EXAMPLE 5 *Using Several Unit Fractions*

Sometimes you may need to use two or three unit fractions in problems like these.

(a) Convert 63 inches to yards.

(b) Convert 2 days to seconds.

(a) Use the unit fraction $\frac{1 \text{ foot}}{12 \text{ inches}}$ to change inches to feet and the unit fraction $\frac{1 \text{ yard}}{3 \text{ feet}}$ to change feet to yards. Notice how all the units cancel except yards, which is the unit you want in the answer.

$$\frac{63 \cancel{\text{ inches}}}{1} \cdot \frac{1 \cancel{\text{ foot}}}{12 \cancel{\text{ inches}}} \cdot \frac{1 \text{ yard}}{3 \cancel{\text{ feet}}} = \frac{63}{36} \text{ yards} = 1\frac{3}{4} \text{ yards}$$

You can also cancel the numbers.

$$\frac{\overset{7}{\cancel{\overset{21}{\cancel{63}}}}}{1} \cdot \frac{1}{\underset{4}{\cancel{12}}} \cdot \frac{1}{\underset{1}{\cancel{3}}} = \frac{7}{4} = 1\frac{3}{4} \text{ yards}$$

(b) Use three unit fractions. The first one changes days to hours, the second one changes hours to minutes, and the third one changes minutes to seconds. All the units cancel except seconds, which is what you want in your answer.

$$\frac{2 \cancel{\text{ days}}}{1} \cdot \frac{24 \cancel{\text{ hours}}}{1 \cancel{\text{ day}}} \cdot \frac{60 \cancel{\text{ minutes}}}{1 \cancel{\text{ hour}}} \cdot \frac{60 \text{ seconds}}{1 \cancel{\text{ minute}}} = 172{,}800 \text{ seconds}$$ ■

WORK PROBLEM 5 AT THE SIDE.

4. Convert using unit fractions.

(a) 16 pints to quarts

(b) 16 quarts to gallons

(c) 3 cups to pints

(d) $2\frac{1}{4}$ gallons to quarts

(e) 48 ounces to pounds

(f) $3\frac{1}{2}$ tons to pounds

(g) $1\frac{3}{4}$ pounds to ounces

(h) 4 ounces to pounds

5. Convert using two or three unit fractions.

(a) 4 tons to ounces

(b) 90 inches to yards

(c) 3 miles to inches

(d) 36 pints to gallons

(e) 4 weeks to minutes

(f) $1\frac{1}{2}$ gallons to cups

ANSWERS

4. (a) 8 quarts (b) 4 gallons (c) $1\frac{1}{2}$ pints (d) 9 quarts (e) 3 pounds (f) 7000 pounds (g) 28 ounces (h) $\frac{1}{4}$ pound or 0.25 pound

5. (a) 128,000 ounces (b) $2\frac{1}{2}$ yards or 2.5 yards (c) 190,080 inches (d) $4\frac{1}{2}$ gallons or 4.5 gallons (e) 40,320 minutes (f) 24 cups

QUEST FOR NUMERACY

Calorie Counting

With a growing interest in maintaining good health, most adults are familiar with the term "calorie." Scientifically, a calorie measures heat energy, but in everyday terms, a calorie relates to the amount and type of food you eat and your level of exercise and activity. Eating too many calories will cause you to gain weight unless you are also burning calories through exercise.

While you may know that eating high-calorie foods is not good for your body, it is very difficult to determine how many calories is a "high" amount. The following chart displays the calories in a variety of foods from McDonald's.

Calorie Content of McDonald's Foods

Item	*Calories*
Hamburger	255
Big Mac	500
Medium French Fries	320
Garden Salad	50
Chocolate Lowfat Shake	320

The next chart estimates the amount of calories a person burns in one hour by participating in various activities. The two columns represent 120- and 180-pound people. Notice that the heavier people burn calories faster. Also notice that common tasks such as food shopping can burn several calories even though they are not considered to be fitness activities.

Calories Burned per Hour

Activity	*120 lb. person*	*180 lb. person*
Moderate Aerobics	340	508
Food Shopping	204	307
Light Office Work	204	308
Stationary Cycling	300	450
Vacuuming	168	248
Walking	260	398
Jogging	444	673

Use the two charts on this page to solve the following problems.

1. How many calories are in a McDonald's lunch including a Hamburger, a Medium French Fries, and a Chocolate Shake? **895 calories**
2. How many more calories are in a Big Mac than a McDonald's Garden Salad? **450 calories**
3. How long must a 180 lb. person vacuum in order to burn 496 calories? **2 hours**
4. Which activity burns calories quicker: moderate aerobics or riding a stationary bicycle? **moderate aerobics**
5. About how long must a 120 lb. person jog in order to burn off the McDonald's lunch described in question #1? **about 2 hours**

EXERCISES

Fill in the blanks with the measurement relationships you have memorized.

Example: 1 hour = ____ minutes **Solution:** 1 hour = **60** minutes

1. 1 yard = 3 feet

2. 1 foot = 12 inches

3. 2 pints = 1 quart

4. 4 quarts = 1 gallon

5. 1 mile = 5280 feet

6. 1 week = 7 days

7. 2000 pounds = 1 ton

8. 16 ounces = 1 pound

9. 1 minute = 60 seconds

10. 1 day = 24 hours

Convert these measurements using unit fractions.

Example:
3000 pounds = ____ tons

Solution:

$$\frac{\overset{3}{\cancel{3000}}\ \cancel{\text{pounds}}}{1} \cdot \frac{1 \text{ ton}}{\underset{2}{\cancel{2000}}\ \cancel{\text{pounds}}} = \frac{3}{2} \text{ ton} = 1\frac{1}{2} \textbf{ tons or 1.5 tons}$$

cancel pounds
cancel numbers

11. 6 feet = 2 yards

12. 36 feet = 12 yards

13. 8 quarts = 2 gallons

14. 24 quarts = 6 gallons

Writing Conceptual Challenging ≈ Estimation

15. An adult gray whale may weigh 40 to 45 tons. How many pounds does it weigh?

80,000 to 90,000 pounds

16. The largest animal that ever lived is the blue whale. It may weigh 150 tons, which equals how many pounds?

300,000 pounds

17. 10 feet = 120 inches

18. 15 feet = 180 inches

19. 7 pounds = 112 ounces

20. 6 pounds = 96 ounces

21. 5 quarts = 10 pints

22. 13 quarts = 26 pints

23. 90 minutes = $1\frac{1}{2}$ or 1.5 hours

24. 45 seconds = $\frac{3}{4}$ or 0.75 minutes

25. 3 inches = $\frac{1}{4}$ or 0.25 feet

26. 30 inches = $2\frac{1}{2}$ or 2.5 feet

27. 24 ounces = $1\frac{1}{2}$ or 1.5 pounds

28. 36 ounces = $2\frac{1}{4}$ or 2.25 pounds

29. 5 cups = $2\frac{1}{2}$ or 2.5 pints

30. 15 quarts = $3\frac{3}{4}$ or 3.75 gallons

31. Mr. Kashpaws worked for 12 hours harvesting wild rice. What part of the day did he work?

$\frac{1}{2}$ day

32. Michelle prepares 4-ounce hamburgers at a fast-food restaurant. Each hamburger is what part of a pound?

$\frac{1}{4}$ pound

33. $2\frac{1}{2}$ tons = ___5000___ pounds

34. $4\frac{1}{2}$ pints = ___9___ cups

35. $4\frac{1}{4}$ gallons = ___17___ quarts

36. $2\frac{1}{4}$ hours = ___135___ minutes

37. Sandra's puppy weighed $1\frac{3}{4}$ pounds. How many ounces did the puppy weigh?

28 ounces

38. Lee has grown to be $5\frac{3}{4}$ feet tall. What is his height in inches?

69 inches

Use two or three unit fractions to convert the following.

Example:
3 days = _____ minutes

Solution:

$$\frac{3\ \cancel{\text{days}}}{1} \cdot \frac{24\ \cancel{\text{hours}}}{1\ \cancel{\text{day}}} \cdot \frac{60\ \text{minutes}}{1\ \cancel{\text{hour}}} = \textbf{4320 minutes}$$

39. 6 yards = ___216___ inches

40. 2 tons = ___64,000___ ounces

41. 112 cups = ___28___ quarts

42. 336 hours = ___2___ weeks

43. 6 days = ___518,400___ seconds

44. 5 gallons = ___80___ cups

45. $1\frac{1}{2}$ tons = ___48,000___ ounces

46. $3\frac{1}{3}$ yards = ___120___ inches

47. The statement 8 = 2 is *not* true. But with appropriate measurement units, it *is* true.

8 *quarts* = 2 *gallons*

Attach measurement units to these numbers to make the statements true.

(a) 1 = 16 pound/ounces
(b) 10 = 20 quarts/pints or pints/cups
(c) 120 = 2 min/hr or sec/min
(d) 2 = 24 feet/inches

48. Explain in your own words why you can add 2 feet + 12 inches to get 3 feet, but you cannot add 2 feet + 12 pounds.

Answer varies.

Convert the following.

49. $2\frac{3}{4}$ miles = 174,240 inches

50. $5\frac{3}{4}$ tons = 184,000 ounces

51. $6\frac{1}{4}$ gallons = 800 fluid ounces

52. $3\frac{1}{2}$ days = 302,400 seconds

PREVIEW EXERCISES

*Place < or > in each blank to make a true statement. (For help, see **Section 3.5.**)*

53. 2 weeks < 15 days

54. 72 hours < 4 days

55. 4 hours > 185 minutes

56. 2 years < 28 months

57. 32 days > 4 weeks

58. 14 minutes > 780 seconds

7.2 THE METRIC SYSTEM—LENGTH

Around 1790, a group of French scientists developed the metric system of measurement. It is an organized system based on multiples of 10, like our number system and our money. After you are familiar with metric units, you will see that they are easier to use than the hodgepodge of English measurement relationships you memorized in Section 7.1.

1 The basic unit of length in the metric system is the **meter** (also spelled *metre*). Use the symbol **m** for meter; do not put a period after it. If you put five of the pages from this textbook side by side, they would measure about 1 meter. Or, look at a yardstick—a meter is just a little longer. (A meter is about 39 inches long.)

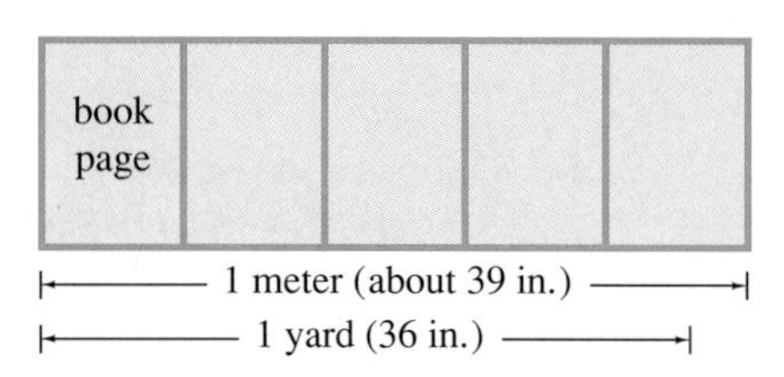

In the metric system you use meters for things like buying fabric for sewing projects, measuring the length of your living room, or talking about heights of buildings.

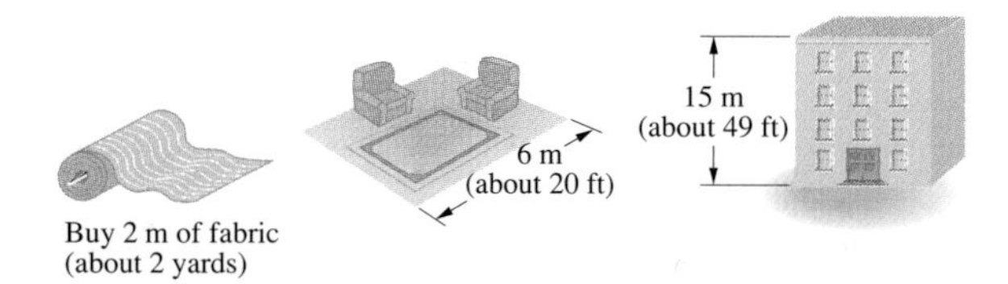

WORK PROBLEM 1 AT THE SIDE.

To make longer or shorter length units in the metric system, **prefixes** are written in front of the word meter. For example, the prefix *kilo* means 1000, so a *kilo*meter is 1000 meters. The table below shows how to use the prefixes for length measurements. It is helpful to memorize the prefixes because they are also used with weight and capacity measurements. The colored boxes are the units you will use most often in daily life.

prefix	*kilo-* meter	*hecto-* meter	*deka-* meter	meter	*deci-* meter	*centi-* meter	*milli-* meter
meaning	1000 meters	100 meters	10 meters	1 meter	$\frac{1}{10}$ of a meter	$\frac{1}{100}$ of a meter	$\frac{1}{1000}$ of a meter
symbol	**km**	**hm**	**dam**	m	**dm**	**cm**	**mm**

OBJECTIVES

1. Know the basic metric units of length.
2. Use unit fractions to convert among units.
3. Move the decimal point to convert among units.

FOR EXTRA HELP

Tape 9 | SSM pp. 210–212 | MAC: A IBM: A

1. Which measurements are about 1 meter?

length of a pencil

length of a baseball bat

height of doorknob from the floor

height of a house

basketball player's arm length

ANSWERS

1. baseball bat, height of door knob, arm length

2. Write the most reasonable metric unit in each blank: km, m, cm, or mm.

(a) The woman's height is 168 _____.

(b) The man's waist is 90 _____ around.

(c) Louise ran the 100 _____ dash in the track meet.

(d) A postage stamp is 22 _____ wide.

(e) Michael paddled his canoe 2 _____ across the lake.

(f) The pencil lead is 1 _____ thick.

(g) A stick of gum is 7 _____ long.

(h) The highway speed limit is 90 _____ per hour.

(i) The classroom was 12 _____ long.

(j) A penny is about 18 _____ across.

ANSWERS

2. (a) cm (b) cm (c) m (d) mm (e) km (f) mm (g) cm (h) km (i) m (j) mm

Here are some comparisons to help you get acquainted with the commonly used length units: km, m, cm, mm. *Kilo*meters will be used instead of miles. A kilometer is 1000 meters. It is about 0.6 mile (a little more than half a mile) or about 6 city blocks. If you participate in a 10 km run, you'll go about 6 miles.

A meter is divided into 100 smaller pieces called *centi*meters. Each centimeter is $\frac{1}{100}$ of a meter. Centimeters will be used instead of inches. A centimeter is a little shorter than $\frac{1}{2}$ inch. The cover of this textbook is 21 cm wide. A nickel is about 2 cm across. Measure the width and length of your little finger on this centimeter ruler. The width of your little finger is probably about 1 centimeter.

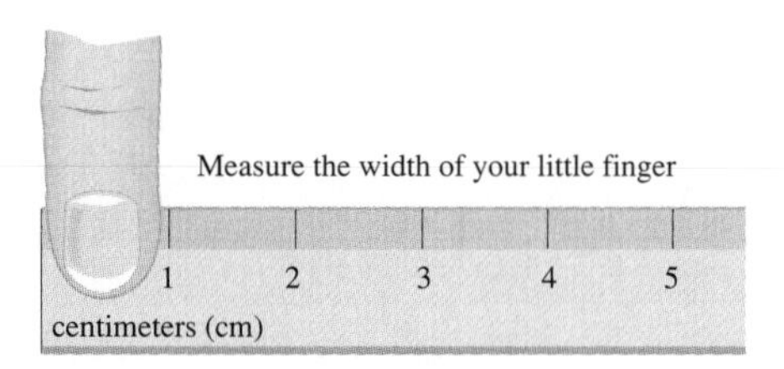

A meter is divided into 1000 smaller pieces called *milli*meters. Each millimeter is $\frac{1}{1000}$ of a meter. It takes 10 mm to equal 1 cm, so it is a very small length. The thickness of a dime is about 1 mm. Measure the width of your pen or pencil and the width of your little finger on this millimeter ruler.

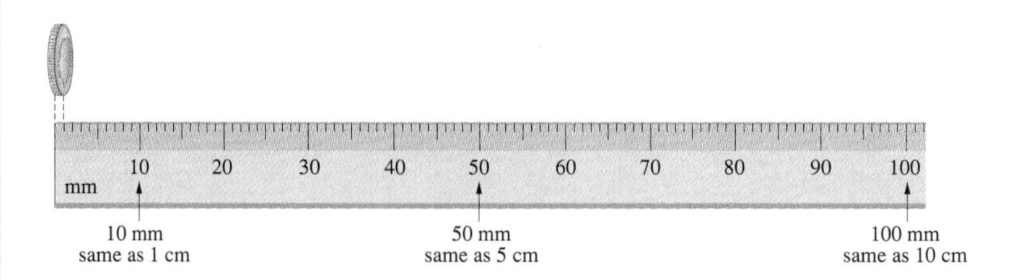

■ **EXAMPLE 1** *Using Metric Length Units*

Write the most reasonable metric unit in each blank. Choose from km, m, cm, and mm.

(a) The distance from home to work is 20 _______.

(b) My wedding ring is 4 _______ wide.

(c) The newborn baby is 50 _______ long.

(a) 20 km because kilometers are used instead of miles. 20 km is about 12 miles.

(b) 4 mm because the width of a ring is very small.

(c) 50 cm, which is half of a meter; a meter is about 39 inches so half a meter is around 20 inches. ■

◀ WORK PROBLEM 2 AT THE SIDE.

2 You can convert among metric length units using unit fractions. Keep these relationships in mind when setting up the unit fractions.

METRIC LENGTH RELATIONSHIPS

1 km = 1000 m
1 m = 1000 mm
1 m = 100 cm
1 cm = 10 mm

EXAMPLE 2 *Using Unit fractions with Length Measurement*
Convert the following.

(a) 5 km to m
(b) 18.6 cm to m

(a) Put the unit for the answer (meters) in the numerator of the unit fraction; the unit you want to change (km) in the denominator.

$$\frac{1000 \text{ m}}{1 \text{ km}} \Big\} \text{ Unit fraction equivalent to 1.}$$

Multiply. Cancel units where possible.

$$\frac{5 \cancel{\text{km}}}{1} \cdot \frac{1000 \text{ m}}{1 \cancel{\text{km}}} = 5000 \text{ m}$$

The answer makes sense because a kilometer is much longer than a meter, so 5 km will contain many meters.

(b) Multiply by a unit fraction that allows you to cancel centimeters.

$$\frac{18.6 \cancel{\text{cm}}}{1} \cdot \overbrace{\frac{1 \text{ m}}{100 \cancel{\text{cm}}}}^{\text{unit fraction}} = \frac{18.6}{100} \text{ m} = 0.186 \text{ m}$$

There are 100 cm in a meter, so 18.6 cm will be a small part of a meter. The answer makes sense. ■

WORK PROBLEM 3 AT THE SIDE.

3 By now you have probably noticed that conversions among metric units are made by multiplying or dividing by 10, by 100, or by 1000. A quick way to multiply by 10 is to move the decimal point one place to the *right*. Move it two places to multiply by 100, three places to multiply by 1000. Division is done by moving the decimal point to the *left*. Use a calculator, or hand calculations, to check the results of the examples at the side of the next page.

3. Convert using unit fractions.

(a) 3.67 m to cm

(b) 92 cm to m

(c) 432.7 cm to m

(d) 65 mm to cm

(e) 0.9 m to mm

(f) 2.5 cm to mm

ANSWERS
3. (a) 367 cm (b) 0.92 m (c) 4.327 m (d) 6.5 cm (e) 900 mm (f) 25 mm

4. Do each multiplication or division by hand or on a calculator. Compare your answer to the one obtained by moving the decimal point.

(a) 43.5 · 10 = ______

43.5 gives 435

(b) 43.5 ÷ 10 = ______

43.5 gives ______

(c) 28 · 100 = ______

28.00 gives ______

(d) 28 ÷ 100 = ______

28. gives ______

(e) 0.7 · 1000 = ______

0.700 gives ______

(f) 0.7 ÷ 1000 = ______

000.7 gives ______

ANSWERS
4. (a) 435 (b) 4.35 (c) 2800 (d) 0.28 (e) 700 (f) 0.0007

WORK PROBLEM 4 AT THE SIDE.

To help you move the decimal point in length conversions, use this **metric conversion line.**

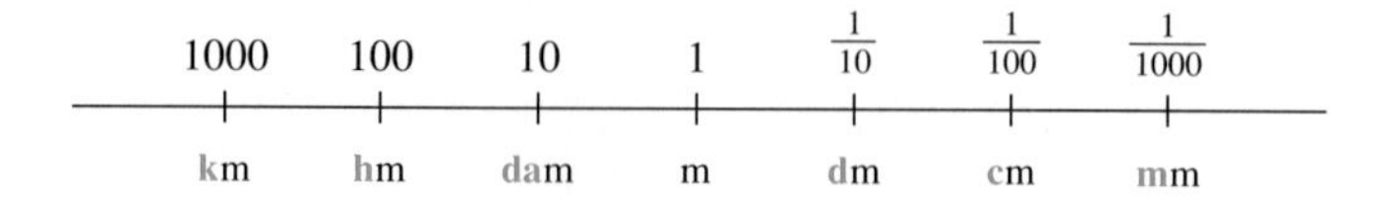

Here are the steps for using the conversion line.

USING THE METRIC CONVERSION LINE

1. Find the unit you are given on the metric conversion line.
2. Count the number of places to get from the unit you are given to the unit you want in the answer.
3. Move the decimal point the *same number of places.* Move in the *same direction* as you did on the conversion line.

EXAMPLE 3 *Using a Metric Conversion Line*

Use the metric conversion line to make the following conversions.

(a) 5.702 km to m

(b) 69.5 cm to m

(c) 8.1 cm to mm

(a) Find **km** on the metric conversion line. To get to **m**, you move *three places* to the *right.* So move the decimal point in 5.702 *three places* to the *right.*

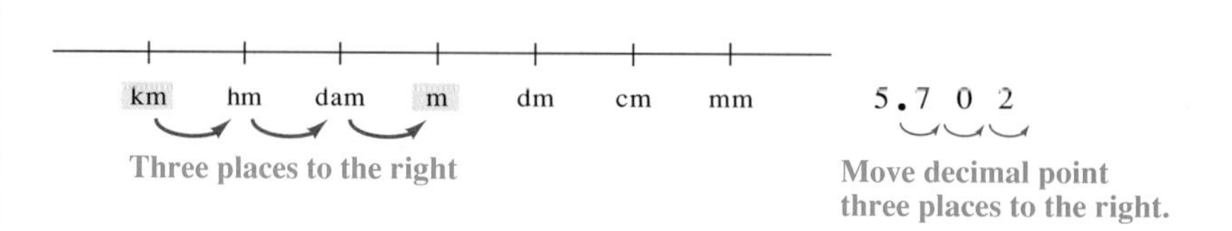

So 5.702 km = 5702 m.

(b) Find **cm** on the conversion line. To get to **m**, move *two places* to the *left.*

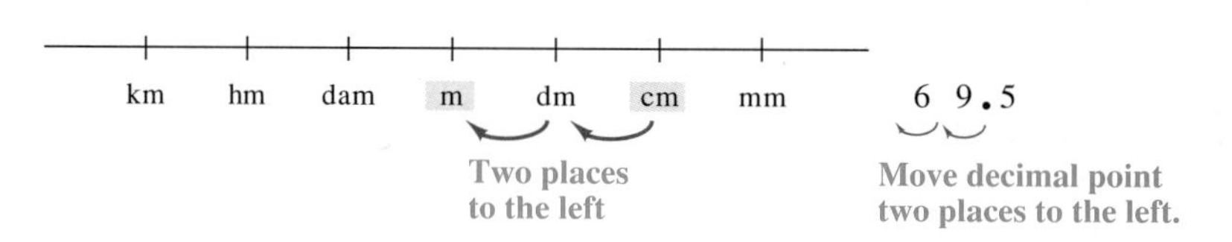

So 69.5 cm = 0.695 m.

(c) From **cm** to **mm** is *one place* to the *right*.

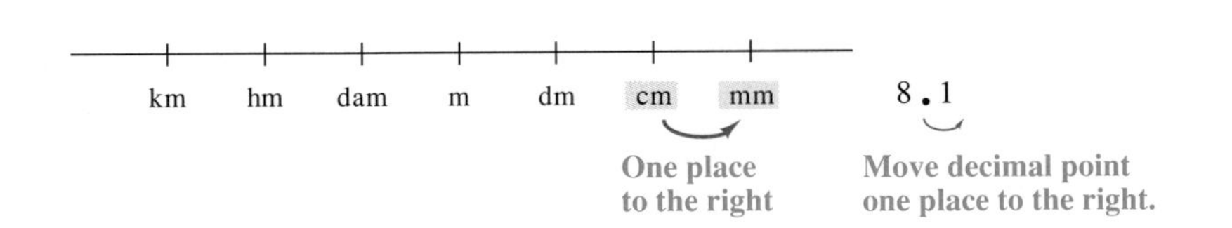

So 8.1 cm = 81 mm. ■

WORK PROBLEM 5 AT THE SIDE.

■ EXAMPLE 4 *Practicing Length Conversions*

Convert using the metric conversion line.

(a) 1.28 m to mm

(b) 60 cm to m

(c) 8 m to km

(a) Moving from **m** to **mm** is going three places to the right. In order to move the decimal point in 1.28 three places to the right, you must write a zero as a placeholder.

1.280 Zero is written in as placeholder.
Move decimal point three places to the right.

1.28 m = 1280 mm

(b) From **cm** to **m** is two places to the left. The decimal point in 60 starts at the *far right side* because 60 is a whole number. Then move it two places to the left.

60. Decimal point starts here.

60. Move decimal point two places. to the left.

60 cm = 0.60 m, which is equivalent to 0.6 m.

(c) From **m** to **km** is three places to the left. The decimal point in 8 starts at the far right side. In order to move it three places to the left, you must write two zeros as placeholders.

8. Decimal point starts here.

008. Two zeros written in as placeholders.
Move decimal point three places to the left.

8 m = 0.008 km. ■

WORK PROBLEM 6 AT THE SIDE.

5. Convert using the metric conversion line.

(a) 12.008 km to m

(b) 561.4 m to km

(c) 20.7 cm to m

(d) 20.7 cm to mm

(e) 4.66 m to cm

(f) 85.6 mm to cm

6. Convert using the metric conversion line.

(a) 9 m to mm

(b) 3 cm to m

(c) 14.6 km to m

(d) 5 mm to cm

(e) 70 m to km

(f) 0.8 m to cm

ANSWERS

5. (a) 12,008 m **(b)** 0.5614 km **(c)** 0.207 m **(d)** 207 mm **(e)** 466 cm **(f)** 8.56 cm

6. (a) 9000 mm **(b)** 0.03 m **(c)** 14,600 m **(d)** 0.5 cm **(e)** 0.07 km **(f)** 80 cm

QUEST FOR NUMERACY

Watts It Worth?

The amount of electricity used by different appliances is measured in watt-hours. You are charged a certain amount to use 1000 watt-hours (a *kilo* watt-hour, abbreviated kwh). Complete this table showing one family's yearly cost to run various appliances. They pay $0.08 per kwh. The cost in your area may be different. Round the yearly cost to the nearest dollar.

Item	*Hours in use per day*	*Hours in use per year (365 days)*	*Watts*	*Watt-hours per year*[1]	*Cost per year at $0.08 per kwh*[2]
Refrigerator[3]	12	4380	350	1,533,000	$123
Color television	5	1825	300	547,500	$44
Microwave oven	0.25	91.25	600	54,750	$4
Clock	24	8760	2	17,520	$1
Radio	3	1095	50	54,750	$4
Electric stove top	0.5	182.5	6000	1,095,000	$88
Toaster	0.1	36.5	1000	36,500	$3
Table lamp	6	2190	100	219,000	$18
Computer	1	365	120	43,800	$4

[1] To find watt-hours, multiply watts times number of hours appliance is in use.
[2] To find number of kwh, divide watt-hours by 1000. "Kilo" means "thousand" and there are 1000 watt-hours in one kwh. Or, using unit fractions, multiply watt-hours by $\frac{1 \text{ kwh}}{1000 \text{ watt-hours}}$.
[3] Although the refrigerator is plugged in 24 hours per day, it is assumed that the motor is running only half the time.

Extend this activity by having students estimate their own hours of usage for different appliances (or even keep a simple diary for one week). Have them find the local cost per kwh and then compute their yearly costs, setting up a table to record the results.

7.2 EXERCISES

NAME DATE HOUR

Use your knowledge of the meaning of metric prefixes to fill in the blanks.

Example: 1 hm = ______ m **Solution:** hecto means 100 so 1 hm = **100** m

1. 1 km = 1000 m

2. 1 dam = 10 m

3. 1 mm = $\frac{1}{1000}$ or 0.001 m

4. 1 dm = $\frac{1}{10}$ or 0.1 m

5. 1 cm = $\frac{1}{100}$ or 0.01 m

6. 1 hm = 100 m

Use this ruler to measure the following.

7. the width of your hand in centimeters

answer varies—about 8 cm

8. the width of your hand in millimeters

answer varies—about 80 mm

9. the width of your thumb in millimeters

answer varies—about 20 mm

10. the width of your thumb in centimeters

answer varies—about 2 cm

Write the most reasonable metric length unit in each blank. Choose from km, m, cm, and mm.

11. The child was 91 cm tall.

12. The cardboard was 3 mm thick.

13. Ming-Na swam in the 200 m backstroke race.

14. The bookcase is 75 cm wide.

15. Adriana drove 400 km on her vacation.

16. The door is 2 m high.

17. An aspirin tablet is 10 mm across.

18. Lamard jogs 4 km every morning.

19. A paper clip is about 3 cm long.

20. My pen is 145 mm long.

21. Dave's truck is 5 m long.

22. Wheelchairs need doorways that are at least 80 cm wide.

23. Describe at least three examples of metric length units that you have come across in your daily life. Answer varies.

24. Explain one reason why the metric system would be easier for a young child to learn than the English system. Answer varies.

Writing Conceptual Challenging Estimation

Convert each measurement. Use unit fractions or the metric conversion line.

Example: 16 mm to m

Solution: $\frac{16 \text{ mm}}{1} \cdot \frac{1 \text{ m}}{1000 \text{ mm}} = \frac{16}{1000} \text{ m} = \mathbf{0.016 \text{ m}}$

or: From mm to m is three places to the left 016. mm = **0.016 m**

25. 7 m to cm

700 cm

26. 18 m to cm

1800 cm

27. 35 mm to m

0.035 m

28. 806 mm to m

0.806 m

29. 9.4 km to m

9400 m

30. 0.7 km to m

700 m

31. 509 cm to m

5.09 m

32. 30 cm to m

0.3 m

33. 400 mm to cm

40 cm

34. 25 mm to cm

2.5 cm

35. 0.91 m to mm

910 mm

36. 4 m to mm

4000 mm

37. Is 82 cm greater than or less than 1 m? What is the difference in the lengths?

less; 18 cm or 0.18 m

38. Is 1022 m greater than or less than 1 km? What is the difference in the lengths?

greater; 22 m or 0.022 km

Convert each measurement.

▲ **39.** 981 km to cm 98,100,000 cm

▲ **40.** 16.5 km to mm 16,500,000 mm

▲ **41.** 5.6 mm to km 0.0000056 km

▲ **42.** 3 cm to km 0.00003 km

PREVIEW EXERCISES

Write each decimal as a fraction in lowest terms. (For help, see ***Section 4.1.****)*

43. 0.875 $\frac{7}{8}$

44. 0.6 $\frac{3}{5}$

45. 0.08 $\frac{2}{25}$

46. 0.075 $\frac{3}{40}$

7.3 THE METRIC SYSTEM—CAPACITY AND WEIGHT (MASS)

We use capacity units to measure liquids, such as the amount of milk in a recipe, the gasoline in our car tank, and the water in a swimming pool. English capacity units are cups, pints, quarts, and gallons. The basic metric unit for capacity is the **liter** (also spelled *litre*). The capital letter **L** is the symbol for liter, to avoid confusion with the numeral 1.

OBJECTIVES

1. Know the basic metric units of capacity.
2. Convert among metric capacity units.
3. Know the basic metric units of weight (mass).
4. Convert among metric weight (mass) units.

FOR EXTRA HELP

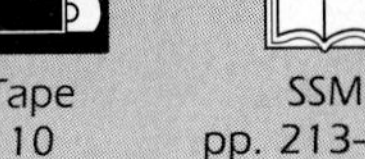

Tape 10 | SSM pp. 213–215 | MAC: A IBM: A

1 The liter is related to metric length in this way: a box that measures 10 cm on every side holds exactly one liter. (The volume of the box is 1000 cubic centimeters. Volume is discussed in **Section 8.7.**) A liter is just a little more than 1 quart.

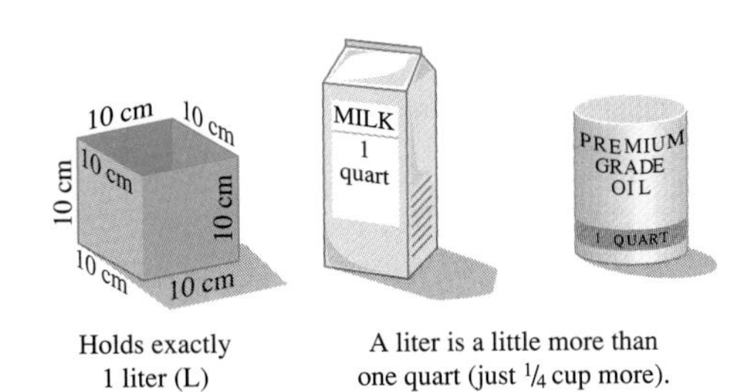

Holds exactly 1 liter (L)

A liter is a little more than one quart (just $\frac{1}{4}$ cup more).

In the metric system you use liters for things like buying milk at the store, filling a pail with water, and describing the size of your home aquarium.

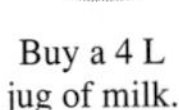

Buy a 4 L jug of milk.

Use a 14 L bucket to wash floors.

Watch the fish in your 40 L aquarium.

WORK PROBLEM 1 AT THE SIDE.

1. Which things would you measure in liters?

amount of water in the bathtub

length of the bathtub

width of your car

amount of gasoline you buy for your car

height of a pail

amount of water in a pail

To make larger or smaller capacity units we use the same **prefixes** as we did with length units. For example, *kilo* means 1000 so a *kilo*meter is 1000 meters. In the same way, a *kilo*liter is 1000 liters.

prefix	*kilo*-liter	*hecto*-liter	*deka*-liter	liter	*deci*-liter	*centi*-liter	*milli*-liter
meaning	1000 liters	100 liters	10 liters	1 liter	$\frac{1}{10}$ of a liter	$\frac{1}{100}$ of a liter	$\frac{1}{1000}$ of a liter
symbol	**kL**	**hL**	**daL**	L	**dL**	**cL**	**mL**

ANSWERS

1. water in bathtub, gasoline, water in a pail

2. Write the most reasonable metric unit in each blank. Choose from L and mL.

(a) I bought 8 _____ of milk at the store.

(b) The nurse gave me 10 _____ of cough syrup.

(c) This is a 100 _____ garbage can.

(d) It took 10 _____ of paint to cover the bedroom walls.

(e) My car's gas tank holds 50 _____.

(f) I added 15 _____ of oil to the pancake mix.

(g) The can of orange soda holds 350 _____.

(h) My boyfriend gave me a 30 _____ bottle of expensive perfume.

ANSWERS
2. (a) L (b) mL (c) L (d) L (e) L (f) mL (g) mL (h) mL

The capacity units you will use most often in daily life are liters (L) and *milli*liters (mL). A tiny box that measures 1 cm on every side holds exactly one milliliter. (In medicine, this small amount is also called 1 cubic centimeter, or 1 cc for short.) It takes 1000 mL to make 1 L. Here are some other useful comparisons.

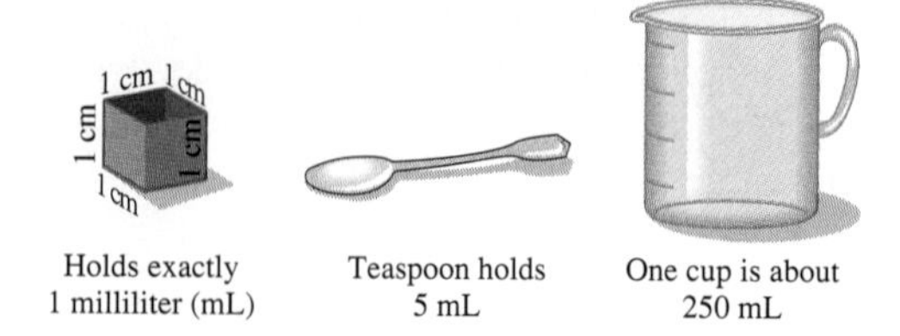

Holds exactly 1 milliliter (mL) | Teaspoon holds 5 mL | One cup is about 250 mL

EXAMPLE 1 *Using Metric Capacity Units*
Write the most reasonable metric unit in each blank. Choose from L and mL.

(a) The bottle of shampoo held 500 _____.

(b) I bought a 2 _____ carton of orange juice.

(a) 500 mL because 500 L would be about 500 quarts, which is too much.

(b) 2 L because 2 mL would be less than a teaspoon. ■

WORK PROBLEM 2 AT THE SIDE.

2 Just as with length units, you can convert between milliliters and liters using unit fractions. The units fractions you need are:

$$\frac{1000 \text{ mL}}{1 \text{ L}} \qquad \frac{1 \text{ L}}{1000 \text{ mL}}$$

Or you can use a metric conversion line to decide how to move the decimal point.

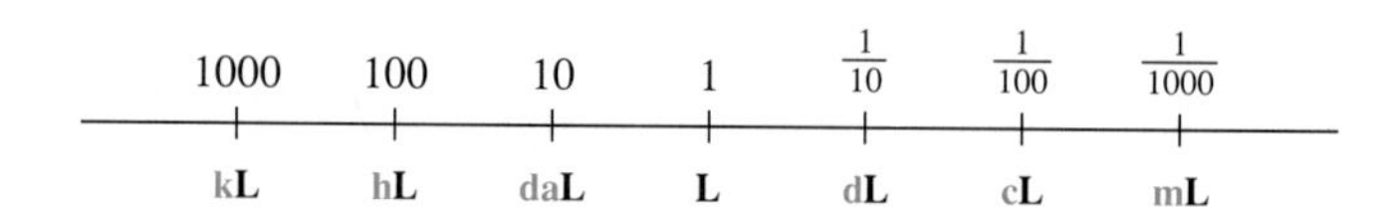

EXAMPLE 2 *Conversions among Metric Capacity Units*
Convert using the metric conversion line or unit fractions.

(a) 2.5 L to mL

Using the metric conversion line:
From **L** to **mL** is *three places* to the *right*.

2.500 — Write two zeros as placeholders.

2.5 L = 2500 mL

Using unit fractions:

Multiply by a unit fraction that allows you to cancel liters.

$$\frac{2.5 \not{L}}{1} \cdot \frac{1000 \text{ mL}}{1 \not{L}} = 2500 \text{ mL}$$

(b) 80 mL to L

Using the metric conversion line:
From **mL** to **L** is *three places* to the *left*.

80. → 080.

Decimal point starts here. Move three places left.

80 mL = 0.080 L or 0.08 L

Using unit fractions:

Multiply by a unit fraction that allows you to cancel mL.

$$\frac{80 \cancel{\text{mL}}}{1} \cdot \frac{1 \text{ L}}{1000 \cancel{\text{mL}}}$$

$$= \frac{80}{1000}\text{L} = 0.08 \text{ L}$$ ■

WORK PROBLEM 3 AT THE SIDE. ▶▶

3 The **gram** is the basic metric unit for mass. Although we often call it "weight," there is a difference. Weight is a measure of the pull of gravity; the farther you are from the center of the earth, the less you weigh. In outer space you become weightless, but your mass, the amount of matter in your body, stays the same regardless of where you are. We will use the word "weight" for everyday purposes.

The gram is related to metric length in this way: the weight of the water in a box measuring 1 cm on every side is 1 gram. This is a very tiny amount of water (1 mL) and a very small weight. One gram is also the weight of a dollar bill or a single raisin. A nickel weighs 5 grams. A regular hamburger weighs from 175 to 200 grams.

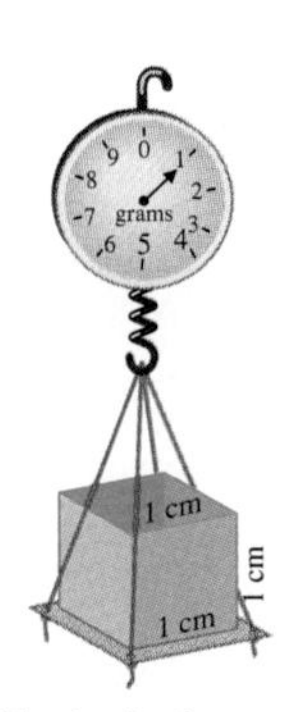

The 1 mL of water in this box weighs 1 gram.

5 grams

A dollar bill weighs 1 gram.

175 to 200 grams

WORK PROBLEM 4 AT THE SIDE. ▶▶

3. Convert.

(a) 9 L to mL

(b) 0.75 L to mL

(c) 500 mL to L

(d) 5 mL to L

(e) 2.07 L to mL

(f) 3275 mL to L

4. Which things would weigh about 1 gram?

a single paperclip

a pair of scissors

one playing card from a deck of cards

a calculator

an average size apple

the check you wrote at the grocery store

ANSWERS
3. (a) 9000 mL (b) 750 mL (c) 0.5 L (d) 0.005 L (e) 2070 mL (f) 3.275 L
4. paperclip, playing card, check

To make larger or smaller weight units, we use the same **prefixes** as we did with length and capacity units. For example, *kilo* means 1000 so a *kilo*meter is 1000 meters, a *kilo*liter is 1000 liters, and a *kilo*gram is 1000 grams.

prefix	*kilo*-gram	*hecto*-gram	*deka*-gram	gram	*deci*-gram	*centi*-gram	*milli*-gram
meaning	1000 grams	100 grams	10 grams	1 gram	$\frac{1}{10}$ of a gram	$\frac{1}{100}$ of a gram	$\frac{1}{1000}$ of a gram
symbol	**kg**	**hg**	**dag**	g	**dg**	cg	**mg**

The units you will use most often in daily life are kilograms (kg), grams (g), and milligrams (mg). *Kilo*grams will be used instead of pounds. A kilogram is 1000 grams. It is about 2.2 pounds. This textbook weighs about 1.7 kg. An average newborn baby weighs 3 to 4 kg; a college football player might weigh 100 to 110 kg.

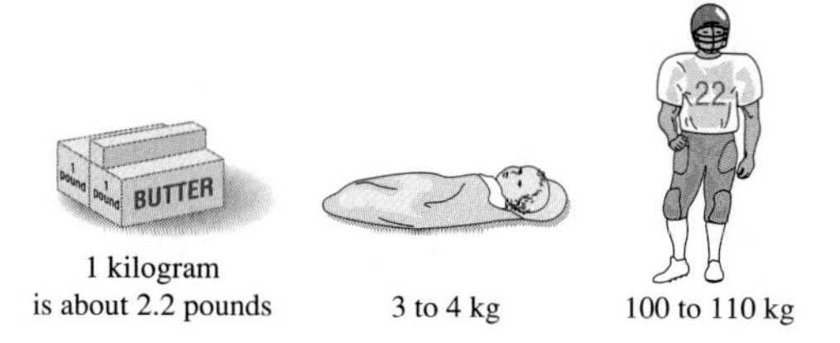

1 kilogram is about 2.2 pounds   3 to 4 kg   100 to 110 kg

Extremely small weights are measured in *milli*grams. It takes 1000 mg to make 1 g. Recall that a dollar bill weighs about 1 g. Think of cutting it into 1000 pieces; the weight of one tiny piece would be 1 mg. Dosages of medicine and vitamins are given in milligrams. You will also use milligrams in science classes.

EXAMPLE 3 *Using Metric Weight Units*

Write the most reasonable metric unit in each blank. Choose from kg, g, and mg.

(a) Ramon's suitcase weighed 20 _____.

(b) LeTia took a 350 _____ aspirin tablet.

(c) Jenny mailed a letter that weighed 30 _____.

(a) 20 kg because kilograms are used instead of pounds. 20 kg is about 44 pounds.

(b) 350 mg because 350 g would be more than the weight of a hamburger, which is too much.

(c) 30 g because 30 kg would be much too heavy and 30 mg is less than the weight of a dollar bill. ■

◀◀ WORK PROBLEM 5 AT THE SIDE.

5. Write the most reasonable metric unit in each blank. Choose from kg, g, and mg.

(a) A thumbtack weighs 800 _____.

(b) A teenager weighs 50 _____.

(c) This large cast-iron frying pan weighs 1 _____.

(d) Jerry's basketball weighed 600 _____.

(e) Tamlyn takes a 500 _____ calcium tablet every morning.

(f) On his diet, Greg can eat 90 _____ of meat for lunch.

(g) One strand of hair weighs 2 _____.

(h) One banana might weigh 150 _____.

ANSWERS

5. (a) mg (b) kg (c) kg (d) g (e) mg (f) g (g) mg (h) g

4 As with length and capacity, you can convert among metric weight units by using unit fractions. The unit fractions you need are:

$$\frac{1000\text{ g}}{1\text{ kg}} \quad \frac{1\text{ kg}}{1000\text{ g}} \quad \frac{1000\text{ mg}}{1\text{ g}} \quad \frac{1\text{ g}}{1000\text{ mg}}$$

Or you can use a metric conversion line to decide how to move the decimal point.

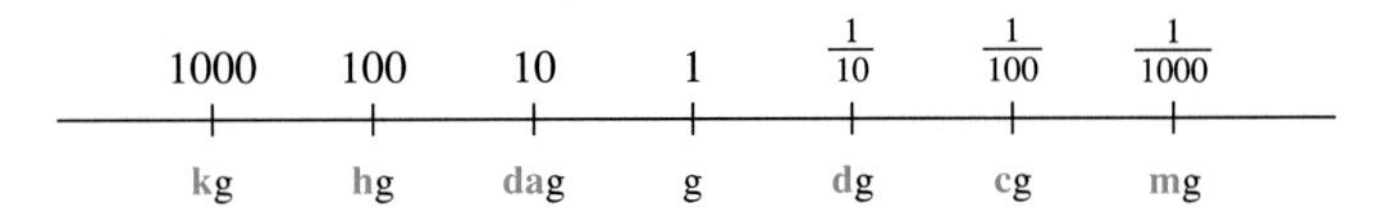

■ **EXAMPLE 4** *Conversions Among Metric Weight Units*

Convert using the metric conversion line or unit fractions.

(a) 7 mg to g

Using the metric conversion line:
From **mg** to **g** is *three places* to the *left*.

7. → 007.

Decimal point starts here. Move three places left.

7 mg = 0.007 g

Using unit fractions:

Multiply by a unit fraction that allows you to cancel mg.

$$\frac{7\ \cancel{\text{mg}}}{1} \cdot \frac{1\text{ g}}{1000\ \cancel{\text{mg}}} = \frac{7}{1000}\text{ g}$$

$$= 0.007\text{ g}$$

(b) 13.72 kg to g

Using the metric conversion line:
From **kg** to **g** is *three* places to the *right*.

13.720 — Decimal point moves three places to the right.

13.72 kg = 13,720 g

a comma (not a decimal point)

Using unit fractions:

Multiply by a unit fraction that allows you to cancel kg.

$$\frac{13.72\ \cancel{\text{kg}}}{1} \cdot \frac{1000\text{ g}}{1\ \cancel{\text{kg}}} = 13{,}720\text{ g}$$

■

WORK PROBLEM 6 AT THE SIDE. ▶▶

6. Convert.

(a) 10 kg to g

(b) 45 mg to g

(c) 6.3 kg to g

(d) 0.077 g to mg

(e) 5630 g to kg

(f) 90 g to kg

ANSWERS

6. (a) 10,000 g (b) 0.045 g (c) 6300 g (d) 77 mg (e) 5.63 kg (f) 0.09 kg

QUEST FOR NUMERACY

Weighing Small Amounts

It is difficult to weigh very light objects, such as a single sheet of paper or a single staple (unless you have a very expensive scientific scale). One way around this problem is to weigh a large number of the items and then divide to find the weight of one item. Of course, before dividing, you must subtract the weight of the box or wrapper that the items are packaged in to find the net weight. Complete this table.

Item	*Total weight*	*Weight of packaging*	*Net weight*	*Weight of one item in grams*	*Weight of one item in milligrams*
Box of 50 envelopes	255 g	40 g	215 g	4.3 g	4300 mg
Box of 1000 staples	350 g	20 g	330 g	0.33 g	330 mg
Bag of 100 cough drops	390 g	30 g	360 g	3.6 g	3600 mg
Roll of 100 stamps	10 g	1 g	9 g	0.09 g	90 mg
Ream of paper (500 sheets)	1.55 kg	50 g	1500 g	3 g	3000 mg
Box of 100 small paper clips	55 g	5 g	50 g	0.5 g	500 mg

Extend the table by asking students to bring in and weigh groups of items, such as a box of toothpicks, stack of 20 nickels, etc. A small inexpensive diet or postage scale is accurate enough for this activity.

7.3 NAME DATE HOUR

EXERCISES

Write the most reasonable metric unit in each blank. Choose from L, mL, kg, g, and mg.

1. The glass held 250 __mL__ of water.

2. Hiromi made 20 __L__ of applesauce from the apples on his trees.

3. Dolores can make 10 __L__ of soup in that pot.

4. Jay gave 2 __mL__ of vitamin drops to the baby.

5. Our labrador dog grew up to weigh 25 __kg__.

6. One dime weighs 2 __g__.

7. Lori caught a small sunfish weighing 150 __g__.

8. A small safety pin weighs 750 __mg__.

9. Andre donated 500 __mL__ of blood today.

10. Barbara bought the large 2 __L__ bottle of cola.

11. The patient received 250 __mg__ of medication each hour.

12. The 8 people on the elevator weighed a total of 500 __kg__.

13. The gas can for the lawn mower holds 4 __L__.

14. Kevin poured 10 __mL__ of vanilla into the bowl.

15. Pam's backpack weighs 5 __kg__ when it is full of books.

16. One grain of salt weighs 2 __mg__.

Today, medical measurements are usually given in the metric system. Since we convert among metric units of measure by moving the decimal point, it is possible that mistakes can be made. Examine the following dosages and indicate whether they are reasonable or unreasonable.

17. Drink 4.1 liters of Kaopectate after each meal.

    unreasonable

18. Drop 1 mL of solution into the eye twice a day.

    reasonable

19. Soak your feet in 3 milligrams of Epsom salts per 4 liters of water.

    unreasonable

20. Inject 0.5 liter of insulin each morning.

    unreasonable

21. Take 15 milliliters of cough syrup every four hours.

    reasonable

22. Take 200 milligrams of vitamin C each day.

    reasonable

23. Take 2 kilograms of aspirin three times a day.

    unreasonable

24. Take an aspirin weighing 0.002 gram.

    unreasonable

25. Describe at least two examples of metric capacity units and two examples of metric weight units that you have come across in your daily life.

Answer varies.

26. Explain in your own words how the meter, liter, and gram are related.

Answer varies.

27. Write out an explanation of each step you would use to convert 20 mg to grams using the metric conversion line.

Answer varies.

28. Describe how you decide which unit fraction to use when converting 6.5 kg to grams.

Answer varies.

Convert each measurement. Use unit fractions or the metric conversion line.

Example: 9 g to kg

Solution: $\frac{9\ \cancel{g}}{1} \cdot \frac{1\ \text{kg}}{1000\ \cancel{g}} = \frac{9}{1000}\ \text{kg} = \mathbf{0.009\ kg}$

or: From g to kg is three places to the left. 009. g = **0.009 kg**

29. 15 L to mL 15,000 mL

30. 6 L to mL 6000 mL

31. 3000 mL to L 3 L

32. 18,000 mL to L 18 L

33. 925 mL to L 0.925 L

34. 200 mL to L 0.2 L

35. 8 mL to L 0.008 L

36. 25 mL to L 0.025 L

37. 4.15 L to mL 4150 mL

38. 11.7 L to mL 11,700 mL

39. 8000 g to kg 8 kg

40. 25,000 g to kg 25 kg

41. 5.2 kg to g 5200 g

42. 12.42 kg to g 12,420 g

43. 0.85 g to mg 850 mg

44. 0.2 g to mg 200 mg

45. 30,000 mg to g 30 g

46. 7500 mg to g 7.5 g

47. 598 mg to g 0.598 g

48. 900 mg to g 0.9 g

49. 60 mL to L 0.06 L

50. 6.007 kg to g 6007 g

51. 3 g to kg 0.003 kg

52. 12 mg to g 0.012 g

53. 0.99 L to mL 990 mL

54. 13,700 mL to L 13.7 L

Solve the following problems.

55. The doctor told Sara to drink two liters of water each day. How many milliliters is that?

2000 mL

56. A juice can holds 1500 mL. How many liters of juice does it hold?

1.5 L

57. The premature infant weighed only 950 grams. How many kilograms did he weigh?

0.95 kg

58. Bill bought three kilograms of potatoes. How many grams did he buy?

3000 g

59. Is 1005 mg greater than or less than 1 g? What is the difference in the weights?

greater; 5 mg or 0.005 g

60. Is 990 mL greater than or less than 1 L? What is the difference in the amounts?

less; 10 mL or 0.01 L

Solve the following word problems.

▲ **61.** One nickel weighs 5 grams. How many nickels are in 1 kilogram of nickels?

200 nickels

▲ **62.** Seawater contains about 3.5 grams of salt per 1000 milliliters of water. How many grams of salt would be in 1 liter of seawater?

3.5 grams

▲ **63.** Helium weighs about 0.0002 grams per milliliter. How much would 1 liter of helium weigh?

0.2 gram

▲ **64.** About 1500 grams of sugar can be dissolved in a liter of warm water. How much sugar could be dissolved in 1 milliliter of warm water?

1.5 grams

Preview Exercises

Name the digit that has the given place value in each of the following. (For help, see ***Section 4.1.****)*

65. 7250.6183
7 thousands
1 hundredths
8 thousandths
2 hundreds

66. 1358.0256
6 ten-thousandths
0 tenths
5 tens
8 ones

7.4 APPLICATIONS OF METRIC MEASUREMENT

OBJECTIVE

1 Solve application problems involving metric measurements.

FOR EXTRA HELP

Tape 10 | SSM pp. 215–216 | MAC: A IBM: A

1 One advantage of the metric system is the ease of comparing measurements in application situations. Just be sure that you are comparing similar units: mg to mg, km to km, and so on.

EXAMPLE 1 *Solving Metric Applications*

(a) Cheddar cheese is on sale at \$8.99 per kg. Jake bought 350 grams of the cheese. How much did he pay, to the nearest cent?

(b) Olivia has 2.5 meters of lace. How many centimeters of lace can she use to trim each of six hair ornaments? Round to the nearest tenth of a centimeter.

(a) The price is \$8.99 per *kilogram*, but the amount purchased is in *grams*. Convert grams to kilograms (the unit in the price). Then multiply the weight times the cost per kilogam.

$$350 \text{ g} = 0.35 \text{ kg}$$

$$\frac{\$8.99}{1 \cancel{\text{kg}}} \cdot \frac{0.35 \cancel{\text{kg}}}{1} = \$3.1465$$

Jake paid \$3.15, to the nearest cent.

(b) The given amount is in *meters*, but the answer must be in *centimeters*, so convert meters to centimeters. Then divide by the number of hair ornaments.

$$2.5 \text{ m} = 250 \text{ cm}$$

$$\frac{250 \text{ cm}}{6} = 41.6666 \text{ cm}$$

Olivia can use 41.7 cm of lace on each ornament (rounded to the nearest tenth). ■

WORK PROBLEM 1 AT THE SIDE.

1. Solve each problem.

(a) Satin ribbon is on sale at \$0.89 per meter. How much will 75 cm cost, to the nearest cent?

(b) Lucinda's doctor wants her to take 1.2 grams of medication each day in three equal doses. How many milligrams should be in each dose?

EXAMPLE 2 *Measurement Applications*

(a) Rubin measured a board and found that the length was 3 meters plus an additional 5 centimeters. He cut off a piece measuring 1 meter 40 centimeters for a shelf. Find the length of the remaining piece in meters.

(b) Amy put a basket of nuts on her scale and saw that they weighed 4 kilograms plus 140 grams. She plans to put the nuts into three gift packs of equal size. Find the number of kilograms in each pack.

ANSWERS
1. (a) \$0.67 (b) 400 mg

2. (a) Andrea has two pieces of fabric. One measures 2 meters 35 centimeters and the other measures 1 meter 85 centimeters. How many meters of fabric does she have in all?

(b) Mr. Green has 9 m 20 cm of rope. He is cutting it into eight pieces so his Scout troop can practice knot tying. How many meters of rope will each Scout get?

(a) The lengths involve two units, m and cm. To make the calculations easier, write each length in terms of meters (the unit called for in the answer). Then subtract to find the leftover length.

Board		Shelf	
3m →	3.00 m	1 m →	1.0 m
plus 5 cm →	+ 0.05 m	plus 40 cm →	+ 0.4 m
	3.05 m		1.4 m

Subtract to find leftover length.	
3.05 m	← board
− 1.40 m	← shelf
1.65 m	← leftover piece

The length of the leftover piece is 1.65 m.

(b) Write 4 kg 140g in terms of kg (the unit for the answer). Then divide by the number of gift packs.

4 kg →	4.00 kg
plus 140 g →	+ 0.14 kg
	4.14 kg

$$\frac{4.14 \text{ kg}}{3} = 1.38 \text{ kg}$$

Each gift pack will have 1.38 kg of nuts. ■

◀◀ WORK PROBLEM 2 AT THE SIDE.

ANSWERS
2. (a) 4.2 m (b) 1.15 m

Solve each problem.

Example: A basket of strawberries weighed 1 kg 80 g. Find the cost of the strawberries if they are priced at $2.19 per kg.

Solution: Write 1 kg 80 g in terms of kg (the unit in the price).

$$\begin{array}{rr} 1 \text{ kg} \rightarrow & 1.00 \text{ kg} \\ \text{plus } 80 \text{ g} \rightarrow & +\ 0.08 \text{ kg} \\ \hline & 1.08 \text{ kg} \end{array} \qquad \frac{\$2.19}{1 \cancel{\text{kg}}} \cdot \frac{1.08 \cancel{\text{kg}}}{1} = \$2.3652$$

In consumer situations, prices are rounded to the nearest cent, so the strawberries cost **$2.37.**

1. Bulk rice is on special at $0.65 per kilogram. Pam scooped some rice into a bag and put it on the scale. How much will she pay for 2 kg 50 g of rice?

$1.33 rounded to the nearest cent

2. Lanh is buying a piece of plastic tubing that measures 3 m 15 cm. The price is $4.75 per meter. How much will Lanh pay?

$14.96 rounded to the nearest cent

3. Kendal works for a garden store. He put 15 grams of fertilizer on each of 650 tomato plants. How many kilograms of fertilizer did he use?

9.75 kg

4. The garden store ordered a 50 liter drum of liquid plant food. They repackaged the plant food into 125 mL bottles. How many bottles were filled?

400 bottles

5. A jug of cider contains 4 L. How many 160 mL servings are in one jug?

25 servings

6. A floor tile measures 30 cm by 30 cm and weighs 185 g. How many kilograms would a carton of 24 tiles weigh?

4.44 kg

7. Rosa is building a bookcase. She has one board that is 2 m 8 cm long and another that is 2 m 95 cm long. How long are the two boards together in meters?

5.03 m

8. Eric's puppy weighed 1 kg 390 g. Three months later it weighed 2 kg 50 g. How much weight had the puppy gained?

0.66 kg or 660 g

9. The apartment building caretaker puts 750 mL of chlorine into the swimming pool every day. How many liters should he order to have a one-month (30-day) supply on hand?

22.5 L

10. Janet has 10 m 30 cm of fabric. She wants to make curtains for three windows that are all the same size. How much fabric is available for each window, to the nearest tenth of a meter?

3.4 m

▲ 11. As a fund raiser, the PTA bought 40 kg of nuts for $113.50. They sold the nuts in 250 g bags for $2.95 each. Find the amount of profit.

$358.50

▲ 12. In chemistry class, each of the 45 students needs 85 mL of acid. How many one-liter bottles of acid need to be ordered?

4 bottles

▲ 13. Which case of shampoo is the better buy: a $16 case that holds 12 1-liter bottles or an $18 case that holds 36 400-mL bottles?

$18 case

▲ 14. James needs 3 m 80 cm of wood molding to frame a picture. The price is $5.89 per meter plus a 7% sales tax. How much will James pay?

$23.95 rounded to the nearest cent

PREVIEW EXERCISES

Multiply. (For help, see **Section 4.5.***)*

15. 0.035×18 = 0.63

16. 28.35×12 = 340.2

17. 6.3×0.91 = 5.733

18. $14.7 \cdot 2.2$

32.34

19. $5.25 \cdot 1.09$

5.7225

20. $8.4 \cdot 3.78$

31.752

7.5 METRIC—ENGLISH CONVERSIONS AND TEMPERATURE

1 Until everyone thinks in the metric system as naturally as they do in the English system, it will be necessary to make conversions from one system to the other. *Approximate* conversions can be made with the help of the following table, in which the values have been rounded to the nearest hundredth or thousandth.

Metric to English			*English to Metric*		
from metric	to English	multiply by	from English	to metric	multiply by
meters	yards	**1.09**	yards	meters	**0.91**
meters	feet	**3.28**	feet	meters	**0.30**
centimeters	inches	**0.39**	inches	centimeters	**2.54**
kilometers	miles	**0.62**	miles	kilometers	**1.61**
grams	ounces	**0.035**	ounces	grams	**28.35**
kilograms	pounds	**2.20**	pounds	kilograms	**0.45**
liters	quarts	**1.06**	quarts	liters	**0.95**
liters	gallons	**0.26**	gallons	liters	**3.78**

EXAMPLE 1 *Converting Metric and English (Length)*
Convert. Round your answers to the nearest hundredth, if necessary.

(a) 10 meters to yards

(b) 24 yards to meters

(c) 6.3 kilometers to miles

(a) In the "Metric to English" part of the table, the number for conversion from meters to yards is 1.09. Multiply 10 meters times 1.09 to get the *approximate* number of yards. Use the "≈" symbol, which means "approximately equal to" instead of the "=" symbol.

$$10 \text{ meters} \approx 10 \cdot 1.09 \approx 10.9 \text{ yards}$$

So 10 meters is approximately equal to 10.9 yards.

(b) In the "English to Metric" part of the table, the number for conversion from yards to meters is 0.91. Multiply 24 yards times 0.91.

$$24 \text{ yards} \approx 24 \cdot 0.91 \approx 21.84 \text{ meters}$$

(c) 6.3 kilometers ≈ 6.3 · 0.62 ≈ 3.91 miles (rounded) ■

Note Because the metric and English systems were developed independently, there are no exact equivalents.

WORK PROBLEM 1 AT THE SIDE.

OBJECTIVES

1. Use a table of conversion factors to convert from metric to English or English to metric units.
2. Know common temperatures on the Celsius scale.
3. Convert temperatures by using the order of operations.

FOR EXTRA HELP

Tape 10 | SSM pp. 216–219 | MAC: A IBM: A

1. Convert. Round your answers to the nearest hundredth.

(a) 23 meters to yards

(b) 40 centimeters to inches

(c) 29.3 meters to feet

(d) 5 kilometers to miles

(e) 3.5 yards to meters

(f) 12 inches to centimeters

ANSWERS
1. (a) 25.07 yd (b) 15.6 in. (c) 96.10 ft (d) 3.1 mi (e) 3.19 m (f) 30.48 cm

2. Convert. Use the table on the previous page.

(a) 17 kilograms to pounds

(b) 5 liters to quarts

(c) 90 grams to ounces

(d) 3.5 gallons to liters

(e) 145 pounds to kilograms

(f) 8 ounces to grams

ANSWERS
2. (a) 37.4 pounds (b) 5.3 quarts (c) 3.15 ounces (d) 13.23 L (e) 65.25 kg (f) 226.8 g

■ **EXAMPLE 2** *Converting Metric and English (Weight and Capacity)*
Convert.

(a) 5 kilograms to pounds

(b) 18 gallons to liters

(c) 300 grams to ounces

Find the numbers for each conversion in the appropriate part of the table on the previous page.

(a) 5 kilograms $\approx 5 \cdot 2.20 \approx$ 11 pounds

(b) 18 gallons $\approx 18 \cdot 3.78 \approx$ 68.04 liters

(c) 300 grams $\approx 300 \cdot 0.035 \approx$ 10.5 ounces ■

WORK PROBLEM 2 AT THE SIDE.

2 In the metric system, temperature is measured on the **Celsius scale.** On the Celsius scale, water freezes at 0° and boils at 100°. The thermometer below shows some typical temperatures in both Celsius and **Fahrenheit** (the English temperature system we now use). For example, comfortable room temperature is about 20°C or 68°F, and body temperature is about 37°C or 98.6°F.

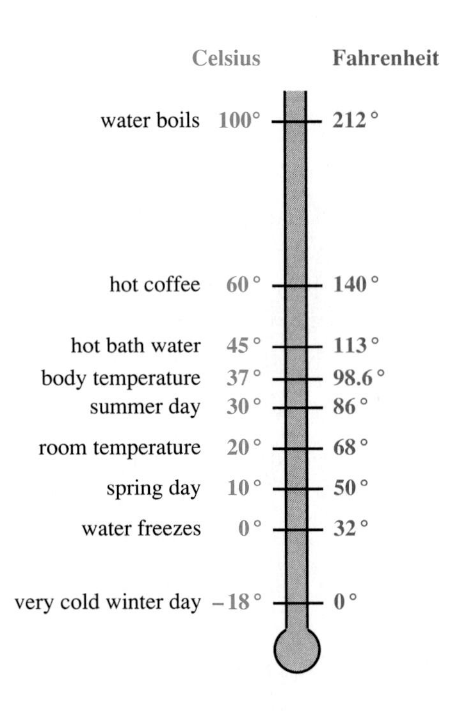

Note The freezing and boiling temperatures are exact. The other temperatures are approximate. Even normal body temperature varies slightly from person to person.

EXAMPLE 3 *Using Celsius Temperatures*

Circle the Celsius temperature that is most reasonable for each situation.

(a) warm summer day 29°C 64°C 90°C

(b) inside a freezer −10°C 3°C 25°C

(a) 29°C is reasonable. 64°C and 90°C are too hot; they're both above the temperature of hot bath water (above 110°F).

(b) −10°C is the reasonable temperature because it is the only one below the freezing point of water (0°C). Your frozen foods would start thawing at 3°C or 25°C. ■

WORK PROBLEM 3 AT THE SIDE.

3 You can convert between Celsius and Fahrenheit temperatures using these formulas.

CELSIUS-FAHRENHEIT CONVERSION FORMULAS

Converting from Fahrenheit (F) to Celsius (C)	Converting from Celsius (C) to Fahrenheit (F)
$C = \frac{5(F - 32)}{9}$	$F = \frac{9 \cdot C}{5} + 32$

As you use these formulas, be sure to use the order of operations from **Sections 1.8** and **3.5.**

1. Do all operations inside parentheses.
2. Simplify any expressions with exponents and find any square roots.
3. Multiply or divide from left to right.
4. Add or subtract from left to right.

EXAMPLE 4 *Converting Fahrenheit to Celsius*

Convert 68°F to Celsius.

Use the formula and the order of operations.

$$C = \frac{5(F - 32)}{9}$$

$$= \frac{5(68 - 32)}{9} \quad \text{Work in parentheses first.}$$

$$= \frac{5(36)}{9}$$

$$= \frac{5(\overset{4}{\cancel{36}})}{\underset{1}{\cancel{9}}} \quad \text{Use cancellation, if possible. Multiply.}$$

$$= 20$$

Thus, 68°F = 20°C. ■

WORK PROBLEM 4 AT THE SIDE.

3. Circle the Celsius temperature that is most reasonable for each situation.

(a) Set the living room thermostat at:
11°C 21°C 71°C

(b) The baby has a fever of:
29°C 39°C 49°C

(c) Wear a sweater outside because it's:
15°C 25°C 50°C

(d) My iced tea is:
−5°C 5°C 30°C

(e) Time to go swimming! It's:
95°C 65°C 35°C

(f) Inside a refrigerator (not the freezer) it's:
−15°C 0°C 3°C

4. Convert to Celsius.

(a) 59°F

(b) 41°F

(c) 212°F

(d) 98.6°F

ANSWERS

3. (a) 21°C (b) 39°C (c) 15°C (d) 5°C (e) 35°C (f) 3°C

4. (a) 15°C (b) 5°C (c) 100°C (d) 37°C

5. Convert to Fahrenheit.

(a) 100°C

(b) 25°C

(c) 80°C

(d) 5°C

ANSWERS
5. (a) 212°F (b) 77°F (c) 176°F (d) 41°F

■ **EXAMPLE 5** *Converting Celsius to Fahrenheit*
Convert 15°C to Fahrenheit.

Use the formula and the order of operations.

$$F = \frac{9 \cdot C}{5} + 32$$

$$= \frac{9 \cdot 15}{5} + 32$$

$$= \frac{9 \cdot \overset{3}{\cancel{15}}}{\underset{1}{\cancel{5}}} + 32 \quad \text{Use cancellation, if possible. Multiply.}$$

$$= 27 + 32 \quad \text{Add.}$$

$$= 59$$

Thus, 15°C = 59°F. ■

◀◀ WORK PROBLEM 5 AT THE SIDE.

EXERCISES

Use the table on page 457 to make approximate conversions from metric to English or English to metric. Round your answers to the nearest tenth.

Example: 36 meters to yards

Solution: Look in the "Metric to English" part of the table.
36 meters ≈ 36 · 1.09 ≈ **39.2 yards** (rounded)

1. 20 meters to yards

21.8 yards

2. 8 kilometers to miles

5.0 miles

3. 80 meters to feet

262.4 feet

4. 85 centimeters to inches

33.2 inches

5. 16 feet to meters

4.8 m

6. 3.2 yards to meters

2.9 m

7. 150 grams to ounces

5.3 ounces

8. 2.5 ounces to grams

70.9 g

9. 248 pounds to kilograms

111.6 kg

10. 7.68 kilograms to pounds

16.9 pounds

11. 28.6 liters to quarts

30.3 quarts

12. 15.75 liters to gallons

4.1 gallons

13. Jeanette bought 18 gallons of gas for her car. How many liters of gas did she buy?

68.0 L

14. The PTA used 16 quart-size containers of orange juice at the pancake breakfast. How many liters of juice were used?

15.2 L

Circle the more reasonable temperature for each of the following.

15. A snowy day

28°C (28°F)

16. Brewing coffee

(80°C) 80°F

17. A high fever

(40°C) 40°F

18. Swimming pool water

78°C (78°F)

19. Oven temperature

(150°C) 150°F

20. Light jacket weather

(10°C) 10°F

21. Would a drop of 20 Celsius degrees be more or less than a drop of 20 Fahrenheit degrees? Explain your answer.

More; explanation varies.

22. Describe one advantage of switching from the Fahrenheit temperature scale to the Celsius scale. Describe one disadvantage.

Answer varies.

Use the conversion formulas on page 459 and the order of operations to convert Fahrenheit temperatures to Celsius and Celsius temperatures to Fahrenheit. Round your answers to the nearest degree, if necessary.

23. 60°F 16°C

24. 80°F 27°C

25. 104°F 40°C

26. 536°F 280°C

27. 8°C 46°F

28. 18°C 64°F

29. 35°C 95°F

30. 0°C 32°F

31. The highest temperature ever recorded on earth was 136°F at Aziza, Libya. Convert this temperature to Celsius.

58°C

32. A recipe for French pastry calls for an oven temperature of 175°C. Convert this to Fahrenheit.

347°F

33. A kiln for firing pottery reaches a temperature of 500°C. Convert this to Fahrenheit.

932°F

34. The temperature of the water in a lake in December is 34°F. Convert this to Celsius.

1°C

Solve the following word problems.

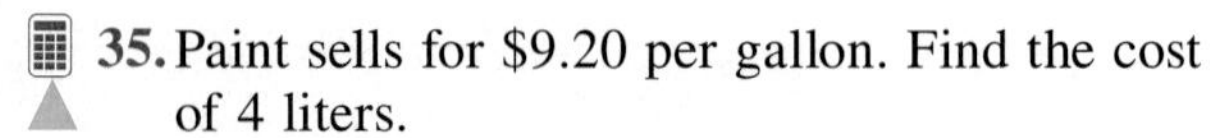

35. Paint sells for $9.20 per gallon. Find the cost of 4 liters.

$9.72 if gallons converted to liters; $9.57 if liters converted to gallons

36. A 3-liter bottle of beverage sells for $2.80. A 1-gallon bottle of the same beverage sells for $3.50. What is the better value?

the 1-gallon bottle

PREVIEW EXERCISES

Work each problem. (For help, see ***Section 1.8.****)*

37. $2 \cdot 8 + 2 \cdot 8$ 32

38. $2 \cdot 12.2 + 2 \cdot 5.6$ 35.6

39. 9^2 81

40. 7^2 49

41. $(5^2) + (4^2)$ 41

42. $(12^2) + (3^2)$ 153

CHAPTER 7 SUMMARY

KEY TERMS

7.1	**English system**	The English system of measurement (American system of units) is the system used for many daily activities in the United States. Common units in this system include quarts, pounds, feet, miles, and degrees Fahrenheit.
	metric system	The metric system of measurement is an international system of measurement used in manufacturing, science, medicine, sports, and other fields. The system uses meters, liters, grams, and degrees Celsius.
	unit fraction	A unit fraction involves measurement units and is equivalent to 1. Unit fractions are used to convert among different measurements.
7.2	**meter**	The meter is the basic unit of length in the metric system. The symbol *m* is used for meter. One meter is a little longer than a yard.
	prefixes	Attaching a prefix to meter, liter, or gram produces larger or smaller units. For example, the prefix *kilo* means 1000 so a *kilo*meter is 1000 meters.
	metric conversion line	The metric conversion line is a line showing the various metric measurement prefixes and their size relationship to each other. See pages 438, 444, and 447.
7.3	**liter**	The liter is the basic unit of capacity in the metric system. The symbol L is used for liter. One liter is a little more than one quart.
	gram	The gram is the basic unit of weight (mass) in the metric system. The symbol *g* is used for gram. One gram is the weight of 1 milliliter of water.
7.5	**Celsius**	The Celsius scale is the scale used to measure temperature in the metric system. Water boils at 100°C and freezes at 0°C.
	Fahrenheit	The Fahrenheit scale is the scale used to measure temperature in the English system. Water boils at 212°F and freezes at 32°F.

QUICK REVIEW

Concepts	Examples
7.1 The English System of Measurement Memorize the basic measurement relationships. Then, to convert units, multiply when changing from a larger unit to a smaller unit; divide when changing from a smaller unit to a larger unit.	Convert each measurement. **(a)** 5 feet to inches $5 \text{ feet} = 5 \cdot 12 = 60 \text{ inches}$ **(b)** 3 pounds to ounces $3 \text{ pounds} = 3 \cdot 16 = 48 \text{ ounces}$ **(c)** 15 quarts to gallons $15 \text{ quarts} = \frac{15}{4} = 3\frac{3}{4} \text{ gallons}$
7.1 Using Unit Fractions Another, more useful, conversion method is multiplying by a unit fraction. The unit you want in the answer should be in the numerator. The unit you want to change should be in the denominator.	Convert 32 ounces to pounds. $\frac{32 \text{ ounces}}{1} \cdot \frac{1 \text{ pound}}{16 \text{ ounces}}$ } unit fraction $= \frac{\overset{2}{\cancel{32}}\ \cancel{\text{ounces}}}{1} \cdot \frac{1 \text{ pound}}{\underset{1}{\cancel{16}}\ \cancel{\text{ounces}}}$ Cancel ounces. Cancel numbers. $= 2 \text{ pounds}$

Concepts	Examples
7.2 Knowing Basic Metric Length Units Use approximate comparisons to judge which units are appropriate: 1 mm is the thickness of a dime. 1 cm is about 1/2 inch. 1 m is a little more than 1 yd. 1 km is about 0.6 mile.	Write the most reasonable metric unit in each blank: km, m, cm, mm. The room is 6 __m__ long. A paper clip is 30 __mm__ long. He drove 20 __km__ to work.
7.2 and 7.3 Converting Within the Metric System **Using Unit Fractions** One method is to multiply by a unit fraction. Use a fraction with the unit you want in the answer in the numerator and the unit you want to change in the denominator.	Convert 9 g to kg $\frac{9 \not{g}}{1} \cdot \frac{1 \text{ kg}}{1000 \not{g}} = \frac{9}{1000} \text{ kg} = 0.009 \text{ kg}$ Convert 3.6 m to cm $\frac{3.6 \not{m}}{1} \cdot \frac{100 \text{ cm}}{1 \not{m}} = 360 \text{ cm}$
Using the Metric Conversion Line Another method is to find the unit you are given on the metric conversion line. Count the number of places to get from the unit you are given to the unit you want. Move the decimal point the same number of places and in the same direction. 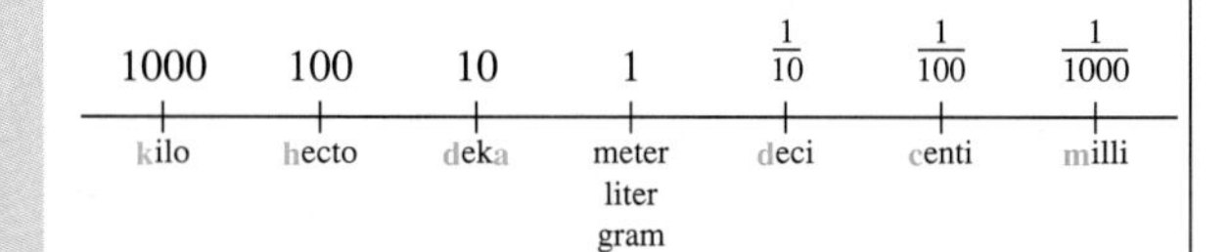 	Convert each of the following. **(a)** 68.2 kg to g From kg to g is three places to the right. 6 8.2 0 0 Decimal point is moved three places to the right. 68.2 kg = 68,200 g **(b)** 300 mL to L From mL to L is three places to the left. 3 0 0. Decimal point is moved three places to the left. 300 ml = 0.3 L **(c)** 825 cm to m From cm to m is two places to the left. 8 2 5. Decimal point is moved two places to the left. 825 cm = 8.25 m

Concepts	Examples
7.3 Knowing Basic Metric Capacity Units	
Use approximate comparisons to judge which units are appropriate: 1 L is a little more than 1 quart. 1 mL is the amount of water in a cube 1 cm on each side. 5 mL is about one teaspoon. 250 mL is about one cup.	Write the most appropriate metric unit in each blank. Choose from L or mL. The pail holds 12 <u>L</u>. The milk carton from the vending machine holds 250 <u>mL</u>.
7.3 Knowing Basic Metric Weight (Mass) Units	
Use approximate comparisons to judge which units are appropriate: 1 kg is about 2.2 pounds. 1 g is the weight of 1 mL of water or one dollar bill. 1 mg is $\frac{1}{1000}$ of a gram; a very tiny amount.	Write the most appropriate metric unit in each blank: kg, g, or mg. The wrestler weighed 95 <u>kg</u>. She took a 500 <u>mg</u> aspirin tablet. One banana weighs 150 <u>g</u>.
7.4 Solving Metric Application Problems	
Convert units so you are comparing kg to kg, cm to cm, and so on. When a measurement involves two units, such as 6 m 20 cm, write it in terms of the unit called for in the answer (6.2 m or 620 cm).	**(a)** Grapes are \$3.95 per kg. How much will 400 g of grapes cost? 400 g = 0.4 kg $\quad \frac{0.4\ \cancel{kg}}{1} \cdot \frac{\$3.95}{1\ \cancel{kg}} = \$1.58$ **(b)** How many meters are left if 1 m 35 cm is cut off a board measuring 3 m? 1 m → 1.00 m; plus 35 cm → + 0.35 m; 1.35 m 3.00 m ← board; − 1.35 m ← cut off; 1.65 m ← left
7.5 Converting from Metric to English and English to Metric	
Multiply the given quantity by the value from the table of conversion factors to get an approximate answer. Use the symbol " ≈ " to mean "approximately equal to."	Convert the following. **(a)** 23 meters to yards $23 \cdot 1.09 \approx 25.07$ yards **(b)** 4 ounces to grams $4 \cdot 28.35 \approx 113.4$ grams **(c)** 5 liters to quarts $5 \cdot 1.06 \approx 5.3$ quarts

Concepts	Examples
7.5 Knowing Common Celsius Temperatures	
Use approximate and exact comparisons to judge which temperatures are appropriate. Exact comparisons 0°C is freezing point (32°F) 100°C is boiling point (212°F) Approximate comparisons 10°C for a spring day (50°F) 20°C for room temperature (68°F) 30°C for summer day (86°F) 37°C for body temperature (98.6°F)	Circle the Celsius temperature that is most reasonable. **(a)** Hot summer day: (35°C) 90°C 110°C **(b)** The first snowy day in winter. −20°C (0°C) 15°C
7.5 Converting between Fahrenheit and Celsius Temperatures	
Use these formulas. $C = \frac{5(F - 32)}{9}$	Convert 176°F to Celsius. $C = \frac{5(176 - 32)}{9}$ $= \frac{5(\overset{16}{\cancel{144}})}{\underset{1}{\cancel{9}}}$ Cancel, if possible. Then multiply. $= 80$ $176°F = 80°C$
$F = \frac{9 \cdot C}{5} + 32$	Convert 80°C to Fahrenheit. $F = \frac{9 \cdot 80}{5} + 32$ $= \frac{9 \cdot \overset{16}{\cancel{80}}}{\underset{1}{\cancel{5}}} + 32$ Cancel, if possible. Then multiply. $= 144 + 32$ Add. $= 176$ $80°C = 176°F$

NAME DATE HOUR

CHAPTER 7 REVIEW EXERCISES

[7.1] *Fill in the blanks with the measurement relationships you have memorized.*

1. 1 pound = 16 ouncesr

2. 3 feet = 1 yard

3. 1 ton = 2000 pounds

4. 24 hours = 1 day

5. 1 hour = 60 minutes

6. 1 quart = 2 pints

7. 4 quarts = 1 gallon

8. 5280 feet = 1 mile

9. 12 inches = 1 foot

10. 1 week = 7 days

11. 60 seconds = 1 minute

12. 1 pint = 2 cups

Convert using unit fractions.

13. 4 feet = 48 inches

14. 3 miles = 15,840 feet

15. 64 ounces = 4 pounds

16. 6000 pounds = 3 tons

17. 150 minutes = $2\frac{1}{2}$ or 2.5 hours

18. 11 cups = $5\frac{1}{2}$ or 5.5 pints

19. 18 hours = $\frac{3}{4}$ or 0.75 day

20. 9 quarts = $2\frac{1}{4}$ or 2.25 gallons

21. 2 miles = 126,720 inches

22. 4 tons = 128,000 ounces

23. 7 gallons = 112 cups

24. 4 days = 345,600 seconds

[7.2] *Write the most reasonable metric length unit in each blank. Choose from km, m, cm, mm.*

25. My thumb is 20 mm wide.

26. Her waist measurement is 66 cm.

27. The two towns are 40 km apart.

28. A basketball court is 30 m long.

29. The height of the picnic bench is 45 cm.

30. The eraser on the end of my pencil is 5 mm long.

Convert using unit fractions or the metric conversion line.

31. 5 m to cm

500 cm

32. 12 km to m

12,000 m

33. 85 mm to cm

8.5 cm

34. 370 cm to m

3.7 m

35. 6.1 km to m

6100 m

36. 0.93 m to mm

930 mm

[7.3] *Write the most reasonable metric unit in each blank. Choose from L , mL , kg, g, and mg.*

37. The eyedropper holds 1 __mL__.

38. I can heat 3 __L__ of water in this pan.

39. Don's paperback book weighed 250 __g__.

40. The crate of dishes weighed 20 __kg__.

41. My fish tank holds 80 __L__ of water.

42. I'll buy the 500 __mL__ bottle of mouthwash.

43. Mara took a 200- __mg__ antibiotic pill.

44. This piece of chicken weighs 100 __g__.

Convert using unit fractions or the metric conversion line.

45. 5000 mL to L

5 L

46. 8 L to mL

8000 mL

47. 4.58 g to mg

4580 mg

48. 0.7 kg to g

700 g

49. 6 mg to g

0.006 g

50. 35 mL to L

0.035 L

[7.4] *Solve each problem.*

51. Each serving of punch at the wedding reception will be 180 mL. How many liters of punch are needed for 175 guests?

31.5 L

52. Loretta is serving a 10-kg turkey to 28 people. How many grams of meat is she allowing for each person? Round to the nearest whole gram.

357 g

53. Yerald weighed 92 kg. Then he lost 4 kg 750 g. What is his weight now in kilograms?

87.25 kg

54. Young-Mi bought 2 kg 20 g of onions. The price was $1.49 per kilogram. How much did she pay, to the nearest cent?

$3.01

[7.5] *Use the table on page 457 to make approximate conversions. Round your answers to the nearest tenth, if necessary.*

55. 10 m to yards

10.9 yards

56. 140 cm to inches

54.6 inches

57. 108 km to miles

67.0 miles

58. 800 miles to km

1288 km

59. 23 quarts to L

21.9 L

60. 41.5 L to quarts

44.0 quarts

Write the appropriate Celsius temperature in each blank.

61. Water freezes at 0°C.

62. Water boils at 100°C.

63. Normal body temperature is about 37°C.

64. Comfortable room temperature is about 20°C.

Use the conversion formulas on page 459 to convert each temperature to Fahrenheit or Celsius. Round to the nearest degree, if necessary.

65. 77°F 25°C

66. 92°F 33°C

67. 12°C 54°F

68. 50°C 122°F

Mixed Review Exercises

Write the most reasonable metric unit in each blank. Choose from km, m, cm, mm, L, mL, kg, g, and mg.

69. I added 1 L of oil to my car.

70. The box of books weighed 15 kg.

71. Larry's shoe is 30 cm long.

72. Jan used 15 mL of shampoo on her hair.

73. My fingernail is 10 mm wide.

74. I walked 2 km to school.

75. The tiny bird weighed 15 g.

76. The new library building is 18 m wide.

77. The cookie recipe uses 250 mL of milk.

78. Renee's pet mouse weighs 30 g.

79. One ant weighs 2 mg.

80. I bought 30 L of gas for my car.

Convert the following using unit fractions, the metric conversion line, or the temperature conversion formulas.

81. 2.75 cm to mm

27.5 mm

82. 45 minutes to hours

$\frac{3}{4}$ hour or 0.75 hour

83. 90 inches to feet

$7\frac{1}{2}$ feet or 7.5 feet

84. 1.3 m to cm

130 cm

85. 25°C to Fahrenheit

77°F

86. 12 gallons to quarts

48 quarts

87. 700 mg to g

0.7 g

88. 0.81 L to mL

810 mL

89. 5 pounds to ounces

80 ounces

90. 60 kg to g

60,000 g

91. 1.8 L to mL

1800 mL

92. 86°F to Celsius

30°C

93. 0.36 m to cm

36 cm

94. 55 mL to L

0.055 L

Solve the following problems.

95. Peggy had a board measuring 2 m 4 cm. She cut off 78 cm. How long is the board now, in meters?

1.26 m

96. Imported wool fabric is $12.99 per meter. What is the cost, to the nearest cent, of a piece that measures 3 m 70 cm?

$48.06

97. Find the length of a football field in meters. (*Hint:* a football field is 100 yards long.)

91 meters

98. A commercial chemical tank is labeled "1000-liter capacity." How many gallons will the tank hold?

260 gallons

NAME CHAPTER 7 TEST DATE HOUR

Convert the following measurements.

1. $3\frac{1}{2}$ feet = ? inches — 1. 42
2. 36 feet = ? yards — 2. 12
3. 228 hours = ? days — 3. 9.5 or $9\frac{1}{2}$
4. 42 pints = ? gallons — 4. 5.25 or $5\frac{1}{4}$
5. 8 pounds = ? ounces — 5. 128
6. 4 hours = ? seconds — 6. 14,400

Write the most reasonable metric unit in each blank. Choose from km, m, cm, mm, L, mL, kg, g, mg.

7. My husband weighs 75 ______. — 7. kg
8. I hiked 5 ______ this morning. — 8. km
9. She bought 125 ______ of cough syrup. — 9. mL
10. This apple weighs 180 ______. — 10. g
11. This page is 21 ______ wide. — 11. cm
12. My little finger is 50 ______ long. — 12. mm
13. I bought 10 ______ of pop for the picnic. — 13. L
14. The bracelet is 18 ______ long. — 14. cm

Convert these measurements.

15. 250 cm to meters — 15. 2.5 m
16. 4.6 km to meters — 16. 4600 m
17. 5 mm to centimeters — 17. 0.5 cm
18. 325 mg to grams — 18. 0.325 g
19. 16 L to milliliters — 19. 16,000 mL
20. 0.4 kg to grams — 20. 400 g

21. 1055 cm

22. 0.095 L

23. 6.32 kg

24. 6.75 m

25. 95°C

26. 0°C

27. 5.5 m

28. 32.4 kg

29. 33.2 ounces

30. 10.3 km

31. 23°C

32. 36°F

33. Answer varies.

34. Answer varies.

21. 10.55 m to centimeters

22. 95 mL to liters

23. Stan's cat weighed 3 kg 740 g. His dog weighed 10 kg 60 g. How much heavier is the dog in kilograms?

24. Denise is making five matching pillows. She needs 1 m 35 cm of braid to trim each pillow. How many meters of braid should she buy?

Pick the Celsius temperature that is most appropriate in each situation.

25. The water is almost boiling.
210°C 155°C 95°C

26. The tomato plants may freeze tonight.
30°C 20°C 0°C

Use the table on page 457 to convert the following measurements. Round your answers to the nearest tenth, if necessary.

27. 18.4 feet to meters

28. 72 pounds to kilograms

29. 948 grams to ounces

30. 6.4 miles to kilometers

Use the conversion formulas to convert each temperature. Round your answers to the nearest degree, if necessary.

31. 74°F to Celsius

32. 2°C to Fahrenheit

33. Describe two benefits the United States would achieve by switching entirely to the metric system.

34. Give two reasons why you think the United States has resisted changing to the metric system.

NAME DATE HOUR

CUMULATIVE REVIEW EXERCISES CHAPTERS 1–7

≈ Round the numbers in each problem so there is only one non-zero digit. Then add, subtract, multiply, or divide the rounded numbers, as indicated, to estimate the answer. Finally, solve for the exact answer.

1. *estimate*: 100 + 3 + 70 = 173; *exact*: 107.5 + 2.548 + 68.79 = 178.838

2. *estimate*: 30,000 − 800 = 29,200; *exact*: 31,007 − 829 = 30,178

3. *estimate*: 90,000 × 200 = 18,000,000; *exact*: 92,075 × 183 = 16,849,725

4. *estimate*: 60 × 5 = 300; *exact*: 56.52 × 4.7 = 265.644

5. *estimate*: $40\overline{)20{,}000}$ = 500; *exact*: $37\overline{)19{,}610}$ = 530

6. *estimate*: $8\overline{)40}$ = 5; *exact*: $8.3\overline{)38.18}$ = 4.6

Add, subtract, multiply, or divide as indicated. Write answers to fraction problems in lowest terms and as whole or mixed numbers when possible.

7. $4\frac{7}{8} + \frac{5}{6}$

$5\frac{17}{24}$

8. 7 + 484,099 + 3939

488,045

9. $2\frac{2}{5} \cdot 2\frac{2}{9}$

$5\frac{1}{3}$

10. 0.86 ÷ 0.066
Round to nearest tenth.

13.0

11. $5\frac{1}{5} - 4\frac{2}{3}$

$\frac{8}{15}$

12. 8 − 0.9207

7.0793

13. $3\frac{3}{4} \div 6$

$\frac{5}{8}$

14. $47\overline{)14{,}467}$

307 R 38

15. (2.54)(0.003)

0.00762

Simplify by using the order of operations.

16. $12 - 6 \div 3 \cdot 4 + (7 - 6)$

5

17. $\sqrt{81} + 2^3 \cdot 5^2$

209

18. Write 307.19 in words.

three hundred seven and nineteen hundredths

19. Write eighty-two ten-thousandths in numbers.

0.0082

20. Arrange in order from smallest to largest.
8.3 0.083 8.03 0.8033

0.083, 0.8033, 8.03, 8.3

Complete this chart.

fraction/mixed number	*decimal*	*percent*
$\frac{7}{8}$	**21.** 0.875	**22.** 87.5% or $87\frac{1}{2}\%$
23. $\frac{1}{25}$	0.04	**24.** 4%
25. $3\frac{1}{2}$	**26.** 3.5	350%

Write each rate or ratio in lowest terms. Change to the same units when necessary.

27. \$44 to \$4

$\frac{11}{1}$

28. 45 seconds to 2 minutes
Compare in seconds.

$\frac{3}{8}$

29. There are 200 students taking a math class and 350 students taking an English class. Find the ratio of English students to math students.

$\frac{7}{4}$

30. Find the best buy on muffins.
package of 4 muffins for \$1.77
package of 6 muffins for \$2.89
package of 10 muffins for \$4.50

4 muffins for \$1.77

Find the missing number in each proportion. Round your answers to hundredths, if necessary.

31. $\frac{x}{16} = \frac{3}{4}$

12

32. $\frac{0.9}{0.75} = \frac{2}{x}$

1.67 (rounded)

Solve each of the following.

33. \$4 is what percent of \$80?

5%

34. 36 hours is 120% of what number of hours?

30 hours

Convert the following measurements. Use the table on page 457 and the temperature conversion formulas when necessary.

35. 6 feet to inches

72 inches

36. 75 minutes to hours

$1\frac{1}{4}$ or 1.25 hours

37. 2.8 m to centimeters

280 cm

38. 65 mg to grams

0.065 g

39. 198 km to miles

122.76 miles

40. 50°F to Celsius

10°C

Write the most reasonable metric unit in each blank. Choose from km, m, cm, mm, L, mL, kg, g, mg.

41. Ron bought the tube of toothpaste weighing 100 __g__.

42. The teacher's desk is 140 __cm__ long.

43. The hallway is 3 __m__ wide.

44. Joe's hammer weighed 1 __kg__.

45. Anne took a 500 __mg__ tablet of vitamin C.

46. Tia added 125 __mL__ of milk to her cereal.

Circle the Celsius temperature that is most appropriate in each situation.

47. John has a slight fever.

(38°C) 70°C 99°C

48. You'll need a light jacket outside.

0°C (12°C) 45°C

Solve the following word problems.

49. Calbert bought an automatic focus camera for \$64.95. He paid $7\frac{1}{2}$% sales tax. Find the amount of tax, to the nearest cent, and the total cost of the camera.

\$4.87; \$69.82

50. Danielle had a roll of 35 mm film developed. She received 24 prints for \$10.25. What is the cost per print, to the nearest cent?

\$0.43

51. Mark bought 650 grams of maple sugar candy on his vacation in Montreal. The candy is priced at $14.98 per kilogram. How much did Mark pay, to the nearest cent?

$9.74

52. On the Illinois map, one centimeter represents 12 kilometers. The center of Springfield is 7.8 cm from the center of Bloomington on the map. What is the actual distance in kilometers?

93.6 km

53. Dimitri took out a $3\frac{1}{2}$ year car loan for $8750 at 9% interest. Find the total amount due on the loan.

$ 11,506.25

54. On a 35-problem math test, Juana solved 31 problems correctly. What percent of the problems were correct? Round to the nearest tenth of a percent.

88.6%

55. An elevator has a weight limit of 1500 pounds. On it are twin boys weighing 88 pounds each and 8 adults weighing this number of pounds each: 240, 189, 127, 165, 143, 219, 116, and 124. The total weight of the people is how much above or below the limit?

1 pound under the limit

56. The Jackson family is making three kinds of holiday cookies that require brown sugar. The recipes call for $2\frac{1}{4}$ cups, $1\frac{1}{2}$ cups, and $\frac{3}{4}$ cup, respectively. They bought two packages of brown sugar, each holding $2\frac{1}{3}$ cups. How much will be left after making the cookies?

$\frac{1}{6}$ cup

57. Leather jackets are on sale at 30% off the regular price. Tracy likes a jacket with a regular price of $189. Find the amount of discount and the sale price she will pay.

$56.70; $132.30

58. Bags of slivered almonds weigh 115 g each. How many kilograms would a carton of 48 bags weigh? The carton alone weighs 450 g.

5.97 kg

59. Akuba is knitting a scarf. Six rows of knitting result in 5 centimeters of scarf. At that rate, how many rows will she knit to make a 100-centimeter scarf?

120 rows

60. A survey of the 5600 students on our campus found that $\frac{3}{8}$ of the students work 20 hours or more per week. How many students is that?

2100 students

Geometry 8

Geometry developed centuries ago when people needed a way to measure land. The name geometry comes from the Greek words *ge,* meaning earth, and *metron,* meaning measure. Today we still use geometry to measure farmland. It is also important in architecture, construction, navigation, art and design, physics, chemistry, and astronomy. You can use it at home, too, when you buy carpet or wallpaper, hang a picture, or build a fence. This chapter discusses the basic terms of geometry and the common geometric shapes that are all around us.

8.1 BASIC GEOMETRIC TERMS

OBJECTIVES

1. Identify lines, line segments, and rays.
2. Identify parallel and intersecting lines.
3. Identify and name an angle.
4. Classify an angle as right, acute, straight, or obtuse.
5. Identify perpendicular lines.

FOR EXTRA HELP

Tape 10	SSM pp. 233–234	MAC: A IBM: A

Geometry starts with the idea of a point. A **point** is a location in space. It has no length or width. A point is represented by a dot and is named by writing a capital letter next to the dot.

Point P

1 A **line** is a straight row of points that goes on forever in both directions. A line is drawn by using arrowheads to show that it never ends. The line is named by using the letters of any two points on the line.

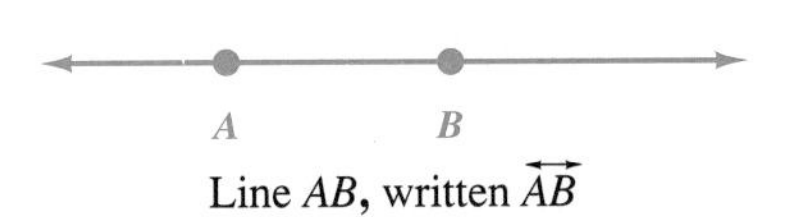

Line AB, written $\overleftrightarrow{AB}$

A piece of a line that has two endpoints is called a **line segment.** A line segment is named for its endpoints. The segment with endpoints P and Q is shown below. It can be named $\overline{PQ}$ or $\overline{QP}$.

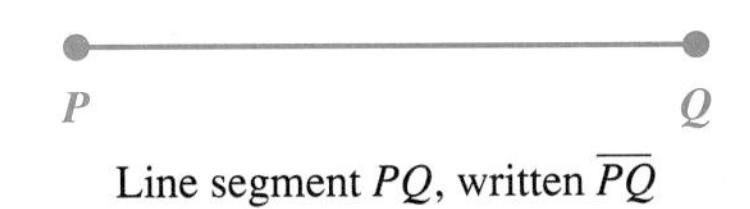

Line segment PQ, written $\overline{PQ}$

A **ray** is a part of a line that has only one endpoint and goes on forever in one direction. A ray is named by using the endpoint and some other point on the ray. The endpoint is always mentioned first.

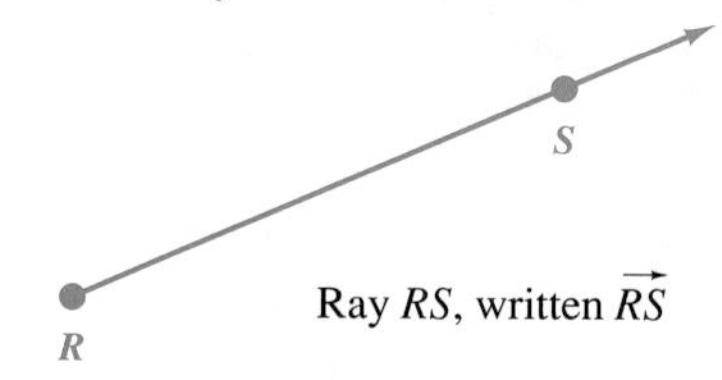

Ray RS, written $\overrightarrow{RS}$

■ **EXAMPLE 1** *Identifying Lines, Rays, and Line Segments*
Identify each of the following as a line, line segment, or ray.

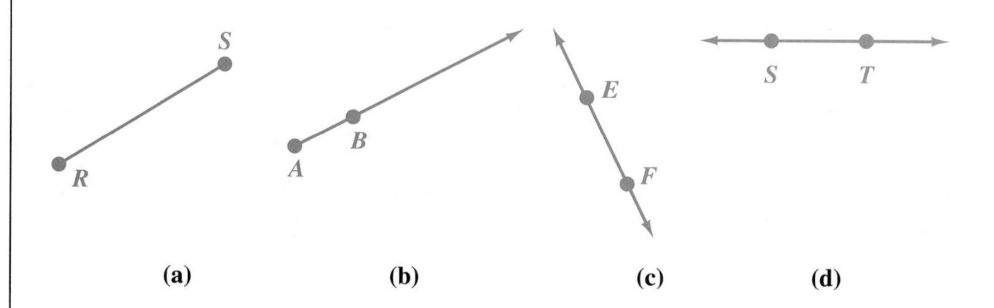

Both **(c)** and **(d)** go on forever in both directions, so they are lines. **(a)** has two endpoints, so it is a line segment. **(b)** starts at point A and goes on forever in one direction, so it is a ray. ■

◀◀ WORK PROBLEM 1 AT THE SIDE.

2 A *plane* is a flat surface, like a floor or a wall. Lines that are in the same plane but that never intersect (cross) are called **parallel lines,** while lines that cross or merge are called **intersecting lines.** (Think of an intersection, where two streets cross each other.)

■ **EXAMPLE 2** *Identifying Parallel and Intersecting Lines*
Label each pair of lines as parallel or intersecting.

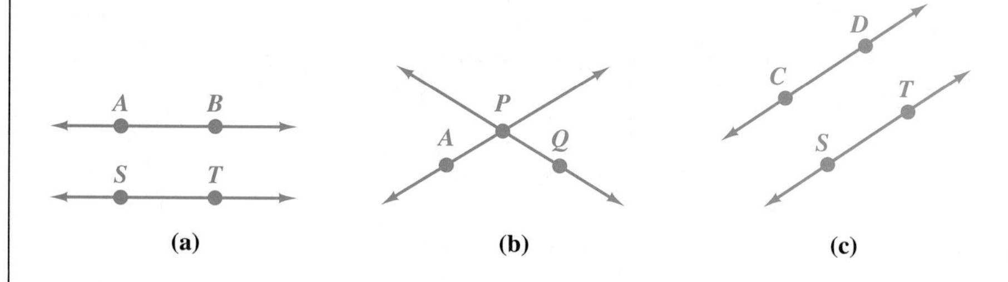

The lines in **(a)** and **(c)** never intersect. They are parallel lines. The lines in **(b)** cross at P, so they are intersecting lines. ■

◀◀ WORK PROBLEM 2 AT THE SIDE.

1. Identify each of the following as a line, line segment, or ray.

(a)

(b)

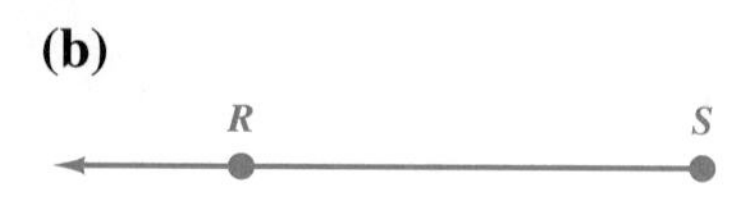

(c)

(d)

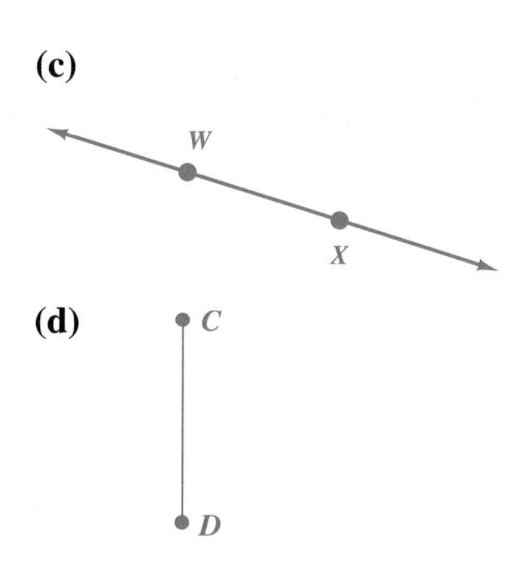

2. Label each pair of lines as parallel or intersecting.

(a)

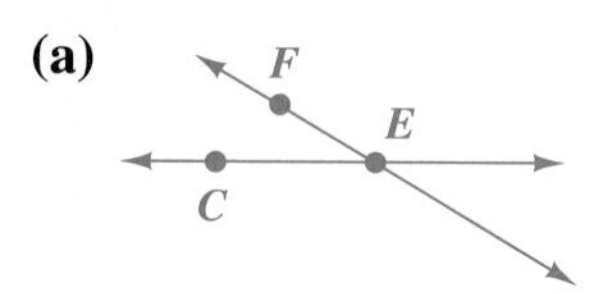

(b)

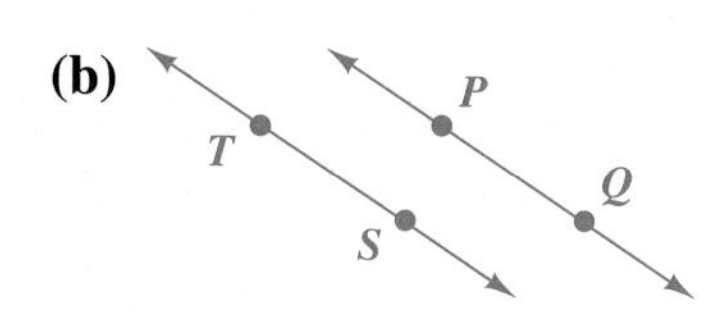

(c)

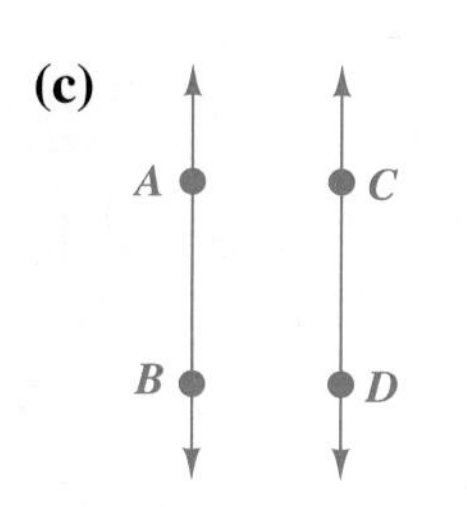

ANSWERS
1. (a) line segment (b) ray (c) line (d) line segment
2. (a) intersecting (b) parallel (c) parallel

3 An **angle** is made up of two rays that start at a common endpoint. This common endpoint is called the *vertex.*

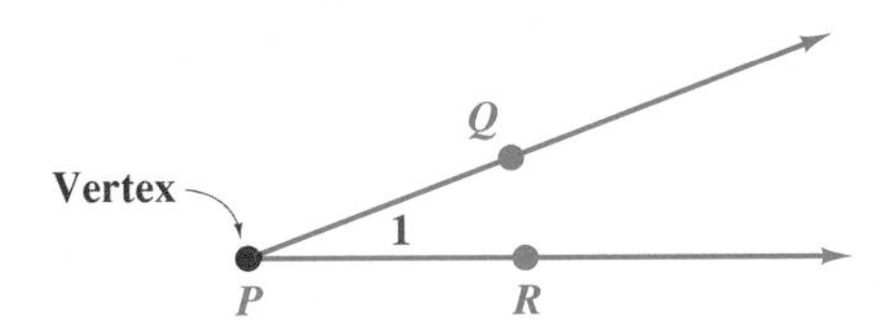

The rays PQ and PR are called *sides.* The angle can be named

$$\angle 1, \quad \angle P, \quad \angle QPR, \quad \text{or} \quad \angle RPQ.$$

NAMING AN ANGLE

When naming an angle, the vertex is written alone or it is written in the middle of two other points. If two or more angles have the same vertex, as in Example 3, do not use the vertex alone to name an angle.

EXAMPLE 3 *Identifying and Naming an Angle*
Identify the highlighted angle.

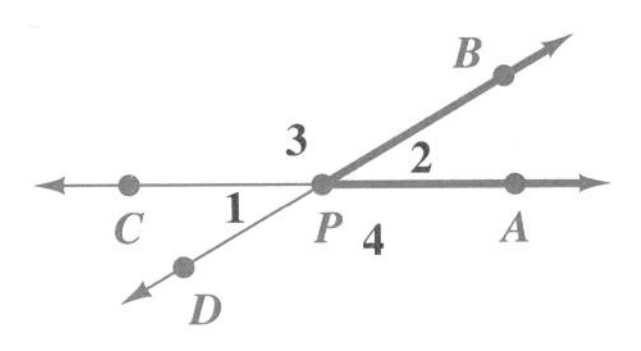

The angle can be named $\angle BPA$, $\angle APB$, or $\angle 2$. It cannot be named $\angle P$, using the vertex alone, because four different angles have P as their vertex. ■

WORK PROBLEM 3 AT THE SIDE.

4 Angles are measured in **degrees.** The symbol for degrees is a small, raised circle °. Think of the minute hand on a clock as a ray in an angle. Suppose it is at 12:00. During one hour of time, the minute hand moves around in a complete circle. It moves 360 *degrees,* or 360°. In half an hour, at 12:30, the minute hand has moved half way around the circle, or 180°. An angle of 180° is called a **straight angle.** (Notice that the two rays in a straight angle form a straight line.)

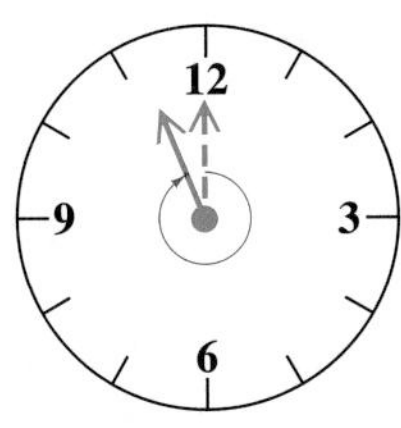

Complete circle
360°

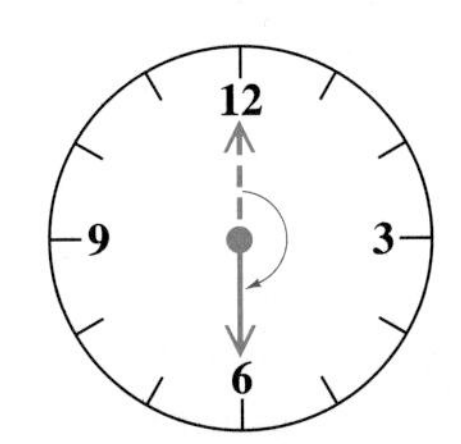

Straight angle
(half a circle)
180°

3. (a) Identify the highlighted angle in three different ways.

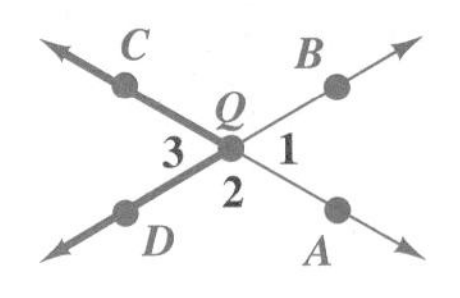

(b) Darken the lines that make up $\angle ZTW$.

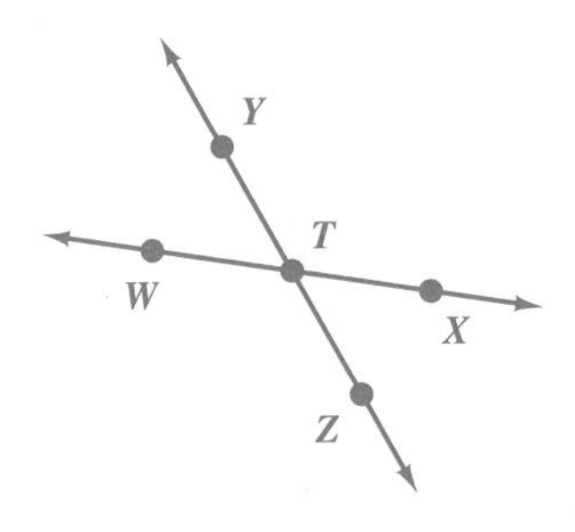

(c) Identify this angle in four different ways.

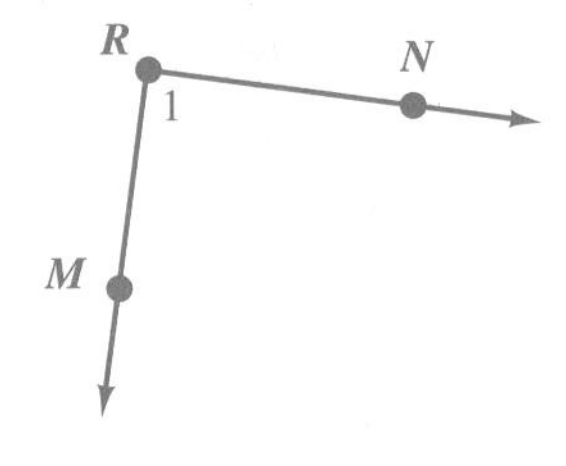

ANSWERS

3. (a) $\angle 3$, $\angle CQD$, $\angle DQC$

(b)

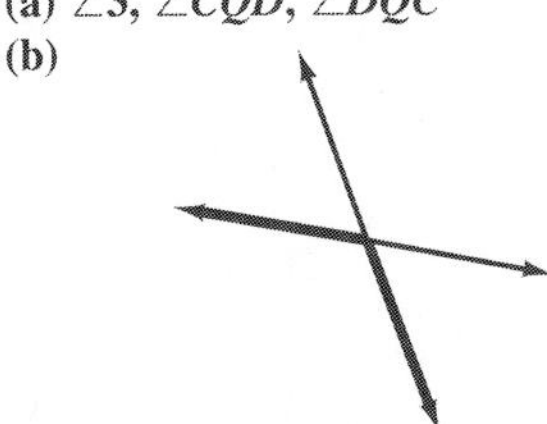

(c) $\angle 1$, $\angle R$, $\angle MRN$, $\angle NRM$

In a quarter of an hour, at 12:15, the minute hand has moved $\frac{1}{4}$ of the way around the circle, or 90°. An angle of 90° is called a **right angle.** Sometimes you hear it called a *square angle.* The minute hands at 12:00 and 12:15 form one corner of a square. So, to show that an angle is a **right angle**, we draw a **small square** at the vertex.

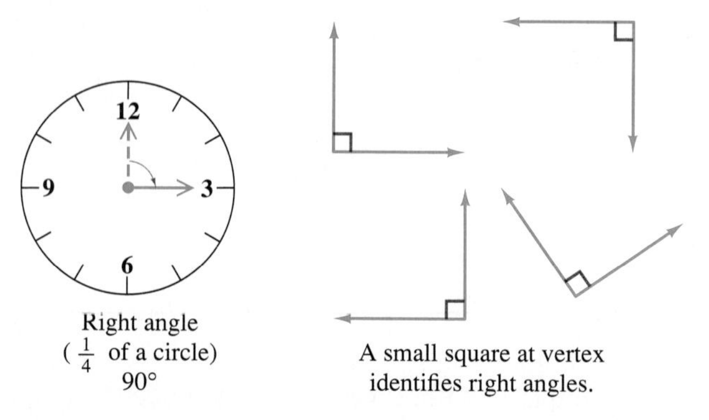

Right angle ($\frac{1}{4}$ of a circle) 90°

A small square at vertex identifies right angles.

You can see that an angle of 1° is very small. To be precise, it is only the distance that the minute hand moves in ten *seconds.* Some other terms used to describe angle size are:

acute angles measure between 0° and 90°
obtuse angles measure between 90° and 180°

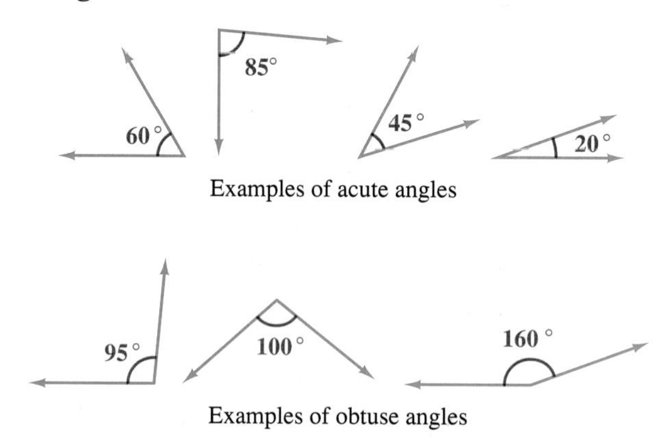

Examples of acute angles

Examples of obtuse angles

Section 10.1 shows you how to use a tool called a *protractor* to measure the number of degrees in an angle.

■ EXAMPLE 4 *Classifying an Angle*

Label each of the following angles as acute, right, obtuse, or straight.

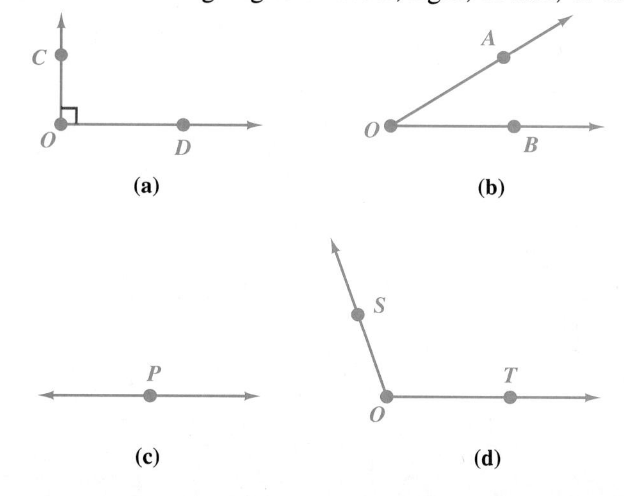

The angle in **(a)** is a right angle (identified by a small square at the vertex). The angle in **(b)** is an acute angle (between 0° and 90°). The angle in **(c)** is a straight angle (180°), and the angle in **(d)** is an obtuse angle (between 90° and 180°). ■

WORK PROBLEM 4 AT THE SIDE. ▶▶

Two lines are called **perpendicular lines** if they intersect to form a right angle.

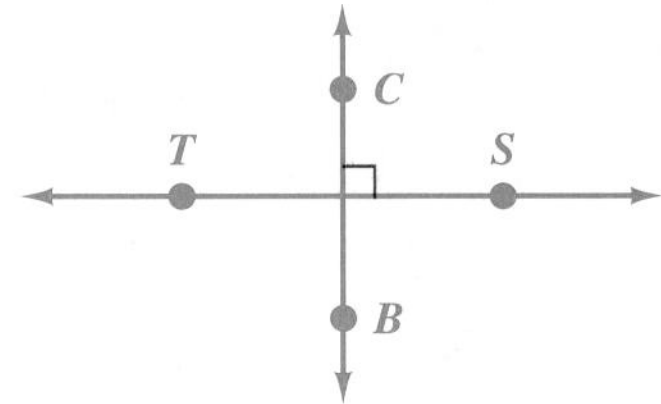

Lines *CB* and *ST* are **perpendicular**, because they intersect at right angles. This can be written in the following way: $\overleftrightarrow{CB} \perp \overleftrightarrow{ST}$.

■ **EXAMPLE 5** *Identifying Perpendicular Lines*

Which of the following pairs of lines are perpendicular?

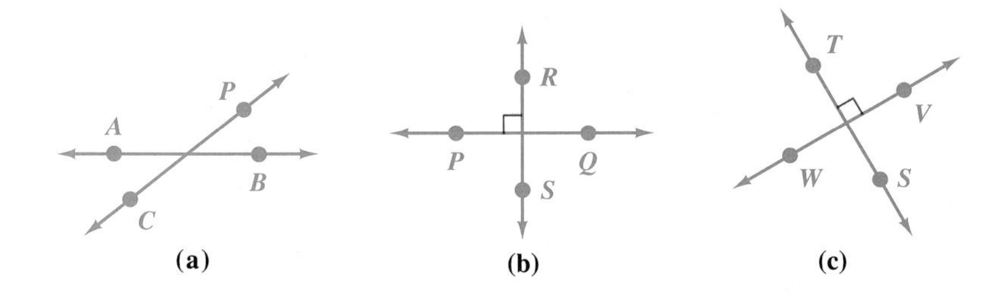

The lines in **(b)** and the lines in **(c)** are perpendicular to each other, because they intersect at right angles. ■

WORK PROBLEM 5 AT THE SIDE. ▶▶

4. Label each of the following as acute, right, obtuse, or straight angles.

(a)

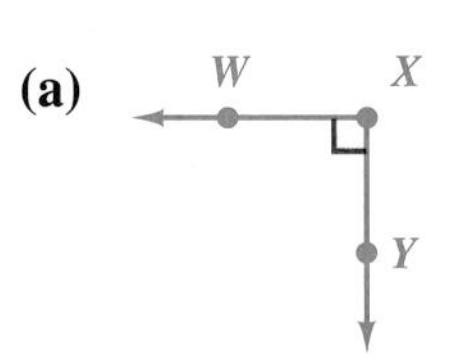

(b) P

(c)

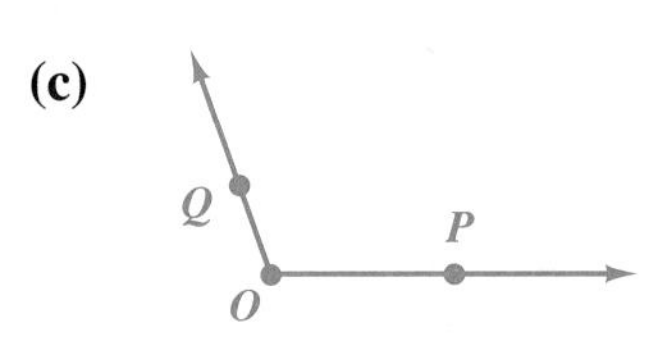

(d) R E F

5. Which pair of lines is perpendicular?

(a)

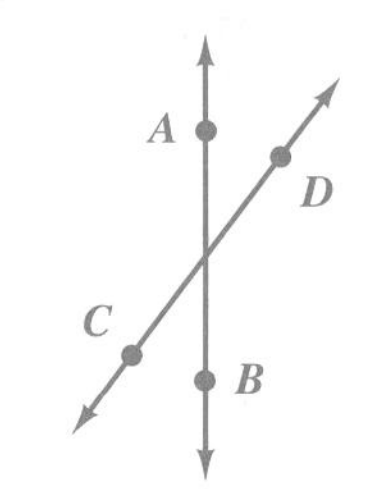

(b) P R S Q

ANSWERS

4. (a) right (b) straight (c) obtuse (d) acute

5. (b)

QUEST FOR NUMERACY

Paycheck Math

The amount an employee earns is called *gross pay*. Nevertheless, certain amounts are withheld for federal and state taxes, social security (FICA), Medicare, and insurance premiums. These amounts are called *deductions*. After subtracting deductions from earnings, the remaining amount is *net pay,* or take-home pay. Fill in the blanks on these paycheck stubs for gross pay and net pay. Use a calculator.

EMPLOYEE NUMBER	PERIOD ENDING	HOURS	
		REGULAR	OVERTIME
66736	1/27	40.00	10.50

EARNINGS		
REGULAR	OVERTIME	GROSS PAY
383.20	150.89	$ 534.09

DEDUCTIONS					NET PAY
FEDERAL TAX	FICA	MEDI-CARE	STATE TAX	HEALTH INS	
80.97	34.72	8.01	39.08	18.78	$ 352.53

What is the employee's regular hourly wage? $ 9.58

What is the employee's overtime hourly wage? $14.37

The overtime hourly wage is what percent of the regular hourly wage? 150%

EMPLOYEE NUMBER	PERIOD ENDING	HOURS	
		REGULAR	OVERTIME
80082	1/27	80.00	.00

EARNINGS		
REGULAR	OVERTIME	GROSS PAY
1427.70	.00	$ 1427.70

DEDUCTIONS					NET PAY
FEDERAL TAX	FICA	MEDI-CARE	STATE TAX	HEALTH INS	
138.89	92.80	21.42	65.89	45.87	$ 1062.83

Round your answers to the nearest tenth of a percent.

Federal and state taxes are what percent of gross pay? 14.3%

Health insurance is what percent of gross pay? 3.2%

What percent of gross pay did the employee actually take home? 74.4%

Extend this activity by asking students to bring in their own paycheck stubs. Or, ask your college business office for help in making a realistic mock-up of a paycheck stub for a mythical student worker on your campus.

8.1 EXERCISES

NAME DATE HOUR

Name each line, line segment, or ray using the appropriate symbol.

1.

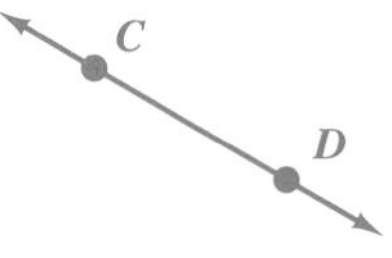

line named $\overleftrightarrow{CD}$ or $\overleftrightarrow{DC}$

2.

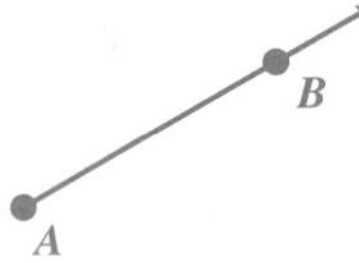

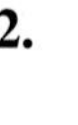

ray named $\overrightarrow{AB}$

3.

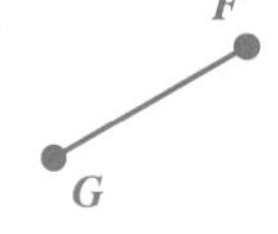

line segment named $\overline{GF}$ or $\overline{FG}$

4.

line segment named $\overline{EF}$ or $\overline{FE}$

5.

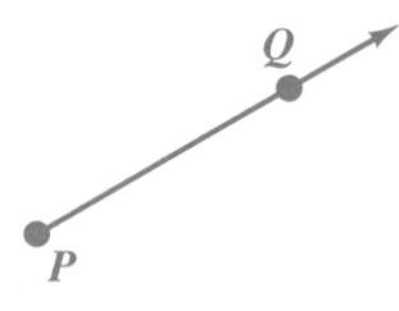

ray named $\overrightarrow{PQ}$

6.

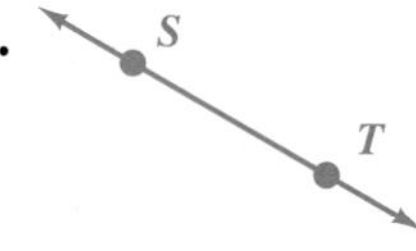

line named $\overleftrightarrow{ST}$ or $\overleftrightarrow{TS}$

Label each pair of lines as parallel, perpendicular, or intersecting.

7.

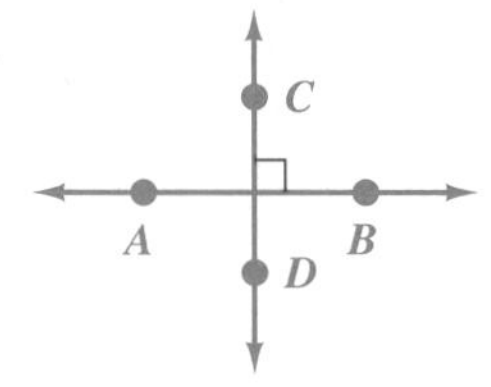

perpendicular

8.

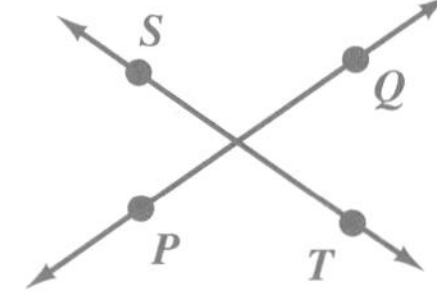

intersecting

9.

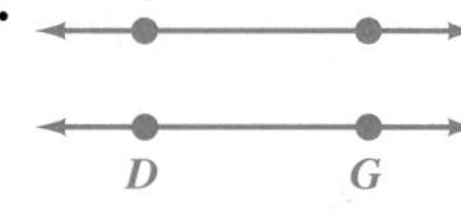

parallel

10.

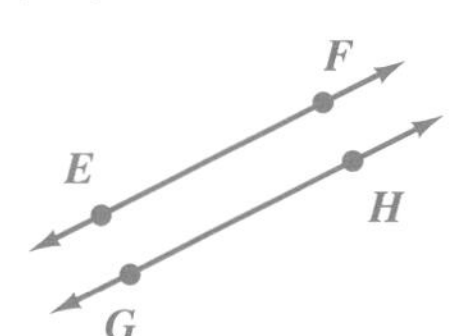

parallel

11.

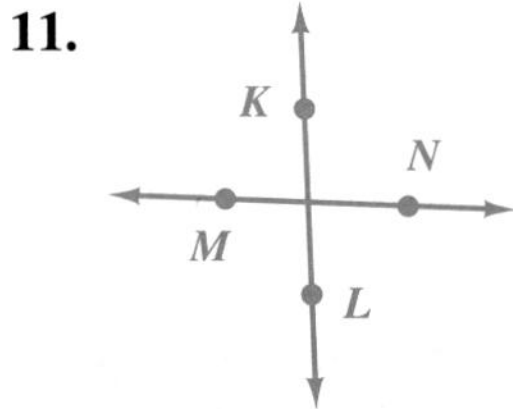

intersecting

12.

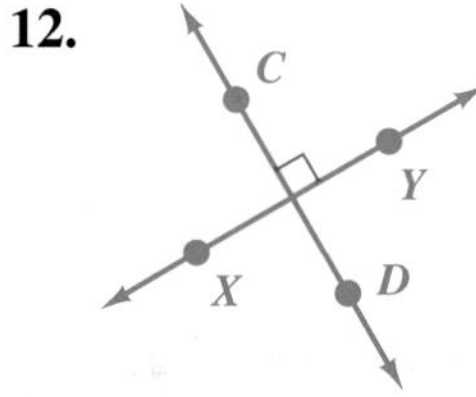

perpendicular

Name each highlighted angle by using the three-letter form of identification.

13.

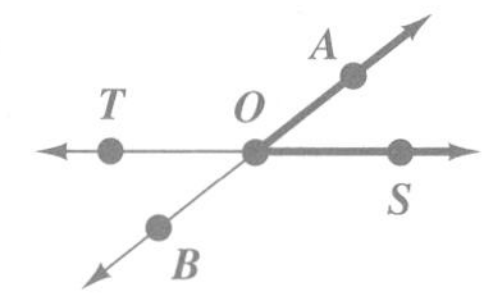

∠*AOS* or ∠*SOA*

14.

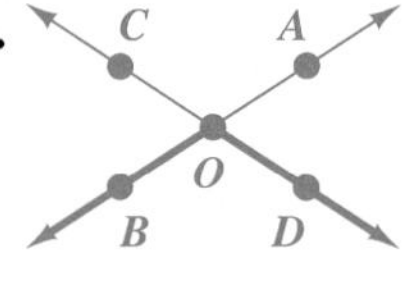

∠*BOD* or ∠*DOB*

15.

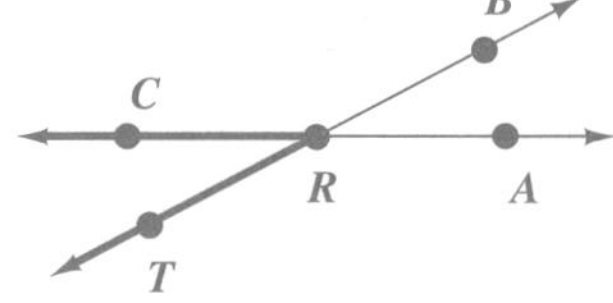

∠*CRT* or ∠*TRC*

16.

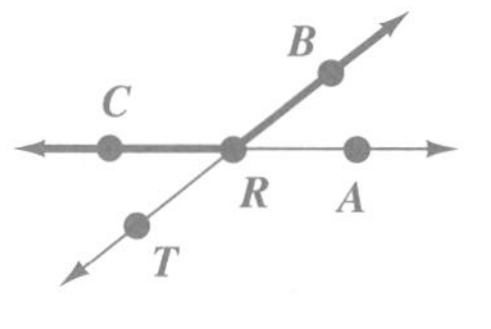

∠*CRB* or ∠*BRC*

17.

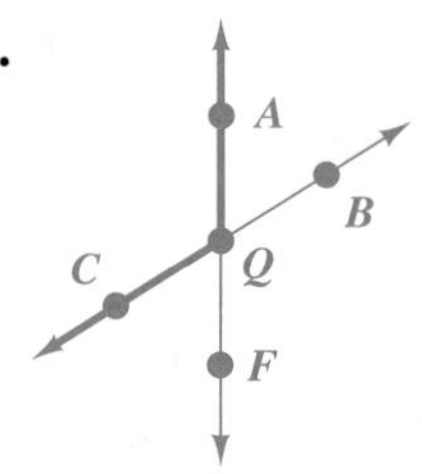

∠*AQC* or ∠*CQA*

18.

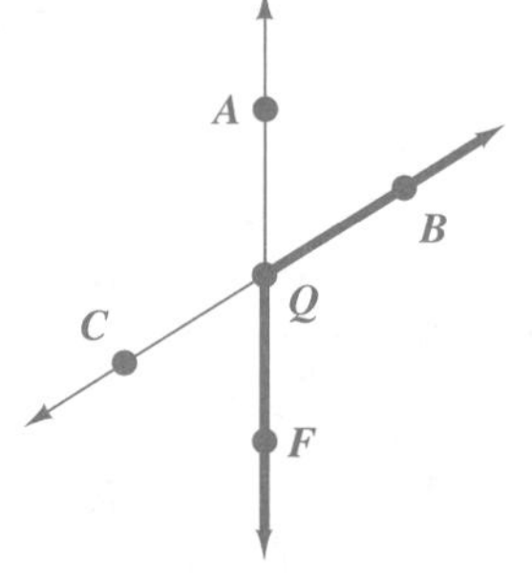

∠*FQB* or ∠*BQF*

Label each of the following as an acute, right, obtuse, or straight angle. For right angles and straight angles, indicate the number of degrees in the angle.

19.

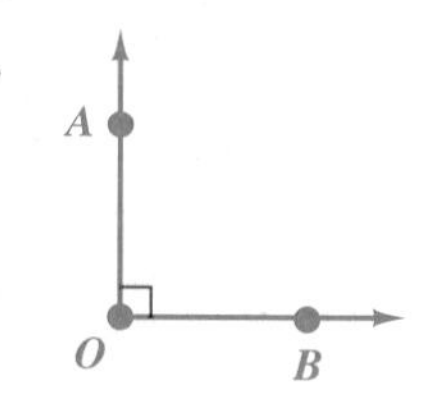

right (90°)

20.

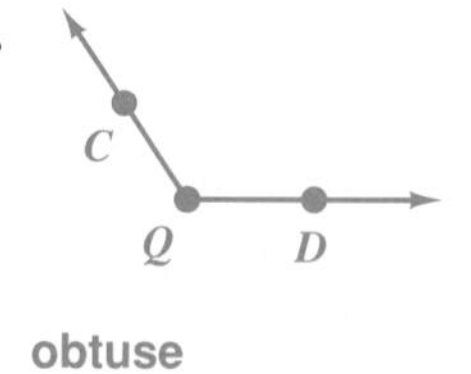

obtuse

21.

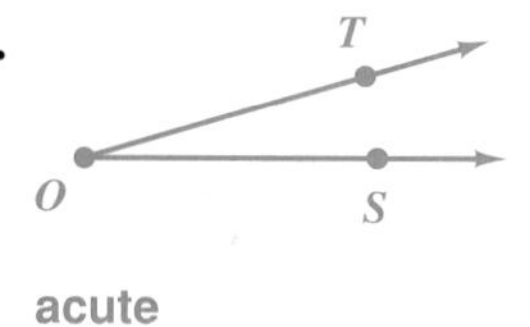

acute

22.

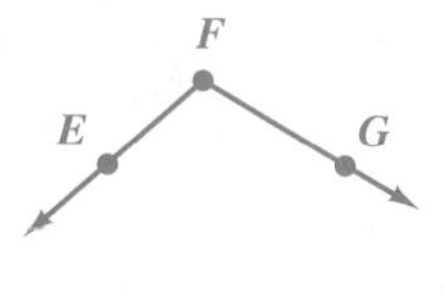

obtuse

23.

straight (180°)

24.

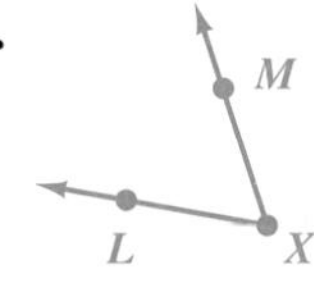

acute

25. Explain what is happening in each sentence.

(a) The road was so slippery that my car did a 360.

(b) After the election, the governor's view on taxes took a 180° turn.

Answer varies.

26. Find at least four examples of right angles in your home, at work, or on the street. Make a sketch of each example and label the right angle.

Answer varies.

Use the diagram to label each statement as true or false.

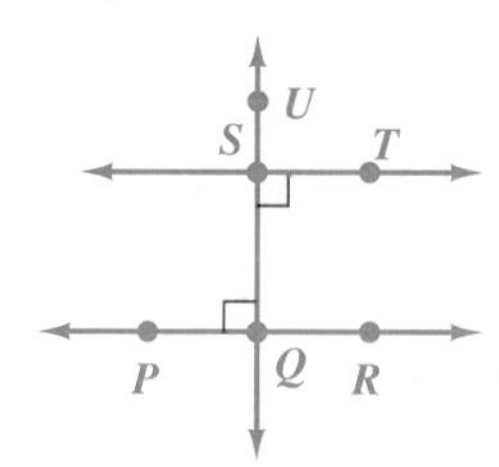

27. $\angle UST$ is 90°. true

28. Lines SQ and PQ are perpendicular. true

29. $\angle USQ$ is smaller than $\angle PQR$. false

30. Lines ST and PR are parallel. true

PREVIEW EXERCISES

Evaluate the following expressions. (For help, see **Section 1.8.***)*

31. $79 + 65$ 144

32. $83 - 26$ 57

33. $(180 - 75) - 15$ 90

34. $180 - (75 - 15)$ 120

35. $(90 - 37) + 15$ 68

36. $90 - (37 + 15)$ 38

OBJECTIVES

1. Identify complementary angles and supplementary angles.
2. Identify congruent angles and vertical angles.

FOR EXTRA HELP

Tape 10 | SSM pp. 234–235 | MAC: A IBM: A

1 Two angles are called **complementary angles** if their sum is 90°. If two angles are complementary, each angle is the *complement* of the other.

EXAMPLE 1 *Identifying Complementary Angles*
Identify each pair of complementary angles.

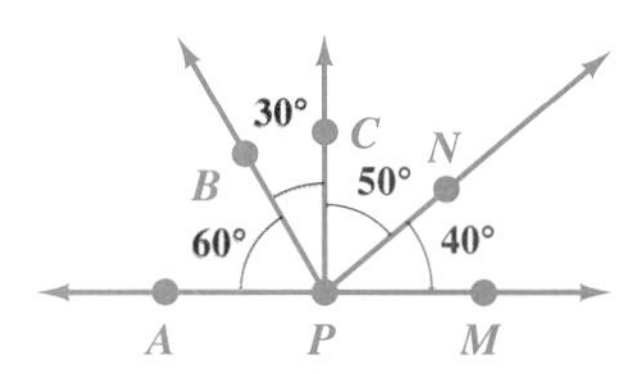

$\angle MPN$ (40°) and $\angle NPC$ (50°) are complementary angles because

$$40° + 50° = 90°.$$

$\angle CPB$ (30°) and $\angle BPA$ (60°) are complementary angles because

$$30° + 60° = 90°.$$ ■

WORK PROBLEM 1 AT THE SIDE.

1. Identify each pair of complementary angles.

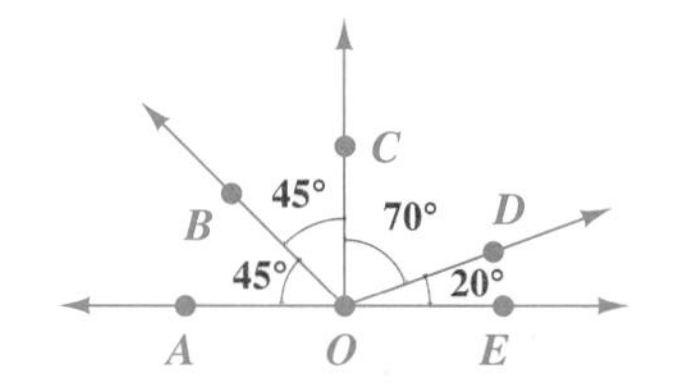

EXAMPLE 2 *Finding the Complement of an Angle*
Find the complement of each angle.

(a) 30° **(b)** 40°

(a) The complement of 30° is 60°, because 90° − 30° = 60°.

(b) The complement of 40° is 50°, because 90° − 40° = 50°. ■

WORK PROBLEM 2 AT THE SIDE.

2. Find the complement of the following angles.

(a) 35°

(b) 80°

Two angles are called **supplementary angles** if their sum is 180°. If two angles are supplementary, each angle is the *supplement* of the other.

EXAMPLE 3 *Identifying Supplementary Angles*
Identify each pair of supplementary angles.

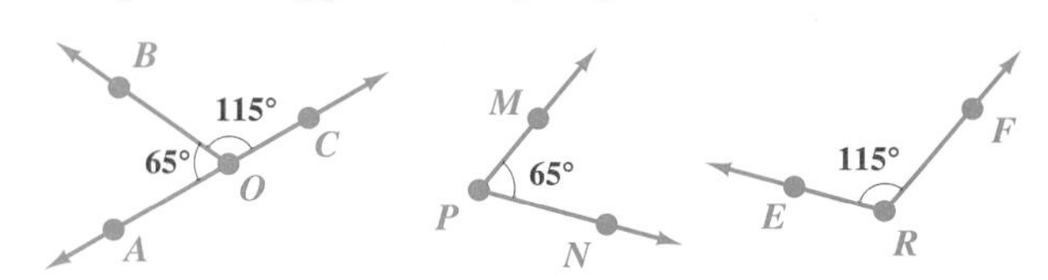

$\angle BOA$ and $\angle BOC$, because 65° + 115° = 180°.

$\angle BOA$ and $\angle ERF$, because 65° + 115° = 180°.

$\angle BOC$ and $\angle MPN$, because 115° + 65° = 180°.

$\angle MPN$ and $\angle ERF$, because 65° + 115° = 180°. ■

WORK PROBLEM 3 AT THE SIDE.

3. Identify each pair of supplementary angles.

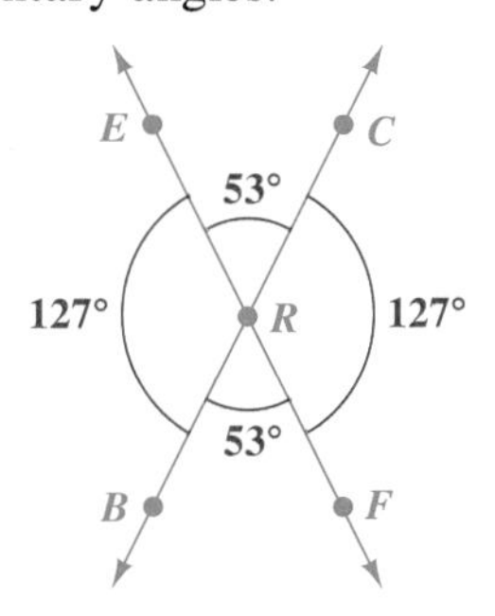

ANSWERS
1. $\angle AOB$ and $\angle BOC$; $\angle COD$ and $\angle DOE$
2. (a) 55° (b) 10°
3. $\angle CRF$ and $\angle BRF$
$\angle CRE$ and $\angle ERB$
$\angle BRF$ and $\angle BRE$
$\angle CRE$ and $\angle CRF$

4. Find the supplement of each angle.

(a) 175°

(b) 30°

5. Identify the angles that are congruent.

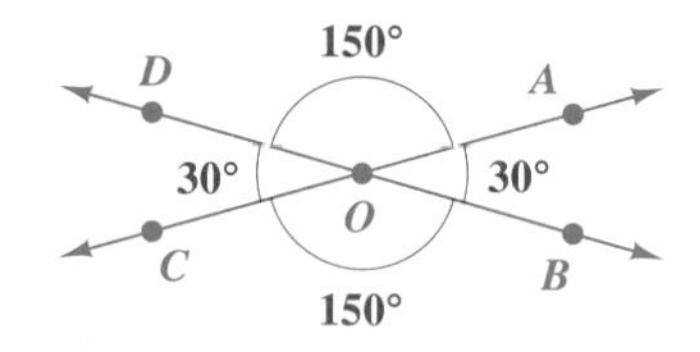

6. Identify the vertical angles.

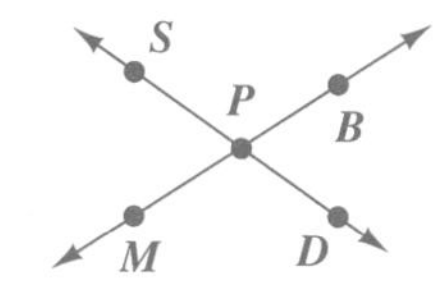

ANSWERS

4. (a) 5° (b) 150°

5. $\angle BOC \cong \angle AOD$, $\angle AOB \cong \angle DOC$

6. $\angle SPB$ and $\angle MPD$, $\angle BPD$ and $\angle SPM$

■ **EXAMPLE 4** *Finding the Supplement of an Angle*

Find the supplement of the following angles.

(a) 70° **(b)** 140°

(a) The supplement of 70° is 110°, because 180° − 70° = 110°.

(b) The supplement of 140° is 40°, because 180° − 140° = 40°. ■

WORK PROBLEM 4 AT THE SIDE.

2 Two angles are called **congruent angles** if they measure the same number of degrees. If two angles are congruent, this is written as

$$\angle A \cong \angle B$$

and read as, "angle A **is congruent to** angle B."

■ **EXAMPLE 5** *Identifying Congruent Angles*

Identify the angles that are congruent.

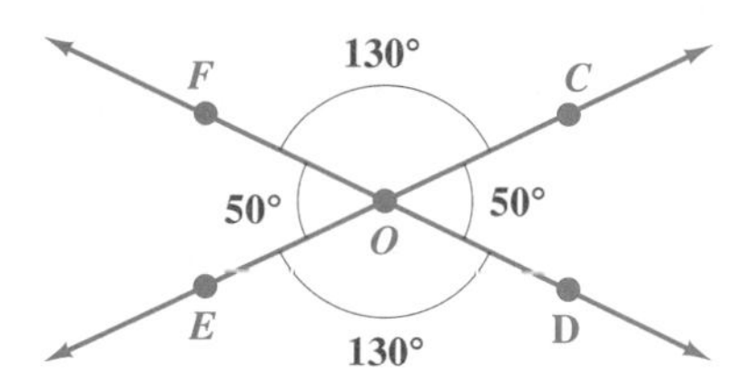

$\angle FOC \cong \angle EOD$ and $\angle COD \cong \angle EOF$ ■

WORK PROBLEM 5 AT THE SIDE.

Angles that do not share a common side are called *nonadjacent* angles. Two nonadjacent angles formed by intersecting lines are called **vertical angles.**

■ **EXAMPLE 6** *Identifying Vertical Angles*

Identify the vertical angles in this figure.

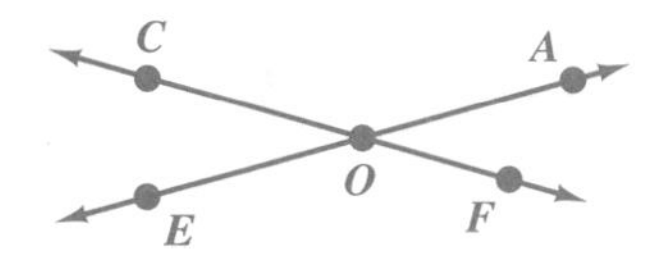

$\angle AOF$ and $\angle COE$ are vertical angles because they are not adjacent to each other and they are formed by two intersecting lines.

$\angle EOF$ and $\angle COA$ also are vertical angles. ■

WORK PROBLEM 6 AT THE SIDE.

Look back at Example 5. Notice that the two congruent angles that measure 130° are also vertical angles. Also, the two congruent angles that measure 50° are vertical angles. This illustrates the following property.

CONGRUENT ANGLES

If two angles are vertical angles, they are congruent, that is, they measure the same number of degrees.

EXAMPLE 7 *Finding the Sizes of Vertical Angles*
In the figure below, find the sizes of the following angles.

(a) $\angle COD$ **(b)** $\angle DOE$ **(c)** $\angle EOF$

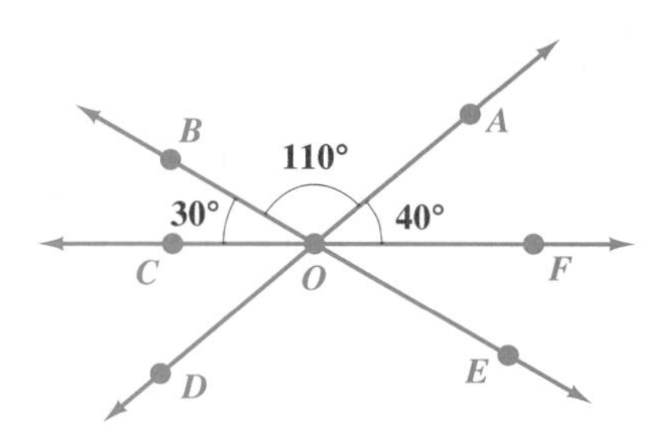

(a) $\angle COD$ and $\angle AOF$ are vertical angles so they are congruent. This means they measure the same number of degrees.

$$\angle AOF = 40° \quad \text{so} \quad \angle COD = 40° \text{ also.}$$

(b) $\angle D$OE and $\angle BOA$ are vertical angles so they are congruent.

$$\angle BOA = 110° \quad \text{so} \quad \angle DOE = 110° \text{ also.}$$

(c) $\angle EO$F and $\angle C$OB are vertical angles so they are congruent.

$$\angle COB = 30° \quad \text{so} \quad \angle EOF = 30° \text{ also.}$$ ■

WORK PROBLEM 7 AT THE SIDE.

7. In the figure below, find the number of degrees in each of the following angles.

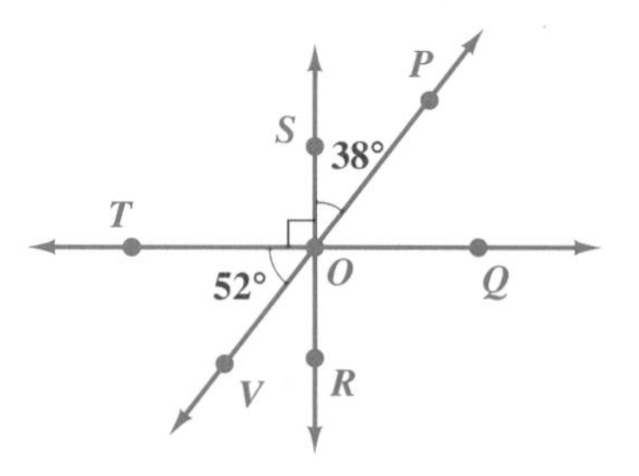

(a) $\angle VOR$

(b) $\angle POQ$

(c) $\angle QOR$

ANSWERS
7. (a) 38° (b) 52° (c) 90°

QUEST FOR NUMERACY

Is 99.9% Accuracy Good Enough?

If you could invest your money in something that was 99.9% safe, would you do it? After all, 99.9% is nearly 100%, so you might think there is no risk of losing your money. But there is always a chance for mistakes when the accuracy rate is less than 100%. Take a look at these possibilities.

If there is 99.9% accuracy, then . . . .

The IRS would lose two million documents this year.

How many documents does the IRS handle in a year?

2,000,000,000 documents

22,000 checks would be deducted from the wrong bank accounts in the next 60 minutes.

How many checks are processed each hour?

22,000,000 checks

18,322 pieces of mail would go to the wrong place in the next hour.

How many pieces per second would go to the wrong place?
5 pieces (rounded)

How many pieces would go to the wrong place in the next year?
≈ 160,500,000 pieces

How many pieces of mail are handled per hour? **18,322,000 pieces**
Per second?
5089 pieces (rounded)

1314 phone calls would be mishandled every minute.

How many calls would be mishandled per day? **1,892,160 calls**

How many calls in all are placed every minute? **1,314,000 calls**

Every day? **1,892,160,000 calls**

Every year?
≈ 690,638,000,000 calls

315 words in Webster's Third New International Dictionary (Unabridged) would be misspelled. How many words would be spelled correctly?

314,685 words

811,000 defective rolls of film would be sold this year. If you buy one roll of film per year, what are your chances of getting a defective roll?

Very small, because only 0.1% of the rolls are defective. You have a 1 in 1000 chance of buying a defective roll.

Extend this activity by asking students to estimate the number of errors in this textbook if the authors were 99.9% accurate. The first step is to estimate the number of problems in the text, perhaps by counting the number in each of several sections, finding an average, multiplying by the number of sections, and so on.

8.2 EXERCISES

NAME DATE HOUR

Identify each pair of complementary angles.

1.

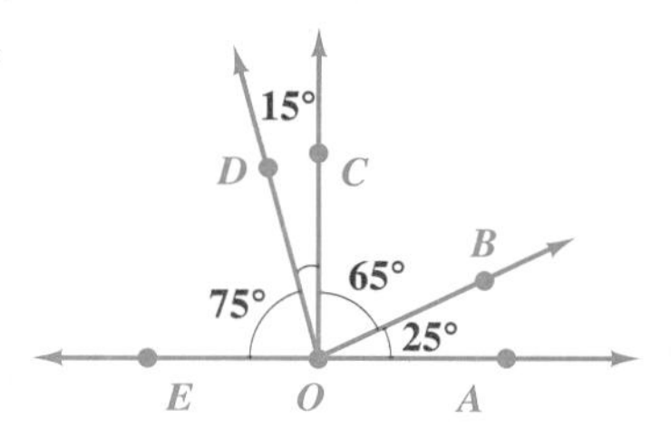

∠*EOD* and ∠*COD*; ∠*AOB* and ∠*BOC*

2.

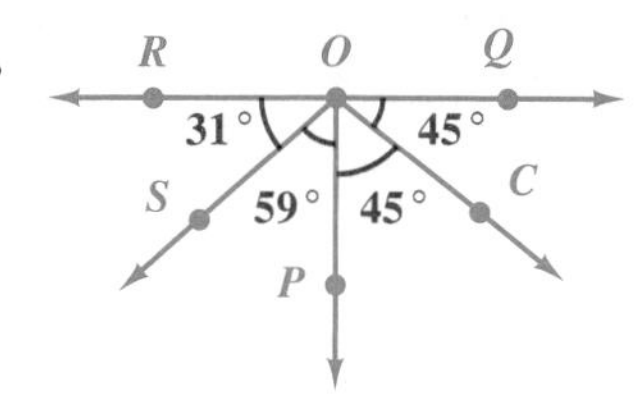

∠*COQ* and ∠*COP*; ∠*SOR* and ∠*SOP*

Identify each pair of supplementary angles.

3.

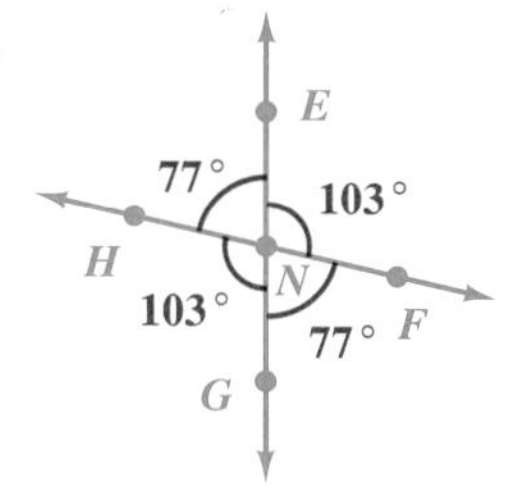

∠*HNE* and ∠*ENF*; ∠*HNG* and ∠*GNF*
∠*HNE* and ∠*HNG*; ∠*ENF* and ∠*GNF*

4.

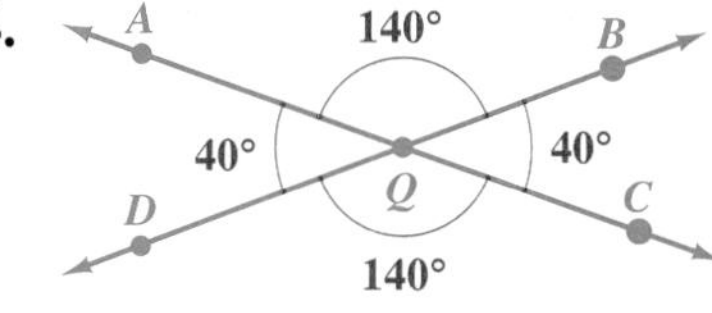

∠*AQB* and ∠*AQD*; ∠*BQC* and ∠*CQD*;
∠*AQB* and ∠*BQC*; ∠*CQD* and ∠*AQD*

Find the complement of each angle.

5. 40°
50°

6. 35°
55°

7. 83°
7°

8. 67°
23°

Find the supplement of each angle.

9. 130°
50°

10. 75°
105°

11. 96°
84°

12. 8°
172°

In each of the following, identify the angles that are congruent.

13.

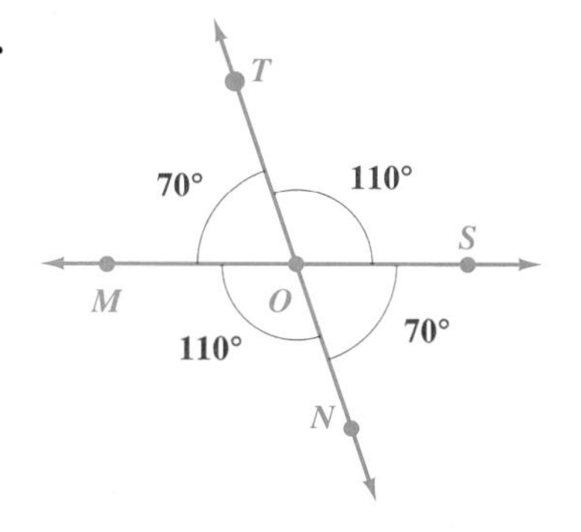

∠*SON* ≅ ∠*TOM*; ∠*TOS* ≅ ∠*MON*

14.

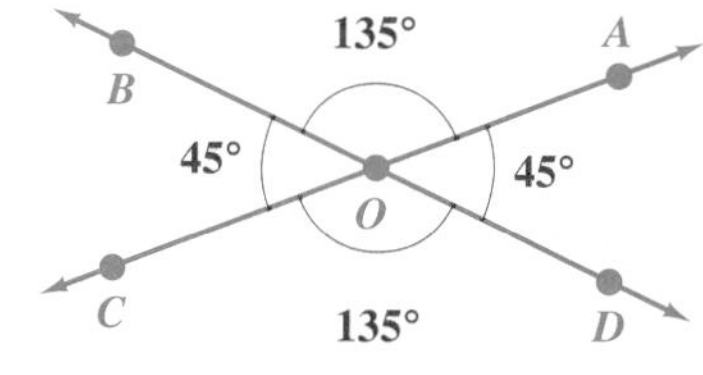

∠*AOB* ≅ ∠*COD*; ∠*AOD* ≅ ∠*BOC*

Writing Conceptual Challenging Estimation

In this figure, $\angle AOH = 37°$ and $\angle COE = 63°$. Find the number of degrees in each angle.

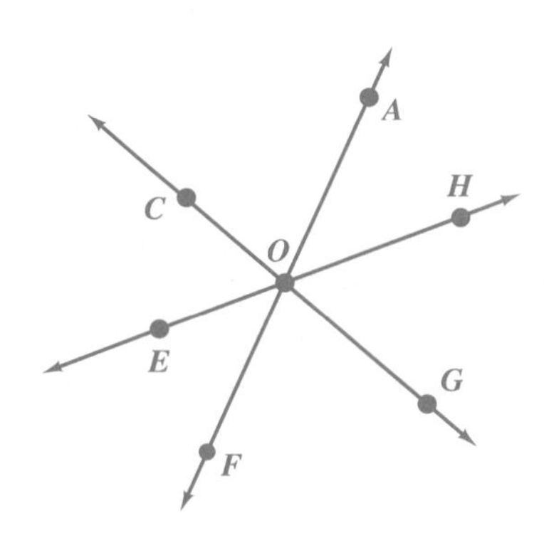

15. $\angle GOH$

63°

16. $\angle AOC$

80°

17. $\angle EOF$

37°

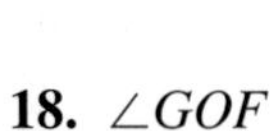

18. $\angle GOF$

80°

19. In your own words, write a definition of complementary angles and a definition of supplementary angles. Draw a picture to illustrate each definition.

Answer varies.

20. Make up a test question in which a student has to use knowledge of vertical angles. Include a drawing with some angles labeled and ask the student to find the size of the remaining angles.

Answer varies.

In each figure, ray AB is parallel to ray CD. Identify two pairs of congruent angles and the number of degrees in each congruent angle.

21.

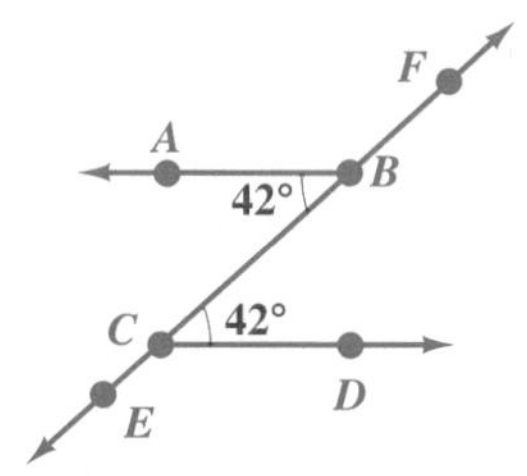

$\angle ABF \cong \angle ECD$ Both are 138°.
$\angle ABC \cong \angle BCD$ Both are 42°.

22.

E
B
133°
D
A
133°
C
G

$\angle ABC \cong \angle BCD$ Both are 133°.
$\angle DCG \cong \angle EBA$ Both are 47°.

PREVIEW EXERCISES

Evaluate the following expressions. (For help, see **Section 1.8.***)*

23. $16 - (3 \cdot 3)$ 7

24. $16 \cdot 4 \div 2$ 32

25. $2 \cdot 3 + 2 \cdot 4$ 14

26. $5 + 2 \cdot 6$ 17

27. $6 \cdot 8 - 5^2$ 23

28. $3^2 + 4^2$ 25

8.3 RECTANGLES AND SQUARES

A **rectangle** is a figure with four sides that intersect to form 90° angles. Each set of opposite sides is parallel and congruent (has the same length).

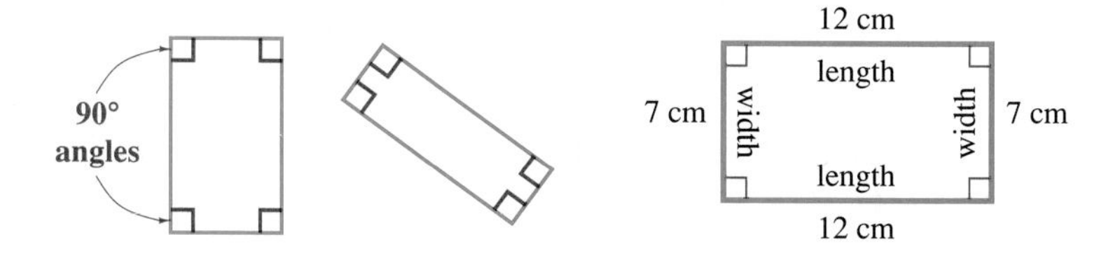

Each longer side of a rectangle is called the length (l) and each shorter side is called the width (w).

OBJECTIVES

1. Find the perimeter and area of a rectangle.
2. Find the perimeter and area of a square.
3. Find the perimeter and area of a composite shape.

FOR EXTRA HELP

Tape 11 | SSM pp. 235–237 | MAC: A IBM: A

WORK PROBLEM 1 AT THE SIDE.

1. Identify all rectangles.

(a)

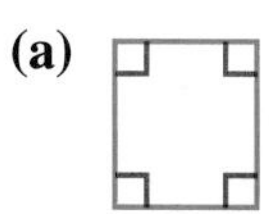

(b)

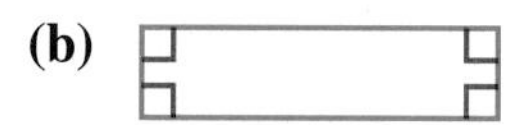

(c)

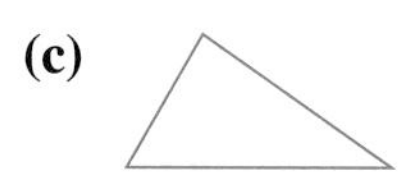

(d)

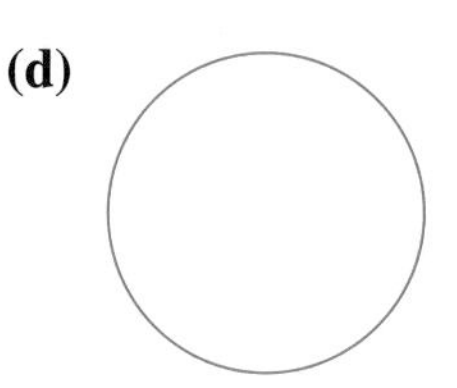

(e)

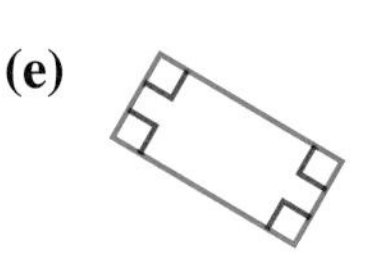

(f)

(g)

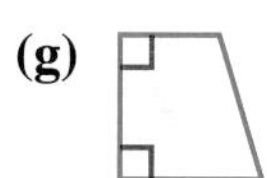

1 The distance around the outside edges of a figure is the **perimeter** of the figure. Think of how much fence you would need to put around the sides of a garden plot, or how far you would walk if you go around the outside edges of your backyard. In either case you would add up the lengths of the sides. In the rectangle on the right above,

$$\text{Perimeter} = 12 \text{ cm} + 7 \text{ cm} + 12 \text{ cm} + 7 \text{ cm} = 38 \text{ cm}$$

Because the two long sides are the same, and the two short sides are the same, you can also use this formula.

FINDING THE PERIMETER OF A RECTANGLE

$$\text{Perimeter of a rectangle} = (2 \cdot \text{length}) + (2 \cdot \text{width})$$

$$P = 2 \cdot l + 2 \cdot w$$

EXAMPLE 1 *Finding the Perimeter of a Rectangle*

Find the perimeter of each rectangle.

(a)

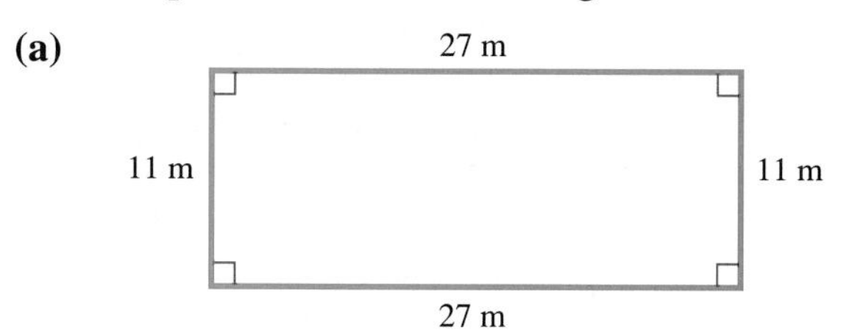

(b) A rectangle 8.9 m by 12.3 m

ANSWERS

1. (a) (b) and (e) are rectangles; (c) (d) (f) and (g) are not.

2. Find the perimeter of each rectangle.

(a)

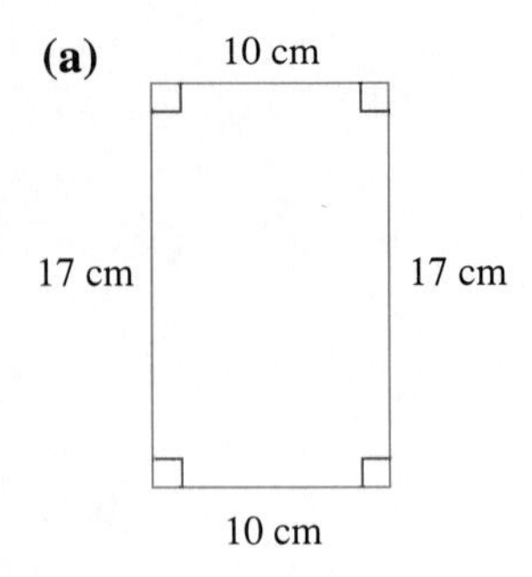

(b)

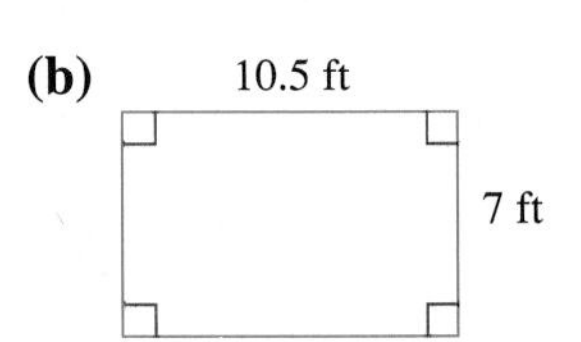

(c) 6 m wide and 11 m long

(d) 0.9 km by 2.8 km

ANSWERS
2. (a) 54 cm (b) 35 ft (c) 34 m (d) 7.4 km

(a) The length is 27 m and the width is 11 m.

$$P = 2 \cdot l + 2 \cdot w$$
$$P = \underbrace{2 \cdot 27 \text{ m}} + \underbrace{2 \cdot \mathbf{11 \text{ m}}}$$
$$P = 54 \text{ m} + 22 \text{ m}$$
$$P = 76 \text{ m}$$

The perimeter of the rectangle (the distance you would walk around the outside edges of the rectangle) is 76 m.

(b)

$$P = 2 \cdot l + 2 \cdot w$$
$$P = \underbrace{2 \cdot 12.3 \text{ m}} + \underbrace{2 \cdot \mathbf{8.9 \text{ m}}}$$
$$P = 24.6 \text{ m} + 17.8 \text{ m}$$
$$P = 42.4 \text{ m} \blacksquare$$

◀ WORK PROBLEM 2 AT THE SIDE.

The perimeter of a rectangle is the distance around the outside edges. The **area** of a rectangle is the amount of surface *inside* the rectangle. We measure area by seeing how many squares of a certain size are needed to cover the surface inside the rectangle. Think of covering the floor of a rectangular living room with carpet. Carpet is measured in square yards, that is, square pieces that measure 1 yard along each side. Here is a drawing of a living room floor.

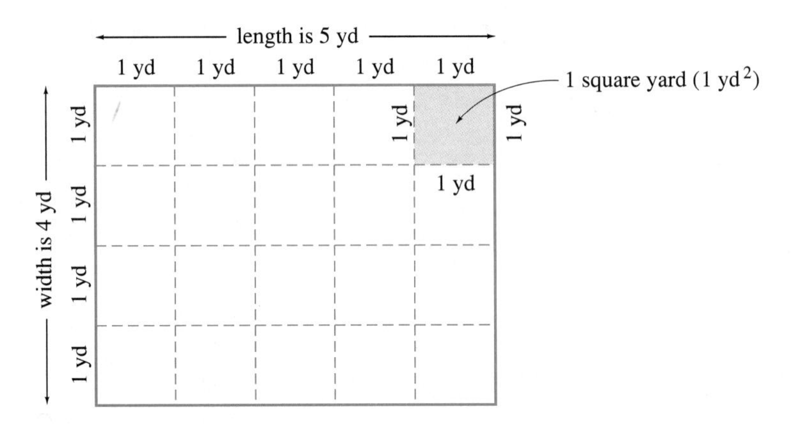

You can see from the drawing that it takes 20 squares to cover the floor. We say that the area of the floor is 20 *square yards.* A shorter way to write square yards is yd^2.

20 **square yards** can be written 20 $\mathbf{yd^2}$

To find the number of squares, you can count them, or you can multiply the number of squares in the length (5) times the number of squares in the width (4) to get 20. The formula is:

FINDING THE AREA OF A RECTANGLE

Area of a rectangle = length · width

$$A = l \cdot w$$

Squares of other sizes can be used to measure area. For smaller areas, you might use these:

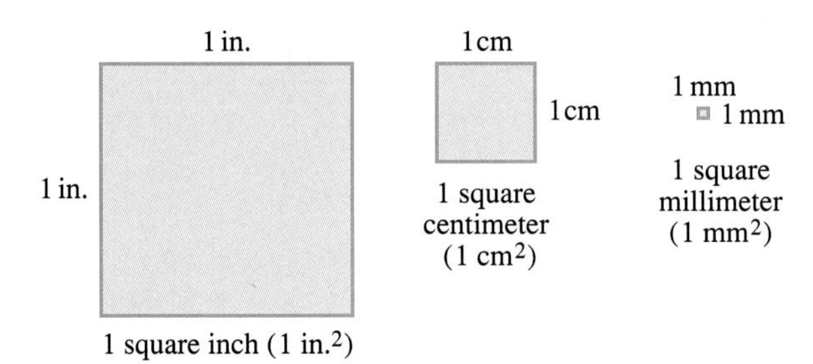

Actual-size drawings

Other sizes of squares that are often used to measure area are listed here, but they are too large to draw on this page.

1 square meter (1 m^2)	1 square foot (1 ft^2)
1 square kilometer (1 km^2)	1 square yard (1 yd^2)
	1 square mile (1 mi^2)

Note The raised 2 in 4^2 means that you multiply $4 \cdot 4$ to get 16. The raised 2 in cm^2 or yd^2 is a short way to write the word "square." When you see 5 cm^2, say "five square centimeters." Do *not* multiply $5 \cdot 5$.

EXAMPLE 2 *Finding the Area of a Rectangle*

Find the area of the each rectangle.

(a)

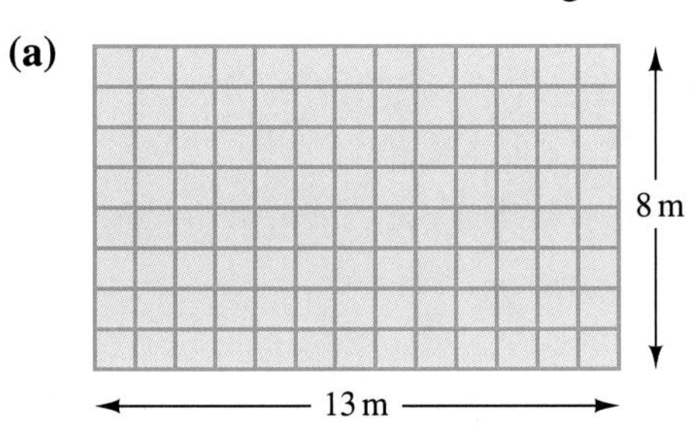

(b) A rectangle measuring 7 cm by 21 cm

(a) The length of this rectangle is 13 m and the width is 8 m.

$$A = l \cdot w$$
$$A = 13 \text{ m} \cdot 8 \text{ m}$$
$$A = 104 \text{ square meters}$$

"Square meters" can be written as m^2, so the area is 104 m^2.

(b) The area is

$$A = 21 \text{ cm} \cdot 7 \text{ cm} = 147 \text{ cm}^2.$$

Note The units for *area* will always be *square* units (cm^2, m^2, yd^2, mi^2, and so on). The units for *perimeter* will be cm, m, yd, mi, and so on (no square units).

WORK PROBLEM 3 AT THE SIDE.

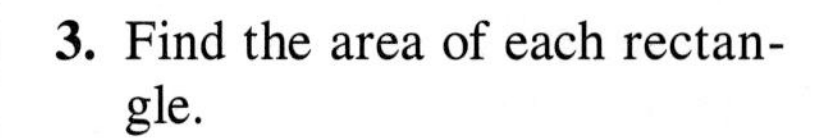

3. Find the area of each rectangle.

(a)

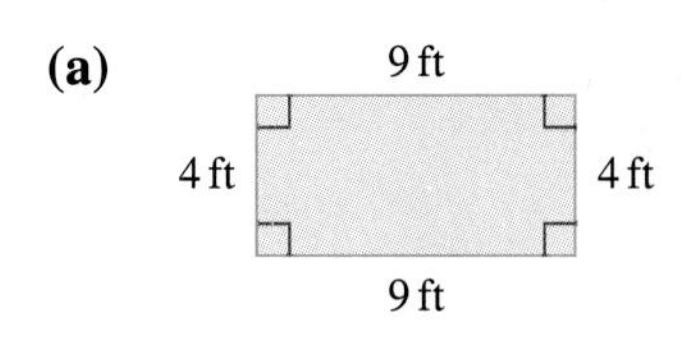

(b) A rectangle that is 6 meters long and 0.5 meter wide.

(c) 8.2 cm by 41.2 cm

ANSWERS

3. (a) 36 ft^2 **(b)** 3 m^2 **(c)** 337.84 cm^2

4. Find the perimeter and area of each square.

(a)

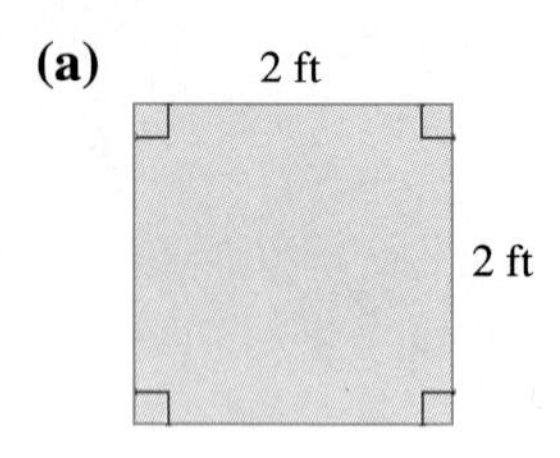

(b) 10.5 cm on each side

(c) 2.1 miles on a side

ANSWERS
4. (a) $P = 8$ ft; $A = 4$ ft^2
(b) $P = 42$ cm; $A = 110.25$ cm^2
(c) $P = 8.4$ mi; $A = 4.41$ mi^2

2 A **square** is a rectangle with all sides the same length. Two squares are shown here. Notice the 90° angles.

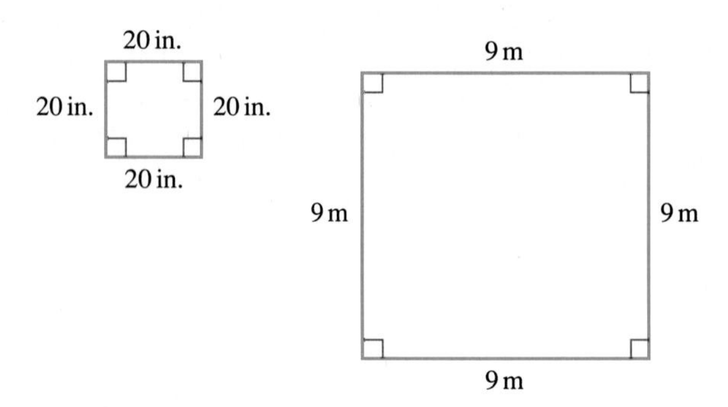

To find the *perimeter* (distance around) of the square on the right, you could add 9 m + 9 m + 9 m + 9 m to get 36 m. A shorter way is to multiply the length of one side times 4, because all 4 sides are the same length.

FINDING THE PERIMETER OF A SQUARE

Perimeter of a square = 4 · side
$$P = 4 \cdot s$$

As with a rectangle, you can multiply length times width to find the area (surface inside) a square. Because the length and the width are the same in a square, the formula is written:

FINDING THE AREA OF A SQUARE

Area of a square = side · side
$$A = s \cdot s$$
$$A = s^2$$

■ **EXAMPLE 3** *Finding the Perimeter and Area of a Square*

(a) Find the perimeter of a square where each side measures 9 m, and

(b) find its area.

(a)
$$P = 4 \cdot s$$
$$P = 4 \cdot 9 \text{ m}$$
$$P = 36 \text{ m}$$

(b)
$$A = s^2$$
$$A = s \cdot s$$
$$A = 9 \text{ m} \cdot 9 \text{ m}$$
$$A = 81 \text{ m}^2 \text{ (square units for area)}$$ ■

Note s^2 does **not** mean $2 \cdot s$. In this example, s is 9 m so s^2 is $9 \cdot 9 = 81$ **not** $2 \cdot 9 = 18$.

◀ WORK PROBLEM 4 AT THE SIDE.

3 As with any other shape, you can find the perimeter (distance around) an irregular shape by adding up the lengths of the sides. To find the area (surface inside the shape), try to break it up into pieces that are squares or rectangles. Find the area of each piece and then add them together.

EXAMPLE 4 *Finding Perimeter and Area of Composite Figures*
A room has the shape shown here.

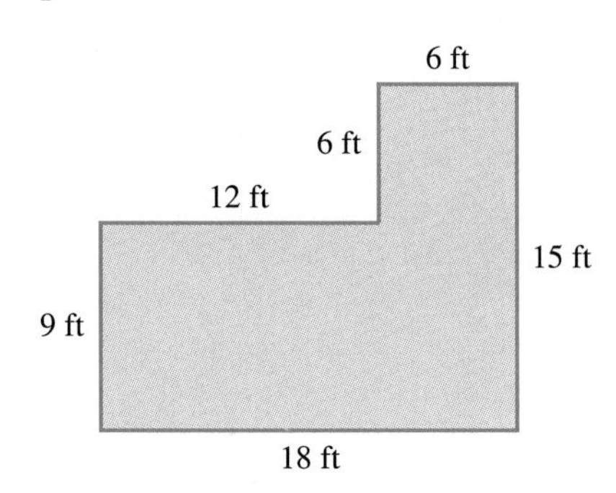

(a) Suppose you want to put a new baseboard (wooden strip) along the base of all the walls. How much material do you need?

(b) The carpet you like costs \$20.50 per square yard. How much will it cost to carpet the room?

(a) Find the perimeter of the room by adding up the lengths of the sides.

$$P = 9 \text{ ft} + 12 \text{ ft} + 6 \text{ ft} + 6 \text{ ft} + 15 \text{ ft} + 18 \text{ ft} = 66 \text{ ft}$$

You need 66 feet of baseboard material.

(b) First change the measurements from feet to yards, because the carpet is sold in square yards. There are 3 feet in 1 yard, so divide by 3.

$$\frac{9 \text{ ft}}{3} = 3 \text{ yd} \quad \frac{12 \text{ ft}}{3} = 4 \text{ yd} \quad \frac{6 \text{ ft}}{3} = 2 \text{ yd} \quad \frac{15 \text{ ft}}{3} = 5 \text{ yd} \quad \frac{18 \text{ ft}}{3} = 6 \text{ yd}$$

Next, break up the room into two pieces. Use just the measurements for the length and width of each piece.

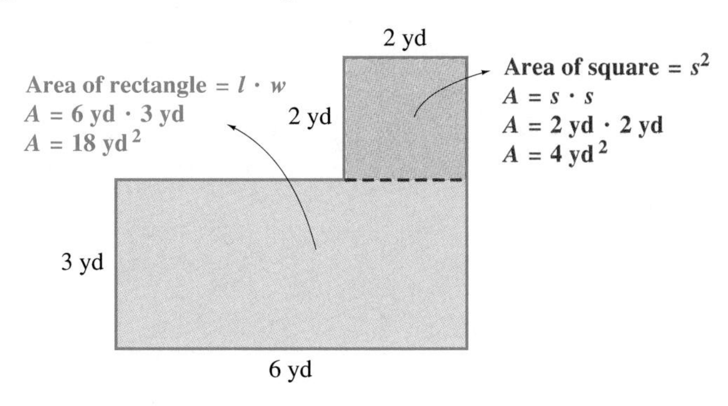

$$\text{Total area} = 18 \text{ yd}^2 + 4 \text{ yd}^2 = 22 \text{ yd}^2$$

so the cost of the carpet is

$$22 \cancel{\text{yd}^2} \cdot \frac{\$20.50}{1 \cancel{\text{yd}^2}} = \$451.00$$

5. Carpet costs \$19.95 per square yard. Find the cost of carpeting the following rooms. Round your answers to the nearest cent.

(a)

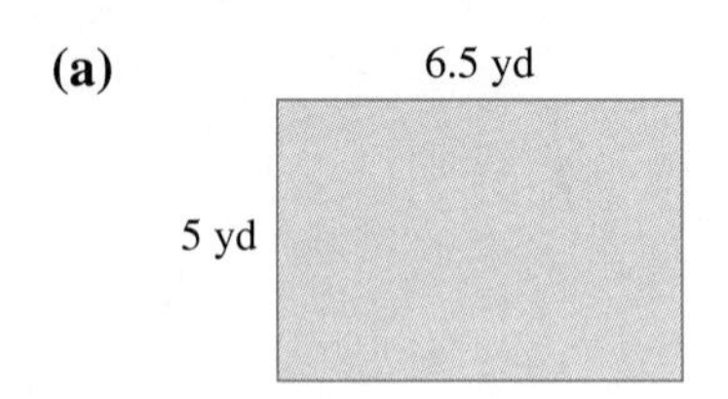

(b)

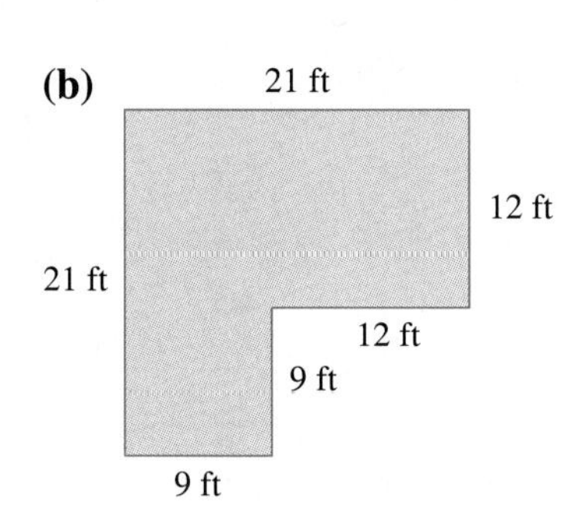

(c) a classroom that is 24 ft long and 18 ft wide

ANSWERS
5. (a) 32.5 yd^2 costs \$648.38
(b) 37 yd^2 costs \$738.15
(c) 48 yd^2 costs \$957.60

You could have cut the room into two rectangles. The final answer is the same.

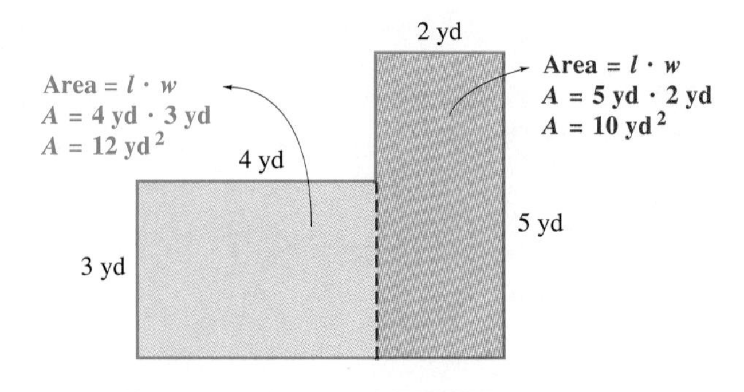

Total area $= 12 \text{ yd}^2 + 10 \text{ yd}^2 = 22 \text{ yd}^2$ ■

WORK PROBLEM 5 AT THE SIDE.

Find the perimeter and area of each rectangle or square.

1.

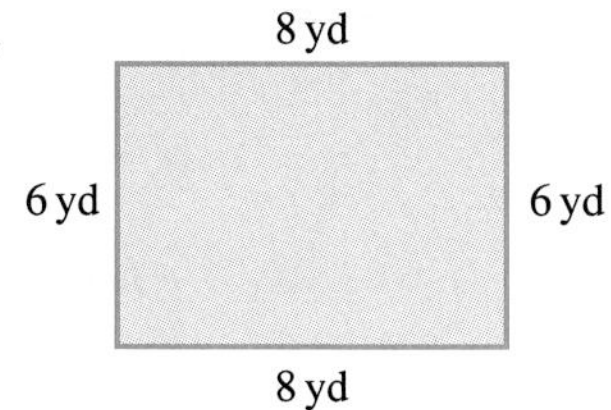

$P = 28$ yd
$A = 48$ yd^2

2.

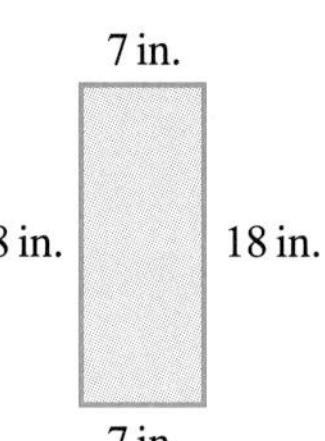

$P = 50$ in.
$A = 126$ in.^{2}

3.

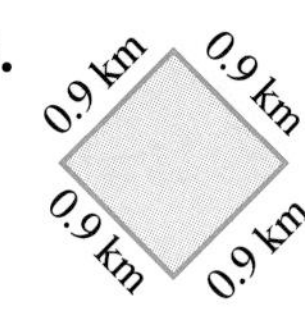

$P = 3.6$ km
$A = 0.81$ km^2

4.

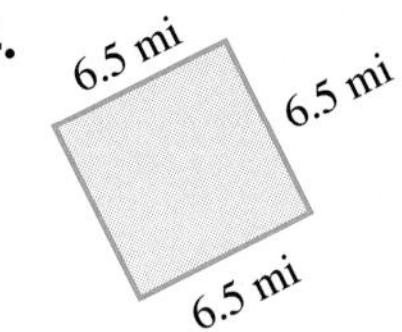

$P = 26$ mi
$A = 42.25$ mi^2

5. 10 ft by 1 ft

$P = 22$ ft
$A = 10$ ft^2

6. 8 cm by 17 cm

$P = 50$ cm
$A = 136$ cm^2

7. 14 m by 0.5 m

$P = 29$ m
$A = 7$ m^2

8. 3.2 km by 4.75 km

$P = 15.9$ km
$A = 15.2$ km^2

9. 76.1 ft by 22 ft

$P = 196.2$ ft
$A = 1674.2$ ft^2

10. 12 m by 12 m

$P = 48$ m
$A = 144$ m^2

11. a square 4 in. wide

$P = 16$ in.
$A = 16$ in.^{2}

12. a square 16.3 cm on a side

$P = 65.2$ cm
$A = 265.69$ cm^2

Find the perimeter and area of each figure.

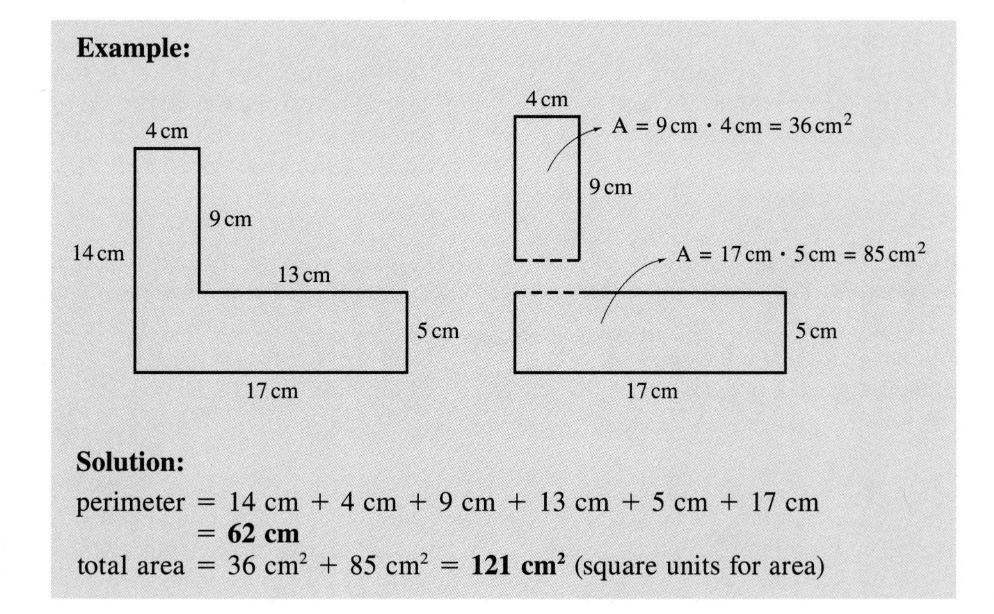

13.

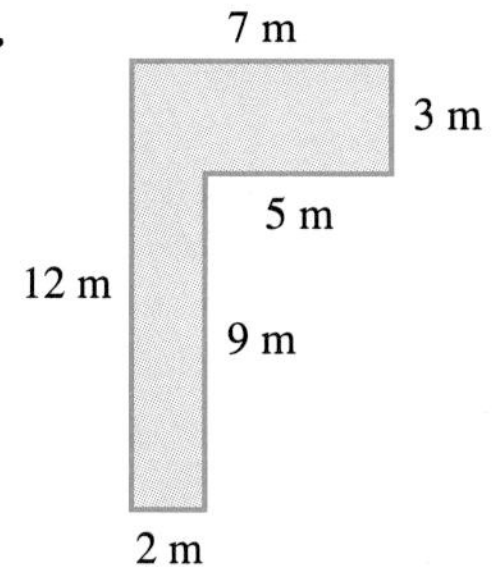

$P = 38$ m
$A = 39$ m^2

14.

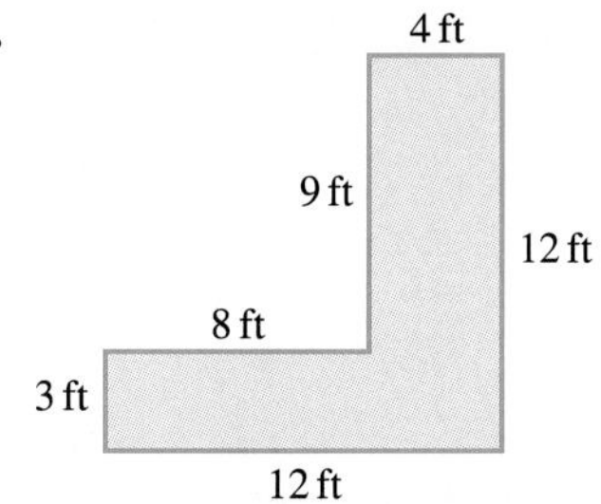

$P = 48$ ft
$A = 72$ ft^2

15.

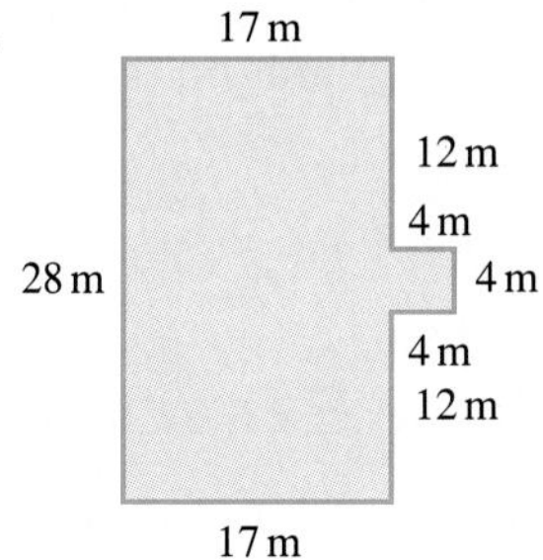

$P = 98$ m
$A = 492$ m^2

16.

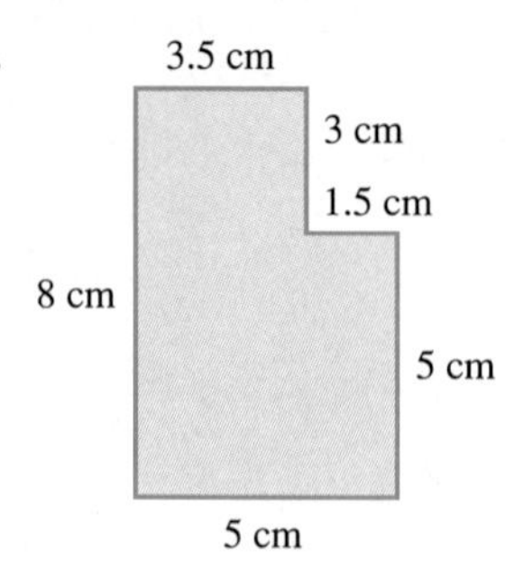

$P = 26$ cm
$A = 35.5$ cm^2

▲ **17.**

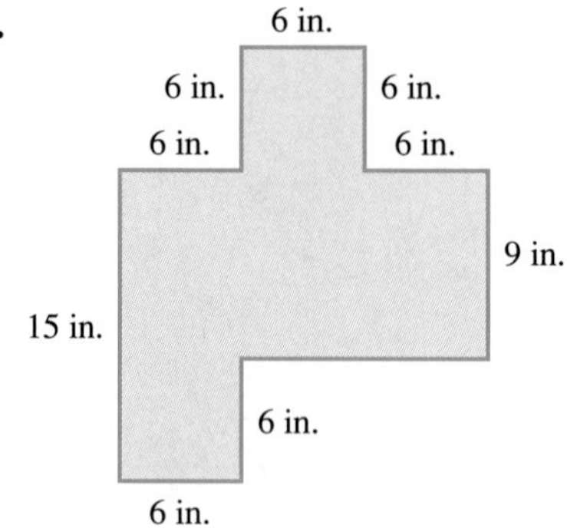

$P = 78$ in.
$A = 234$ in.^{2}

▲ **18.**

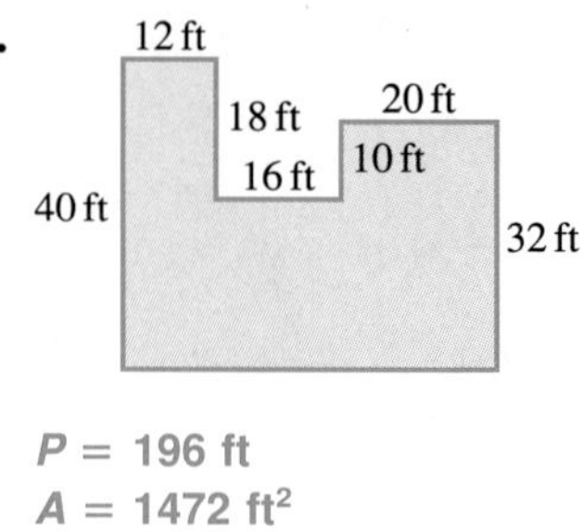

$P = 196$ ft
$A = 1472$ ft^2

Solve each word problem.

19. The Wang's family room measures 20 feet by 25 feet. They are covering the floor with square tiles that measure 1 foot on a side and cost \$0.72 each. How much will they spend on tile?

\$360

20. A page in this book measures 27.5 centimeters from top to bottom and 20 centimeters wide. Find the perimeter and the area of the page.

$P = 95$ cm
$A = 550$ cm^2

21. Tyra's kitchen is 4.4 meters wide and 5.1 meters long. She is pasting a decorative strip that costs \$4.99 per meter around the top edge of all the walls. How much will she spend?

\$94.81

22. Mr. and Mrs. Gomez are buying carpet for their square-shaped bedroom that is 15 feet wide. The carpet is \$23 per square yard and padding and installation is another \$6 per square yard. How much will they spend in all?

\$725

23. In your own words, describe the difference between perimeter and area. Then make a drawing of a square and a rectangle, label the sides with measurements, and show the steps in finding the perimeter and area of each figure.

Answer varies.

24. Suppose you had 16 feet of fencing. Draw three different square or rectangular garden plots that would use exactly 16 feet of fencing; label the lengths of the sides. What shape plot would have the greatest area?

Answer varies.

▲ **25.** A lot is 124 feet by 172 feet. County rules require that nothing be built on land within 12 feet of any edge of the lot. Draw a sketch of the lot, showing the land that cannot be built on. What is the area of the land that cannot be built on?

Sketches vary.
$A = 6528$ ft^2

▲ **26.** Find the cost of fencing needed for this rectangular field. Fencing along the highway costs \$4.25 per foot. Fencing for the other three sides costs \$2.75 per foot.

\$1333

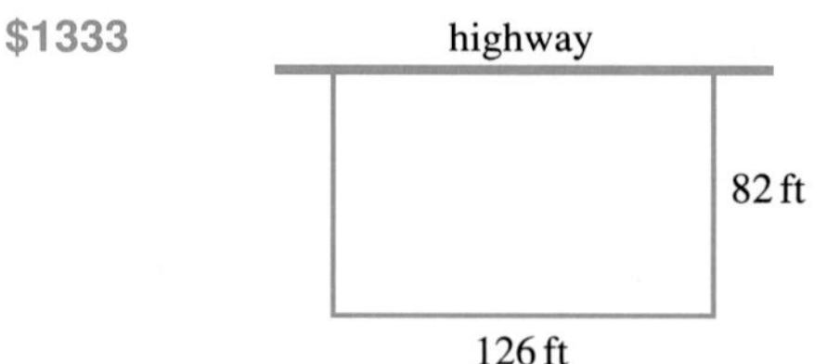

PREVIEW EXERCISES

Convert the following measurements. (For help, see ***Sections 7.1 and 7.2.****)*

27. 50 cm to m 0.5 m

28. 8 km to m 8000 m

29. 12 ft to in. 144 in.

30. 75 ft to yd 25 yd

31. 2700 in. to yd 75 yd

32. 400 mm to cm 40 cm

A **parallelogram** is a four-sided figure with opposite sides parallel, such as these. Notice that opposite sides have the same length.

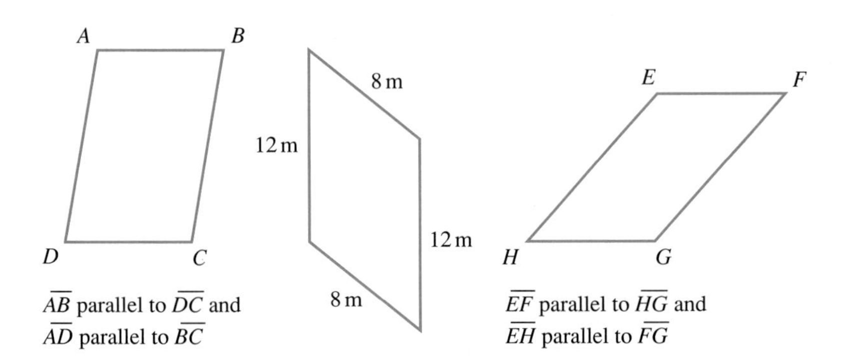

$\overline{AB}$ parallel to $\overline{DC}$ and $\overline{AD}$ parallel to $\overline{BC}$

$\overline{EF}$ parallel to $\overline{HG}$ and $\overline{EH}$ parallel to $\overline{FG}$

1 Perimeter is the distance around a figure, so the easiest way to find the perimeter of a parallelogram is to add the lengths of the four sides.

EXAMPLE 1 *Finding the Perimeter of a Parallelogram*

Find the perimeter of the middle parallelogram above.

$$P = 12 \text{ m} + 8 \text{ m} + 12 \text{ m} + 8 \text{ m} = 40 \text{ m}$$

WORK PROBLEM 1 AT THE SIDE.

To find the area of a parallelogram, first draw a dashed line as shown here.

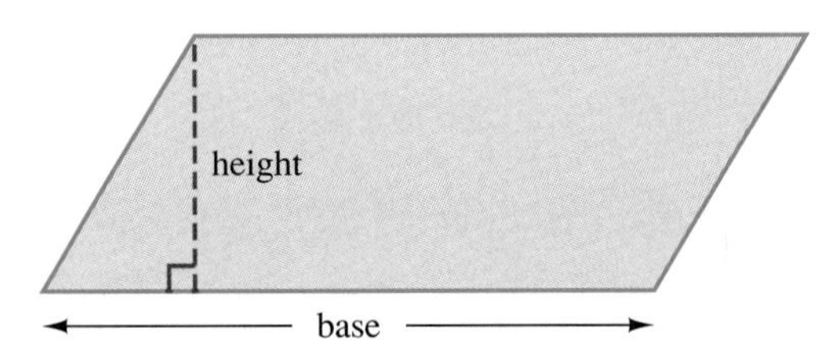

The length of the dashed line is the *height* of the parallelogram. It forms a *right angle* with the base. The height is the shortest distance between the base and the opposite side.

Now cut off the triangle created on the left side of the parallelogram and move it to the right side, as shown below.

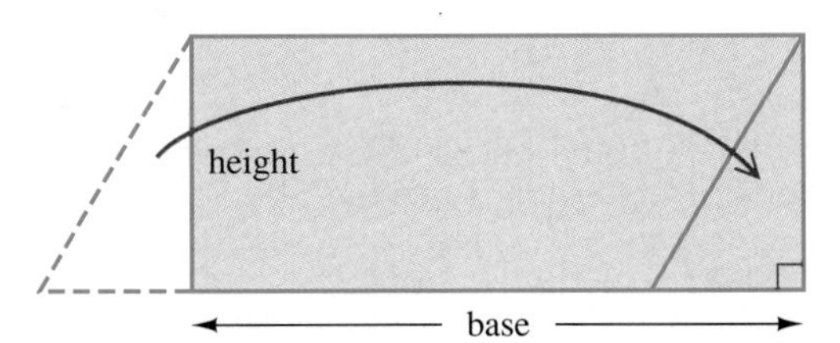

The parallelogram has been made into a rectangle. You can see that the area of the parallelogram and the rectangle are the same. The area of the rectangle is length times width. In the parallelogram, this translates into base times height.

OBJECTIVES

1. Find the perimeter and area of parallelograms.
2. Find the perimeter and area of trapezoids.

FOR EXTRA HELP

Tape 11 | SSM pp. 238–240 | MAC: A IBM: A

1. Find the perimeter of each parallelogram.

(a)

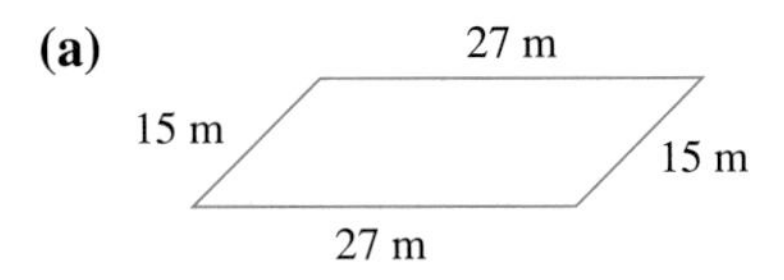

(b)

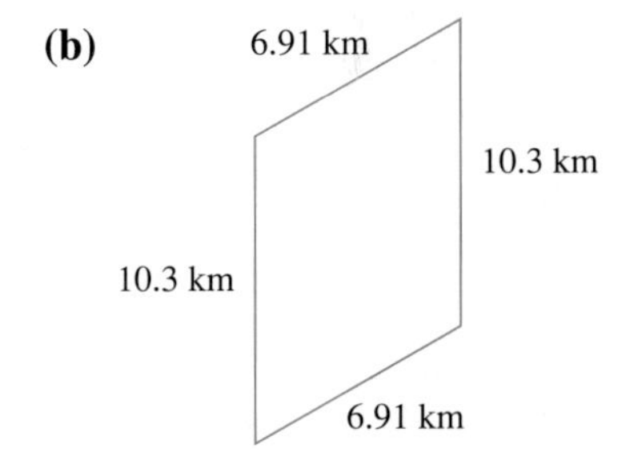

ANSWERS

1. (a) 84 m (b) 34.42 km

FINDING THE AREA OF A PARALLELOGRAM

Area of parallelogram = base · height

$$A = b \cdot h$$

■ EXAMPLE 2 ***Finding the Area of a Parallelogram***

Find the area of each parallelogram.

(a)

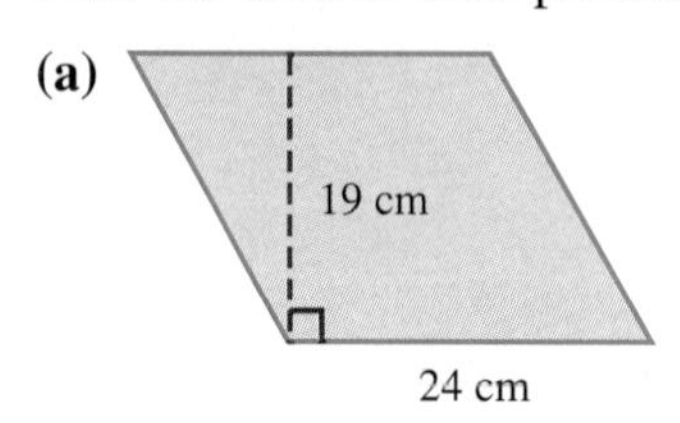

(b)

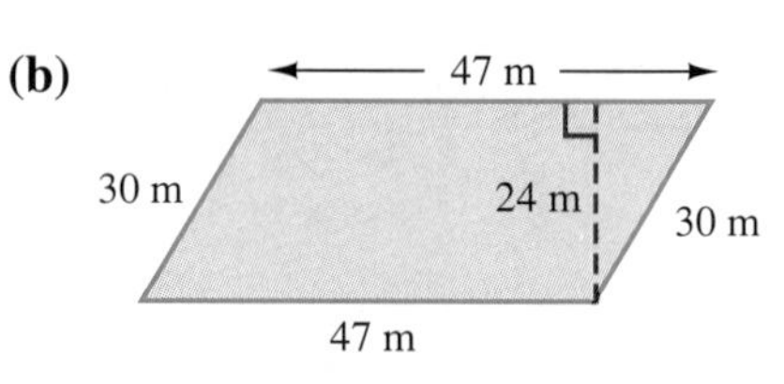

(a) The base is 24 cm and the height is 19 cm. The area is

$$A = b \cdot h$$
$$A = 24 \text{ cm} \cdot 19 \text{ cm}$$
$$A = 456 \text{ cm}^2$$

(b) $A = 47 \text{ m} \cdot 24 \text{ m} = 1128 \text{ m}^2$ (square units for area)

Notice that the 30-m sides are not used in finding the area. ■

◀◀ **WORK PROBLEM 2 AT THE SIDE.**

2 A **trapezoid** is a four-sided figure with at least one pair of parallel sides, such as the ones shown here. Opposite sides may *not* have the same length, as in parallelograms.

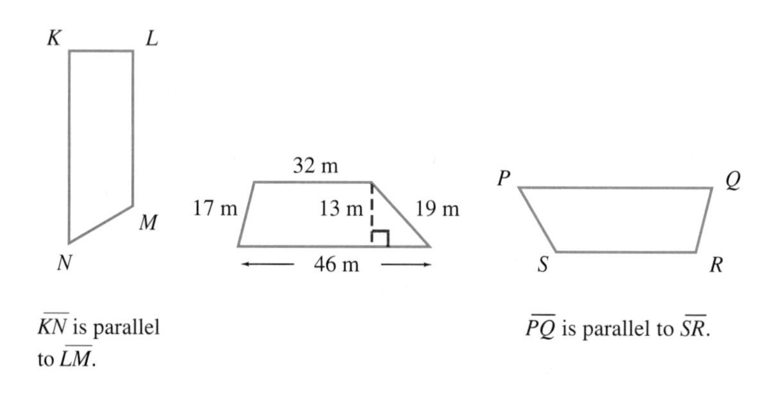

$\overline{KN}$ is parallel to $\overline{LM}$.

$\overline{PQ}$ is parallel to $\overline{SR}$.

■ EXAMPLE 3 ***Finding the Perimeter of a Trapezoid***

Find the perimeter of the middle trapezoid above. Add the lengths of the sides.

$$P = 17 \text{ m} + 32 \text{ m} + 19 \text{ m} + 46 \text{ m} = 114 \text{ m}$$

Notice that the height (13 m) is *not* part of the perimeter. ■

◀◀ **WORK PROBLEM 3 AT THE SIDE.**

Use the following formula to find the area of a trapezoid.

2. Find the area of each parallelogram.

(a)

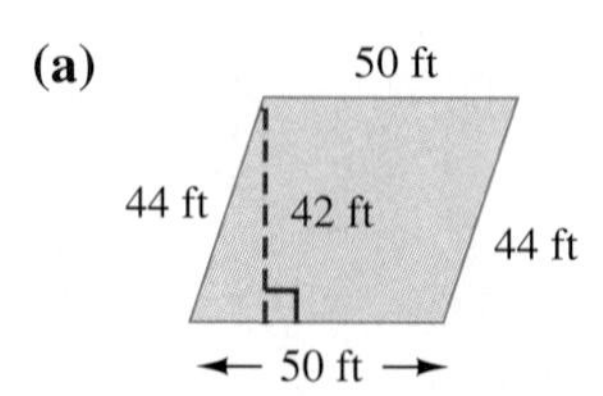

(b)

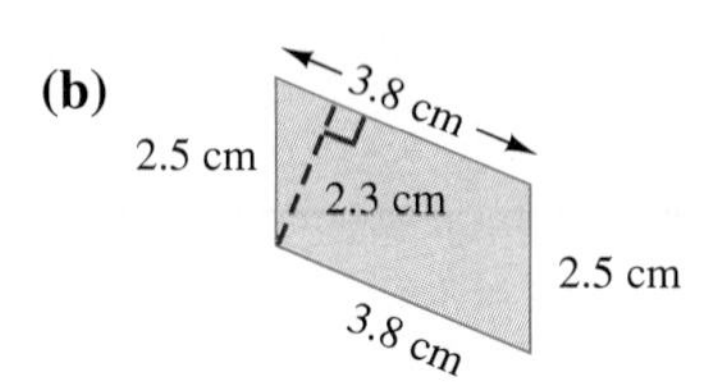

(c) a parallelogram with base $12\frac{1}{2}$ m and height $4\frac{3}{4}$ m (Hint: Write $12\frac{1}{2}$ as 12.5 and $4\frac{3}{4}$ as 4.75.)

3. Find the perimeter of each trapezoid.

(a)

9.8 in.

5.6 in. 5.6 in. 6.7 in.

6.9 in.

(b)

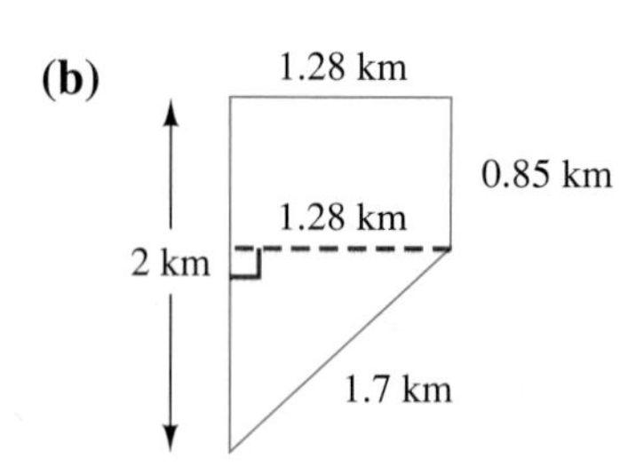

(c) a trapezoid with sides 39.7 cm, 29.2 cm, 74.9 cm, and 16.4 cm

ANSWERS

2. (a) 2100 ft² (b) 8.74 cm²

(c) $59\frac{3}{8}$ m² or 59.375 m²

3. (a) 29 in. (b) 5.83 km (c) 160.2 cm

FINDING THE AREA OF A TRAPEZOID

$$\text{Area} = \frac{1}{2} \cdot \text{height} \cdot (\text{short base} + \text{long base})$$

$$A = \frac{1}{2} \cdot h \cdot (b + B)$$

$$\text{or} \quad A = 0.5 \cdot h \cdot (b + B)$$

■ **EXAMPLE 4** *Finding the Areas of Trapezoids*

Find the area of this trapezoid. The short base and long base are the parallel sides.

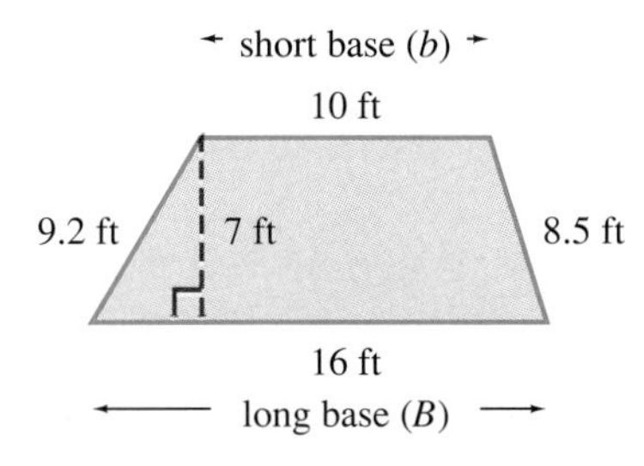

The height (h) is 7 ft, the short base (b) is 10 ft, and the long base (B) is 16 ft. You do not need the 9.2 ft or 8.5 ft sides to find the area.

$$A = \frac{1}{2} \cdot h \cdot (b + B)$$

$$A = \frac{1}{2} \cdot 7 \text{ ft} \cdot (\mathbf{10 \text{ ft}} + \mathbf{16 \text{ ft}})$$

$$A = \frac{1}{\underset{1}{\not 2}} \cdot 7 \text{ ft} \cdot (\overset{13}{\not 26} \text{ ft})$$

$$A = 91 \text{ ft}^2 \quad \text{(square units for area)}$$

You can also solve the problem by using 0.5, the decimal equivalent for $\frac{1}{2}$, in the formula.

$$A = 0.5 \cdot h \cdot (\mathbf{b} + \mathbf{B})$$

$$A = 0.5 \cdot 7 \cdot (\mathbf{10} + \mathbf{16}) = 0.5 \cdot 7 \cdot 26 = 91 \text{ ft}^2 \quad ■$$

WORK PROBLEM 4 AT THE SIDE. ▶▶

■ **EXAMPLE 5** *Finding the Area of Composite Figures*

Find the area of this figure.

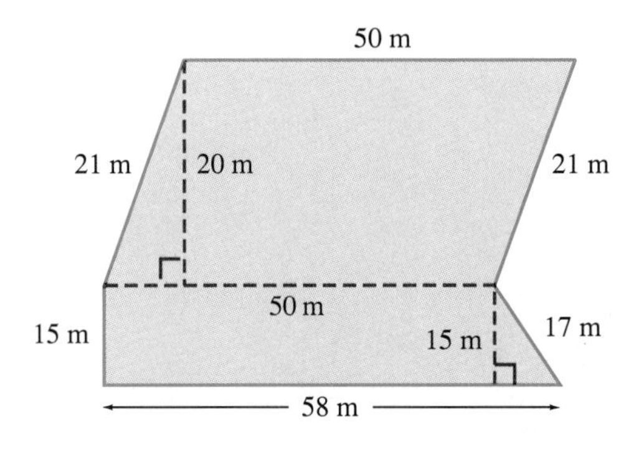

4. Find the area of each trapezoid.

(a)

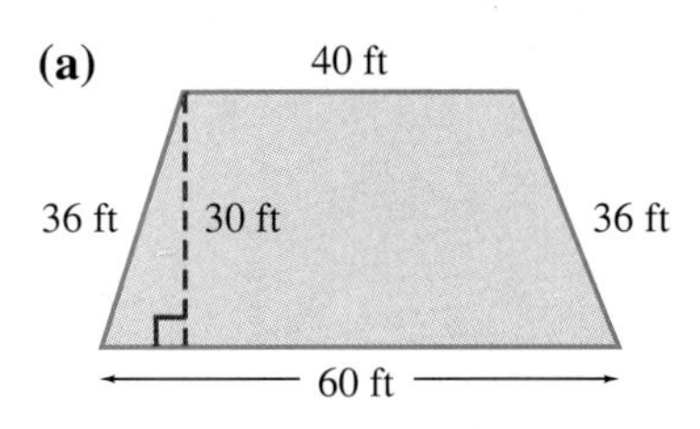

(b)

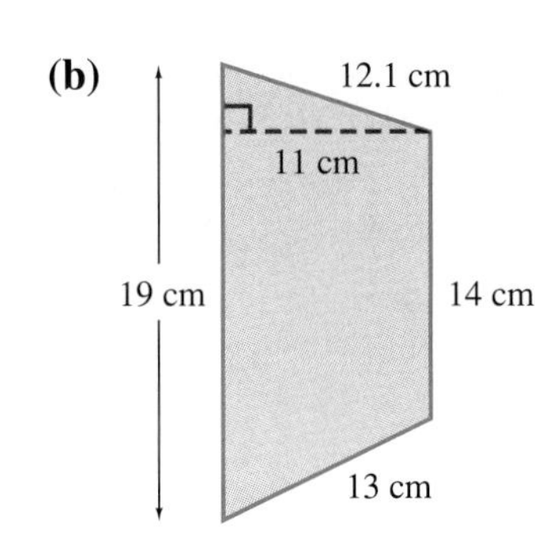

(c) a trapezoid with height 4.7 m, short base 9 m, and long base 10.5 m.

ANSWERS
4. (a) 1500 ft^2 (b) 181.5 cm^2 (c) 45.825 m^2

5. Find the area of each floor.

(a)

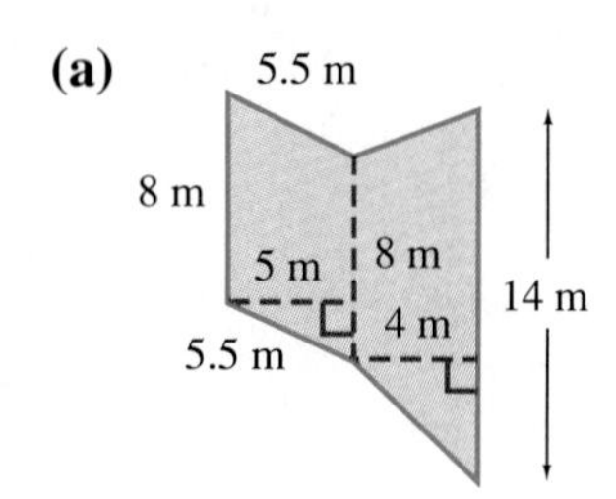

(b)

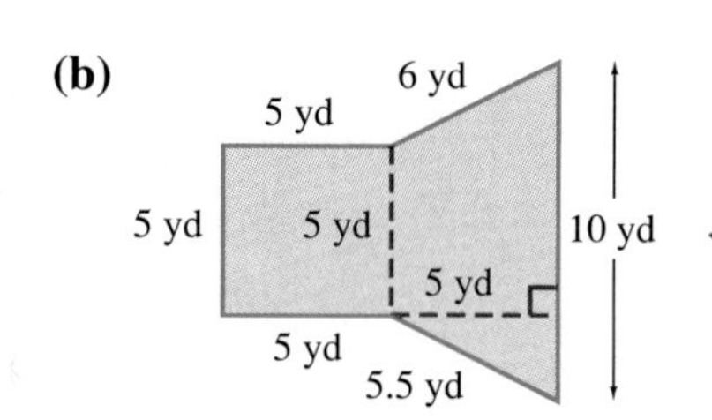

6. Find the cost of carpeting the floors in Problem 5. The cost of carpet is as follows:

(a) Floor (a), \$18.50 per square meter.

(b) Floor (b), \$28 per square yard.

ANSWERS
5. (a) $84\ m^2$ **(b)** $62.5\ yd^2$
6. (a) \$1554 **(b)** \$1750

Break the figure into two pieces, a parallelogram and a trapezoid.

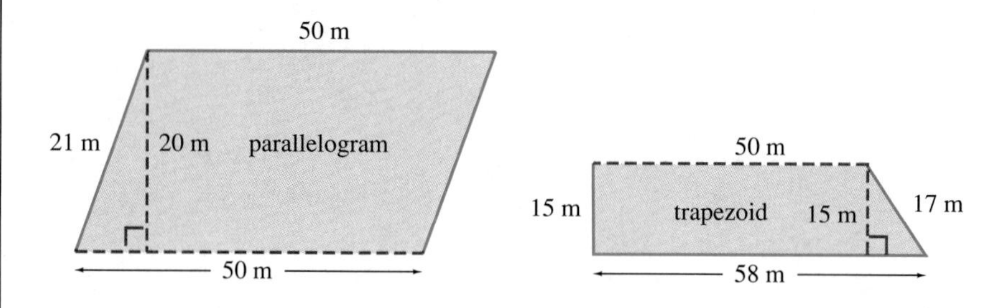

$$A = b \cdot h \qquad\qquad A = \frac{1}{2} \cdot h \cdot (b + B)$$

$$A = 50\text{ m} \cdot 20\text{ m} \qquad\qquad A = 0.5 \cdot 15\text{ m} \cdot (50\text{ m} + 58\text{ m})$$

$$A = \mathbf{1000\ m^2} \qquad\qquad A = \mathbf{810\ m^2}$$

$$\text{Total area} = \mathbf{1000\ m^2} + \mathbf{810\ m^2} = 1810\ m^2$$ ■

WORK PROBLEM 5 AT THE SIDE.

■ **EXAMPLE 6** *Applying Knowledge of Area*
Suppose the figure in Example 5 represents the floor plan of a hotel lobby. What is the cost of labor to install tile on the floor if the labor charge is \$35.11 per square meter?

From Example 5, the floor area is $1810\ m^2$. To find the labor cost, multiply the number of square meters times the cost of labor per square meter.

$$\text{Cost} = 1810\ \cancel{m^2} \cdot \frac{\$35.11}{1\ \cancel{m^2}}$$

$$\text{Cost} = \$63{,}549.10$$

The cost of the labor is \$63,549.10. ■

WORK PROBLEM 6 AT THE SIDE.

8.4 EXERCISES

NAME DATE HOUR

Find the perimeter of each figure.

1.

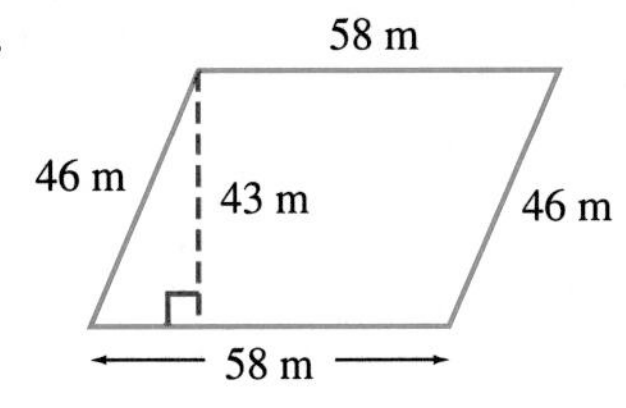

208 m

2.

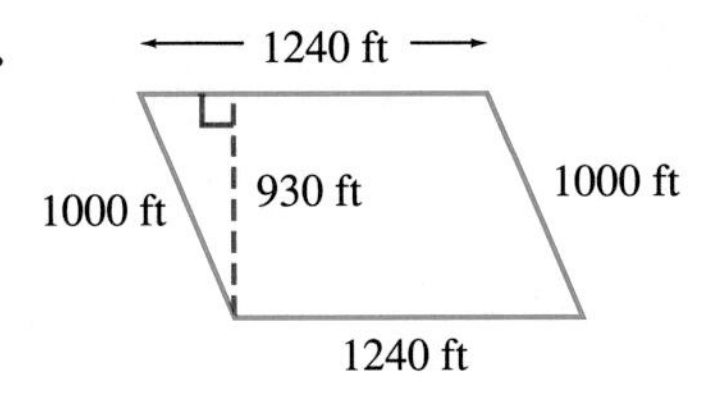

4480 ft

3.

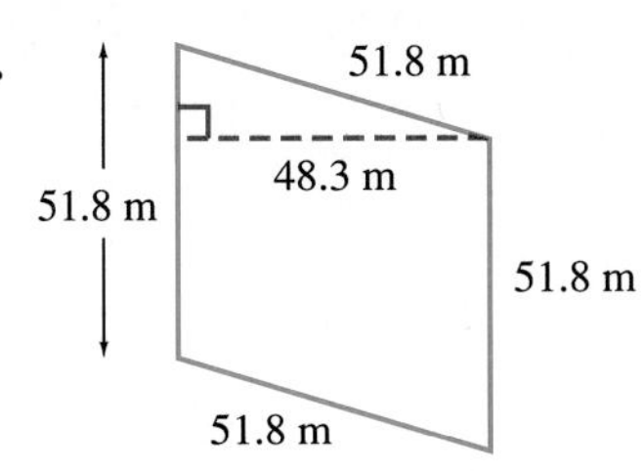

207.2 m

4.

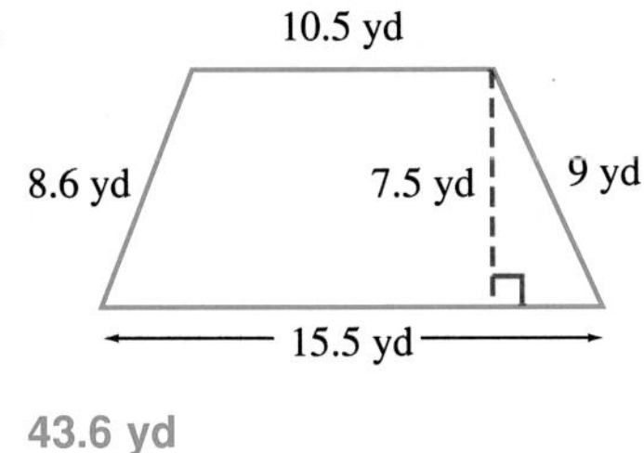

43.6 yd

5.

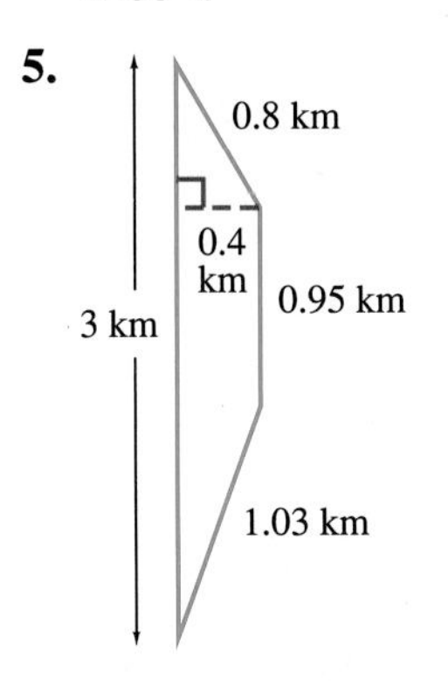

5.78 km

6.

7.33 cm
2.8 cm
3 cm
4.3 cm
4.17 cm

18.6 cm

Find the area of each figure.

7.

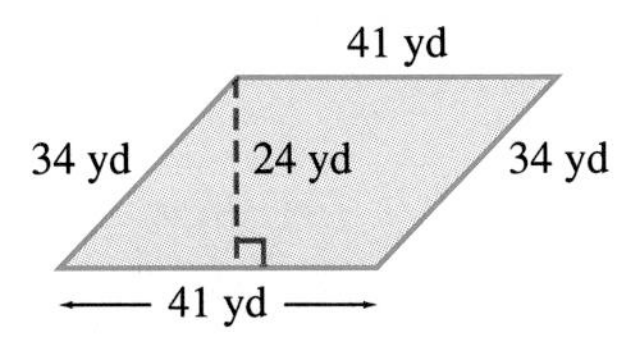

984 yd^2

8.

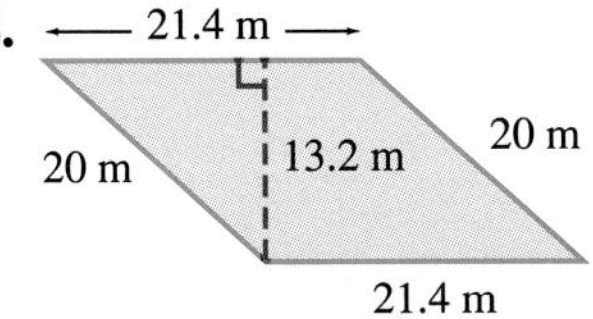

282.48 m^2

9.

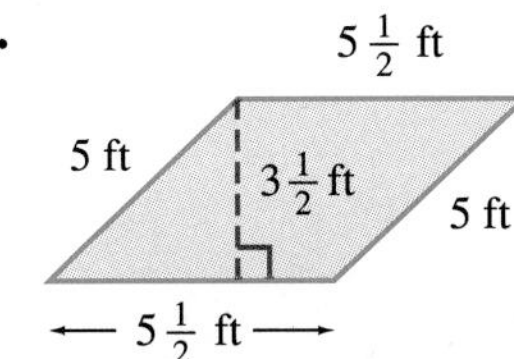

19.25 ft^2

10.

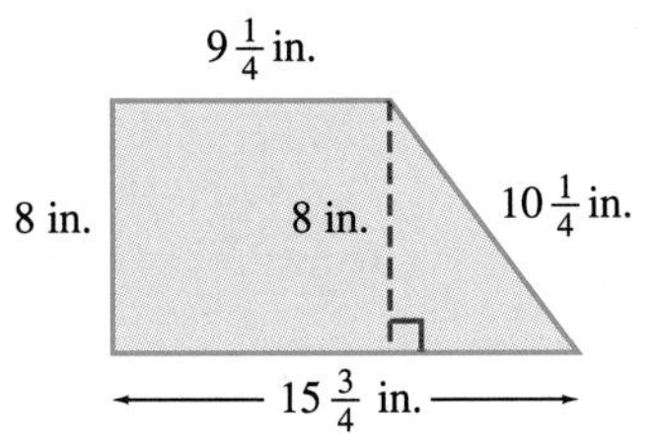

100 in.^{2}

11.

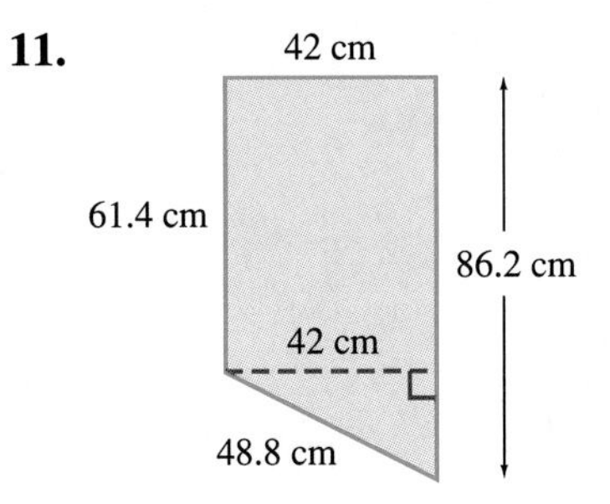

3099.6 cm^2

12.

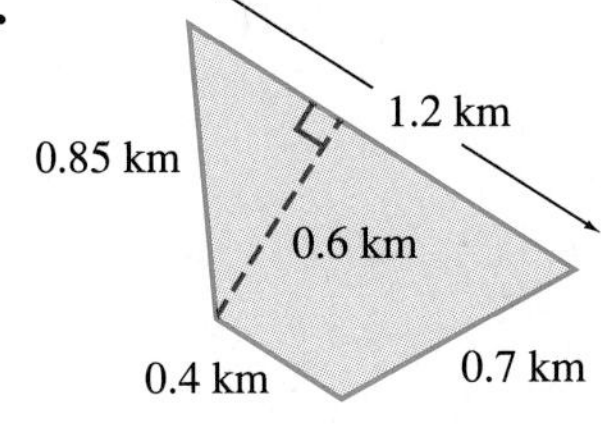

0.48 km^2

Solve each word problem.

13. The backyard of a new home is shaped like a trapezoid with a height of 45 ft and bases of 80 ft and 110 ft. What is the cost of putting sod on the yard, if the landscaper charges \$0.22 per square foot for sod?

\$940.50

14. A swimming pool is in the shape of a parallelogram with a height of 9.6 m and base of 12.4 m. Find the labor cost to make a custom solar cover for the pool at a cost of \$4.92 per square meter.

\$585.68

Find two errors in each student's solution below. Write a sentence explaining each error. Then show how to work the problem correctly.

15.

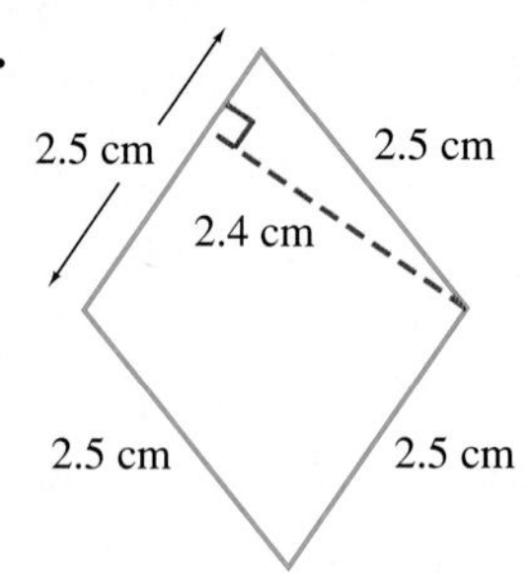

$P = 2.5 \text{ cm} + 2.4 \text{ cm} + 2.5 \text{ cm} + 2.5 \text{ cm} + 2.5 \text{ cm}$
$P = 12.4 \text{ cm}^2$

Answer varies.

16.

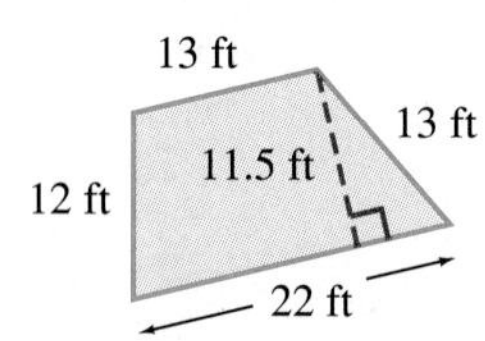

$A = 0.5 \cdot 11.5 \text{ ft} \cdot (12 \text{ ft} + 13 \text{ ft})$
$A = 143.75 \text{ ft}$

Answer varies.

Find the area of each figure.

17.

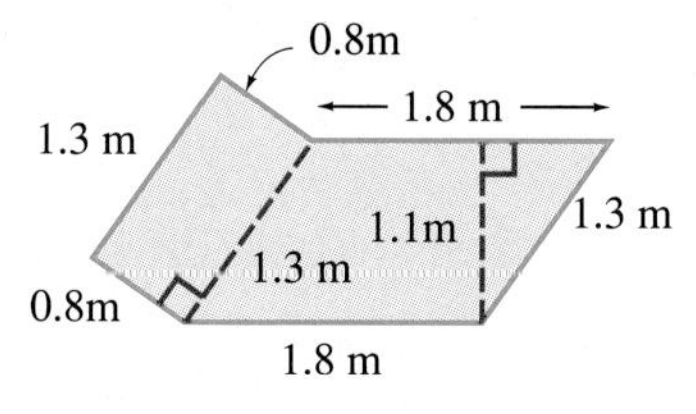

3.02 m²

18.

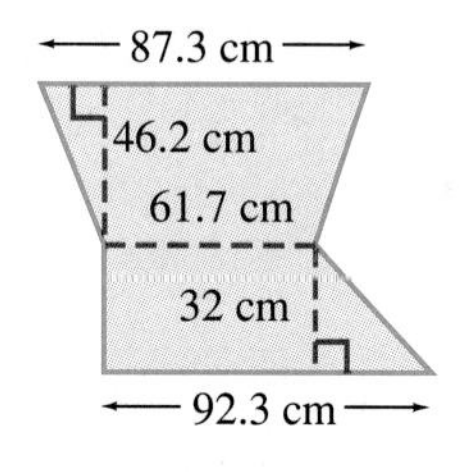

5905.9 cm²

19.

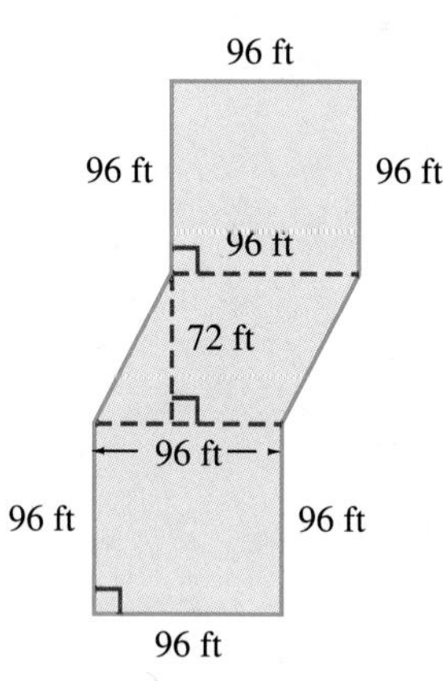

25,344 ft²

20. Find the area of this figure.

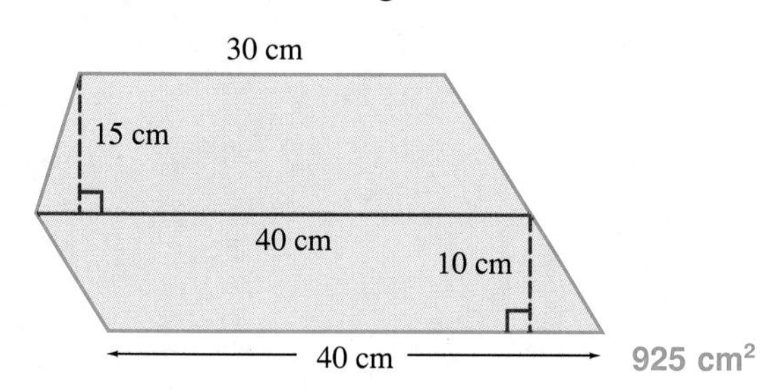

925 cm²

PREVIEW EXERCISES

Multiply the following. Write all answers as whole numbers or mixed numbers in lowest terms. (For help, see ***Sections 2.5 and 2.8.****)*

21. $6\frac{1}{2} \cdot 8$ 52

22. $12 \cdot \frac{5}{6}$ 10

23. $25 \cdot 1\frac{3}{7} \cdot \frac{28}{75}$ $13\frac{1}{3}$

24. $\frac{15}{34} \cdot 3\frac{2}{5} \cdot 1\frac{1}{3}$ 2

8.5 TRIANGLES

A **triangle** is a figure with exactly three sides, as shown below.

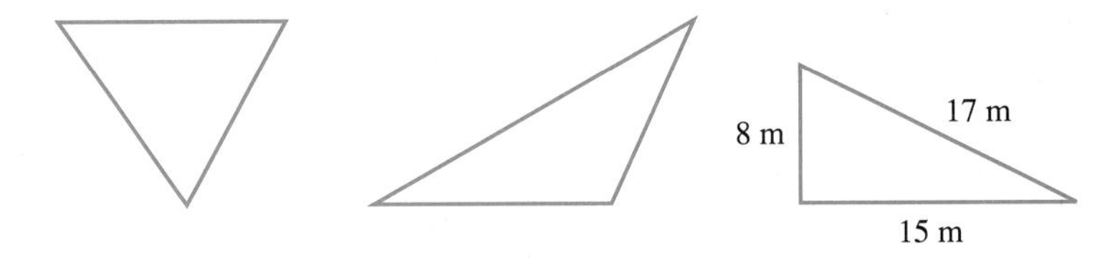

1 To find the perimeter of a triangle (the distance around the edges), add the lengths of the three sides.

EXAMPLE 1 *Finding the Perimeter of a Triangle*
The perimeter of the above triangle on the right is

$$P = 8 \text{ m} + 15 \text{ m} + 17 \text{ m} = 40 \text{ m}.$$

WORK PROBLEM 1 AT THE SIDE.

As with parallelograms, you can find the *height* of a triangle by measuring the distance from one corner of the triangle to the opposite side (the base). The height line must be *perpendicular* to the base, that is, it must form a right angle with the base. Sometimes you have to extend the base line in order to draw the height perpendicular to it.

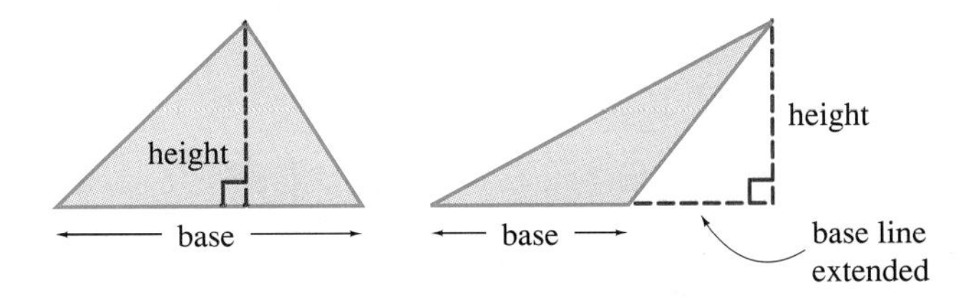

If you cut out two identical triangles and turn one upside down, you can fit them together to form a parallelogram.

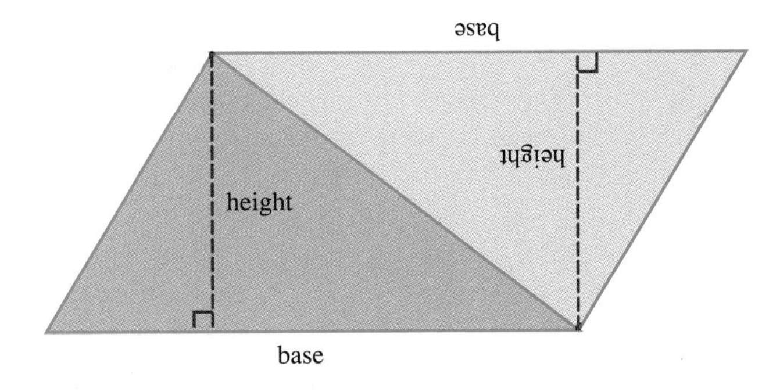

The area of the parallelogram is base times height. Because each triangle is *half* of the parallelogram, the area of one triangle is

$$\frac{1}{2} \text{ of base times height.}$$

OBJECTIVES

1. Find the perimeter of a triangle.
2. Find the area of a triangle.
3. Given two angles in a triangle, find the third angle.

FOR EXTRA HELP

Tape 11 | SSM pp. 240–241 | MAC: A IBM: A

1. Find the perimeter of each triangle.

(a)

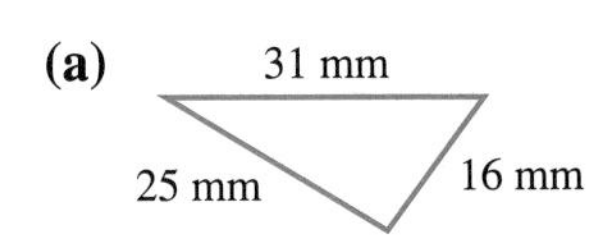

(b)

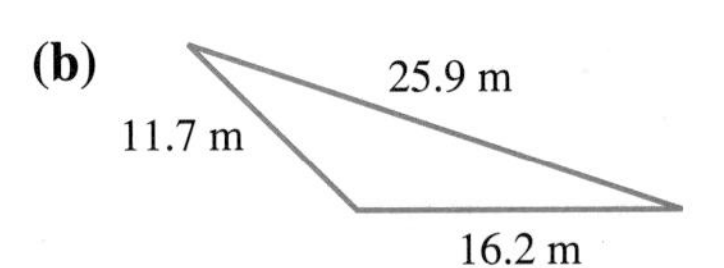

(c) a triangle with sides $6\frac{1}{2}$ yd, $9\frac{3}{4}$ yd, and $11\frac{1}{4}$ yd

ANSWERS
1. (a) 72 mm (b) 53.8 m (c) $27\frac{1}{2}$ yd or 27.5 yd

2. Find the area of each triangle.

(a)

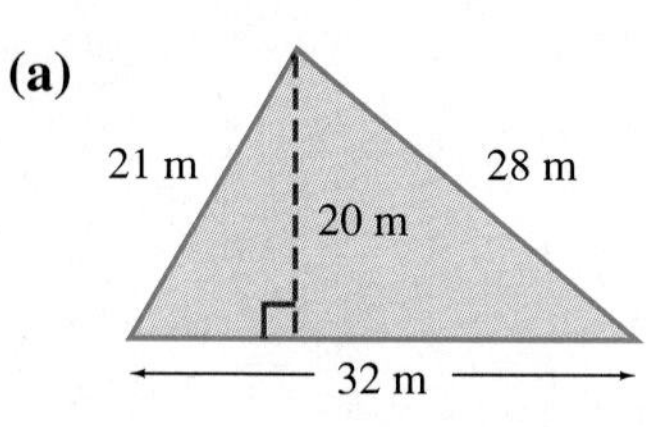

(b)

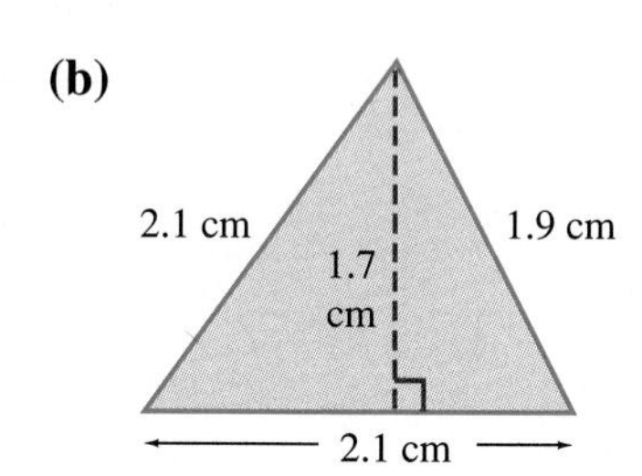

(c)

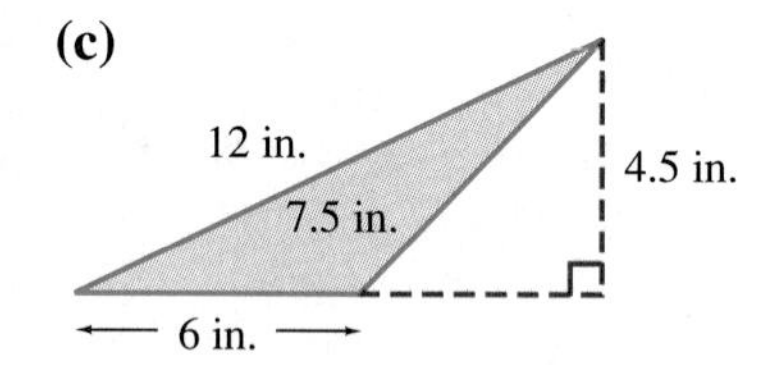

(d)

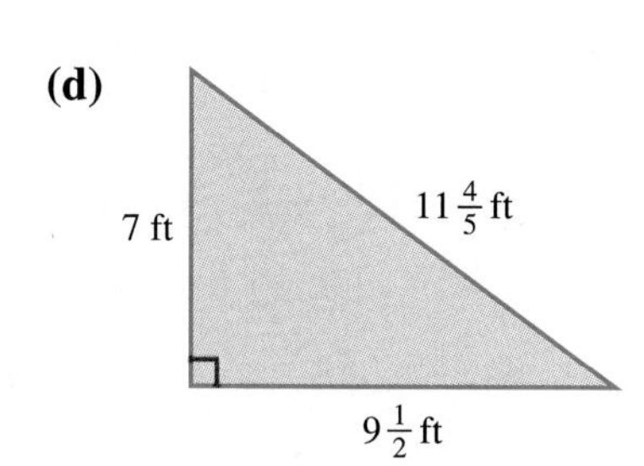

ANSWERS

2. **(a)** 320 m² **(b)** 1.785 cm² **(c)** 13.5 in.² **(d)** 33.25 ft² or $33\frac{1}{4}$ ft²

2 Use the following formula to find the area of a triangle.

FINDING THE AREA OF A TRIANGLE

$$\text{Area of triangle} = \frac{1}{2} \cdot \text{base} \cdot \text{height}$$

$$A = \frac{1}{2} \cdot b \cdot h$$

$$\text{or} \quad A = 0.5 \cdot b \cdot h$$

EXAMPLE 2 *Finding the Area of a Triangle*

Find the area of each triangle.

(a)

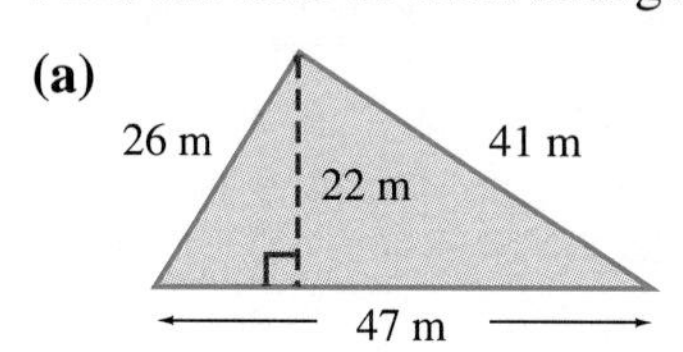

(b)

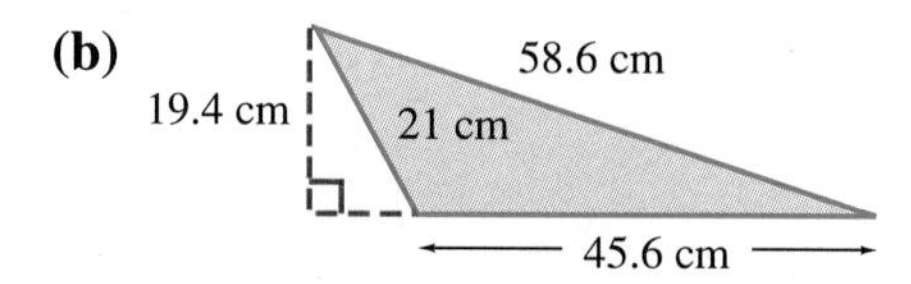

(c)

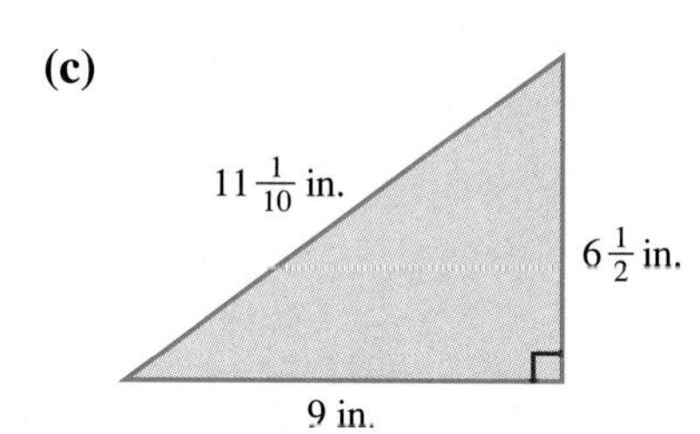

(a) The base is 47 m and the height is 22 m. You do not need the 26 m or 41 m sides to find the area.

$$A = \frac{1}{2} \cdot b \cdot h$$

$$A = \frac{1}{\cancel{2}_1} \cdot 47 \text{ m} \cdot \overset{11}{\cancel{22}} \text{ m}$$

$$A = 517 \text{ m}^2 \quad \text{(square units for area)}$$

(b) The base line must be extended to draw the height. However, still use 45.6 cm for b in the formula. Because the measurements are decimal numbers, it is easier to use 0.5 in the formula (the decimal equivalent of $\frac{1}{2}$).

$$A = 0.5 \cdot 45.6 \text{ cm} \cdot 19.4 \text{ cm}$$

$$A = 442.32 \text{ cm}^2$$

(c) Because two sides of the triangle are perpendicular to each other, use those sides as the base and the height.

$$A = \frac{1}{2} \cdot 9 \text{ in.} \cdot 6\frac{1}{2} \text{ in.} \quad \text{or} \quad A = 0.5 \cdot 9 \text{ in.} \cdot 6.5 \text{ in.}$$

$$A = 29\frac{1}{4} \text{ in.}^2 \quad \text{or} \quad 29.25 \text{ in.}^2$$ ■

◀ WORK PROBLEM 2 AT THE SIDE.

■ **EXAMPLE 3** *Using the Concept of Area*
Find the area of the shaded part in this figure.

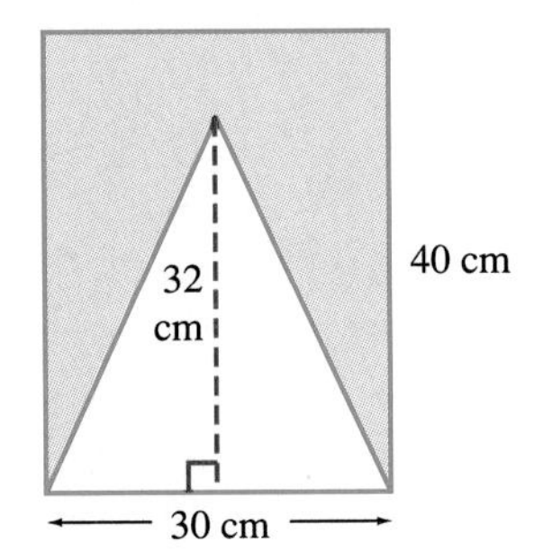

The *entire* figure is a rectangle.

$$A = l \cdot w$$
$$A = 30 \text{ cm} \cdot 40 \text{ cm} = 1200 \text{ cm}^2$$

The *un*shaded part is a triangle.

$$A = \frac{1}{2} \cdot 30 \text{ cm} \cdot 32 \text{ cm}$$
$$A = 480 \text{ cm}^2$$

Subtract to find the area of the shaded part.

$$A = \underbrace{1200 \text{ cm}^2}_{\text{entire area}} - \underbrace{480 \text{ cm}^2}_{\text{unshaded part}} = \underbrace{720 \text{ cm}^2}_{\text{shaded part}} \quad ■$$

WORK PROBLEM 3 AT THE SIDE. ▶▶

■ **EXAMPLE 4** *Applying the Concept of Area*
The Department of Transportation cuts triangular signs out of rectangular pieces of metal using the measurements shown above in Example 3. If the metal costs $0.02 per square centimeter, how much does the metal for the sign cost? What is the cost of the metal that is *not* used?

From Example 3, the area of the triangle (the sign) is 480 cm². Multiply that times the cost per square centimeter.

$$\text{Cost of sign} = 480 \cancel{\text{cm}^2} \cdot \frac{\$0.02}{1 \cancel{\text{cm}^2}} = \$9.60$$

The metal that is *not* used is the *shaded* part from Example 3. The area is 720 cm².

$$\text{Cost of unused metal} = 720 \cancel{\text{cm}^2} \cdot \frac{\$0.02}{1 \cancel{\text{cm}^2}}$$
$$= \$14.40 \quad ■$$

WORK PROBLEM 4 AT THE SIDE. ▶▶

3. Find the area of the shaded part in this figure.

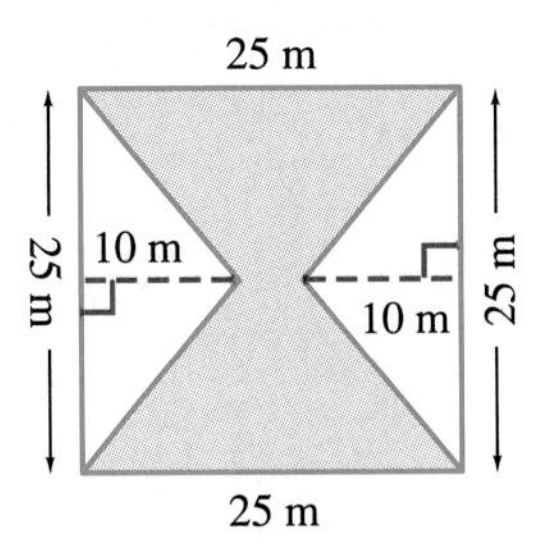

4. Suppose the figure directly above is an auditorium floor plan. The shaded part will be covered with carpet costing $27 per square meter. The rest will be covered with vinyl floor covering costing $18 per square meter. What is the total cost of covering the floor?

ANSWERS
3. $625 \text{ m}^2 - 125 \text{ m}^2 - 125 \text{ m}^2 = 375 \text{ m}^2$
4. $10,125 + $2250 + $2250 = $14,625

5. Find the size of the missing angle.

(a)

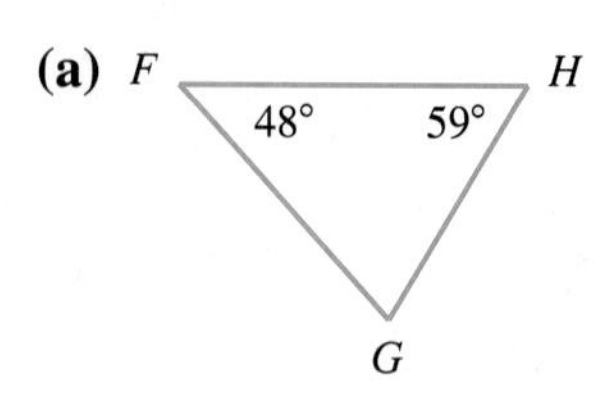

(b)

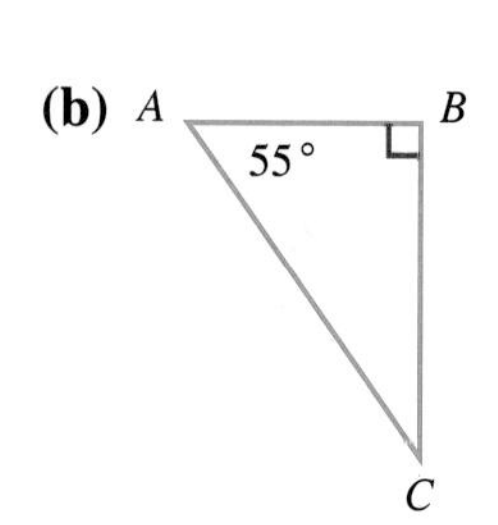

(c)

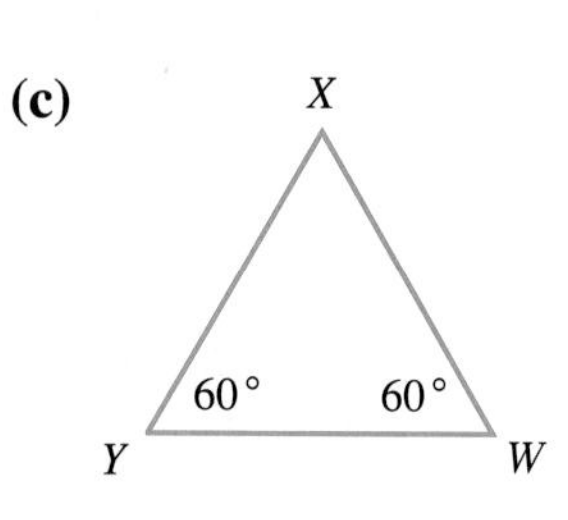

ANSWERS
5. (a) $\angle G = 73°$ (b) $\angle C = 35°$
(c) $\angle X = 60°$

3 The *tri* in *tri*angle means *three*. So the name tells you that a triangle has three angles. The sum of the three angles in any triangle is always 180° (a straight angle). You can see it by drawing a triangle, cutting off the three angles, and rearranging them to make a straight angle.

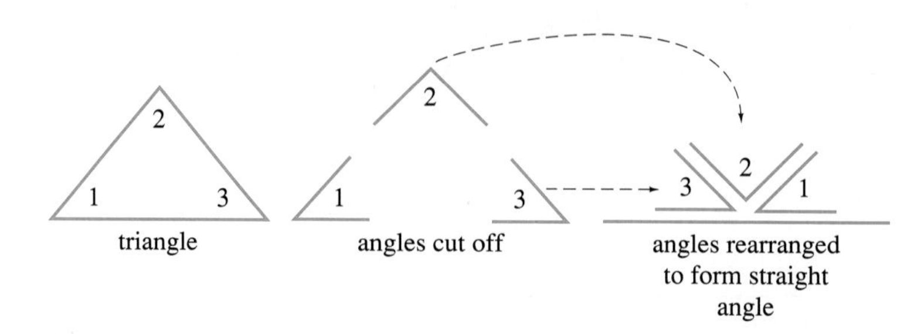

FINDING THE MISSING ANGLE IN A TRIANGLE

Step 1 Add the number of degrees in the two angles you are given.
Step 2 Subtract the sum from 180°.

EXAMPLE 5

How many degrees are in:

(a) angle R

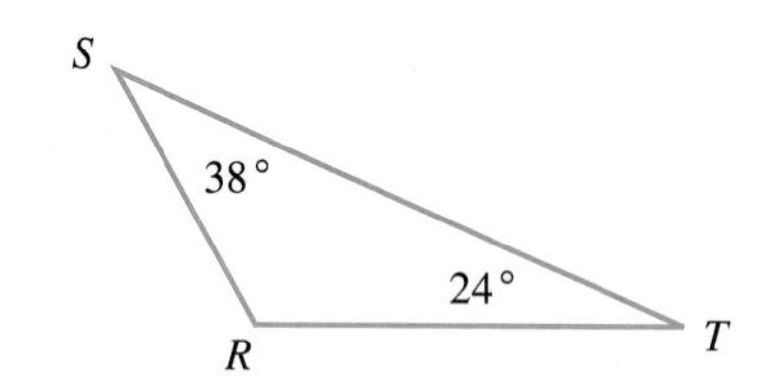

(b) angle F

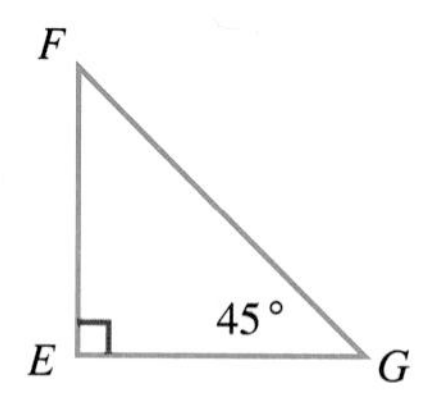

(a) *Step 1* Add the two angles you are given,

$$38° + 24° = 62°$$

Step 2 Subtract the sum from 180°.

$$180° - 62° = 118°$$

$\angle R$ is 118°.

(b) $\angle E$ is a right angle, which is 90°.

Step 1 $90° + 45° = 135°$

Step 2 $180° - 135° = 45°$

$\angle F$ is 45°. ■

WORK PROBLEM 5 AT THE RIGHT.

8.5 EXERCISES

NAME DATE HOUR

Find the perimeter of each triangle.

1.

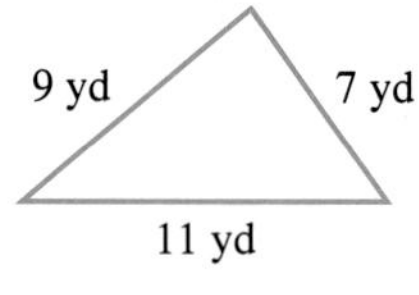

27 yd

2.

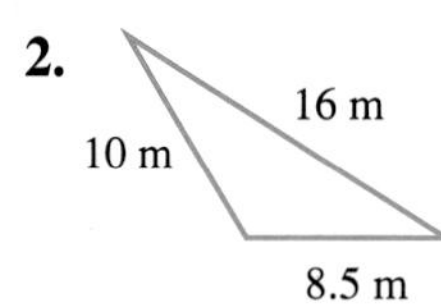

34.5 m

3.

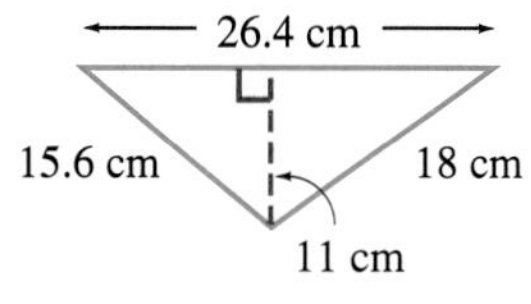

60 cm

4.

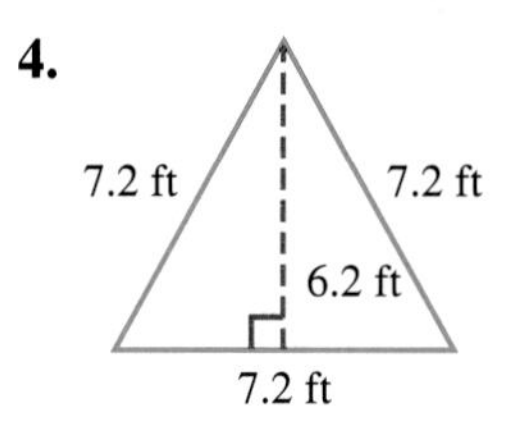

21.6 ft

Find the area of each triangle.

5.

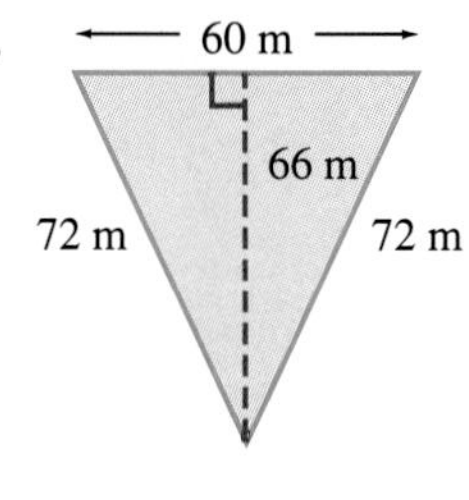

1980 m^2

6.

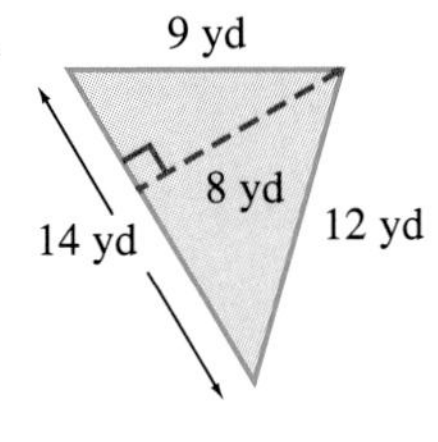

56 yd^2

7.

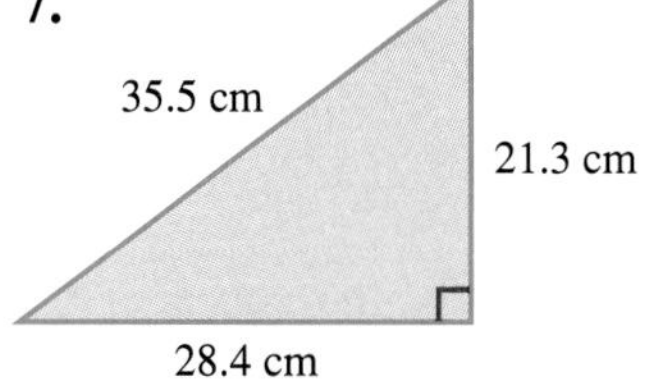

302.46 cm^2

8.

$6\frac{1}{4}$ ft

18 ft

9 ft

$28\frac{1}{8}$ ft^2 or 28.125 ft^2

Find the shaded area in each figure.

9.

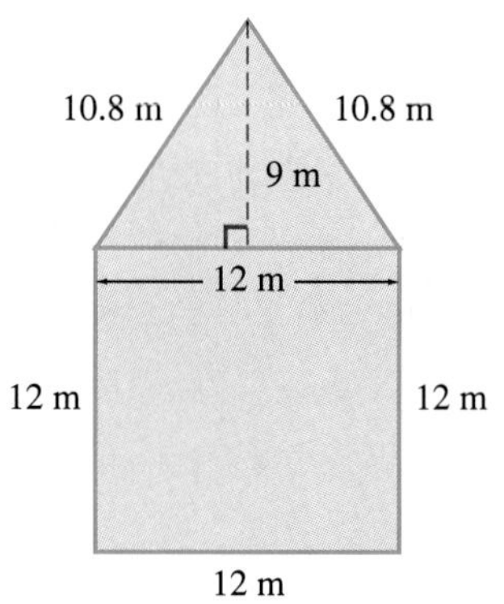

198 m^2

10.

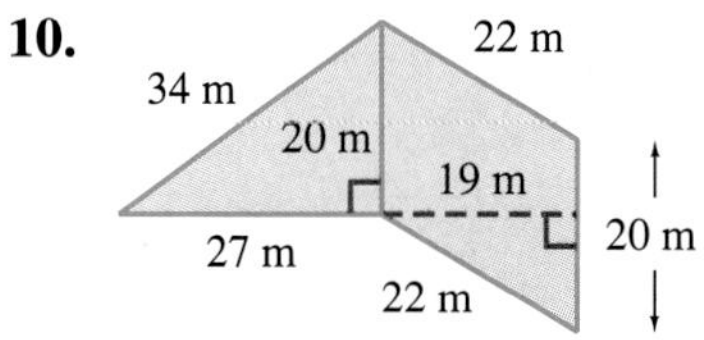

650 m^2

11.

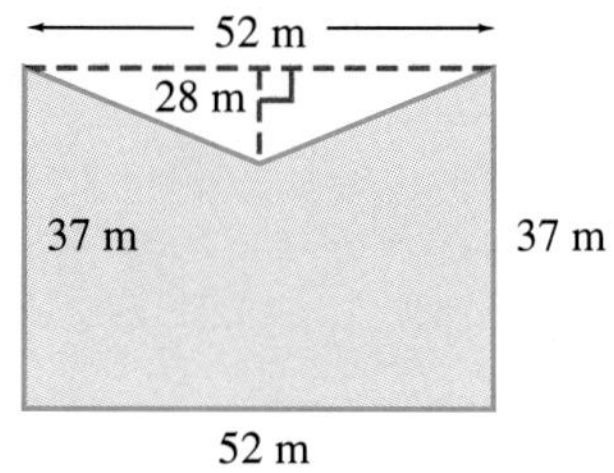

1196 m^2

12.

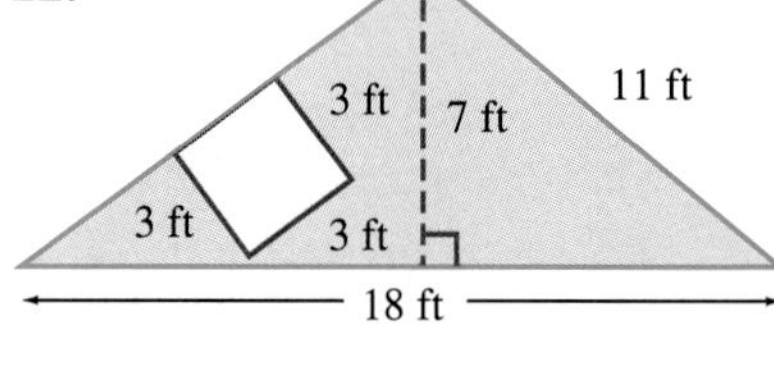

54 ft^2

Find the size of the missing angle.

13.

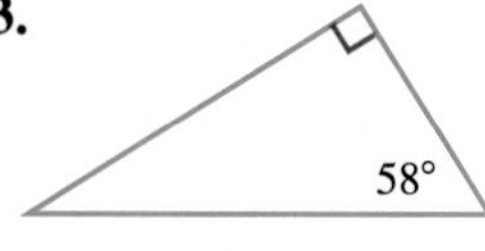

32°

14.

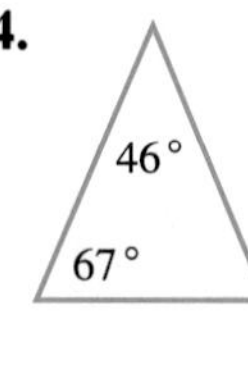

67°

15.

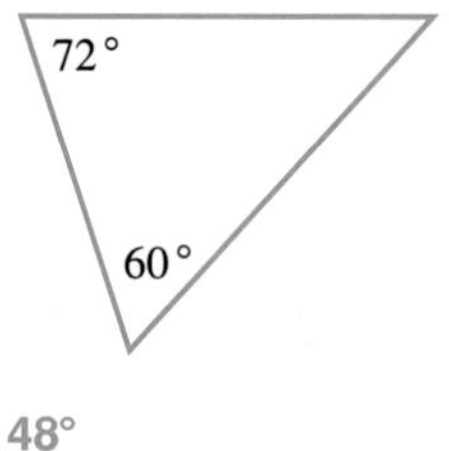

48°

16.

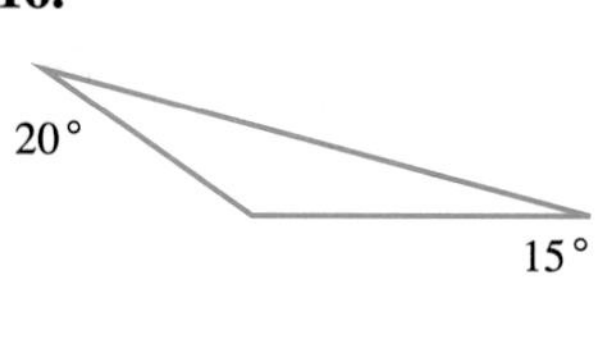

145°

17. Can a triangle have two right angles? Explain your answer.

Answer varies.

18. In your own words, explain where the $\frac{1}{2}$ comes from in the formula for area of a triangle.

Answer varies.

Solve each word problem.

19. **(a)** Find the area of one side of the house.
(b) Find the area of one roof section.

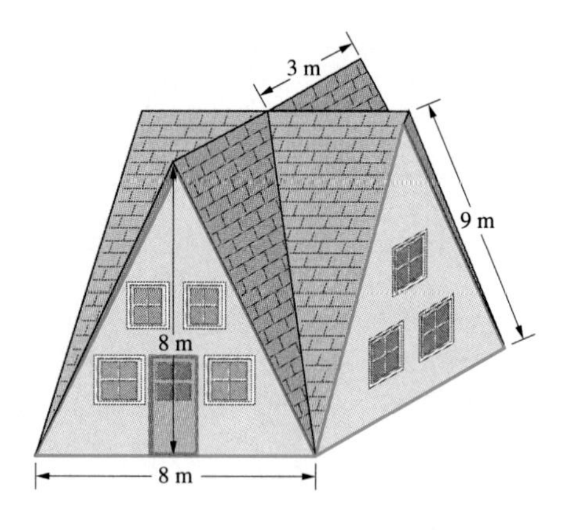

(a) 32 m^2 (b) 13.5 m^2

20. The sketch shows the plan for an office building. The shaded area will be a parking lot. What is the cost of building the parking lot, if the contractor charges $28.00 per square yard?

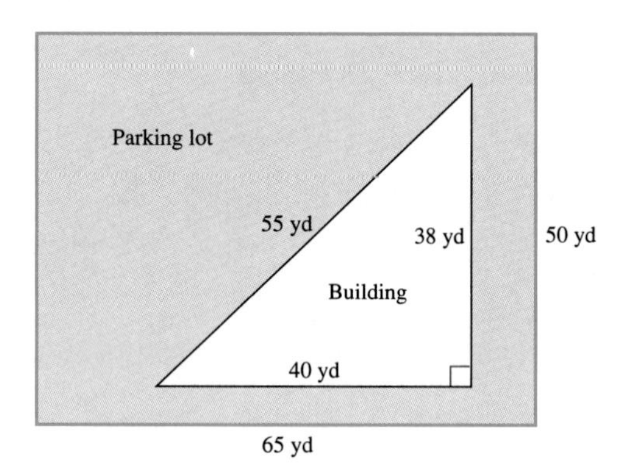

$69,720

PREVIEW EXERCISES

Multiply the following decimals. (For help, see **Section 4.5.**)

21. $3.14 \cdot 16$ 50.24

22. $2.13 \cdot 4.65$ 9.9045

23. $0.8 \cdot 0.8$ 0.64

24. $15 \cdot 0.7$ 10.5

8.6 CIRCLES

1 Suppose you start with one dot on a piece of paper. Then you draw a bunch of dots that are each 2 cm away from the first dot. If you draw enough dots (points) you'll end up with a **circle.** Each point on the circle is exactly 2 cm away from the *center* of the circle. The 2 cm distance is called the **radius,** r, of the circle. The distance across the circle (passing through the center) is called the **diameter**, d, of the circle.

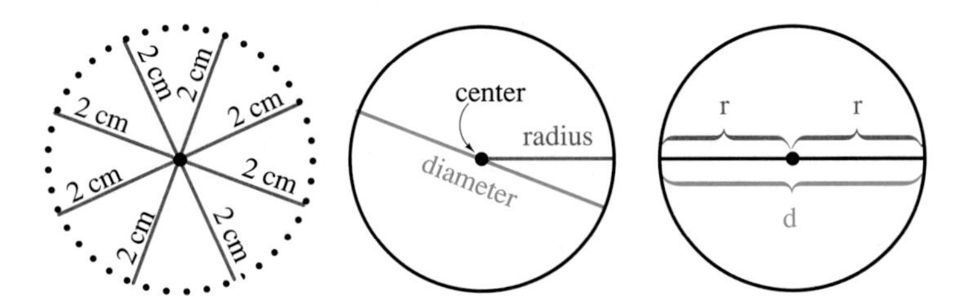

As the circle on the right shows,

FINDING THE DIAMETER AND RADIUS OF A CIRCLE

$$\text{diameter} = 2 \cdot \text{radius}$$
$$d = 2 \cdot r$$
and $$r = \frac{d}{2}$$

EXAMPLE 1 *Finding the Diameter and Radius of a Circle*
Find the missing diameter or radius in each circle.

(a)

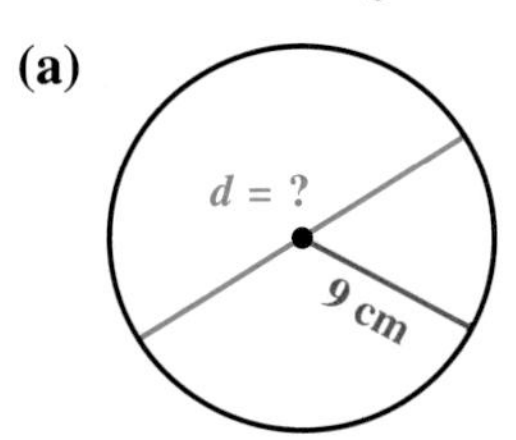

(b)

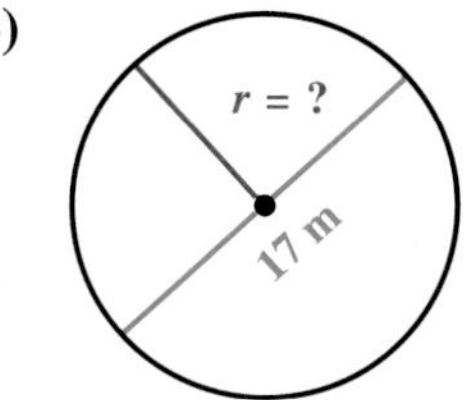

(a) Because the radius is 9 cm, the diameter is twice as long.

$$d = 2 \cdot r$$
$$d = 2 \cdot 9 \text{ cm}$$
$$d = 18 \text{ cm}$$

(b) The radius is half the diameter.

$$r = \frac{d}{2}$$
$$r = \frac{17 \text{ m}}{2}$$
$$r = 8.5 \text{ m} \quad \text{or} \quad 8\frac{1}{2} \text{ m}$$ ■

WORK PROBLEM 1 AT THE SIDE.

OBJECTIVES

1. Find the radius and diameter of a circle.
2. Find the circumference of a circle.
3. Find the area of a circle.

FOR EXTRA HELP

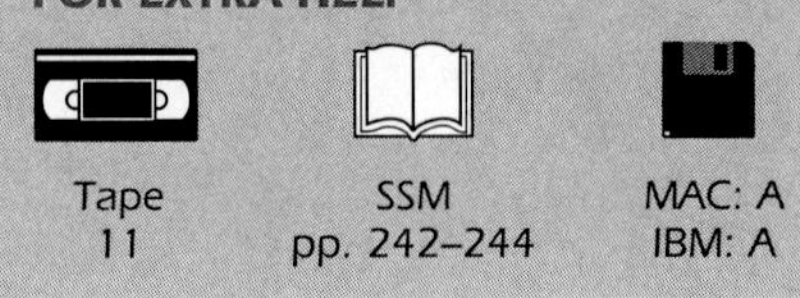

Tape 11 | SSM pp. 242–244 | MAC: A IBM: A

1. Find the missing diameter or radius in each circle.

(a)

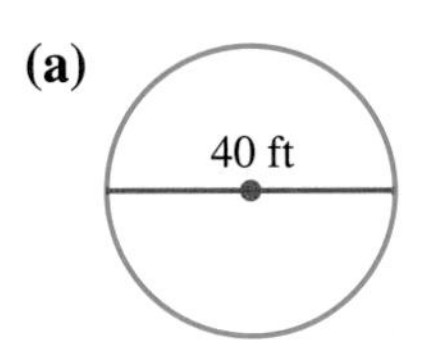

(b)

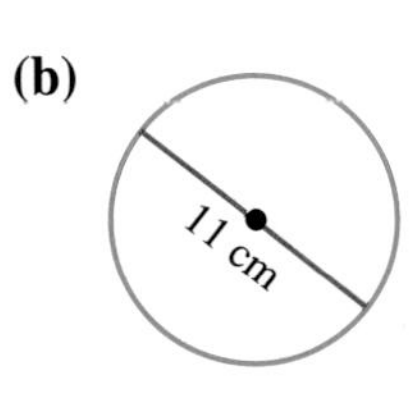

(c)

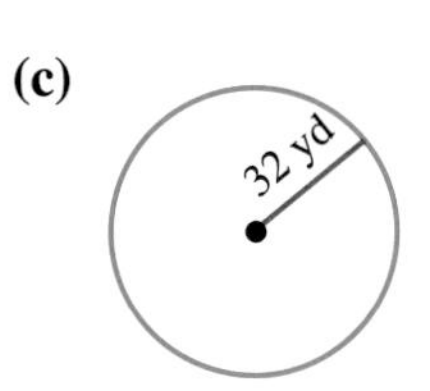

(d)

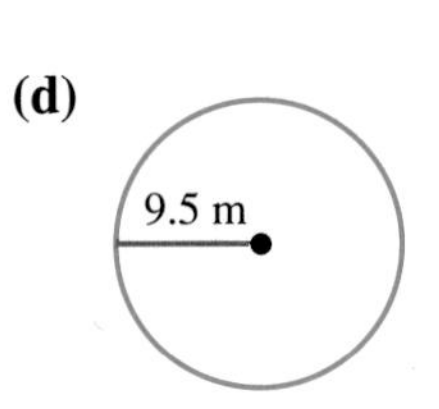

ANSWERS
1. (a) $r = 20$ ft
(b) $r = 5.5$ cm
(c) $d = 64$ yd
(d) $d = 19$ m

2 The perimeter of a circle is called its **circumference.** Circumference is the distance around the edge of a circle.

The diameter of the can in the drawing is about 10.6 cm, and the circumference of the can is about 33.3 cm. Dividing the circumference of the circle by the diameter gives an interesting result.

$$\frac{\text{circumference}}{\text{diameter}} = \frac{33.3}{10.6} = 3.14 \quad \text{(rounded)}$$

Dividing the circumference of *any* circle by its diameter *always* gives an answer close to 3.14. This means that going around the edge of any circle is a little more than 3 times as far as going straight across the circle.

This ratio of circumference to diameter is called π (the Greek letter **pi**). There is no decimal that is exactly equal to π, but approximately:

$$\pi = 3.14159265359$$

ROUNDING THE VALUE OF Pi (π)

We usually round π to 3.14.

Use the following formulas to find the circumference of a circle.

FINDING THE DISTANCE AROUND A CIRCLE

$$\text{Circumference} = \pi \cdot \text{diameter}$$

$$C = \pi \cdot d$$

or, because $d = 2 \cdot r$ $\quad C = \pi \cdot 2 \cdot r \quad$ usually written $2 \cdot \pi \cdot r$

■ **EXAMPLE 2** *Finding the Circumference of a Circle*

Find the circumference of each circle. Use 3.14 as the approximate value for π. Round answers to the nearest tenth.

(a)

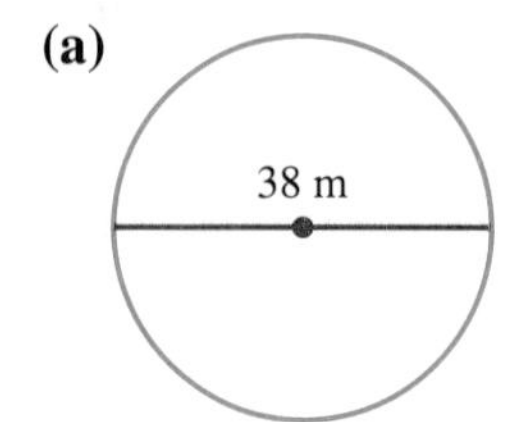

(b)

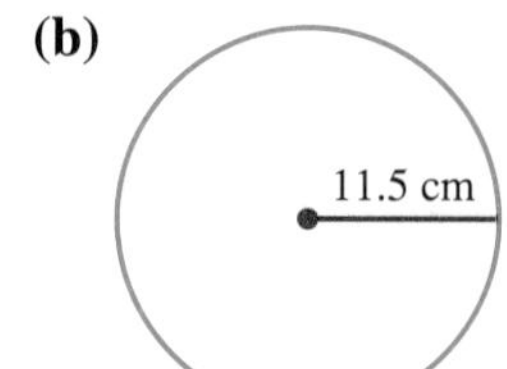

(a) The diameter is 38 m, so use the formula with d in it.

$C = \pi \cdot d$

$C = \mathbf{3.14} \cdot 38 \text{ m}$

$C = 119.3 \text{ m}$ (rounded)

(b) In this example, r is known, so it is easier to use the formula

$C = 2 \cdot \pi \cdot r$

$C = 2 \cdot \mathbf{3.14} \cdot 11.5 \text{ cm}$

$C = 72.2 \text{ cm}$ (rounded) ■

WORK PROBLEM 2 AT THE SIDE.

3 To find the formula for the area of a circle, start by cutting two circles into many pie-shaped pieces.

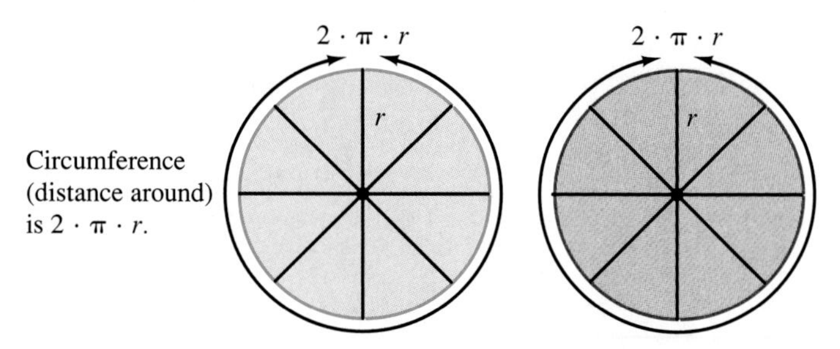

Unfold the circles, much as you might "unfold" a peeled orange, and put them together as shown here.

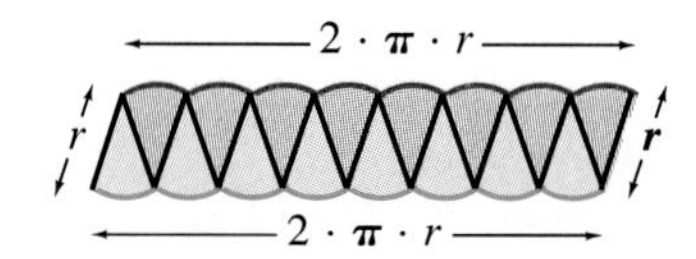

The figure is approximately a rectangle with width r (the radius of the original circle) and length $2 \cdot \pi \cdot r$ (the circumference of the original circle). The area of the "rectangle" is length times width.

$$\text{Area} = l \cdot w$$
$$\text{Area} = 2 \cdot \pi \cdot r \cdot r$$
$$\text{Area} = 2 \cdot \pi \cdot r^2$$

Because the "rectangle" was formed from *two* circles, the area of *one* circle is half as much.

$$\frac{1}{\underset{1}{\cancel{2}}} \cdot \overset{1}{\cancel{2}} \cdot \pi \cdot r^2 = 1 \cdot \pi \cdot r^2 \quad \text{or simply} \quad \pi \cdot r^2$$

FINDING THE AREA OF A CIRCLE

$$\text{Area of a circle} = \pi \cdot \text{radius} \cdot \text{radius}$$
$$A = \pi \cdot r^2$$

EXAMPLE 3 *Finding the Area of a Circle*

Find the area of each circle. Use 3.14 for π. Round your answers to the nearest tenth.

(a) A circle with radius 8.2 cm.

(b)

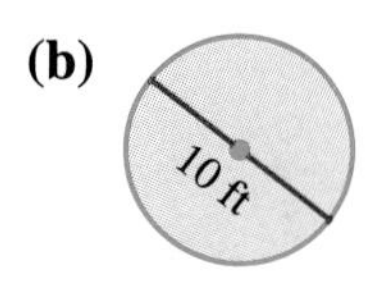

2. Find the circumference of each circle. Use 3.14 as the approximate value for π. Round answers to the nearest tenth.

(a)

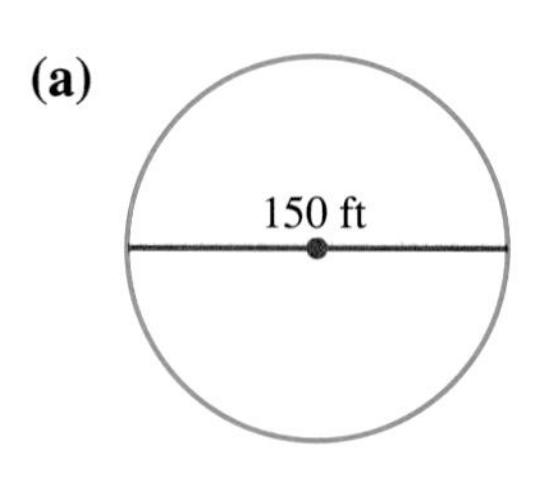

(b)

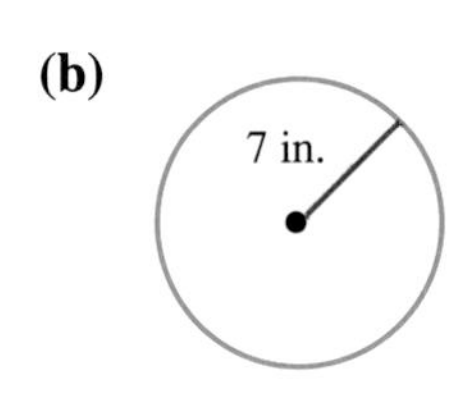

(c) diameter 0.9 km

(d) radius 4.6 m

ANSWERS

2. (a) $C = 471$ **ft (b)** $C = 44.0$ **in. (c)** $C = 2.8$ **km (d)** $C = 28.9$ **m**

3. Find the area of each circle. Use 3.14 for π. Round your answers to the nearest tenth.

(a)

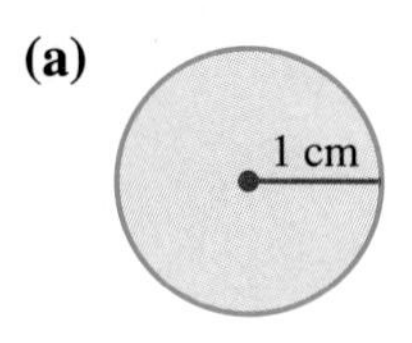

(b)

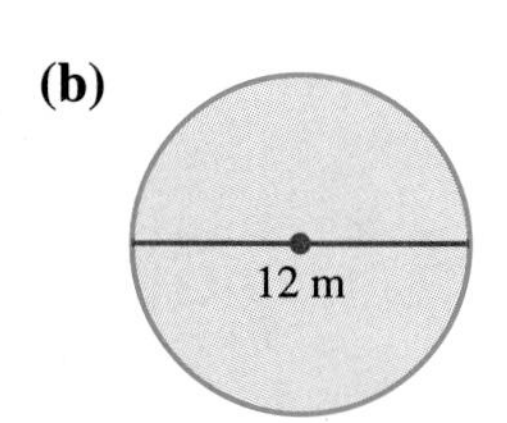

(Hint: The diameter is 12 m so r = ___ m)

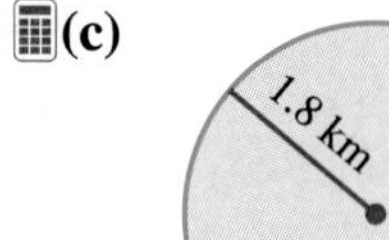

(c)

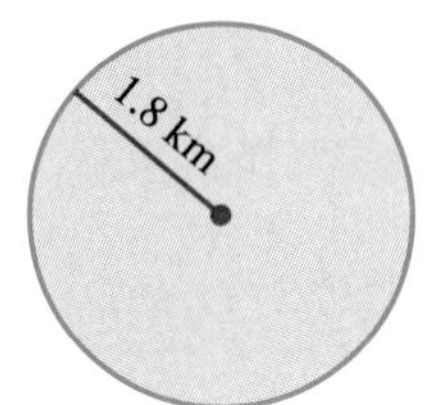

(d)

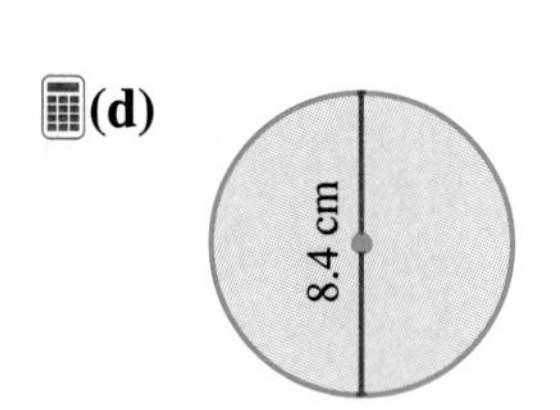

ANSWERS

3. **(a)** 3.1 cm^2 **(b)** 113.0 m^2 **(c)** 10.2 km^2 **(d)** 55.4 cm^2

(a) Use the formula $A = \pi \cdot r^2$, which means $\pi \cdot r \cdot r$.

$$A = \pi \cdot r \cdot r$$
$$A = \mathbf{3.14} \cdot 8.2\text{ cm} \cdot 8.2\text{ cm}$$
$$A = 211.1\text{ cm}^2 \text{ (square units for area)}$$

(b) To use the formula, you need to know the radius (r). In this circle, the diameter is 10 ft. First find the radius.

$$r = \frac{d}{2}$$
$$r = \frac{10\text{ ft}}{2} = 5\text{ ft}$$

Now find the area.

$$A = 3.14 \cdot 5\text{ ft} \cdot 5\text{ ft}$$
$$A = 78.5\text{ ft}^2 \quad ■$$

Note When finding *circumference,* you can start with either the radius or the diameter. When finding *area,* you must use the *radius.* If you are given the diameter, divide it by 2 to find the radius. Then find the area.

◀◀ WORK PROBLEM 3 AT THE SIDE.

■ **EXAMPLE 4** *Finding the Area of a Semicircle*

Find the area of the semicircle. Use 3.14 for π. Round your answer to the nearest tenth.

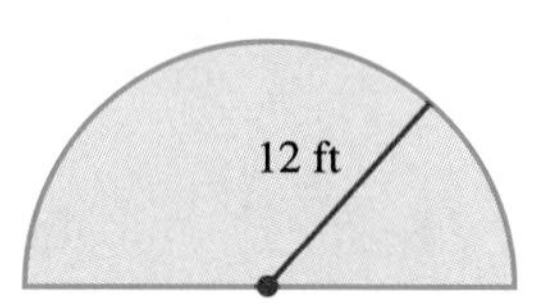

First, find the area of an entire circle with a radius of 12 ft.

$$A = \pi \cdot r \cdot r$$
$$A = \mathbf{3.14} \cdot 12\text{ ft} \cdot 12\text{ ft}$$
$$A = 452.2\text{ ft}^2 \quad \text{(rounded)}$$

Divide the area of the whole circle by 2 to find the area of the semicircle.

$$\frac{452.2\text{ ft}^2}{2} = 226.1\text{ ft}^2 \quad ■$$

WORK PROBLEM 4 AT THE SIDE.

EXAMPLE 5 *Applying the Concept of Circumference*
A circular rug is 8 feet in diameter. The cost of fringe for the edge is \$2.25 per foot. What will it cost to add fringe to the rug? Use 3.14 for π.

$$\text{Circumference} = \pi \cdot d$$
$$C = 3.14 \cdot 8 \text{ ft}$$
$$C = 25.12 \text{ ft}$$

$$\text{Cost} = \text{Cost per foot} \cdot \text{Circumference}$$
$$\text{Cost} = \frac{\$2.25}{1 \text{ ft}} \cdot \frac{25.12 \text{ ft}}{1}$$
$$\text{Cost} = \$56.52$$ ■

WORK PROBLEM 5 AT THE SIDE.

EXAMPLE 6 *Applying the Concept of Area*
Find the cost of covering the rug in Example 5 with a plastic cover. The material for the cover costs \$1.50 per square foot. Use 3.14 for π.

First find the radius.

$$r = \frac{d}{2} = \frac{8 \text{ ft}}{2} = 4 \text{ ft}$$

Then, $A = \pi \cdot r^2$
$$A = 3.14 \cdot 4 \text{ ft} \cdot 4 \text{ ft}$$
$$A = 50.24 \text{ ft}^2$$

And, $\text{cost} = \frac{\$1.50}{1 \text{ ft}^2} \cdot \frac{50.24 \text{ ft}^2}{1} = \$75.36$ ■

WORK PROBLEM 6 AT THE SIDE.

4. Find the area of each semicircle. Use 3.14 for π. Round your answers to the nearest tenth.

(a)

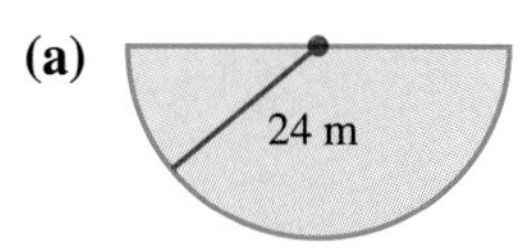

(b)

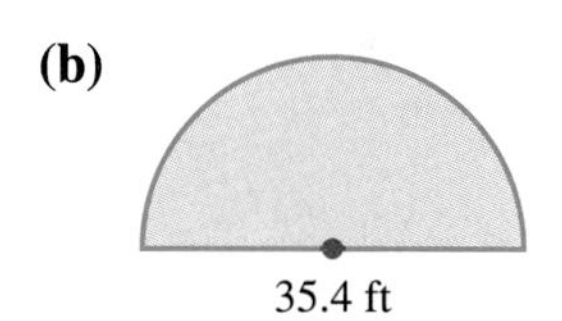

(c)

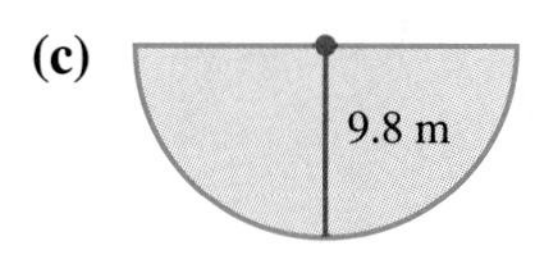

5. Find the cost of binding around the edge of a circular rug that is 3 meters in diameter. The binder charges \$4.50 per meter. Use 3.14 for π.

6. Find the cost of covering the rug in problem 5 above with a nonslip rubber backing. The rubber backing costs \$2 per square meter.

ANSWERS
4. (a) 904.3 m² (b) 491.9 ft² (c) 150.8 m²
5. \$42.39
6. \$14.13

QUEST FOR NUMERACY

Best Buy on Pizza

PIZZA MENU

	Small $7\frac{1}{2}''$	*Medium* $13''$	*Large* $16''$	*Family* $24''$
Cheese only	$2.80	$6.50	$9.30	$21.69
"The works"	3.70	9.95	14.30	31.25
Deep dish combo	4.35	10.95	15.65	36.00

Find the "best buy" for each type of pizza. The "best buy" is the lowest cost per square inch of pizza. All the pizzas are circular in shape, and the measurement given on the menu board is the diameter of the pizza in inches. Use 3.14 as the approximate value of π. Round the area to the nearest tenth.

As you work on this problem, organize your results in the form of a table. Show the cost per square inch for each type of pizza and circle the "best buy." (Round the cost per square inch to the nearest thousandth, rather than hundredth, in order to compare more easily.) Here is the outline of a table to get you started.

	Small	*Medium*	*Large*	*Family*
Area in square inches	44.2	132.7	201.0	452.2
Cost per in.^{2} Cheese	$0.063	$0.049	$0.046 Best buy	$0.048
Cost per in.^{2} The Works	$0.084	$0.075	$0.071	$0.069 Best buy
Cost per in.^{2} Deep dish combo	$0.098	$0.083	$0.078 Best buy	$0.080

Extend this activity by bringing in the menus from several local pizza places or ask if any students work there. Find the "best buy" for different combinations at different restaurants. Discuss what other factors could influence your decision to buy a particular size pizza besides lowest cost per square inch.

8.6 EXERCISES

NAME DATE HOUR

Find the missing value in each circle.

1.

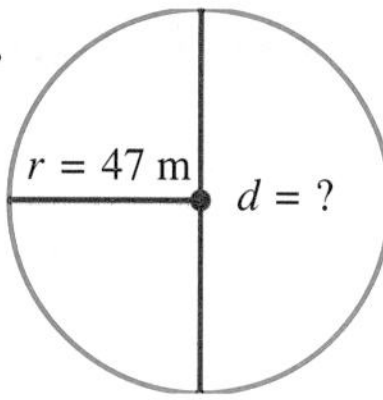

94 m

2.

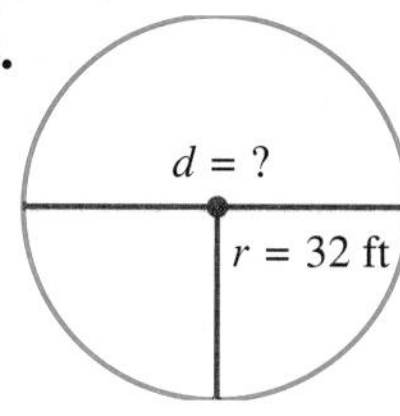

64 ft

3.

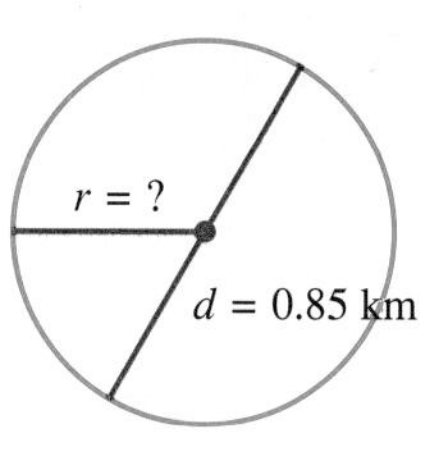

0.425 km

4.

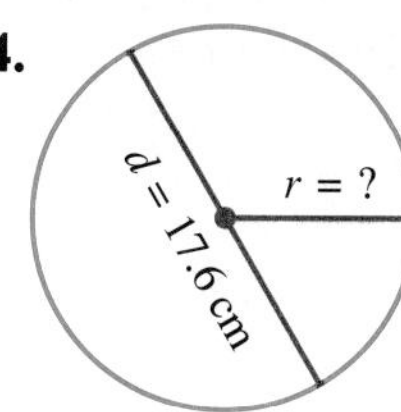

8.8 cm

Find the circumference and area of each circle. Use 3.14 as the approximate value for π. Round your answers to the nearest tenth.

5.

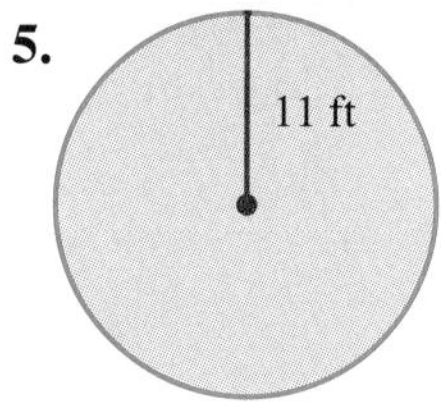

$C = 69.1$ ft
$A = 379.9\ \text{ft}^2$

6.

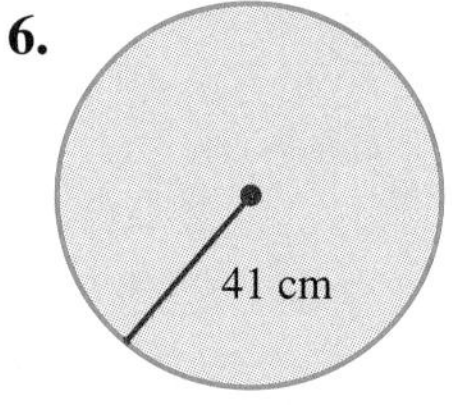

$C = 257.5$ cm
$A = 5278.3\ \text{cm}^2$

7.

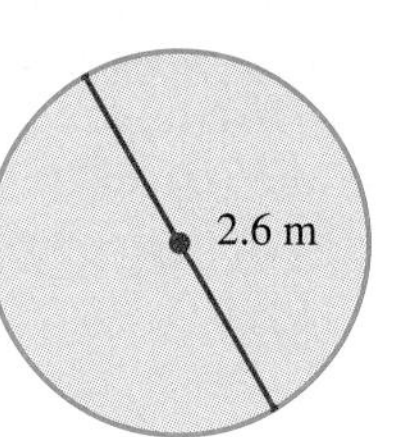

$C = 8.2$ m
$A = 5.3\ \text{m}^2$

8.

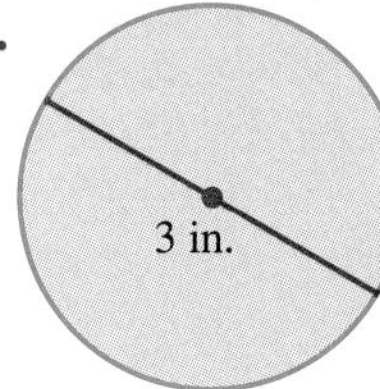

$C = 9.4$ in.
$A = 7.1\ \text{in.}^2$

Find the circumference and area of circles having the following diameters. Use 3.14 for π. Round your answers to the nearest tenth.

9. $d = 15$ cm

$C = 47.1$ cm
$A = 176.6\ \text{cm}^2$

10. $d = 39$ ft

$C = 122.5$ ft
$A = 1194.0\ \text{ft}^2$

11. $d = 7\frac{1}{2}\ m$

$C = 23.6$ m
$A = 44.2\ \text{m}^2$

12. $d = 4\frac{1}{2}$ yd

$C = 14.1$ yd
$A = 15.9\ \text{yd}^2$

13. $d = 8.65$ km

$C = 27.2$ km
$A = 58.7\ \text{km}^2$

14. $d = 19.5$ mm

$C = 61.2$ mm
$A = 298.5\ \text{mm}^2$

Solve each problem.

15. How far does a point on the tread of a tire move in one turn, if the diameter of the tire is 70 cm?

219.8 cm

16. If you swing a ball held at the end of a string 2 m long, how far will the ball travel on each turn?

12.6 m

17. A wave energy extraction device is a huge undersea dome used to harness the power of ocean waves. The dome is 250 ft in diameter. Find its circumference.

785.0 ft

18. Find the area of the base of the dome in Exercise 17.

49,062.5 ft^2

19. A radio station can be heard 150 miles in all directions during evening hours. How many square miles are in the station's broadcast area?

70,650 mi^2

20. An earthquake was felt by people 900 km away in all directions from the epicenter (the source of the earthquake). How much area was affected by the quake?

2,543,400 km^2

21. How would you explain π to a friend who is not in your math class? Write an explanation. Then make up a test question which requires the use of π, and show how to solve it.

Answer varies.

22. Explain how circumference and perimeter are alike. How are they different? Make up two problems, one involving perimeter, the other circumference. Show how to solve your problems.

Answer varies.

Find each shaded area in Exercises 23–26. Use 3.14 as the approximate value of π. Round your answers to the nearest tenth.

Example:

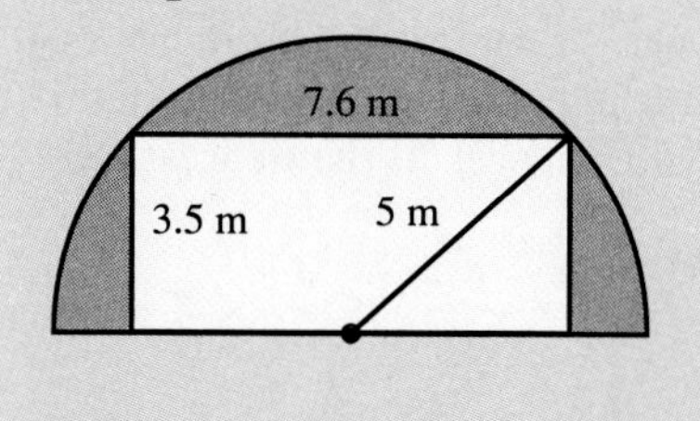

Solution:

First, find the area of the entire circle.

$$A = \pi \cdot r^2 = 3.14 \cdot 5 \text{ m} \cdot 5 \text{ m} = 78.5 \text{ m}^2$$

Next, find the area of the semicircle.

$$\frac{78.5 \text{ m}^2}{2} = 39.3 \text{ m}^2 \quad \text{(rounded)}$$

Now, find the area of the rectangle.

$$3.5 \text{ m} \cdot 7.6 \text{ m} = 26.6 \text{ m}^2$$

Finally, subtract to find the shaded area.

$$39.3 \text{ m}^2 - 26.6 \text{ m}^2 = \mathbf{12.7 \text{ m}^2}$$

23.

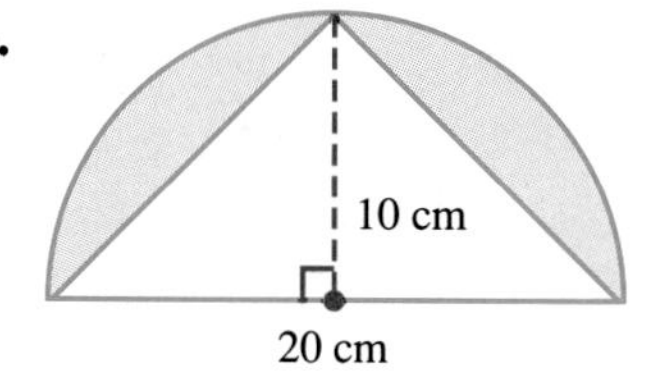

57 cm^2

24.

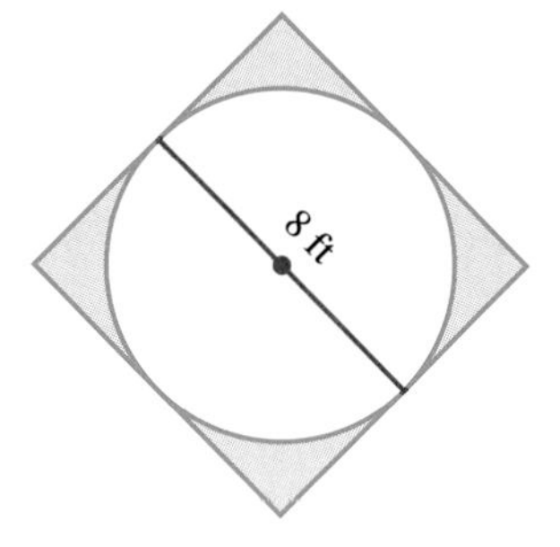

13.8 ft^2

25.

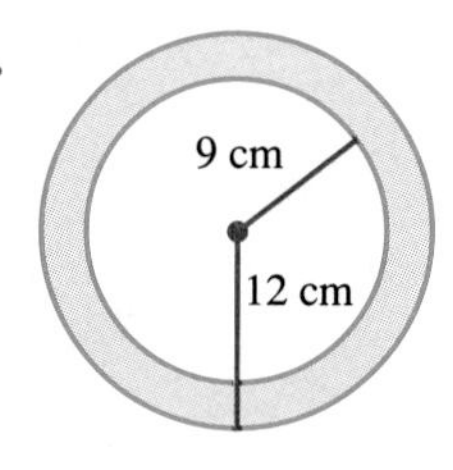

197.8 cm^2

26.

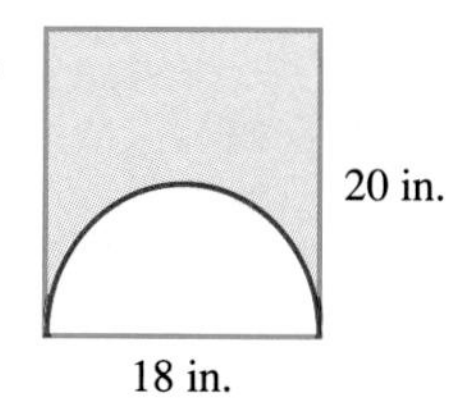

232.8 in.^2

Solve each word problem.

▲ **27.** Find the cost of sod, at \$1.76 per square foot, for the following playing field.

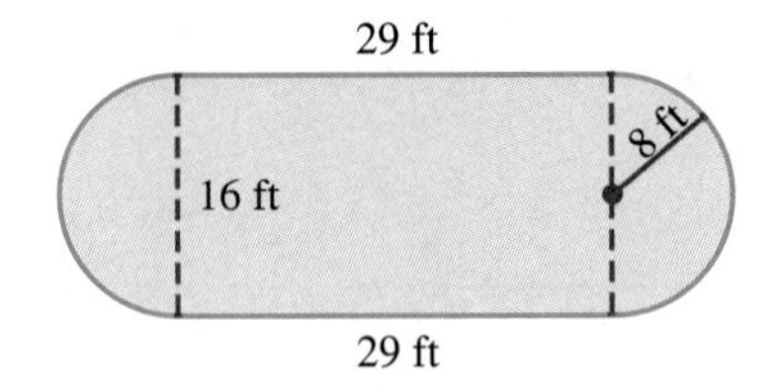

\$1170.33 (rounded)

▲ **28.** A forest ranger measured 56 ft around the base of a giant sequoia tree. The ranger wanted to find the diameter of the tree without cutting it down. Find the diameter. Round your answer to the nearest tenth.

17.8 ft

▲ **29.** The circumference of a circular swimming pool is 22 meters. What is the radius of the pool, to the nearest tenth of a meter?

3.5 m

▲ **30.** Find the area of this skating rink.

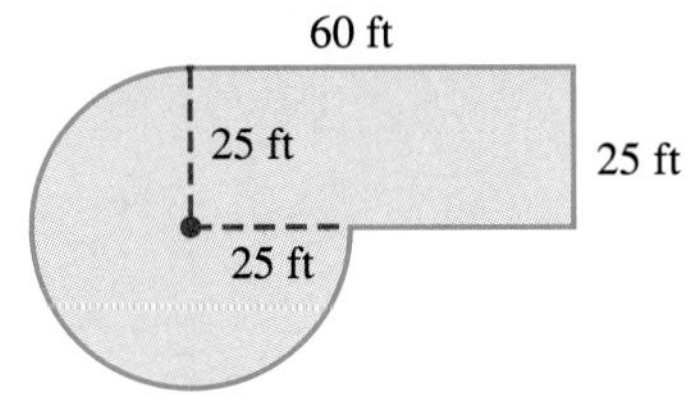

2971.9 ft^2

PREVIEW EXERCISES

Write each fraction or mixed number as a decimal. Round to the nearest thousandth, if necessary. (For help, see ***Section 4.7.****)*

31. $\frac{1}{2}$ 0.5

32. $\frac{1}{3}$ 0.333

33. $2\frac{3}{4}$ 2.75

34. $4\frac{1}{4}$ 4.25

35. $\frac{2}{3}$ 0.667

36. $\frac{4}{3}$ 1.333

37. $10\frac{1}{2}$ 10.5

38. $18\frac{3}{4}$ 18.75

8.7 VOLUME

OBJECTIVES

Find the volume of a

1. rectangular solid;
2. sphere;
3. cylinder;
4. cone and pyramid.

FOR EXTRA HELP

Tape 11

SSM pp. 244–246

MAC: A
IBM: A

1 A shoe box and a cereal box are examples of three-dimensional (or solid) figures. The three dimensions are length, width, and height. (A rectangle or square is a two-dimensional figure. The two dimensions are length and width.) If you want to know how much the shoe box will hold, you find its **volume.** We measure volume by seeing how many cubes of a certain size will fill the space inside the box. Three sizes of *cubic units* are shown here.

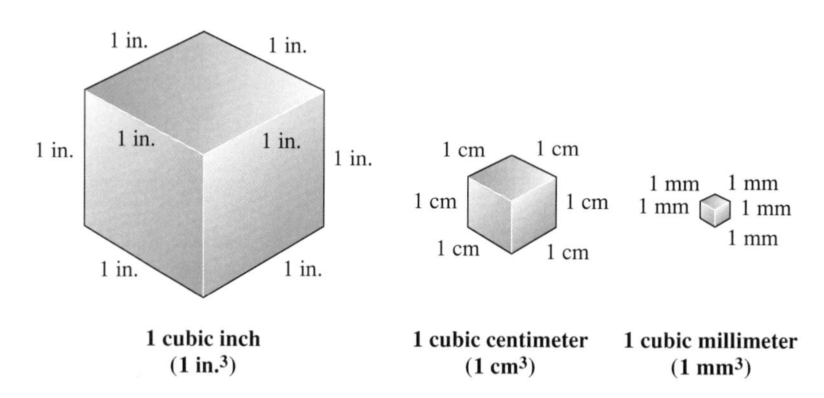

1 cubic inch (1 in.^3) **1 cubic centimeter (1 cm^3)** **1 cubic millimeter (1 mm^3)**

Some other sizes of cubes that are used to measure volume are 1 cubic foot (1 ft^3), 1 cubic yard (1 yd^3), and 1 cubic meter (1 m^3).

> *Note* The raised 3 in 4^3 means that you multiply $4 \cdot 4 \cdot 4$ to get 64. The raised 3 in cm^3 or ft^3 is a short way to write the word "cubic." When you see 5 cm^3, say "five cubic centimeters." Do *not* multiply $5 \cdot 5 \cdot 5$.

Use the following formula for finding the volume of rectangular solids (box-like shapes).

FINDING THE VOLUME OF BOX-LIKE SHAPES

Volume of rectangular solid = length · width · height

$$V = l \cdot w \cdot h$$

EXAMPLE 1 *Finding the Volume of a Rectangular Solid*

Find the volume of each box.

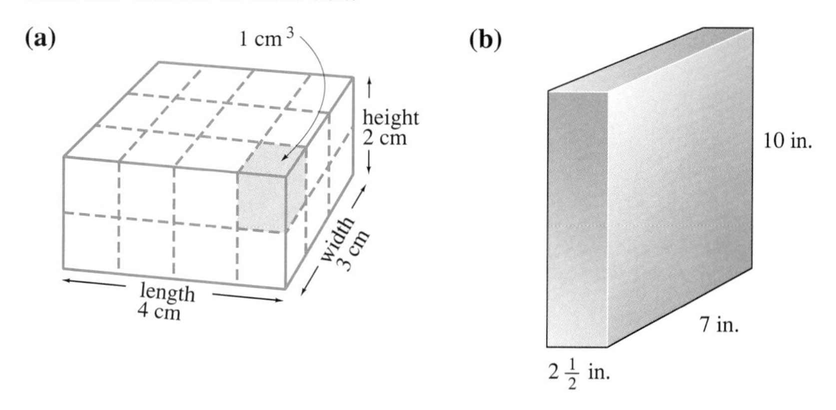

(a) Each cube that fits in the box is 1 cubic centimeter (1 cm^3). To find the volume, you can count the number of cubes.

bottom layer has 12 cubes, top layer has 12 cubes } total of 24 cubes (24 cm^3)

1. Find the volume of each box. Round your answers to the nearest tenth, if necessary.

(a)

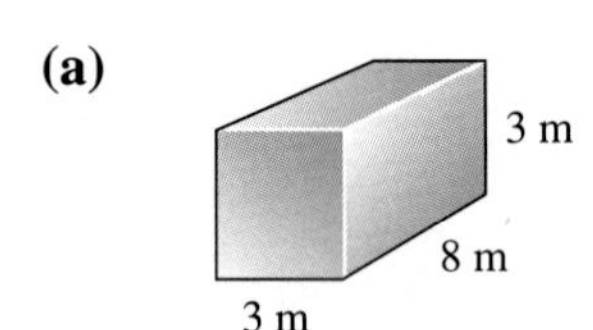

(b)

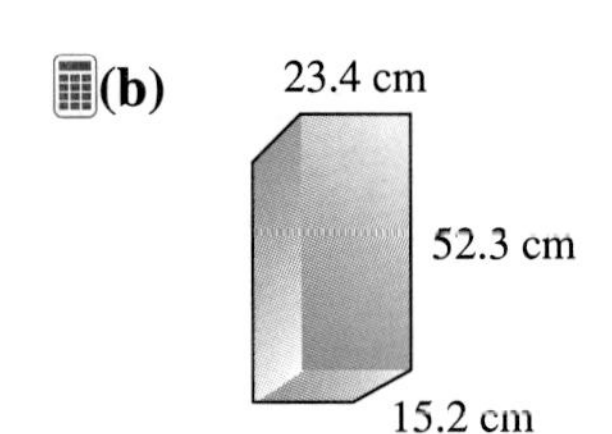

(c) length $6\frac{1}{4}$ ft, width $3\frac{1}{2}$ ft, height 2 ft

ANSWERS

1. (a) 72 m³ **(b)** 18,602.1 cm³

(c) $43\frac{3}{4}$ ft³ or 43.8 ft³

Or you can use the formula for rectangular solids.

$$V = l \cdot w \cdot h$$
$$V = 4 \text{ cm} \cdot 3 \text{ cm} \cdot 2 \text{ cm}$$
$$V = 24 \text{ cm}^3 \quad \text{(cubic units for volume)}$$

(b) Use the formula. If you like, use 2.5 inches, the decimal equivalent of $2\frac{1}{2}$ inches, for the width.

$$V = 7 \text{ in.} \cdot 2\frac{1}{2} \text{ in.} \cdot 10 \text{ in.} \quad \text{or} \quad V = 7 \text{ in.} \cdot 2.5 \text{ in.} \cdot 10 \text{ in.}$$
$$V = 175 \text{ in.}^3 \qquad\qquad V = 175 \text{ in.}^3$$ ■

◀ **WORK PROBLEM 1 AT THE SIDE.**

2 A *sphere* is shown here. Examples of spheres include baseballs, oranges, and the earth. (The last two aren't perfect spheres, but they're close.)

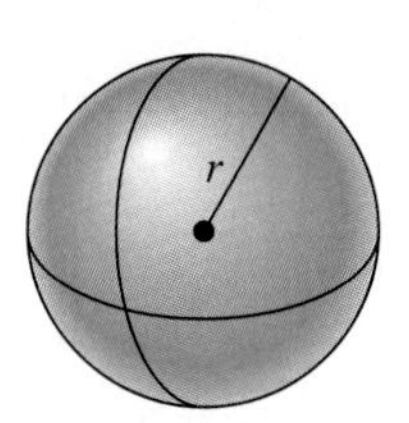

As with circles, the *radius* of a sphere is the distance from the center to the edge of the sphere. Use the following formula to find the volume of a sphere.

FINDING THE VOLUME OF A SPHERE

$$\text{Volume of sphere} = \frac{4}{3} \cdot \pi \cdot r \cdot r \cdot r$$
$$V = \frac{4}{3} \cdot \pi \cdot r^3 \quad \text{or} \quad \frac{4 \cdot \pi \cdot r^3}{3}$$

■ **EXAMPLE 2** *Finding the Volume of a Sphere*

Find the volume of each sphere with the help of a calculator. Use 3.14 as the approximate value of π. Round your answers to the nearest tenth.

(a)

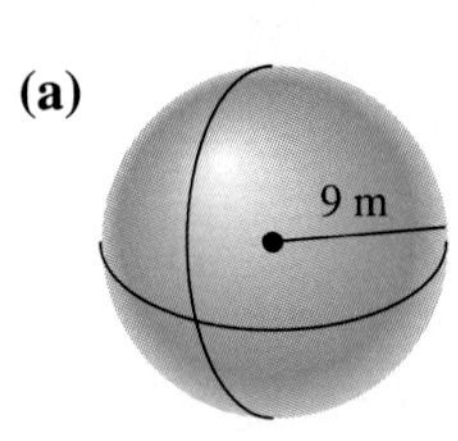

(b)

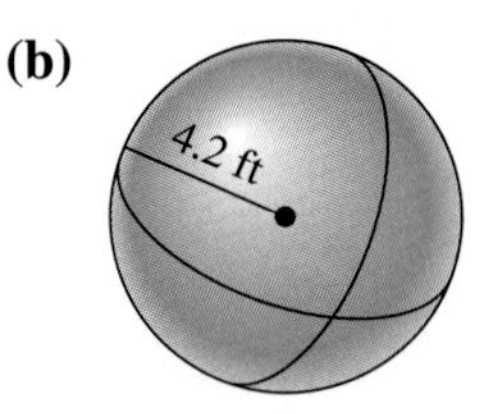

(a) $V = \frac{4}{3} \cdot \pi \cdot r^3$

$$V = \frac{4 \cdot 3.14 \cdot 9 \text{ m} \cdot 9 \text{ m} \cdot 9 \text{ m}}{3}$$

$$V = 3052.1 \text{ m}^3 \quad \text{(rounded)}$$

(b) $V = \dfrac{4 \cdot 3.14 \cdot 4.2 \text{ ft} \cdot 4.2 \text{ ft} \cdot 4.2 \text{ ft}}{3}$

$V = 310.2 \text{ ft}^3$ (rounded) ■

WORK PROBLEM 2 AT THE SIDE. ▶▶

Half a sphere is called a *hemisphere.* The volume of a hemisphere is *half* the volume of a sphere. Use the following formula to find the volume of a hemisphere.

FINDING THE VOLUME OF A HEMISPHERE

$$\text{Volume of hemisphere} = \frac{\overset{1}{\cancel{1}}}{\underset{1}{\cancel{2}}} \cdot \frac{\overset{2}{\cancel{4}}}{3} \cdot \pi \cdot r \cdot r \cdot r$$

$$V = \frac{2}{3} \cdot \pi \cdot r^3 \quad \text{or} \quad \frac{2 \cdot \pi \cdot r^3}{3}$$

■ **EXAMPLE 3** *Finding the Volume of a Hemisphere*
Find the volume of the hemisphere with the help of a calculator. Use 3.14 for π. Round your answer to the nearest tenth.

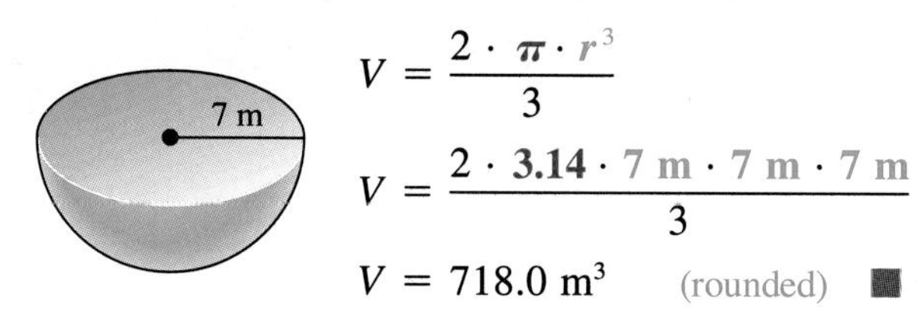

$V = \dfrac{2 \cdot \pi \cdot r^3}{3}$

$V = \dfrac{2 \cdot 3.14 \cdot 7 \text{ m} \cdot 7 \text{ m} \cdot 7 \text{ m}}{3}$

$V = 718.0 \text{ m}^3$ (rounded) ■

WORK PROBLEM 3 AT THE SIDE. ▶▶

3 Several *cylinders* are shown here.

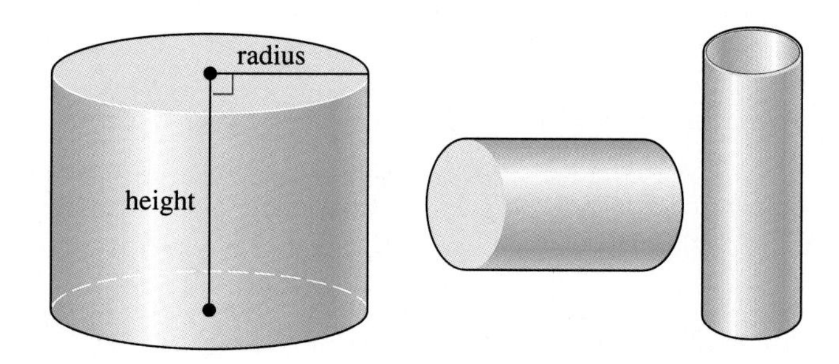

These are called *right circular cylinders* because the top and bottom are circles, and the side makes a right angle with the top and bottom. Examples of cylinders are a soup can, a home water heater, and a piece of pipe.

Use the following formula to find the volume of a cylinder. Notice that the first part of the formula, $\pi \cdot r \cdot r$, is the area of the circular base.

FINDING THE VOLUME OF A CYLINDER

$$\text{Volume of cylinder} = \pi \cdot r \cdot r \cdot h$$
$$V = \pi \cdot r^2 \cdot h$$

2. Find the volume of each sphere. Use 3.14 for π. Round your answers to the nearest tenth.

(a)

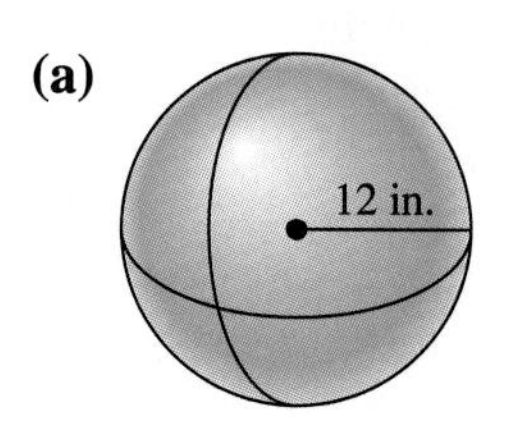

(b)

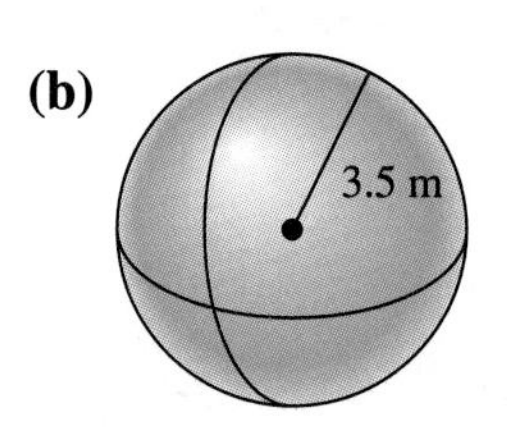

(c) radius 2.7 cm

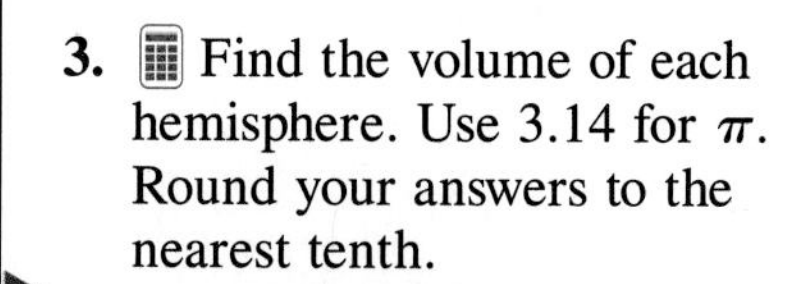

3. Find the volume of each hemisphere. Use 3.14 for π. Round your answers to the nearest tenth.

(a)

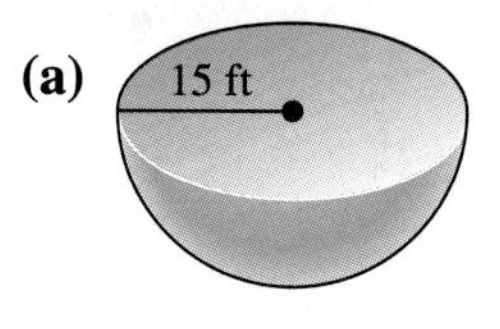

(b)

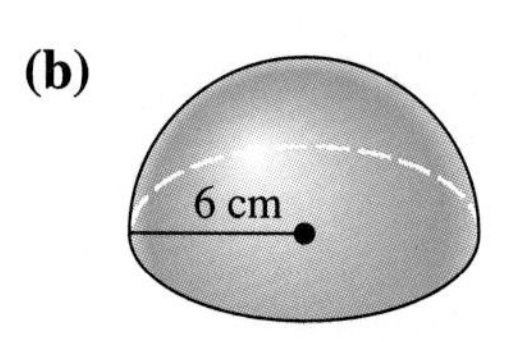

(c) radius 3.7 mm

ANSWERS
2. (a) 7234.6 in.^3 **(b)** 179.5 m^3 **(c)** 82.4 cm^3
3. (a) 7065 ft^3 **(b)** 452.2 cm^3 **(c)** 106.0 mm^3

4. Find the volume of each cylinder. Use 3.14 for π. Round your answers to the nearest tenth. (A calculator is helpful on these problems.)

(a)

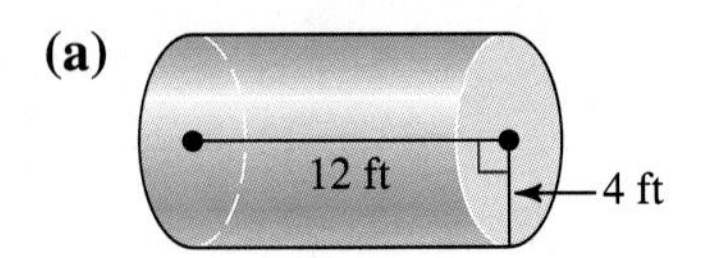

(b)

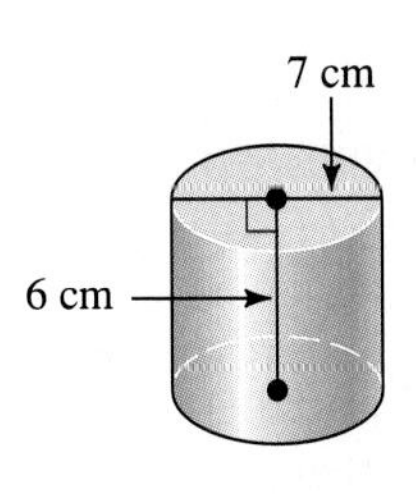

(c) radius 14.5 yd, height 3.2 yd

■ **EXAMPLE 4** *Finding the Volume of a Cylinder*

Find the volume of each cylinder. Use 3.14 as the approximate value of π. Round your answers to the nearest tenth.

(a)

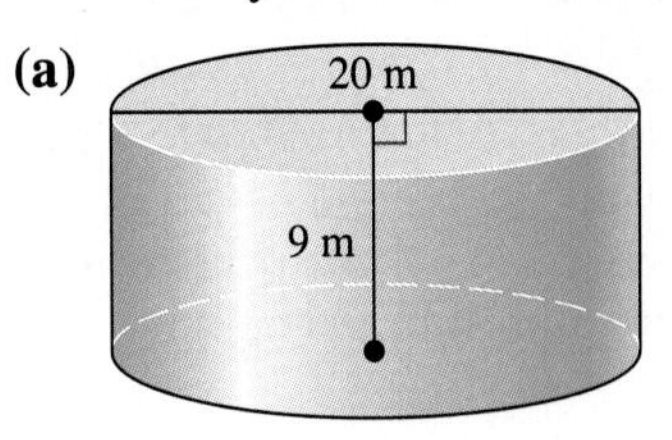

(b)

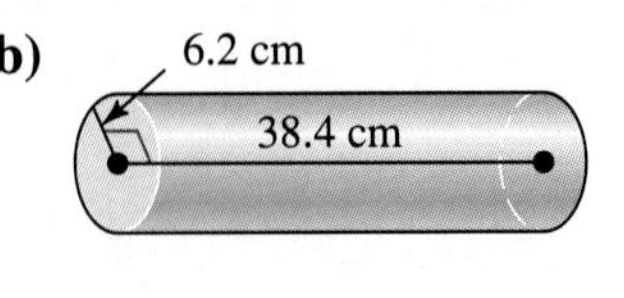

(a) The diameter is 20 m so the radius is $\frac{20 \text{ m}}{2} = 10$ m. The height is 9 m. Use the formula to find the volume.

$$V = \pi \cdot r \cdot r \cdot h$$
$$V = 3.14 \cdot 10 \text{ m} \cdot 10 \text{ m} \cdot 9 \text{ m}$$
$$V = 2826 \text{ m}^3$$

(b) $V = 3.14 \cdot 6.2 \text{ cm} \cdot 6.2 \text{ cm} \cdot 38.4 \text{ cm}$

$V = 4634.9 \text{ cm}^3$ (rounded) ■

◀ WORK PROBLEM 4 AT THE SIDE.

4 A cone and a pyramid are shown here.

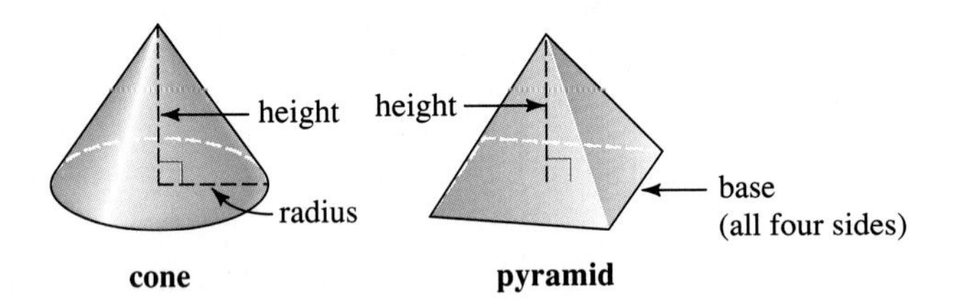

Use the following formula to find the volume of a cone.

FINDING THE VOLUME OF A CONE

$$\text{Volume of cone} = \frac{1}{3} \cdot \pi \cdot r^2 \cdot h$$

$$\text{or} \quad V = \frac{\pi \cdot r^2 \cdot h}{3}$$

■ **EXAMPLE 5** *Finding the Volume of a Cone*

Find the volume of the cone. Use 3.14 for π. Round your answer to the nearest tenth.

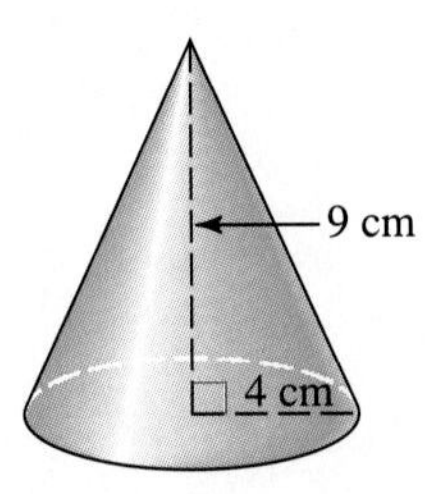

ANSWERS

4. (a) 602.9 ft³ (b) 230.8 cm³ (c) 2112.6 yd³

The radius is 4 cm, and the height is 9 cm.

$$V = \frac{\pi \cdot r \cdot r \cdot h}{3}$$

$$V = \frac{3.14 \cdot 4 \text{ cm} \cdot 4 \text{ cm} \cdot 9 \text{ cm}}{3}$$

$$V = 150.7 \text{ cm}^3 \quad \text{(rounded)}$$ ■

WORK PROBLEM 5 AT THE SIDE. ▶▶

5. Find the volume of a cone with base radius 2 ft and height 11 ft. Use 3.14 for π. Round your answer to the nearest tenth.

Use the following formula to find the volume of a pyramid.

FINDING THE VOLUME OF A PYRAMID

$$\text{Volume of pyramid} = \frac{1}{3} \cdot B \cdot h$$

$$\text{or} \quad V = \frac{B \cdot h}{3}$$

where B is the area of the base of the pyramid and h is the height of the pyramid.

■ **EXAMPLE 6** *Finding the Volume of a Pyramid*

Find the volume of the pyramid. Round your answer to the nearest tenth.

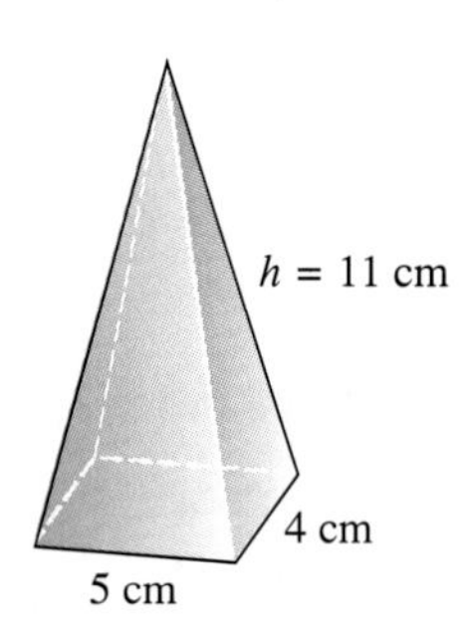

6. Find the volume of a pyramid with base 10 m by 10 m and height 8 m. Round your answer to the nearest tenth.

First find the area of the base.

$$4 \text{ cm} \cdot 5 \text{ cm} = 20 \text{ cm}^2$$

$$B = 20 \text{ cm}^2$$

Next, find the volume.

$$V = \frac{B \cdot h}{3}$$

$$V = \frac{20 \text{ cm}^2 \cdot 11 \text{ cm}}{3}$$

$$V = 73.3 \text{ cm}^3 \quad \text{(rounded)}$$ ■

WORK PROBLEM 6 AT THE SIDE. ▶▶

ANSWERS
5. 46.1 ft^3
6. 266.7 m^3

QUEST FOR NUMERACY

More Vocabulary Roots

In Section 5.1 you read about the prefixes that help you understand number names: *bi* means 2, *tri* means 3, *quad* means 4, and so on. Here are some more Latin and Greek root words and prefixes with their meanings in parentheses. You've seen math terms in this textbook that use these prefixes. List at least one math term and one non-mathematical word that use each prefix or root word. **There are many answers. Some possibilities are listed.**

circum (around): **circumference; circumvent**

de (down): **denominator; deduction**

dia (through): **diameter; diagonal**

equ (equal): **equation; equinox**

fract (break): **fraction; fracture**

hemi (half): **hemisphere; hemitrope**

lateral (side): **quadrilateral; bilateral**

par (beside): **parallel; paramedic**

per (divide): **percent; per capita**

peri (around): **perimeter; periscope**

rad (ray): **radius; radiate**

re (back or again): **reciprocal; reduce**

rect (right): **rectangle; rectify**

sub (below): **subtract; submarine**

How could you use your knowledge of roots and prefixes in these situations?

1. You have trouble remembering which part of a fraction is the denominator.
   The *de* prefix in denominator means down so the denominator is the number down below the fraction bar.
2. You get area and perimeter confused.
   The *peri* prefix in perimeter means around, so perimeter is the distance around the edges of a shape.
3. You can't remember the difference between radius, diameter, and circumference.
   The *ra* prefix tells you that the radius is a ray from the center to the edge of a circle. The *dia* prefix means the diameter goes through the circle, and *circum* means the circumference is the distance around.
4. You get perpendicular and parallel lines confused.
   The *par* prefix means beside, so parallel lines are beside each other.
5. You aren't sure how to change a percent to a decimal.
   ***Per* means divide and *cent* means 100, so divide by 100 to change a percent to a decimal.**

8.7 EXERCISES

NAME DATE HOUR

Find the volume of each figure. Use 3.14 as the approximate value of π. Round your answers to the nearest tenth.

1. 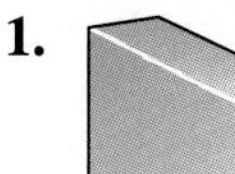
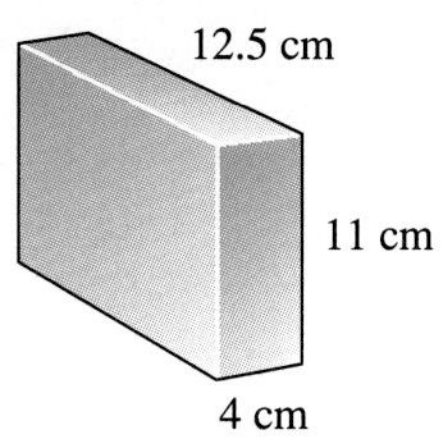

550 cm³

2. 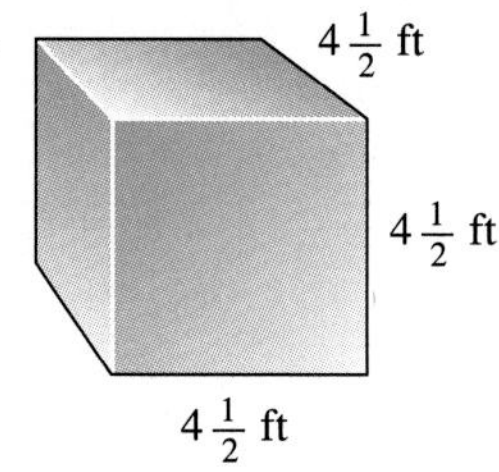

91.1 ft³

3. A cereal box that measures $7\frac{1}{2}$ inches by 2 inches by 10 inches.

150 in.³

4. A shoe box that is 31 cm long, 15 cm wide, and 10.5 cm high.

4882.5 cm³

 **5.** 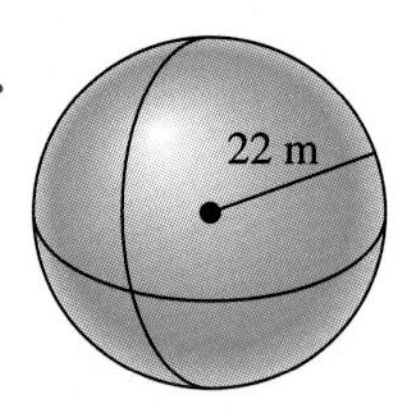

44,579.6 m³

6. 
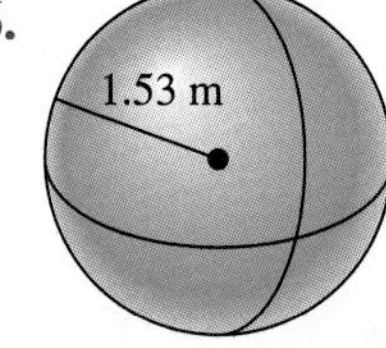

15.0 m³

7. 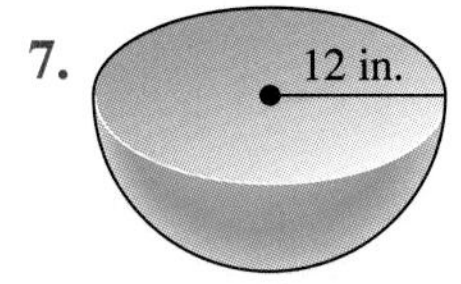

3617.3 in.³

8. 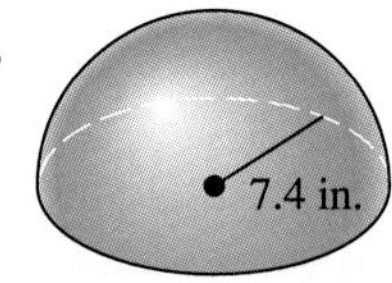

848.3 in.³

9. 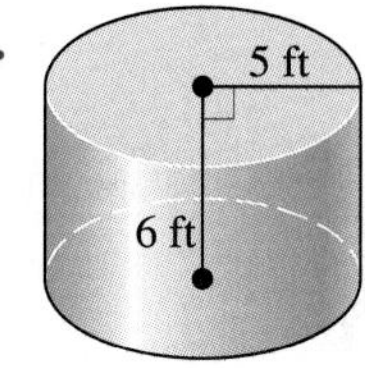

471 ft³

10. 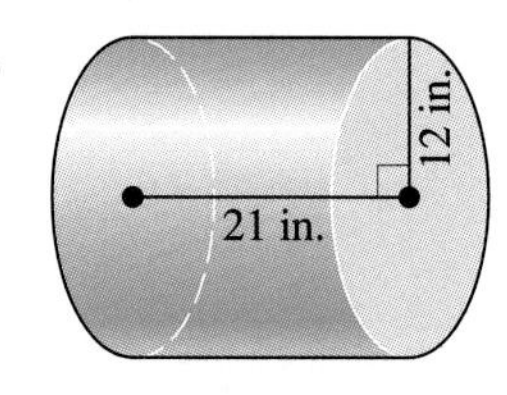

9495.4 in.³

11. A juice can with diameter 6.6 cm and height 12 cm.

410.3 cm³

12. A farm silo with diameter 12 ft and height 40 ft.

4521.6 ft³

13.

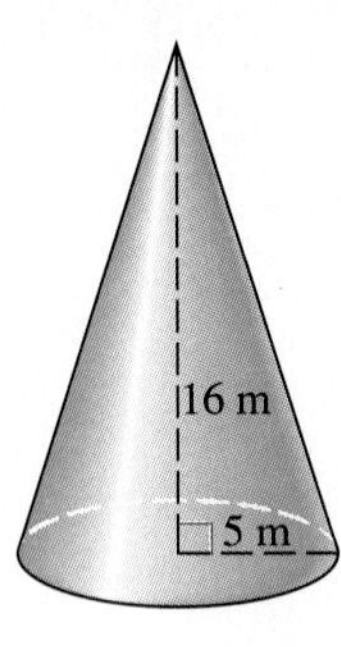

418.7 m^3

14.

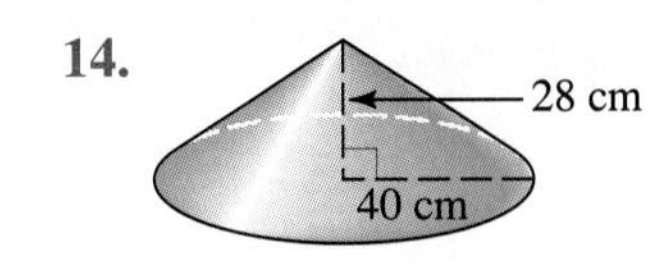

46,890.7 cm^3

15.

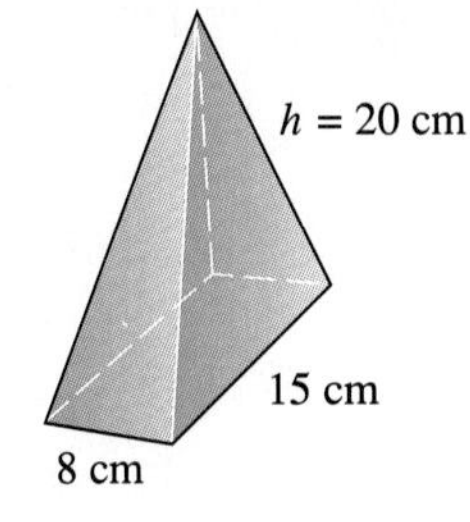

800 cm^3

16.

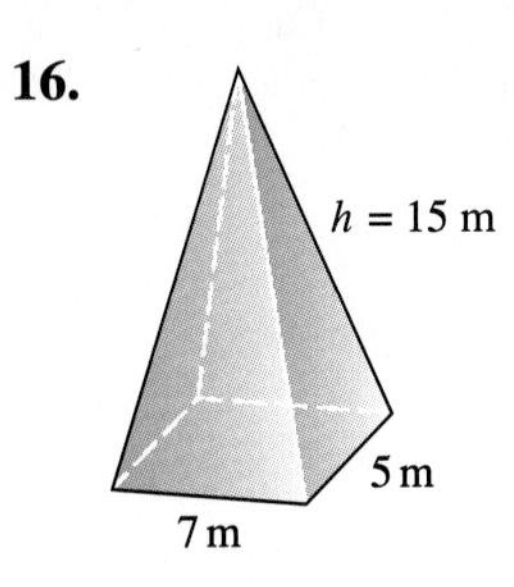

175 m^3

17. A tennis ball has a diameter of 6.8 cm. What is the volume of the ball?

164.6 cm^3

18. An ice cream cone has a diameter of 2 inches and a height of 4 inches. Find its volume.

4.2 $in.^3$

19. Explain the two errors made by this student in finding the volume of a cylinder with a diameter of 7 cm and a height of 5 cm. Find the correct answer.

$V = 3.14 \cdot 7 \cdot 7 \cdot 5$
$V = 769.3 \text{ cm}^2$

Explanations vary.
Correct answer is 192.3 cm^3.

20. Compare the formulas for volume of a cylinder and a cone. How are they similar? Suppose you know the volume of a cylinder. How can you find the volume of a cone with the same radius and height by doing just a one-step calculation?

Answer varies.

21. Find the volume.

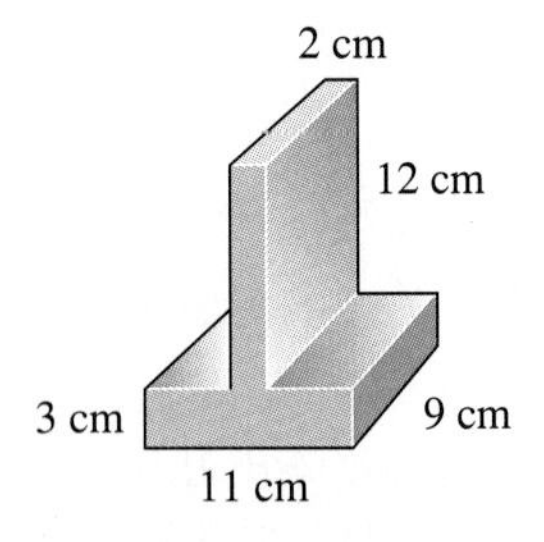

513 cm^3

22. Find the volume of the shaded part. (Hint: Notice the hole in the center of the shape.)

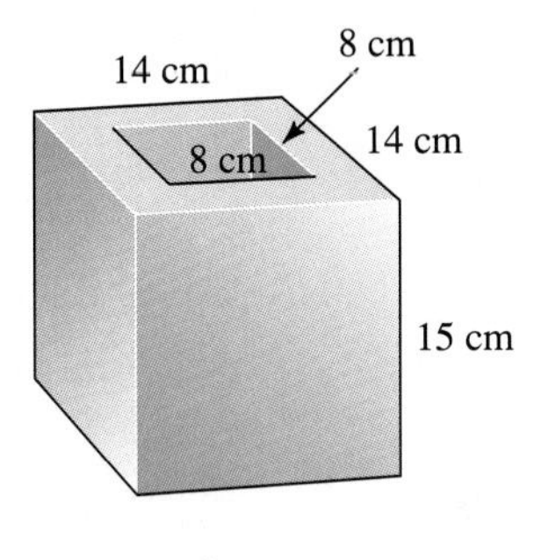

1980 cm^3

PREVIEW EXERCISES

Simplify each expression. (For help, see ***Section 1.8.****)*

23. 8^2 64

24. 14^2 196

25. $\sqrt{16}$ 4

26. $\sqrt{144}$ 12

27. $\sqrt{64}$ 8

28. $\sqrt{4}$ 2

8.8 PYTHAGOREAN THEOREM

Recall the formula for area of a square, $A = s^2$. The square on the left has an area of 25 cm².

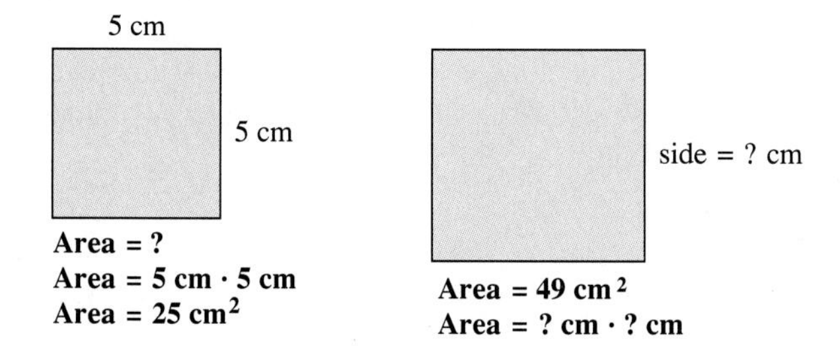

The square on the right has an area of 49 cm². To find the length of a side, ask yourself, "What number can be multiplied by itself to give 49?" Because $7 \cdot 7 = 49$, the length of each side is 7 cm.

Remember: $7 \cdot 7 = 49$, so 7 is the **square root** of 49, or $\sqrt{49} = 7$. Also, $\sqrt{81} = 9$, since $9 \cdot 9 = 81$.

WORK PROBLEM 1 AT THE SIDE.

A number that has a whole number as its square root is called a *perfect square*. For example, 9 is a perfect square because $\sqrt{9} = 3$, and 3 is a whole number.

The first few perfect squares are listed here.

$\sqrt{1} = 1$	$\sqrt{16} = 4$	$\sqrt{49} = 7$	$\sqrt{100} = 10$
$\sqrt{4} = 2$	$\sqrt{25} = 5$	$\sqrt{64} = 8$	$\sqrt{121} = 11$
$\sqrt{9} = 3$	$\sqrt{36} = 6$	$\sqrt{81} = 9$	$\sqrt{144} = 12$

1 If a number is not a perfect square, then you can find its approximate square root by using a calculator with a square root key.

EXAMPLE 1 *Finding the Square Root of a Number*

Use a calculator to find each square root. Round your answers to the nearest thousandth.

(a) $\sqrt{7}$ **(b)** $\sqrt{35}$ **(c)** $\sqrt{124}$ **(d)** $\sqrt{200}$

(a) To use a calculator, press 7 and the $\sqrt{\ }$ key to get 2.6457513 as the square root. Round the answer.

$$\sqrt{7} = 2.646 \quad \text{(rounded to nearest thousandth)}$$

The number 2.646 is an *approximate* square root of 7. Check this by multiplying 2.646 and 2.646. The result is 7.001316 which is very close to 7. Even the original calculator result is approximate. Multiplying 2.6457513 and 2.6457513 gives a result of 6.9999999.

(b) $\sqrt{35} = 5.916$

(c) $\sqrt{124} = 11.136$

(d) $\sqrt{200} = 14.142$ ■

OBJECTIVES

1. Use the square root key on a calculator.
2. Find the missing length in a right triangle.
3. Solve application problems involving right triangles.

FOR EXTRA HELP

1. Find each square root.

(a) $\sqrt{36}$

(b) $\sqrt{25}$

(c) $\sqrt{9}$

(d) $\sqrt{100}$

(e) $\sqrt{121}$

ANSWERS
1. (a) 6 (b) 5 (c) 3 (d) 10 (e) 11

2. Use a calculator with a square root key to find each square root. Round to the nearest thousandth if necessary.

(a) $\sqrt{11}$

(b) $\sqrt{40}$

(c) $\sqrt{56}$

(d) $\sqrt{196}$

(e) $\sqrt{147}$

ANSWERS
2. (a) 3.317 (b) 6.325 (c) 7.483 (d) 14 (e) 12.124

◀ WORK PROBLEM 2 AT THE SIDE.

2 One place you will use square roots is when working with the *Pythagorean Theorem.* This theorem applies only to *right* triangles (triangles with a 90° angle). The longest side of a right triangle is called the **hypotenuse.** It is opposite the right angle. The other two sides are called *legs.* The legs form the right angle.

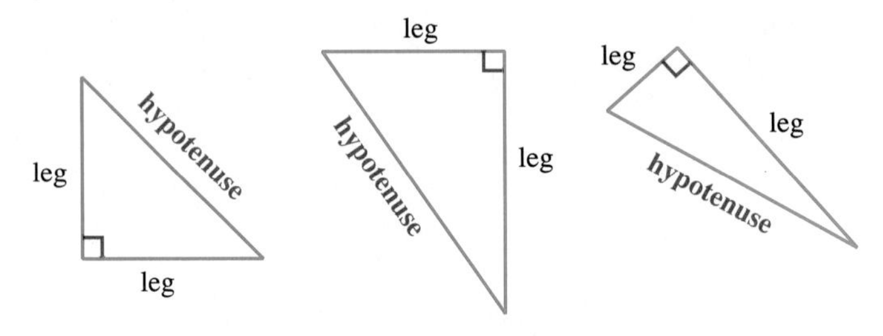

Examples of right triangles

PYTHAGOREAN THEOREM

$$(\text{hypotenuse})^2 = (\text{leg})^2 + (\text{leg})^2$$

In other words, square the length of each side. After you have squared all the sides, the sum of the squares of the two legs will equal the square of the hypotenuse.

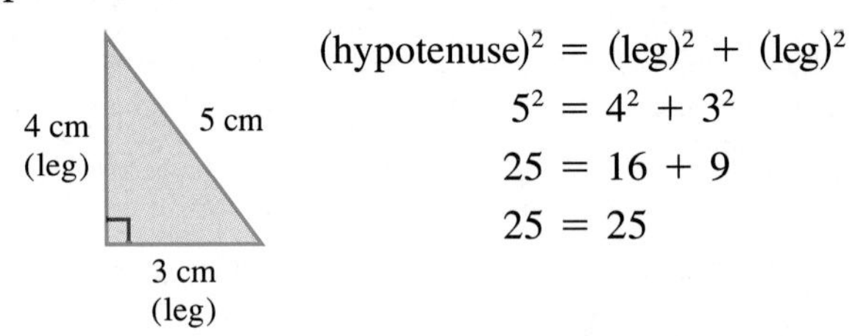

$$(\text{hypotenuse})^2 = (\text{leg})^2 + (\text{leg})^2$$
$$5^2 = 4^2 + 3^2$$
$$25 = 16 + 9$$
$$25 = 25$$

The theorem is named after Pythagoras, a Greek mathematician who lived about 2500 years ago. He and his followers may have used floor tiles to prove the theorem, as shown here.

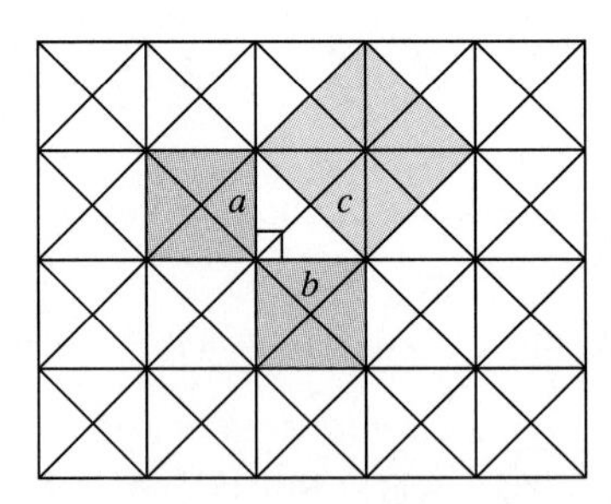

The right triangle in the center of the drawing has sides a, b, and c. The square drawn on side a contains four triangles. The square on side b contains four triangles. The square on side c contains eight triangles. The number of triangles in the square on side c equals the sum of the number of triangles in the squares on sides a and b ($8 = 4 + 4$).

If you know the lengths of any two sides in a right triangle, you can use the Pythagorean Theorem to find the third side.

USING THE PYTHAGOREAN THEOREM

To find the hypotenuse, use this formula:

$$\textbf{hypotenuse} = \sqrt{(\textbf{leg})^2 + (\textbf{leg})^2}$$

To find a leg, use this formula:

$$\textbf{leg} = \sqrt{(\textbf{hypotenuse})^2 - (\textbf{leg})^2}$$

Note *Remember:* A small square drawn in one angle of a triangle indicates a right angle. You can use the Pythagorean Theorem *only* on triangles that have a right angle.

EXAMPLE 2 *Finding the Missing Length in a Right Triangle*
Find the missing length in each right triangle.

(a)

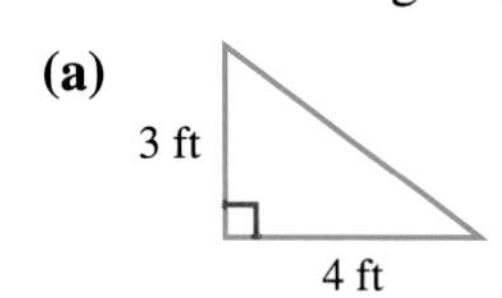

(b)

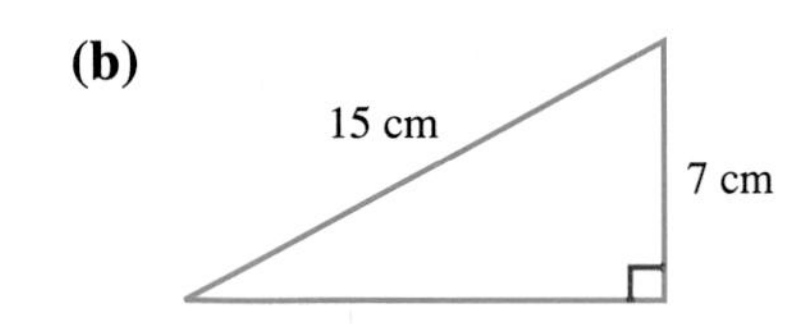

(a) The side opposite the right angle is missing. That side is the hypotenuse, so use this formula.

$$\text{hypotenuse} = \sqrt{(\text{leg})^2 + (\text{leg})^2}$$

$$\text{hypotenuse} = \sqrt{(3)^2 + (4)^2} \quad \text{Legs are 3 and 4.}$$

$$= \sqrt{9 + 16} \quad 3 \cdot 3 \text{ is } 9 \text{ and } 4 \cdot 4 \text{ is } 16$$

$$= \sqrt{25} = 5$$

The hypotenuse is 5 ft long.

Note You use the Pythagorean Theorem to find the *length* of one side, *not* the area of the triangle. Your answer will be in linear units, such as ft, yd, cm, m, and so on (*not* ft^2, cm^2, m^2).

(b) You *do* know the length of the hypotenuse (15 cm), so it is one of the legs that is unknown. Use this formula.

$$\text{leg} = \sqrt{(\text{hypotenuse})^2 - (\text{leg})^2} \quad \text{Find a leg.}$$

$$\text{leg} = \sqrt{(15)^2 - (7)^2} \quad \text{Hypotenuse is 15, one leg is 7.}$$

$$= \sqrt{225 - 49} \quad 15 \cdot 15 \text{ is } 225 \text{ and } 7 \cdot 7 \text{ is } 49$$

$$= \sqrt{176}$$

Use the $\sqrt{\ }$ key on your calculator to find that $\sqrt{176} = 13.266$ (rounded to nearest thousandth). The length of the leg is approximately 13.266 cm. ■

WORK PROBLEM 3 AT THE SIDE.

3. Find the missing length in each right triangle. Round your answers to the nearest thousandth, if necessary.

(a) 5 in. 12 in.

(b) 7 cm 25 cm 90°

(c) 13 m 17 m

(d) 18 ft 20 ft

(e) 8 mm 5 mm

ANSWERS
3. (a) 13 in. **(b)** 24 cm **(c)** 21.401 m **(d)** 8.718 ft **(e)** 9.434 mm

4. These problems show ladders leaning against buildings. Find the missing lengths. Round to the nearest thousandth of a foot.

(a)

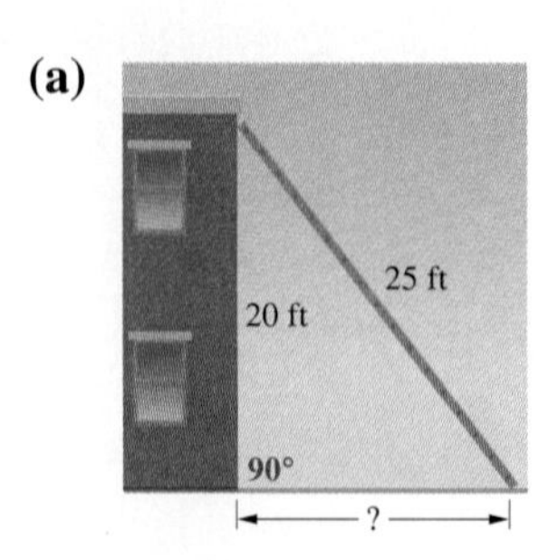

How far away from the building is the bottom of the ladder?

(b)

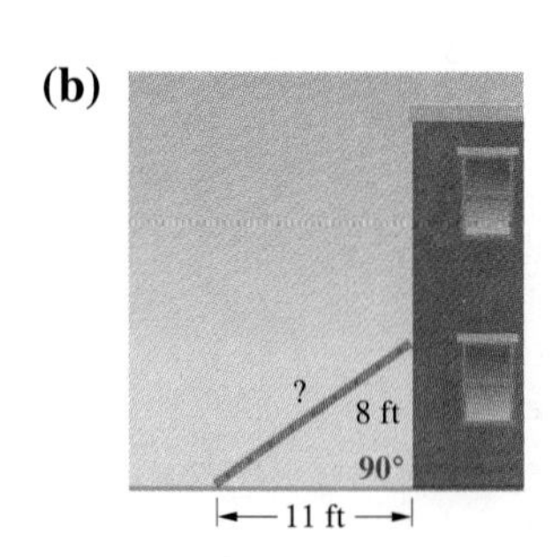

How long is the ladder?

(c) A 17-foot ladder is leaning against a building. The bottom of the ladder is 10 ft from the building. How high up on the building will the ladder reach? Hint: Start by drawing the building and the ladder.

ANSWERS
4. (a) 15 ft (b) $\sqrt{185} = 13.601$ ft
(c) $\sqrt{189} = 13.748$ ft

3 The next example shows an application of the Pythagorean theorem.

EXAMPLE 3 *Using the Pythagorean Theorem*

A television antenna is on the roof of a house, as shown. Find the length of the support wire.

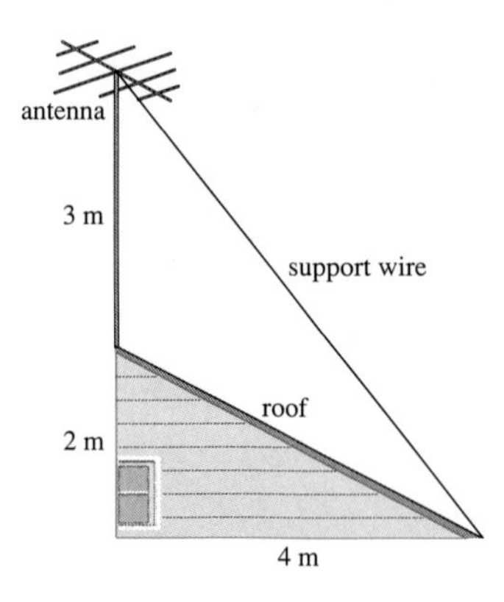

The total length of the side at the left is 3 m + 2 m = 5 m.

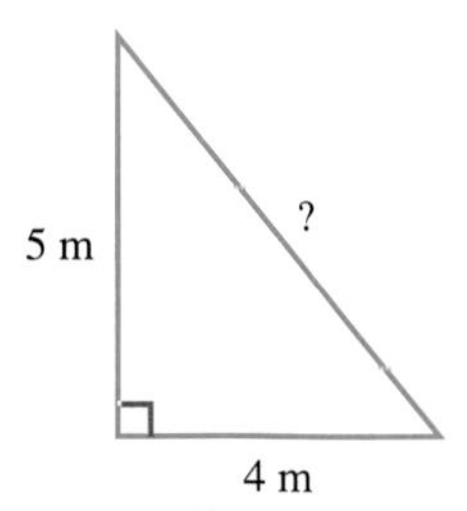

Find the length of the support wire with the formula for the hypotenuse.

$$\text{hypotenuse} = \sqrt{(\text{leg})^2 + (\text{leg})^2}$$

$$\text{hypotenuse} = \sqrt{(5)^2 + (4)^2} \quad \text{Legs are 4 and 5.}$$

$$= \sqrt{25 + 16} \quad 5^2 \text{ is 25 and } 4^2 \text{ is 16.}$$

$$= \sqrt{41} = 6.403 \quad \text{Use } \sqrt{\ } \text{ key on a calculator.}$$

The support wire has a length of about 6.403 m. ■

WORK PROBLEM 4 AT THE SIDE.

8.8 EXERCISES

NAME DATE HOUR

Find each square root. Starting with Exercise 5, use the $\sqrt{\ }$ *key on a calculator. Round your answers to the nearest thousandth, when necessary.*

1. $\sqrt{49}$ 7

2. $\sqrt{144}$ 12

3. $\sqrt{64}$ 8

4. $\sqrt{81}$ 9

5. $\sqrt{11}$ 3.317

6. $\sqrt{23}$ 4.796

7. $\sqrt{8}$ 2.828

8. $\sqrt{6}$ 2.449

9. $\sqrt{73}$ 8.544

10. $\sqrt{80}$ 8.944

11. $\sqrt{101}$ 10.050

12. $\sqrt{125}$ 11.180

13. $\sqrt{190}$ 13.784

14. $\sqrt{160}$ 12.649

15. $\sqrt{175}$ 13.229

16. $\sqrt{200}$ 14.142

Find the areas of the squares on the sides of the right triangles in Exercises 17 and 18. Check to see if the Pythagorean Theorem holds true.

17.

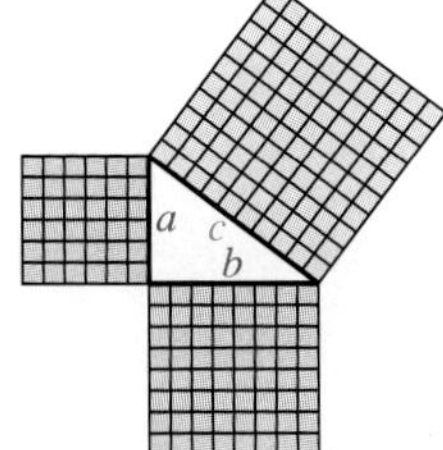

$a^2 = 36$
$b^2 = 64$
$c^2 = 100$
side $c = 10$; true

18.

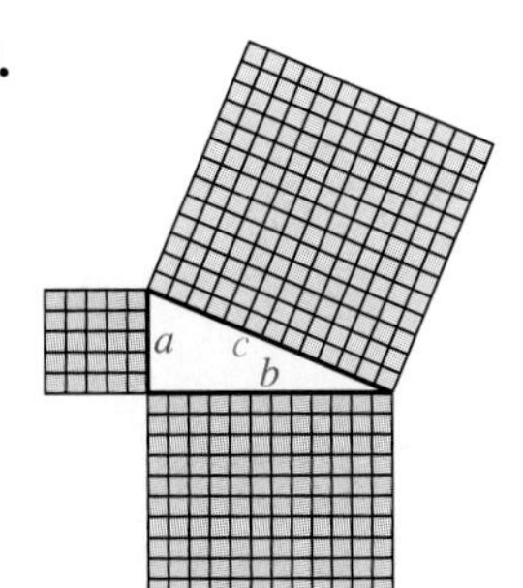

$a^2 = 25$
$b^2 = 144$
$c^2 = 169$
side $c = 13$; true

Find the missing length in each right triangle. Use a calculator to find square roots. Round your answers to the nearest thousandth.

19.

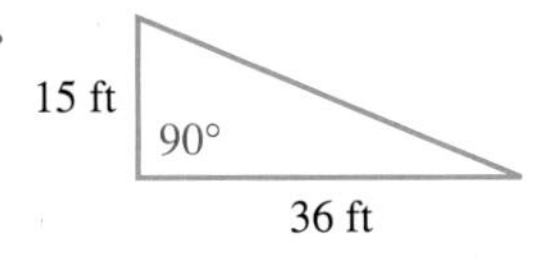

39 ft

20.

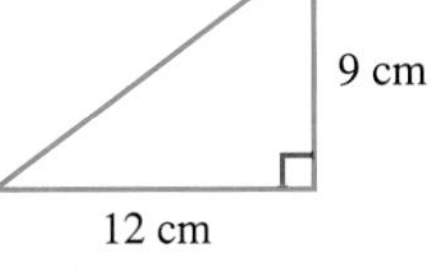

15 cm

21.

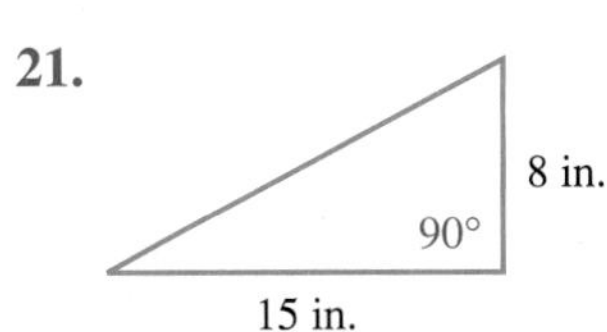

17 in.

22.

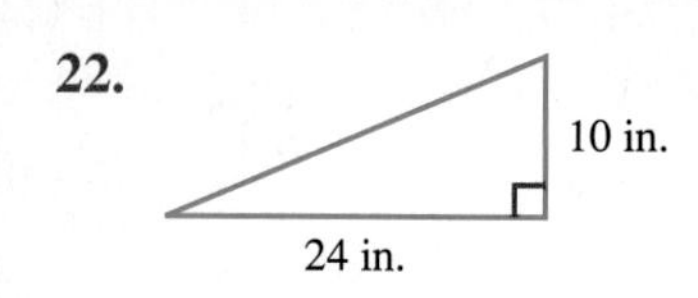

26 in.

23.

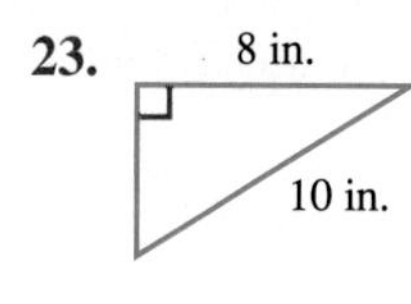

6 in.

24.

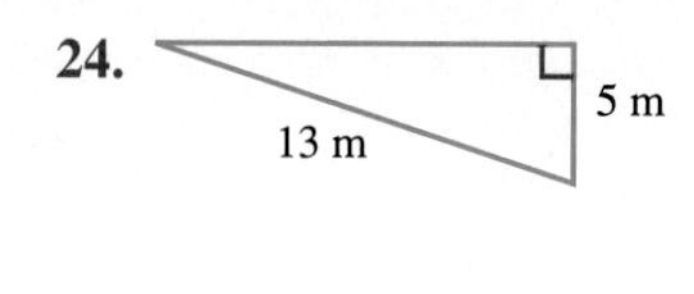

12 m

25.

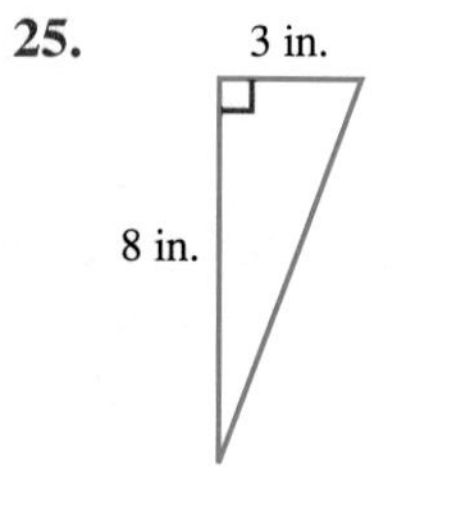

$\sqrt{73} = 8.544$ in.

26.

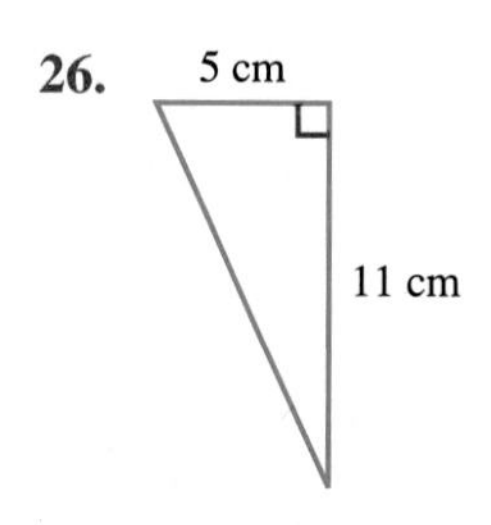

$\sqrt{146} = 12.083$ cm

27.

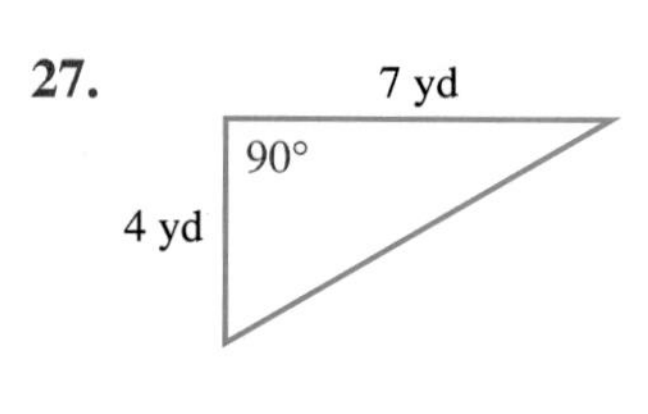

$\sqrt{65} = 8.062$ yd

28.

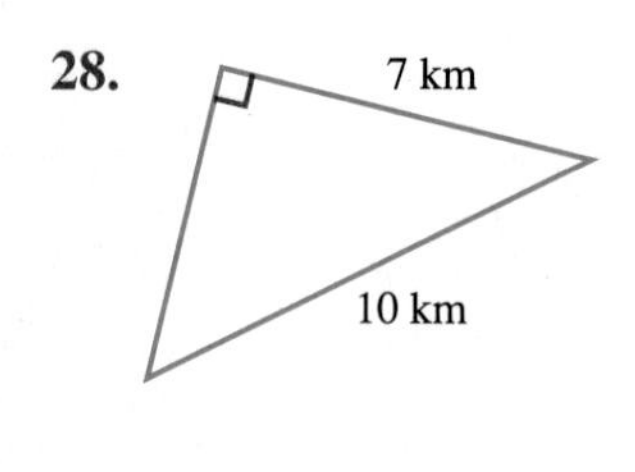

$\sqrt{51} = 7.141$ km

29.

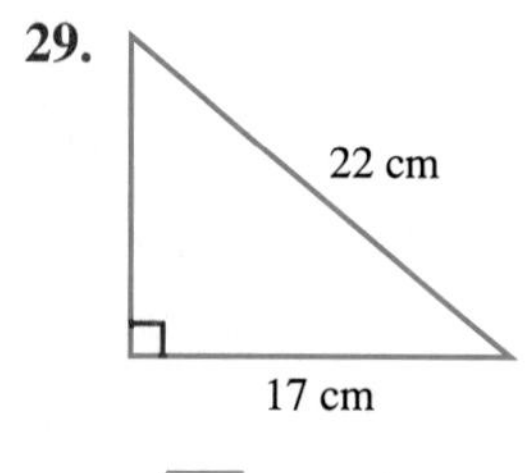

$\sqrt{195} = 13.964$ cm

30.

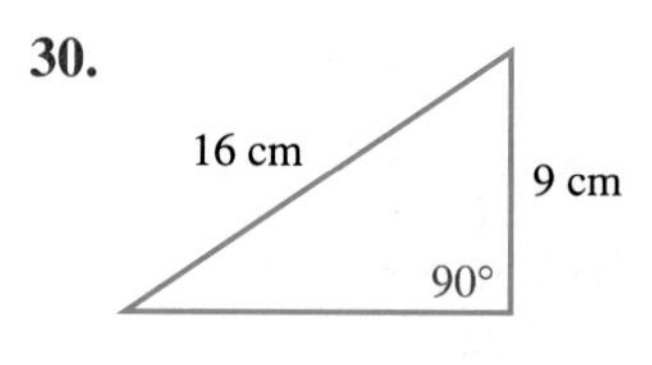

$\sqrt{175} = 13.229$ cm

31.

1.3 m

90°

2.5 m

$\sqrt{7.94} = 2.818$ m

32.

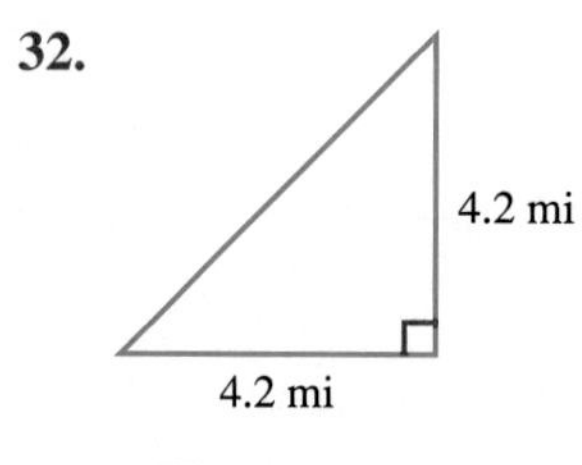

$\sqrt{35.28} = 5.940$ mi

33.

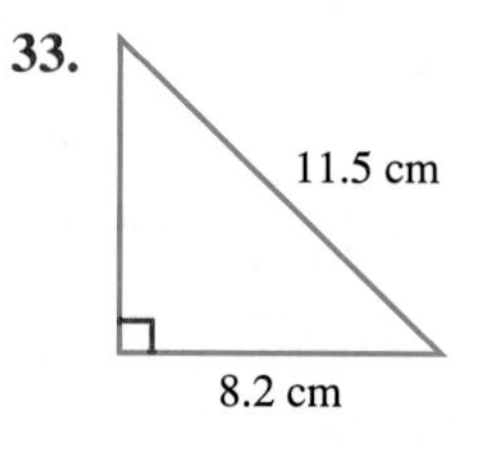

$\sqrt{65.01} = 8.063$ cm

34.

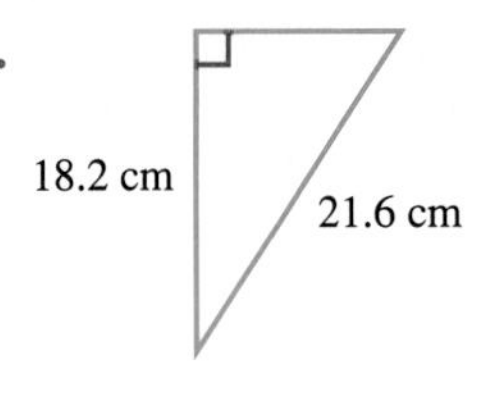

$\sqrt{135.32} = 11.633$ cm

35.

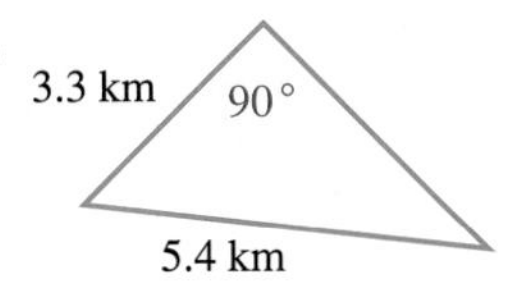

$\sqrt{18.27} = 4.274$ km

36.

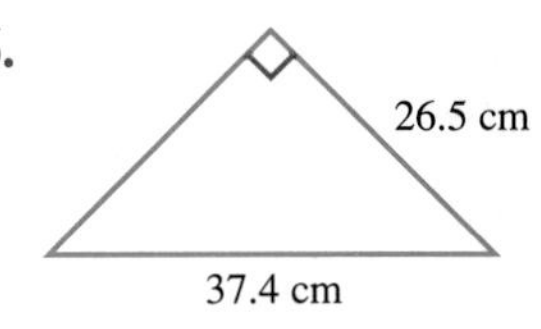

$\sqrt{696.51} = 26.391$ cm

Solve each word problem. Round your answers to the nearest tenth.

37. Find the length of this loading dock.

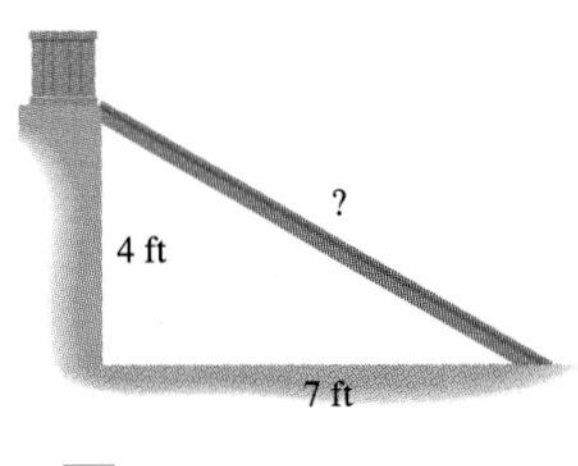

$\sqrt{65} = 8.1$ ft

38. Find the marked length in this roof plan.

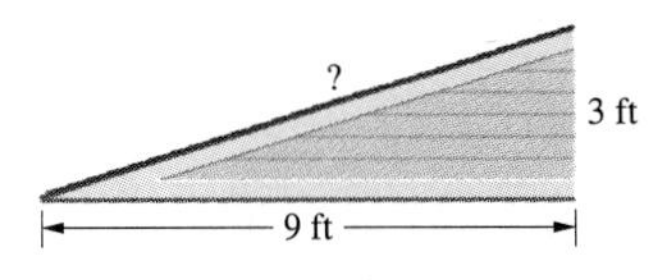

$\sqrt{90} = 9.5$ ft

39. How high is the kite above the ground?

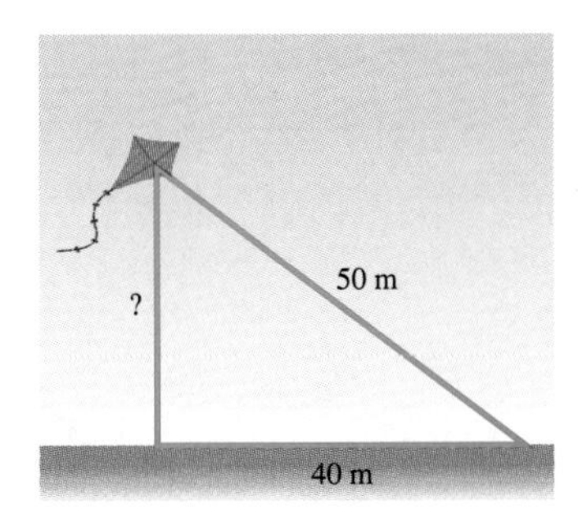

$\sqrt{900} = 30$ m

40. Find the height of this telephone pole.

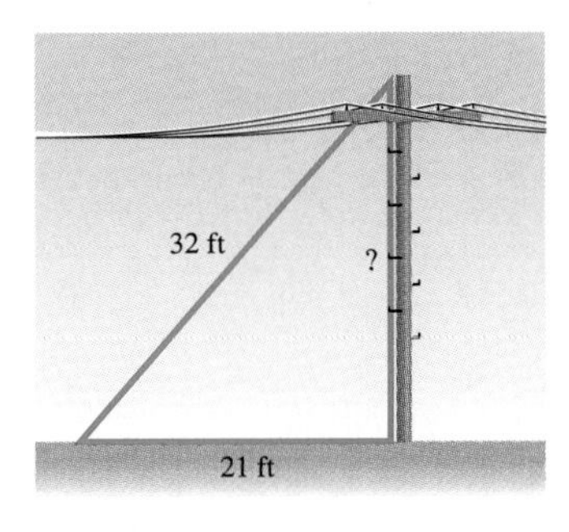

$\sqrt{583} = 24.1$ ft

41. A 12-foot ladder is leaning against a building, with the base of the ladder 3 feet from the building. How high on the building will the top of the ladder reach? Draw a sketch of the building and ladder and solve the problem.

Sketches vary.
$\sqrt{135} = 11.6$ ft

42. William drove his car 15 miles north, then made a right turn and drove 7 miles east. How far is he, in a straight line, from his starting point? Draw a sketch to illustrate the problem and solve it.

Sketches vary.
$\sqrt{274} = 16.6$ miles

43. You know that $\sqrt{25} = 5$ and $\sqrt{36} = 6$. Using just that information (no calculator), describe how you could estimate $\sqrt{30}$. How would you estimate $\sqrt{26}$ or $\sqrt{35}$? Now check your estimates using a calculator.

Answer varies.

44. Describe the two errors made by a student in solving this problem. Also find the correct answer.

$\sqrt{(13)^2 + (20)^2}$

$\sqrt{169 + 400}$

$\sqrt{569} = 23.9 \text{ m}^2$

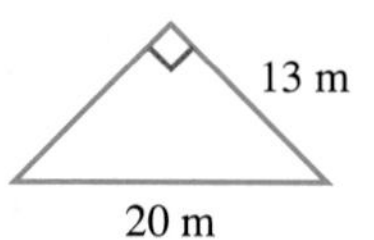

Descriptions vary.
Correct answer is 15.2 m.

45. Find lengths BC and BD.

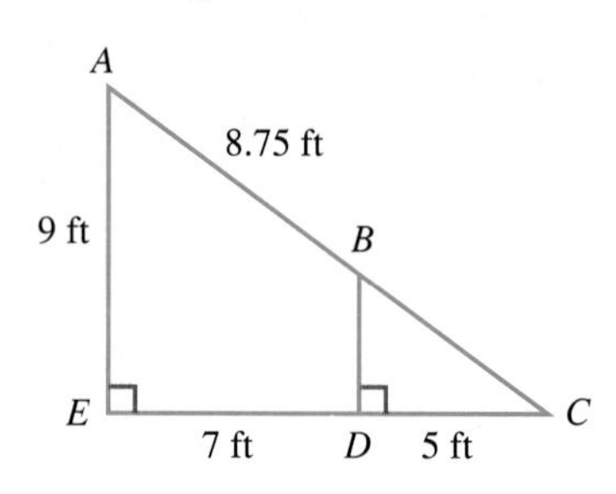

$BC = 6.25$ ft
$BD = 3.75$ ft

46. Find the length of CD and DB. Round your answers to the nearest tenth.

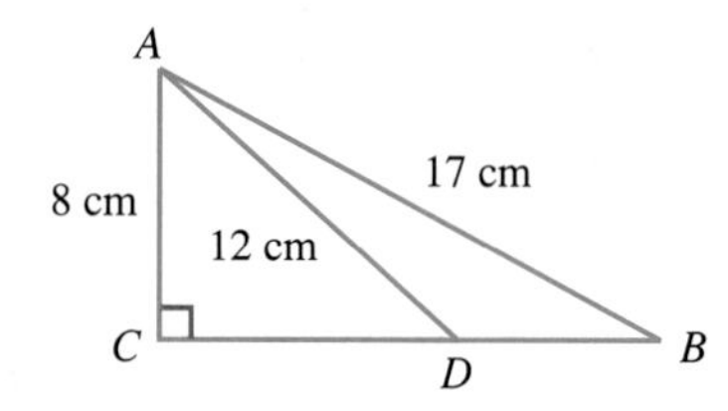

$CD = 8.9$ cm
$DB = 6.1$ cm

PREVIEW EXERCISES

Find the missing number in each of the following proportions. (For help, see **Section 5.4.**)

47. $\frac{2}{9} = \frac{x}{36}$ 8

48. $\frac{7}{x} = \frac{21}{24}$ 8

49. $\frac{x}{9.2} = \frac{15.6}{7.8}$ 18.4

50. $\frac{0.8}{5} = \frac{12.4}{x}$ 77.5

8.9 SIMILAR TRIANGLES

OBJECTIVES

1. Identify corresponding parts in similar triangles.
2. Find the lengths of unknown sides in similar triangles.
3. Solve problems with similar triangles.

FOR EXTRA HELP

Tape 12

SSM pp. 248–250

MAC: A
IBM: A

Two triangles with the same shape (but not necessarily the same size) are called **similar triangles.** Three pairs of similar triangles are shown here.

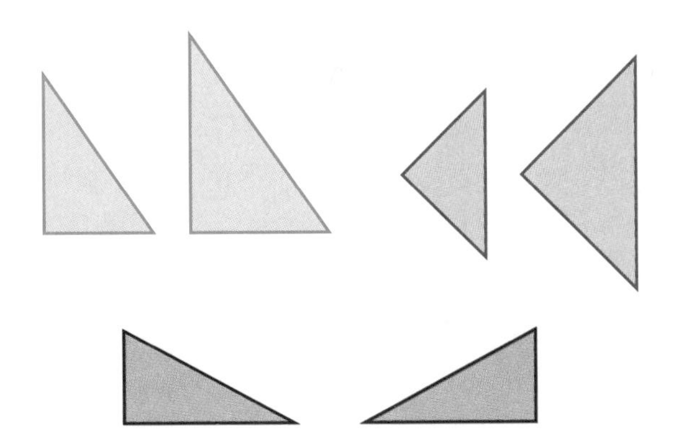

1 The two triangles shown below are different sizes but have the same shape, so they are similar triangles. Angles A and P measure the same number of degrees and are called *corresponding angles.* Angles B and Q are corresponding angles, as are angles C and R.

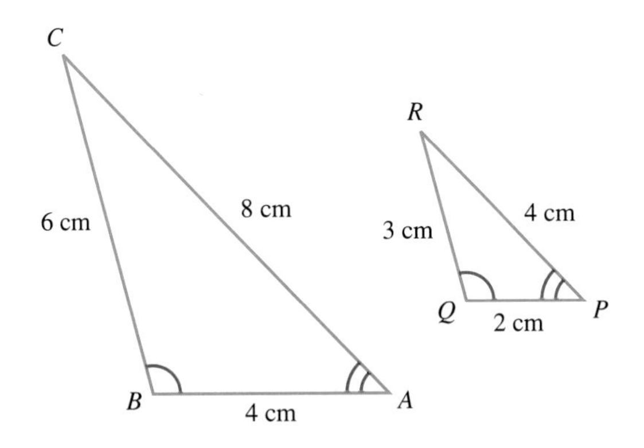

Sides PR and AC are called *corresponding sides,* since they are *opposite* corresponding angles. Also, QR and BC are corresponding sides, as are PQ and AB. Although corresponding angles measure the same number of degrees, corresponding sides do *not* need to be the same in length. In the triangles here, each side in the smaller triangle is half the length of the corresponding side in the larger triangle.

WORK PROBLEM 1 AT THE SIDE.

2 Similar triangles are useful because of the following property.

SIMILAR TRIANGLES

In similar triangles, the ratios of the lengths of corresponding sides are equal.

1. Identify corresponding angles and sides in these similar triangles.

(a)

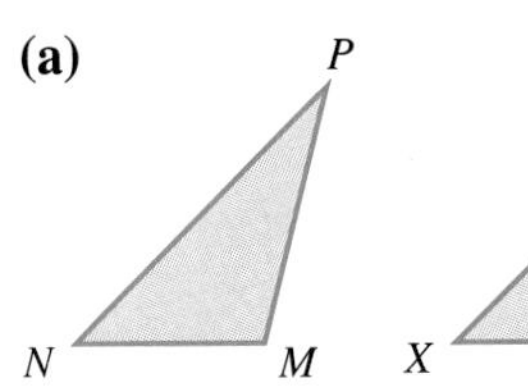

Angles:
P and ____
N and ____
M and ____
Sides:
PN and ____
PM and ____
NM and ____

(b)

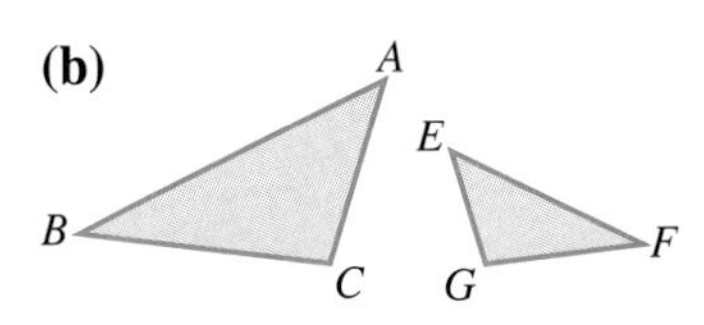

Angles:
A and ____
B and ____
C and ____
Sides:
AB and ____
BC and ____
AC and ____

ANSWERS
1. (a) Z; X; Y; ZX; ZY; XY
(b) E; F; G; EF; GF; EG

2. Use the same triangles as in Example 1, but find EF.

■ **EXAMPLE 1** *Finding the Lengths of Unknown Sides in Similar Triangles*

Find the length of y in the smaller triangle. Assume the triangles are similar.

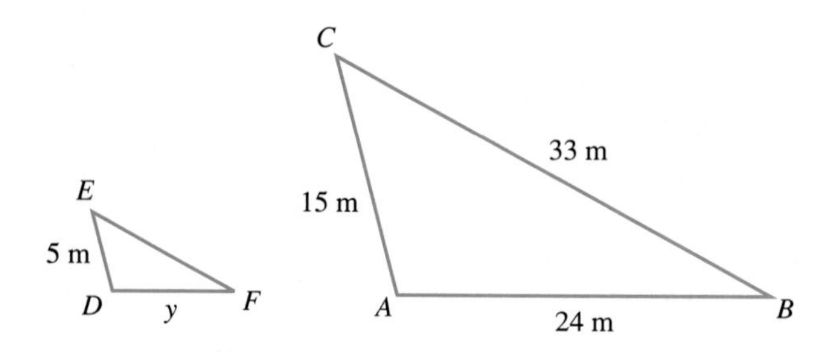

Sides ED and CA are corresponding sides. The ratio of the lengths of these sides can be written as a fraction in lowest terms.

$$\frac{5\ \cancel{m}}{15\ \cancel{m}} = \frac{1}{3} \qquad \text{(lowest terms)}$$

As mentioned above, the ratios of the lengths of corresponding sides are equal. Side DF in the smaller triangle corresponds to side AB in the larger triangle. Since the ratios of corresponding sides are equal,

$$\frac{DF}{AB} = \frac{1}{3}$$

Replace DF with y and AB with 24 to get the proportion

$$\frac{y}{24} = \frac{1}{3}.$$

Find cross products.

$$24 \cdot 1 = 24$$

$$\frac{y}{24} = \frac{1}{3}$$

$$y \cdot 3$$

Show that cross products are equal.

$$y \cdot 3 = 24$$

Divide both sides by 3.

$$\frac{y \cdot \overset{1}{\cancel{3}}}{\underset{1}{\cancel{3}}} = \frac{24}{3}$$

$$y = 8$$

Side DF has a length of 8 m. ■

WORK PROBLEM 2 AT THE SIDE.

ANSWER
2. 11 m

■ EXAMPLE 2 *Using a Ratio to Find an Unknown Side*
Find x in the smaller triangle. Assume the triangles are similar.

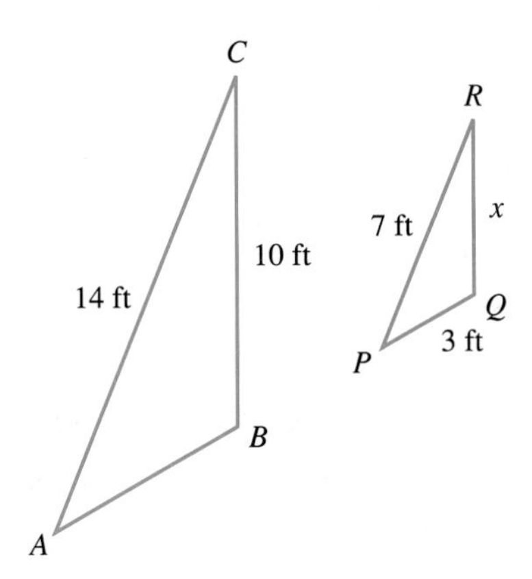

Sides PR and AC are corresponding sides. The ratio of their lengths can be written as a fraction in lowest terms.

$$\frac{7\ \cancel{ft}}{14\ \cancel{ft}} = \frac{1}{2} \quad \text{(lowest terms)}$$

The two triangles are similar, so the ratio of any pair of corresponding sides will also equal $\frac{1}{2}$. Because sides RQ and CB are corresponding sides,

$$\frac{RQ}{CB} = \frac{1}{2}.$$

Replace RQ with x and CB with 10 to make a proportion.

$$\frac{x}{10} = \frac{1}{2}$$

Find cross products.

$$10 \cdot 1 = 10$$

$$\frac{x}{10} = \frac{1}{2}$$

$$x \cdot 2$$

Show that cross products are equal.

$$x \cdot 2 = 10$$

Divide both sides by 2.

$$\frac{x \cdot \overset{1}{\cancel{2}}}{\underset{1}{\cancel{2}}} = \frac{10}{2}$$

$$x = 5$$

Side RQ has a length of 5 ft. ■

WORK PROBLEM 3 AT THE SIDE. ▶

3. (a) Find the length of side AB in Example 2.

(b) Find x and y if the triangles are similar.

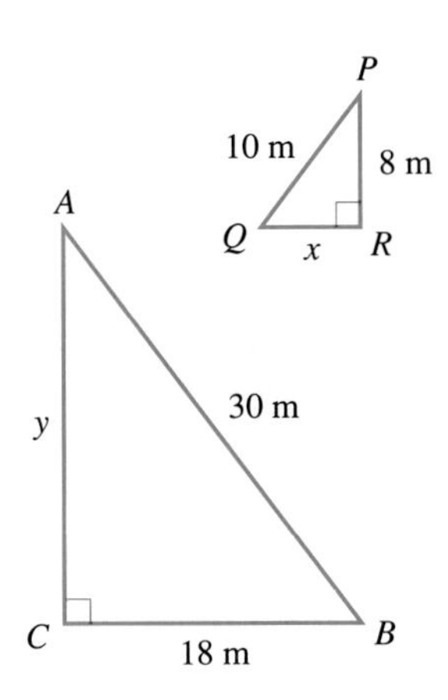

ANSWERS
3. (a) 6 ft (b) $x = 6$ m, $y = 24$ m

4. Find the height of each flagpole.

(a)

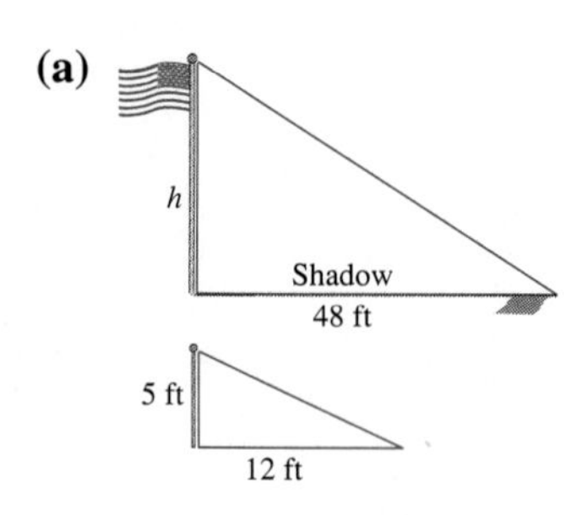

(b)

12.5 m

7.2 m

5m

ANSWER

4. (a) 20 ft (b) 18 m

3 The next example shows an application of similar triangles.

■ EXAMPLE 3 *Using Similar Triangles in Applications*

A flagpole casts a shadow 99 m long at the same time that a pole 10 m tall casts a shadow 18 m long. Find the height of the flagpole.

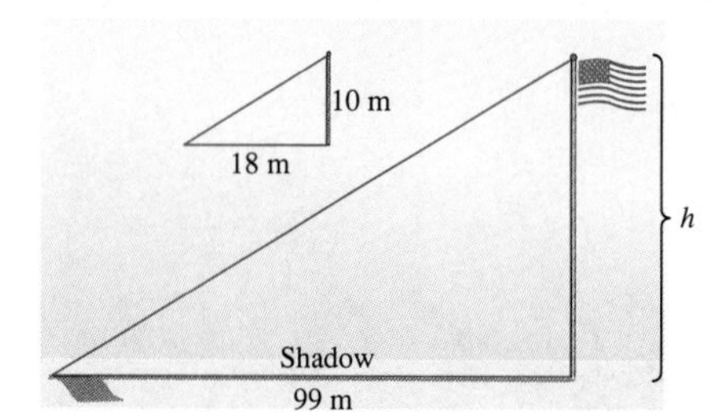

The triangles shown are similar, so write a proportion to find h.

$$\frac{h}{10} = \frac{99}{18}$$

Find cross products and place them equal to each other.

$$h \cdot 18 = 10 \cdot 99$$
$$h \cdot 18 = 990$$

Divide both sides by 18.

$$\frac{h \cdot \overset{1}{\cancel{18}}}{\underset{1}{\cancel{18}}} = \frac{990}{18}$$
$$h = 55$$

The flagpole is 55 m high. ■

◀ **WORK PROBLEM 4 AT THE SIDE.**

Write similar *or* not similar *for each pair of triangles.*

1. 

similar

2. 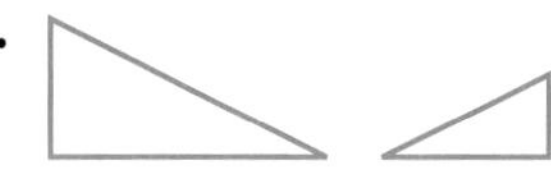

similar

3. 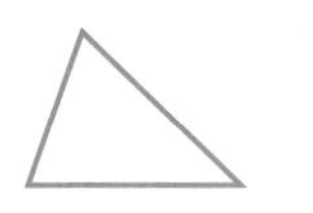

not similar

4. 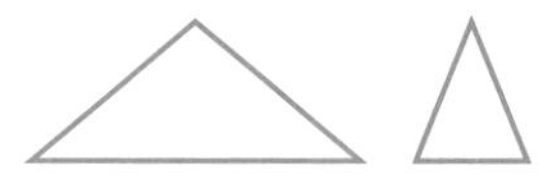

not similar

5. 

similar

6. 

not similar

Name the corresponding angles and the corresponding sides in each pair of similar triangles.

7. 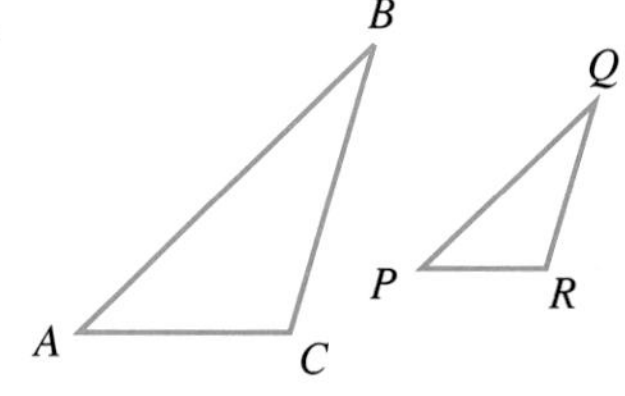

B and *Q*
C and *R*
A and *P*
AB and *PQ*
BC and *QR*
AC and *PR*

8. 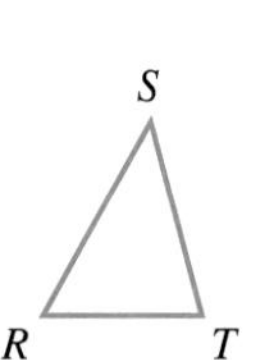

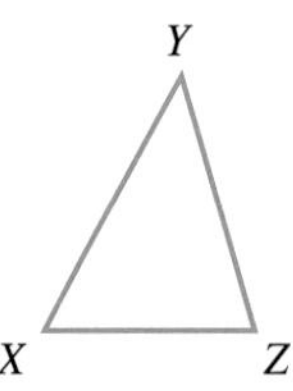

S and *Y*
R and *X*
T and *Z*
SR and *YX*
ST and *YZ*
RT and *XZ*

9. 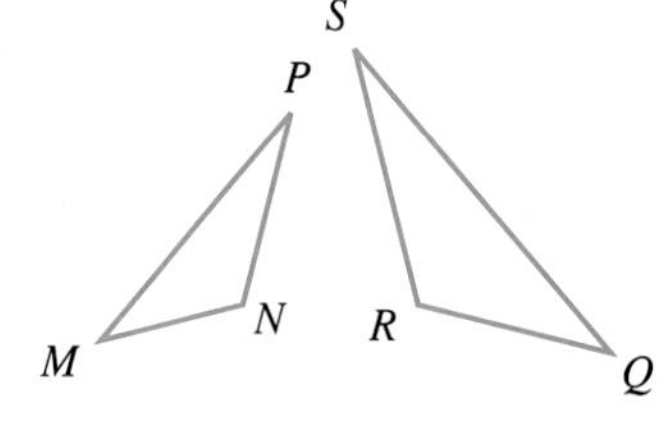

P and *S*
N and *R*
M and *Q*
MP and *QS*
MN and *QR*
NP and *RS*

10. 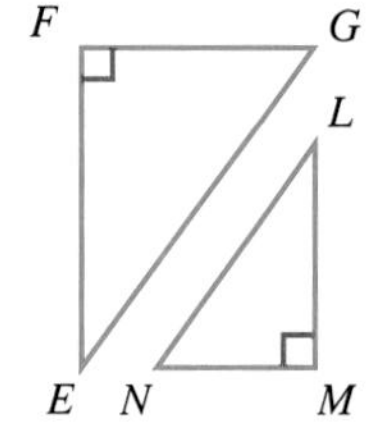

G and *N*
F and *M*
E and *L*
FG and *NM*
FE and *LM*
EG and *LN*

Find all the ratios for the triangles shown below. Write the ratios as fractions in lowest terms.

11. $\frac{AB}{PQ}; \frac{AC}{PR}; \frac{BC}{QR}$

$\frac{3}{2}; \frac{3}{2}; \frac{3}{2}$

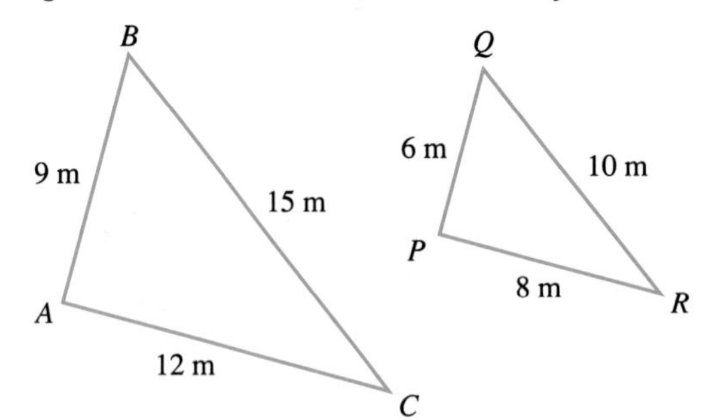

12. $\frac{AB}{PQ}; \frac{AC}{PR}; \frac{BC}{QR}$

$\frac{2}{3}; \frac{2}{3}; \frac{2}{3}$

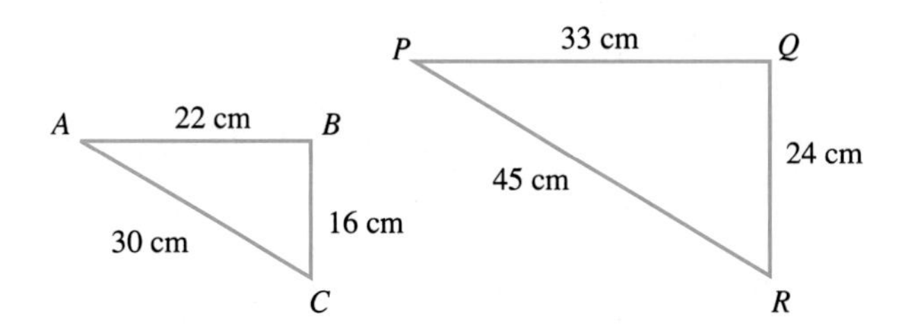

Find the unknown lengths in each pair of similar triangles.

13.

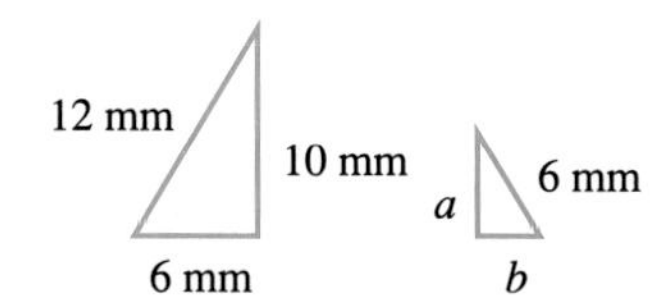

$a = 5$ mm
$b = 3$ mm

14.

75 m
b
a
10 m
20 m
25 m

$a = 30$ m
$b = 60$ m

15.

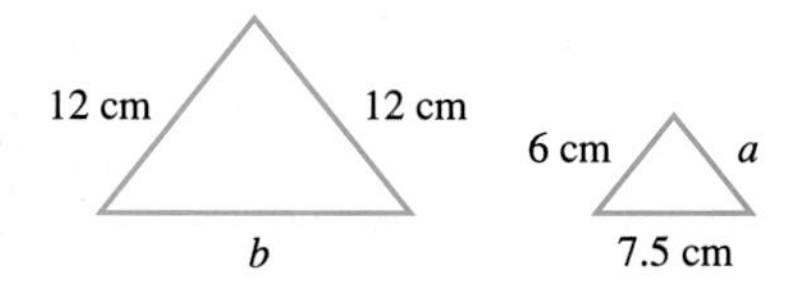

$a = 6$ cm
$b = 15$ cm

16.

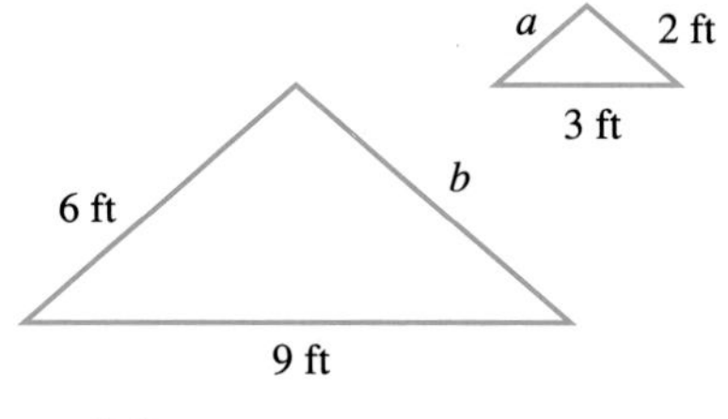

$a = 2$ ft
$b = 6$ ft

17.

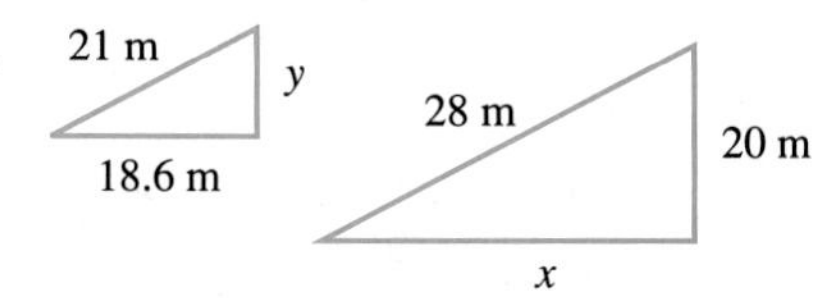

$x = 24.8$ m
$y = 15$ m

18.

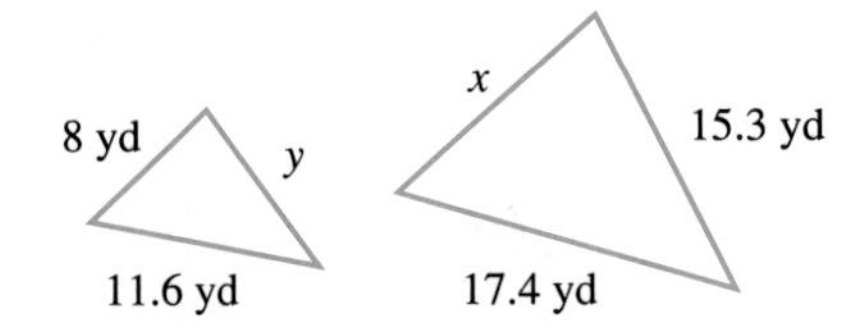

$x = 12$ yd
$y = 10.2$ yd

Solve the following problems.

19. The height of the house shown here can be found by using similar triangles and a proportion. Find the height of the house by writing a proportion and solving it.

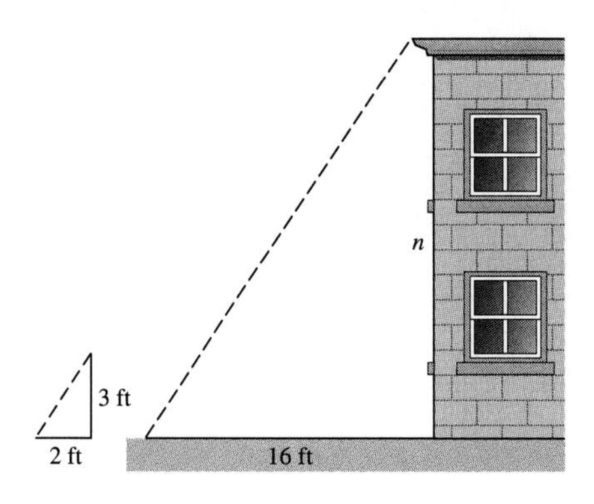

$n = 24$ ft

20. A sailor on the USS *Ramapo* saw one of the highest waves ever recorded. He used the height of the ship's mast, the length of the deck, and similar triangles to find the height of the wave. Using the information in the figure, write a proportion and then find the height of the wave.

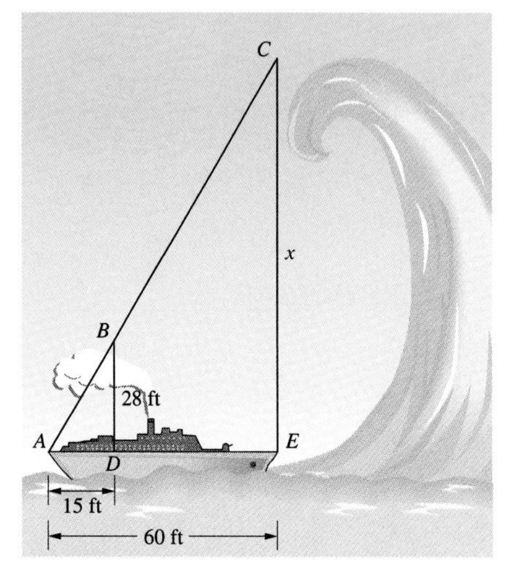

$x = 112$ ft

21. Look up the word *similar* in a dictionary. What is the non-mathematical definition of this word? Find two examples of similar objects at home or school.

Answer varies.

22. *Congruent* objects have the same shape and the same size. Sketch a pair of congruent triangles. Find two examples of congruent objects at home or school.

Answer varies.

23. Use similar triangles and a proportion to find the length of the lake shown here. (*Hint:* The side 100 m long in the smaller triangle corresponds to a side of 100 + 120 = 220 m in the larger triangle.)

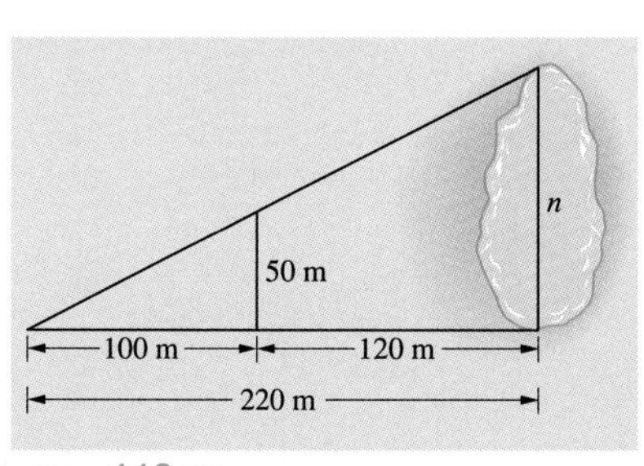

$n = 110$ m

24. To find the height of the tree, find y and then add $5\frac{1}{2}$ feet for the distance from the ground to eye level.

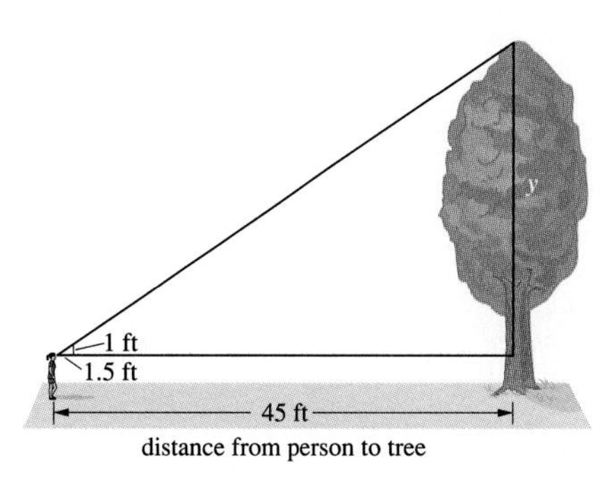

35.5 ft

Find the unknown length. Round your answers to the nearest tenth.
Note: When a line is drawn parallel to one side of a triangle, the smaller triangle that is formed will be similar to the original triangle.

25.

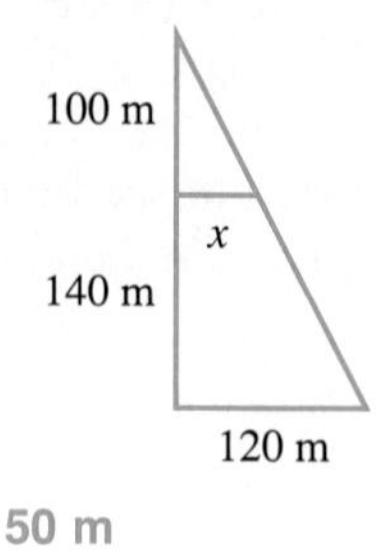

50 m

26.

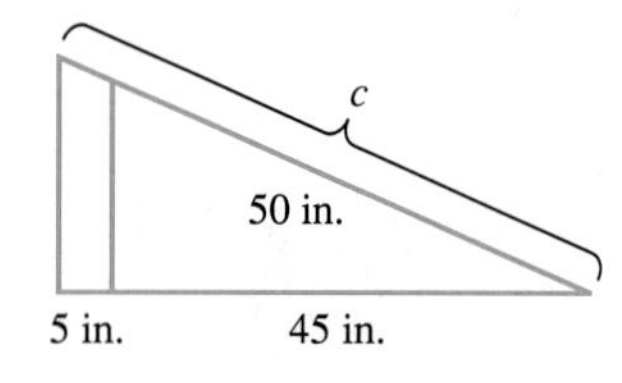

55.6 in.

PREVIEW EXERCISES

Work each problem by using the order of operations. (For help, see **Section 1.8.**)

27. $8 \div 4 + 3 \cdot 2$ 8

28. $16 + 4 \div 2$ 18

29. $13 - 2 \cdot 5 + 7$ 10

30. $18 + 7 \cdot 2 - 33 \div 3$ 21

31. $16 - 5 + 2(6 - 4)$ 15

32. $3(2 + 2) \div 12 \cdot 4$ 4

KEY TERMS

8.1	**point**	A point is a location in space.
	line	A line is a straight row of points that goes on forever in both directions.
	line segment	A line segment is a piece of a line with two endpoints.
	ray	A ray is a part of a line that has one endpoint and extends forever in one direction.
	angle	An angle is made up of two rays that have a common endpoint called the vertex.
	degrees	A system used to measure angles in which a complete circle is 360 degrees, written 360°.
	right angle	A right angle is an angle that measures 90°.
	acute angle	An acute angle is an angle that measures between 0° and 90°.
	obtuse angle	An obtuse angle is an angle that measures between 90° and 180°.
	straight angle	A straight angle is an angle that measures 180°; its sides form a straight line.
	intersecting lines	Intersecting lines cross or merge.
	perpendicular lines	Perpendicular lines are two lines that intersect to form a right angle.
	parallel lines	Parallel lines are two lines that never intersect and are equidistant from each other.
8.2	**complementary angles**	Complementary angles are two angles with a sum of 90°.
	supplementary angles	Supplementary angles are two angles with a sum of 180°.
	congruent angles	Congruent angles are angles that measure the same number of degrees.
	vertical angles	Vertical angles are two nonadjacent angles formed by intersecting lines.
8.3, 8.4, 8.5	**perimeter**	Perimeter is the distance around the outside edges of a figure. It is measured in linear units such as ft, yd, cm, m, km, and so on.
8.3, 8.4, 8.5, 8.6	**area**	Area is the surface covered by a two-dimensional (flat) figure. It is measured by seeing how many squares of a certain size are needed to cover the surface inside the figure. Commonly used units are square inches (in.^{2}); square feet (ft^2); square yards (yd^2); square centimeters (cm^2); and square meters (m^2).
8.3	**rectangle**	A rectangle is a four-sided figure with all sides meeting at 90° angles.
	square	A square is a rectangle with all four sides the same length.
8.4	**parallelogram**	A parallelogram is a four-sided figure with both pairs of opposite sides parallel.
	trapezoid	A trapezoid is a four-sided figure with one pair of opposite sides parallel.
8.5	**triangle**	A triangle is a figure with exactly three sides.
8.6	**circle**	A circle is a figure with all points the same distance from a fixed center point.
	radius	Radius is the distance from the center to any point on the circle.
	diameter	Diameter is the distance across the circle, passing through the center.
	circumference	Circumference is the distance around a circle.
	π (pi)	π is the ratio of the circumference to the diameter of any circle. It is approximately equal to 3.14.
8.7	**volume**	Volume is the space inside a three-dimensional (solid) figure.

8.8	**square root**	A square root is one of two equal factors of a number.
	hypotenuse	The hypotenuse is the side of a right triangle opposite the 90° angle.
8.9	**similar triangles**	Similar triangles are two triangles with the same shape but not necessarily the same size.

QUICK REVIEW

Concepts	Examples
8.1 Lines	
If a line has one endpoint, it is a ray. If it has two endpoints, it is a line segment.	Identify each of the following as a line, line segment, or ray. (a) (b) (c) **(a)** is a ray, **(b)** is a line, and **(c)** is a line segment.
If two lines intersect at right angles, they are perpendicular. If two lines never intersect, they are parallel.	Label each pair of lines as parallel or perpendicular. (a) (b) **(a)** shows two perpendicular lines (they intersect at 90°). **(b)** shows two parallel lines (they never intersect).
8.2 Angles	
If the sum of two angles is 90°, they are complementary. If the sum of two angles is 180°, they are supplementary.	Find the complement and supplement of a 35° angle. $90° - 35° = 55°$ (the complement) $180° - 35° = 145°$ (the supplement)
If two angles measure the same number of degrees, the angles are congruent. The symbol for congruent is ≅. Two nonadjacent angles formed by intersecting lines are called vertical angles. Vertical angles are congruent.	Identify the congruent and vertical angles in the following figure. $\angle 1 \cong \angle 3$ and $\angle 2 \cong \angle 4$ $\angle 1$ and $\angle 3$ are vertical angles. $\angle 2$ and $\angle 4$ are vertical angles.

Concepts	Examples
8.3 Rectangles and Squares	
Use this formula to find the perimeter of a **rectangle.** $P = 2 \cdot \text{length} + 2 \cdot \text{width}$ Use this formula to find area. $A = \text{length} \cdot \text{width}$ Area is measured in **square units.**	Find the perimeter and area of the rectangle. 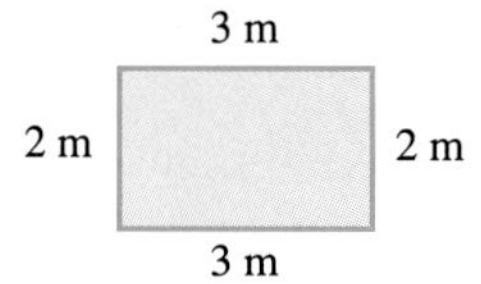 $P = 2 \cdot l + 2 \cdot w = 2 \cdot 3\text{ m} + 2 \cdot 2\text{ m}$ $= 6\text{ m} + 4\text{ m} = 10\text{ m}$ $A = l \cdot w = 3\text{ m} \cdot 2\text{ m} = 6\text{ m}^2$
Use these formulas to find the perimeter and area of a **square.** $P = 4 \cdot \text{side}$ $A = (\text{side})^2$ Area is measured in **square units.**	Find the perimeter and area of a square with a side of 6 meters. $P = 4 \cdot s = 4 \cdot 6\text{ m} = 24\text{ m}$ $A = s^2 = s \cdot s = 6\text{ m} \cdot 6\text{ m} = 36\text{ m}^2$
8.4 Parallelograms	
Use these formulas to find the perimeter and area. $P = \text{sum of the lengths of the sides}$ $A = \text{base} \cdot \text{height}.$ Area is measured in **square units.**	Find the perimeter and area of the parallelogram. 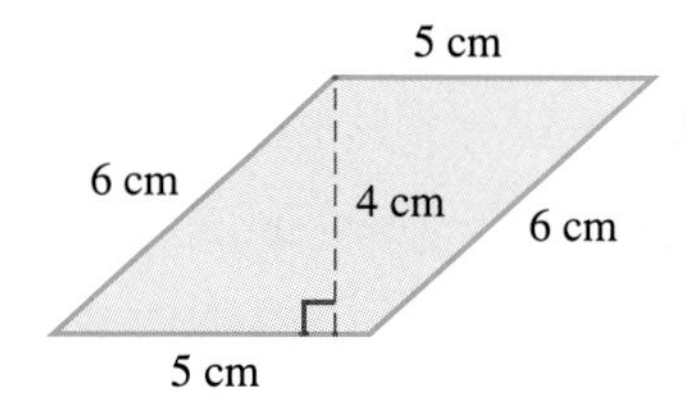 $P = 5\text{ cm} + 6\text{ cm} + 5\text{ cm} + 6\text{ cm} = 22\text{ cm}$ $A = 5\text{ cm} \cdot 4\text{ cm} = 20\text{ cm}^2$
8.4 Trapezoids	
Use these formulas to find the perimeter and area. $P = \text{sum of the lengths of the sides}$ $A = \frac{1}{2} \cdot \text{height} \cdot (b + B)$ where b is the short base and B is the long base. Area is measured in **square units.**	Find the perimeter and area of the trapezoid. 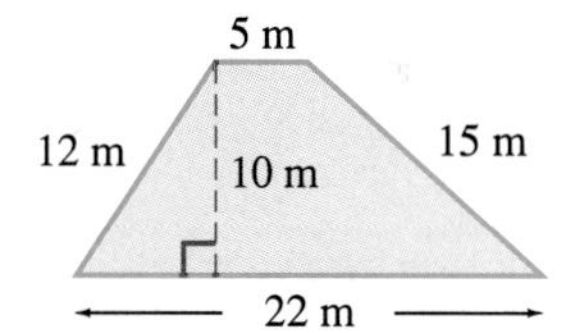 $P = 5\text{ m} + 15\text{ m} + 22\text{ m} + 12\text{ m} = 54\text{ m}$ $A = \frac{1}{\cancel{2}_1} \cdot \overset{5}{\cancel{10}}\text{ m} \cdot (5\text{ m} + 22\text{ m})$ $= 5\text{ m} \cdot (27\text{ m}) = 135\text{ m}^2$
8.5 Triangles	
Use these formulas to find the perimeter and area. $P = \text{sum of the lengths of the sides}$ $A = \frac{1}{2} \cdot \text{base} \cdot \text{height}.$ or $A = 0.5 \cdot \text{base} \cdot \text{height}$ Area is measured in **square units.**	Find the perimeter and area of the triangle. 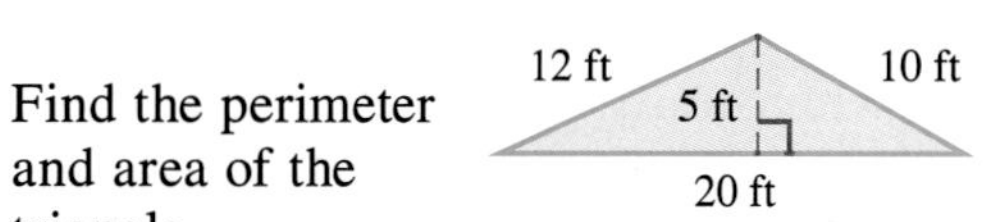 $P = 12\text{ ft} + 10\text{ ft} + 20\text{ ft} = 42\text{ ft}$ $A = \frac{1}{2} \cdot b \cdot h$ $A = \frac{1}{\cancel{2}_1} \cdot \overset{10}{\cancel{20}}\text{ ft} \cdot 5\text{ ft} = 50\text{ ft}^2$

Concepts	Examples
8.6 Circles	
Use this formula to find the **diameter,** given the radius. diameter = 2 · radius	Find the diameter of a circle if the radius is 3 meters. $d = 2 \cdot r = 2 \cdot 3 \text{ m} = 6 \text{ m}$
Use this formula to find the **radius,** given the diameter. radius $= \frac{1}{2} \cdot$ diameter or radius $= \frac{\text{diameter}}{2}$	Find the radius of a circle if the diameter is 5 cm. $r = \frac{d}{2} = \frac{5 \text{ cm}}{2} = 2.5 \text{ cm}$
Use these formulas for **circumference.** $C = 2 \cdot \pi \cdot \text{radius}$ or $C = \pi \cdot \text{diameter}$	Find the circumference and area of a circle with a radius of 3 cm. Circumference $= 2 \cdot \pi \cdot r = 2 \cdot \mathbf{3.14} \cdot 3 \text{ cm}$; $C = 18.84 \text{ cm}$
Use this formula to find **area.** $A = \pi \cdot (\text{radius})^2$ Area is measured in **square units.**	Area $= \pi \cdot r^2 = \mathbf{3.14} \cdot 3 \text{ cm} \cdot 3 \text{ cm}$; $A = 28.26 \text{ cm}^2$
8.7 Volume of a Rectangular Solid	
Use this formula. Volume = length · width · height. Volume is measured in **cubic units.**	Find the volume of this box. 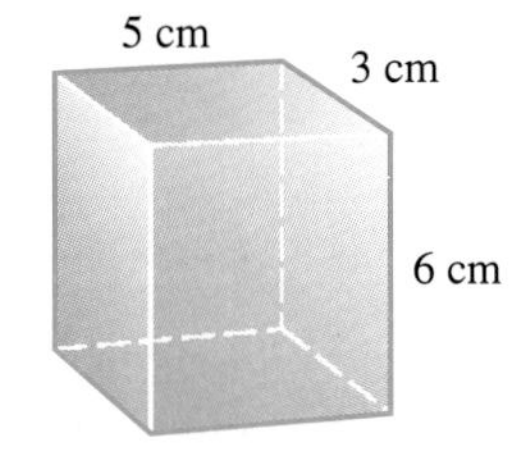 $V = l \cdot w \cdot h$; $V = 5 \text{ cm} \cdot 3 \text{ cm} \cdot 6 \text{ cm}$; $V = 90 \text{ cm}^3$
8.7 Volume of a Sphere	
Use this formula to find the volume of a **sphere.** Volume $= \frac{4}{3} \cdot \pi \cdot r^3$ or $V = \frac{4 \cdot \pi \cdot r^3}{3}$ where r is the radius of the sphere. Volume is measured in **cubic units.**	Find the volume of a sphere with a radius of 5 m. $V = \frac{4 \cdot \pi \cdot (\text{radius})^3}{3}$; $V = \frac{4 \cdot 3.14 \cdot 5 \text{ m} \cdot 5 \text{ m} \cdot 5 \text{ m}}{3}$; $V = 523.33 \text{ m}^3$ (rounded)
Use this formula to find the volume of a **hemisphere.** Volume $= \frac{2}{3} \cdot \pi \cdot r^3$ or $V = \frac{2 \cdot \pi \cdot r^3}{3}$ where r is the radius of the hemisphere. Volume is measured in **cubic units.**	Find the volume of a hemisphere with a radius of 20 cm. $V = \frac{2 \cdot \pi \cdot (\text{radius})^3}{3}$; $V = \frac{2 \cdot 3.14 \cdot 20 \text{ cm} \cdot 20 \text{ cm} \cdot 20 \text{ cm}}{3}$; $V = 16{,}746.67 \text{ cm}^3$ (rounded)

Concepts	Examples
8.7 Volume of a Cylinder Use this formula. $\text{Volume} = \pi \cdot r^2 \cdot h$ where r is the radius of the base and h is the height. Volume is measured in **cubic units.**	Find the volume of a cylinder that is 10 m high with a radius of 4 m. $V = \pi \cdot r^2 \cdot h$ $V = 3.14 \cdot 4 \text{ m} \cdot 4 \text{ m} \cdot 10 \text{ m}$ $V = 502.4 \text{ m}^3$
8.7 Volume of a Cone Use this formula. $\text{Volume} = \frac{1}{3} \cdot \pi \cdot r^2 \cdot h$ or $V = \frac{\pi \cdot r^2 \cdot h}{3}$ where r is the radius of the base and h is the height of the cone. Volume is measured in **cubic units.**	Find the volume of a cone, with a height of 9 inches and a base with a radius of 4 inches. $V = \frac{\pi \cdot r^2 \cdot h}{3}$ $V = \frac{3.14 \cdot 4 \text{ in.} \cdot 4 \text{ in.} \cdot 9 \text{ in.}}{3}$ $V = 150.72 \text{ in.}^3$
8.7 Volume of a Pyramid Use this formula. $V = \frac{1}{3} \cdot B \cdot h$ or $V = \frac{B \cdot h}{3}$ where B is the area of the base and h is the height of the pyramid. Volume is measured in **cubic units.**	Find the volume of a pyramid with a square base 2 cm by 2 cm and a height of 6 cm. $V = \frac{B \cdot h}{3}$ $\quad B = 2 \text{ cm} \cdot 2 \text{ cm} = 4 \text{ cm}^2$ $V = \frac{4 \text{ cm}^2 \cdot 6 \text{ cm}}{3}$ $V = 8 \text{ cm}^3$
8.8 Finding the Square Root of a Number Use the $\sqrt{\ }$ key on a calculator.	$\sqrt{43} = 6.557$ (rounded) $\sqrt{64} = 8$

Concepts	Examples

8.8 Finding the Unknown Length in a Right Triangle

To find the **hypotenuse,** use the formula

$$\text{hypotenuse} = \sqrt{(\text{leg})^2 + (\text{leg})^2}$$

The hypotenuse is the side opposite the right angle.

Find the hypotenuse. Round to thousandths.

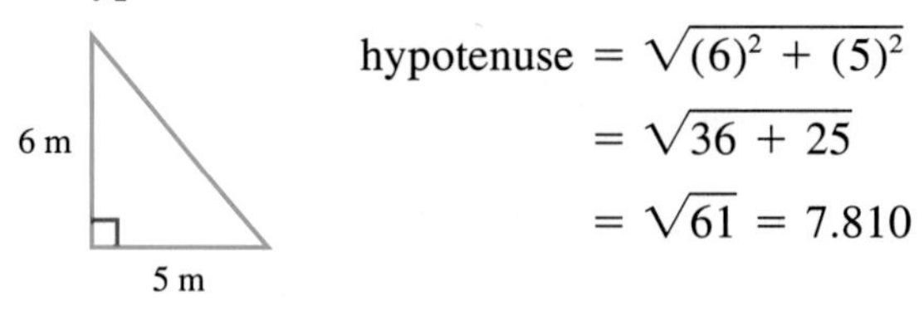

$$\begin{aligned}\text{hypotenuse} &= \sqrt{(6)^2 + (5)^2}\\ &= \sqrt{36 + 25}\\ &= \sqrt{61} = 7.810\end{aligned}$$

The hypotenuse is about 7.810 m long.

To find a **leg,** use the formula

$$\text{leg} = \sqrt{(\text{hypotenuse})^2 - (\text{leg})^2}$$

The legs are the sides that form the right angle.

Find the missing leg of the following triangle. Round to thousandths.

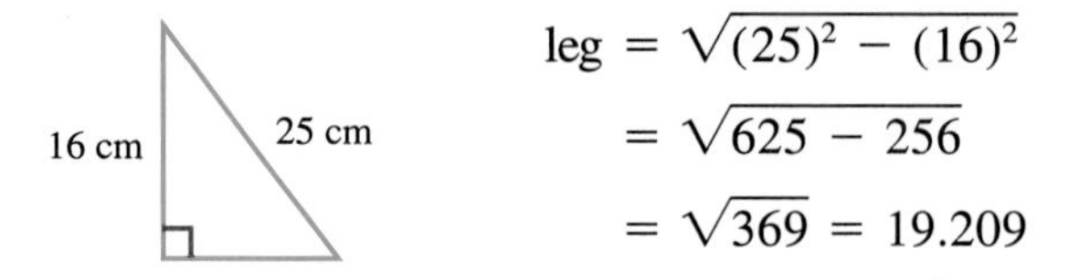

$$\begin{aligned}\text{leg} &= \sqrt{(25)^2 - (16)^2}\\ &= \sqrt{625 - 256}\\ &= \sqrt{369} = 19.209\end{aligned}$$

The leg is about 19.209 cm long.

8.9 Finding the Unknown Sides of Similar Triangles

Use the fact that in similar triangles, the ratios of the lengths of corresponding sides are equal. Write a proportion. Then find cross products and set them equal to each other. Finish solving for the unknown side.

Find x and y if the triangles are similar.

$$\frac{x}{8} = \frac{5}{10}$$

$$x \cdot 10 = 8 \cdot 5$$

$$\frac{x \cdot \overset{1}{\cancel{10}}}{\underset{1}{\cancel{10}}} = \frac{40}{10}$$

$$x = 4 \text{ m}$$

$$\frac{y}{12} = \frac{5}{10}$$

$$y \cdot 10 = 12 \cdot 5$$

$$\frac{y \cdot \overset{1}{\cancel{10}}}{\underset{1}{\cancel{10}}} = \frac{60}{10}$$

$$y = 6 \text{ m}$$

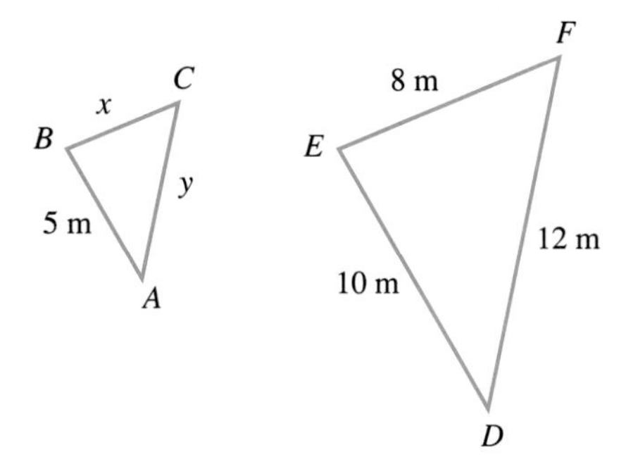

NAME DATE HOUR

CHAPTER 8 REVIEW EXERCISES

[8.1] *Name each line, line segment, or ray.*

1.

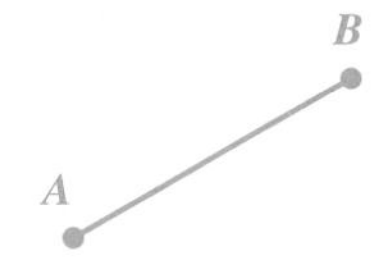

line segment named
$\overline{AB}$ or $\overline{BA}$

2.

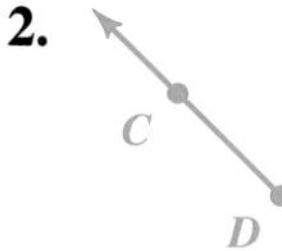

line named
$\overleftrightarrow{CD}$ or $\overleftrightarrow{DC}$

3.

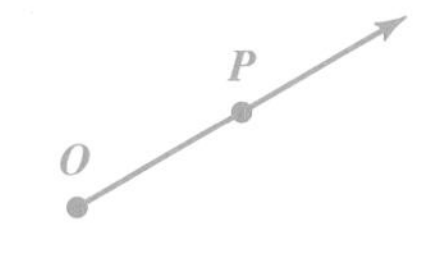

ray named
$\overrightarrow{OP}$

[8.1] *Label each pair of lines as parallel, perpendicular, or intersecting.*

4.

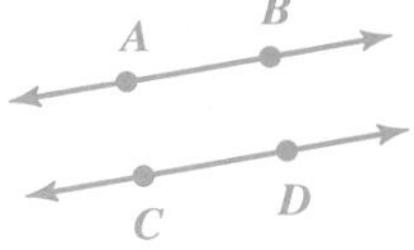

parallel

5.

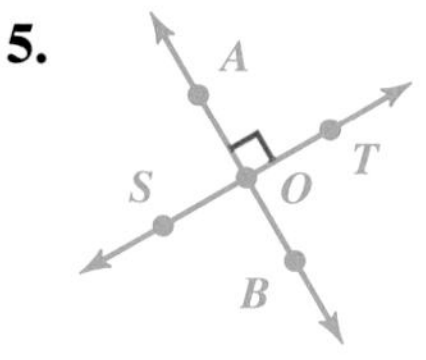

perpendicular

6.

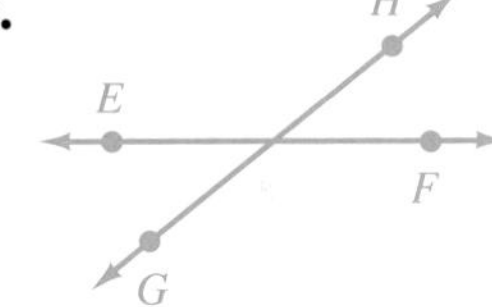

intersecting

[8.1] *Label each angle as an acute, right, obtuse, or straight angle. For right and straight angles, indicate the number of degrees in the angle.*

7.

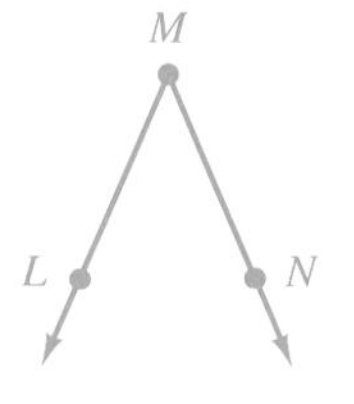

acute

8.

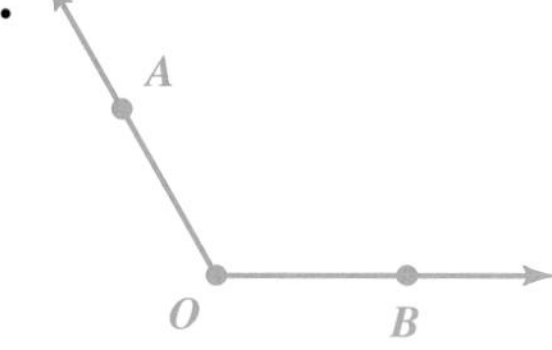

obtuse

9.

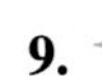

straight; 180°

10.

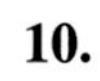
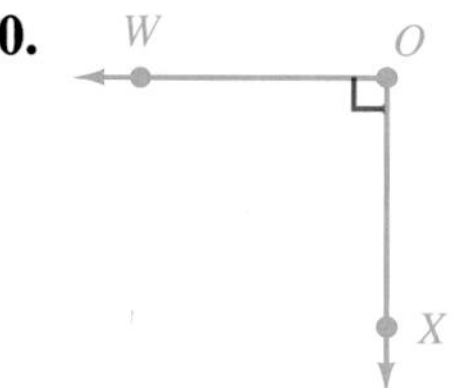

right; 90°

[8.2] *First identify the congruent angles in each figure. Then find all pairs of complementary angles.*

11.

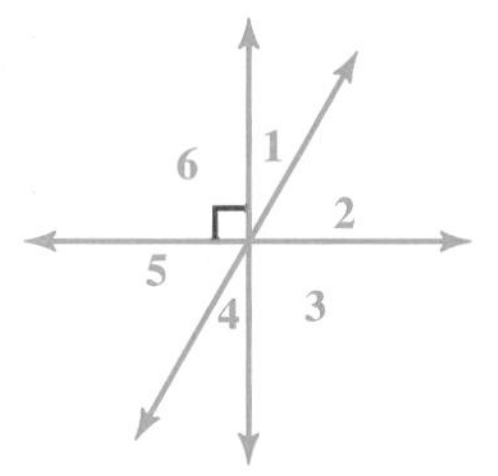

$\angle 1 \cong \angle 4$; $\angle 2 \cong \angle 5$; $\angle 3 \cong \angle 6$
$\angle 1$ and $\angle 2$; $\angle 4$ and $\angle 5$
$\angle 5$ and $\angle 1$; $\angle 4$ and $\angle 2$

12.

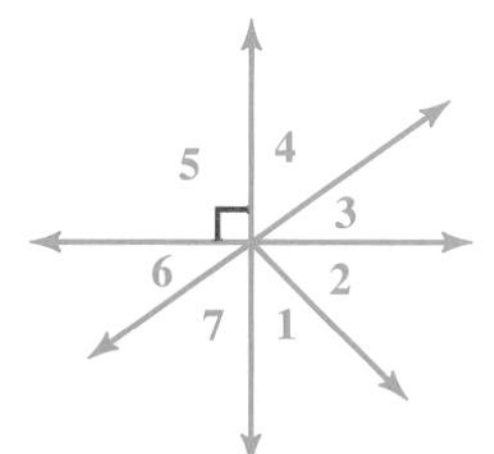

$\angle 4 \cong \angle 7$; $\angle 3 \cong \angle 6$
$\angle 1$ and $\angle 2$; $\angle 3$ and $\angle 4$
$\angle 6$ and $\angle 7$; $\angle 6$ and $\angle 4$
$\angle 7$ and $\angle 3$

[8.2] *Find the supplementary angles in each of the following.*

13.

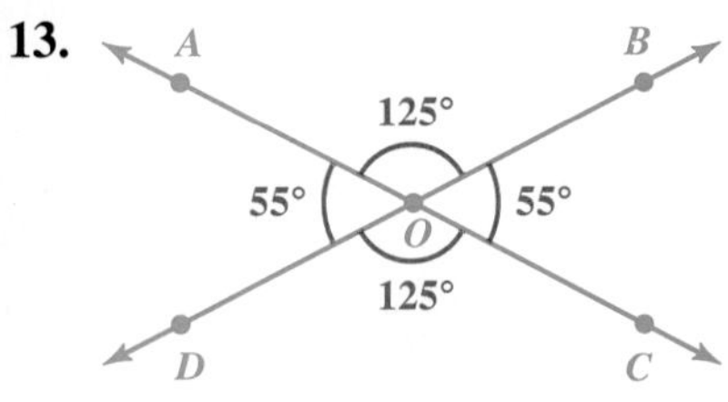

$\angle AOB$ and $\angle BOC$; $\angle BOC$ and $\angle COD$
$\angle COD$ and $\angle DOA$; $\angle DOA$ and $\angle AOB$

14.

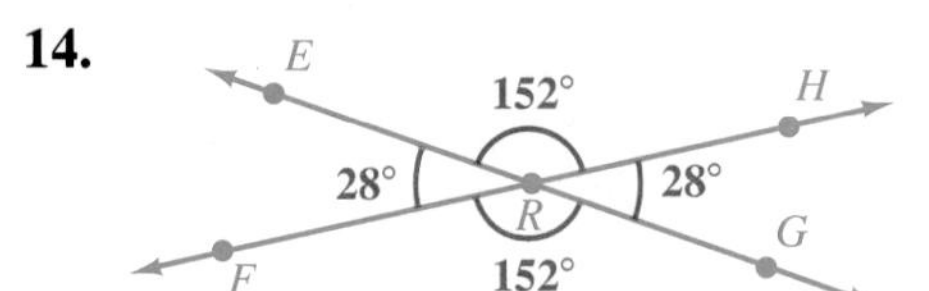

$\angle ERH$ and $\angle HRG$; $\angle HRG$ and $\angle GRF$
$\angle FRG$ and $\angle FRE$; $\angle FRE$ and $\angle ERH$

[8.2] *Find the complement or supplement of each angle.*

15. Find the complement of:
(a) 5° 85°
(b) 45° 45°
(c) 77° 13°

16. Find the supplement of:
(a) 9° 171°
(b) 140° 40°
(c) 90° 90°

[8.3] *Find each perimeter.*

17.

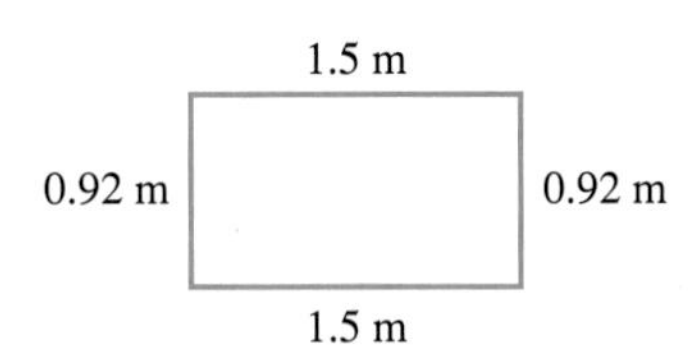

4.84 m

18.

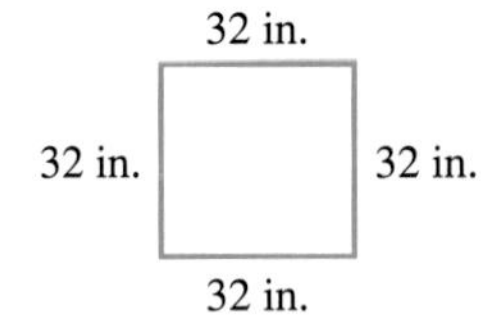

128 in.

19. A square park that is 2.4 km along each side.

9.6 km

20. A rectangular living room that is $12\frac{1}{2}$ feet wide and 18 feet long.

61 ft

Find the area of each rectangle or square. Round your answers to the nearest tenth when necessary.

21.

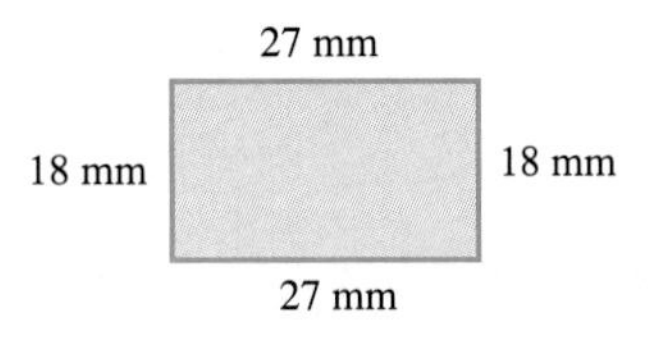

486 mm^2

22.

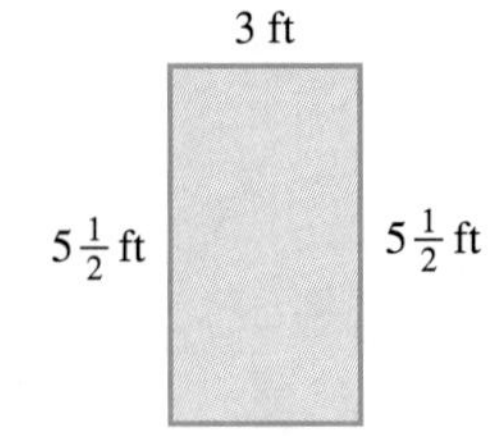

16.5 ft^2 or $16\frac{1}{2}$ ft^2

23.

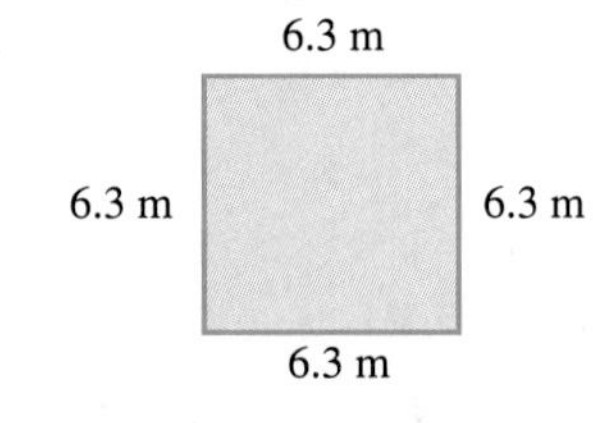

39.7 m^2

[8.4] *Find the perimeter and area of each parallelogram or trapezoid. Round your answers to the nearest tenth when necessary.*

24.

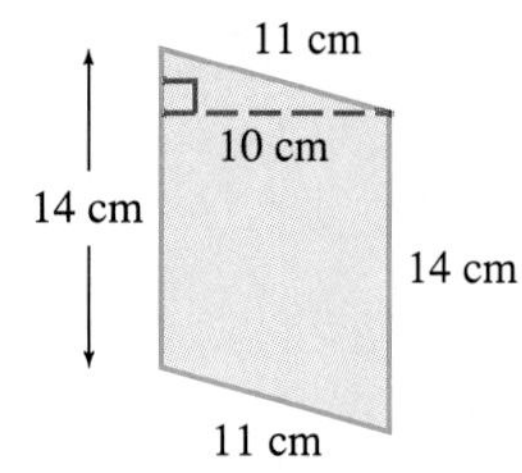

$P = 50$ cm
$A = 140$ cm^2

25.

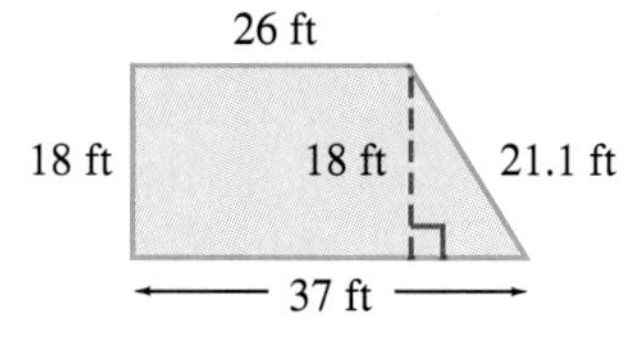

$P = 102.1$ ft
$A = 567$ ft^2

26.

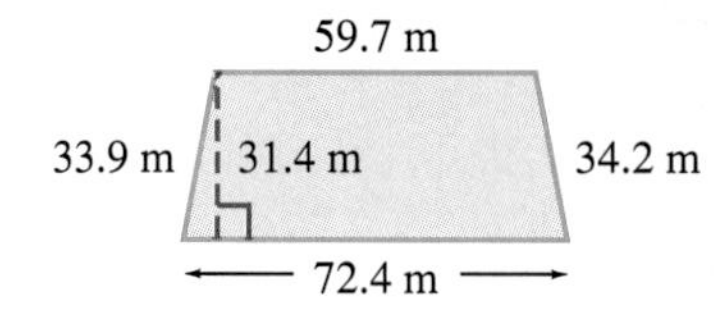

$P = 200.2$ m
$A = 2074.0$ m^2

[8.5] *Find the perimeter and area of each triangle.*

27.

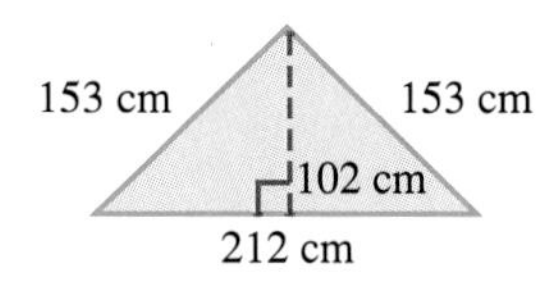

$P = 518$ cm
$A = 10{,}812$ cm^2

28.

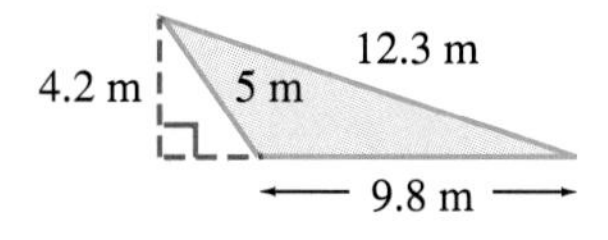

$P = 27.1$ m
$A = 20.58$ m^2

29.

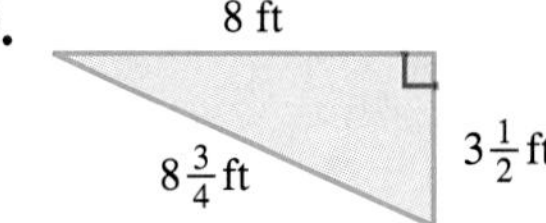

$P = 20\frac{1}{4}$ ft or 20.25 ft
$A = 14$ ft^2

Find the size of the missing angle.

30.

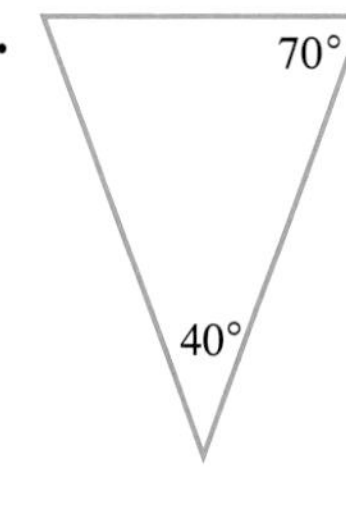

70°

31.

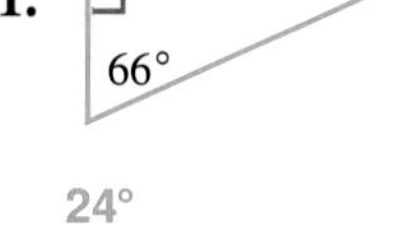

24°

[8.6] *Find the missing value for each circle.*

32. radius is 7.8 m, find the diameter

15.6 m

33. diameter is 23 ft, find the radius

$11\frac{1}{2}$ ft or 11.5 ft

Find the circumference and area of each circle. Use 3.14 as the approximate value for π. Round your answers to the nearest tenth.

34.

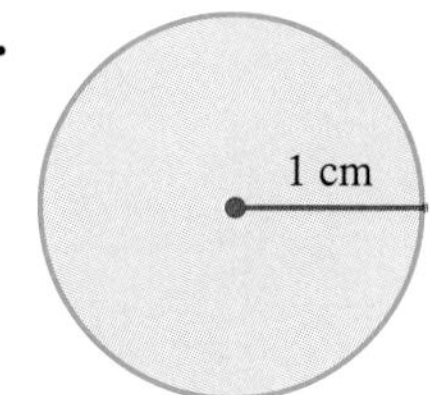

$C = 6.3$ cm
$A = 3.1$ cm^2

35.

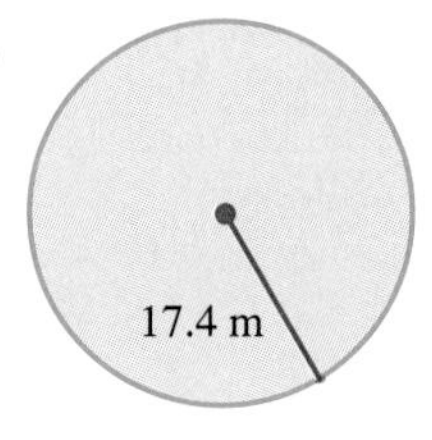

$C = 109.3$ m
$A = 950.7$ m^2

36.

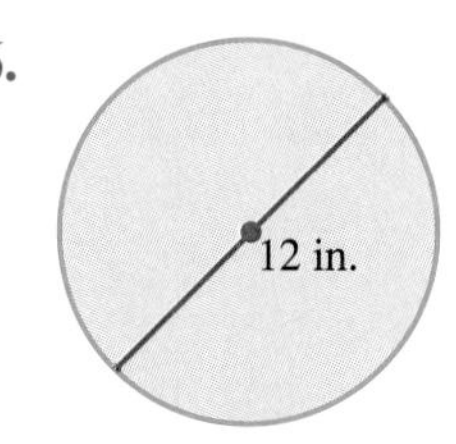

$C = 37.7$ in.
$A = 113.0$ in.^{2}

[8.3–8.6] *Find each shaded area. Use 3.14 for π. Round your answers to the nearest tenth when necessary.*

37.

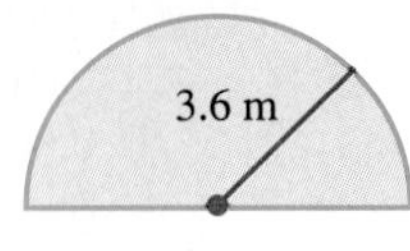

20.3 m²

38.

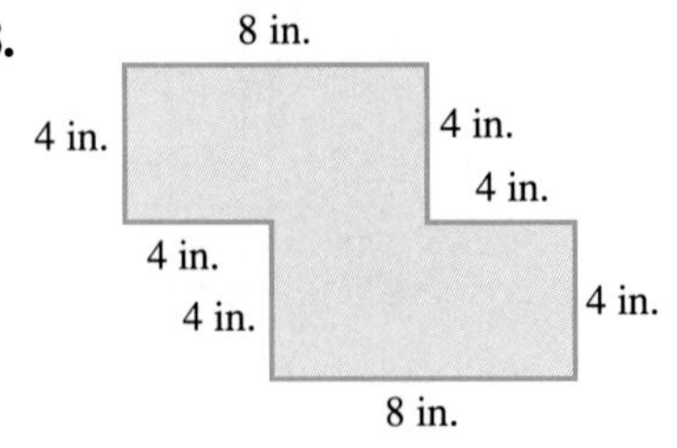

64 in.²

39.

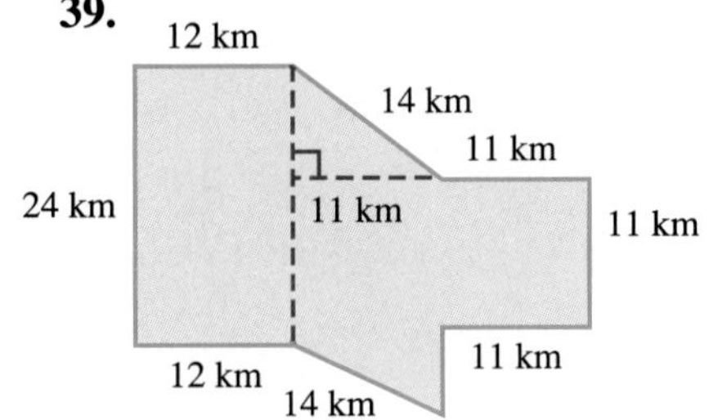

673 km²

40.

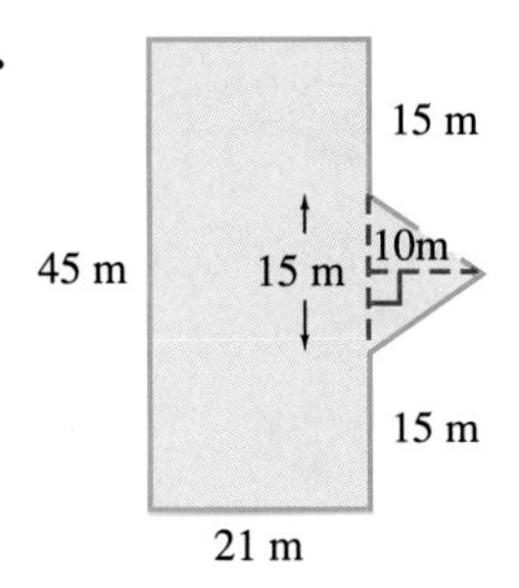

1020 m²

41.

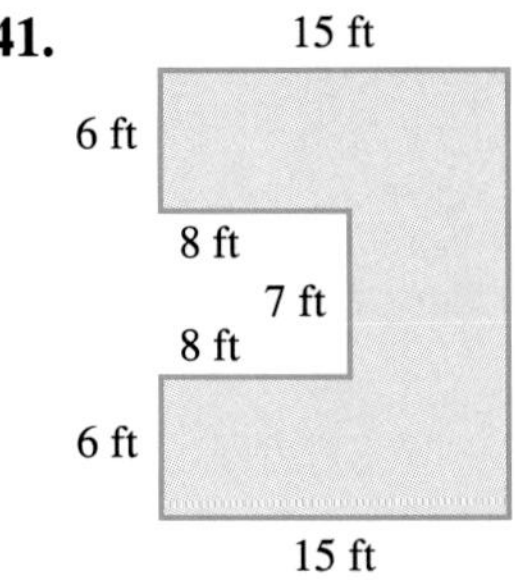

229 ft²

42.

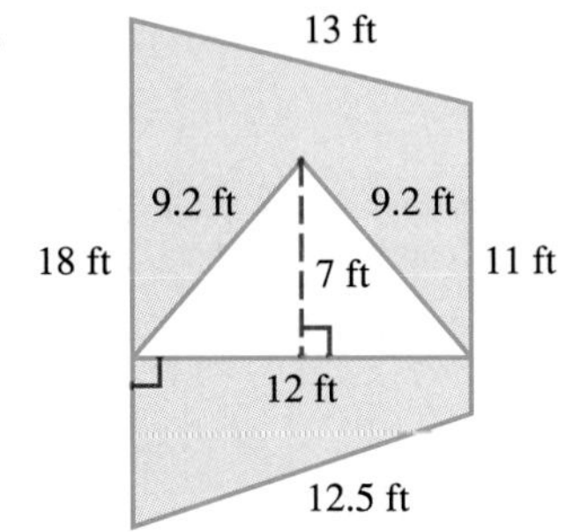

132 ft²

43.

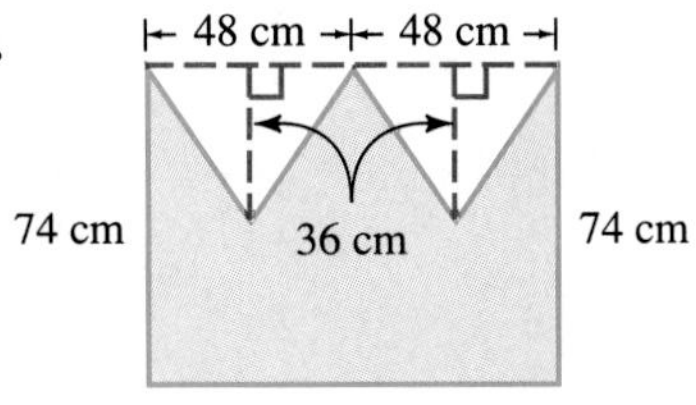

5376 cm²

44.

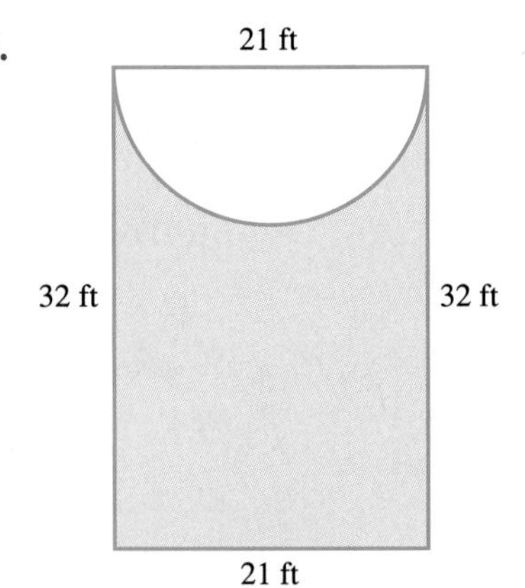

498.9 ft²

45.

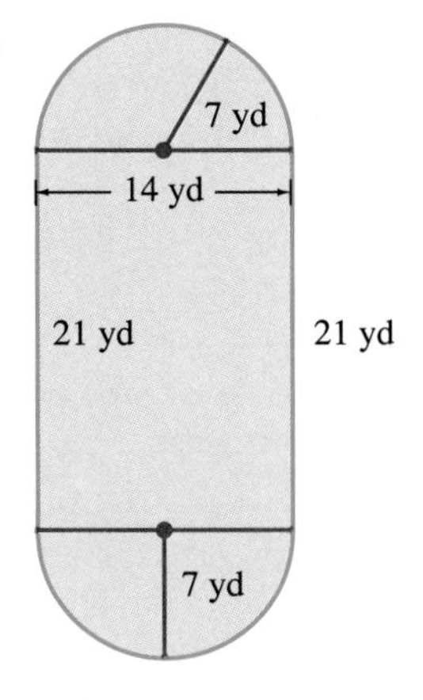

447.9 yd²

[8.7] *Find each volume. Use 3.14 for π. Round your answers to the nearest tenth when necessary.*

46.

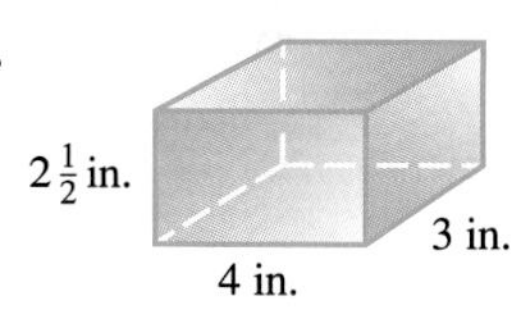

30 in.^{3}

47.

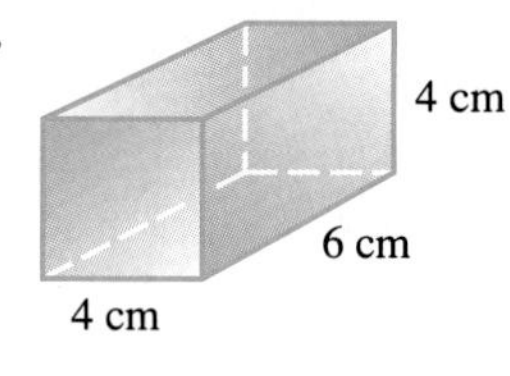

96 cm^3

48.

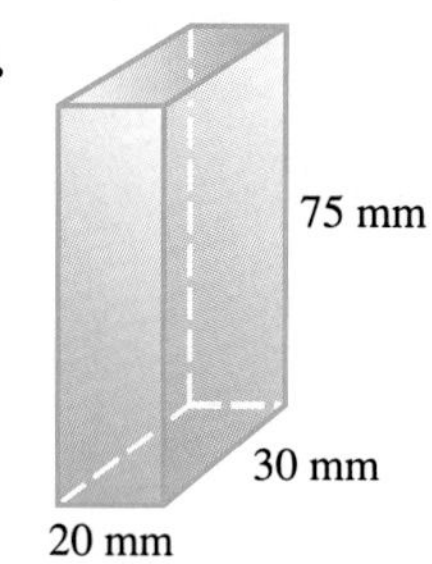

45,000 mm^3

49.

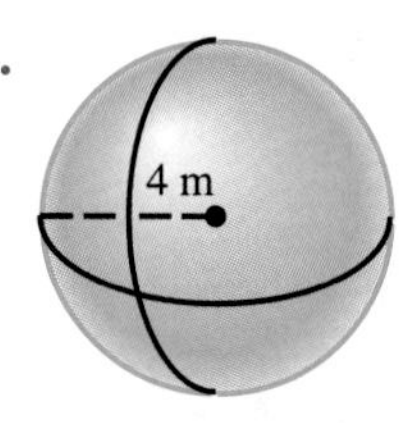

267.9 m^3

50.

452.2 ft^3

51.

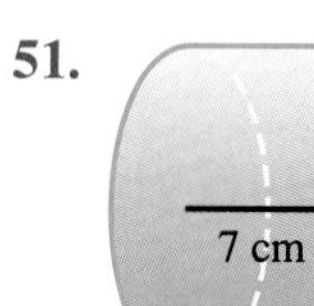

549.5 cm^3

52.

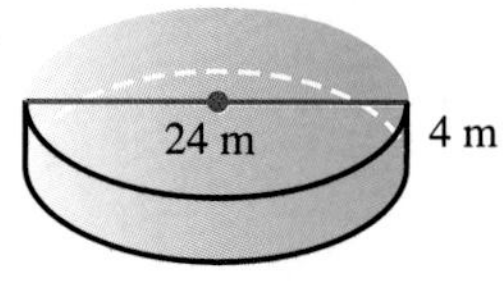

1808.6 m^3

53.

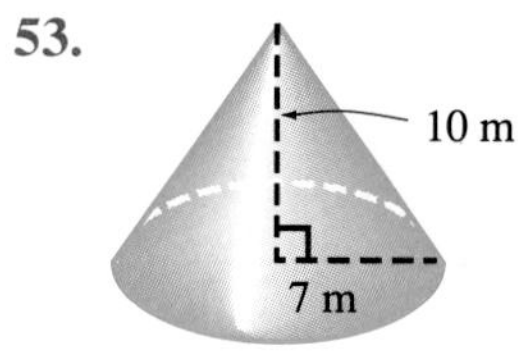

512.9 m^3

54.

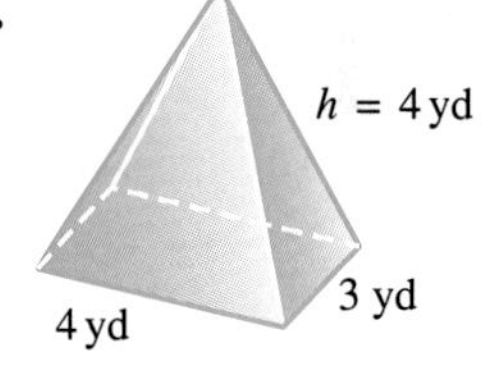

16 yd^3

[8.8] *Find each square root. Round your answers to the nearest thousandth when necessary.*

55. $\sqrt{7}$

2.646

56. $\sqrt{19}$

4.359

57. $\sqrt{16}$

4

58. $\sqrt{196}$

14

59. $\sqrt{58}$

7.616

60. $\sqrt{121}$

11

61. $\sqrt{105}$

10.247

62. $\sqrt{80}$

8.944

Find the missing length in each right triangle. Use a calculator to find square roots. Round your answers to the nearest thousandth.

63.

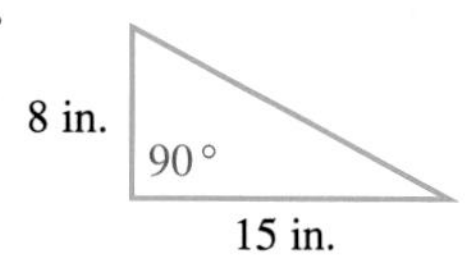

17 in

64.

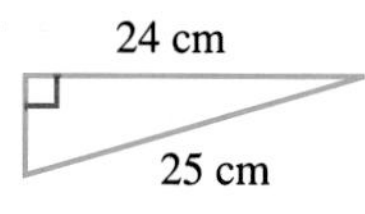

7 cm

65.

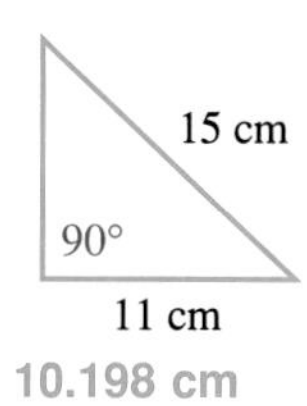

10.198 cm

66.

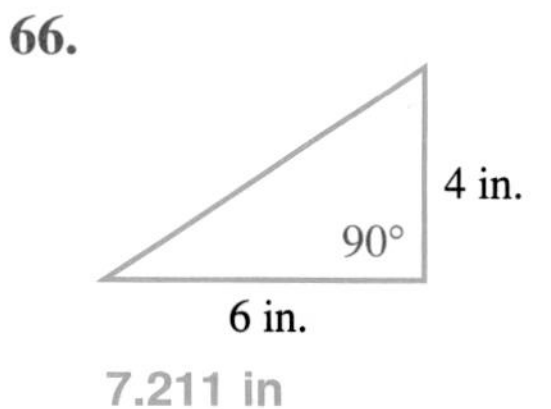

7.211 in

67.

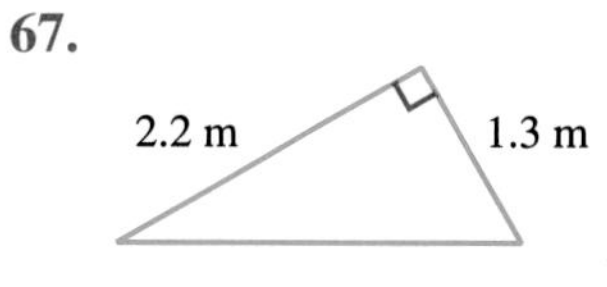

2.556 m

68.

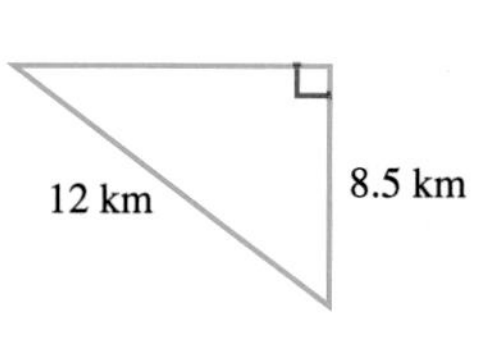

8.471 km

[8.9] *Find the lengths of the missing sides in each pair of similar triangles.*

69.

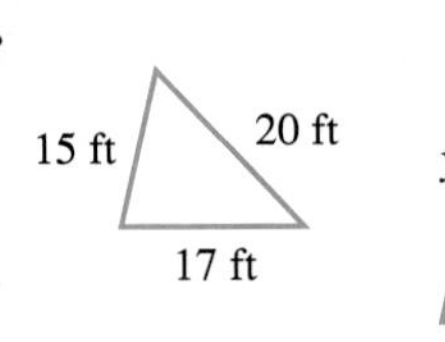

$y = 30$ ft
$x = 34$ ft

70.

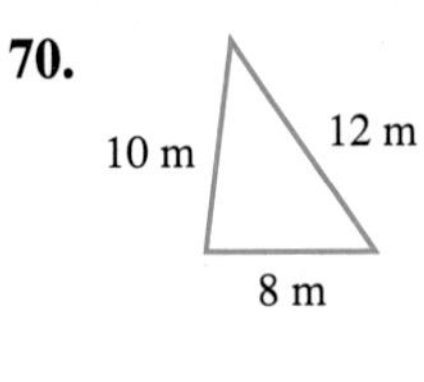

$y = 15$ m
$x = 18$ m

71.

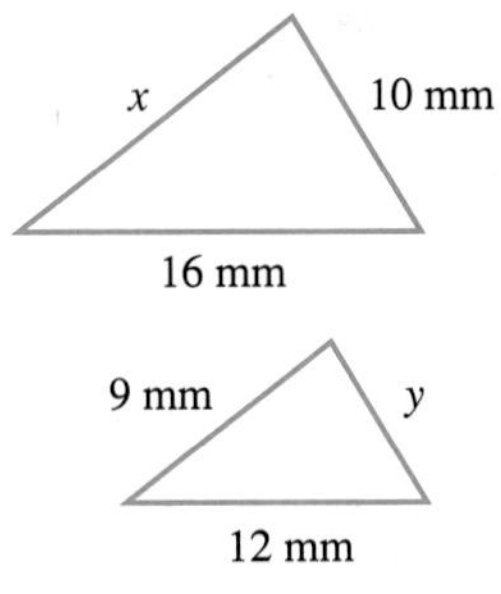

$x = 12$ mm
$y = 7.5$ mm

MIXED REVIEW EXERCISES

Find the perimeter (or circumference) and area of each figure. Use 3.14 for π.

72.

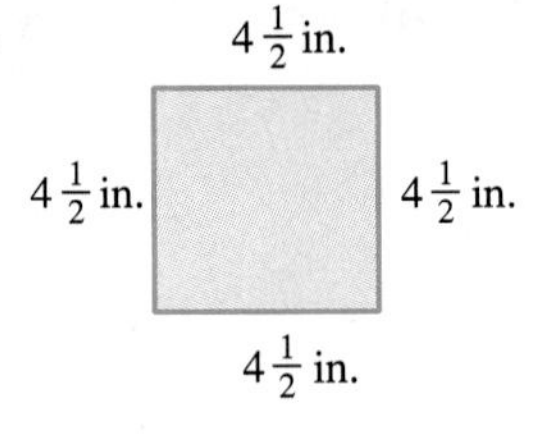

$P = 18$ in.
$A = 20.25$ in.^{2}
or $20\frac{1}{4}$ in^2

73.

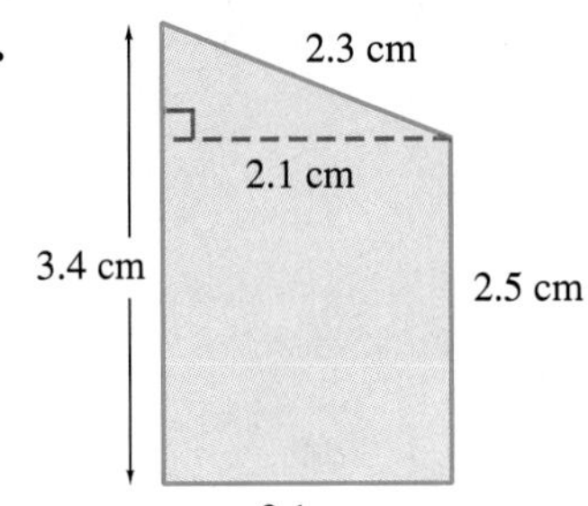

$P = 10.3$ cm
$A = 6.195$ cm^2

74.

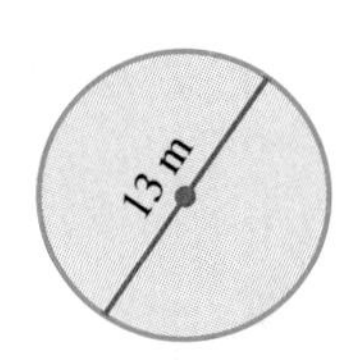

$C = 40.82$ m
$A = 132.665$ m^2

75.

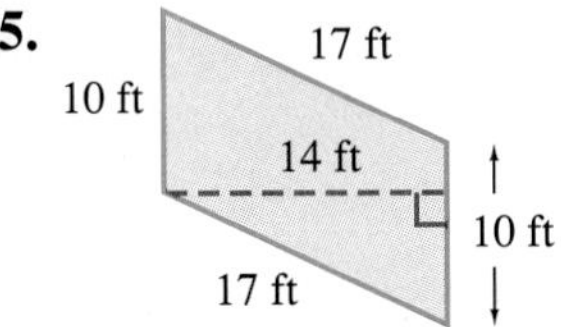

$P = 54$ ft
$A = 140$ ft^2

76.

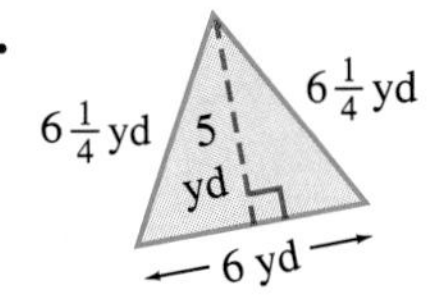

$P = 18.5$ yd or $18\frac{1}{2}$ yd
$A = 15$ yd^2

77.

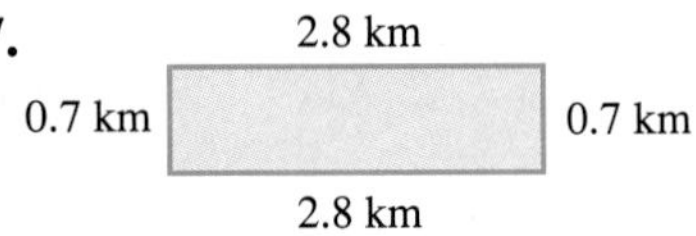

$P = 7$ km
$A = 1.96$ km^2

78.

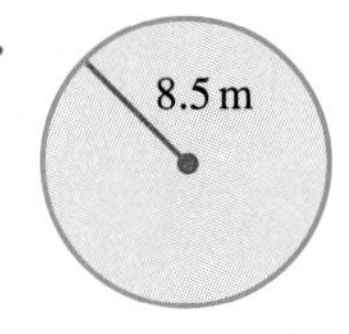

$C = 53.38$ m
$A = 226.865$ m^2

79.

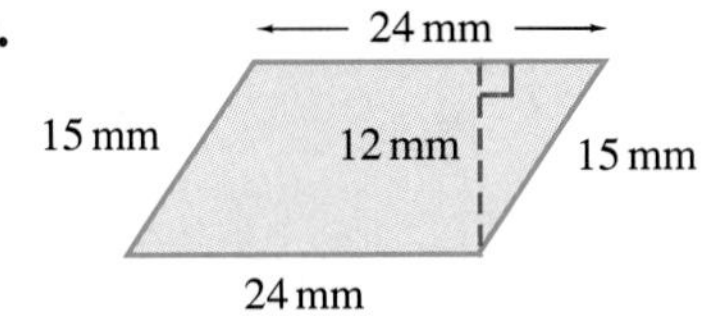

$P = 78$ mm
$A = 288$ mm^2

80.

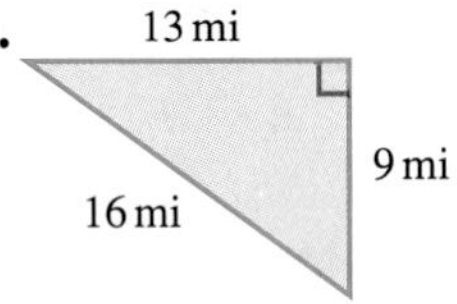

$P = 38$ mi
$A = 58.5$ mi^2

Label each figure. Choose from these labels: line, line segment, ray, parallel lines, perpendicular lines, intersecting lines, acute angle, right angle, straight angle, obtuse angle.

81. 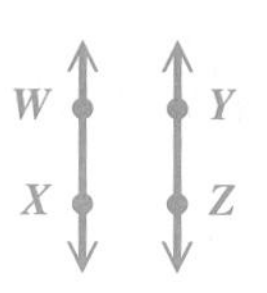

parallel lines

82. 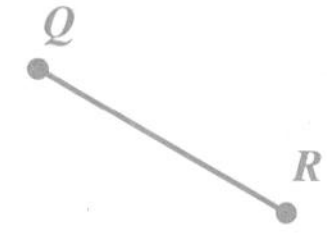

line segment

83. 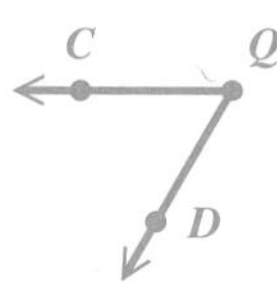

acute angle

84. 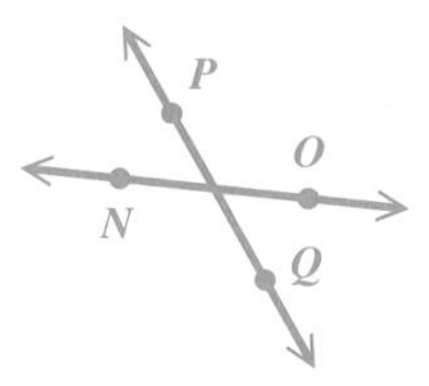

intersecting lines

85. 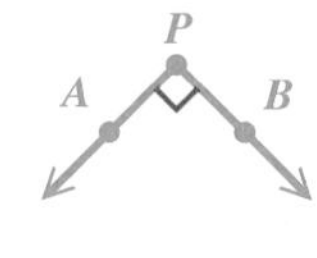

right angle

86. 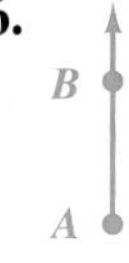

ray

87. 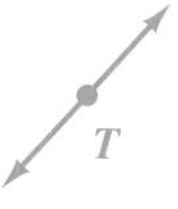

straight angle

88. 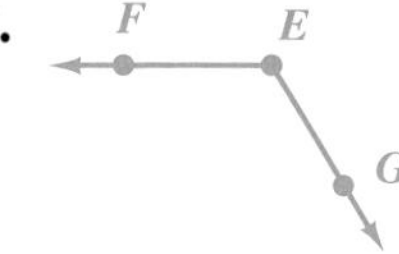

obtuse angle

89. 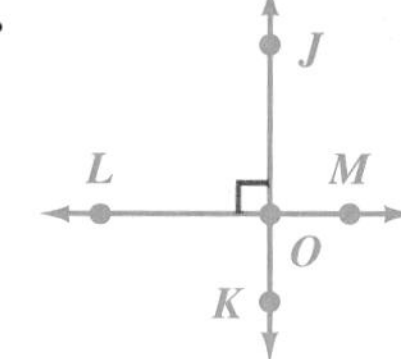

perpendicular lines

90. What is the complement of 61°?

29°

91. What is the supplement of 85°?

95°

Find the perimeter and area of the following figures.

92. 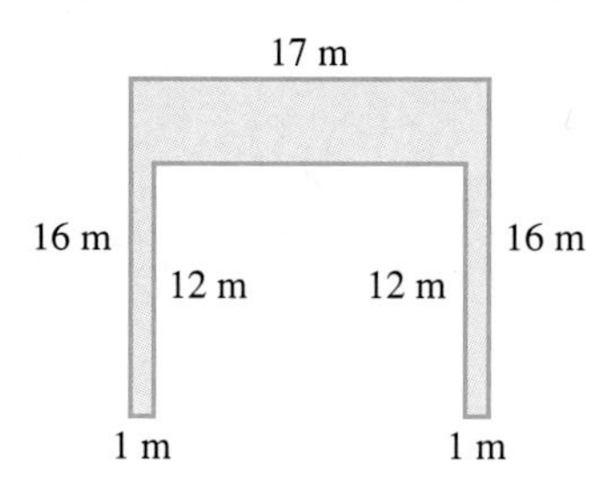

$P = 90$ m
$A = 92$ m^2

93. 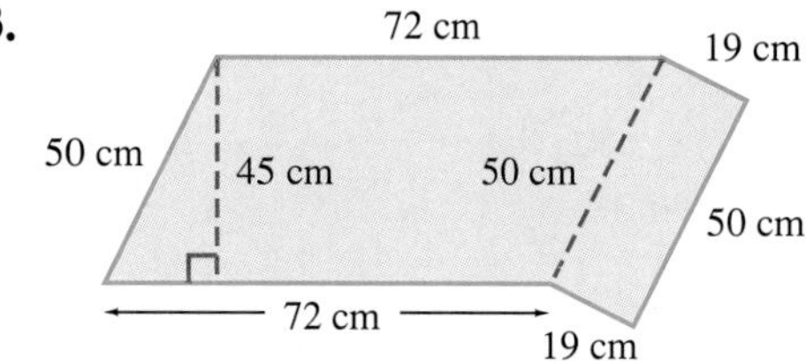

$P = 282$ cm
$A = 4190$ cm^2

Find the volume of each of the following figures. Use 3.14 for π. Round your answers to the nearest tenth when necessary.

94.

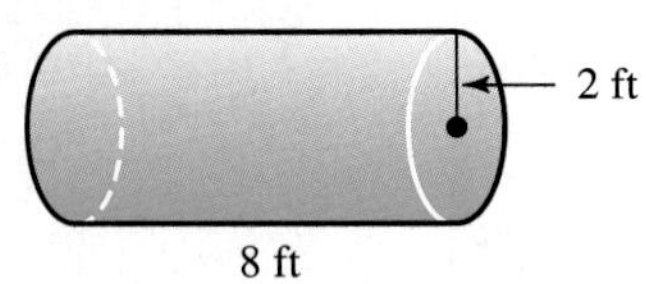

100.5 ft^3

95.

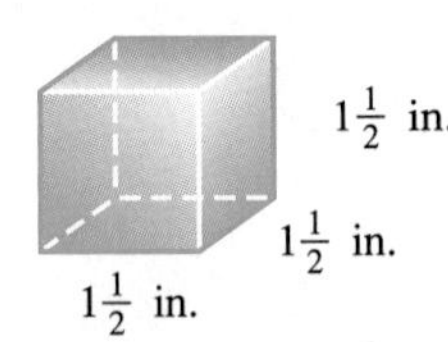

3.4 in.^{3} or $3\frac{3}{8}$ in.^{3}

96.

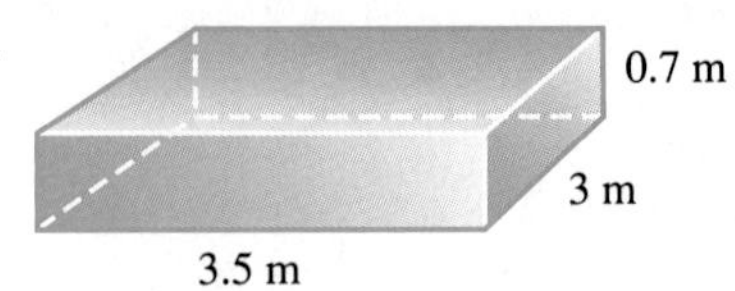

7.4 m^3

97.

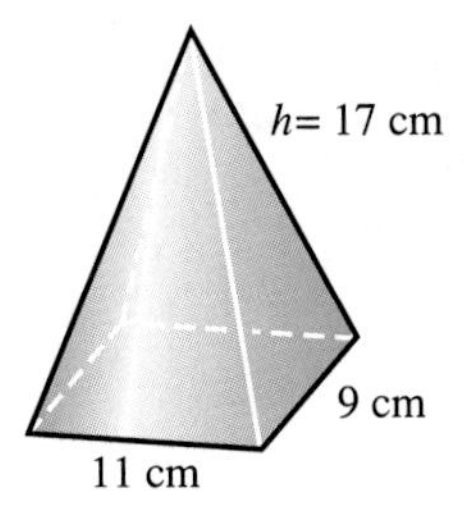

561 cm^3

98.

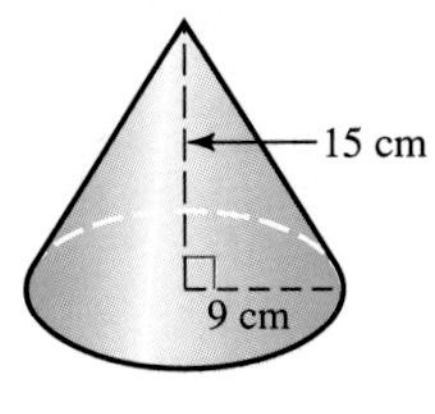

1271.7 cm^3

99.

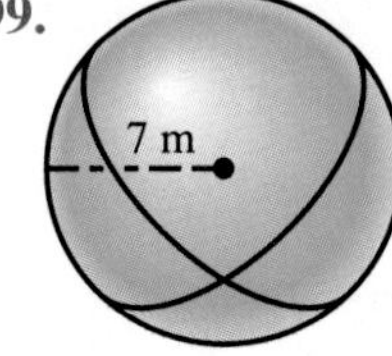

1436.0 m^3

Find the missing angle or side. Round your answers to the nearest thousandth when necessary.

100.

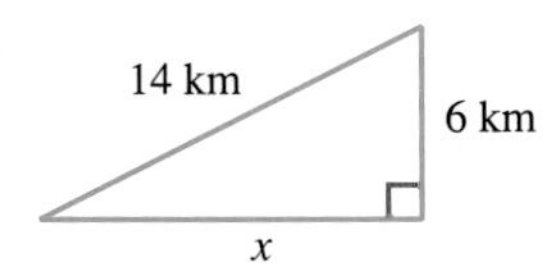

$x = 12.649$ km

101.

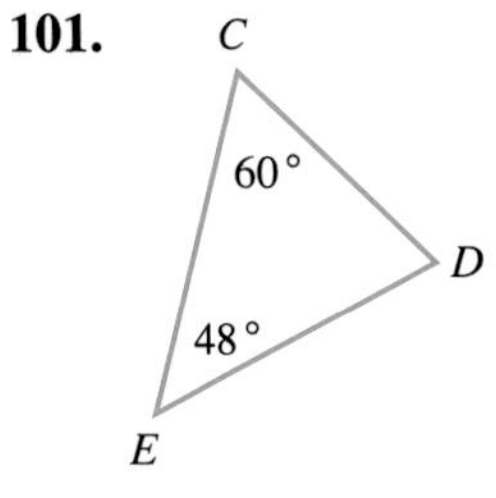

$\angle D = 72°$

102. similar triangles

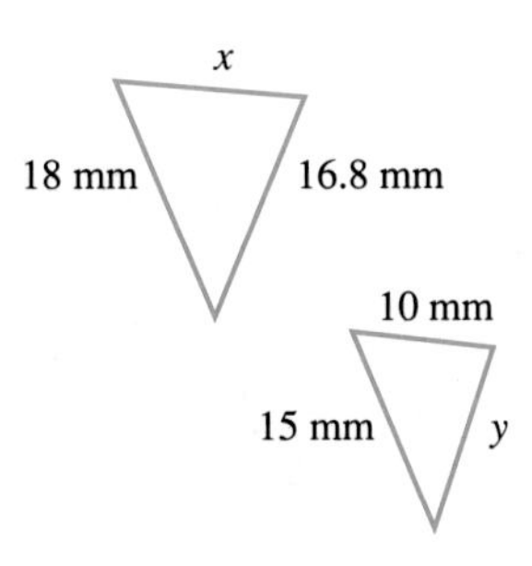

$x = 12$ mm
$y = 14$ mm

NAME DATE HOUR

CHAPTER 8 TEST

Choose the figure that matches each label.

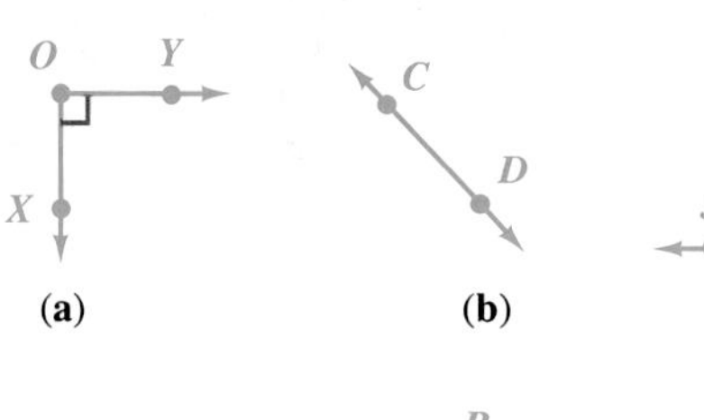

J K Q (c)

G H (d)

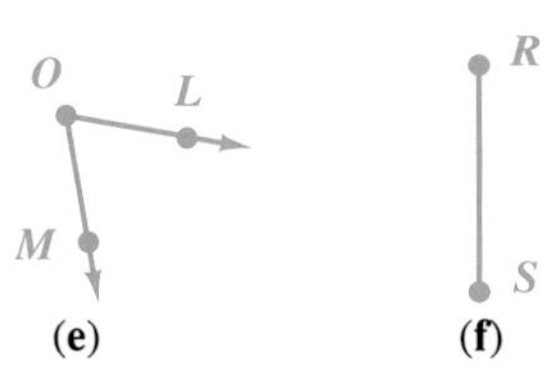

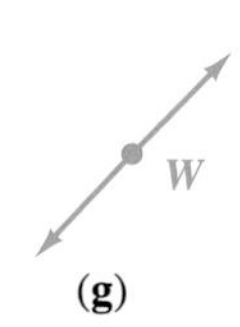

1. Acute angle is figure ____
2. Right angle is figure ____
3. Line segment is figure ____
4. Straight angle is figure ____
5. Ray is figure ____

6. Write a definition of parallel lines and a definition of perpendicular lines. Make a sketch to illustrate each definition.

7. Find the complement of a 16° angle.
8. Find the supplement of a 75° angle.
9. Identify all pairs of congruent angles.

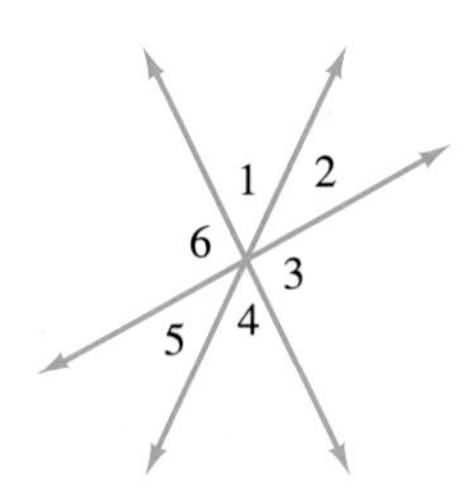

Find the perimeter and area of each figure.

10.

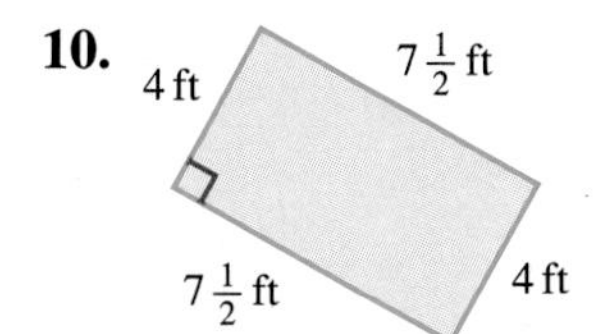

11.

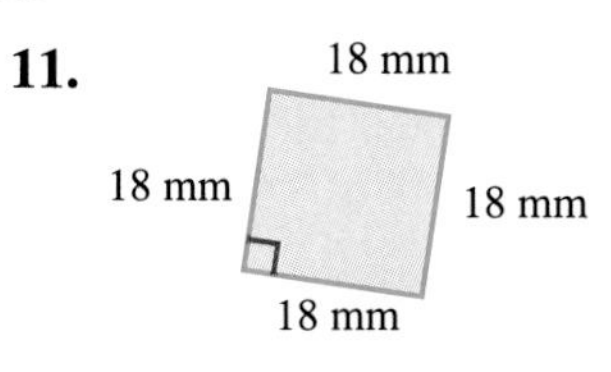

12.

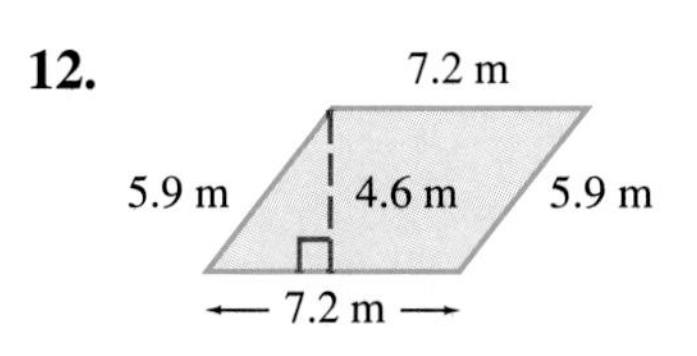

13.

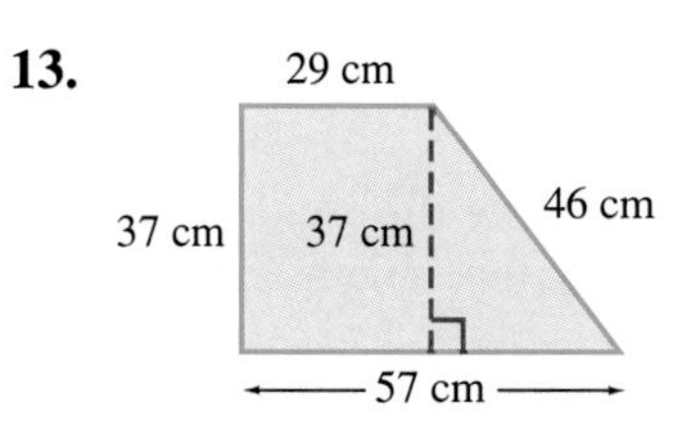

1. e
2. a
3. f
4. g
5. d
6. Answer varies.
7. 74°
8. 105°
9. ∠1 and ∠4; ∠2 and ∠5; ∠3 and ∠6
10. $P = 23$ ft; $A = 30$ ft^2
11. $P = 72$ mm; $A = 324$ mm^2
12. $P = 26.2$ m; $A = 33.12$ m^2
13. $P = 169$ cm; $A = 1591$ cm^2

14. $P = 32.05$ cm
$A = 48\text{ m}^2$

15. $P = 37.8$ yd or $37\frac{4}{5}$ yd
$A = 58.5\text{ yd}^2$

16. 107°

17. 12.5 in. or $12\frac{1}{2}$ in.

18. 5.7 km

19. 206.0 cm^2

20. 39.3 m^2

21. 6480 m^3

22. 33.5 ft^3

23. 5086.8 ft^3

24. 9.220 cm

25. $y = 12$ cm; $z = 6$ cm

26. Answer varies.

Find the perimeter and area of each triangle.

14.

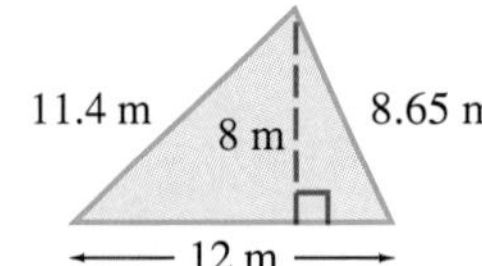

15.

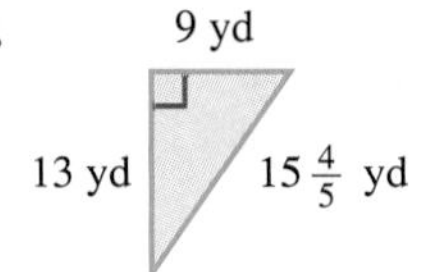

16. A triangle has angles that measure 18° and 55°. What does the third angle measure?

In problems 17–23, use 3.14 for π. Round your answers to the nearest tenth when necessary.

17. Find the radius.

18. Find the circumference.

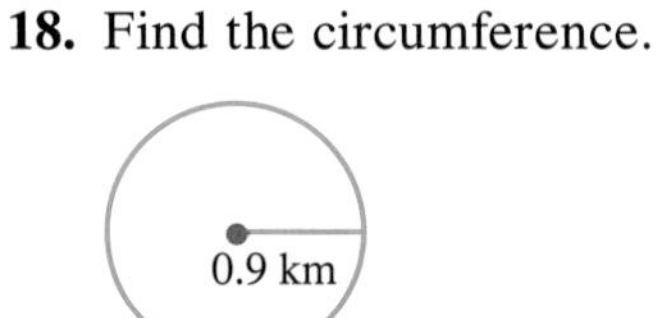

Find the area of each figure.

19.

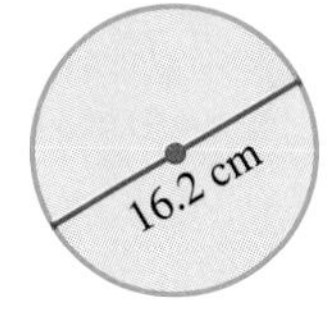

20.

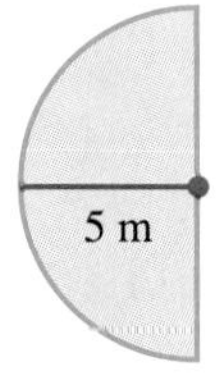

Find the volume of each figure.

21.

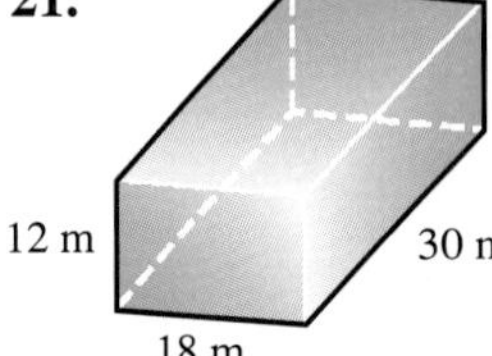

22.

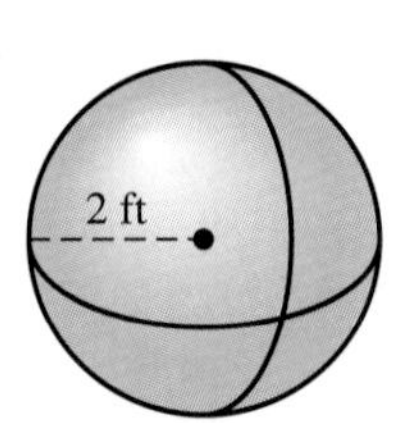

23.

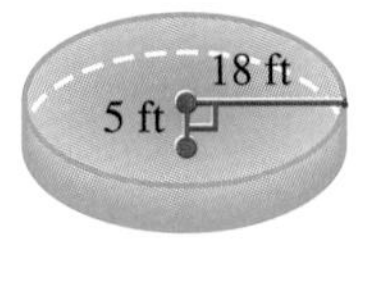

Find the mising lengths. Round your answers to the nearest thousandth when necessary.

24.

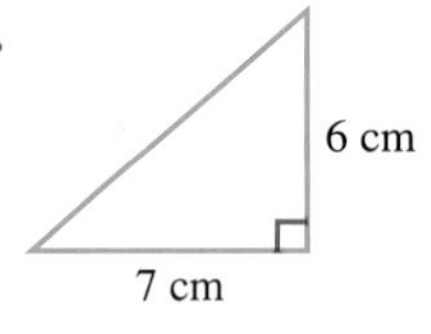

25. similar triangles

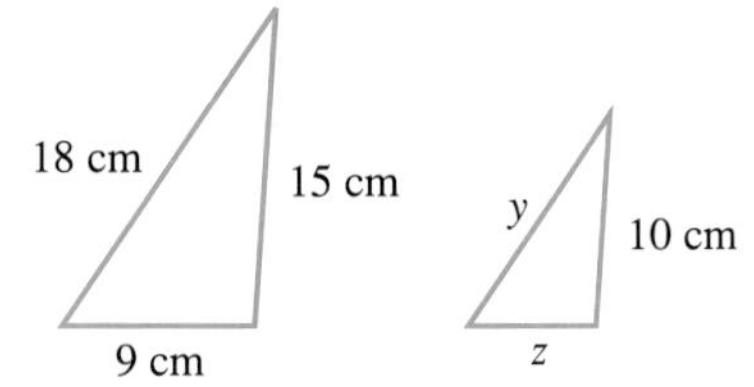

26. Explain the difference between cm, cm^2, and cm^3. In what types of geometry problems might you use each of these units?

NAME DATE HOUR

CUMULATIVE REVIEW EXERCISES CHAPTERS 1–8

≈ *Round the numbers in each problem so there is only one non-zero digit. Then add, subtract, multiply, or divide the rounded numbers, as indicated, to estimate the answer. Finally, solve for the exact answer.*

1. *estimate*

$$\begin{array}{r} 300 \\ 60{,}000 \\ +\ 6000 \\ \hline 66{,}300 \end{array}$$

exact

$$\begin{array}{r} 319 \\ 58{,}028 \\ +\ 6{,}227 \\ \hline 64{,}574 \end{array}$$

2. *estimate*

$$\begin{array}{r} 20 \\ -\ 10 \\ \hline 10 \end{array}$$

exact

$$\begin{array}{r} 20.07 \\ -\ 9.828 \\ \hline 10.242 \end{array}$$

3. *estimate*

$$\begin{array}{r} 4 \\ \times\ 7 \\ \hline 28 \end{array}$$

exact

$$\begin{array}{r} 3.664 \\ \times\ 7.3 \\ \hline 26.7472 \end{array}$$

4. *estimate*

$$\begin{array}{r} 30{,}000 \\ \times\ 70 \\ \hline 2{,}100{,}000 \end{array}$$

exact

$$\begin{array}{r} 28{,}419 \\ \times\ 73 \\ \hline 2{,}074{,}587 \end{array}$$

5. *estimate* $3\overline{)600}$ = 200 *exact* $2.8\overline{)562.24}$ = 200.8

6. *estimate* $50\overline{)5000}$ = 100 *exact* $52\overline{)4888}$ = 94

Add, subtract, multiply, or divide as indicated. Write answers to fraction problems in lowest terms and as whole or mixed numbers when possible.

7. $8 \div 3\frac{3}{5}$

$2\frac{2}{9}$

8. $1 - 0.0868$

0.9132

9. $81\overline{)5749}$

70 R79

10. $1\frac{7}{10} + 1\frac{7}{8}$

$3\frac{23}{40}$

11. $(0.006)(0.013)$

0.000078

12. $4\frac{2}{3} \cdot 2\frac{4}{7}$

12

13. $0.8 \div 3.64$ Round to nearest thousandth.

0.220

14. $6\frac{1}{6} - 1\frac{3}{4}$

$4\frac{5}{12}$

15. $752.6 + 83 + 0.485$

836.085

Simplify by using order of operations.

16. $16 - (10 - 2) \div 2 \cdot 3 + 5$

9

17. $2^4 \div \sqrt{64} + 6^2$

38

18. Write 0.0208 in words.

two hundred eight ten-thousandths

19. Write six hundred sixty-five and fifty-one hundredths in numbers.

665.51

20. Arrange in order from smallest to largest.
2.55 2.505 2.055 2.5005

2.055; 2.5005; 2.505; 2.55

Complete this chart.

fraction/mixed number	*decimal*	*percent*
21. $\frac{1}{20}$	0.05	**22.** 5%
$1\frac{3}{4}$	**23.** 1.75	**24.** 175%
25. $\frac{2}{5}$	**26.** 0.4	40%

Write each rate or ratio in lowest terms. Change to the same units when necessary.

27. 3 hours to 30 minutes; compare in minutes.

$\frac{6}{1}$

28. Last month there were 9 cloudy days and 21 sunny days. What was the ratio of sunny days to cloudy days?

$\frac{7}{3}$

Find the missing number in each proportion. Round your answer to hundredths, if necessary.

29. $\frac{5}{13} = \frac{x}{91}$

35

30. $\frac{207}{69} = \frac{300}{x}$

100

31. $\frac{4.5}{x} = \frac{6.7}{3}$

2.01 (rounded)

Solve each of the following.

32. 72 patients is what percent of 45 patients?

160%

33. $18 is 3% of what number of dollars?

$600

Convert the following measurements.

34. 5 minutes to seconds

300 seconds

35. 40 ounces to pounds

$2\frac{1}{2}$ or 2.5 pounds

36. 5 mL to liters

0.005 L

37. 1.25 kg to grams

1250 g

Write the most reasonable metric unit in each blank. Choose from km, m, cm, mm, L, mL, kg, g, mg.

38. Her wristwatch strap is 15 mm wide.

39. Jon added 2 L of oil to his car.

40. The child weighs 15 kg.

41. The bookcase is 90 cm high.

42. At what metric temperature does water freeze? 0°C

Find the perimeter or circumference and area of each figure. Use 3.14 for π.

43.

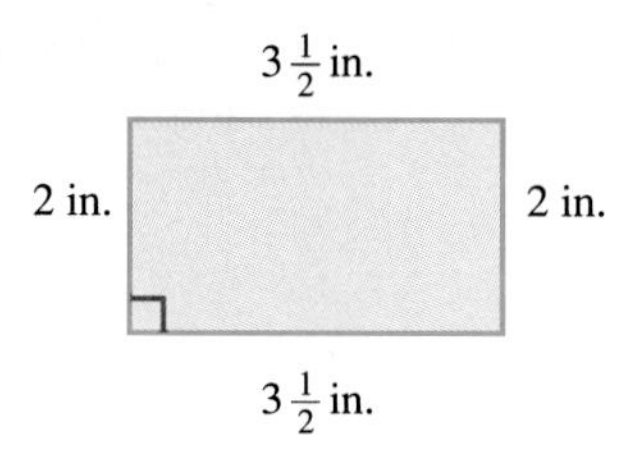

$P = 11$ in.
$A = 7$ in.^{2}

44.

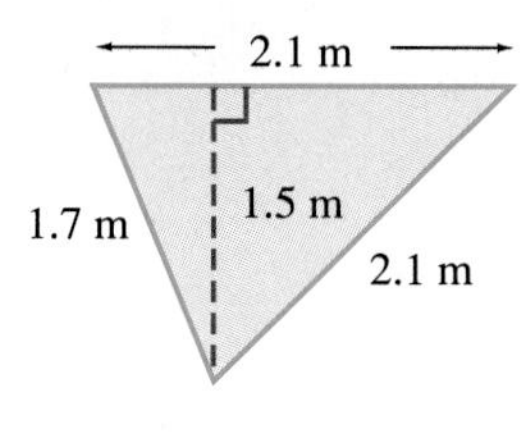

$P = 5.9$ m
$A = 1.575$ m^2

45.

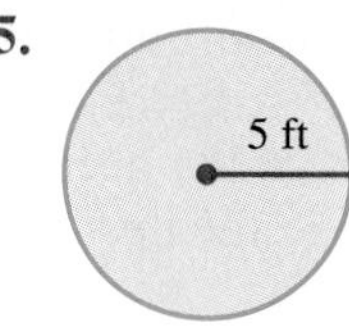

$C = 31.4$ ft
$A = 78.5$ ft^2

46.

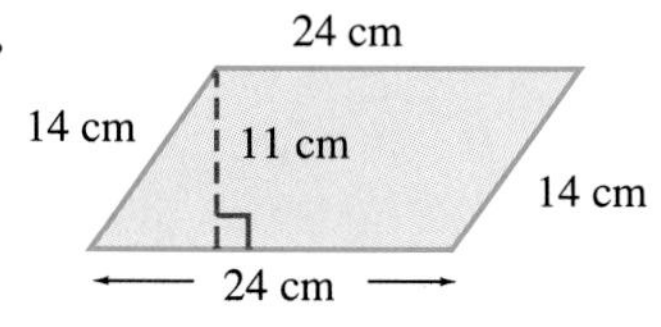

$P = 76$ cm
$A = 264$ cm^2

47.

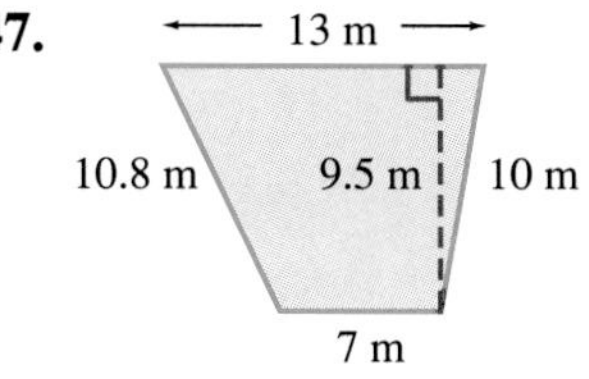

$P = 40.8$ m
$A = 95$ m^2

48.

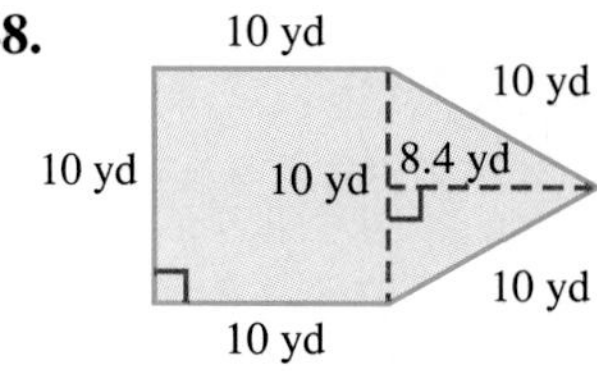

$P = 50$ yd
$A = 142$ yd^2

Find the length of the missing side in each figure. Round your answers to the nearest tenth.

49.

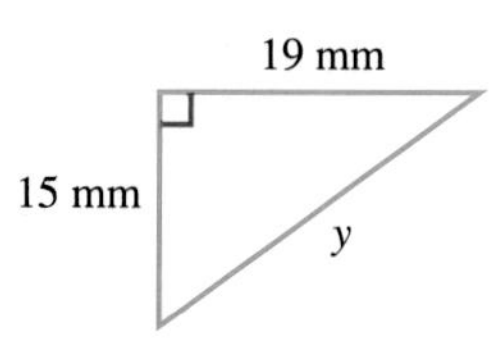

$y = 24.2$ mm

50. similar triangles

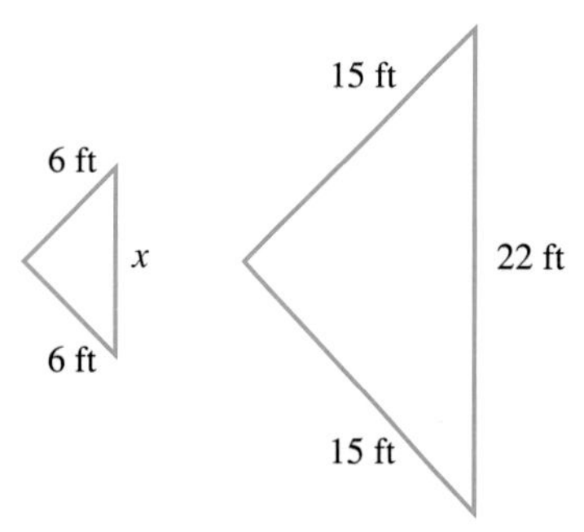

$x = 8.8$ ft

Solve the following word problems.

51. Mei Ling must earn 90 credits to receive an associate of arts degree. She has 53 credits. What percent of the necessary credits does she have? Round to the nearest whole percent.

59%

52. Which box of facial tissues is the best buy: 250 tissues for $2.99, 175 tissues for $1.89, or 95 tissues for $1.19?

175 tissues for $1.89

53. A coffee can has a diameter of 13 cm and a height of 17 cm. Find the volume of the can. Use 3.14 for π and round your answer to the nearest tenth.

2255.3 cm^3

54. Swimsuits are on sale in August at 65% off the regular price. How much will Lanece pay for a suit that has a regular price of $44?

$15.40

55. Steven bought $4\frac{1}{2}$ yards of canvas material to repair the tents used by the scout troop. He used $1\frac{2}{3}$ yards on one tent and $1\frac{3}{4}$ yards on another. How much material is left?

$1\frac{1}{12}$ yard

56. The cooks at a homeless shelter used 30 pounds of meat to make stew for 140 people. At that rate, how much meat is needed for stew to feed 200 people? Round to the nearest tenth.

42.9 pounds

57. Graciela needs 85 cm of yarn to make a tassel for one corner of a pillow. How many meters of yarn does she need to put a tassel on each corner of a square-shaped pillow?

3.4 m

58. A photograph measures 8 inches wide and 10 inches long. Earl put it in a frame that is 2 inches wide. Find the perimeter of the frame. Draw a sketch of the frame to help solve the problem.

52 inches

9 Basic Algebra

9.1 SIGNED NUMBERS

All the numbers you have studied so far in this book have been either 0 or greater than 0. Numbers greater than 0 are called *positive numbers.* For example, you have worked with these positive numbers:

salary of \$800
temperature of 98.6°F
length of $3\frac{1}{2}$ feet

1 Not all numbers are positive. For example, "15 degrees below 0" or "a loss of \$500" are expressed with numbers less than 0. Numbers less than 0 are called **negative numbers.** Zero is neither positive nor negative.

WRITING NEGATIVE NUMBERS

Write negative numbers with a *negative sign,* $-$.

For example, "15 degrees below 0" is written with a negative sign, as $-15°$. And "a loss of \$500" is written $-\$500$.

WORK PROBLEM 1 AT THE SIDE.

2 In **Section 3.5** you graphed positive numbers on a number line. Negative numbers can also be shown on a number line. Zero separates the positive numbers from the negative numbers on the number line. The number -5 is read "negative five."

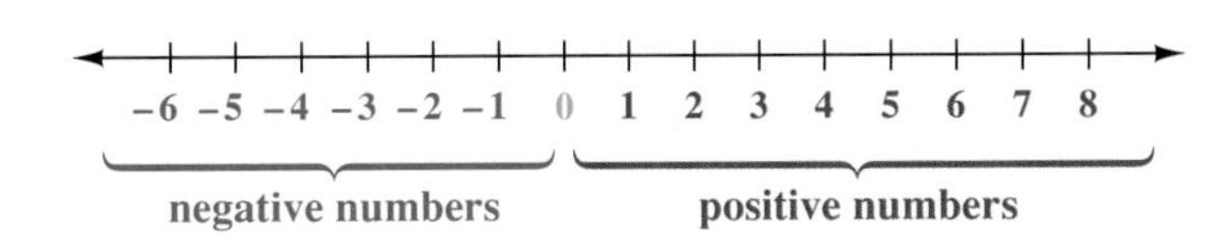

Note For every positive number there is a corresponding negative number on the opposite side of 0 on a number line.

When you work with both positive and negative numbers (and zero) we say you are working with **signed numbers.**

OBJECTIVES

1. Write negative numbers.
2. Use number lines.
3. Graph numbers.
4. Use the < and > symbols.
5. Find absolute value.
6. Find the opposite of a number.

FOR EXTRA HELP

Tape 12	SSM pp. 268–269	MAC: A IBM: A

1. Write each number.

(a) A temperature at the North Pole of 70 degrees below 0.

(b) Your checking account is overdrawn by 15 dollars.

(c) The altitude of a place 284 feet below sea level.

ANSWERS
1. (a) $-70°$ **(b)** $-\$15$ **(c)** -284 **ft**

WRITING POSITIVE NUMBERS

Positive numbers can be written in two ways:

1. Use a "+" sign. For example, +2 is "positive two."
2. Do not write any sign. For example, 3 is assumed to be "positive three."

WORK PROBLEM 2 AT THE SIDE.

3 The next example shows you how to graph signed numbers.

■ EXAMPLE 1 *Graphing Signed Numbers*

Graph **(a)** -4 **(b)** 3 **(c)** -1 **(d)** 0 **(e)** $1\frac{1}{4}$

Place a dot at the correct location for each number. ■

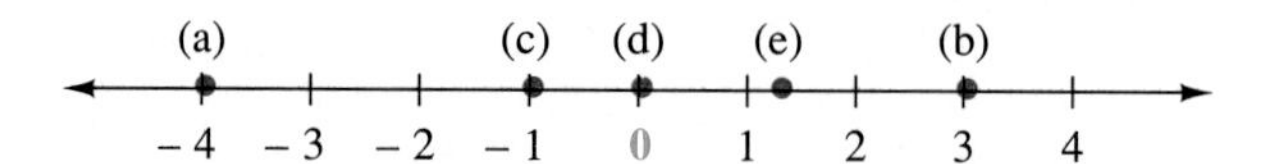

WORK PROBLEM 3 AT THE SIDE.

4 As shown on the following number line, 3 is to the left of 5.

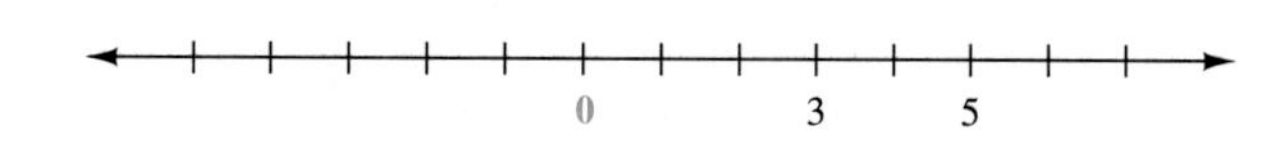

Also, 3 is *less than* 5.

Recall the following symbols for comparing two numbers.

$<$ means "is less than"

$>$ means "is greater than"

Use these symbols to write "3 is less than 5" as follows.

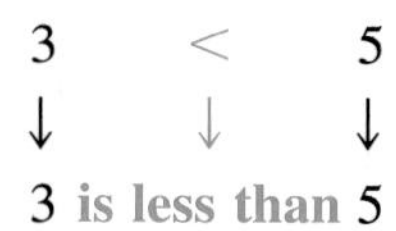

As this example suggests,

> The lesser of two numbers is the one farther to the *left* on a number line.

■ EXAMPLE 2 *Using the Symbols < and >*

Use this number line and $>$ or $<$ to make true statements.

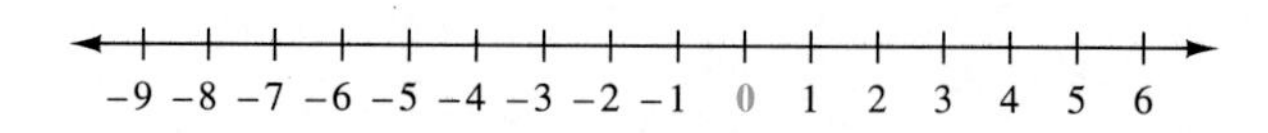

---

2. Write *positive, negative,* or *neither* for each number.

(a) -8

(b) $-\frac{3}{4}$

(c) 1

(d) 0

3. Graph each list of numbers.

(a) $-1, 1, -3, 3$

−4 −3 −2 −1 0 1 2 3 4

(b) $-2, 4, 0, -1, -4$

−4 −3 −2 −1 0 1 2 3 4

ANSWERS

2. (a) negative **(b)** negative **(c)** positive **(d)** neither

3. (a) −4 −3 −2 −1 0 1 2 3 4

(b) −4 −3 −2 −1 0 1 2 3 4

(a) On the number line, 2 is to the *left* of 6, so 2 is less than 6, or $2 < 6$.

(b) -9 is to the *left* of -4, so $-9 < -4$.

(c) 2 is to the *right* of -1, so $2 > -1$.

(d) $-4 < 0$ ■

Note When using $>$ and $<$, the *small* pointed end of the symbol points to the *smaller* (lesser) number.

WORK PROBLEM 4 AT THE SIDE. ▶

4. Place $<$ or $>$ in each blank to make a true statement.

(a) 4 _____ 0

(b) -1 _____ 0

(c) -3 _____ -1

(d) -8 _____ -9

(e) 0 _____ -3

5 In order to graph a number on the number line, you need to know two things:

1. Which *direction* it is from 0. It can be in a *positive* direction or a *negative* direction. You can tell the direction by looking for a positive or negative sign.
2. How *far* it is from zero. The *distance* from zero is the **absolute value** of a number.

Absolute value is indicated by two vertical bars. For example, $|6|$ is read "the absolute value of 6."

Note Absolute value is never negative, because it is a distance and distance is never negative.

■ **EXAMPLE 3** *Finding Absolute Value*

Find each of the following.

(a) $|8|$ **(b)** $|-8|$ **(c)** $|0|$ **(d)** $-|-3|$

(a) The distance from 0 to 8 is 8, so $|8| = 8$.

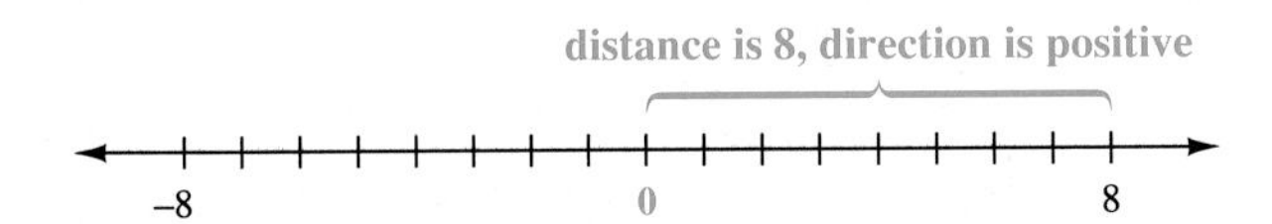

(b) The distance from 0 to -8 is also 8, so $|-8| = 8$.

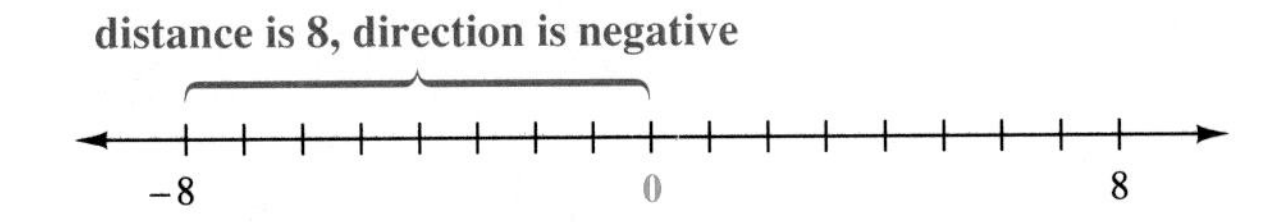

(c) $|0| = 0$

(d) First, $|-3| = 3$. Then,

$$-(3) = -3. \quad ■$$

Note A negative sign *outside* the absolute value bars is *not* affected by the absolute value bars. Therefore, your final answer is negative, as in Example 3(d) above.

WORK PROBLEM 5 AT THE SIDE. ▶

5. Find each of the following.

(a) $|5|$

(b) $|-5|$

(c) $|-17|$

(d) $-|-9|$

(e) $-|2|$

ANSWERS

4. (a) $>$ (b) $<$ (c) $<$ (d) $>$ (e) $>$

5. (a) 5 (b) 5 (c) 17 (d) -9 (e) -2

6. Give the opposite of each number.

(a) 4

(b) 10

(c) 49

(d) $\frac{2}{5}$

7. Find the opposite of each number.

(a) -4

(b) -10

(c) -25

(d) -1.9

(e) -0.85

(f) $-\frac{3}{4}$

ANSWERS

6. (a) -4 (b) -10 (c) -49 (d) $-\frac{2}{5}$

7. (a) 4 (b) 10 (c) 25 (d) 1.9 (e) 0.85 (f) $\frac{3}{4}$

6 Two numbers that are the same distance from 0 on a number line, but on opposite sides of 0, are called **opposites** of each other. As this number line shows, -3 and 3 are opposites of each other.

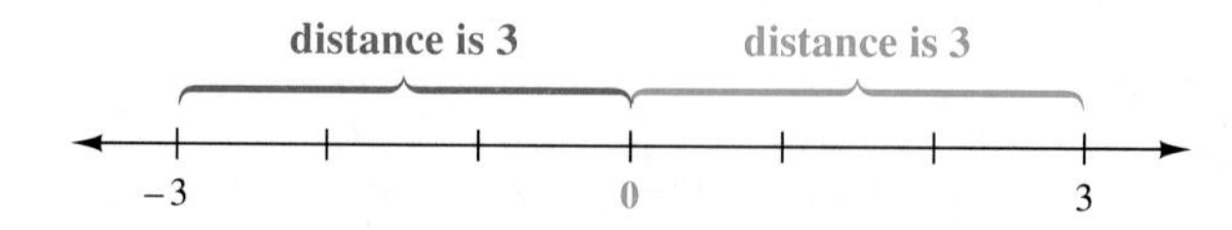

To indicate the opposite of a number, write a negative sign in front of the number.

EXAMPLE 4 *Finding Opposites*

Give the opposite of each number.

number	*opposite*
5	-5 ← Write a negative sign.
9	-9
$\frac{4}{5}$	$-\frac{4}{5}$
0	0

The opposite of 0 is 0. Zero is neither positive nor negative. ■

WORK PROBLEM 6 AT THE SIDE.

Some numbers have two negative signs, such as

$$-(-3).$$

The negative sign in front of -3 means the *opposite* of -3. The opposite of -3 is 3, so

$$-(-3) = 3.$$

Use the following rule to find the opposite of a negative number.

DOUBLE NEGATIVE RULE

$$-(-x) = x$$

The opposite of a negative number is positive.

EXAMPLE 5 *Finding Opposites*

Find the opposite of each number.

number	*opposite*	
-2	$-(-2) = 2$	by double negative rule
-9	$-(-9) = 9$	
$-\frac{1}{2}$	$-\left(-\frac{1}{2}\right) = \frac{1}{2}$ ■	

WORK PROBLEM 7 AT THE SIDE.

9.1 EXERCISES

NAME DATE HOUR

Write a signed number for each of the following.

1. The temperature is 12 degrees above zero

+12

2. She made a profit of \$920

+920

3. The price of the stock fell \$12.

−12

4. His checking account is overdrawn by \$30.

−30

5. The plane is 18,000 feet above sea level.

+18,000

6. The team lost 6 yards on that play.

−6

Write positive, negative, *or* neither *for each of the following numbers.*

7. 24

positive

8. −8

negative

9. $-\frac{7}{10}$

negative

10. $2\frac{1}{3}$

positive

11. 0

neither

12. +6

positive

13. −6.3

negative

14. −0.25

negative

Graph each of the following lists of numbers.

Example: $-3, -\frac{2}{3}, -5, 1, 2, \frac{3}{4}$

Solution: Place a dot on a number line for each number.

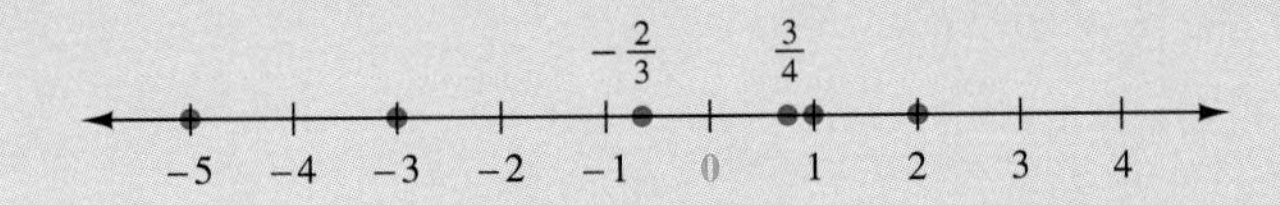

15. 4, −1, 2, 3, 0, −2

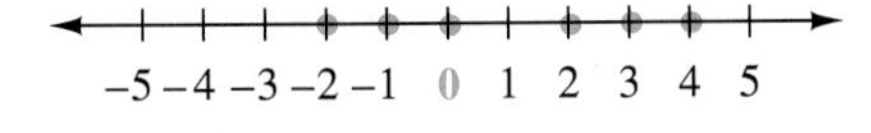

16. −5, −3, 1, 4, 0

−5 −4 −3 −2 −1 0 1 2 3 4 5

17. $-\frac{1}{2}, -3, -5, \frac{1}{2}, 1\frac{3}{4}, 3$

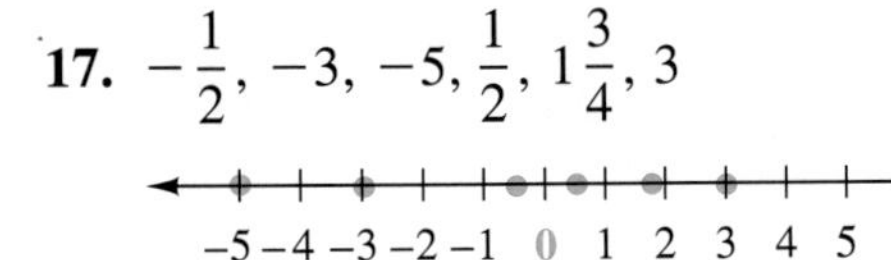

18. $-4, -\frac{3}{4}, -2, 4, 1, 2\frac{1}{2}$

−5 −4 −3 −2 −1 0 1 2 3 4 5

19. $-2, -4, -3\frac{1}{5}, -\frac{5}{8}, 1, 2$

−5 −4 −3 −2 −1 0 1 2 3 4 5

20. $-5, -3, -2, -4\frac{2}{3}, -1\frac{1}{2}, 0, 1$

−5 −4 −3 −2 −1 0 1 2 3 4 5

21. $3, 4.5, -1.5, 2.2, 0.1$

−5 −4 −3 −2 −1 0 1 2 3 4 5

22. $3.25, -1, 4.5, 1.25, 2$

−5 −4 −3 −2 −1 0 1 2 3 4 5

23. $-14, -11, -10.5, -13, -7.3$

−14 −12 −10 −8 −6

24. $-10, -7, -14.8, -9.25, -13.75$

−14 −12 −10 −8 −6

Write $<$ or $>$ in each of the following to make a true statement.

Examples:

$-4 \quad 2$ $\qquad$ $-5 \quad -9$

Solutions:

Because -4 is to the *left* of 2 on a number line, -4 is less than 2.

$-4 < 2$

Because -5 is to the *right* of -9 on a number line, -5 is greater than -9.

$-5 > -9$

25. $9 < 14$ **26.** $6 < 11$ **27.** $0 > -2$ **28.** $0 < 2$

29. $-6 < 3$ **30.** $-4 < 7$ **31.** $1 > 0$ **32.** $-1 < 0$

33. $-11 < -2$ **34.** $-5 < -1$ **35.** $-7 < -4$ **36.** $-5 > -6$

37. $2 > -1$ **38.** $4 > -9$ **39.** $-5 > -10$ **40.** $-15 > -20$

Find the absolute value of the following.

Examples: $|8|, |-7|, -|-2|$
Solutions: $|8| = \mathbf{8}$ $\quad |-7| = \mathbf{7}$ $\quad -|-2| = \mathbf{-2}$

41. $|5|$ 5

42. $|12|$ 12

43. $|-5|$ 5

44. $|-16|$ 16

45. $|-1|$ 1

46. $|-14|$ 14

47. $|251|$ 251

48. $|397|$ 397

49. $|0|$ 0

50. $|-199|$ 199

51. $\left|-\frac{1}{2}\right|$ $\frac{1}{2}$

52. $\left|-\frac{9}{5}\right|$ $\frac{9}{5}$

53. $|-9.5|$ 9.5

54. $|-0.72|$ 0.72

55. $|0.618|$ 0.618

56. $|4.7|$ 4.7

57. $\left|\frac{3}{4}\right|$ $\frac{3}{4}$

58. $\left|\frac{7}{3}\right|$ $\frac{7}{3}$

59. $-\left|-\frac{5}{2}\right|$ $-\frac{5}{2}$

60. $-\left|-\frac{1}{3}\right|$ $-\frac{1}{3}$

61. $-|-9|$ -9

62. $-|-12|$ -12

63. $-|4|$ -4

64. $-|20|$ -20

Give the opposite of each number.

Examples: **Solutions:**

number	*opposite*
8	**−8**
−5	$-(-5) = \mathbf{5}$
0	**0**

65. 2 -2

66. 7 -7

67. -54 54

68. -75 75

69. -11 11

70. -24 24

71. 163 -163

72. 502 -502

73. $\frac{4}{3}$ $-\frac{4}{3}$

74. $\frac{5}{8}$ $-\frac{5}{8}$

75. $-4\frac{1}{2}$ $4\frac{1}{2}$

76. $-1\frac{2}{3}$ $1\frac{2}{3}$

77. 5.2 −5.2

78. 3.7 −3.7

79. −1.4 1.4

80. −0.65 0.65

81. In your own words, explain opposite numbers. Include an example of opposite numbers and draw a number line to illustrate your example.

Answer varies.

82. Explain why the opposite of zero is zero.

Answer varies.

83. Describe three different situations at home, work, or school where you have used negative numbers.

Answer varies.

84. Explain in your own words why absolute value is never negative.

Answer varies.

Write true or false for each statement.

85. $|-5| > 0$

true

86. $|-12| > |-15|$

false

87. $0 < -(-6)$

true

88. $-9 < -(-6)$

true

89. $-|-4| < -|-7|$

false

90. $-|-0| > 0$

false

PREVIEW EXERCISES

Add or subtract the following numbers. (For help, see ***Section 3.3.****)*

91. $\frac{3}{4} + \frac{1}{5}$ $\frac{19}{20}$

92. $\frac{3}{10} + \frac{3}{8}$ $\frac{27}{40}$

93. $\frac{5}{6} - \frac{1}{4}$ $\frac{7}{12}$

94. $\frac{2}{3} - \frac{5}{9}$ $\frac{1}{9}$

9.2 ADDITION AND SUBTRACTION OF SIGNED NUMBERS

You can show a positive number on a number line by drawing an arrow pointing to the right. In the following examples both arrows represent positive 4 units.

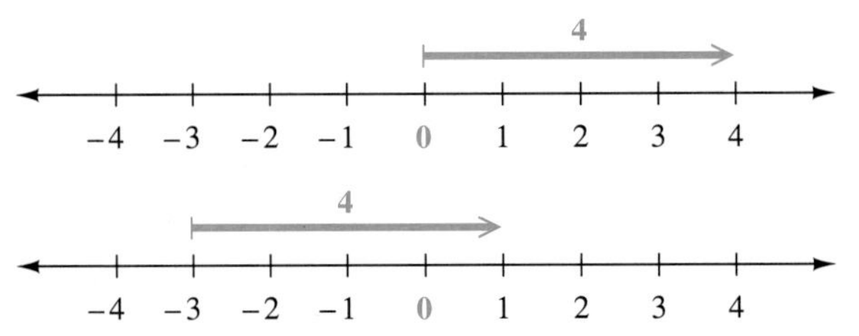

Draw arrows pointing to the left to show negative numbers. Both of the following arrows represent −3 units.

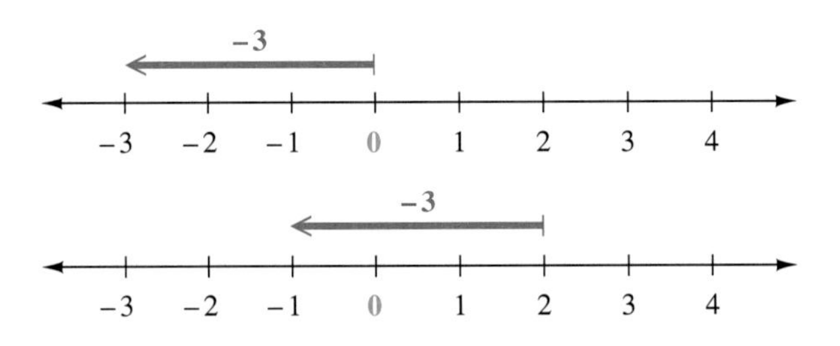

WORK PROBLEM 1 AT THE SIDE.

1 You can use a number line to add signed numbers. For example, the next number line shows how to add 2 and 3.

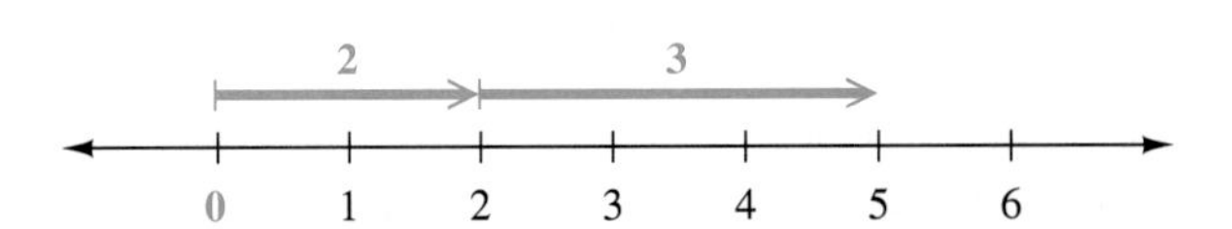

Add 2 and 3 by starting at zero and drawing an arrow 2 units to the right. From the end of this arrow, draw another arrow 3 units to the right. This second arrow ends at 5, showing that

$$2 + 3 = 5.$$

EXAMPLE 1 *Adding Signed Numbers by Using a Number Line*

Add by using a number line.

(a) $4 + (-1)$ **(b)** $-6 + 2$ **(c)** $-3 + (-5)$

(a) $4 + (-1)$

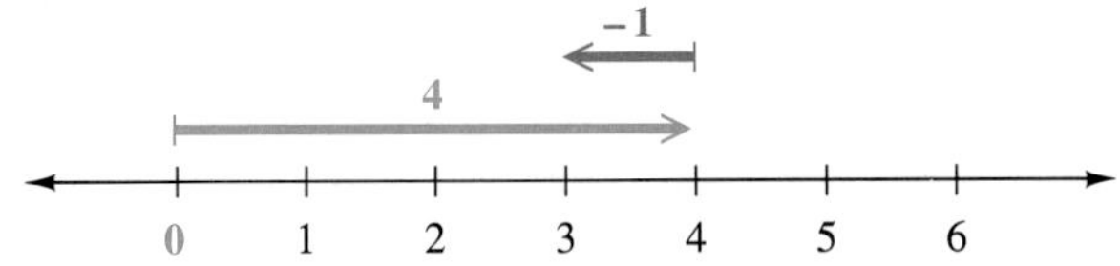

(*Note:* Always start at 0 when adding on a number line.)

Start at zero and draw an arrow 4 units to the right. From the end of this arrow, draw an arrow 1 unit to the *left*. (Remember to go to the left for a negative number.) This second arrow ends at 3, so

$$4 + (-1) = 3.$$

OBJECTIVES

1. Add signed numbers by using a number line.
2. Add signed numbers without using a number line.
3. Find the additive inverse of a number.
4. Subtract signed numbers.
5. Add or subtract a series of signed numbers.

FOR EXTRA HELP

Tape 12 | SSM pp. 270–273 | MAC: A IBM: A

1. Complete each arrow so it represents the indicated number of units.

(a)

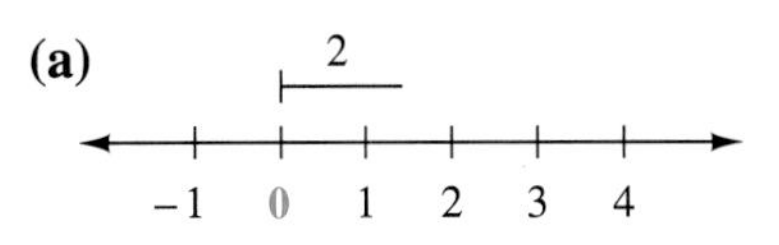

(b)

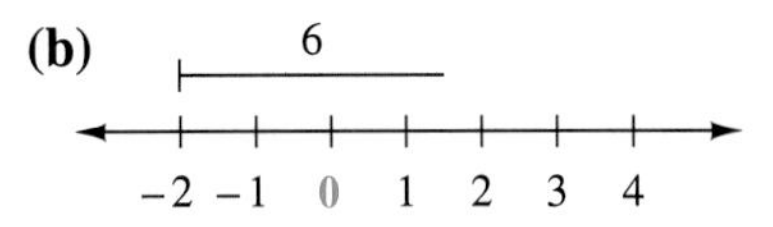

(c)

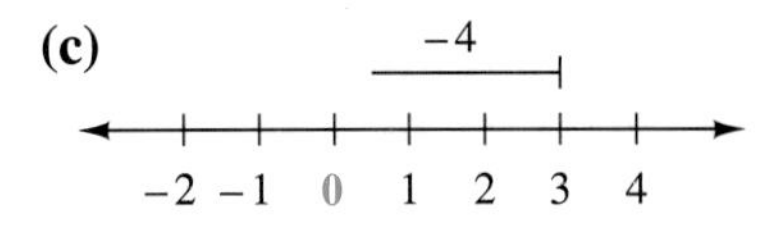

(d)

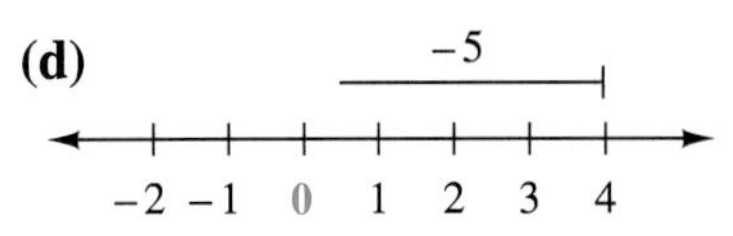

ANSWERS

1. (a)

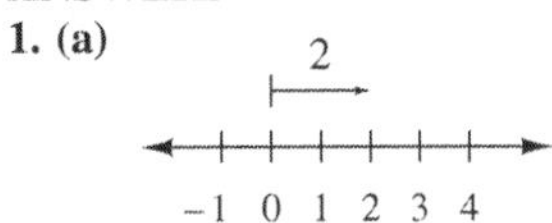

(b)

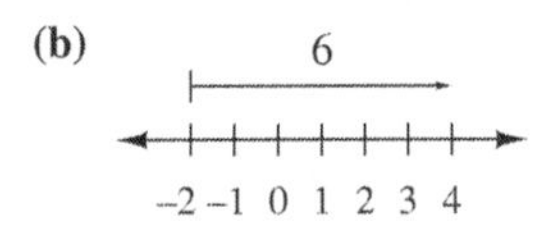

(c)

−4

−2 −1 0 1 2 3 4

(d)

−5

−2 −1 0 1 2 3 4

(b) $-6 + 2$

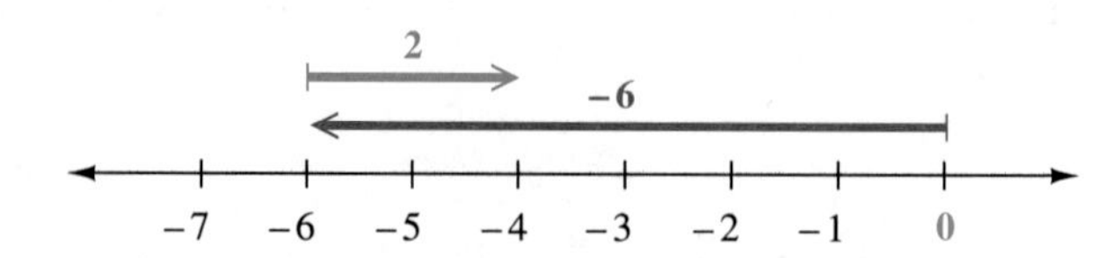

Draw an arrow from zero going 6 units to the left. From the end of this arrow, draw an arrow 2 units to the right. This second arrow ends at -4, so

$$-6 + 2 = -4.$$

(c) $-3 + (-5)$

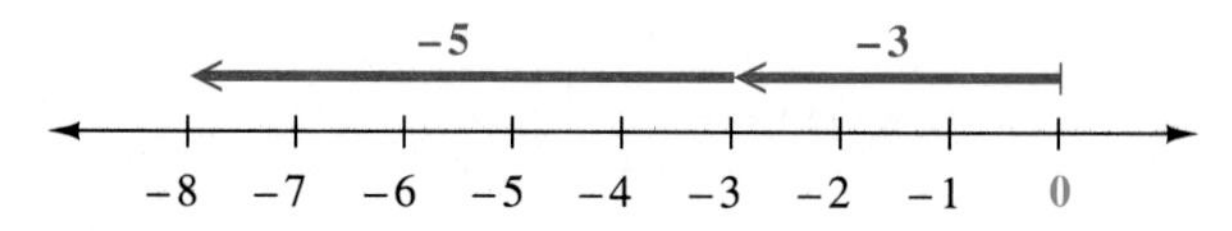

As the arrows along the number line show,

$$-3 + (-5) = -8. \blacksquare$$

WORK PROBLEM 2 AT THE SIDE.

2 After working with number lines for awhile, you will see ways to add signed numbers without drawing arrows. You already know how to add two positive numbers (from Chapter 1). Here are the steps for adding two negative numbers.

ADDING NEGATIVE NUMBERS

Step 1 Add the absolute values of the numbers.
Step 2 Write a negative sign in front of the sum.

■ **EXAMPLE 2** *Adding Negative Numbers*
Add (without number lines).

(a) $-4 + (-12)$ **(b)** $-5 + (-25)$ **(c)** $-11 + (-46)$

(d) $-\frac{3}{4} + \left(-\frac{1}{2}\right)$

(a) $-4 + (-12)$
The absolute value of -4 is 4.
The absolute value of -12 is 12.
Add the absolute values.

$$4 + 12 = 16$$

Write a negative sign in front of the sum.

$$-4 + (-12) = -16$$ Write a negative sign in front of 16.

2. Draw arrows to find each of the following.

(a) $3 + (-2)$

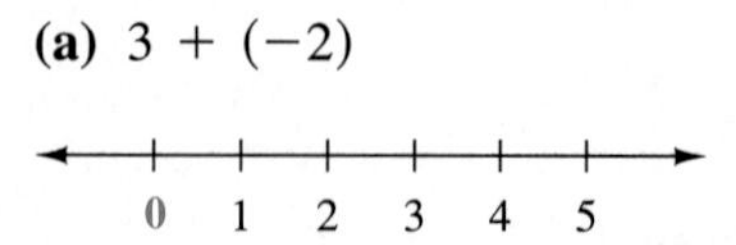

(b) $-4 + 1$

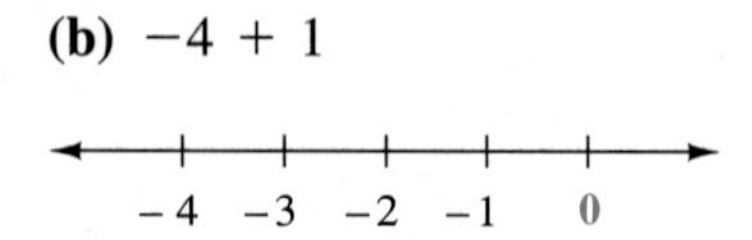

(c) $-3 + 7$

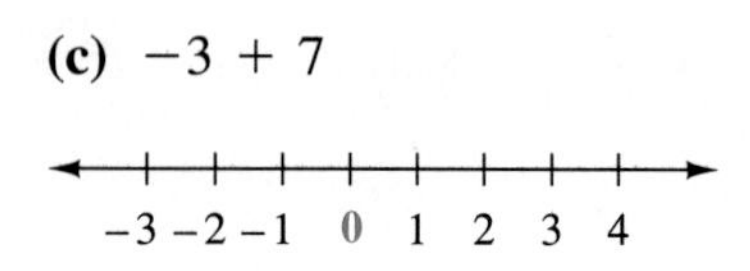

(d) $-1 + (-4)$

−6 −5 −4 −3 −2 −1 0

ANSWERS

2. (a) $3 + (-2) = 1$

(b) $-4 + 1 = -3$

(c) $-3 + 7 = 4$

(d) $-1 + (-4) = -5$

(b) $-5 + (-25) = -30$ ← Sum of absolute values, with a negative sign written in front of 30.

(c) $-11 + (-46) = -57$

(d) The absolute value of $-\frac{3}{4}$ is $\frac{3}{4}$, and the absolute value of $-\frac{1}{2}$ is $\frac{1}{2}$. Add the absolute values. Check that the answer is in lowest terms.

$$\frac{3}{4} + \frac{1}{2} = \frac{3}{4} + \frac{2}{4} = \frac{5}{4} \quad \text{Lowest terms.}$$

Write a negative sign in front of the sum.

$$-\frac{3}{4} + \left(-\frac{1}{2}\right) = -\frac{5}{4} \quad \text{Write a negative sign.}$$ ■

WORK PROBLEM 3 AT THE SIDE. ▶▶

Use the following steps to add two numbers with *different* signs.

ADDING TWO NUMBERS WITH DIFFERENT SIGNS

Step 1 *Subtract* the smaller absolute value from the larger absolute value.

Step 2 Write the sign of the number with the *larger* absolute value in front of the answer.

■ **EXAMPLE 3** *Adding Two Numbers with Different Signs*

Add:

(a) $8 + (-3)$ **(b)** $-12 + 4$ **(c)** $15 + (-21)$

(d) $-13 + 9$ **(e)** $-\frac{1}{2} + \frac{2}{3}$.

(a) $8 + (-3)$

Find this sum with a number line as follows.

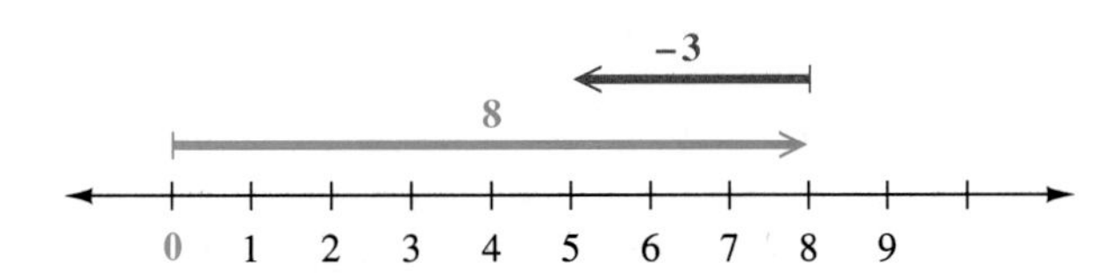

Because the top arrow ends at 5,

$$8 + (-3) = 5.$$

Find the sum by using the rule as follows. First, find the absolute value of each number.

$$|8| = 8 \qquad |-3| = 3$$

3. Add.

(a) $-4 + (-4)$

(b) $-3 + (-20)$

(c) $-31 + (-5)$

(d) $-10 + (-8)$

(e) $-\frac{9}{10} + \left(-\frac{3}{5}\right)$

ANSWERS

3. (a) -8 (b) -23 (c) -36 (d) -18 (e) $-\frac{3}{2}$

4. Add.

(a) $10 + (-2)$

(b) $-7 + 8$

(c) $-11 + 11$

(d) $23 + (-32)$

(e) $-\frac{7}{8} + \frac{1}{4}$

ANSWERS

4. (a) 8 (b) 1 (c) 0 (d) −9 (e) $-\frac{5}{8}$

Subtract the absolute value of the smaller number from the larger number.

$$8 - 3 = 5$$

Here the positive number 8 has the larger absolute value, so the answer is positive.

$$8 + (-3) = 5 \leftarrow \text{positive answer}$$

(b) $-12 + 4$
First, find absolute values.

$$|-12| = 12 \qquad |4| = 4$$

Subtract:

$$12 - 4 = 8$$

The negative number -12 has the larger absolute value, so the answer is negative.

$$-12 + 4 = -8$$

Write negative sign in front of the answer because -12 has the larger absolute value.

(c) $15 + (-21) = -6$

(d) $-13 + 9 = -4$

(e) $-\frac{1}{2} + \frac{2}{3}$

The absolute value of $-\frac{1}{2}$ is $\frac{1}{2}$, and the absolute value of $\frac{2}{3}$ is $\frac{2}{3}$. Subtract the absolute value of the smaller number from the larger.

$$\frac{2}{3} - \frac{1}{2} = \frac{4}{6} - \frac{3}{6} = \frac{1}{6}$$

Because the positive number $\frac{2}{3}$ has the larger absolute value, the answer is positive.

$$-\frac{1}{2} + \frac{2}{3} = \frac{1}{6} \leftarrow \text{positive answer}$$ ■

WORK PROBLEM 4 AT THE SIDE.

3 Recall that the opposite of 9 is -9, and the opposite of -4 is $-(-4)$, or 4. Add these opposites as follows.

$$9 + (-9) = 0 \quad \textit{and} \quad -4 + 4 = 0$$

The sum of a number and its opposite is always 0. For this reason, opposites are also called **additive inverses** of each other.

ADDITIVE INVERSE

The opposite of a number is called its **additive inverse.** The sum of a number and its opposite is zero.

EXAMPLE 4 *Finding the Additive Inverse*
This chart shows you several numbers and the additive inverse of each.

number	*additive inverse*	*sum of number and inverse*
6	−6	$6 + (-6) = 0$
−8	−(−8) or 8	$(-8) + 8 = 0$
4	−4	$4 + (-4) = 0$
−3	−(−3) or 3	$-3 + 3 = 0$
$\frac{5}{8}$	$-\frac{5}{8}$	$\frac{5}{8} + \left(-\frac{5}{8}\right) = 0$
0	0	$0 + 0 = 0$ ■

WORK PROBLEM 5 AT THE SIDE.

4 You may have noticed that negative numbers are often written with parentheses, like (−8). This is especially helpful when subtracting because the − sign is used both to indicate a **negative number** and to indicate **subtraction.**

Example	*How to say it*
(−5)	**negative five**
8	positive eight
−3 − 2	**negative three minus** positive two
6 − (−4)	positive six **minus negative four**
−7 − (−1)	**negative seven minus negative one**

WORK PROBLEM 6 AT THE SIDE.

Recall from **Section 1.3** that a subtraction problem can be written as an addition problem. That is helpful when working with signed numbers. For example, you know that $6 - 4 = 2$. But you get the same result by *adding* 6 and the *opposite* of 4, that is, $6 + (-4)$.

$$6 - 4 = 2$$
$$6 + (-4) = 2$$
same result

This suggests the following definition of subtraction.

DEFINING SUBTRACTION

The difference of two numbers, a and b, is

$$a - b = a + (-b).$$

Subtract two numbers by adding the first number and the opposite (additive inverse) of the second.

5. Give the additive inverse of each number.

(a) 12

(b) −9

(c) −3.5

(d) $-\frac{7}{10}$

(e) 0

6. Write each example in words.

(a) −7 − 2

(b) −10

(c) 3 − (−5)

(d) 4

(e) −8 − (−6)

ANSWERS

5. (a) −12 (b) 9 (c) 3.5 (d) $\frac{7}{10}$ (e) 0

6. (a) negative seven minus positive two
(b) negative ten
(c) positive three minus negative five
(d) positive four
(e) negative eight minus negative six

7. Subtract.

(a) $-6 - 5$

(b) $3 - (-10)$

(c) $-8 - (-2)$

(d) $4 - 9$

(e) $-7 - (-15)$

(f) $-\frac{2}{3} - \left(-\frac{5}{12}\right)$

ANSWERS

7. (a) -11 (b) 13 (c) -6 (d) -5 (e) 8 (f) $-\frac{1}{4}$

SUBTRACTING SIGNED NUMBERS

To subtract two signed numbers, add the opposite of the second number to the first number. Use these steps.

Step 1 Change the subtraction sign to addition.
Step 2 Change the sign of the second number to its opposite.
Step 3 Proceed as in addition.

■ EXAMPLE 5 *Subtracting Signed Numbers*

Subtract:

(a) $8 - 11$ **(b)** $-9 - 15$ **(c)** $-5 - (-7)$

(d) $7.6 - (-8.3)$ **(e)** $\frac{5}{8} - \left(-\frac{1}{2}\right)$.

(a) $8 - 11$

The first number, 8, stays the same. Change the subtraction sign to addition. Change the sign of the second number to its opposite.

positive 8 stays the same

$$\begin{array}{ccc} 8 & - & 11 \\ \downarrow & & \downarrow \\ 8 & + & (-11) \end{array}$$

positive 11 changed to its opposite -11

subtraction changed to addition

Now add.

$$8 + (-11) = -3$$

$$\text{So } 8 - 11 = -3 \text{ also.}$$

(b)
$$\begin{array}{l} -9 - 15 \\ \downarrow \quad \downarrow \\ -9 + (-15) = -24 \end{array}$$

positive 15 is changed to its opposite (-15)

(c)
$$\begin{array}{l} -5 - (-7) \\ \downarrow \quad \downarrow \\ -5 + (+7) = 2 \end{array}$$

negative 7 is changed to its opposite $(+7)$

(d)
$$\begin{array}{l} 7.6 - (-8.3) \\ \downarrow \quad \downarrow \\ 7.6 + (+8.3) = 15.9 \end{array}$$

(e) $\frac{5}{8} - \left(-\frac{1}{2}\right) = \frac{5}{8} + \left(\frac{1}{2}\right) = \frac{9}{8} = 1\frac{1}{8}$ ■

Note The pattern in each example is:

1st number − 2nd number
= 1st number + opposite of 2nd number

◀ WORK PROBLEM 7 AT THE SIDE.

5 If a problem involves both addition and subtraction, use the order of operations and work from left to right.

EXAMPLE 6 *Combining Addition and Subtraction of Signed Numbers*

Perform the operations from left to right.

(a) $-6 + (-11) - (-5)$ **(b)** $4 - (-3) + (-9)$

(a) $\underbrace{-6 + (-11)} - (-5)$

$-17 - (-5)$

$\underbrace{-17 + (+5)}$

-12

(b) $\underbrace{4 - (-3)} + (-9)$

$\underbrace{7 + (-9)}$

-2 ■

WORK PROBLEM 8 AT THE SIDE.

EXAMPLE 7 *Using the Order of Operations to Combine More Than Two Numbers*

Find each sum.

(a) $\underbrace{-7 + 12} + (-3)$

$\underbrace{5 + (-3)}$

2

(b) $\underbrace{14 + (-9)} - (-8) + 10$

$\underbrace{5 - (-8)} + 10$

$\underbrace{13 + 10}$

23

(c)

$$\begin{array}{r} -6.3 \\ -14.9 \\ 8.5 \\ -7.4 \\ \underline{5.2} \end{array}$$

Start at the top.

$$\begin{array}{r} -6.3 \\ -14.9 \\ 8.5 \\ -7.4 \\ \underline{5.2} \end{array} \rightarrow \begin{array}{r} -21.2 \\ 8.5 \\ -7.4 \\ \underline{5.2} \end{array} \rightarrow \begin{array}{r} -12.7 \\ -7.4 \\ \underline{5.2} \end{array} \rightarrow \begin{array}{r} -20.1 \\ \underline{5.2} \\ -14.9 \end{array}$$ ■

WORK PROBLEM 9 AT THE SIDE.

8. Perform the operations from left to right.

(a) $6 - 7 + (-3)$

(b) $-2 + (-3) - (-5)$

(c) $-3 - (-9) - (-5)$

(d) $8 - (-2) + (-6)$

9. Add.

(a) $-1 - 2 + 3 - 4$

(b) $7 - 6 - 5 + (-4)$

(c) $-6 + (-15) - (-19) + (-25)$

(d)
$$\begin{array}{r} -19.2 \\ -6.7 \\ 15.8 \\ 17.1 \\ \underline{-\ 5.4} \end{array}$$

ANSWERS

8. (a) −4 (b) 0 (c) 11 (d) 4

9. (a) −4 (b) −8 (c) −27 (d) 1.6

QUEST FOR NUMERACY

Phone Home

This chart shows the times during the week when you get a discount on long distance telephone calls.

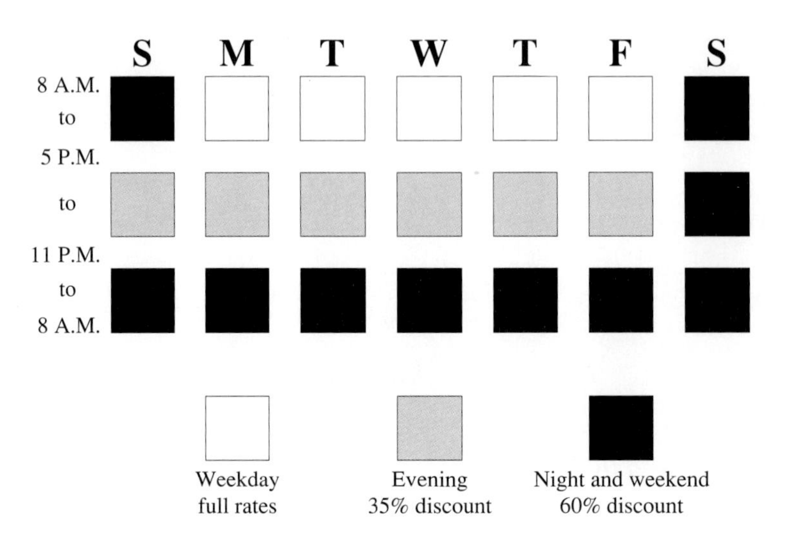

Use the information on discounts to complete the table below. Round all costs to the nearest cent. Use your calculator.

Time Call is Started	*Cost of Call at Full Rate*	*Percent Discount*	*Discounted Cost of Call*
6:30 A.M. Friday	$8.45	60%	$ 3.38
6:30 P.M. Sunday	$15.03	35%	$ 9.77
11:01 A.M. Wednesday	$18.88	0%	$18.88
8:15 P.M. Saturday	$11.25	60%	$ 4.50
5:00 P.M. Tuesday	$23.95	35%	$15.57

Extend this activity by asking students to check the front of their local phone book for information on long distance discounts. Or have them bring in their own phone bills. Are the discount periods similar or different from the ones shown here? What happens if you start the call during one period and hang up in another period? Why might the company choose these particular times for discounts?

Add by using the number line.

1. $-2 + 5$ 3

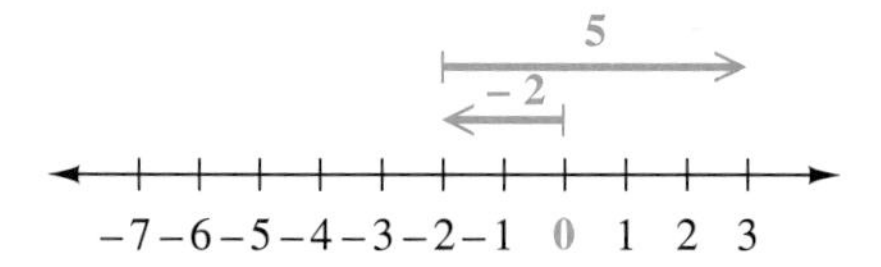

2. $-3 + 4$ 1

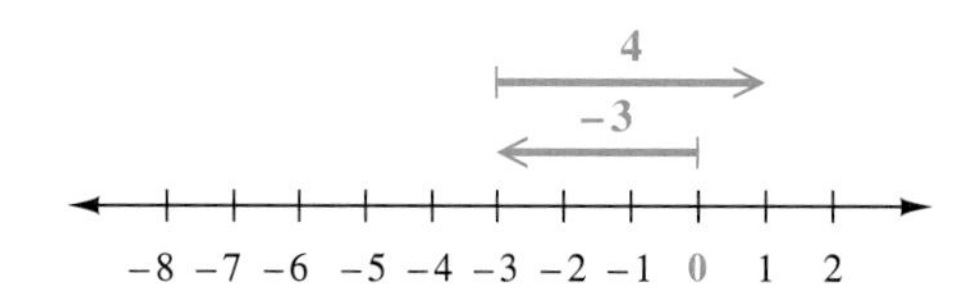

3. $-5 + (-2)$ -7

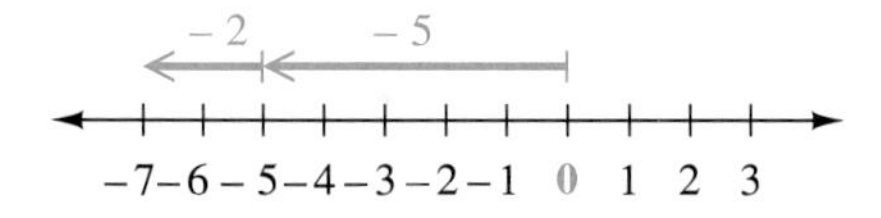

4. $-2 + (-2)$ -4

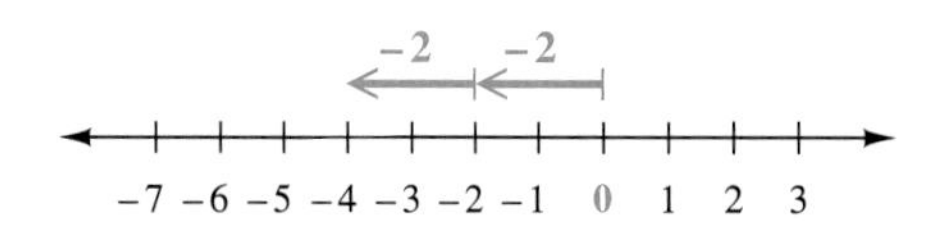

5. $3 + (-4)$ -1

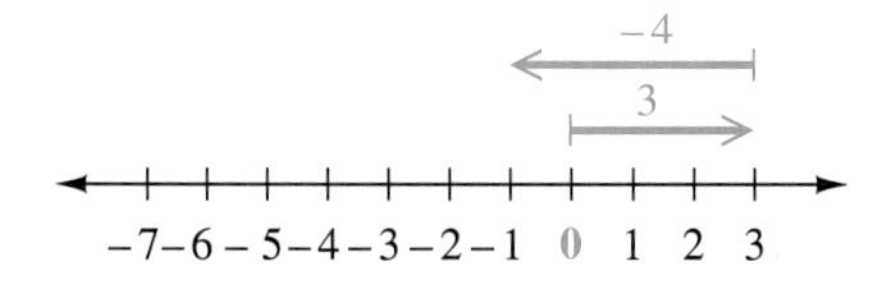

6. $5 + (-1)$ 4

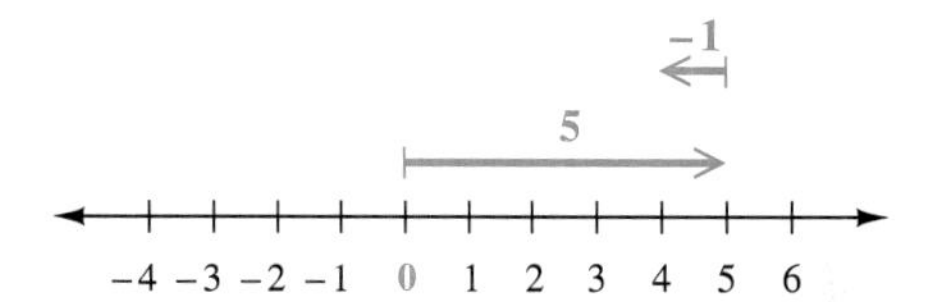

Add.

Example:
$-4 + (-11)$

Solution:
Add absolute values.
$|-4| = 4 \quad |-11| = 11$
$4 + 11 = 15$

Write a negative sign in front of the sum, because both numbers are negative.

$-4 + (-11) = \mathbf{-15}$

7. $-8 + 5$ -3

8. $-3 + 2$ -1

9. $-1 + 8$ 7

10. $-4 + 10$ 6

11. $-2 + (-5)$ -7

12. $-7 + (-3)$ -10

13. $6 + (-5)$ 1

14. $11 + (-3)$ 8

Write and solve an addition problem for each situation.

15. The football team lost 8 yards on the first play and lost 9 yards on the second play.

$-8 + (-9) = -17$

16. Carl lost 7 pounds last month on his diet and another 4 pounds this month.

$-7 + (-4) = -11$

17. Nicole's checking account was overdrawn by \$52.50. She deposited \$50 in the account.

$-\$52.50 + \$50 = -\$2.50$

18. \$48.40 was stolen from Jay's car. He got \$30 of it back.

$-\$48.40 + \$30 = -\$18.40$

Add.

19. $7.8 + (-14.6)$

-6.8

20. $4.9 + (-8.1)$

-3.2

21. $-\frac{1}{2} + \frac{3}{4}$

$\frac{1}{4}$

22. $-\frac{2}{3} + \frac{5}{6}$

$\frac{1}{6}$

23. $-\frac{7}{10} + \frac{2}{5}$

$-\frac{3}{10}$

24. $-\frac{3}{4} + \frac{3}{8}$

$-\frac{3}{8}$

25. $-\frac{7}{3} + \left(-\frac{5}{9}\right)$

$-\frac{26}{9}$ or $-2\frac{8}{9}$

26. $-\frac{8}{5} + \left(-\frac{3}{10}\right)$

$-\frac{19}{10}$ or $-1\frac{9}{10}$

27. Explain in your own words how to add two numbers with different signs. Include two examples in your explanation, one that has a positive answer and one that has a negative answer.

Answer varies.

28. Work these two examples:
(a) $-5 + 3$ **(b)** $3 + (-5)$
Explain why the answers are the same.

Answer varies.

Give the additive inverse of each number.

	number	*additive inverse*		*number*	*additive inverse*		*number*	*additive inverse*		*number*	*additive inverse*
29.	3	-3	**30.**	4	-4	**31.**	-9	9	**32.**	-14	14
33.	$\frac{1}{2}$	$-\frac{1}{2}$	**34.**	$\frac{7}{8}$	$-\frac{7}{8}$	**35.**	-6.2	6.2	**36.**	-0.5	0.5

Subtract by changing subtraction to addition.

Examples: $-8 - (-2)$ $\quad$ $7 - 11$

Solutions: Change subraction to addition. Change the sign of the second number to its opposite.

$-8 - (-2)$ $\quad$ $7 - 11$
↓ ↓ $\quad$ ↓ ↓
$-8 + (+2) = \mathbf{-6}$ $\quad$ $7 + (-11) = \mathbf{-4}$

37. $19 - 5$ $\quad 14$

38. $24 - 11$ $\quad 13$

39. $10 - 12$ $\quad -2$

40. $1 - 8$ $\quad -7$

41. $7 - 19$ $\quad -12$

42. $2 - 17$ $\quad -15$

43. $-15 - 10$ $\quad -25$

44. $-10 - 4$ $\quad -14$

45. $-9 - 14$ -23

46. $-3 - 11$ -14

47. $-3 - (-8)$ 5

48. $-1 - (-4)$ 3

49. $6 - (-14)$ 20

50. $8 - (-1)$ 9

51. $1 - (-10)$ 11

52. $6 - (-1)$ 7

53. $-\frac{7}{10} - \frac{4}{5}$

$-\frac{3}{2}$ or $-1\frac{1}{2}$

54. $-\frac{8}{15} - \frac{3}{10}$

$-\frac{5}{6}$

55. $-10 - (-4)$

-6

56. $-9 - (-5)$

-4

57. $-6.4 - (-2.8)$

-3.6

58. $-2 - (-3.9)$

1.9

59. Explain the purpose of the "−" sign in each of these examples.
(a) $6 - 9$ **(b)** (-9) **(c)** $-(-2)$

Answer varies.

60. Solve these two examples.
(a) $8 - 3$ **(b)** $3 - 8$
How are the answers similar? How are they different? Write a rule that explains what happens when you switch the order of the numbers in a subtraction problem.

Answer varies.

Follow the order of operations to work each problem.

Example:	**Solution:**
$-6 + (-9) - (-10)$	$= -15 - (-10)$
The order of operations	$= -15 + 10$
tells you to add and subtract	$= \mathbf{-5}$
from left to right.	

61. $-2 + (-11) - (-3)$

-10

62. $-5 - (-2) + (-6)$

-9

63. $4 - (-13) + (-5)$

12

64. $6 - (-1) + (-10)$

-3

65. $-12 - (-3) - (-2)$

-7

66. $-1 - (-7) - (-4)$

10

67. $4 - (-4) - 3$ 5

68. $5 - (-2) - 8$ −1

69. $\frac{1}{2} - \frac{2}{3} + \left(-\frac{5}{6}\right)$

−1

70. $\frac{2}{5} - \frac{7}{10} + \left(-\frac{3}{2}\right)$

$-\frac{9}{5}$ or $-1\frac{4}{5}$

71. $-5.7 - (-9.4) - 8.1$ −4.4

72. $-6.5 - (-11.2) - 1.4$ 3.3

Add or subtract the following.

▲ **73.** $-2 + (-11) + |-2|$ −11

▲ **74.** $|-7 + 2| + (-2) + 4$ 7

▲ **75.** $-3 - (-2 + 4) + (-5)$ −10
Hint: Work inside parentheses first.

▲ **76.** $5 - 8 - (6 - 7) + 1$ −1

▲ **77.** $2\frac{1}{2} + 3\frac{1}{4} - \left(-1\frac{3}{8}\right) - 2\frac{3}{8}$ $\frac{19}{4}$ or $4\frac{3}{4}$

▲ **78.** $\frac{5}{8} - \left(-\frac{1}{2} - \frac{3}{4}\right)$ $\frac{15}{8}$ or $1\frac{7}{8}$

PREVIEW EXERCISES

Multiply or divide the following. (For help, see ***Sections 1.4, 1.5, 2.5, 2.8, or 4.5.****)*

79. $23 \cdot 46$

1058

80. $\frac{8}{11} \cdot \frac{3}{5}$

$\frac{24}{55}$

81. $71.20 \cdot 21.25$

1513

82. $2\frac{2}{3} \cdot 2\frac{3}{4}$

$\frac{22}{3}$ or $7\frac{1}{3}$

83. $1235 \div 5$ 247

84. $1\frac{2}{3} \div 2\frac{7}{9}$ $\frac{3}{5}$

85. $\frac{7}{9} \div \frac{14}{27}$ $\frac{3}{2}$ or $1\frac{1}{2}$

9.3 MULTIPLICATION AND DIVISION OF SIGNED NUMBERS

How do you multiply two numbers with different signs? Look for a pattern in the following list of products.

$$4 \cdot 2 = 8$$
$$3 \cdot 2 = 6$$
$$2 \cdot 2 = 4$$
$$1 \cdot 2 = 2$$
$$0 \cdot 2 = 0$$
$$-1 \cdot 2 = ?$$

As the numbers in blue decrease by 1, the numbers in red decrease by 2. You can continue the pattern by replacing ? with a number 2 *less than* 0, which is -2. Therefore:

$$-1 \cdot 2 = -2$$

1 The pattern above suggests a rule for multiplying numbers with different signs.

MULTIPLYING NUMBERS WITH DIFFERENT SIGNS

The product of two numbers with *different* signs is *negative.*

EXAMPLE 1 *Mutiplying Numbers with Different Signs*
Multiply.

(a) $-8 \cdot 4$ **(b)** $6 \cdot (-3)$ **(c)** $-5 \cdot (11)$ **(d)** $12 \cdot (-7)$

(a) $-8 \cdot 4 = -32$ ← The product is negative. **(b)** $6 \cdot (-3) = -18$

(c) $-5 \cdot (11) = -55$ **(d)** $12 \cdot (-7) = -84$ ■

WORK PROBLEM 1 AT THE SIDE.

For two numbers with the same signs, look at this pattern.

$$4 \cdot (-2) = -8$$
$$3 \cdot (-2) = -6$$
$$2 \cdot (-2) = -4$$
$$1 \cdot (-2) = -2$$
$$0 \cdot (-2) = 0$$
$$-1 \cdot (-2) = ?$$

This time, as the numbers in blue decrease by 1, the products increase by 2. You can continue the pattern by replacing ? with a number 2 *greater than 0,* which is positive 2. Therefore:

$$-1 \cdot (-2) = 2$$

2 In the pattern above, a negative number times a negative number gave a positive result.

OBJECTIVES

1. Multiply or divide two numbers with opposite signs.
2. Multiply or divide two numbers with the same sign.

FOR EXTRA HELP

Tape 12

SSM pp. 273–275

MAC: A IBM: A

1. Multiply.

(a) $5 \cdot (-4)$

(b) $-9 \cdot (15)$

(c) $12 \cdot (-1)$

(d) $-6 \cdot (6)$

(e) $\left(-\frac{7}{8}\right)\left(\frac{4}{3}\right)$

ANSWERS
1. (a) -20 **(b)** -135 **(c)** -12 **(d)** -36 **(e)** $-\frac{7}{6}$ or $-1\frac{1}{6}$

2. Multiply.

(a) $(-5) \cdot (-5)$

(b) $(-14)(-1)$

(c) $-7 \cdot (-8)$

(d) $3 \cdot 12$

(e) $\left(-\frac{2}{3}\right)\left(-\frac{6}{5}\right)$

3. Divide.

(a) $\frac{-20}{4}$

(b) $\frac{-50}{-5}$

(c) $\frac{44}{2}$

(d) $\frac{6}{-6}$

(e) $\frac{-15}{-1}$

(f) $\frac{-\frac{3}{5}}{\frac{9}{10}}$

ANSWERS

2. **(a)** 25 **(b)** 14 **(c)** 56 **(d)** 36 **(e)** $\frac{4}{5}$

3. **(a)** -5 **(b)** 10 **(c)** 22 **(d)** -1 **(e)** 15 **(f)** $-\frac{2}{3}$

MULTIPLYING NUMBERS WITH THE SAME SIGN

The product of two numbers with the *same* sign is *positive.*

EXAMPLE 2 *Multiplying Two Numbers with the Same Sign*

Multiply.

(a) $(-9)(-2)$ The numbers have the same sign (both are negative).

$(-9)(-2) = 18$ ← The product is positive.

(b) $-7 \cdot (-4) = 28$ **(c)** $(-6)(-2) = 12$

(d) $(-10)(-5) = 50$ **(e)** $7 \cdot 5 = 35$ ■

WORK PROBLEM 2 AT THE SIDE.

You can use the same rules for dividing signed numbers as you use for multiplying signed numbers.

DIVIDING SIGNED NUMBERS

When two nonzero numbers with *different* signs are divided, the result is *negative.*

When two nonzero numbers with the *same* sign are divided, the result is *positive.*

EXAMPLE 3 *Dividing Signed Numbers*

Divide.

(a) $\frac{-15}{5}$ Numbers have different signs so the answer is negative. $\frac{-15}{5} = -3$

(b) $\frac{-8}{-4}$ Numbers have same sign (both negative) so answer is positive. $\frac{-8}{-4} = 2$

(c) $\frac{-75}{-25} = 3$ **(d)** $\frac{-6}{3} = -2$

(e) $\frac{15}{5} = 3$ **(f)** $\frac{-90}{9} = -10$

(g) $\frac{-\frac{2}{3}}{-\frac{5}{9}} = -\frac{2}{3} \cdot \left(-\frac{9}{5}\right)$ Invert the divisor: $-\frac{5}{9}$ becomes $-\frac{9}{5}$.

$= -\frac{2}{\cancel{3}_1} \cdot \left(-\frac{\cancel{9}^3}{5}\right)$ Cancel when possible. Then multiply.

$= \frac{6}{5}$ or $1\frac{1}{5}$ ■

WORK PROBLEM 3 AT THE SIDE.

EXERCISES

Multiply.

Examples: $-6 \cdot 5$ $\quad (-4)(-3) \quad -\frac{3}{4} \cdot \left(-\frac{2}{3}\right)$

Solutions:

$-6 \cdot 5 = \mathbf{-30} \quad (-4)(-3) = \mathbf{12} \quad -\frac{\overset{1}{3}}{\underset{2}{4}} \cdot \left(-\frac{\overset{1}{2}}{\underset{1}{3}}\right) = \mathbf{\frac{1}{2}}$

1. $-5 \cdot 7$
-35

2. $-10 \cdot 2$
-20

3. $-9 \cdot 3$
-27

4. $-4 \cdot 9$
-36

5. $3 \cdot (-6)$
-18

6. $8 \cdot (-6)$
-48

7. $10 \cdot (-5)$
-50

8. $5 \cdot (-11)$
-55

9. $-1 \cdot 12$
-12

10. $9 \cdot (-1)$
-9

11. $-8 \cdot (-4)$
32

12. $-3 \cdot (-9)$
27

13. $11 \cdot 7$
77

14. $4 \cdot 25$
100

15. $-19 \cdot (-7)$
133

16. $-21 \cdot (-3)$
63

17. $-13 \cdot (-1)$
13

18. $-1 \cdot (-31)$
31

19. $0 \cdot (-3)$
0

20. $-7 \cdot 0$
0

21. $-\frac{1}{2} \cdot (-8)$
4

22. $\frac{1}{3} \cdot (-15)$
-5

23. $-10 \cdot \left(\frac{2}{5}\right)$
-4

24. $-25 \cdot \left(-\frac{7}{10}\right)$
$\frac{35}{2}$ or $17\frac{1}{2}$

25. $\frac{3}{5} \cdot \left(-\frac{1}{6}\right)$
$-\frac{1}{10}$

26. $-\frac{7}{9} \cdot \left(-\frac{3}{4}\right)$
$\frac{7}{12}$

27. $-\frac{7}{5} \cdot \left(-\frac{10}{3}\right)$
$\frac{14}{3}$ or $4\frac{2}{3}$

28. $-\frac{9}{10} \cdot \frac{5}{4}$
$-\frac{9}{8}$ or $-1\frac{1}{8}$

29. $-\frac{7}{15} \cdot \frac{25}{14}$

$-\frac{5}{6}$

30. $-\frac{5}{9} \cdot \frac{18}{25}$

$-\frac{2}{5}$

31. $-\frac{5}{2} \cdot \left(-\frac{7}{10}\right)$

$\frac{7}{4}$ or $1\frac{3}{4}$

32. $-\frac{8}{5} \cdot \left(-\frac{15}{16}\right)$

$\frac{3}{2}$ or $1\frac{1}{2}$

33. $9 \cdot (-4.7)$

-42.3

34. $15 \cdot (-6.3)$

-94.5

35. $-0.5 \cdot (-12)$

6

36. $-3.15 \cdot (-5)$

15.75

37. $-6.2 \cdot (5.1)$

-31.62

38. $-4.3 \cdot (9.7)$

-41.71

39. $-1.25 \cdot (-3.6)$

4.5

40. $6.33 \cdot 0.2$

1.266

Divide.

Examples: $\frac{-10}{2}$ $\frac{32}{-8}$ $\frac{-21}{-3}$

Solutions: $\frac{-10}{2} = \mathbf{-5}$ $\frac{32}{-8} = \mathbf{-4}$ $\frac{-21}{-3} = \mathbf{7}$

41. $\frac{-14}{7}$ -2

42. $\frac{-8}{2}$ -4

43. $\frac{30}{-6}$ -5

44. $\frac{21}{-7}$ -3

45. $\frac{-28}{4}$ -7

46. $\frac{-40}{4}$ -10

47. $\frac{14}{-1}$ -14

48. $\frac{25}{-1}$ -25

49. $\frac{-20}{-2}$ 10

50. $\frac{-80}{-4}$ 20

51. $\frac{-48}{-12}$ 4

52. $\frac{-30}{-15}$ 2

53. $\frac{-18}{18}$ -1

54. $\frac{50}{-50}$ -1

55. $\frac{-573}{-3}$ 191

56. $\frac{-580}{-5}$ 116

57. $\frac{-\frac{5}{7}}{-\frac{15}{14}}$ $\frac{2}{3}$

58. $\frac{-\frac{3}{4}}{-\frac{9}{16}}$ $\frac{4}{3}$ or $1\frac{1}{3}$

59. $-\frac{2}{3} \div (-2)$ $\frac{1}{3}$

60. $-\frac{3}{4} \div (-9)$ $\frac{1}{12}$

61. $5 \div \left(-\frac{5}{8}\right)$ -8

62. $7 \div \left(-\frac{14}{15}\right)$ $-\frac{15}{2}$ or $-7\frac{1}{2}$

63. $-\frac{7}{5} \div \frac{3}{10}$ $-\frac{14}{3}$ or $-4\frac{2}{3}$

64. $-\frac{4}{9} \div \frac{8}{3}$ $-\frac{1}{6}$

65. $\frac{-18.92}{-4}$ 4.73

66. $\frac{-22.75}{-7}$ 3.25

67. $\frac{-7.05}{1.5}$ -4.7

68. $\frac{-17.02}{7.4}$ -2.3

69. $\frac{45.58}{-8.6}$ -5.3

70. $\frac{6.27}{-0.3}$ -20.9

Following the order of operations, work from left to right in each of the following.

71. $(-4) \cdot (-6) \cdot \frac{1}{2}$ 12

72. $(-9) \cdot (-3) \cdot \frac{2}{3}$ 18

73. $(-0.6)(-0.2)(-3)$ -0.36

74. $(-4)(-1.2)(-0.7)$ -3.36

75. $\left(-\frac{1}{2}\right) \cdot \left(\frac{2}{5}\right) \cdot \left(\frac{7}{8}\right)$ $-\frac{7}{40}$

76. $\left(\frac{3}{4}\right) \cdot \left(-\frac{5}{6}\right) \cdot \left(\frac{2}{3}\right)$ $-\frac{5}{12}$

77. Write three examples for each of these situations:
(a) a positive number multiplied by -1
(b) a negative number multiplied by -1
Now write a rule that explains what happens when you multiply a signed number by -1.

Answer varies.

78. Write three examples for each of these situations:
(a) a negative number divided by -1
(b) a positive number divided by -1
(c) a negative number divided by itself
Now write a rule that explains what happens when you divide a signed number by -1. Write another rule for a negative number divided by itself.

Answer varies.

79. Explain what is different and what is similar between multiplying and dividing signed numbers.

Answer varies.

80. Explain why $\frac{0}{-3}$ and $\frac{-3}{0}$ do **not** give the same result.

Answer varies.

Simplify the following.

81. $-36 \div (-2) \div (-3) \div (-3) \div (-1)$

-2

82. $-48 \div (-8) \cdot (-4) \div (-4) \div (-3)$

-2

83. $|-8| \div (-4) \cdot |-5|$

-10

84. $-6 \cdot |-3| \div |9| \cdot (-2)$

4

PREVIEW EXERCISES

Simplify the following. (For help, see ***Section 1.8.****)*

85. $8 + 4 \cdot 2 \div 8$ 9

86. $16 \div 2 \cdot 4 + (15 - 2 \cdot 3)$ 41

87. $7 + 6 \div 2 \cdot 4 - 9$ 10

88. $9 \div 3 + 8 \div 4 \cdot 2 - 7$ 0

9.4 ORDER OF OPERATIONS

In the last two sections you worked examples that mixed either addition and subtraction or multiplication and division. In those situations you worked from left to right. Here are two more examples.

$$-8 - (-6) + (-11) \quad\to\quad -2 + (-11) \quad\to\quad -13$$

$$(-15) \div (-3) \cdot 6 \quad\to\quad 5 \cdot 6 \quad\to\quad 30$$

WORK PROBLEM 1 AT THE SIDE.

1 Before working examples that mix division with addition or include parentheses, let's review the order of operations from **Section 1.8**.

ORDER OF OPERATIONS

1. Work inside **parentheses.**
2. Simplify expressions with **exponents,** and find any **square roots.**
3. Multiply or divide from **left to right.**
4. Add or subtract from **left to right.**

EXAMPLE 1 *Using the Order of Operations*

Use the order of operations to simplify each of the following.

(a) $6 \div 3 + 4$

Steps 1 and 2 in the order of operations do not apply, so start with Step 3. Divide first; then add.

$$6 \div 3 + 4 \quad\to\quad 2 + 4 \quad\to\quad 6$$

(b) $-12 + (-3) \cdot (-6)$

Start by multiplying. Then add.

$$-12 + (-3) \cdot (-6) \quad\to\quad -12 + 18 \quad\to\quad 6$$

(c) $8 \cdot (5 - 7) - 9$

Work inside parentheses first.

$$8 \cdot (5 - 7) - 9$$

Now multiply. $8 \cdot (-2) - 9$

Subtract last. $-16 - 9$

Change subtraction to addition. $-16 + (-9)$

$$-25$$ ■

WORK PROBLEM 2 AT THE SIDE.

OBJECTIVES

1 Use the order of operations.

2 Use the order of operations with exponents.

FOR EXTRA HELP

Tape 13 | SSM pp. 275–278 | MAC: A IBM: A

1. Simplify.

(a) $-9 + (-15) + (-3)$

(b) $-8 + (-2) + (-6)$

(c) $-2 - (-7) - (-4)$

(d) $3 \cdot (-4) \div (-6)$

(e) $-18 \div 9 \cdot (-4)$

2. Use the order of operations to simplify each of the following.

(a) $10 + 8 \div 2$

(b) $4 - 6 \cdot (-2)$

(c) $-3 + (-5) \cdot 2$

(d) $-2 \cdot (4 - 7)$

(e) $(9 - 17) \cdot (-3 - 1)$

ANSWERS

1. (a) -27 **(b)** -16 **(c)** 9 **(d)** 2 **(e)** 8

2. (a) 14 **(b)** 16 **(c)** -13 **(d)** 6 **(e)** 32

3. Simplify each of the following.

(a) $2^3 - 3^2$

(b) $4^2 - 3^2 \cdot (5 - 2)$

(c) $-18 \div (-3) \cdot 2^3$

(d) $(-3)^3 + (-5)^2$

(e) $\frac{3}{8} + \left(-\frac{1}{2}\right)^2 \div \frac{1}{4}$

ANSWERS
3. (a) -1 (b) -11 (c) 48 (d) -2 (e) $\frac{11}{8}$ or $1\frac{3}{8}$

2 Remember that 2^3 means 2 is used as a factor 3 times:

$$2^3 = 2 \cdot 2 \cdot 2 = 8.$$

The 3 is called an *exponent*. Exponents are also used with signed numbers. For example:

$$(-3)^2 = (-3) \cdot (-3) = 9$$
$$(-4)^3 = (-4) \cdot (-4) \cdot (-4) = -64$$
$$\left(-\frac{1}{2}\right)^4 = \left(-\frac{1}{2}\right) \cdot \left(-\frac{1}{2}\right) \cdot \left(-\frac{1}{2}\right) \cdot \left(-\frac{1}{2}\right) = \frac{1}{16}$$
$$(0.6)^2 = (0.6) \cdot (0.6) = 0.36$$

Be very careful with exponents and signed numbers. Although

$$(-3)^2 = (-3) \cdot (-3) = 9,$$

the expression -3^2, with no parentheses, is different.

$$-3^2 = -(3 \cdot 3) = -9$$

Note $(-3)^2$ is **not** the same as -3^2
$(-3)^2 = (-3) \cdot (-3) = 9$ *but* $-3^2 = -(3 \cdot 3) = -9$

EXAMPLE 2 *Using the Order of Operations*

Simplify each of the following.

(a) $4^2 - (-3)^2$

Because $4^2 = 4 \cdot 4 = 16$ and $(-3)^2 = (-3) \cdot (-3) = 9$,

$$4^2 - (-3)^2 = 16 - 9 = 7.$$

(b)

$-5^2 - (4 - 6)^2 \cdot (-3)$ Work inside parentheses first.

$-5^2 - (-2)^2 \cdot (-3)$ Use the exponents next.

$-25 - 4 \cdot (-3)$ Multiply.

$-25 - (-12)$ Change subtraction to addition.

$-25 + (+12)$ Add.

-13

(c)

$\left(\frac{2}{3} - \frac{1}{6}\right)^2 \div \left(-\frac{3}{8}\right)$ Inside parentheses $\frac{2}{3} - \frac{1}{6} = \frac{4}{6} - \frac{1}{6} = \frac{3}{6} = \frac{1}{2}$

$\left(\frac{1}{2}\right)^2 \div \left(-\frac{3}{8}\right)$ Use the exponent: $(\frac{1}{2})^2 = \frac{1}{2} \cdot \frac{1}{2} = \frac{1}{4}$

$\frac{1}{4} \div \left(-\frac{3}{8}\right)$ Invert.

$\frac{1}{4} \cdot \left(-\frac{8}{3}\right)$ Multiply.

$-\frac{2}{3}$ ■

WORK PROBLEM 3 AT THE SIDE.

9.4 EXERCISES

NAME DATE HOUR

Simplify each of the following.

Examples: $-6 - 3^2$ $\qquad$ $4 \cdot (6 - 11)^2 - (-8)$

Solutions:

$= -6 - 9$
$= -6 + (-9)$
$= \mathbf{-15}$

$= 4 \cdot (-5)^2 - (-8)$
$= 4 \cdot 25 - (-8)$
$= 100 - (-8)$
$= 100 + (+8)$
$= \mathbf{108}$

1. $20 \div 5 + 10$

14

2. $30 \div 6 + 4$

9

3. $-1 + 15 + (-7) \cdot 2$

0

4. $9 + (-5) + 2 \cdot (-2)$

0

5. $6^2 + 4^2$

52

6. $3^2 + 8^2$

73

7. $10 - 7^2$

−39

8. $5 - 5^2$

−20

9. $(-2)^5 + 2$

−30

10. $(-2)^4 - 7$

9

11. $4^2 + 3^2 + (-8)$

17

12. $5^2 + 2^2 + (-12)$

17

13. $2 - (-5) + 3^2$

16

14. $6 - (-9) + 2^3$

23

15. $(-4)^2 + (-3)^2 + 5$

30

16. $(-5)^2 + (-6)^2 + 12$

73

17. $3 + 5 \cdot (6 - 2)$

23

18. $4 + 3 \cdot (8 - 3)$

19

19. $-7 + 6 \cdot (8 - 14)$

−43

20. $-3 + 5 \cdot (9 - 12)$

−18

21. $-6 + (-5) \cdot (9 - 14)$

19

22. $-5 + (-3) \cdot (6 - 7)$

-2

23. $(-5) \cdot (7 - 13) \div (-10)$

-3

24. $(-4) \cdot (9 - 17) \div (-8)$

-4

25. $9 \div (-3)^2 + (-1)$

0

26. $-48 \div (-4)^2 + 3$

0

27. $2 - (-5) \cdot (-3)^2$

47

28. $1 - (-10) \cdot (-2)^3$

-79

29. $(-2) \cdot (-7) + 3 \cdot 9$

41

30. $4 \cdot (-2) + (-3) \cdot (-5)$

7

31. $30 \div (-5) - 36 \div (-9)$

-2

32. $8 \div (-4) - 42 \div (-7)$

4

33. $2 \cdot 5 - 3 \cdot 4 + 5 \cdot 3$

13

34. $9 \cdot 3 - 6 \cdot 4 + 3 \cdot 7$

24

35. $4 \cdot 3^2 + 7 \cdot (3 + 9) - (-6)$

126

36. $5 \cdot 4^2 - 6 \cdot (1 + 4) - (-3)$

53

37. $-12 \cdot (-1) + 5^2 - (-3)$

40

38. $-6 \cdot (-5) + 3^2 - (-7)$

46

39. $2 \cdot 4^2 - 4 \cdot (6 - 2) - 6^2$

-20

40. $3 \cdot 4^2 - 5 \cdot (9 - 2) - 6^2$

-23

41. $3^2 \cdot (2 - 5) \div (4 + 5) - (-6)$

3

42. $8^2 \cdot (9 - 14) \div (3 - 13) - (-2)$

34

43. $(-4)^2 \cdot (7 - 9)^2 \div 2^3$

8

44. $(-5)^2 \cdot (9 - 17)^2 \div (-10)^2$

16

45. $(-0.3)^2 + (-0.5)^2 + 0.9$

1.24

46. $(0.2)^3 - (-0.4)^2 + 3.02$

2.868

47. $(-0.75) \cdot (3.6 - 5)^2$

-1.47

48. $(-0.3) \cdot (4 - 6.8)^2$

-2.352

49. $(0.5)^2 \cdot (-8) - (0.31)$

-2.31

50. $(0.3)^3 \cdot (-5) - (-2.8)$

2.665

51. $\frac{2}{3} \div \left(-\frac{5}{6}\right) - \frac{1}{2}$

$-\frac{13}{10}$ or $-1\frac{3}{10}$

52. $\frac{5}{8} \div \left(-\frac{10}{3}\right) - \frac{3}{4}$

$-\frac{15}{16}$

53. $\left(-\frac{1}{2}\right)^2 - \left(\frac{3}{4} - \frac{7}{4}\right)$

$\frac{5}{4}$ or $1\frac{1}{4}$

54. $\left(-\frac{2}{3}\right)^2 - \left(\frac{1}{6} - \frac{11}{6}\right)$

$\frac{19}{9}$ or $2\frac{1}{9}$

55. $\frac{3}{5} \cdot \left(-\frac{7}{6}\right) - \left(\frac{1}{6} - \frac{5}{3}\right)$

$\frac{4}{5}$

56. $\frac{2}{7} \cdot \left(-\frac{14}{5}\right) - \left(\frac{4}{3} - \frac{13}{9}\right)$

$-\frac{31}{45}$

57. $5^2 \cdot (9 - 11) \cdot (-3) \cdot (-2)^3$

-1200

58. $4^2 \cdot (13 - 17) \cdot (-2) \cdot (-3)^2$

1152

59. $1.6 \cdot (-0.8) \div (-0.32) \div 2^2$

1

60. $6.5 \cdot (-4.8) \div (-0.3) \div (-2)^3$

-13

61. Explain the difference between -5^2 and $(-5)^2$.

Answer varies.

62. Solve this series of examples.

$(-2)^2 =$
$(-2)^3 =$
$(-2)^4 =$
$(-2)^5 =$

What pattern do you see in the sign of the answers?

Answer varies.

Simplify each of the following.

63. $5 - 4 \cdot 12 \div 3 \cdot 2$

-27

64. $4 + 27 \div 3 \cdot 2 - 6$

16

65. $-7 \cdot \left(6 - \frac{5}{8} \cdot 24 + 3 \cdot \frac{8}{3}\right)$

7

66. $(-0.3)^2 \cdot (-5 \cdot 3) + (6 \div 2 \cdot 0.4)$

-0.15

67. $|-12| \div 4 + 2 \cdot 3^2 \div 6$

6

68. $6 - (2 - 3 \cdot 4) + 5^2 \div \left(-2 \cdot \frac{5}{2}\right) + (2)^2$

15

PREVIEW EXERCISES

Evaluate the following. (For help, see ***Section 3.5.****)*

69. $\frac{6}{7} \cdot \frac{14}{9}$

$\frac{4}{3}$ or $1\frac{1}{3}$

70. $\frac{17}{15} \div \frac{34}{5}$

$\frac{1}{6}$

71. $\frac{5}{3} - \frac{3}{8} \cdot \frac{4}{9}$

$\frac{3}{2}$ or $1\frac{1}{2}$

72. $\frac{7}{12} \div \frac{21}{10} \cdot \frac{4}{5}$

$\frac{2}{9}$

9.5 EVALUATING EXPRESSIONS AND FORMULAS

In formulas you have seen that numbers can be represented by letters. For example, you used this formula for finding simple interest in **Section 6.8**.

$$I = p \cdot r \cdot t$$

In this formula, p (principal) represents the amount of money borrowed, r is the rate of interest, and t is the time in years. In algebra, we often write multiplication without the multiplication dots. If there is no operation sign written between two letters, or between a letter and a number, you assume it is multiplication.

SHOWING MULTIPLICATION IN ALGEBRA

If there is no operation sign, it is understood to be multiplication. Here is an example.

$$I = p \cdot r \cdot t \quad \text{is written} \quad I = prt$$

1 Letters (such as the I, p, r, or t used here) that represent numbers are called **variables.** A combination of letters and numbers is an **expression.** Three examples of expressions are shown here.

$$9 + p \qquad 8r \qquad 7k - 2m$$

2 The value of an expression changes depending upon the value of each variable. To find the value of an expression, replace the variables with their values.

EXAMPLE 1 *Finding the Value of an Expression*
Find the value of $5x - 3y$, if $x = 2$ and $y = 7$.

Replace x with 2 and y with 7.

$$5x - 3y = 5(2) - 3(7)$$

(*Note:* Here, $5x$ means 5 times x and $3y$ means 3 times y.) Next, multiply, and then add, using the order of operations.

$$= 10 - 21$$
$$= -11$$

WORK PROBLEM 1 AT THE SIDE.

EXAMPLE 2 *Finding the Value of an Expression*
Find the value of $7m - 8n + p$, if $m = -2$, $n = 4$, and $p = 3$.

Replace m with -2, n with 4, and p with 3.

$$7m - 8n + p = 7(-2) - 8(4) + 3$$

Multiply.

$$= -14 - 32 + 3$$
$$= -46 + 3$$
$$= -43$$

WORK PROBLEMS 2 AND 3 AT THE SIDE.

OBJECTIVES

1. Define variable and expression.
2. Find the value of an expression when values of the variables are given.

FOR EXTRA HELP

Tape 13 | SSM pp. 278–280 | MAC: A IBM: A

1. Find the value of $5x - 3y$ if

(a) $x = 1$, $y = 2$.

(b) $x = 3$, $y = -4$.

(c) $x = 0$, $y = 6$.

2. Find the value of $7m - 8n + p$ if

(a) $m = 1$, $n = 2$, $p = 5$.

(b) $m = -4$, $n = -3$, $p = -7$.

(c) $m = -5$, $n = 0$, $p = -1$.

3. Find the value of $x + 6y$ if

(a) $x = 9$, $y = -3$.

(b) $x = -2$, $y = 1$.

(c) $x = 6$, $y = -1$.

ANSWERS
1. (a) -1 (b) 27 (c) -18
2. (a) -4 (b) -11 (c) -36
3. (a) -9 (b) 4 (c) 0

4. Find the value of $\frac{3k + r}{2s}$ if

(a) $k = 1, r = 1, s = 2.$

(b) $k = 8, r = -2,$ $s = -4.$

(c) $k = -3, r = 1,$ $s = -2.$

5. Find the value of A, P, d, and C in these formulas.

(a) $A = \frac{1}{2}bh;\ b = 6$ yd $h = 12$ yd

(b) $P = 2l + 2w;\ l = 10,$ $w = 8$

(c) $d = rt;\ r = 4,$ $t = 8$

(d) $C = 2\pi r;\ \pi = 3.14,$ $r = 6$

ANSWERS

4. (a) $\frac{4}{4} = 1$ **(b)** $\frac{22}{-8} = -\frac{11}{4}$ **(c)** $\frac{-8}{-4} = 2$

5. (a) $A = 36$ yd² **(b)** $P = 36$ **(c)** $d = 320$ **(d)** $C = 37.68$

■ **EXAMPLE 3** *Finding the Value of an Expression*

What is the value of $5x - y$, if $x = 2$ and $y = -3$?

Replace x with 2 and y with -3.

$$5x - y = 5(2) - (-3)$$
$$= 10 + 3$$
$$= 13 \quad ■$$

■ **EXAMPLE 4** *Finding the Value of an Expression*

What is the value of $\frac{6k + 2r}{5s}$, if $k = -2$, $r = 5$ and $s = -1$?

Replace k with -2, r with 5, and s with -1.

$$\frac{6k + 2r}{5s} = \frac{6(-2) + 2(5)}{5(-1)}$$

$$= \frac{-12 + 10}{-5} \quad \text{Multiply.}$$

$$= \frac{-2}{-5} \quad \text{Add in numerator.}$$

$$= \frac{2}{5} \quad \text{Quotient of two numbers with the same sign is positive.} \quad ■$$

◀ WORK PROBLEM 4 AT THE SIDE.

■ **EXAMPLE 5** *Finding the Value of an Expression*

The formula you used in Chapter 8 for the area of a triangle can now be written without the multiplication dots.

$$A = \frac{1}{2} \cdot b \cdot h \quad \text{is written} \quad A = \frac{1}{2}bh$$

In this formula, b is the length of the base and h is the height. What is the area if $b = 9$ cm and $h = 24$ cm?

Replace b with 9 cm and h with 24 cm.

$$A = \frac{1}{2}bh$$

$$A = \frac{1}{2}(9 \text{ cm})(24 \text{ cm})$$

$$A = \frac{1}{\underset{1}{\cancel{2}}}(9 \text{ cm})(\overset{12}{\cancel{24}} \text{ cm})$$

$$A = 108 \text{ cm}^2$$

The area of the triangle is 108 cm². ■

Note Area is measured in square units. The short way to write square centimeters is cm².

◀ WORK PROBLEM 5 AT THE SIDE.

EXERCISES

Find the value of the expression $2r + 4s$ for each of the following values of r and s.

Example: $r = 3, s = -5$

Solution: Replace r with 3 and s with -5.

$$2r + 4s = 2(3) + 4(-5)$$
$$= 6 + (-20)$$
$$= \mathbf{-14}$$

1. $r = 2, s = 6$

28

2. $r = 6, s = 1$

16

3. $r = 1, s = -3$

-10

4. $r = 7, s = -2$

6

5. $r = -4, s = 4$

8

6. $r = -3, s = 5$

14

7. $r = -1, s = -7$

-30

8. $r = -3, s = -5$

-26

9. $r = 0, s = -2$

-8

10. $r = -7, s = 0$

-14

Use the given values of the variables to find the value of each expression.

11. $8x - y; x = 1, y = 8$

0

12. $a - 5b; a = 10, b = 2$

0

13. $6k + 2s; k = 1, s = -2$

2

14. $7p + 7q; p = -4, q = 1$

-21

15. $\dfrac{-m + 5n}{2s + 2}; m = 4, n = -8, s = 0$

-22

16. $\dfrac{2y - z}{x - 2}; y = 0, z = 5, x = 1$

5

17. $-m - 3n; m = \dfrac{1}{2}, n = \dfrac{3}{8}$

$-\dfrac{13}{8}$ or $-1\dfrac{5}{8}$

18. $7k - 3r; k = \dfrac{2}{3}, r = -\dfrac{1}{3}$

$\dfrac{17}{3}$ or $5\dfrac{2}{3}$

In each of the following, use the given formula and values of the variables to find the value of the remaining variable.

Example: $P = 2l + 2w; l = 33, w = 16$

Solution: Replace l with 33 and w with 16.

$$P = 2l + 2w$$
$$P = 2 \cdot (33) + 2 \cdot (16)$$
$$P = 66 + 32$$
$$P = \mathbf{98}$$

19. $P = 4s; s = 7.5$

30

20. $P = 4s; s = 0.8$

3.2

21. $P = 21 + 2w; l = 9, w = 5$

28

22. $P = 2l + 2w; l = 12, w = 2$

28

23. $A = \frac{1}{2}bh$; $b = 14$, $h = 20$

140

24. $A = \frac{1}{2}bh$; $b = 23$, $h = 12$

138

25. $A = \frac{1}{2}bh$; $b = 15$, $h = 3$

$\frac{45}{2}$ or $22\frac{1}{2}$

26. $A = \frac{1}{2}bh$; $b = 5$, $h = 11$

$\frac{55}{2}$ or $27\frac{1}{2}$

27. $V = \frac{1}{3}bh$; $b = 30$, $h = 60$

600

28. $V = \frac{1}{3}bh$; $b = 105$, $h = 5$

175

29. $d = rt$; $r = 53$, $t = 6$

318

30. $d = rt$; $r = 180$, $t = 5$

900

31. $C = 2\pi r$; $\pi = 3.14$, $r = 4$

25.12

32. $C = 2\pi r$; $\pi = 3.14$, $r = 18$

113.04

33. Find and correct the error made by the student who solved this example:

Find the value of $-x - 4y$ if $x = -3$ and $y = -1$.

$$-x - 4y$$
$$-3 - 4(-1)$$
$$-3 - (-4)$$
$$-3 + (+4)$$
$$1$$

Also write a sentence next to each step, explaining what is being done in that step.

Answer varies.

34. Go back to Chapter 8 and find one of the formulas listed below. Pick values for the variables indicated and find the value of A. Then pick different values for the variables and again find the value of A.

Area of a trapezoid; pick value for h, a, and b.

Volume of a rectangular solid; pick values for l, w, and h.

Answer varies.

Use the given formula and values of the variables to find the value of the remaining variable.

▲ **35.** $F = \frac{9C}{5} + 32;\ C = -40$

−40

▲ **36.** $C = \frac{5(F - 32)}{9};\ F = -4$

−20

▲ **37.** $V = \frac{4\pi r^3}{3};\ \pi = 3.14,\ r = 3$

113.04

▲ **38.** $c^2 = a^2 + b^2;\ a = 3,\ b = 4$

5

PREVIEW EXERCISES

Simplify the following. (For help, see ***Sections 2.5 and 9.3.****)*

39. $-\frac{1}{9} \cdot 9$

−1

40. $4 \cdot \left(-\frac{3}{2}\right)$

−6

41. $-\frac{4}{3} \cdot \left(-\frac{3}{4}\right)$

1

42. $-\frac{2}{5} \cdot \left(-\frac{5}{2}\right)$

1

9.6 SOLVING EQUATIONS

OBJECTIVES

1. Tell whether a number is a solution of an equation.
2. Solve equations using the addition property of equations.
3. Solve equations using the multiplication property of equations.

FOR EXTRA HELP

Tape 13 | SSM pp. 280–284 | MAC: A IBM: A

An **equation** is a statement that says two expressions are equal. Examples of equations include

$$x + 1 = 9, \qquad 20 = 5k \qquad \text{and} \qquad 6r - 1 = 17.$$

The **equal sign** in an equation divides the equation into two parts, the *left side* and the *right side.* In $6r - 1 = 17$, the left side is $6r - 1$, and the right side is 17.

$$\underbrace{6r - 1}_{\text{left side}} = \underbrace{17}_{\text{right side}}$$

You solve an equation by finding all numbers that can be substituted for the variable to make the equation true. These numbers are called **solutions** of the equation.

1 To tell whether a number is a solution of the equation, substitute the number in the equation to see whether the result is true.

■ **EXAMPLE 1** *Determining If a Number Is a Solution of an Equation*
Is 7 a solution of the following equations?

(a) $12 = x + 5$ **(b)** $2y + 1 = 16$

(a) Replace x with 7.

$$12 = x + 5$$
$$12 = 7 + 5 \qquad \text{Let } x = 7.$$
$$12 = 12 \qquad \text{true}$$

Because the statement is true, 7 is a solution of $12 = x + 5$.

(b) Replace y with 7.

$$2y + 1 = 16$$
$$2(7) + 1 = 16$$
$$14 + 1 = 16$$
$$15 = 16 \qquad \text{false}$$

The false statement shows that 7 is *not* a solution of $2y + 1 = 16$. ■

WORK PROBLEM 1 AT THE SIDE.

1. Decide if the given number is a solution of the equation.

(a) $p + 1 = 8; \quad 7$

(b) $30 = 5r; \quad 6$

(c) $3k - 2 = 4; \quad 3$

(d) $23 = 4y + 3; \quad 5$

2 If the equation $a = b$ is true, and if a number c is added to both a and b, the new equation is also true. This rule is called the **addition property of equations.** It means that you can add the same number to both sides of an equation and still have a true equation.

ADDITION PROPERTY OF EQUATIONS

If $a = b$, then $a + c = b + c$. In other words, you may add the same number to both sides of an equation.

You can use the addition property to solve equations. The idea is to get the variable (the letter) by itself on one side of the equal sign and a number by itself on the other side.

ANSWERS
1. (a) solution (b) solution (c) not a solution (d) solution

2. Solve each equation. Check each solution.

(a) $n - 5 = 8$

(b) $5 = r - 10$

(c) $3 = z + 1$

(d) $k + 17 = 9$

(e) $-2 = y + 9$

(f) $x - 2 = -6$

ANSWERS
2. (a) 13 (b) 15 (c) 2 (d) −8 (e) −11 (f) −4

EXAMPLE 2 *Solving an Equation by Using the Addition Property*
Solve each equation.

(a) $k - 4 = 6$

To get k by itself on the left side, add 4 to the left side, because $k - 4 + 4$ gives $k + 0$. You must then add 4 to the right side also.

$$k - 4 = 6 \quad \leftarrow \text{original equation}$$
$$k - \mathbf{4 + 4} = 6 + \mathbf{4} \quad \text{Add 4 to both sides.}$$
$$k + 0 = 10$$
$$k = 10$$

The solution is 10. Check by replacing k with 10 in the original equation.

$$k - 4 = 6$$
$$10 - 4 = 6 \quad \text{Let } k = 10.$$
$$6 = 6 \quad \text{true}$$

This result is true, so 10 is the solution.

(b) $2 = z + 8$

To get z by itself on the right side, add -8 to both sides.

$$2 = z + 8 \quad \leftarrow \text{original equation}$$
$$2 + (\mathbf{-8}) = z + 8 + (\mathbf{-8}) \quad \text{Add } (-8) \text{ to both sides.}$$
$$-6 = z + 0$$
$$-6 = z$$

Check the solution by replacing z with -6 in the original equation.

$$2 = z + 8$$
$$2 = -6 + 8$$
$$2 = 2 \quad \text{true, so } -6 \text{ is the solution}$$

Notice, that we *added* -8 to both sides to get z by itself. We can accomplish the same thing by *subtracting* 8 from both sides. Recall from **Section 9.2** that subtraction is defined in terms of addition. On the left side of the equation above

$$2 - 8 \text{ gives the same result as } 2 + (-8).$$ ■

Note You may add *or subtract* the same number on both sides of an equation.

WORK PROBLEM 2 AT THE SIDE.

Here is a summary of the rules you can use to solve equations using the addition property. In these rules, x is the variable and a and b are numbers.

SOLVING EQUATIONS USING THE ADDITION PROPERTY

Solve $x - a = b$ or $b = x - a$ by adding a to both sides.

Solve $x + a = b$ or $b = x + a$ by subtracting a from both sides.

MULTIPLICATION PROPERTY OF EQUATIONS

If $a = b$ and c does not equal 0, then

$$a \cdot c = b \cdot c \quad \text{and} \quad \frac{a}{c} = \frac{b}{c}.$$

In other words, you can multiply or divide both sides of an equation by the same number. (The only exception is you cannot divide by zero.)

■ EXAMPLE 3 *Solving an Equation by Using the Multiplication Property*

Solve each equation.

(a) $9p = 63$

You want to get the variable, p, by itself on the left side. The expression $9p$ means $9 \cdot p$. To get p by itself, *divide* both sides by 9.

$$9p = 63$$

$$\frac{\overset{1}{\cancel{9}} \cdot p}{\underset{1}{\cancel{9}}} = \frac{63}{9} \qquad \text{Divide both sides by 9.}$$

$$p = 7$$

Check.

$$9p = 63$$

$$9 \cdot 7 = 63 \qquad \text{Let } p = 7.$$

$$63 = 63 \qquad \text{true}$$

The result is true, so 7 is the solution.

(b) $-4r = 24$

Divide both sides by -4 to get r by itself on the left.

$$\frac{\overset{1}{-\cancel{4}} \cdot r}{\underset{1}{-\cancel{4}}} = \frac{24}{-4} \qquad \text{Divide both sides by } -4.$$

$$r = -6$$

Check this solution: $-4 \cdot (-6) = 24$ is true.

(c) $-55 = -11m$

Divide by -11.

$$\frac{-55}{-11} = \frac{\overset{1}{-\cancel{11}} \cdot m}{\underset{1}{-\cancel{11}}}$$

$$5 = m$$

Check this solution: $-55 = -11 \cdot (5)$ is true, so 5 is the solution. ■

WORK PROBLEM 3 AT THE SIDE.

3. Solve each equation. Check each solution.

(a) $2y = 14$

(b) $42 = 7p$

(c) $-8a = 32$

(d) $-3r = -15$

(e) $-60 = -6k$

ANSWERS

3. (a) 7 (b) 6 (c) −4 (d) 5 (e) 10

4. Solve each equation. Check each solution.

(a) $\frac{a}{4} = 2$

(b) $\frac{y}{7} = -3$

(c) $-8 = \frac{k}{6}$

(d) $8 = -\frac{4}{5}z$

(e) $-\frac{5}{8}p = -10$

ANSWER
4. (a) 8 (b) −21 (c) −48 (d) −10 (e) 16

■ EXAMPLE 4 *Solving an Equation by Using the Muliplication Property*

Solve each equation.

(a) $\frac{x}{2} = 9$

Dividing by 2 is the same as multiplying by $\frac{1}{2}$.

$$\frac{x}{2} = 9 \quad \text{means} \quad \frac{1}{2}x = 9$$

To get x by itself on the left, multiply both sides by $\frac{2}{1}$.

$$\frac{1}{2}x = 9$$

$$\frac{\overset{1}{\cancel{2}}}{1} \cdot \frac{1}{\underset{1}{\cancel{2}}}x = 2 \cdot 9 \qquad \text{Multiply both sides by } \tfrac{2}{1} \text{ (which equals 2)}$$

$$1x = 18$$

$$x = 18$$

Check. $\frac{x}{2} = 9$

$$\frac{1}{2}x = 9$$

$$\frac{1}{\underset{1}{\cancel{2}}} \cdot \overset{9}{\cancel{18}} = 9 \qquad \text{Let } x = 18.$$

$$9 = 9 \qquad \text{true}$$

The solution is 18.

(b) $-\frac{2}{3}r = 4$

Multiply both sides by $-\frac{3}{2}$ (because the product of $-\frac{3}{2}$ and $-\frac{2}{3}$ is 1).

$$-\frac{2}{3}r = 4$$

$$-\frac{\overset{1}{\cancel{3}}}{\underset{1}{\cancel{2}}} \cdot \left(-\frac{\overset{1}{\cancel{2}}}{\underset{1}{\cancel{3}}}r\right) = -\frac{3}{\underset{1}{\cancel{2}}} \cdot \frac{\overset{2}{\cancel{4}}}{1}$$

$$r = -6$$

Check this solution. ■

◀◀ WORK PROBLEM 4 AT THE SIDE.

Here is a summary of the rules for using the multiplication property. In these rules, x is the variable and a, b, and c are numbers.

SOLVING EQUATIONS USING THE MULTIPLICATION PROPERTY

Solve the equation $ax = b$ by dividing both sides by a.

Solve the equation $\frac{a}{b}x = c$ by multiplying both sides by $\frac{b}{a}$.

9.6 EXERCISES

NAME DATE HOUR

Decide whether the given number is a solution of the equation.

1. $x + 7 = 11$; 4

yes

2. $k - 2 = 7$; 9

yes

3. $4y = 28$; 7

yes

4. $5p = 30$; 6

yes

5. $2z - 1 = -15$; -8

no

6. $6r - 3 = -14$; -2

no

Solve each equation by using the addition property. Check each solution.

Example: $m - 2 = 7$

Solution:

Add 2 to both sides.

$$m - 2 + 2 = 7 + 2$$
$$m = \mathbf{9}$$

Check:

$$m - 2 = 7$$
$$9 - 2 = 7$$
$$7 = 7 \quad \text{true}$$

7. $p + 5 = 9$

4

8. $a + 3 = 12$

9

9. $k + 10 = 40$

30

10. $y + 11 = 15$

4

11. $z - 5 = 3$

8

12. $x - 9 = 4$

13

13. $8 = r - 2$

10

14. $3 = b - 5$

8

15. $-5 = n + 3$

-8

16. $-1 = a + 8$

-9

17. $7 = r + 13$

-6

18. $12 = z + 7$

5

19. $-4 + k = 14$

18

20. $-9 + y = 7$

16

21. $-8 + x = 1$

9

22. $-3 + m = -9$

-6

23. $-5 = -2 + r$

-3

24. $-1 = -10 + y$

9

25. $d + \frac{2}{3} = 3$

$\frac{7}{3}$ or $2\frac{1}{3}$

26. $x + \frac{1}{2} = 4$

$\frac{7}{2}$ or $3\frac{1}{2}$

27. $z - \frac{7}{8} = 10$

$10\frac{7}{8}$

28. $m - \frac{3}{4} = 21$

$21\frac{3}{4}$

29. $k - 2 = \frac{1}{2}$

$2\frac{1}{2}$

30. $t - 1 = \frac{3}{5}$

$1\frac{3}{5}$

31. $m - 1\frac{4}{5} = 2\frac{1}{10}$

$3\frac{9}{10}$

32. $z - 5\frac{1}{3} = 3\frac{5}{9}$

$8\frac{8}{9}$

33. $x - 0.8 = 5.07$

5.87

34. $a - 3.82 = 7.9$

11.72

35. $4.76 + r = 3.25$

−1.51

36. $10.5 + b = 8.9$

−1.6

Solve each equation. Check each solution.

Example: $5k = 60$

Solution:

Divide each side by 5.

$$\frac{\overset{1}{\cancel{5}} \cdot k}{\underset{1}{\cancel{5}}} = \frac{60}{5}$$

$$k = \mathbf{12}$$

Check:

$$5k = 60$$
$$5 \cdot 12 = 60$$
$$60 = 60 \quad \text{true}$$

37. $6z = 12$

2

38. $8k = 24$

3

39. $48 = 12r$

4

40. $99 = 11m$

9

41. $3y = -24$

−8

42. $5a = -25$

−5

43. $-6k = 36$

−6

44. $-7y = 70$

−10

45. $-36 = -4p$

9

46. $-54 = -9r$

6

47. $-1.2m = 8.4$

−7

48. $-5.4z = 27$

−5

49. $-8.4p = -9.24$

1.1

50. $-3.2y = -16.64$

5.2

Solve each equation. Check each solution.

Example: $\frac{p}{4} = -3$

Solution:

Multiply each side by 4.

$$\frac{\overset{1}{\cancel{4}}}{1} \cdot \frac{p}{\underset{1}{\cancel{4}}} = 4 \cdot (-3)$$

$$p = \mathbf{-12}$$

Check:

$$\frac{p}{4} = -3$$

$$\frac{-12}{4} = -3$$

$$-3 = -3 \quad \text{true}$$

51. $\frac{k}{2} = 17$ 34

52. $\frac{y}{3} = 5$ 15

53. $11 = \frac{a}{6}$ 66

54. $5 = \frac{m}{8}$ 40

55. $\frac{r}{3} = -12$ −36

56. $\frac{z}{9} = -3$ −27

57. $-\frac{2}{5}p = 8$ −20

58. $-\frac{5}{6}k = 15$ −18

59. $-\frac{3}{4}m = -3$ 4

60. $-\frac{9}{10}b = -18$ 20

61. $\frac{3}{8}x = 6$ 16

62. $\frac{2}{3}a = 4$ 6

63. $\frac{y}{2.6} = 0.5$

1.3

64. $\frac{k}{0.7} = 3.2$

2.24

65. $\frac{z}{-3.8} = 1.3$

−4.94

66. $\frac{m}{-5.2} = 2.1$

−10.92

67. Explain the addition property of equations. Then show an example of an equation where you would use the addition property to solve it. Make the equation so it has -3 as the solution.

Answer varies.

68. Explain the multiplication property of equations. Then show an example of an equation where you would use the multiplication property to solve it. Make the equation so it has $+6$ as the solution.

Answer varies.

Solve the following equations.

69. $x - 17 = 5 - 3$

19

70. $y + 4 = 10 - 9$

-3

71. $3 = x + 9 - 15$

9

72. $-1 = y + 7 - 9$

1

73. $\frac{7}{2}x = \frac{4}{3}$

$\frac{8}{21}$

74. $\frac{3}{4}x = \frac{5}{3}$

$\frac{20}{9}$ or $2\frac{2}{9}$

PREVIEW EXERCISES

Use the order of operations to simplify the following. (For help, see **Sections 3.5 and 9.4.)**

75. $2\frac{1}{5} \div \left(3\frac{1}{3} - 4\frac{1}{5}\right)$

$-\frac{33}{13}$ or $-2\frac{7}{13}$

76. $\frac{2}{3} + \frac{5}{6} \cdot \left(-\frac{3}{4}\right)$

$\frac{1}{24}$

77. $\frac{1}{2} + \frac{3}{4} \cdot \frac{8}{9} - \frac{1}{6}$

1

78. $2 - \left(3 \cdot \frac{1}{2} \div \frac{2}{3}\right) + \frac{7}{4}$

$\frac{3}{2}$ or $1\frac{1}{2}$

9.7 SOLVING EQUATIONS WITH SEVERAL STEPS

You cannot solve the equation $5m + 1 = 16$ by just adding the same number to both sides, nor by just dividing both sides by the same number.

1 Instead, you will use a combination of both operations. Here are the steps.

SOLVING EQUATIONS USING THE ADDITION AND MULTIPLICATION PROPERTIES

Step 1 Add or subtract the same thing on both sides of the equation so that the variable term ends up by itself on one side.
Step 2 Multiply or divide both sides by the same number to find the solution.
Step 3 Check the solution.

EXAMPLE 1 *Solving Equations with Several Steps*
Solve $5m + 1 = 16$.
First subtract 1 from both sides so that $5m$ will be by itself on the left side.

$$5m + 1 - 1 = 16 - 1$$
$$5m = 15$$

Next, divide both sides by 5.

$$\frac{\overset{1}{\cancel{5}} \cdot m}{\underset{1}{\cancel{5}}} = \frac{15}{5}$$
$$m = 3$$

Check.

$$5m + 1 = 16$$
$$5 \cdot 3 + 1 = 16 \quad \text{Let } m = 3.$$
$$15 + 1 = 16$$
$$16 = 16 \quad \text{true}$$

The solution is 3. ■

WORK PROBLEM 1 AT THE SIDE. ▶

2 We can use the order of operations to simplify these expressions:

$$2(6 + 8) \quad \text{and} \quad 2 \cdot 6 + 2 \cdot 8$$

On the left $\quad 2(6 + 8) = 2(14) = 28$
On the right $\quad 2 \cdot 6 + 2 \cdot 8 = 12 + 16 = 28$

Because both answers are the same, the two expressions are equivalent.

$$2(6 + 8) = 2 \cdot 6 + 2 \cdot 8$$

This is an example of the **distributive property.**

DISTRIBUTIVE PROPERTY

$$a(b + c) = ab + ac$$

OBJECTIVES

1. Solve equations with several steps.
2. Use the distributive property.
3. Combine like terms.
4. Solve more difficult equations.

FOR EXTRA HELP

Tape 13	SSM pp. 285–288	MAC: A IBM: A

1. Solve each equation. Check each solution.

(a) $2r + 7 = 13$

(b) $20 = 6y - 4$

(c) $7m + 9 = 44$

(d) $-2 = 4p + 10$

(e) $-10z - 9 = 11$

ANSWER
1. (a) 3 (b) 4 (c) 5 (d) −3 (e) −2

2. Use the distributive property.

(a) $3(2 + 6)$

(b) $8(k - 3)$

(c) $-6(r + 5)$

(d) $-9(s - 8)$

3. Combine like terms.

(a) $5y + 11y$

(b) $10a - 28a$

(c) $3x + 3x - 9x$

(d) $k + k$

(e) $6b - b - 7b$

ANSWERS

2. (a) $3 \cdot 2 + 3 \cdot 6 = 24$ (b) $8k - 24$ (c) $-6r + (-6) \cdot 5 = -6r + (-30) = -6r - 30$ (d) $-9s - (-9) \cdot 8 = -9s - (-72) = -9s + 72$

3. (a) $16y$ (b) $-18a$ (c) $-3x$ (d) $2k$ (e) $-2b$

EXAMPLE 2 *Using the Distributive Property*

Simplify each expression by using the distributive property.

(a) $9(4 + 2) = 9 \cdot 4 + 9 \cdot 2 = 36 + 18 = 54$

The 9 on the outside of the parentheses is *distributed* over the 4 and the 2 on the inside of the parentheses. That means that each number inside the parentheses is multiplied by 9.

(b) $-3(k + 9) = -3 \cdot k + (-3) \cdot 9 = -3k + (-27) = -3k - 27$

(c) $6(y - 5) = 6 \cdot y - 6 \cdot 5 = 6y - 30$

(d) $-2(x - 3) = -2 \cdot x - (-2) \cdot 3 = -2x - (-6) = -2x + 6$ ■

WORK PROBLEM 2 AT THE SIDE.

3 A single letter or number, or the product of a variable and a number, makes up a *term.* Here are six examples of terms.

$$3y \qquad 5 \qquad -9 \qquad 8r \qquad 10r^2 \qquad x$$

Terms with exactly the same variable and the same exponent are called ***like terms.***

$5x$	and	$3x$	like terms
$5x$	and	$3m$	not like terms; variables are different
$5x^2$	and	$5x^3$	not like terms; exponents are different

The distributive property can be used to simplify a sum of like terms such as $6r + 3r$.

$$6r + 3r = (6 + 3)r = 9r$$

This process is called *combining like terms.*

EXAMPLE 3 *Combining Like Terms*

Combine like terms.

(a) $5k + 11k$ **(b)** $10m - 14m + 2m$ **(c)** $-5x + x$

(a) $5k + 11k = (5 + 11)k = 16k$

(b) $10m - 14m + 2m = (10 - 14 + 2)m = -2m$

(c) $-5x + x$ can be written $-5x + 1x = (-5 + 1)x = -4x$ ■

WORK PROBLEM 3 AT THE SIDE.

4 The next examples show you how to solve more difficult equations using the addition, multiplication, and distributive properties.

EXAMPLE 4 *Solving Equations*
Solve each equation.

(a) $6r + 3r = 36$
Because $6r + 3r = 9r$, the equation becomes

$$9r = 36.$$

Next, divide both sides by 9.

$$\frac{\overset{1}{\cancel{9}} \cdot r}{\underset{1}{\cancel{9}}} = \frac{36}{9}$$

$$r = 4$$

Check.

$$6r + 3r = 36$$
$$6 \cdot 4 + 3 \cdot 4 = 36 \quad \text{Let } r = 4.$$
$$24 + 12 = 36$$
$$36 = 36 \quad \text{true}$$

The solution is 4.

(b) $2k - 2 = 5k - 11$
First, to get the variable term on one side, subtract $5k$ from both sides.

$$2k - 2 - 5k = 5k - 11 - 5k$$
$$2k - 5k - 2 = 5k - 5k - 11$$
$$-3k - 2 = -11$$

Next, add 2 to both sides.

$$-3k - 2 + 2 = -11 + 2$$
$$-3k = -9$$

Finally, divide both sides by -3.

$$\frac{-3k}{-3} = \frac{-9}{-3}$$
$$k = 3$$

Check.

$$2k - 2 = 5k - 11$$
$$2(3) - 2 = 5(3) - 11 \quad \text{Let } k = 3$$
$$6 - 2 = 15 - 11$$
$$4 = 4 \quad \text{true}$$

The solution is 3. ■

WORK PROBLEM 4 AT THE SIDE. ▶▶

4. Solve each equation. Check each solution.

(a) $3y - 1 = 2y + 7$

(b) $5a + 7 = 3a - 9$

(c) $3p - 2 = p - 6$

ANSWER
4. (a) $y = 8$ **(b)** $a = -8$ **(c)** $p = -2$

5. Solve each equation. Check each solution.

(a) $-12 = 4(y - 1)$

(b) $5(m + 4) = -10$

(c) $6(t - 2) = 18$

ANSWER
5. (a) $y = -2$ (b) $m = -6$ (c) $t = 5$

■ **EXAMPLE 5** *Solving Equations Using the Distributive Property*

Solve $-18 = 3(y - 2)$

Use the distributive property on the right side of the equation.

$$3(y - 2) \text{ becomes } 3 \cdot y - 3 \cdot 2 \quad \text{or} \quad 3y - 6$$

Now the equation looks like this.

$$-18 = 3y - 6$$

Add 6 to both sides in order to get the variable term by itself on the right side.

$$-18 + 6 = 3y - 6 + 6$$
$$-12 = 3y$$

Finally, divide both sides by 3.

$$\frac{-12}{3} = \frac{\overset{1}{\cancel{3}} \cdot y}{\underset{1}{\cancel{3}}}$$
$$-4 = y$$

Check.

$$-18 = 3(y - 2)$$
$$-18 = 3(-4 - 2) \qquad \text{Let } y = -4.$$
$$-18 = 3(-6)$$
$$-18 = -18 \qquad \text{true}$$

The solution is -4. ■

◀ WORK PROBLEM 5 AT THE SIDE.

9.7 EXERCISES

NAME DATE HOUR

Solve each equation. Check each solution.

Examples:

$9p - 7 = 11$

Solutions:

Add 7 to both sides.

$9p - 7 + 7 = 11 + 7$

$9p = 18$

Divide both sides by 9.

$\frac{\overset{1}{\cancel{9}} \cdot p}{\underset{1}{\cancel{9}}} = \frac{18}{9}$

$p = \mathbf{2}$

Check:

$9 \cdot 2 - 7 = 11$ Let $p = 2$.

$18 - 7 = 11$

$11 = 11$ true

The solution is **2.**

$-3m + 2 = 8$

Subtract 2 from both sides.

$-3m + 2 - 2 = 8 - 2$

$-3m = 6$

Divide both sides by -3.

$\frac{\overset{1}{\cancel{-3}} \cdot m}{\underset{1}{\cancel{-3}}} = \frac{6}{-3}$

$m = \mathbf{-2}$

Check:

$-3(-2) + 2 = 8$ Let $m = -2$.

$6 + 2 = 8$

$8 = 8$ true

The solution is **−2.**

1. $7p + 5 = 12$

1

2. $6k + 3 = 15$

2

3. $2 = 8y - 6$

1

4. $10 = 11p - 12$

2

5. $-3m + 1 = -5$

2

6. $-4k + 5 = -7$

3

7. $28 = -9a + 10$

−2

8. $5 = -10p + 25$

2

9. $-5x - 4 = 16$

−4

10. $-12a - 3 = 21$

−2

11. $-\frac{1}{2}z + 2 = -1$

6

12. $-\frac{5}{8}r + 4 = -6$

16

Writing Conceptual Challenging Estimation

Use the distributive property to simplify.

13. $6(x + 4)$

$6x + 24$

14. $8(k + 5)$

$8k + 40$

15. $7(p - 8)$

$7p - 56$

16. $9(t - 4)$

$9t - 36$

17. $-3(m + 6)$

$-3m - 18$

18. $-5(a + 2)$

$-5a - 10$

19. $-2(y - 3)$

$-2y + 6$

20. $-4(r - 7)$

$-4r + 28$

21. $-5(z - 9)$

$-5z + 45$

Combine like terms.

22. $11r + 6r$

$17r$

23. $2m + 5m$

$7m$

24. $8z + 7z$

$15z$

25. $10x - 2x$

$8x$

26. $9y - 3y$

$6y$

27. $4t - 9t + 3t$

$-2t$

28. $3p + 3p - 10p$

$-4p$

29. $-5a + a$

$-4a$

30. $-6p + p$

$-5p$

Solve each equation. Check each solution.

31. $4k + 6k = 50$

5

32. $3a + 2a = 15$

3

33. $45 = 13m - 8m$

9

34. $21 = 14z - 7z$

3

35. $2b - 6b = 24$

-6

36. $3r - 9r = 18$

-3

37. $-12 = 6y - 18y$

1

38. $-5 = 10z - 15z$

1

39. $6p - 2 = 4p + 6$

4

40. $5y - 5 = 2y + 10$

5

41. $9 + 7z = 9z + 13$

-2

42. $8 + 4a = 2a + 2$

-3

43. $-2y + 6 = 6y - 10$

2

44. $5x - 4 = -3x + 4$

1

45. $-3.6m + 1 = 2.4m + 7$

-1

46. $-0.2p + 7 = 1.8p - 3$

5

47. $\frac{y}{2} - 2 = \frac{y}{4} + 3$

20

48. $\frac{z}{3} + 1 = \frac{z}{2} - 3$

24

49. $-10 = 2(y + 4)$

-9

50. $-3 = 3(x + 6)$

-7

51. $-4(t + 2) = 12$

-5

52. $-5(k + 3) = 25$

-8

53. $6(x - 1) = 42$

8

54. $7(r - 5) = 7$

6

55. Solve $-2t - 10 = 3t + 5$. Show each step you take in solving it. Next to each step, write a sentence that explains what you did in that step. Be sure to tell when you used the addition property of equations and when you used the multiplication property of equations.

Answer varies.

56. Here is one student's solution to an equation.

$$3(2x + 5) = -7$$
$$6x + 5 = -7$$
$$6x + 5 - 5 = -7 - 5$$
$$6x = -12$$
$$x = -2$$

Show how to check the solution. If the solution does not check, find and correct the error.

Answer varies.

Solve each equation.

▲ **57.** $30 - 40 = -2x + 7x - 4x$

−10

▲ **58.** $-6 - 5 + 14 = -50a + 51a$

3

▲ **59.** $0 = -2(y - 2)$

2

▲ **60.** $0 = -9(b - 1)$

1

▲ **61.** $3 + 4r = 10 - 7$

0

▲ **62.** $18 - 8 = -2k + 10$

0

PREVIEW EXERCISES

Solve each word problem. (For help, see **Section 6.7**.)

63. Swimsuits are on sale in September at 60% off the regular price. How much will Beyanjeru pay for a suit regularly priced at $38?

$15.20

64. A "going out of business" sale at the Great Goods store promises 75% off on all items. Batteries for electric wheelchairs are regularly priced at $99. Find the amount Will paid for a battery during the sale.

$24.75

65. If the sales tax rate is 6% and a refrigerator costs $420, what is the amount of sales tax?

$25.20

66. A VCR sells for $450 plus 4% sales tax. Find the price of the VCR including sales tax.

$468

9.8 APPLICATIONS

It is rare for an application problem to be presented as an equation. Usually, the problem is given in words. You need to *translate* these words into an equation that you can solve.

OBJECTIVES

1. Translate word phrases by using variables.
2. Solve word problems.

FOR EXTRA HELP

Tape 13 | SSM pp. 288–291 | MAC: A IBM: A

1 The following examples show you how to translate word phrases into algebra.

EXAMPLE 1 *Translating Word Phrases by Using Variables*
Write in symbols by using x as the variable.

words	*algebra*
a number **plus** 2	$x + 2$
the **sum** of a number and 8	$x + 8$
5 **more than** a number	$x + 5$
-35 **added to** a number	$x + (-35)$
9 **less than** a number	$x - 9$
3 **subtracted from** a number	$x - 3$
a number **decreased by** 4	$x - 4$ ■

WORK PROBLEM 1 AT THE SIDE.

1. Write in symbols by using x as the variable.

(a) 15 less than a number

(b) 12 more than a number

(c) a number increased by 13

(d) a number minus 8

(e) -10 plus a number

EXAMPLE 2 *Translating Word Phrases by Using Variables*
Write in symbols by using x as the variable.

words	*algebra*
8 **times** a number	$8x$
the **product** of 12 and a number	$12x$
double a number (meaning "2 times")	$2x$
the **quotient** of 6 and a number	$\frac{6}{x}$
a number **divided by** 10	$\frac{x}{10}$
one-third of a number	$\frac{1}{3}x$ or $\frac{x}{3}$
the result **is**	$=$ ■

WORK PROBLEM 2 AT THE SIDE.

2. Write in symbols by using x as the variable.

(a) double a number

(b) the product of -8 and a number

(c) the quotient of 15 and a number

(d) one-half of a number

2 The next examples show you how to solve word problems. Notice that you begin each solution by selecting a variable to represent the unknown.

ANSWERS
1. (a) $x - 15$ **(b)** $x + 12$ **(c)** $x + 13$ **(d)** $x - 8$ **(e)** $-10 + x$ or $x + (-10)$
2. (a) $2x$ **(b)** $-8x$ **(c)** $\frac{15}{x}$ **(d)** $\frac{1}{2}x$ or $\frac{x}{2}$

3. (a) If 3 times a number is added to 4, the result is 19. Find the number.

(b) If -6 times a number is added to 5, the result is -13. Find the number.

(c) Susan donated $10 more than twice what LuAnn donated. If Susan donated $22, how much did LuAnn donate?

EXAMPLE 3 *Solving Word Problems*

If 5 times a number is added to 11, the result is 26. Find the number.

Let x represent the unknown number.
Use the information in the problem to write an equation.

$$\underbrace{\text{5 times a number}}\ \underbrace{\text{added to}}\ \text{11 is 26.}$$

$$5x + 11 = 26$$ ■

Note The phrase "the result is" translates to " =."

Next, solve the equation. First subtract 11 from both sides.

$$5x + 11 - \mathbf{11} = 26 - \mathbf{11}$$
$$5x = 15$$

Divide both sides by 5.

$$\frac{5x}{5} = \frac{15}{5}$$
$$x = 3$$

The number is 3. To check the solution, go back to the words of the original problem. First take five times the solution $(5 \cdot 3)$, which is 15. Then, 15 added to 11 gives 26, the result stated in the problem. The solution checks.

EXAMPLE 4 *Solving Word Problems*

Michael has 5 less than three times as many lab experiments completed as David. If Michael has completed 13 experiments, how many lab experiments has David completed?

Let x represent the number of experiments David has completed.

$$\underbrace{\text{3 times a number}}\ \text{minus 5 is 13}$$

$$3x - 5 = 13$$

Next, solve the equation. First add 5 to both sides.

$$3x - 5 + 5 = 13 + 5$$
$$3x = 18$$

Divide both sides by 3.

$$\frac{3x}{3} = \frac{18}{3}$$
$$x = 6$$

David has completed 6 lab experiments. Check by using the words of the original problem. First take three times the solution $(3 \cdot 6)$, which is 18. Then, 18 decreased by 5 is 13, which matches the 13 experiments completed by Michael. ■

WORK PROBLEM 3 AT THE SIDE.

ANSWERS

3. (a) $3x + 4 = 19$, $x = 5$
(b) $-6x + 5 = -13$, $x = 3$
(c) $2x + 10 = 22$, $x = 6$; LuAnn donated $6.

The steps in solving a word problem are summarized for you here.

SOLVING ALGEBRA WORD PROBLEMS

Step 1 Choose a variable to represent the unknown.

Step 2 Use the information in the problem to write an equation that relates known information to the unknown. Make a sketch or drawing, if possible.

Step 3 Solve the equation.

Step 4 Answer the question raised in the problem.

Step 5 Check the solution with the original words of the problem.

EXAMPLE 5 *Solving Word Problems*

During the day, Sheila drove 72 km more than Russell. The total distance traveled by them both was 232 km. Find the distance traveled by Russell.

Let x be the distance traveled by Russell.

Since Sheila drove 72 km more than Russell, the distance she traveled is $x + 72$ km. Next, write an equation.

distance for Russell plus distance for Sheila is total distance

$$x + x + 72 = 232$$

Because $x = 1x$, the sum $x + x$ is $1x + 1x = 2x$. The equation becomes

$$2x + 72 = 232.$$

Subtract 72 from both sides.

$$2x + 72 - 72 = 232 - 72$$

$$2x = 160 \qquad \text{Divide both sides by 2.}$$

$$x = 80$$

Russell traveled 80 km. To check, use the words of the original problem. If Russell drove 80 km, Sheila drove 72 km more than Russell, or $80 + 72 = 152$ km. The total distance traveled by both was $152 + 80 = 232$ km. This checks with the statement of the problem. ■

WORK PROBLEM 4 AT THE SIDE. ▶▶

4. (a) In a day of work, Keonda made $12 more than her daughter. Together they made $182. Find the amount made by Keonda's daughter.

(b) A rope is 21 m long. Marcos cut it into two pieces, so that one piece is 3 m longer than the other. Find the length of the shorter piece.

ANSWERS

4. (a) daughter made x
Keonda made $x + 12$
$x + x + 12 = 182$
daughter made $85
(b) shorter piece is x
longer piece is $x + 3$
$x + x + 3 = 21$
shorter piece is 9 m

5. Make a drawing to help solve this problem. The length of Ann's rectangular garden plot is 3 m more than the width. She used 22 m of fencing around the edge. Find the length and the width of the garden.

■ **EXAMPLE 6** *Solving a Geometry Problem*

The length of a rectangle is 2 cm more than the width. The perimeter is 68 cm. Find the length.

Let x be the width of the rectangle.

Since the length is 2 cm more than the width, the length is $x + 2$. A drawing of the rectangle will help you see these relationships.

$$\text{width} = x$$
$$\text{length} = x + 2$$
$$\text{perimeter} = 68$$

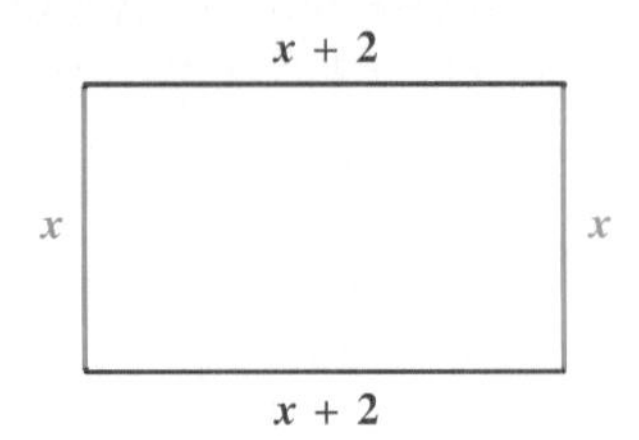

Use the formula for perimeter of a rectangle, $P = 2 \cdot l + 2 \cdot w$.

$$P = 2 \cdot l + 2 \cdot w$$
$$68 = \underbrace{2(x + 2)} + \underbrace{2 \cdot x}$$
$$68 = 2x + 4 + 2x \quad \text{Use distributive property.}$$
$$68 = 4x + 4 \quad \text{Combine like terms.}$$
$$68 - 4 = 4x + 4 - 4 \quad \text{Subtract 4 from both sides.}$$
$$64 = 4x$$
$$\frac{64}{4} = \frac{\overset{1}{\cancel{4}} \cdot x}{\underset{1}{\cancel{4}}} \quad \text{Divide both sides by 4.}$$
$$16 = x$$

This does **not** answer the question in the problem, because x represents the *width* of the rectangle and the problem asks for the *length.*

width $= x$ so width is 16 cm

length $= x + 2$ so length is $16 + 2$ or **18 cm**

Check using the words of the original problem. It says the length is 2 cm more than the width. 18 cm is 2 cm more than 16 cm, so that part checks. The original problem also says the perimeter is 68 cm. Use 18 cm and 16 cm to find the perimeter.

$$68 = 2 \cdot l + 2 \cdot w \quad \text{Let } l = 18 \text{ and } w = 16.$$
$$68 = 2 \cdot \mathbf{18} + 2 \cdot \mathbf{16}$$
$$68 = 36 + 32$$
$$68 = 68 \quad \text{true}$$

The solution is length $= 18$ cm. ■

◀◀ **WORK PROBLEM 5 AT THE SIDE.**

ANSWER

5. width $= x$
length $= x + 3$
$22 = 2(x + 3) + 2 \cdot x$
width is 4 m
length is 7 m

9.8 NAME DATE HOUR

EXERCISES

Write in symbols by using x as the variable.

Examples:	Solutions:
the sum of 7 and a number	$7 + x$
12 added to a number	$x + 12$
a number subtracted from 57	$57 - x$
three times a number	$3x$
the sum of 15 and twice a number	$15 + 2x$

1. 14 plus a number

$14 + x$

2. the sum of 9 and a number

$9 + x$

3. -5 added to a number

$-5 + x$

4. a number increased by -10

$x + (-10)$

5. the sum of a number and 6

$x + 6$

6. the total of a number and 18

$x + 18$

7. 9 less than a number

$x - 9$

8. a number subtracted from 2

$2 - x$

9. subtract 4 from a number

$x - 4$

10. 3 fewer than a number

$x - 3$

11. six times a number

$6x$

12. double a number

$2x$

13. triple a number

$3x$

14. half a number

$\frac{x}{2}$ or $\frac{1}{2}x$

15. a number divided by 2

$\frac{x}{2}$

16. 4 divided by a number

$\frac{4}{x}$

Writing Conceptual Challenging Estimation

17. twice a number added to 8

$8 + 2x$

18. five times a number plus 5

$5x + 5$

19. 10 fewer than seven times a number

$7x - 10$

20. 12 less than six times a number

$6x - 12$

21. the sum of twice a number and the number

$2x + x$

22. triple a number subtracted from the number

$x - 3x$

23. In your own words, write a definition for each of these words: variable, expression, equation. Give three examples to illustrate each definition.

Answer varies.

24. "You can use any letter to represent the unknown in a word problem." Is this statement true or false? Explain your answer.

Answer varies.

Solve each word problem. Use the five steps for solving algebra word problems.

Example: If a number is multiplied by 5 and the product is added to 2, the result is -13. Find the number.

Solution:

Let x represent the unknown number.

$$\underbrace{\text{a number multiplied by 5}}_{5x}\ \underbrace{\text{added to}}_{+}\ \underset{2}{\overset{\downarrow}{2}}\ \underbrace{\text{result is}}_{=}\ \underset{-13}{\overset{\downarrow}{-13}}$$

Solve the equation.

$$5x + 2 - 2 = -13 - 2 \quad \text{Subtract 2 from both sides.}$$
$$5x = -15 \quad \text{Divide both sides by 5.}$$
$$x = -3$$

The number is **-3.**

Check with the words of the original problem: $-3 \cdot 5 + 2 = -13$

$-13 = -13$ true

25. If four times a number is decreased by 2, the result is 26. Find the number.

7

26. The sum of 8 and five times a number is 53. Find the number.

9

27. If twice a number is added to the number, the result is -15. What is the number?

-5

28. If a number is subtracted from three times the number, the result is -8. What is the number?

-4

29. If half a number is added to twice the number, the answer is 50. Find the number.

20

30. If one-third of a number is added to three times the number, the result is 30. Find the number.

9

31. A board is 78 cm long. Rosa cut the board into two pieces, with one piece 10 cm longer than the other. Find the length of both pieces.

34 cm and 44 cm

32. Ed and Marge were candidates for city council. Marge won, with 93 more votes than Ed. The total number of votes cast in the election was 587. Find the number of votes received by each candidate.

247 votes for Ed; 340 votes for Marge

33. Kerwin rented a chain saw for a one-time $9 sharpening fee plus $16 a day rental. His total bill was $89. For how many days did Kerwin rent the saw?

5 days

34. Mrs. Chao made a $50 down payment on a sofa. Her monthly payments were $158. She paid $998 in all. For how many months did she make payments?

6 months

In the next exercises, use the formula for the perimeter of a rectangle, $P = 2l + 2w$. Make a drawing to help you solve each problem.

35. The perimeter of a rectangle is 48 m. The width is 5 m. Find the length.

19 m

36. The length of a rectangle is 27 cm, and the perimeter is 74 cm. Find the width of the rectangle.

10 cm

37. A rectangular dog pen is twice as long as it is wide. The perimeter of the pen is 36 ft. Find the length and the width of the pen.

12 ft, 6 ft

38. A new city park is a rectangular shape. The length is triple the width. It will take 240 meters of fencing to go around the park. Find the length and width of the park.

90 m, 30 m

▲ **39.** When 75 is subtracted from four times Tamu's age, the result is Tamu's age. How old is Tamu?

25 years old

▲ **40.** If three times Linda's age is decreased by 36, the result is twice Linda's age. How old is Linda?

36 years old

▲ **41.** A fence is 706 m long. It is to be cut into three parts. Two parts are the same length, and the third part is 25 m longer than the other two. Find the length of each part.

227 m, 227 m, 252 m

▲ **42.** A wooden railing is 82 m long. It is to be divided into four pieces. Three pieces will be the same length, and the fourth piece will be 2 m longer than each of the other three. Find the length of each piece.

20 m, 20 m, 20 m, 22 m

In the following exercises, use the formula for interest, $I = prt$.

▲ **43.** For how long must $800 be deposited at 12% per year to earn $480 interest?

5 years

▲ **44.** How much money must be deposited at 12% per year for 7 years to earn $1008 interest?

$1200

PREVIEW EXERCISES

Solve the following problems. (For help, see **Section 6.5** *or* **Section 6.6.**)

45. 80% of $2900 is how much money?

$2320

46. What is 15% of $360? $54

47. Jeffrey puts 5% of his $1830 monthly salary into a retirement plan. How much goes into his plan each month?

$91.50

48. Marshall College expects 9% of incoming students to need emergency loans to buy books. How many of the 2200 incoming students are expected to need loans?

198 students

CHAPTER 9 SUMMARY

KEY TERMS

9.1	**negative numbers**	Negative numbers are numbers that are less than zero.
	signed numbers	Signed numbers are positive numbers, negative numbers, and zero.
	absolute value	Absolute value is the distance of a number from zero on a number line. Absolute value is never negative.
	opposite of a number	The opposite of a number is a number the same distance from zero on a number line as the original number but on the opposite side of zero.
9.2	**additive inverse**	The additive inverse is the opposite of a number. The sum of a number and its additive inverse is always 0.
9.5	**variables**	Variables are letters that represent numbers.
	expression	An expression is a combination of letters and numbers.
9.6	**equation**	An equation is a statement that says two expressions are equal.
	solution	The solution is a number that can be substituted for the variable in an equation, so that the equation is true.
	addition property of equations	The addition property of equations states that the same number can be added or subtracted on both sides of an equation.
	multiplication property of equations	The multiplication property of equations states that the same nonzero number can be multiplied or divided on both sides of an equation.
9.7	**distributive property**	If a, b, and c are three numbers, the distributive property says that $a(b + c) = ab + ac$
	like terms	Like terms are terms with exactly the same variable and the same exponent.

QUICK REVIEW

Concepts	Examples
9.1 Graphing Signed Numbers Place a dot at the correct location on the number line.	Graph -2, 1, 0, and $2\frac{1}{2}$. 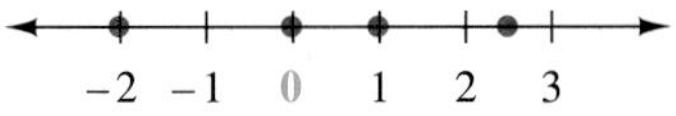
9.1 Identifying the Smaller of Two Numbers Place the symbols < (less than) or > (greater than) between two numbers to make the statement true. The small pointed end of the symbol points to the smaller number.	Use the symbol < or > to make the following statements true. 2 __>__ 1 -3 __>__ -5 -6 __<__ 2
9.1 Finding the Absolute Value of a Number Determine the distance from 0 to the given number on the number line.	Find each of the following. **(a)** $\lvert 8 \rvert$ $\lvert 8 \rvert = 8$ **(b)** $\lvert -7 \rvert$ $\lvert -7 \rvert = 7$

Concepts	Examples
9.1 Finding the Opposite of a Number Determine the number that is the same distance from 0 as the given number, but on the opposite side of 0 on a number line.	Find the opposite of each of the following. **(a)** -6 $-(-6) = 6$ **(b)** $+9$ $-(+9) = -9$
9.2 Adding Two Signed Numbers Case 1: *Two positive numbers* Add the numerical values. Case 2: *Two negative numbers* Add the absolute values and write a negative sign in front of the sum.	Add the following. **(a)** $8 + 6 = 14$ **(b)** $-8 + (-6)$ Find absolute values. $\|-8\| = 8 \quad \|-6\| = 6$ Add absolute values. $8 + 6 = 14$ Write a negative sign: -14. So, $-8 + (-6) = -14$
Case 3: *Two numbers with different signs* Subtract the absolute values and write the sign of the number with the larger absolute value in front of the answer.	**(c)** $5 + (-7)$ Find absolute values. $\|5\| = 5 \quad \|-7\| = 7$ Subtract the smaller absolute value from the larger. $7 - 5 = 2$ The number with the larger absolute value is -7. Its sign is negative, so write a negative sign in front of the answer. $5 + (-7) = -2$
9.2 Subtracting Two Signed Numbers Follow these steps: Change the subtraction sign to addition. Change the sign of the second number to its opposite. Proceed as in addition.	Subtract. **(a)** $-6 - 5 = -6 + (-5) = -11$ **(b)** $5 - (-8) = 5 + (+8) = 13$ **(c)** $7 - 9 = 7 + (-9) = -2.$
9.3 Multiplying Signed Numbers Use these rules: The product of two numbers with the same sign is positive. The product of two numbers with different signs is negative.	Multiply the following. **(a)** $7 \cdot 3 = 21$ **(b)** $(-3) \cdot 4 = -12$ **(c)** $(-9) \cdot (-6) = 54$
9.3 Dividing Signed Numbers Use the same rules as for multiplying signed numbers: When two numbers have the same sign, the quotient is positive. When two numbers have different signs, the quotient is negative.	Divide. **(a)** $\frac{8}{4} = 2$ **(b)** $\frac{-20}{5} = -4$ **(c)** $\frac{50}{-5} = -10$ **(d)** $\frac{-12}{-6} = 2$

Concepts	Examples
9.4 Using the Order of Operations to Evaluate Numerical Expressions Use the following order of operations to evaluate numerical expressions: Work inside parentheses first. Simplify expressions with exponents and find any square roots. Multiply or divide from left to right. Add or subtract from left to right.	Simplify the following. **(a)** $-4 + 6 \div (-2) = -4 + (-3)$ $= -7$ **(b)** $3^2 \cdot 4 + 3 \cdot (8 \div 2)$ $= \mathbf{3^2} \cdot 4 + 3 \cdot (4)$ $= \mathbf{9} \cdot 4 + 3 \cdot (4)$ $= 36 + 12$ $= 48$ **(c)** $9 \cdot 2^3 - (-4 \cdot 3)$ $= 9 \cdot \mathbf{8} - (-12)$ $= 72 + 12$ $= 84$
9.5 Evaluating Expressions Replace the variables in the expression with the numerical values. Use the order of operations to evaluate.	What is the value of $6p - 5s$, if $p = -3$ and $s = -4$? $6p - 5s = 6(-3) - 5(-4)$ $= -18 - (-20)$ $= 2$
9.6 Determining if a Number is a Solution of an Equation Substitute the number for the variable in the equation. If the equation is true, the number is a solution.	Is the number 4 a solution of the following equation? $3x - 5 = 7$ Replace x with 4. $3(4) - 5 = 7$ $12 - 5 = 7$ $7 = 7$ true 4 is the solution.
9.6 Using the Addition Property of Equations to Solve an Equation Add or subtract the same number on both sides of the equation, so that you get the variable by itself on one side.	Solve each equation. **(a)** $x - 6 = 9$ $x - 6 + \mathbf{6} = 9 + \mathbf{6}$ Add 6 to both sides. $x + 0 = 15$ $x = 15$ **(b)** $-7 = x + 9$ $-7 - \mathbf{9} = x + 9 - \mathbf{9}$ Subtract 9 from both sides. $-16 = x + 0$ $-16 = x$

Concepts	Examples
9.6 Using the Multiplication Property of Equations to Solve an Equation Multiply or divide both sides of the original equation by the same nonzero number, so that you get the variable by itself on one side.	Solve each equation. **(a)** $-54 = 6x$ $\frac{-54}{6} = \frac{\overset{1}{\cancel{6}} \cdot x}{\underset{1}{\cancel{6}}}$ $-9 = x$ **(b)** $\frac{1}{3}x = 8$ $\overset{1}{\cancel{3}} \cdot \frac{1}{\underset{1}{\cancel{3}}}x = 3 \cdot 8$ $1x = 24$ $x = 24$
9.7 Solving Equations with Several Steps Use the following steps: *Step 1* Add or subtract the same expression on both sides so that the variable ends up by itself on one side. *Step 2* Multiply or divide both sides by the same number to find the solution. *Step 3* Check the solution.	Solve: $2p - 3 = 9$ $2p - 3 + 3 = 9 + 3$ Add 3 to both sides. $2p = 12$ $\frac{2p}{2} = \frac{12}{2}$ Divide both sides by 2. $p = 6$ Check: $2p - 3 = 9$ $2 \cdot 6 - 3 = 9$ Let $p = 6$. $12 - 3 = 9$ $9 = 9$ true The solution is 6.
9.7 Using the Distributive Property To simplify expressions, use the distributive property: $a(b + c) = ab + ac.$	Simplify: $-2(x + 4)$ $= -2 \cdot x + (-2) \cdot 4$ $= -2x - 8$
9.7 Combining Like Terms If terms are like, combine the numbers that multiply each variable.	Combine like terms in the following. **(a)** $6p + 7p$ $6p + 7p = (6 + 7)p = 13p$ **(b)** $8m - 11m$ $8m - 11m = (8 - 11)m = -3m$
9.8 Translating Word Phrases by Using Variables Use x as a variable and symbolize the operations described by the words of the problem.	Write the following word phrases in symbols using x as the variable. **(a)** Two more than a number $x + 2$ **(b)** A number decreased by 8 $x - 8$ **(c)** The product of a number and 15 $15x$ **(d)** A number divided by 9 $\frac{x}{9}$

NAME DATE HOUR

CHAPTER 9 REVIEW EXERCISES

[9.1] *Graph the following lists of numbers.*

1. $2, -3, 4, 1, 0, -5$

−7 −6 −5 −4 −3 −2 −1 0 1 2 3 4

2. $-2, 5, -4, -1, 3, -6$

−6 −5 −4 −3 −2 −1 0 1 2 3 4 5

3. $-1\frac{1}{4}, -\frac{5}{8}, -3\frac{3}{4}, 2\frac{1}{8}, 1\frac{1}{2}, -2\frac{1}{8}$

−8 −7 −6 −5 −4 −3 −2 −1 0 1 2 3

4. $0, -\frac{3}{4}, 1\frac{1}{4}, -4\frac{1}{2}, \frac{7}{8}, -7\frac{2}{3}$

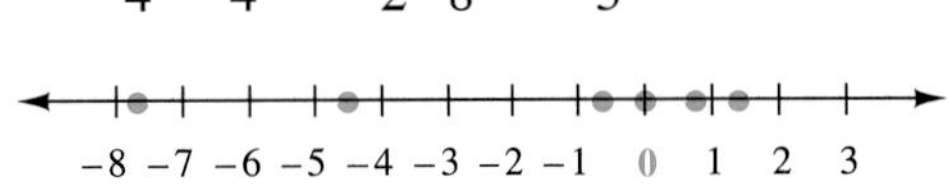

Place < or > in each of the following to get a true statement.

5. $0 > -2$

6. $-5 < 0$

7. $-1 > -4$

8. $-9 < -6$

Find each of the following.

9. $|8|$ 8

10. $|-19|$ 19

11. $-|-7|$ −7

12. $-|15|$ −15

[9.2] *Add.*

13. $-4 + 6$ 2

14. $-3 + 12$ 9

15. $-11 + (-8)$ −19

16. $-9 + (-24)$ −33

17. $8 + (-15)$ −7

18. $1 + (-20)$ −19

19. $\frac{9}{10} + \left(-\frac{3}{5}\right)$ $\frac{3}{10}$

20. $-\frac{7}{8} + \frac{1}{2}$ $-\frac{3}{8}$

21. $-6.7 + 1.5$ −5.2

22. $3.2 + (-2.9)$ 0.3

[9.2] *Give the additive inverse of each number.*

23. 6

−6

24. −14

14

25. $-\frac{5}{8}$

$\frac{5}{8}$

26. 3.75

−3.75

Subtract.

27. $4 - 10$

−6

28. $7 - 15$

−8

29. $-6 - 1$

−7

30. $-12 - 5$

−17

31. $8 - (-3)$

11

32. $2 - (-9)$

11

33. $-1 - (-14)$

13

34. $-10 - (-4)$

−6

[9.3] *Multiply or divide.*

35. $-4 \cdot 6$

−24

36. $5 \cdot (-4)$

−20

37. $-3 \cdot (-5)$

15

38. $-8 \cdot (-8)$

64

39. $\frac{80}{-10}$

−8

40. $\frac{-9}{3}$

−3

41. $\frac{-25}{-5}$

5

42. $\frac{-120}{-6}$

20

43. $\frac{2}{3} \cdot \left(-\frac{6}{7}\right)$

$-\frac{4}{7}$

44. $-\frac{5}{4} \cdot \left(-\frac{2}{15}\right)$

$\frac{1}{6}$

45. $-0.5 \cdot (-2.8)$

1.4

46. $-5.1 \cdot 2.9$

−14.79

[9.4] *Use the order of operations to simplify each of the following.*

47. $2 \cdot (-5) - 11$

−21

48. $(-4) \cdot (-8) - 9$

23

49. $48 \div (-2)^3 - (-5)$

−1

50. $-36 \div (-3)^2 - (-2)$

-2

51. $5 \cdot 4 - 7 \cdot 6 + 3 \cdot (-4)$

-34

52. $2 \cdot 8 - 4 \cdot 9 + 2 \cdot (-6)$

-32

53. $-4 \cdot 3^3 - 2 \cdot (5 - 9)$

-100

54. $6 \cdot (-4)^2 - 3 \cdot (7 - 14)$

117

55. $5^2 \cdot (0.9 - 1.6) \div (0.5 - 0.3)$

-87.5

56. $(-0.8)^2 \cdot (0.2) - (-1.2)$

1.328

57. $\left(-\frac{1}{3}\right)^2 + \frac{1}{4} \cdot \left(-\frac{4}{9}\right)$

0

58. $\frac{2}{9} \cdot \left(-\frac{6}{7}\right) - \left(-\frac{1}{3}\right)$

$\frac{1}{7}$

[9.5] *Find the value of each expression with the given values of the variables.*

59. $3k + 5m$
$k = 4, m = 3$

27

60. $3k + 5m$
$k = -6, m = 2$

-8

61. $2p - q$
$p = -5, q = -10$

0

62. $2p - q$
$p = 6, q = -7$

19

63. $\dfrac{5a - 7y}{2 + m}$
$a = 1, y = 4, m = -3$

23

64. $\dfrac{5a - 7y}{2 + m}$
$a = 2, \quad y = -2, \quad m = -26$

-1

In each of the following, use the formula and the values of the variables to find the value of the remaining variable.

65. $P = a + b + c; \quad a = 9, b = 12, c = 14$

35

66. $A = \frac{1}{2}bh; \quad b = 6, h = 9$

27

[9.6–9.7] *Solve each equation. Check each solution.*

67. $y + 2 = 11$

9

68. $a - 8 = 8$

16

69. $-5 = z - 6$

1

70. $-8 = -9 + r$

1

71. $-\frac{3}{4} + x = -2$

$-\frac{5}{4}$ or $-1\frac{1}{4}$

72. $12.92 + k = 4.87$

-8.05

73. $-8r = 56$

-7

74. $3p = 24$

8

75. $\frac{z}{4} = 5$

20

76. $\frac{a}{5} = -11$

-55

77. $20 = 3y - 7$

9

78. $-5 = 2b + 3$

-4

Use the distributive property to simplify.

79. $6(r - 5)$

$6r - 30$

80. $11(p + 7)$

$11p + 77$

81. $-9(z - 3)$

$-9z + 27$

82. $-8(x + 4)$

$-8x - 32$

Combine like terms.

83. $3r + 8r$

$11r$

84. $10z - 15z$

$-5z$

85. $3p - 12p + p$

$-8p$

86. $-6x - x + 9x$

$2x$

Solve each equation. Check each solution.

87. $-4z + 2z = 18$

-9

88. $9k - 2k = -35$

-5

89. $4y - 3 = 7y + 12$

-5

90. $b + 6 = 3b - 8$

7

91. $-14 = 2(a - 3)$

-4

92. $30 = 5(t + 4)$

2

[9.8] *Write in symbols by using x to represent the variable.*

93. 18 plus a number

$18 + x$

94. half a number

$\frac{1}{2}x$

95. the sum of four times a number and 6

$4x + 6$

96. five times a number decreased by 10

$5x - 10$

Solve each word problem.

97. If eight times a number is subtracted from eleven times the number, the result is -9. Find the number.

-3

98. A snowmobile can be rented for $45 for the first day and $35 for each additional day. The bill for Scott's rental was $255. For how many days did he rent it?

7 days

99. The perimeter of a rectangle is 124 cm. The width is 25 cm. Find the length. (Use the formula for the perimeter of a rectangle, $P = 2l + 2w$.)

37 cm

100. My sister is 9 years older than I am. The sum of our ages is 51. Find our ages.

21 and 30 years old

Mixed Review Exercises

Add, subtract, multiply, or divide as indicated.

101. $-6 - (-9)$

3

102. $-8 \cdot (-5)$

40

103. $-12 + 11$

-1

104. $\frac{-70}{10}$

-7

105. $-4 \cdot 4$

-16

106. $5 - 14$

-9

107. $\frac{-42}{-7}$

6

108. $16 + (-11)$

5

109. $|-6| + 2 - 3 \cdot (-8) - 5^2$

7

110. $9 \div |-3| + 6 \cdot (-5) + 2^3$

-19

Solve each equation.

111. $-45 = -5y$

9

112. $b - 8 = -12$

-4

113. $6z - 3 = 3z + 9$

4

114. $-5 = r + 5$

-10

115. $-3x = 33$

-11

116. $2z - 7z = -15$

3

117. $3(k - 6) = 6 - 12$

4

118. $4(t + 2) = -3 + 7$

-1

119. $-10 = \frac{a}{5} - 2$

-40

120. $4 + 8p = 4p + 16$

3

Solve each word problem.

121. When twice a number is decreased by 8, the result is the number increased by 7. Find the number.

15

122. The length of a rectangle is 3 inches more than twice the width. The perimeter is 36 inches. Find the length. (Use $P = 2l + 2w$.)

13 inches

123. Sam weighs 25 pounds more than Pam. Together they weigh 299 pounds. Find each person's weight.

Pam weighs 137 pounds; Sam weighs 162 pounds.

124. A 90-centimeter pipe is cut into two pieces so that one piece is 6 centimeters shorter than the other. Find the length of each piece.

48 cm and 42 cm

NAME DATE HOUR

CHAPTER 9 TEST

Work each of the following problems.

1. Graph the numbers $-4, -1, 1\frac{1}{2}, 3, 0$

1. −5 −4 −3 −2 −1 0 1 2 3 4 5

2. Place < or > in the blanks to make true statements:
-3 ______ 0 -4 ______ -8

2. <, >

3. Find $|-7|$ and $|15|$

3. 7, 15

Add or subtract.

4. $-8 + 7$

4. -1

5. $-11 + (-2)$

5. -13

6. $6.7 + (-1.4)$

6. 5.3

7. $8 - 15$

7. -7

8. $4 - (-12)$

8. 16

9. $-\frac{1}{2} - \left(-\frac{3}{4}\right)$

9. $\frac{1}{4}$

Multiply or divide.

10. $8(-4)$

10. -32

11. $-7(-12)$

11. 84

12. $\frac{-100}{4}$

12. -25

Use the order of operations to simplify each of the following.

13. $-5 + 3 \cdot (-2) - (-12)$

13. 1

14. $2 - (6 - 8) - (-5)^2$

14. -21

Find the value of 8k − 3m, given the following.

15. $k = -4, \quad m = 2$

15. -38

16. $k = 7, \quad m = 9$

16. 29

17. Answer varies.

17. In Exercises 15 and 16, you were evaluating an expression. Explain the difference between evaluating an expression and solving an equation.

18. 110

18. The formula for the area of a triangle is $A = \frac{1}{2}bh$. Find A, if $b = 20$ and $h = 11$.

Solve each equation.

19. 5

20. 31

21. −11

22. −15

23. 5

24. 2

25. −3

19. $x - 9 = -4$

20. $30 = -1 + r$

21. $77 = -7y$

22. $\frac{p}{5} = -3$

23. $3t - 8t = -25$

24. $3m - 5 = 7m - 13$

25. $-15 = 3(a - 2)$

Solve each word problem.

26. 10

26. If seven times a number is decreased by 23, the result is 47. Find the number.

27. 57 cm; 61 cm

27. A board is 118 cm long. Karin cut it into two pieces, with one piece 4 cm longer than the other. Find the length of both pieces.

28. Answer varies.

28. Make up your own word problem in which your age and the age of a friend are to be found. Include the sum of your ages and the difference between your ages in the problem. Then show the steps for solving your problem and explain each step.

NAME DATE HOUR

CUMULATIVE REVIEW EXERCISES CHAPTERS 1–9

$\approx$ *Round the numbers in each problem so there is only one non-zero digit. Then add, subtract, multiply, or divide the rounded numbers, as indicated, to estimate the answer. Finally, solve for the exact answer.*

1. *estimate*

$$\begin{array}{r} 9 \\ 1 \\ +\ 40 \\ \hline 50 \end{array}$$

exact

$$\begin{array}{r} 8.7 \\ 0.902 \\ +\ 41 \\ \hline 50.602 \end{array}$$

2. *estimate*

$$\begin{array}{r} 6 \\ \times\ 50 \\ \hline 300 \end{array}$$

exact

$$\begin{array}{r} 6.27 \\ \times\ 49.2 \\ \hline 308.484 \end{array}$$

3. *estimate*

$$80\overline{)40{,}000} \quad 500$$

exact

$$78\overline{)39{,}234} \quad 503$$

Add, subtract, multiply, or divide as indicated. Write answers to fraction problems in lowest terms and as whole or mixed numbers when possible.

4. $17 - 8.094$

8.906

5. $3\frac{2}{5} - 2\frac{3}{4}$

$\frac{13}{20}$

6. $4.06 \div 0.072$ Round to nearest tenth.

56.4

7. $1\frac{2}{3} \div 4\frac{1}{6}$

$\frac{2}{5}$

8. $(1309)(408)$

534,072

9. $5\frac{5}{6} \cdot \frac{9}{10}$

$5\frac{1}{4}$

10. $-12 + 7$ -5

11. $-5(-8)$ 40

12. $-3 - (-7)$ 4

13. $3.2 + (-4.5)$ -1.3

14. $\frac{30}{-6}$ -5

15. $\frac{1}{4} - \frac{3}{4}$ $-\frac{1}{2}$

Simplify by using the order of operations.

16. $45 \div \sqrt{25} - 2 \cdot 3 + (10 \div 5)$

5

17. $-6 - (4 - 5) + (-3)^2$

4

Write < or > in the blanks to make true statements.

18. $\frac{3}{10}$ __>__ $\frac{4}{15}$

19. 0.7072 __<__ 0.72

20. -5 __<__ -2

21. Write 6% as a decimal and as a fraction in lowest terms.

$0.06; \frac{3}{50}$

22. Write $2\frac{3}{10}$ as a decimal and as a percent.

2.3; 230%

23. There are 18 infants and 48 toddlers in the day care center. Write the ratio of toddlers to infants as a fraction in lowest terms.

$\frac{8}{3}$

Find the missing number in each proportion. Round your answer to hundredths, if necessary.

24. $\frac{x}{12} = \frac{1.5}{45}$

0.4

25. $\frac{350}{x} = \frac{3}{2}$

233.33 (rounded)

26. $\frac{38}{190} = \frac{9}{x}$

45

Solve each of the following.

27. 90 cars is 180% of what number of cars?

50 cars

28. \$5.80 is what percent of \$145?

4%

Convert the following measurements.

29. $3\frac{1}{2}$ gallons to quarts

14 quarts

30. 72 hours to days

3 days

31. 3.7 L to milliliters

3700 mL

32. 40 cm to meters

0.4 m

Write the most reasonable metric unit in each blank. Choose from km, m, cm, mm, L, mL, kg, g, mg.

33. The building is 15 __m__ high.

34. Rita took 15 __mL__ of cough syrup.

35. Bruce walked 2 __km__ to work.

36. The robin weighs 100 __g__.

37. At what metric temperature does water boil? __100°C__

Find the perimeter or circumference and the area of each figure. Use 3.14 for π.

38.

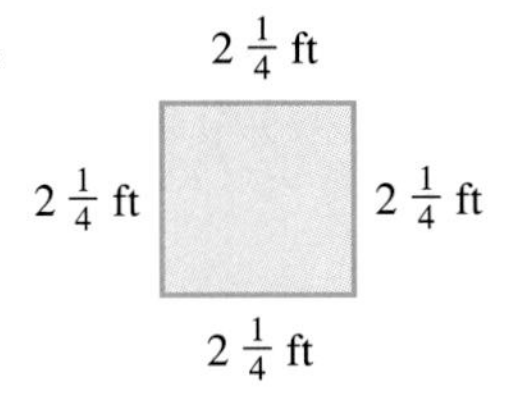

$P = 9$ ft

$A = 5\frac{1}{16}$ or $5.0625\ \text{ft}^2$

39.

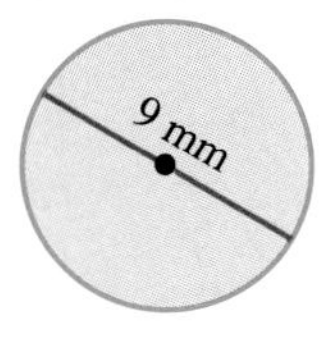

$C = 28.26$ mm

$A = 63.585\ \text{mm}^2$

40.

24 miles

7 miles

25 miles

$P = 56$ mi

$A = 84\ \text{mi}^2$

41.

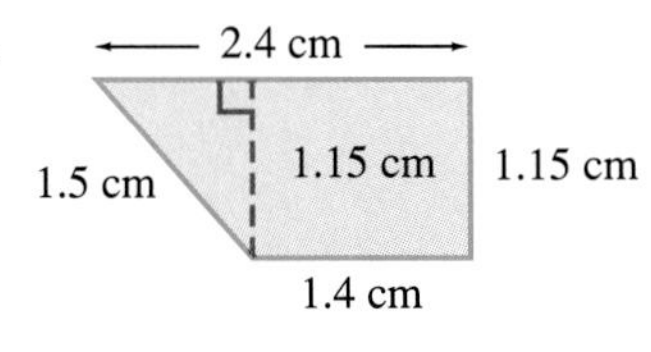

$P = 6.45$ cm

$A = 2.185\ \text{cm}^2$

42.

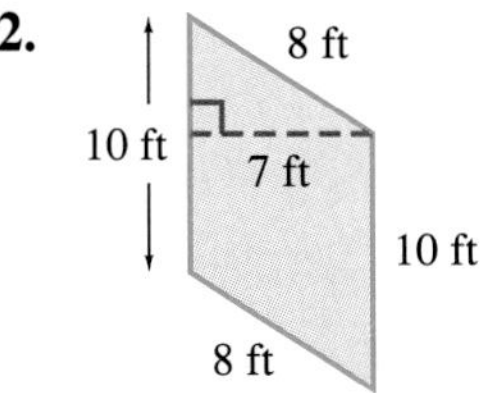

$P = 36$ ft

$A = 70\ \text{ft}^2$

43.

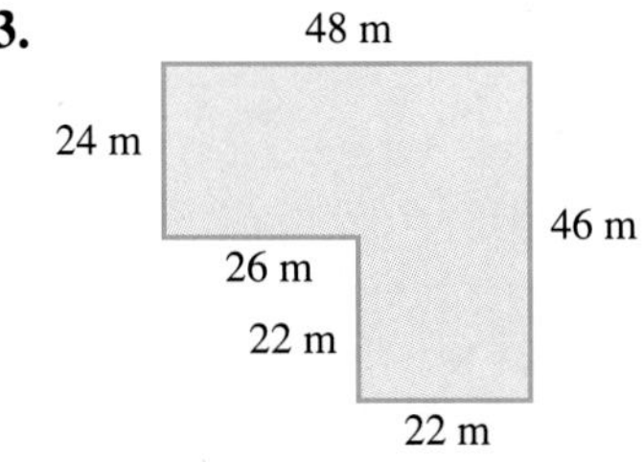

$P = 188$ m

$A = 1636\ \text{m}^2$

Find the length of the missing side in each figure. Round your answers to the nearest tenth.

44.

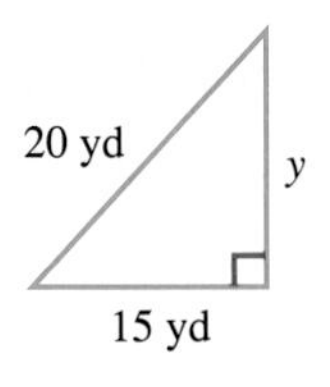

$y = 13.2$ yd

45. similar triangles

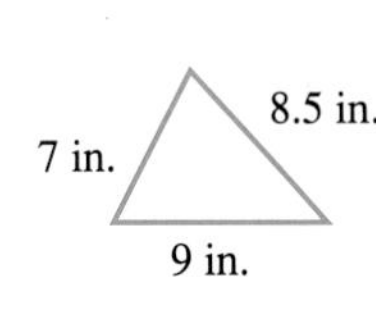

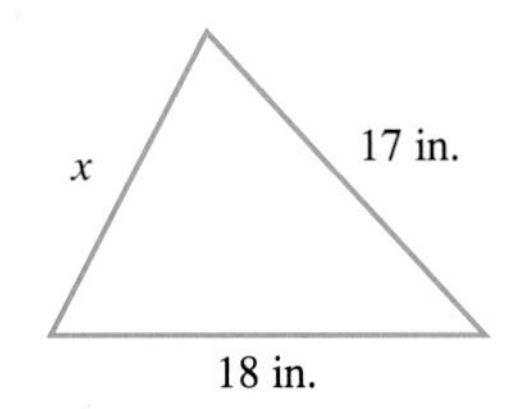

$x = 14$ in.

Solve each equation.

46. $-20 = 6 + y$

-26

47. $-2t - 6t = 40$

-5

48. $3x + 5 = 5x - 11$

8

49. $6(p + 3) = -6$

-4

Write an equation to solve each of the following word problems.

50. If 40 is added to four times a number, the result is zero. Find the number.

-10

51. $1000 in prize money is being split between Reggie and Donald. Donald should get $300 more than Reggie. How much will each man receive?

$350 for Reggie, $650 for Donald

52. Make a drawing to help solve this problem. The length of a picture frame is 5 cm more than the width. The perimeter of the frame is 82 cm. Find the length and the width.

23 cm, 18 cm

Solve the following word problems.

53. Portia bought two CDs at $11.98 each. The sales tax rate is $6\frac{1}{2}\%$. Find the total amount charged to Portia's credit card, to the nearest cent.

$25.52

54. Brian's spaghetti sauce recipe calls for $3\frac{1}{3}$ cups of tomato sauce. He wants to make $2\frac{1}{2}$ times the usual amount. How much tomato sauce does he need?

$8\frac{1}{3}$ cups

55. The local food shelf received 2480 pounds of food this month. Their goal was 2000 pounds. What percent of their goal was received?

124%

56. Sayoko bought 720 g of chicken priced at $5.97 per kilogram. How much did she pay for the chicken, to the nearest cent?

$4.30

57. A packing crate measures 2.4 m long, 1.2 m wide, and 1.2 m high. A trucking company wants crates that hold 4 m^3. The crate's volume is how much more or less than 4 m^3?

0.544 m^3 less

58. The Mercado family has 35 feet of fencing to put around a garden plot. The plot is rectangular in shape. If it is $6\frac{1}{2}$ feet wide, find the length of the plot.

11 feet

59. Rich spent 25 minutes reading 14 pages in his sociology textbook. At that rate, how long will it take him to read 30 pages? Round to the nearest whole number of minutes.

54 minutes

60. Jackie drove her car 365 miles on 14.5 gallons of gas. Maya used 16.3 gallons to drive 406 miles. Naomi drove 300 miles on 11.9 gallons. Which car had the highest number of miles per gallon?

Naomi's car

Statistics 10

median % circle graph mode

The word *statistics* originally came from words that mean *state numbers.* State numbers refer to numerical information, or **data,** gathered by the government such as the number of births, deaths, or marriages in a population. Today the word *statistics* has a much broader application; data from the fields of economics, social science, science, and business can all be organized and studied under the branch of mathematics called **statistics.**

10.1 CIRCLE GRAPHS

OBJECTIVES

1. Read a circle graph.
2. Use a circle graph.
3. Draw a circle graph.

FOR EXTRA HELP

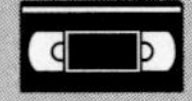

Tape 14 | SSM pp. 310–312 | MAC: A IBM: A

1 It can be hard to understand a large collection of data. The graphs described in this section can be used to help you make sense of such data. The **circle graph** is used to show how a total amount is divided into parts. The circle graph below shows you how 24 hours in the life of a college student are divided among different activities.

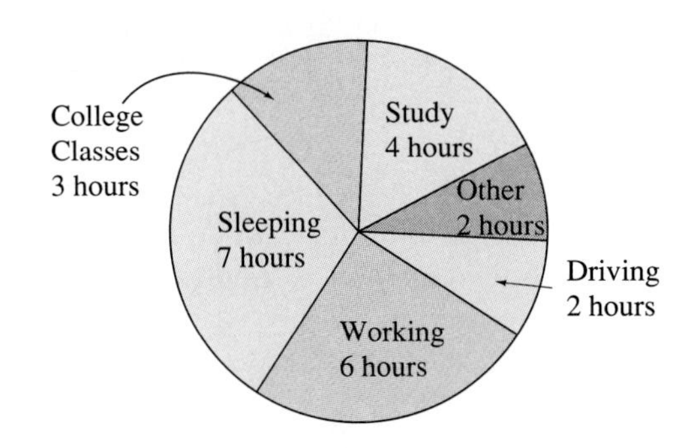

2 The circle graph uses pie-shaped pieces called **sectors** to show the amount of time spent on each activity (the total must be 24 hours); a circle graph can therefore be used to compare the time spent on one activity to the total number of hours in the day.

■ **EXAMPLE 1** *Using a Circle Graph*

Find the ratio of time spent in college classes to the total number of hours in a day. Write the ratio as a fraction in lowest terms.

The circle graph shows that 3 of the 24 hours in a day are spent in class. The ratio of class time to the hours in a day is

$$\frac{3 \text{ hours (college classes)}}{24 \text{ hours (whole day)}} = \frac{3 \text{ hours}}{24 \text{ hours}} = \frac{1}{8}. \quad \text{lowest terms}$$ ■

1. Use the circle graph to find the following ratios. Write the ratios as fractions in lowest terms.

(a) hours in class to whole day

(b) hours spent studying to whole day

(c) hours spent sleeping to whole day

(d) hours spent working to whole day

2. Use the circle graph to find the following ratios. Write the ratios as fractions in lowest terms.

(a) hours spent studying to hours spent working

(b) hours spent working to hours spent sleeping

(c) hours spent studying to hours spent driving

(d) hours spent in class to hours spent for other

ANSWERS

1. (a) $\frac{1}{8}$ (b) $\frac{1}{6}$ (c) $\frac{7}{24}$ (d) $\frac{1}{4}$

2. (a) $\frac{2}{3}$ (b) $\frac{6}{7}$ (c) $\frac{2}{1}$ (d) $\frac{3}{2}$

WORK PROBLEM 1 AT THE SIDE.

This circle graph can also be used to find the ratio of the time spent on one activity to the time spent on any other activity.

EXAMPLE 2 *Finding a Ratio from a Circle Graph*

Find the ratio of working time to class time.

The circle graph shows 6 hours spent working and 3 hours spent in class. The ratio of working time to class time is

$$\frac{6 \text{ hours (working)}}{3 \text{ hours (class)}} = \frac{6 \text{ hours}}{3 \text{ hours}} = \frac{2}{1}$$ ■

WORK PROBLEM 2 AT THE SIDE.

A circle graph often shows data as percents. For example, a family with an annual income of \$36,000 kept track of expenses for a year. The next circle graph shows how their income was divided among different types of expenses. The circle represents all the income, and each sector represents an expense as a percent of the annual income (the total must be 100%).

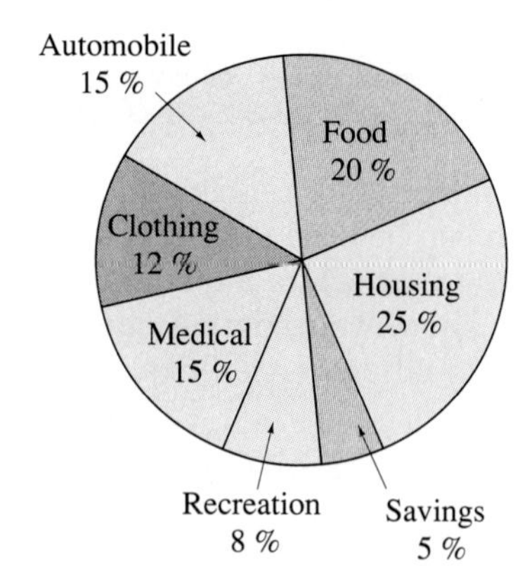

EXAMPLE 3 *Calculating Amounts Using a Circle Graph*

Use the circle graph on family expenses to find the amount spent on housing for the year.

Recall the percent equation:

$$\text{amount} = \text{percent} \cdot \text{base}$$

or

$$a = p \cdot b$$

The total income is \$36,000, so $b = 36{,}000$. The percent is 25, so $p = 0.25$. Find a.

$$\begin{array}{ccccc} \text{amount} & = & \text{percent} & \cdot & \text{base} \\ \downarrow & & \downarrow & & \downarrow \\ a & = & 0.25 & \cdot & 36{,}000 = 9000 \end{array}$$

The amount spent on housing is \$9000. ■

WORK PROBLEM 3 AT THE SIDE. ▶▶

3 The coordinator of the Fair Oaks Youth Soccer League organizes teams in five age groups. She places the players in various age groups as follows.

Age Group	*Percent of Total*
Under 8 years	20%
Under 10 years	15%
Under 12 years	25%
Under 14 years	25%
Under 16 years	15%
Total	100%

The league coordinator can show these percents by using a circle graph. A circle has 360 degrees (written 360°). The 360° represents the entire swimming class.

EXAMPLE 4 *Drawing a Circle Graph*

Using the data on *age groups*, find the number of degrees in the sector that would represent the "Under 8" group, and begin constructing a circle graph.

Because the "Under 8" group makes up 20% of the total number of players, the number of degrees needed for the "Under 8" sector of the circle graph is

$$360^\circ \times 20\% = 360^\circ \times 0.2 = 72^\circ.$$

Use a tool called a **protractor** to make a circle graph. First, draw a line from the center of a circle to the left edge. Place the hole in the protractor over the center of the circle, making sure that zero on the protractor lines up with the line that was drawn. Find 72° and make a mark as shown in the illustration. Then draw a line from the center of the circle to the 72° mark at the edge of the circle.

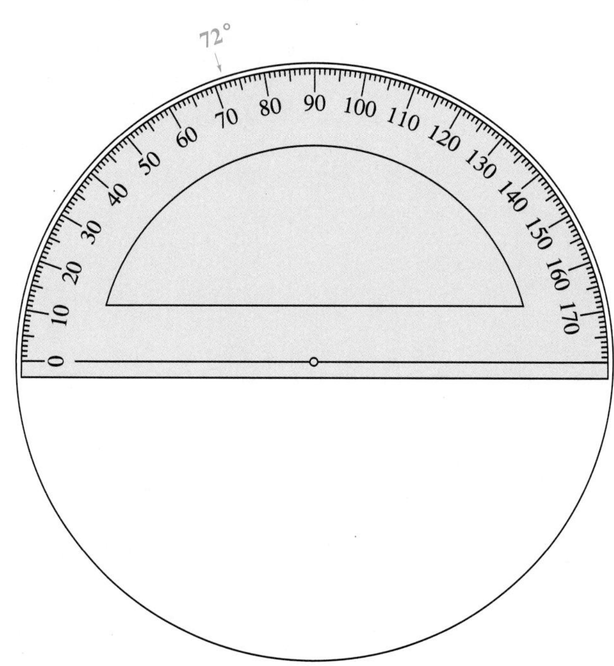

3. Use the circle graph on family expenses to find the following:

(a) the amount spent on clothing

(b) the amount spent on medical costs

(c) the amount spent on food

(d) the amount saved

ANSWERS

3. (a) $4320 (b) $5400 (c) $7200 (d) $1800

4. Using the information on the soccer age groups in the table, find the number of degrees needed for each of the following and complete the circle graph.

(a) Under 12 group

(b) Under 14 group

(c) Under 16 group

ANSWERS
4. (a) 90° (b) 90° (c) 54°

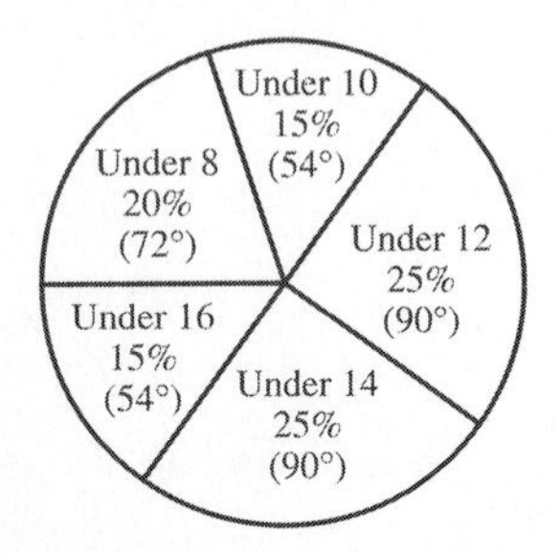

To draw the "Under 10" sector, begin by finding the number of degrees in the sector.

$$360° \times 15\% = 360° \times 0.15 = 54°$$

Again, place the hole of the protractor at the center of the circle, but this time align zero on the second line that was drawn. Make a mark at 54° and draw a line as before. This sector is 54° and represents the "Under 10" group. ■

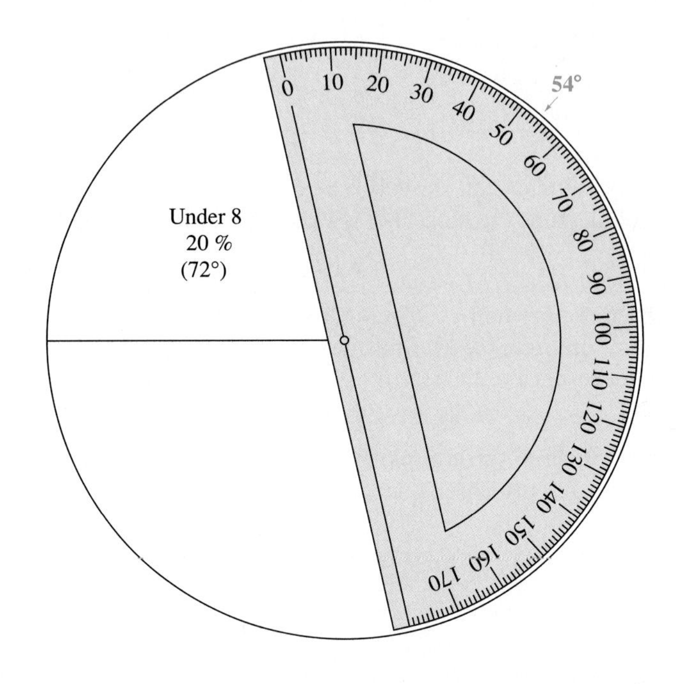

Note You must be certain that the hole in the protractor is placed on the exact center of the circle each time you measure the size of a sector.

◀◀ WORK PROBLEM 4 AT THE SIDE.

10.1 EXERCISES

NAME DATE HOUR

Use the circle graph below, which shows the cost of adding an art studio to an existing building, to find each of the following. Write ratios as fractions in lowest terms.

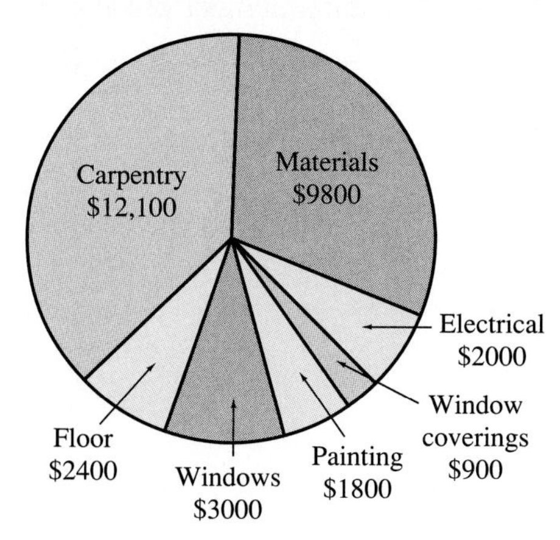

1. Find the total cost of adding the art studio.

$32,000

2. What is the largest single expense in adding the studio?

carpentry

3. Find the ratio of the cost of materials to the total remodeling cost.

$\frac{9800}{32{,}000} = \frac{49}{160}$

4. Find the ratio of the cost of painting to the total remodeling cost.

$\frac{1800}{32{,}000} = \frac{9}{160}$

5. Find the ratio of the cost of carpentry to the cost of window coverings.

$\frac{12{,}100}{900} = \frac{121}{9}$

6. Find the ratio of the cost of windows to the cost of the floor.

$\frac{3000}{2400} = \frac{5}{4}$

7. The circle graph at the right shows the number of students at Rockfield College who are enrolled in various majors. In which major are the least number of students enrolled?

history

8. Using the circle graph at the right, which major has the second highest number of students enrolled?

social science

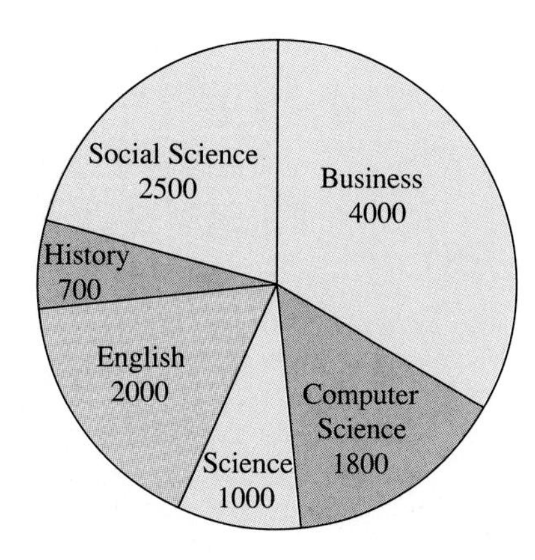

Use the circle graph in Exercise 7 to find each ratio. Write the ratios as fractions in lowest terms.

9. business majors to the total number of students

$\frac{4000}{12,000} = \frac{1}{3}$

10. English majors to the total number of students

$\frac{2000}{12,000} = \frac{1}{6}$

11. computer science majors to the number of English majors

$\frac{1800}{2000} = \frac{9}{10}$

12. history majors to the number of social science majors

$\frac{700}{2500} = \frac{7}{25}$

13. business majors to the number of science majors

$\frac{4000}{1000} = \frac{4}{1}$

14. science majors to the number of history majors

$\frac{1000}{700} = \frac{10}{7}$

The following circle graph shows the expenses necessary to comply with the Americans with Disabilities Act (ADA) at the Dos Pueblos College. Each expense item is expressed as a percent of the total cost of $1,740,000. Use the graph to find the dollar amount spent for each of the following.

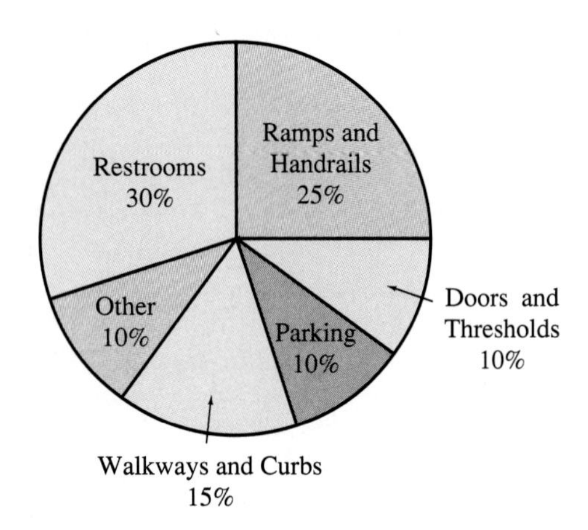

15. restrooms

$522,000

16. ramps and handrails

$435,000

17. doors and thresholds

$174,000

18. parking

$174,000

19. walkways and curbs

$261,000

20. other

$174,000

The circle graph below shows how the Ecology Club's income of $19,600 is budgeted. Use the graph to find the dollar amount budgeted for each of the following.

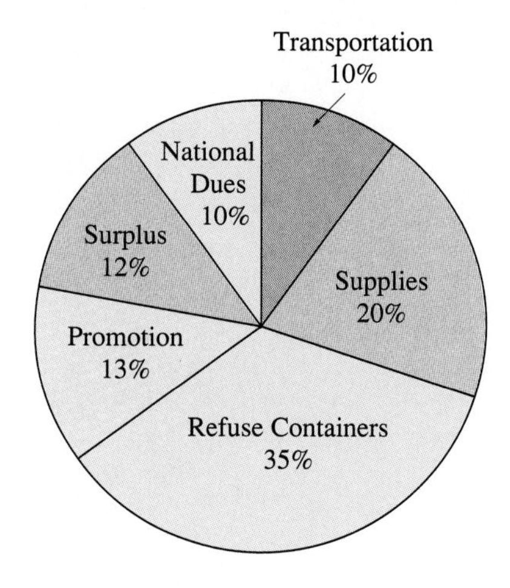

21. supplies

$3920

22. refuse containers

$6860

23. promotion

$2548

24. surplus

$2352

25. national dues

$1960

26. transportation

$1960

27. Describe the procedure for determining how large each sector must be to represent each of the items in a circle graph.

Answer varies.

28. A protractor is the tool used to draw a circle graph. Give a brief explanation of what the protractor does and how you would use it to measure and draw each sector in the circle graph.

Answer varies.

During one semester Zoë Werner spent $4200 for school expenses as shown in the following chart. Find all numbers missing from the chart.

item	*dollar amount*	*percent of total*	*degrees of a circle*
29. rent	\$1050	25%	90°
30. food	\$840	20%	72°
31. clothing	\$420	10%	36°
32. books	\$420	10%	36°
33. entertainment	\$630	15%	54°
34. savings	\$210	5%	18°
35. other	\$630	15%	54°

36. Draw a circle graph by using the above information.

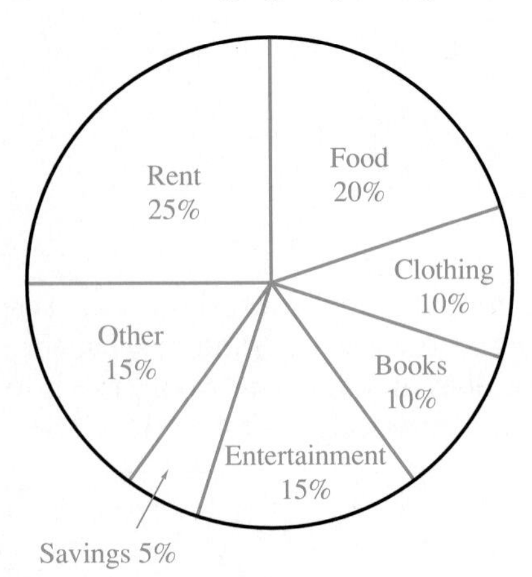

37. White Water Rafting Company divides its annual sales into five categories, as follows.

Category	*Annual Sales*
Adventure classes	\$12,500
Grocery and provision sales	\$40,000
Equipment rentals	\$60,000
Rafting tours	\$50,000
Equipment sales	\$37,500

(a) Find the total sales for the year.

\$200,000

(b) Find the number of degrees in a circle graph for each item.

22.5°; 72°; 108°; 90°; 67.5°

(c) Make a circle graph showing this information.

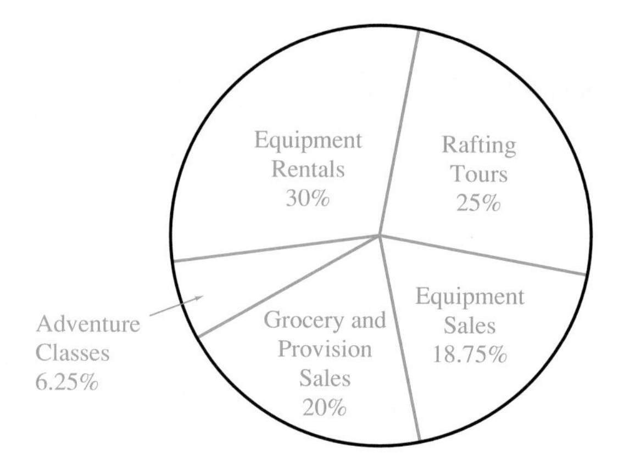

38. A book publisher had 25% of total sales in mysteries, 10% in biographies, 15% in cookbooks, 15% in romantic novels, 20% in science, and the rest in business books.

(a) Find the number of degrees in a circle graph for each type of book.

90°; 36°; 54°; 54°; 72°; 54°

(b) Draw a circle graph.

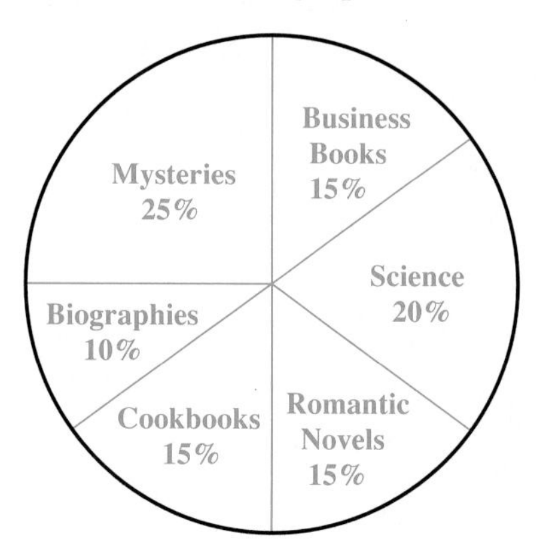

▲ **39.** A family kept track of its expenses for a year and recorded the following results. Complete the chart and draw a circle graph.

Item	*Amount*	*Percent of Total*	*Number of Degrees*
Housing	\$9600	30%	108°
Food	\$6400	20%	72°
Automobile	\$4800	15%	54°
Clothing	\$3200	10%	36°
Medical	\$1600	5%	18°
Savings	\$1600	5%	18°
Other	\$4800	15%	54°
Total	\$32,000		

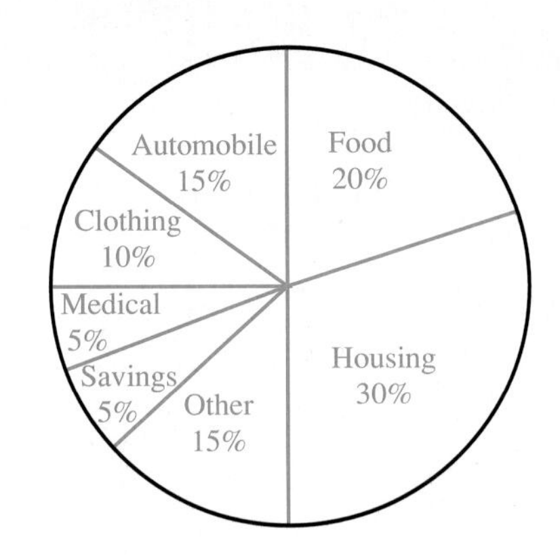

PREVIEW EXERCISES

Write $<$ or $>$ to make a true statement. (For help, see ***Sections 3.5, 4.7, and 6.1.****)*

40. $\frac{1}{4} < \frac{3}{8}$

41. $\frac{4}{5} < \frac{7}{8}$

42. $0.4219 < 0.422$

43. $0.0118 > 0.01$

44. $38.25\% < 38.29\%$

45. $25.9\% < 26.01\%$

46. $60{,}500 > 60{,}498$

47. $44{,}272.68 > 44{,}272.098$

48. $799{,}802 < 799{,}899$

10.2 BAR GRAPHS AND LINE GRAPHS

OBJECTIVES

Read and understand

1. a bar graph;
2. a double-bar graph;
3. a line graph;
4. a comparison line graph.

FOR EXTRA HELP

Tape 14 | SSM pp. 312–313 | MAC: A IBM: A

1 **Bar graphs** are useful when showing comparisons. For example, the bar graph below compares the number of college graduates who continued taking advanced courses in their major field during each of five years.

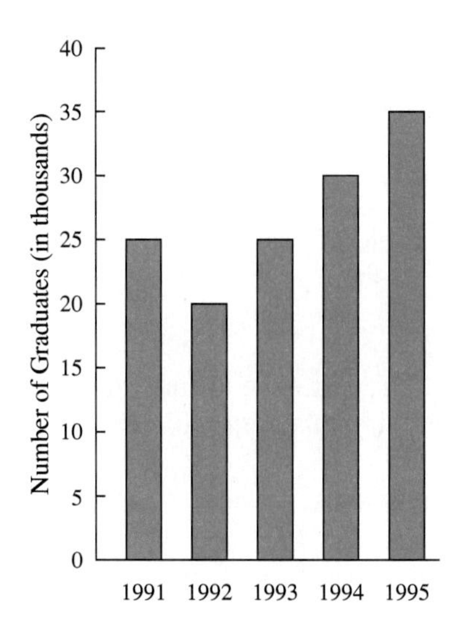

■ **EXAMPLE 1** *Using a Bar Graph*

How many college graduates took advanced classes in their major field in 1993?

The bar for 1993 rises to 25. Notice the label along the left side of the graph that says "Number of Graduates (in thousands)." The phrase *in thousands* means you have to multiply 25 by 1000 to get 25,000. So, 25,000 (not 25) graduates took advanced classes in their major field in 1993. ■

WORK PROBLEM 1 AT THE SIDE.

1. Use the bar graph in the text to find the number of college graduates who took advanced classes in their major field in each of these years.

(a) 1991

(b) 1992

(c) 1994

(d) 1995

2 A **double-bar graph** can be used to compare two sets of data. This graph shows the number of new cable television installations each quarter for two different years.

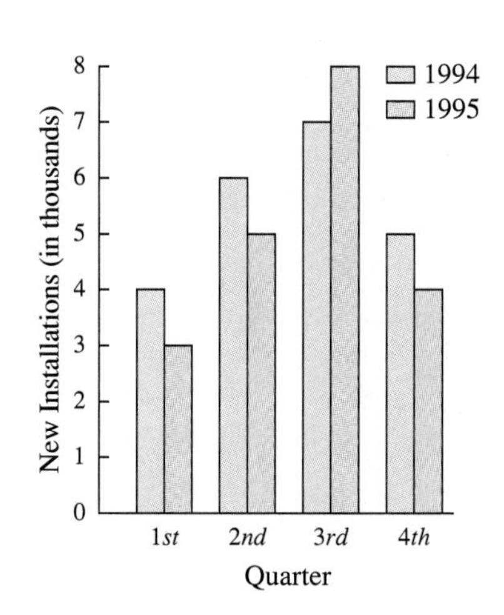

ANSWERS

1. (a) 25,000 (b) 20,000 (c) 30,000 (d) 35,000

2. Use the double-bar graph to find the number of new cable television installations in 1994 and 1995 for each of the following quarters.

(a) 1st quarter

(b) 3rd quarter

(c) 4th quarter

(d) Find the greatest number of installations. Identify the quarter and the year in which they occurred.

3. Use the line graph in the text to find the number of trout stocked in each of the following months.

(a) June

(b) August

(c) April

(d) July

ANSWERS
2. (a) 4000; 3000 (b) 7000; 8000 (c) 5000; 4000 (d) 8000; 3rd quarter of 1995
3. (a) 55,000 (b) 25,000 (c) 40,000 (d) 60,000

■ **EXAMPLE 2** *Reading a Double-Bar Graph*

Use the double-bar graph to find each of the following.

(a) the number of new cable television installations in the second quarter of 1994

There are two bars for the second quarter. The one on the *left* is for the 2nd quarter of *1994*. It rises to 6. Multiply 6 by 1000 because the label on the left side of the graph says *in thousands.* So there were 6000 new installations for the second quarter in 1994.

(b) the number of new cable television installations in the second quarter of 1995

In the second quarter of 1995, there were 5000 new installations.

Note Use a ruler or straight edge to line up the top of the bar with the number on the left side of the graph. ■

WORK PROBLEM 2 AT THE SIDE.

3 A **line graph** is often useful for showing a trend. The line graph that follows shows the number of trout stocked along the Feather River over a five-month period. Each dot indicates the number of trout stocked during the month directly below that dot.

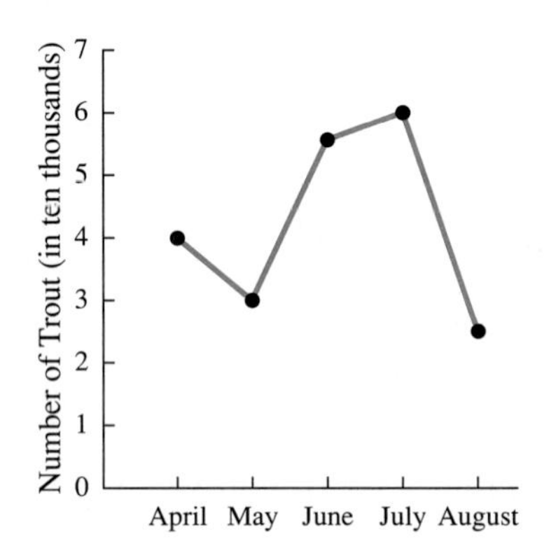

■ **EXAMPLE 3** *Understanding a Line Graph*

Use the line graph to find the following.

(a) In which month were the least number of trout stocked?

The least number of trout were stocked in August.

(b) How many trout were stocked in May?

There were 30,000 $(3 \cdot 10{,}000)$ trout stocked in May. ■

WORK PROBLEM 3 AT THE SIDE.

4 Two sets of data can also be compared by drawing two line graphs together as a **comparison line graph.** For example, the following line graph compares the number of installed cellular phones and the number of transportable cellular phones sold during each of five years.

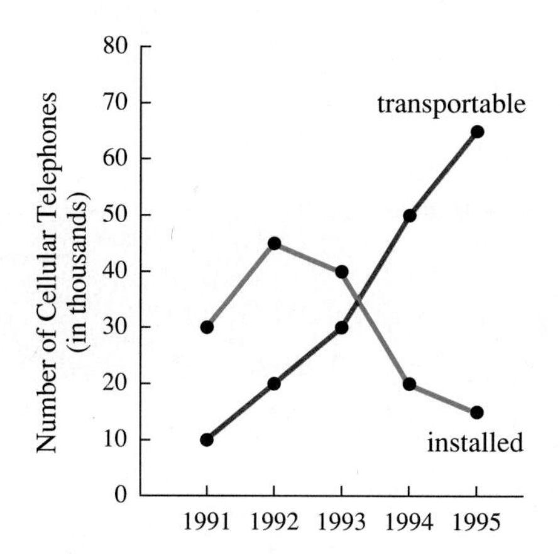

EXAMPLE 4 *Interpreting a Comparison Line Graph*
Use the comparison line graph above to find the following.

(a) the number of installed cellular phones sold in 1992.

The blue line on the graph shows that 45,000 ($45 \cdot 1000$) installed cellular phones were sold in 1992.

(b) the number of transportable cellular phones in 1995.

The red line on the graph shows that 65,000 transportable cellular phones were sold in 1995. ■

Note Both the double bar graph and the comparison line graph are used to compare two or more sets of data.

WORK PROBLEM 4 AT THE SIDE.

4. Use the comparison line graph in the text to find the following.

(a) the number of installed cellular phones sold in 1991, 1993, 1994, and 1995.

(b) the number of transportable cellular phones sold in 1991, 1992, 1993, and 1994.

(c) the first full year in which the number of transportable cellular phones sold was greater than the number of installed cellular phones sold.

ANSWERS
4. (a) 30,000; 40,000; 20,000; 15,000
(b) 10,000; 20,000; 30,000; 50,000
(c) 1994

QUEST FOR NUMERACY

Baseball Card Ups and Downs

Many people collect baseball cards as a hobby, some collect them as an investment, and a few people support themselves through college by buying and selling cards. A 1952 Jackie Robinson card recently sold for $4675. (The one pictured here is from 1953.)

In the past year some cards have gone up in value while others have gone down.

National Baseball Library and Archive, Cooperstown, New York

THE UPS AND DOWNS OF LAST YEAR

ROOKIE CARD	CARD PRICE*	% CHANGE FROM LAST YEAR
Frank Thomas	$ 6.00	+33
Cal Ripkin Jr.	65.00	Unch.
Dave Justice	3.00	−25
Nolan Ryan	1620.00	+35
Ken Griffey Jr.	5.00	−45
José Canseco	3.50	−65
Will Clark	10.00	−33
Mark McGwire	4.00	−67

*Average prices for rookie cards

Find the price of each of the cards *last* year for each of the players listed. (Round to the nearest half dollar.)

Frank Thomas $4.50

Cal Ripkin Jr. $65

Dave Justice $4

Nolan Ryan $1200

Ken Griffey Jr. $9

José Canseco $10

Will Clark $15

Mark McGwire $12

This bar graph shows the attendance at the Folsom Rodeo for each of five days in July.

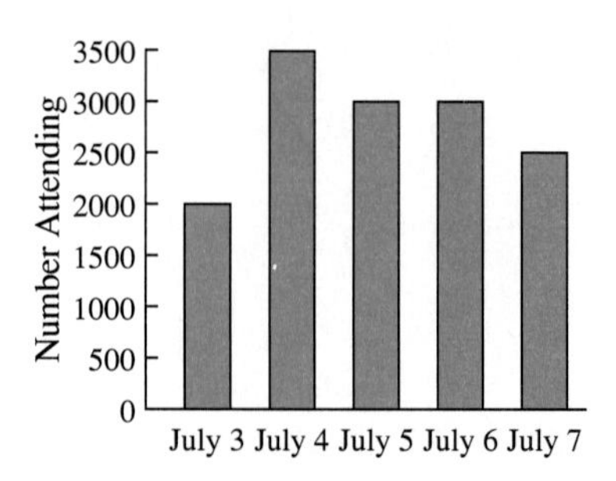

1. Find the attendance on July 5. 3000

2. Which day had the lowest attendance? July 3

3. Find the attendance on July 3. 2000

4. Find the attendance on July 4. 3500

5. Which day had the greatest attendance? July 4

6. Find the attendance on July 7. 2500

This double-bar graph shows the number of workers who were unemployed in a city during the first six months of 1994 and 1995.

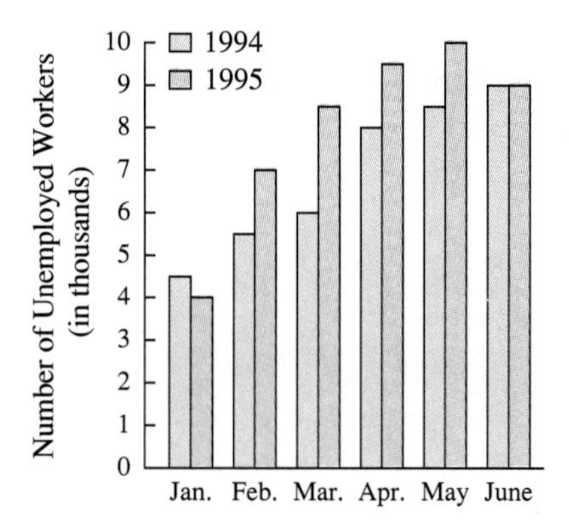

7. In which month in 1995 were the greatest number of workers unemployed? What was the total number unemployed in that month?

May; 10,000

8. How many workers were unemployed in January of 1994?

4500

9. How many more workers were unemployed in February of 1995 than in February of 1994?

1500

10. How many fewer workers were unemployed in March of 1994 than in March of 1995?

2500

11. Find the increase in the number of unemployed workers from February 1994 to April 1994.

2500

12. Find the increase in the number of unemployed workers from January 1995 to June 1995.

5000

This double-bar graph shows sales of super unleaded and supreme unleaded gasoline at a service station for each of the five years.

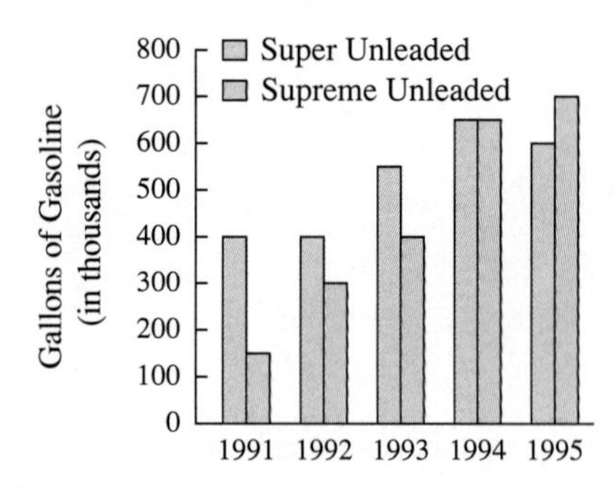

13. How many gallons of supreme unleaded gasoline were sold in 1991?

150,000 gallons

14. How many gallons of super unleaded gasoline were sold in 1994?

650,000 gallons

15. In which year did the greatest difference in sales between super unleaded and supreme unleaded gasoline occur?

1991

16. In which year did the sales of supreme unleaded gasoline surpass the sales of super unleaded gasoline?

1995

17. Find the increase in supreme unleaded gasoline sales from 1991 to 1995.

550,000 gallons

18. Find the increase in super unleaded gasoline sales from 1991 to 1995.

200,000 gallons

This line graph shows the number of burglaries in a community during the first six months of last year.

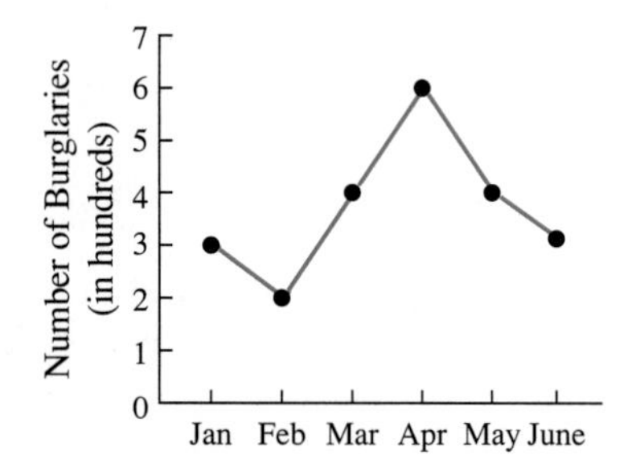

19. In which month did the greatest number of burglaries occur?

April

20. In which month did the least number of burglaries occur?

February

21. Find the number of burglaries in June.

300

22. Find the number of burglaries in March.

400

23. Find the increase in the number of burglaries from March to April.

200

24. Find the decrease in the number of burglaries from April to May.

200

This comparison line graph below shows the number of compact discs (CD's) sold by two different chain stores during each of five years. Find the annual number of CD's sold in each of the following years.

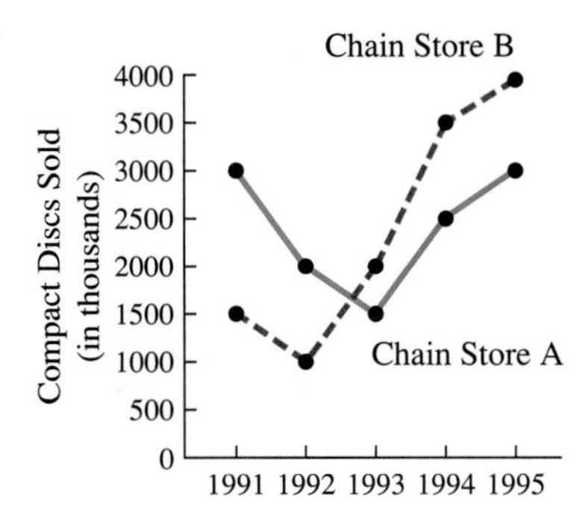

25. Chain Store A in 1995

3,000,000

26. Chain Store A in 1994

2,500,000

27. Chain Store A in 1993

1,500,000

28. Chain Store B in 1995

4,000,000

29. Chain Store B in 1994

3,500,000

30. Chain Store B in 1993

2,000,000

31. Explain in your own words why a bar graph or a line graph (not a double-bar graph or comparison line graph) can be used to show only one set of data.

Answer varies.

32. The double-bar graph and the comparison line graph are both useful for comparing two sets of data. Explain how this works and give your own example.

Answer varies.

This comparison line graph shows the sales and profits of Tacos-To-Go for each of four years. Use the graph to answer the following questions.

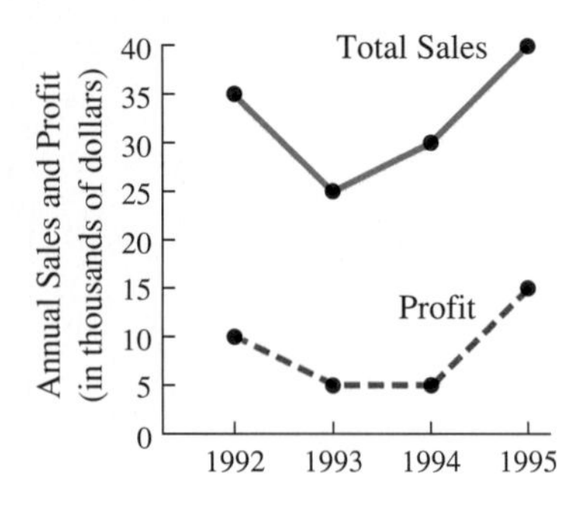

▲ **33.** total sales in 1995

$40,000

▲ **34.** total sales in 1994

$30,000

▲ **35.** total sales in 1993

$25,000

▲ **36.** profit in 1995

$15,000

▲ **37.** profit in 1994

$5000

▲ **38.** profit in 1993

$5000

PREVIEW EXERCISES

The circle graph below shows where the money collected for parking tickets on campus actually goes. The amount collected is $16,800. Use the graph to find the amount going to each place. (For help, see ***Section 10.1.****)*

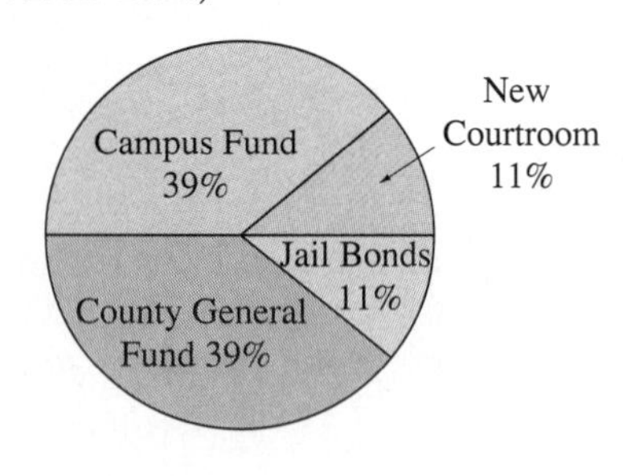

39. new courtroom $1848

40. campus fund $6552

41. county general fund $6552

42. jail bonds $1848

10.3 HISTOGRAMS AND FREQUENCY POLYGONS

An employer asked her 30 employees how many college credits each had completed. The list of responses looked like this:

74	133	4	127	20	30
103	27	139	118	138	121
149	132	64	141	130	76
42	50	95	56	65	104
4	140	12	88	119	64

OBJECTIVES

1. Arrange data in class intervals.
2. Understand class frequency.
3. Read and understand a histogram.
4. Read, understand, and construct a frequency polygon.

FOR EXTRA HELP

Tape 14 | SSM pp. 313–314 | MAC: A IBM: A

1 A long list of numbers can be confusing. You can make the data easier to read by dividing the numbers into smaller groups, or *class intervals,* such as 0 to 24 credits, or 25–49 credits, 50–74 credits, and so on.

2 Next, use a **tally** column, as shown below. Make a mark for each employee whose number of credits falls in that interval. Then count the number of marks to find the **class frequency** (number of employees) for that interval.

Class Interval (Number of Credits)	*Tally*	*Class Frequency (Number of Employees)*
0–24	||||	4
25–49	|||	3
50–74	~~||||~~ |	6
75–99	|||	3
100–124	~~||||~~	5
125–149	~~||||~~ ||||	9

3 The results in the table above have been used to draw this special bar graph, called a **histogram.** In a histogram, the width of each bar represents a range of numbers (*class interval*). The height of each bar in a histogram gives the *class frequency,* that is the number of occurrences in each class interval.

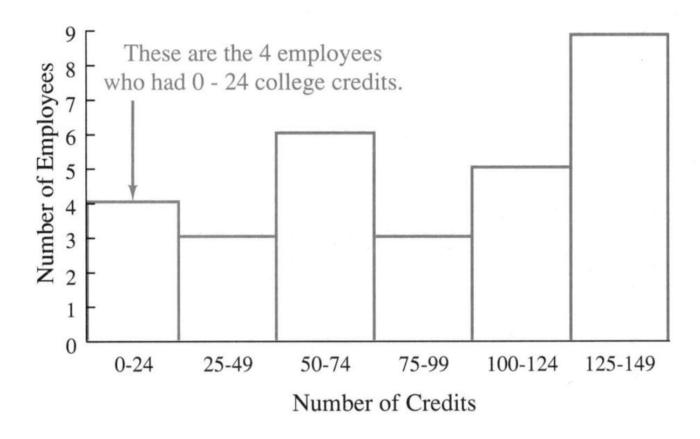

■ EXAMPLE 1 *Using a Histogram*

Use the histogram to find the number of employees who have completed fewer than 50 college credits.

Because 4 employees have completed 0–24 credits and 3 employees have completed 25–49 credits, the number of employees who have completed fewer than 50 credits is 4 + 3 = 7. ■

1. Use the histogram in the text to find the following.

(a) the number of employees who have completed 100 credits or more

(b) the number of employees who have completed 25 to 99 credits

(c) the number of employees who have completed fewer than 75 credits

2. Use the frequency polgyon to find the number of customers who used their charge accounts:

(a) 60 or fewer times in the year

(b) more than 20 times in the year

(c) from 21 to 60 times in the year.

ANSWERS
1. (a) 14 (b) 12 (c) 13
2. (a) $30 + 35 + 20 = 85$
(b) $35 + 30 + 10 + 5 = 80$
(c) $35 + 30 = 65$

Note The number of class intervals can vary. There are usually 5 to 15 intervals.

WORK PROBLEM 1 AT THE SIDE.

4 Sometimes the results of dividing data into class intervals are shown on a graph different from a histogram. For example, suppose the number of times that 100 credit card customers used their charge accounts was recorded. The results are listed in the following table. The middle of each class interval, called the **class midpoint,** is included in the table.

Class Interval (Number of Charge Account Uses)	*Class Midpoint*	*Class Frequency*
1–20	10.5	20
21–40	30.5	35
41–60	50.5	30
61–80	70.5	10
81–100	90.5	5

First a histogram is drawn as shown below. Then each class midpoint is located, and a dot is placed at the corresponding point at the top of the bar. These dots are connected to form a **frequency polygon.**

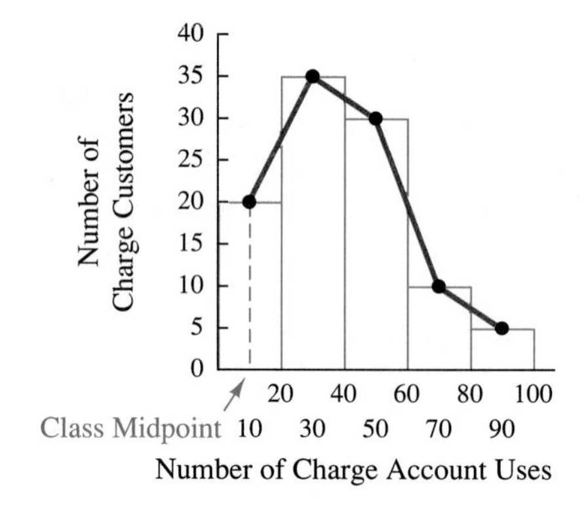

■ **EXAMPLE 2** *Using a Frequency Polygon*

Use the frequency polygon above to find the number of customers who used their charge accounts more than 40 times in the year.

The number of customers who used their charge accounts more than 40 times was $30 + 10 + 5 = 45$

WORK PROBLEM 2 AT THE SIDE.

10.3 EXERCISES

NAME DATE HOUR

The Toy Train Collectors' Club recorded the ages of its members and used the results to construct the histogram below. Use the histogram to find each of the following.

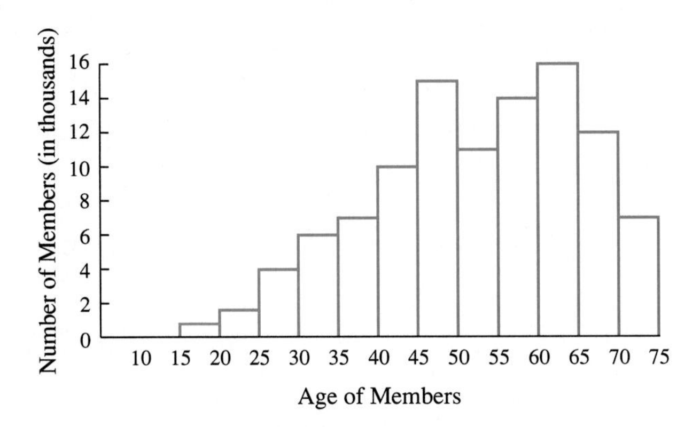

1. The greatest number of members are in which age group?

60–65 years

2. The least number of members are in which age group?

15–20 years

3. Find the number of members under 35 years of age.

13,000 members

4. Find the number of members 60 years of age and older.

35,000 members

5. How many members are 40 to 60 years of age?

50,000 members

6. How many members are 45 to 55 years of age?

26,000 members

This frequency polygon shows the annual salaries for part-time and full-time employees of the Tickle Time Pickle Factory.

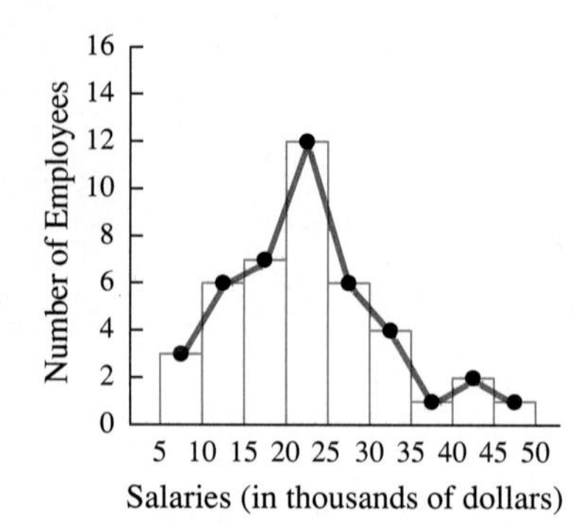

7. The greatest number of employees are in which salary group?

$20,000–$25,000

8. The least number of employees are in which salary groups?

$35,000–$40,000 and $45,000–$50,000

9. Find the number of employees who earn $15–$20 thousand.

7 employees

10. Find the number of employees who earn $45–$50 thousand.

1 employee

11. How many employees earn $25,000 or less?

28 employees

12. How many employees earn $30,000 or more?

8 employees

13. Describe class interval and class frequency. How are these both used when preparing a histogram?

Answer varies.

14. What is the class midpoint? How is the class midpoint used when preparing a frequency polygon?

Answer varies.

This list shows the number of sets of encyclopedias sold annually by the members of the local sales staff. Use it to complete the table.

120	130	144	132	147	158	174
135	142	155	174	162	151	178
145	151	139	128	147	134	146

	number of sets	*tally*	*frequency*
15.	120–129	II	2
16.	130–139	~~IIII~~	5
17.	140–149	~~IIII~~ I	6
18.	150–159	IIII	4
19.	160–169	I	1
20.	170–179	III	3

Here is a list of the daily high temperatures in Fahrenheit degrees for Phoenix, Arizona, during June. Use these numbers to complete the following table.

79°	84°	88°	96°	102°	104°	110°	108°	106°	106°
104°	99°	97°	92°	94°	90°	82°	74°	72°	83°
85°	92°	100°	99°	101°	107°	111°	102°	97°	94°

	temperature	*tally*	*frequency*
21.	70°–74°	II	2
22.	75°–79°	I	1
23.	80°–84°	III	3
24.	85°–89°	II	2
25.	90°–94°	~~IIII~~	5
26.	95°–99°	~~IIII~~	5
27.	100°–104°	~~IIII~~ I	6
28.	105°–109°	IIII	4
29.	110°–114°	II	2

30. Construct a histogram by using the data in Exercises 21–29.

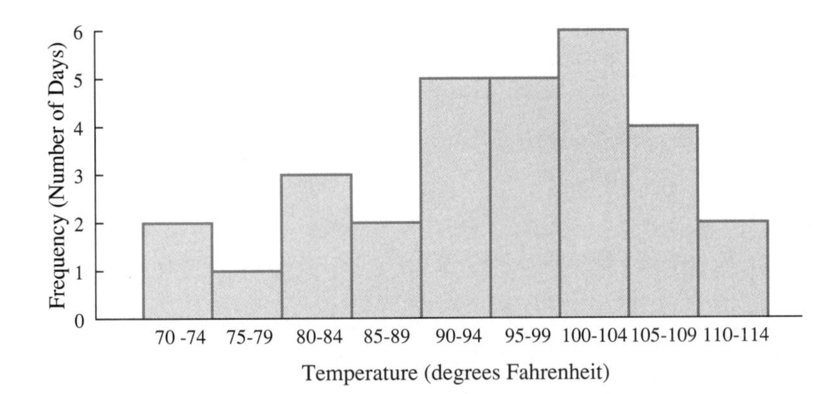

Southside Real Estate has 60 salespeople spread over its five offices. The number of new homes sold by each of these salespeople during the past year is shown below. Use these numbers to complete the following table.

9	33	14	8	17	10	25	11	4	16	3	9
15	24	19	30	16	31	21	20	30	2	6	6
3	8	5	11	15	26	7	18	29	10	7	3
11	6	10	4	2	35	10	25	5	19	34	2
8	13	25	15	23	26	12	4	22	12	21	12

	new homes sold	*tally*	*frequency*
▲ **31.**	1–5	𝍸 𝍸 I	11
▲ **32.**	6–10	𝍸 𝍸 IIII	14
▲ **33.**	11–15	𝍸 𝍸 I	11
▲ **34.**	16–20	𝍸 II	7
▲ **35.**	21–25	𝍸 III	8
▲ **36.**	26–30	𝍸	5
▲ **37.**	31–35	IIII	4

▲ **38.** Make a bar graph showing the results in the table.

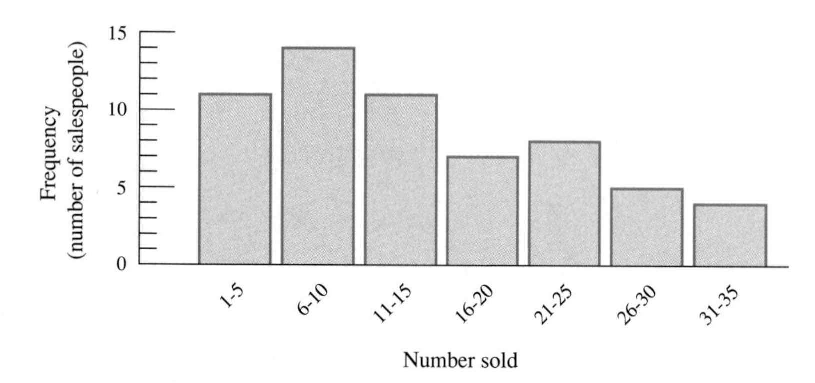

PREVIEW EXERCISES

Solve each problem by using the order of operations. (For help, see **Section 1.8.**)

39. $(18 + 53 + 31 + 26) \div 5$

25.6

40. $(17 + 23 + 46 + 36 + 29) \div 4$

37.75

41. $(8 \cdot 6) + (3 \cdot 8) \div 5$

52.8

42. $(9 \cdot 2) + (12 \cdot 4) \div 6$

26

43. $(4 \cdot 3) + (3 \cdot 9) + (2 \cdot 3) \div 15$

39.4

44. $(4 \cdot 3) + (3 \cdot 6) + (2 \cdot 6) + (3 \cdot 3) \div 18$

42.5

10.4 MEAN, MEDIAN, AND MODE

OBJECTIVES

1. Find the mean of a list of numbers.
2. Find a weighted mean.
3. Find the median.
4. Find the mode.

FOR EXTRA HELP

Tape	SSM	MAC: A
14	pp. 314–316	IBM: A

1 When analyzing data, one of the first things to look for is a *measure of central tendency*—a single number that we can use to represent the entire list of numbers. One such measure is the *average* or **mean.** The mean can be found with the following formula.

FINDING THE MEAN (AVERAGE)

$$\text{mean} = \frac{\text{sum of all values}}{\text{number of values}}$$

■ **EXAMPLE 1** *Finding the Mean*
David had test scores of 84, 90, 95, 98, and 88. Find his average or mean of these scores.

Use the formula for finding mean. Add up all the test scores and then divide by the number of tests.

$$\text{mean} = \frac{84 + 90 + 95 + 98 + 88 \leftarrow \text{sum of test scores}}{5 \leftarrow \text{number of tests}}$$

$$= \frac{455}{5} \quad \text{Divide.}$$

$$= 91$$

The mean score is 91. ■

WORK PROBLEM 1 AT THE SIDE.

■ **EXAMPLE 2** *Applying the Average or Mean*
The sales of photo albums at Sarah's Card Shop for each day last week were

$86, $103, $118, $117, $126, $158, and $149.

For Sarah's Card Shop, the mean (rounded to the nearest cent) is

$$\text{mean} = \frac{86 + 103 + 118 + 117 + 126 + 158 + 149}{7}$$

$$= \frac{857}{7}$$

$$= \$122.43. \; ■$$

WORK PROBLEM 2 AT THE SIDE.

2 Some items in a list might appear more than once. In this case, we find a **weighted mean,** in which each value is "weighted" by multiplying it by the number of times it occurs.

1. Tanya had test scores of 96, 98, 84, 88, 82, and 92. Find her average or mean score.

2. Find the mean for each list of numbers.

(a) Monthly phone bills of $25.12, $42.58, $76.19, $32, $81.11, $26.41, $19.76, $59.32, $71.18, and $21.03.

(b) A list of the sales for one year at eight different office supply stores: $749,820; $765,480 $643,744; $824,222 $485,886; $668,178 $702,294; $525,800

ANSWERS
1. 90
2. (a) $\frac{454.70}{10} = \$45.47$
(b) $\frac{5{,}365{,}424}{8} = \$670{,}678$

3. Find the weighted mean for the numbers given in the following table.

Value	*Frequency*
$ 6	2
$ 7	3
$ 8	3
$ 9	4
$10	6

EXAMPLE 3 *Understanding the Weighted Mean*

The following table shows the amount of contribution and the number of times the amount was given (frequency) to a food pantry. Find the weighted mean.

Contribution Value	*Frequency*
$ 3	4
$ 5	2
$ 7	1
$ 8	5
$ 9	3
$10	2
$12	1
$13	2

The same amount was given by more than one person: for example, $5 was given twice and $8 was given five times. Other amounts, such as $12, were given once. To find the mean, multiply each contribution value by its frequency. Then add the products. Next, add the numbers in the *frequency* column to find the total number of values.

Value	*Frequency*	*Product*
$ 3	4	$(3 \cdot 4) = \$12$
$ 5	2	$(5 \cdot 2) = \$10$
$ 7	1	$(7 \cdot 1) = \$\ 7$
$ 8	5	$(8 \cdot 5) = \$40$
$ 9	3	$(9 \cdot 3) = \$27$
$10	2	$(10 \cdot 2) = \$20$
$12	1	$(12 \cdot 1) = \$12$
$13	2	$(13 \cdot 2) = \$26$
Totals	20	$154

Finally, divide the totals.

$$\text{mean} = \frac{154}{20} = \$7.70. \quad \blacksquare$$

The mean contribution to the food pantry was $7.70.

WORK PROBLEM 3 AT THE SIDE.

A common use of the weighted mean is to find a student's *grade point average,* as shown by the next example.

ANSWER
3. $8.50

EXAMPLE 4 *Applying the Weighted Mean*
Find the grade point average for a student earning the following grades. Assume A = 4, B = 3, C = 2, D = 1, and F = 0.

Course	*Credits*	*Grade*	*Credits · Grade*
Mathematics	3	A (= 4)	$3 \cdot 4 = 12$
Speech	3	C (= 2)	$3 \cdot 2 = 6$
English	3	B (= 3)	$3 \cdot 3 = 9$
Computer Science	3	A (= 4)	$3 \cdot 4 = 12$
Lab for Computer Science	2	D (= 1)	$2 \cdot 1 = 2$
Totals	14		41

It is common to round grade point averages to the nearest hundredth. So the grade point average for this student is

$$\frac{41}{14} = 2.93 \text{ (rounded).}$$

WORK PROBLEM 4 AT THE SIDE.

3 Because it can be affected by extremely high or low numbers, the mean is often a poor indicator of central tendency for a list of numbers. In cases like this, another measure of central tendency, called the **median,** can be used. The *median* divides a group of numbers in half; half the numbers lie above the median, and half lie below the median.

Find the median by listing the numbers *in order* from *smallest* to *largest*. If the list contains an *odd* number of items, the median is the *middle number*.

EXAMPLE 5 *Using the Median*
Find the median for the following list of prices.

$7, $23, $15, $6, $18, $12, $24

First arrange the numbers in numerical order from smallest to largest.

smallest → 6, 7, 12, 15, 18, 23, 24 ← largest

Next, find the middle number in the list.

$$\underbrace{6, 7, 12}_{\text{three are below}},\ 15,\ \underbrace{18, 23, 24}_{\text{three are above}}$$

↓ middle number

The median price is $15.

WORK PROBLEM 5 AT THE SIDE.

If a list contains an *even* number of items, there is no single middle number. In this case, the median is defined as the mean (average) of the *middle two* numbers.

4. Find the grade point average for a student earning the following grades.

Course	*Credits*	*Grade*
Mathematics	3	A (= 4)
P.E.	1	C (= 2)
English	3	C (= 2)
Keyboarding	3	B (= 3)
Recreation	3	B (= 3)

5. Find the median for the following list of quiz scores.
35, 33, 27, 31, 39, 50, 59, 25, 30

ANSWERS
4. 2.92 rounded to the nearest hundredth
5. 33 (the middle number when the scores are arranged from smallest to largest)

6. Find the median for the following list of measurements.
178 ft, 261 ft, 126 ft, 189 ft, 121 ft, 195 ft

7. Find the mode for each list of numbers.

(a) Ages of part-time employees (in years): 28, 16, 22, 28, 34

(b) Total points on a screening exam of 312, 219, 782, 312, 219, 426

(c) Monthly commissions of sales people: $1706, $1289, $1653, $1892, $1301, $1782

ANSWERS
6. 183.5 ft
7. (a) 28 years
(b) bimodal, 219 points and 312 points (this list has two modes)
(c) no mode (no number occurs more than once)

■ **EXAMPLE 6** *Finding the Median*
Find the median for the following list of ages.

7, 13, 15, 25, 28, 32, 47, 59, 68, 74

The numbers are already in numerical order, so find the middle two numbers.

smallest → 7, 13, 15, 25, 28, 32, 47, 59, 68, 74 ← largest
(28, 32: middle two numbers)

The median age is the mean of these two numbers.

$$\text{median} = \frac{28 + 32}{2} = \frac{60}{2} = 30 \text{ years}$$ ■

WORK PROBLEM 6 AT THE SIDE.

4 The last important statistical measure is the **mode,** the number that occurs most often in a list of numbers. For example, if the test scores for 10 students were

74, 81, 39, 74, 82, 80, 100, 92, 74, and 85

then the mode is 74. Three students earned a score of 74, so 74 appears more times on the list than any other score.

A list can have two modes; such a list is sometimes called **bimodal.** If no number occurs more frequently than any other number in a list, the list has *no mode.*

■ **EXAMPLE 7** *Finding the Mode*
Find the mode for each list of numbers.

(a) 51, 32, 49, 73, 49, 90

The number 49 occurs more often than any other number; therefore, 49 is the mode. (It is not necessary to place the numbers in numerical order when looking for the mode.)

(b) 482, 485, 483, 485, 487, 487, 489

Because both 485 and 487 occur twice, each is a mode. This list is bimodal.

(c) 10,708; 11,519; 10,972; 12,546; 13,905; 12,182

No number occurs more than once. This list has no mode. ■

MEASURES OF CENTRAL TENDENCY

The **mean** is the sum of all the values divided by the number of values. It is the mathematical average.

The **median** is the middle number in a group of values that are listed from smallest to largest. It divides a group of numbers in half.

The **mode** is the value that occurs most often in a group of values.

WORK PROBLEM 7 AT THE SIDE.

10.4 EXERCISES

NAME DATE HOUR

Find the mean for each list of numbers. Round answers to the nearest tenth.

1. Children's ages (in years) of 4, 7, 15, 18, 21

13 years

2. Monthly utility bills of $49, $50, $30, $39, $78, $54

$50

3. Scores on an exam of 40, 51, 59, 62, 68, 73, 49, 80

60.3

4. Quiz scores of 31, 37, 43, 51, 58, 64, 79, 83

55.8

5. Annual salaries of $21,900, $22,850, $24,930, $29,710, $28,340, $40,000

$27,955

6. Numbers of people attending baseball games: 27,500; 18,250; 17,357; 14,298; 33,110

22,103 people

Solve the following word problems.

7. The Athletic Shoe Store sold shoes at the following prices: $75.52, $36.15, $58.24, $21.86, $47.68, $106.57, $82.72, $52.14, $28.60, $72.92. Find the average (mean) shoe sales amount.

$58.24

8. In one evening, a waitress collected the following checks from her dinner customers: $30.10, $42.80, $91.60, $51.20, $88.30, $21.90, $43.70, $51.20. Find the average (mean) dinner check amount.

$52.60

9. The table below shows the face value (policy amount) of life insurance policies sold and the number of policies sold for each amount by the New World Life Company during one week. Find the weighted mean amount for the policies sold.

Policy Amount	*Number of Policies Sold*
$10,000	6
$20,000	24
$25,000	12
$30,000	8
$50,000	5
$100,000	3
$250,000	2

$35,500

10. Detroit Metro-Sales Company prepares the following table showing the gasoline mileage obtained by each of the cars in their automobile fleet. Find the weighted mean to determine the miles per gallon for the fleet of cars.

Miles per Gallon	*Number of Autos*
15	5
20	6
24	10
30	14
32	5
35	6
40	4

27.7 miles per gallon

Find the weighted mean. Round answers to the nearest tenth.

11.

Quiz Scores	*Frequency*
3	4
5	2
9	1
12	3

6.7

12.

Value	*Frequency*
9	3
12	5
15	1
18	1

12

13.

Value	*Frequency*
12	4
13	2
15	5
19	3
22	1
23	5

17.2

14.

Students per Class	*Frequency*
25	1
26	2
29	5
30	4
32	3
33	5

30.2 students

Find the median for the following lists of numbers.

15. Number of meassages received:
8, 11, 13, 16, 26, 26, 33

16 messages

16. Miles driven by a courier:
114, 114, 125, 135, 150, 172

130 miles

17. Customers served each day:
328, 516, 420, 592, 715

516 customers

18. Number of books loaned each day:
412, 298, 501, 346, 1275, 521, 515, 528

508 books

Find the mode or modes for each list of numbers.

19. Number of samples taken each hour:
3, 8, 5, 1, 7, 6, 8, 4

8 samples

20. Water bills of
$21, $32, $46, $32, $49, $32, $49

$32

21. Ages of retirees (in years) at the village:
74, 68, 68, 68, 75, 75, 74, 74, 70

68 and 74 years

22. Tires balanced by different employees:
30, 19, 25, 78, 36, 20, 45, 85, 38

no mode

23. Can the mean be a poor indicator of the central tendency? When can this happen?

Answer varies.

24. What is the purpose of the weighted mean? Give an example of where it is used.

Answer varies.

25. When is the median a better average to use than the mean to describe a set of data? Make up a list of numbers to illustrate your explanation. Calculate both the mean and the median.

Answer varies.

26. Suppose you own a hat shop and can order hats in only one size. You look at last year's sales to decide on the size to order. Should you find the mean, median, or mode for these sales? Explain your answer.

Answer varies.

Find the grade point average for students earning the following grades. Assume A = 4, B = 3, C = 2, D = 1, and F = 0. Round answers to the nearest tenth.

27.

Credits	*Grade*
4	B
2	A
5	C
1	F
3	B

2.6

28.

Credits	*Grade*
3	A
3	B
4	B
2	C
3	C

2.9

Find the median for the following lists of numbers.

29. Test scores of
32, 58, 97, 21, 49, 38, 97, 46, 53

49

30. Number of gallons of paint sold per week:
1072, 1068, 1093, 1042, 1056, 205, 1009

1056 gallons

Find the mode or modes for each list of numbers.

31. The number of boxes of candy sold by each child:
5, 9, 17, 3, 2, 8, 19, 1, 4, 20

no mode

32. The weights of soccer players (in pounds);
158, 161, 165, 162, 165, 157, 163

165 pounds

CHAPTER 10 SUMMARY

KEY TERMS

10.1 circle graph — A circle graph shows how a total amount is divided into parts or sectors. It is based on percents of 360°.

protractor — A protractor is a device (usually in the shape of a half-circle) used to measure the number of degrees in an angle or parts of a circle.

10.2 bar graph — A bar graph uses bars of various heights to show quantity or frequency.

double-bar graph — A double-bar graph compares two sets of data by showing two sets of bars.

line graph — A line graph uses dots connected by lines to show trends.

comparison line graph — A comparison line graph shows how several different items relate to each other.

10.3 histogram — A histogram is a bar graph in which the width of each bar represents a range of numbers (class interval) and the height represents the quantity or frequency of items that fall within the interval.

frequency polygon — A frequency polygon results when a dot is placed at the midpoint at the top of each bar in a histogram and the dots are connected with lines.

10.4 mean — The mean is the sum of all the values divided by the number of values. It is often called the *average.*

weighted mean — The weighted mean is a mean calculated so that each value is multiplied by its frequency.

median — The median is the middle number in a group of values that are listed from smallest to largest. It divides a group of values in half. If there are an even number of values, the median is the mean (average) of the two middle values.

mode — The mode is the value that occurs most often in a group of values.

QUICK REVIEW

Concepts	Examples
10.1 Constructing a Circle Graph	
1. Determine the percent of the total for each item.	Construct a circle graph for the following table, which lists expenses for a business trip.
2. Find the number of degrees out of 360° that each percent represents.	
3. Use a protractor to measure the number of degrees for each item in the circle.	

Item	*Amount*
Transportation	\$200
Lodging	\$300
Food	\$250
Entertainment	\$150
Other	\$100
Total	\$1000

Item	*Amount*	*Percent of Total*	*Sector Size*
Transportation	\$200	$\frac{\$200}{\$1000} = \frac{1}{5} = 20\%$ so $360° \cdot 20\%$ $= 360 \cdot 0.20$	$= 72°$
Lodging	\$300	$\frac{\$300}{\$1000} = \frac{3}{10} = 30\%$ so $360° \cdot 30\%$ $= 360 \cdot 0.30$	$= 108°$
Food	\$250	$\frac{\$250}{\$1000} = \frac{1}{4} = 25\%$ so $360° \cdot 25\%$ $= 360 \cdot 0.25$	$= 90°$
Entertainment	\$150	$\frac{\$150}{\$1000} = \frac{3}{20} = 15\%$ so $360° \cdot 15\%$ $= 360 \cdot 0.15$	$= 54°$
Other	\$100	$\frac{\$100}{\$1000} = \frac{1}{10} = 10\%$ so $360° \cdot 10\%$ $= 360 \cdot 0.10$	$= 36°$

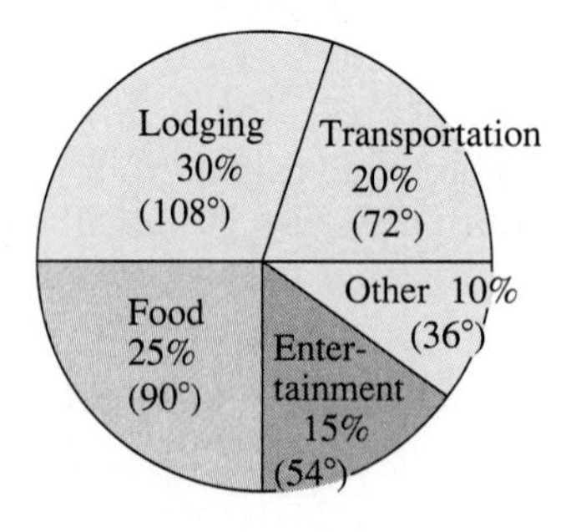

Concepts	Examples

10.2 Reading a Bar Graph

The height of the bar is used to show the quantity or frequency (number) in a specific category.

Use the bar graph below to determine the number of students who earned each letter grade.

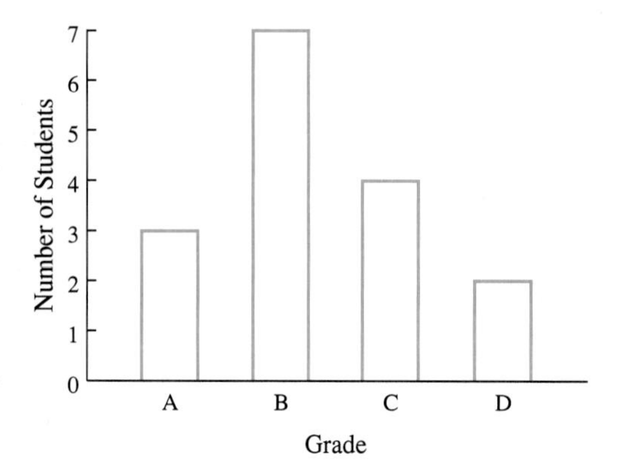

Grade	*Number of Students*
A	3
B	7
C	4
D	2

10.2 Reading a Line Graph

A point is used to show the number or quantity in a specific class. The points are connected with lines. This kind of graph is used to show a trend.

The line graph below shows the sales volume for each of four years.

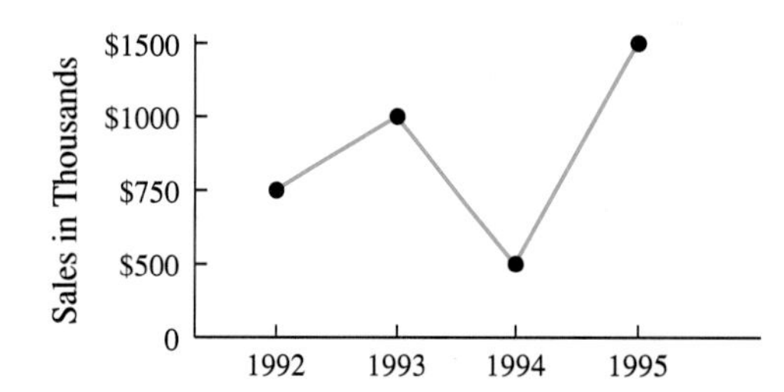

Find the sales in each year.

Year	*Total Sales*
1992	$\$750 \cdot 1000 = \$750{,}000$
1993	$\$1000 \cdot 1000 = \$1{,}000{,}000$
1994	$\$500 \cdot 1000 = \$500{,}000$
1995	$\$1500 \cdot 1000 = \$1{,}500{,}000$

Concepts	Examples
10.3 Constructing a Histogram and a Frequency Polygon from Raw Data	
1. Construct a table listing each value, and the number of times this value occurs.	Draw a frequency polygon for the following list of student quiz scores.
2. Divide the data into groups, categories, or classes.	
3. Draw bars representing these groups to make a histogram.	
4. Connect the midpoints at the tops of the bars with lines to make a frequency polygon.	

12	15	15	14
13	20	10	12
11	9	10	12
17	20	16	17
14	18	19	13

Quiz Score	*Tally*	*Frequency*	
9	I	1	**1st class interval**
10	II	2	
11	I	1	
12	III	3	**2nd class interval**
13	II	2	
14	II	2	
15	II	2	**3rd class interval**
16	I	1	
17	II	2	
18	I	1	**4th class interval**
19	I	1	
20	II	2	

Class Interval (Quiz Scores)	*Frequency*
9–11	4
12–14	7
15–17	5
18–20	4

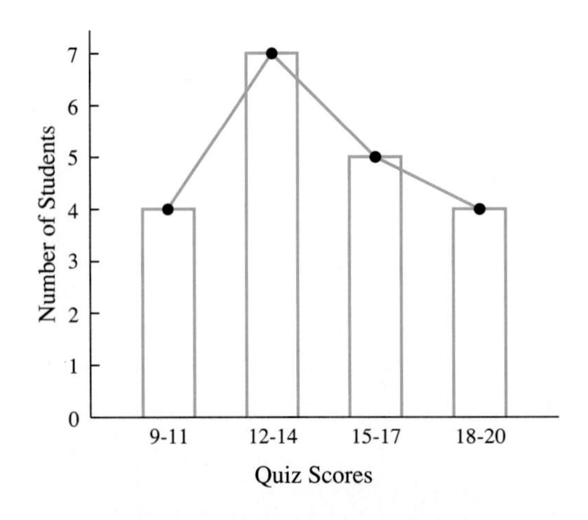

Concepts	Examples
10.4 Finding the Mean (average) of a Set of Numbers	
1. Add all values to obtain a total. **2.** Divide the total by the number of values.	The test scores for Heather Hall in her business math course were as follows: 85 76 93 91 78 82 87 85 Find Heather's test average (mean) to the nearest tenth. $\text{Mean} = \frac{85 + 76 + 93 + 91 + 78 + 82 + 87 + 85}{8}$ $= \frac{677}{8} = 84.6$ (rounded)
10.4 Finding the Weighted Mean	
1. Multiply frequency by value. **2.** Add all the products from Step 1. **3.** Divide the sum in Step 2 by the total number of pieces of data.	This table shows the distribution of the number of school-age children in a survey of 30 families.

Number of School-Age Children	*Frequency (Number of Families)*
0	12
1	6
2	7
3	3
4	2
Total of 30 Families	

Find the mean number of school-age children per family.

Number	*Frequency*	*Product*
0	12	$(0 \cdot 12) = 0$
1	6	$(1 \cdot 6) = 6$
2	7	$(2 \cdot 7) = 14$
3	3	$(3 \cdot 3) = 9$
4	2	$(4 \cdot 2) = 8$
	37	

$$\text{Mean} = \frac{37}{30} = 1.23 \quad \text{(rounded)}$$

The mean number of school-age children per family is 1.23.

Concepts	Examples
10.4 Finding the Median of a Set of Numbers	
1. Arrange the data from smallest to largest. 2. Select the middle value or the average of the two middle values, if there is an even number of values.	Find the median for Heather Hall's grades from the previous page. The data arranged from smallest to largest is as follows: 76 78 82 85 85 87 91 93 (85 85: middle values) The middle two values are 85 and 85. The average of these two values is $\frac{85 + 85}{2} = 85$
10.4 Determining the Mode of a Set of Values	
Find the value that appears most often in the list of values.	Find the mode for Heather's grades in the previous example. The most frequently occurring score is 85 (it occurs twice). Therefore, the mode is 85.

CHAPTER 10 REVIEW EXERCISES

[10.1]

1. This circle graph shows the cost of a family vacation. What is the largest single expense of the vacation? Lodging

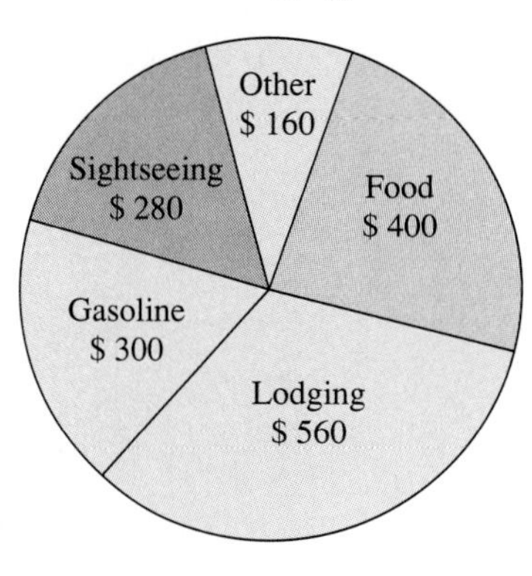

Using the circle graph in Exercise 1, find each of the following ratios. Write ratios as fractions in lowest terms.

2. cost of the food to the total cost of the vacation

$\frac{400}{1700} = \frac{4}{17}$

3. cost of the gasoline to the total cost of the vacation

$\frac{300}{1700} = \frac{3}{17}$

4. cost of sightseeing to the total cost of the vacation

$\frac{280}{1700} = \frac{14}{85}$

5. cost of gasoline to the cost of the *other* category

$\frac{300}{160} = \frac{15}{8}$

6. cost of the lodging to the cost of the food

$\frac{560}{400} = \frac{7}{5}$

This circle graph shows the inventory of books, supplies, and equipment for a school. Find the dollar value of each category listed. The total value of the inventory is $265,500.

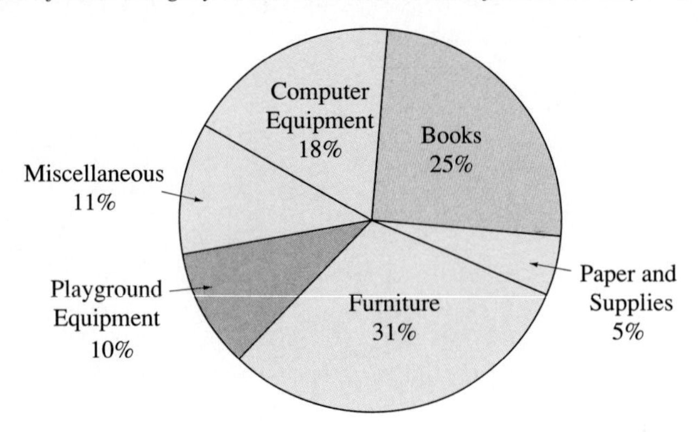

7. books

$66,375

8. paper and supplies

$13,275

9. furniture

$82,305

10. playground equipment

$26,550

11. miscellaneous

$29,205

12. computer equipment

$47,790

[10.2] *This double-bar graph shows the number of acre-feet of water in Lake Natoma for each of the first six months of 1994 and 1995.*

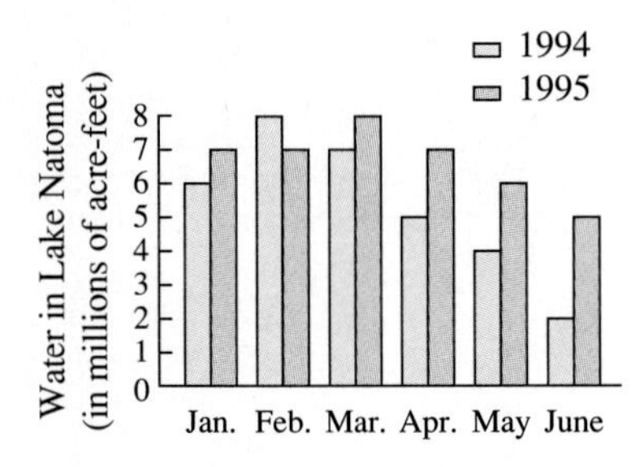

13. During which month in 1995 was the greatest amount of water in the lake?

March

14. During which month in 1994 was the least amount of water in the lake?

June

15. How many acre-feet of water were in the lake in March of 1995?

8 million acre-feet

16. How many acre-feet of water were in the lake in May of 1994?

4 million acre-feet

17. Find the decrease in the amount of water in the lake from March 1994 to June 1994.

5 million acre-feet

18. Find the decrease in the amount of water in the lake from April 1995 to June 1995.

2 million acre-feet

This comparison line graph shows the annual grocery purchases of two different childcare centers during each of five years. Find the amount of annual grocery purchases in each of the following years.

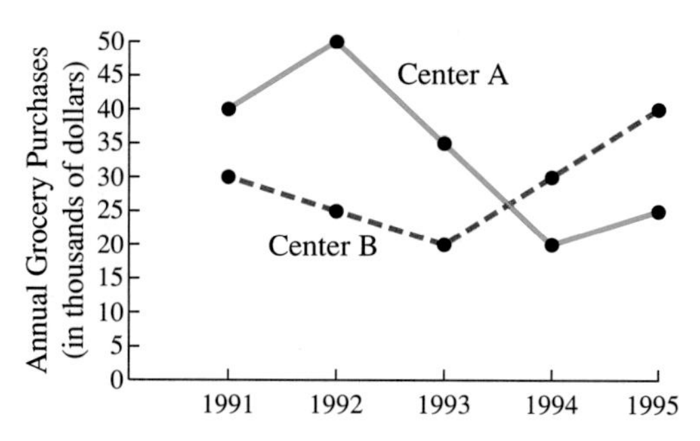

19. center A in 1995

$25,000

20. center A in 1994

$20,000

21. center A in 1993

$35,000

22. center B in 1995

$40,000

23. center B in 1994

$30,000

24. center B in 1993

$20,000

[10.4] *Find the mean for each list of numbers. Round to the nearest tenth.*

25. Computers sold:
18, 12, 15, 24, 9, 42, 54, 87, 21, 3

28.5 computers

26. Number of speeders ticketed:
31, 9, 8, 22, 46, 51, 48, 42, 53, 42,

35.2 speeders

Find the weighted mean for each of the following. Round to the nearest tenth if necessary.

27.

Dollar Value	*Frequency*
$42	3
$47	7
$53	2
$55	3
$59	5

$51.10

28.

Total Points	*Frequency*
243	1
247	3
251	5
255	7
263	4
271	2
279	2

257.3 points

Find the median for each list of numbers.

29. Number of charge customers:
54, 28, 35, 43, 13, 37, 68, 75, 39

39 customers

30. Commissions of \$576, \$578, \$542, \$151, \$559, \$565, \$525, \$590

\$562

Find the mode or modes for each list of numbers.

31. Western boots priced at \$80, \$72, \$64, \$64, \$72, \$53, \$64

\$64

32. Boat launchings: 18, 25, 63, 32, 28, 37, 32, 26, 18

18 and 32 launchings (bimodal)

MIXED REVIEW EXERCISES

The Broadway Hair Salon spent \$22,400 to open a new shop. This amount was spent as shown in the following chart. Find all the missing numbers in Exercises 33–37.

item	*dollar amount*	*percent of total*	*degrees of circle*
33. plumbing and electrical changes	\$2240	10%	36°
34. work stations	\$7840	35%	126°
35. small appliances	\$4480	20%	72°
36. interior decoration	\$5600	25%	90°
37. supplies	\$2240	10%	36°

38. Draw a circle graph by using the information in Exercises 33–37.

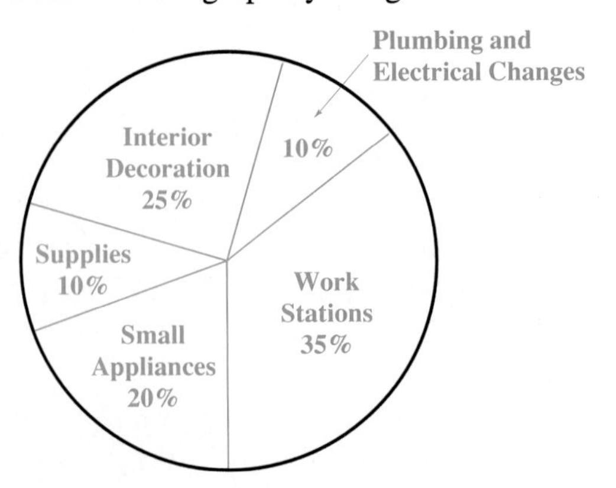

Find the mean for each list of numbers. Round answers to the nearest tenth.

39. Contestant ages (in years) of 24, 36, 26, 74, 90

50 years

40. Number of tacks in a handful:
122, 135, 146, 159, 128, 147, 168, 139, 158

144.7 tacks

Find the mode or modes for each list of numbers.

41. Scores earned by "A" students:
97, 95, 94, 95, 94, 97, 97

97

42. Living units in each complex:
26, 31, 31, 37, 43, 51, 31, 43, 43

31 and 43 living units (bimodal)

Find the median for each list of numbers.

43. Hours worked: 4.7, 3.2, 2.9, 5.3, 7.1, 8.2, 9.4, 1.0

5.0 hours

44. Number of yard sales each Saturday:
7, 15, 28, 3, 14, 18, 46, 59, 1, 2, 9, 21

14.5 yard sales

Here are the scores of 40 students on a computer science exam. Complete the table.

78	89	36	59	78	99	92	86
73	78	85	57	99	95	82	76
63	93	53	76	92	79	72	62
74	81	77	76	59	84	76	94
58	37	76	54	80	30	45	38

score	*tally*	*frequency*
45. 30–39	IIII	4
46. 40–49	I	1
47. 50–59	~~IIII~~ I	6
48. 60–69	II	2
49. 70–79	~~IIII~~ ~~IIII~~ III	13
50. 80–89	~~IIII~~ II	7
51. 90–99	~~IIII~~ II	7

52. Construct a frequency polygon by using the data in Exercises 45–51.

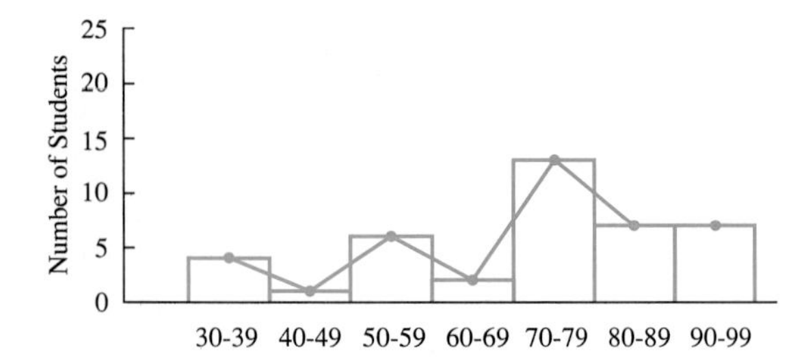

Find the weighted mean for each of the following. Round answers to the nearest tenth, if necessary.

53.

Test Score	*Frequency*
23	2
27	5
31	4
35	6
39	5

32.3

54.

Dollar Value	*Frequency*
$104	6
$112	14
$115	21
$119	13
$123	22
$127	6
$132	9

$118.80

NAME DATE HOUR

CHAPTER 10 TEST

This circle graph shows the development costs for the Shady Brook Public Housing Subdivision. Find the dollar amount budgeted for each category. The total budget of the subdivision is $5,600,000.

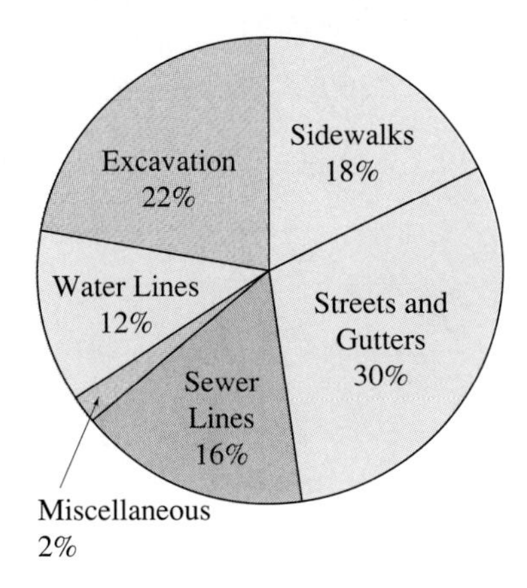

1. excavation

1. $1,232,000

2. sidewalks

2. $1,008,000

3. streets and gutters

3. $1,680,000

4. sewer lines

4. $896,000

5. miscellaneous

5. $112,000

6. water lines

6. $672,000

During a one-year period, the Daily Blat newspaper had the following expenses. Find all numbers missing from the chart.

	item	*dollar amount*	*percent of total*	*degrees of a circle*
7.	newsprint	\$12,000	20%	________
8.	ink	\$6000	________	36°
9.	wire service	\$18,000	30%	________
10.	salaries	\$18,000	30%	________
11.	other	\$6000	10%	________

7. 72°

8. 10%

9. 108°

10. 108°

11. 36°

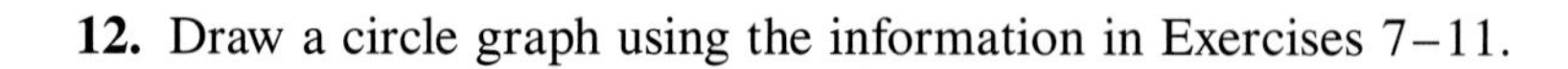

12. Draw a circle graph using the information in Exercises 7–11.

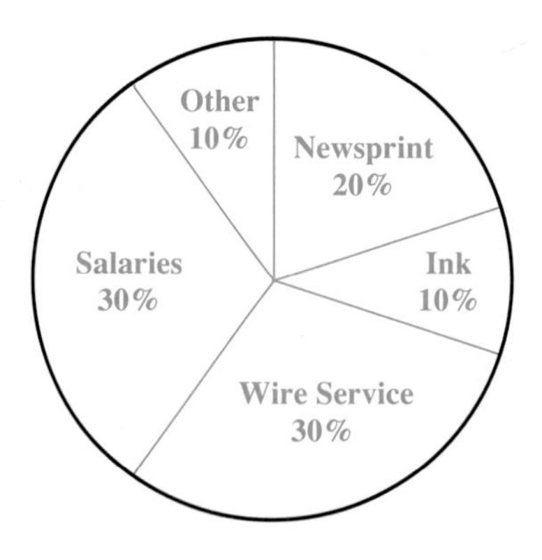

Here are the profits for each of the past 20 weeks from the snack bar vending machines. Complete the following table.

$142 $137 $125 $132 $147 $129 $151 $172 $175 $129
$159 $148 $173 $160 $152 $174 $169 $163 $149 $173

	profit	***number of weeks***
13.	$120–129	________
14.	$130–139	________
15.	$140–149	________
16.	$150–159	________
17.	$160–169	________
18.	$170–179	________

13. 3

14. 2

15. 4

16. 3

17. 3

18. 5

19. Use the numbers found in Exercises 13–18 to draw a histogram.

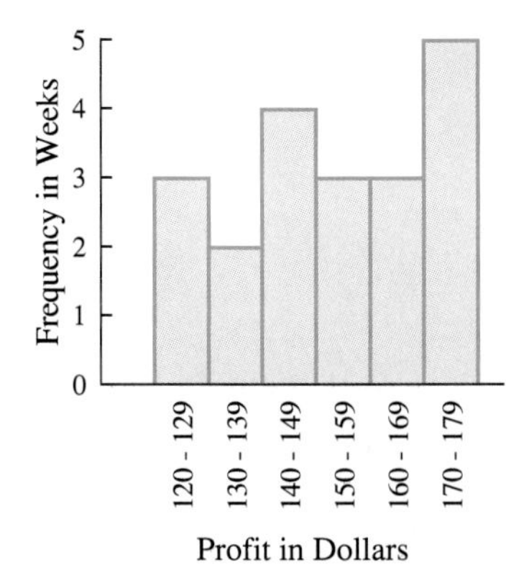

Find the mean for each of the following. Round answers to the nearest tenth if necessary.

20. First-time bowling scores of 52, 61, 68, 69, 73, 75, 79, 84, 91, 98

20. 75

21. Number of miles per gallon by older cars: 22, 28, 24, 27, 29, 32, 33, 35

21. 28.8 miles per gallon

22. Total points of students in a history class: 458, 432, 496, 491, 500, 508, 512, 396, 492, 504

22. 478.9 points

23. Answer varies.

24. Answer varies.

25. $11.30

26. 173.7

27. 31.5 calls

28. 9.3 liters

29. 57 meters

30. 103° and 104°

23. Explain why a weighted mean must be used to determine a student grade point average. Calculate your own grade point average for last semester or quarter. If you are a new student, make up a grade point average problem of your own and solve it.

24. Explain in your own words the procedure for finding the median when there are an odd number of numbers in a list. Make up a problem with a list of five numbers and solve for the median.

Find the weighted mean for each of the following. Round answers to the nearest tenth, if necessary.

25.

Cost	*Frequency*
$ 6	7
$10	3
$11	4
$14	2
$19	3
$24	1

26.

Value	*Frequency*
150	15
160	17
170	21
180	28
190	19
200	7

Find the median for each list of numbers.

27. Listener phone calls to a radio talk show (each hour) of 32, 41, 28, 28, 37, 35, 16, 31

28. Number of liters of water lost in evaporation: 9.3, 10.0, 8.1, 6.3, 1.2, 11.4, 22.8, 10.3, 8.6

Find the mode or modes for each list of numbers:

29. Drilling depth (in meters) of 61, 57, 58, 42, 81, 92, 57

30. Hot tub temperatures (Fahrenheit) of 96°, 104°, 103°, 104°, 103°, 104°, 91°, 74°, 103°

NAME DATE HOUR

CUMULATIVE REVIEW EXERCISES CHAPTERS 1–10

Round each number to the place shown.

1. \$65.736 to the nearest cent. \$65.74

2. \$475.499 to the nearest dollar. \$475

3. 75,696 to the nearest ten. 75,700

4. 983,168 to the nearest ten thousand. 980,000

Simplify each of the following by using the order of operations.

5. $7 + 6 \div 2 + 5 \cdot 4$ 30

6. $\sqrt{81} - 4 \cdot 2 + 9$ 10

Solve each of the following.

7. $3^2 \cdot 2^4$ 144

8. $6^2 \cdot 3^2$ 324

≈ *Round the numbers in each problem so that there is only one non-zero digit. Then add, subtract, multiply, or divide the rounded numbers as indicated to estimate the answer. Finally, solve for the exact answer.*

9.

estimate: $\begin{array}{r} 60{,}000 \\ 200{,}000 \\ 90 \\ +\ 20{,}000 \\ \hline 280{,}090 \end{array}$ exact: $\begin{array}{r} 62{,}318 \\ 159{,}680 \\ 89 \\ +\ 22{,}308 \\ \hline 244{,}395 \end{array}$

10.

estimate: $\begin{array}{r} 3 \\ 800 \\ 40 \\ 50 \\ +\ 8 \\ \hline 901 \end{array}$ exact: $\begin{array}{r} 2.607 \\ 796.2 \\ 37.96 \\ 53.72 \\ +\ 8.06 \\ \hline 898.547 \end{array}$

11.

estimate: $\begin{array}{r} 300{,}000 \\ -\ 100{,}000 \\ \hline 200{,}000 \end{array}$ exact: $\begin{array}{r} 321{,}508 \\ -\ 147{,}725 \\ \hline 173{,}783 \end{array}$

12.

estimate: $\begin{array}{r} 900 \\ -\ 60 \\ \hline 840 \end{array}$ exact: $\begin{array}{r} 875.62 \\ -\ 63.757 \\ \hline 811.863 \end{array}$

13.

estimate: $\begin{array}{r} 7000 \\ \times\ 600 \\ \hline 4{,}200{,}000 \end{array}$ exact: $\begin{array}{r} 7064 \\ \times\ 635 \\ \hline 4{,}485{,}640 \end{array}$

14.

estimate: $\begin{array}{r} 90 \\ \times\ 2 \\ \hline 180 \end{array}$ exact: $\begin{array}{r} 86.34 \\ \times\ 1.715 \\ \hline 148.0731 \end{array}$

15.

estimate: $50\overline{)40{,}000}$ = 800 exact: $48\overline{)41{,}088}$ = 856

16.

estimate: $4\overline{)60}$ = 15 exact: $4.25\overline{)62.56}$ = 14.72

Add, subtract, multiply, or divide as indicated. Write answers in lowest terms and as whole or mixed numbers when possible.

17. $\frac{5}{6} + \frac{2}{3}$ $1\frac{1}{2}$

18. $\frac{3}{4} + \frac{5}{8} + \frac{1}{2}$ $1\frac{7}{8}$

19. $2\frac{3}{5} + 8\frac{2}{3}$ $11\frac{4}{15}$

20. $\frac{5}{8} - \frac{1}{4}$ $\frac{3}{8}$

21. $5\frac{1}{3} - 2\frac{3}{4}$ $2\frac{7}{12}$

22. $46\frac{3}{4} - 15\frac{4}{5}$ $30\frac{19}{20}$

23. $\frac{7}{8} \cdot \frac{4}{5}$ $\frac{7}{10}$

24. $9\frac{3}{5} \cdot 4\frac{5}{8}$ $44\frac{2}{5}$

25. $22 \cdot \frac{2}{5}$ $8\frac{4}{5}$

26. $\frac{5}{6} \div \frac{5}{8}$ $1\frac{1}{3}$

27. $12 \div \frac{2}{3}$ 18

28. $3\frac{1}{3} \div 8\frac{3}{4}$ $\frac{8}{21}$

Simplify each of the following. Use the order of operations.

29. $\left(\frac{7}{8} - \frac{3}{4}\right) \cdot \frac{2}{3}$ $\frac{1}{12}$

30. $\left(\frac{5}{6} - \frac{1}{3}\right) + \left(\frac{1}{2}\right)^2 \cdot \frac{3}{4}$ $\frac{11}{16}$

Write each fraction in decimal form. Round to the nearest thousandth, if necessary.

31. $\frac{3}{4}$ 0.75

32. $\frac{5}{8}$ 0.625

33. $\frac{3}{5}$ 0.6

34. $\frac{11}{20}$ 0.55

Write in order, from smallest to largest.

35. 0.218, 0.22, 0.199, 0.207, 0.2215

0.199, 0.207, 0.218, 0.22, 0.2215

36. 0.6319, $\frac{5}{8}$, 0.608, $\frac{13}{20}$, 0.58

0.58, 0.608, $\frac{5}{8}$, 0.6319, $\frac{13}{20}$

Write each of the following ratios in lowest terms. Be sure to make all necessary conversions.

37. 3 hours to 45 minutes
Compare in minutes. $\frac{4}{1}$

38. $1\frac{1}{4}$ inches to 10 inches $\frac{1}{8}$

Use cross-multiplication to decide whether the following proportions are true or false.

39. $\frac{6}{15} = \frac{18}{45}$

(True) False

40. $\frac{52}{180} = \frac{36}{120}$

True (False)

Find the missing numbers in each proportion.

41. $\frac{1}{4} = \frac{x}{12}$ 3

42. $\frac{14}{x} = \frac{364}{104}$ 4

43. $\frac{200}{135} = \frac{24}{x}$ 16.2

44. $\frac{x}{208} = \frac{6.5}{26}$ 52

Write the percents as decimals. Write the decimals as percents.

45. 35% 0.35

46. 0.025 2.5%

47. 250% 2.50 or 2.5

48. 4.35% 0.0435

Write each percent as a fraction or mixed number in lowest terms. Write each fraction as a percent.

49. 2% $\frac{1}{50}$

50. $62\frac{1}{2}\%$ $\frac{5}{8}$

51. $\frac{3}{20}$ 15%

52. $3\frac{1}{4}$ 325%

Solve these percent problems.

53. 25% of 384 miles

96 miles

54. 5.4% of $900 is how much?

$48.60

55. $8\frac{1}{2}\%$ of what number of people is 238 people?

2800 people

56. 48 is 15% of what number?

320

57. What percent of 520 is 182?

35%

58. 13 weeks is what percent of 52 weeks?

25%

Convert the following.

59. __3__ feet = 1 yard

60. 16 quarts = __4__ gallons

61. 5 days = __120__ hours

62. __12,000__ pounds = 6 tons

Convert each of the following measures as indicated.

63. 1 m to cm 100 cm

64. 2182 mm to m 2.182 m

65. 8.3 g to mg 8300 mg

66. 230 g to kg 0.23 kg

67. 4 mL to L 0.004 L

68. 0.28 L to mL 280 mL

Write the most reasonable metric unit in each blank. Choose from L, mL, kg, g, mg, km, cm, m, and mm.

69. The fuel tank on the chain saw has a capacity of 750 __mL__ of fuel.

70. A nickel weighs 5 __g__.

71. The distance of the run this Saturday is 10 __km__.

72. The heaviest player on the team weighs 108 __kg__.

Find the area of each figure. Use 3.14 as the approximate value of π. Round answers to the nearest tenth.

73. a rectangle 5.6 m by 8.7 m 48.7 m^2

74. a trapezoid with bases of 6.2 cm and 8.4 cm and height 5.3 cm. 38.7 cm^2

75. a triangle with base 8.5 ft and height 9 ft. 38.3 ft^2

76. a circle with diameter of 13 cm 132.7 cm^2

Solve the following. Use 3.14 for π. Round answers to the nearest tenth.

77. Find the volume of a cylinder with radius 8.6 cm and height 3.8 cm. 882.5 cm^3

78. Find the volume of a rectangular solid with length $5\frac{1}{2}$ m, width 2m, and height 9 m. 99 m^3

Find the length of the missing side in each right triangle.

79. 

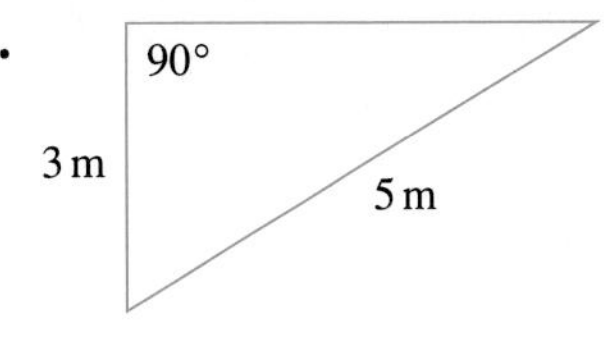

4 m

80. 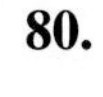

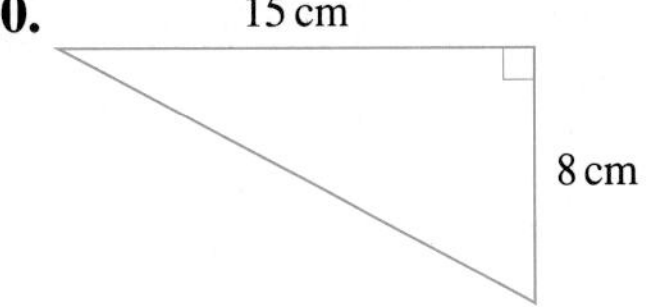

17 cm

Add, subtract, multiply, or divide as indicated.

81. $-6 + (-9)$ −15

82. $-5.7 - (-12.6)$ 6.9

83. $6 \cdot (-5)$ −30

84. $-14.6 \cdot (-5.7)$ 83.22

85. $\dfrac{-45}{-9}$ 5

86. $\dfrac{-34.04}{14.8}$ −2.3

Solve each equation.

87. $4x - 3 = 17$ $x = 5$

88. $-12 = 3(x + 2)$ $x = -6$

89. $19x - 12x = 14$ $x = 2$

90. $3.4x + 6 = 1.4x - 8$ $x = -7$

Find the mean, the median, and the mode for each of the following.

91. Repairs per service representative: 16, 37, 27, 31, 19, 25, 15, 38, 43, 19 27; 26; 19

92. Number of tons of beets processed each hour: 20.6, 8.6, 3.3, 5.7, 10.6, 11.4, 4.6, 8.7, 5.7 8.8; 8.6; 5.7

93. The Bel Air Market sold 3696 gallons of milk in a recent week. If 462 of these cartons were low-fat milk, find the percent that were low-fat.

12.5%

94. In one state the sales tax is 7%. On a recent purchase the amount of sales tax was $78.68. Find the cost of the item purchased.

$1124

95. A gasoline additive is used at the rate of $2\frac{3}{4}$ liters for each storage tank. If $280\frac{1}{2}$ liters of additive are available, how many storage tanks can receive the additive?

102 tanks

96. A survey found that 19 out of every 25 adults are nonsmokers. If the Food Club has 2850 employees, how many would be expected to be nonsmokers.

2166 employees

97. Marja Strutz had sales of $29,580 in building supplies last month. If her commission rate is 12.5%, find the amount of her commission.

$3697.50

98. The sketch below shows the plans for a lobby in a large commercial complex. What is the cost of carpeting the lobby, excluding the atrium, if the contractor charges $43.50 per square yard? Use 3.14 for π.

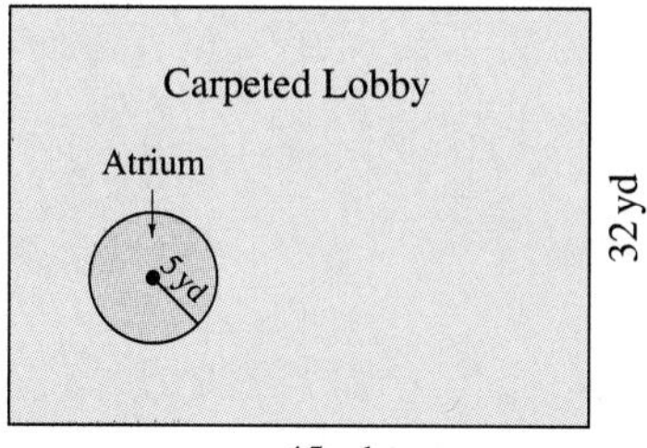

$59,225.25

99. The Spa Service Center services 140 spas. If each spa will need 125 ml of muriatic acid, how many liters of acid will be needed to service the spas?

17.5 L

100. A loan of $3500 will be paid back with $7\frac{1}{2}$% interest at the end of 6 months. Find the total amount due.

$3631.25

Appendices

APPENDIX A: CALCULATORS

Calculators are among the more popular inventions of the last two decades. Each year better calculators are developed and costs drop. A machine that cost $200 a quarter-century ago could add, subtract, multiply, and divide decimals (but could not locate the decimal point in a division problem). Today, these same calculations are performed quite well on a calculator costing less than $10. And today's $200 pocket calculators have more ability to solve problems than some of the early computers.

Many colleges allow students to use calculators in basic mathematics courses. There are many types of calculators available, from the inexpensive basic calculator to the more complex financial and programmable models. The discussion here is confined to the common 8-digit four-function (add, subtract, multiply, and divide), percent key, square root key, and memory function models. Any explanation needed for specific calculator models or special function keys is best gained by referring to the booklet supplied with the calculator.

OBJECTIVES

1. Learn the basic calculator keys.
2. Understand the [C] and [CE] keys.
3. Understand the floating decimal point.
4. Use the [%] and [√] keys.
5. Solve problems with negative numbers.
6. Use the calculator memory function.
7. Use the calculator for problem solving.

1 Most calculators use **algebraic logic.** These calculators can be recognized by the

[+] and [=]

keys. On these calculators, the problem 9 + 8 would be entered as

9 [+] 8 [=]

and 17 would appear as the answer. Enter 17 − 8 as

17 [−] 8 [=]

and 9 appears as the answer. If your calculator does not work problems in this way, check its instruction book to see how to proceed.

2 All calculators have a

[C]

key. Pressing this key erases everything in the calculator and prepares the calculator to begin a new problem. Some calculators also have a

[CE]

key. Pressing this key erases *only* the number displayed and allows the person using the calculator to correct a mistake without having to start the problem over.

Many calculators combine the [C] key and [CE] key and use an [ON/C] key. This key turns the calculator on and is also used to erase the calculator display. If the [ON/C] is pressed after the [=] or one of the operation keys ([+], [−], [×], [÷]), everything in the calculator is erased. If the wrong operation key is pressed, you press the correct key and the error is corrected. For example, 7 [+] [−] 3 [=] 4. Pressing the [−] key cancels out the previous [+] key entry.

3 Most calculators have a **floating decimal** which locates the decimal point in the final result. For example, to buy 55.75 square yards of vinyl floor covering at \$18.99 per square yard, proceed as follows.

55.75 × 18.99 = 1058.6925

The decimal point is automatically placed in the answer. You should **round** money answers to the nearest cent.

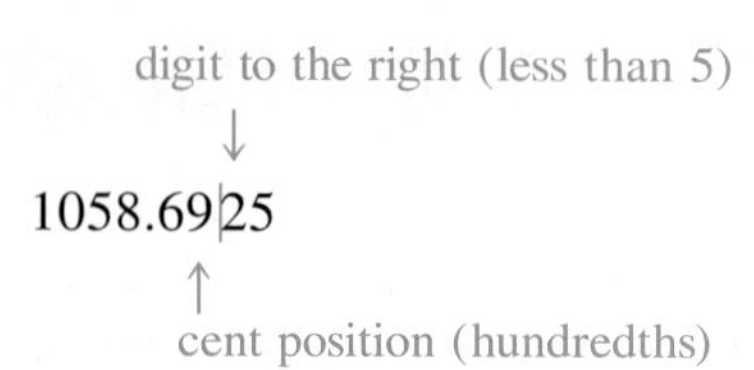

Because the digit to the right of the position being rounded is smaller than 5 (1, 2, 3, or 4) the position being rounded remains the same and everything to the right is dropped. If the digit to the right had been 5 or greater (5, 6, 7, 8, 9), we would have rounded up by adding 1 to the cent position. The answer is rounded to \$1058.69.

4 When using a calculator with a floating decimal, enter the decimal point as needed. For example, enter \$47 as

47

with no decimal point, but enter 95¢ as

. 95

One problem using a floating decimal is shown by the following example (adding \$21.38 and \$1.22).

21.38 + 1.22 = 22.6

The calculator does not show the final 0. You must remember that the problem dealt with money and write the final 0 making the answer \$22.60.

The % moves the decimal point two places to the left when used following multiplication or division. The problem, 8% of \$4205 is solved as follows.

4205 × 8 % = 336.4 = \$336.40

The √ calculates the square root of the number that appears in the display. For example, the $\sqrt{144}$ is found by entering 144 and the square root key.

144 √ 12

The square root of 144 is 12.

The square root of 20 is

20 √ 4.4721359

which may be rounded to the desired position.

5 Negative numbers may be entered by using the − before entering the number. For example, solve $\frac{-10 + 22}{3}$ as follows.

− 10 + 22 ÷ 3 = 4

6 Many calculators feature memory keys, which are a sort of electronic scratch paper. These memory keys are used to store intermediate steps in a calculation. On some calculators, a key labeled M is used to store the numbers in the display, with MR used to recall the numbers from memory.

Other calculators have M+ and M− keys. The M+ key adds the number displayed to the number already in memory. For example, if the memory contains the number 0 at the beginning of a problem, and the calculator display contains the number 29.4, then pushing M+ will cause 29.4 to be stored in the memory (the result of adding 0 and 29.4). If 57.8 is then entered into the display, pushing M+ will cause

$$29.4 + 57.8 = 87.2$$

to be stored. If 11.9 is then entered into the display, with M− pushed, the memory will contain

$$87.2 - 11.9 = 75.3.$$

The MR key is used to recall the number in memory as needed, with MC used to clear the memory.

Note Always clear the memory before starting a problem—forgetting to do so is a very common error.

Memory keys are very useful when working long calculations, called **chain calculations.** For example, to find

$$\frac{8 \times 19.4}{15.7 + 11.8 \times 4.6}$$

calculate the denominator first and follow the order of operations.

Start by calculating 11.8×4.6.

11.8 × 4.6 = 54.28

Next, add 15.7.

\+ 15.7 = 69.98

Store this result in memory, using M or M+, depending on the calculator. Then find the numerator.

8 × 19.4 = 155.2

To divide, push ÷ and then MR =, giving the final quotient.

$$\frac{8 \times 19.4}{15.7 + 11.8 \times 4.6} = 2.218 \quad \text{(rounded)}$$

7 A calculator is especially helpful when multiplying and dividing large numbers. Example 4(a) in Section 4.6 is $27.69 \div 0.003$. To solve this problem by using a calculator, proceed as follows.

27.69 ÷ 0.003 = 9230.

This is much faster than using long division.

The calculator can be used to solve Example 5 in Section 6.7:

A furniture store has a sofa with an original price of \$470 on sale at 15% off. Find the sale price of the sofa.

470 [×] 15 [%] [=] 70.5

Store this result in memory by using [M] or [M+], depending on the calculator. Next, find the actual price.

470 [−] [MR] [=] 399.5 = \$399.50

Example 3 in Section 6.9 involves compound interest:

Find the compounded amount and the compound interest earned on a deposit of \$850 at 6% compounded semiannually for 8 years. Multiply \$850 by the figure from the table on page 401.

850 [×] 1.6047 [=] 1363.995 = \$1364 compound amount (rounded)

Finally, subtract the deposit from the compound amount.

1364 [−] 850 [=] 514 = \$514 interest

OBJECTIVES

1. Use inductive reasoning to analyze patterns.
2. Use deductive reasoning to analyze arguments.
3. Use deductive reasoning to solve problems.

1 In many scientific experiments, conclusions are drawn from specific outcomes. After many repetitions and similar outcomes, the findings are generalized into statements that appear to be true. When general conclusions are drawn from specific observations, we are using a type of reasoning called **inductive reasoning.** In the next several examples, this type of reasoning will be illustrated.

■ **EXAMPLE 1** *Using Inductive Reasoning*
Find the next number in the sequence 3, 7, 11, 15, ...

To discover a pattern, calculate the difference between each pair of successive numbers.

$$7 - 3 = 4$$
$$11 - 7 = 4$$
$$15 - 11 = 4$$

As shown, the difference is 4. Each number is 4 greater than the previous one. Thus, the next number in the pattern is 15 + 4, or 19. ■

WORK PROBLEM 1 AT THE SIDE.

1. Find the next number in the sequence 2, 8, 14, 20, ...

■ **EXAMPLE 2** *Using Inductive Reasoning*
Find the number that comes next in the sequence.

7, 11, 8, 12, 9, 13, ...

The pattern in this example can be determined as follows.

$$7 + 4 = 11$$
$$11 - 3 = 8$$
$$8 + 4 = 12$$
$$12 - 3 = 9$$
$$9 + 4 = 13$$

To get the second number, we add 4 to the first number. To get the third number, we subtract 3 from the second number. To obtain subsequent numbers, this pattern is continued. The next number is 13 − 3, or 10. ■

WORK PROBLEM 2 AT THE SIDE.

2. Find the next number in the sequence
6, 11, 7, 12, 8, 13, ...

■ **EXAMPLE 3** *Using Inductive Reasoning*
Find the next number in the sequence 1, 2, 4, 8, 16, ...

Each number after the first is obtained by multiplying the previous number by 2. So the next number would be $16 \cdot 2 = 32$. ■

WORK PROBLEM 3 AT THE SIDE.

3. Find the next number in the sequence 2, 6, 18, 54, ...

ANSWERS
1. 26
2. 9
3. 162

4. Find the next shape in the following sequence.

■ **EXAMPLE 4** *Using Inductive Reasoning*
Find the next geometric shape in the following sequence.

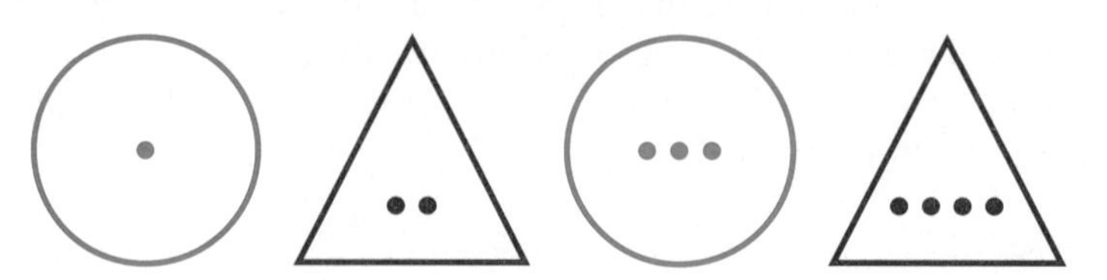

In this sequence, the figures alternate between a circle and a triangle. In addition, the number of dots increases by 1 in each subsequent figure. Thus, the next figure should be a circle with five dots contained in it, or

■

■ **EXAMPLE 5** *Using Inductive Reasoning*
Find the next geometric shape in the following sequence.

The first two shapes consist of vertical lines with horizontal lines at the bottom facing left and right. The third shape is a vertical line with a horizontal line at the top facing to the left. The fourth shape should be a vertical line with a horizontal line at the top facing to the right, or

■

◀◀ WORK PROBLEM 4 AT THE SIDE.

2 In the previous discussion, specific cases were used to find patterns and predict the next event. There is another type of reasoning called **deductive reasoning,** which moves from general cases to specific conclusions.

ANSWERS
4.

EXAMPLE 6 *Using Deductive Reasoning*

Does the conclusion follow from the premises in this argument?

All Buicks are automobiles.
All automobiles have horns.
∴ All Buicks have horns.

In this example, the first two statements are called *premises* and the third statement (below the line) is called a conclusion. The symbol ∴ is a mathematical symbol meaning "therefore." The entire set of statements is called an *argument*. The focus of deductive reasoning is to determine if the conclusion follows (is valid) from the premises. A series of circles called **Euler circles** is used to analyze the argument. In Example 6, the statement "All Buicks are automobiles" can be represented by two circles, one for Buicks and one for automobiles.

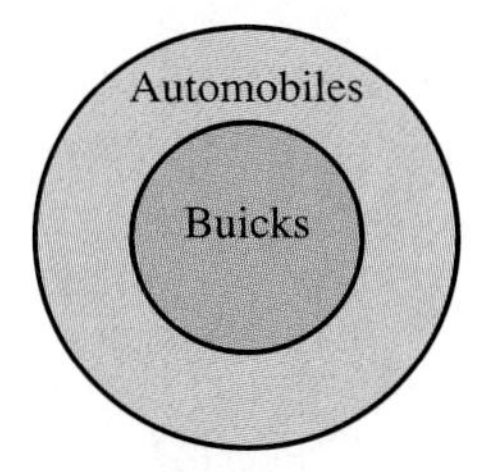

Note that the circle representing Buicks is totally inside the circle representing automobiles.

If a circle representing the second statement is added, a circle representing vehicles with horns must surround the circle representing automobiles.

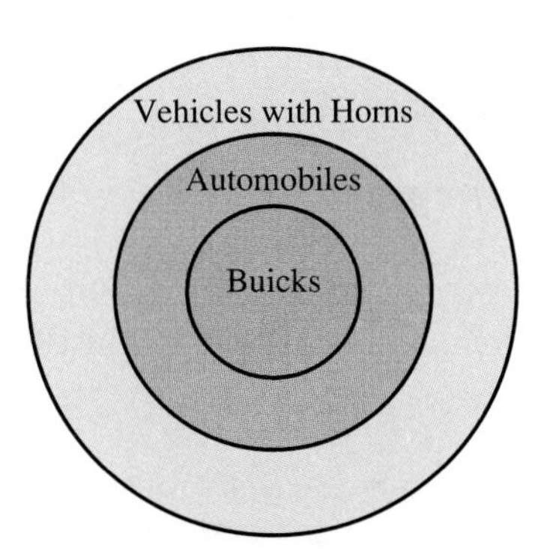

Notice that the circle representing Buicks is completely inside the circle representing vehicles with horns. It must follow that

all Buicks have horns. ■

WORK PROBLEM 5 AT THE SIDE.

5. Does the conclusion follow from the premises in the following argument?

All cars have four wheels.
All Fords are cars.
∴ All Fords have four wheels.

ANSWERS

5. The conclusion follows from the premises.

6. Does each conclusion follow from the premises?

(a) All animals are wild.
All cats are animals.
∴ All cats are wild.

(b) All students use math.
All adults use math.
∴ All adults are students.

ANSWERS
6. (a) The conclusion follows from the premises.
(b) The conclusion does not follow from the premises.

■ EXAMPLE 7 *Using Deductive Reasoning*

Does the conclusion follow from the premises in this argument?

All tables are round.
All glasses are round.
∴ All glasses are tables.

Using Euler circles, a circle representing tables is drawn inside a circle representing round objects.

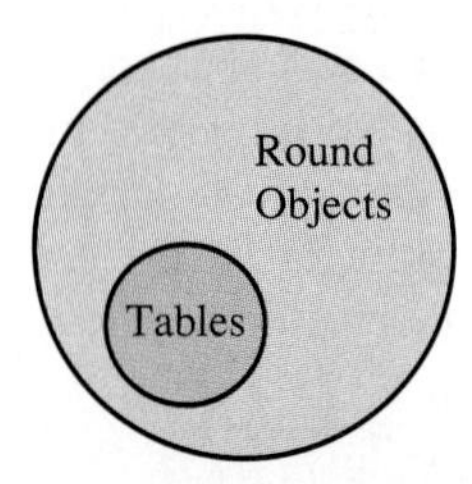

The second statement requires that a circle representing glasses must now be drawn inside the circle representing round objects but not necessarily inside the circle representing tables.

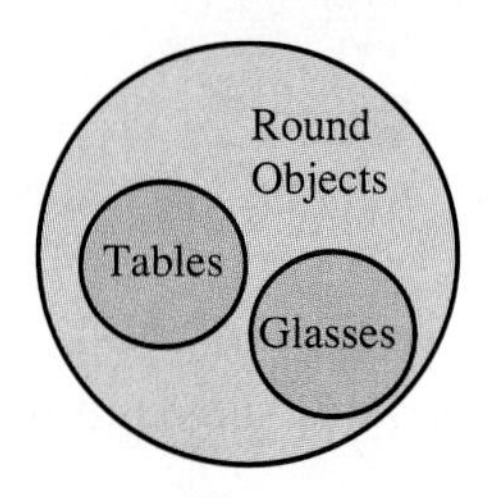

The conclusion does not follow from the premises. This means that the conclusion is invalid or untrue. ■

◀◀ WORK PROBLEM 6 AT THE SIDE.

3 Another type of deductive reasoning problem occurs when a set of facts is given in a problem and a conclusion must be drawn by using these facts.

■ EXAMPLE 8 *Using Deductive Reasoning*

There were 25 students enrolled in a ceramics class. During the class, 10 of the students made a bowl and 8 students made a birdbath. Three students made both a bowl and a birdbath. How many students did not make either a bowl or a birdbath?

This type of problem is best solved by organizing the data by using a device called a *Venn diagram.* Two overlapping circles are drawn, with each circle representing one item made by students.

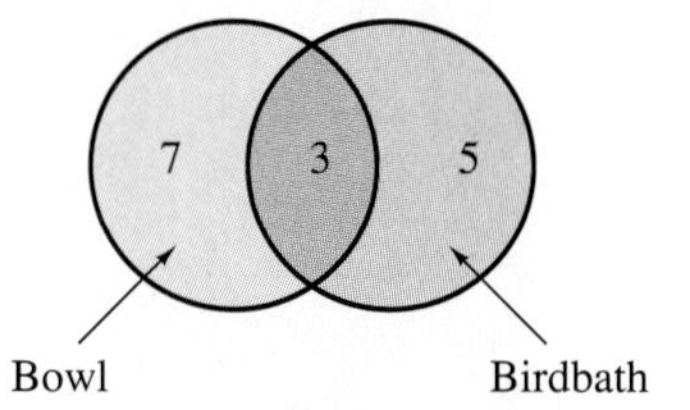

In the region where the circles overlap, place the number that represents the number of students who made both items, namely 3. In the remaining portion of the birdbath circle, write the number 5, which when added to 3 will give the total number of students who made a birdbath, namely 8. In a similar manner, write 7 in the remaining portion of the bowl circle, since $7 + 3 = 10$, the total number of students who made a bowl. The three numbers that have been written in the regions total 15. Since there are 25 students in the class, this means $25 - 15$ or 10 students did not make either a birdbath or a bowl. ■

WORK PROBLEM 7 AT THE SIDE. ▶▶

7. In a college class of 100 students, 35 take both math and history, 50 take history, and 40 take math. How many take neither math nor history?

■ **EXAMPLE 9** *Using Deductive Reasoning*

Four cars in a race finish first, second, third, and fourth. The following facts are known.

(a) Car A beat Car C.

(b) Car D finished between Cars C and B.

(c) Car C beat Car B.

In which order did the cars finish?

To solve this type of problem, it is helpful to use a line diagram.

1. *Write A before C,* since Car A beat Car C (fact a).

A C

2. *Write B after C,* since Car C beat Car B (fact c).

A C B

3. *Write D between C and B,* since Car D finished between Cars C and B (fact b).

So

A C D B

is the correct order of finish. ■

WORK PROBLEM 8 AT THE SIDE. ▶▶

8. A Chevy, BMW, Cadillac, and Oldsmobile are parked side by side.

(a) The Oldsmobile is on the right end.

(b) The BMW is next to the Cadillac.

(c) The Chevy is between the Oldsmobile and the Cadillac.

Which car is parked on the left end?

ANSWERS
7. 45 **8.** BMW

NAME DATE HOUR

APPENDIX B: EXERCISES

Find the next number in each of the following sequences.

1. 2, 9, 16, 23, 30, ... 37

2. 5, 8, 11, 14, 17, ... 20

3. 1, 6, 11, 16, 21, ... 26

4. 3, 5, 7, 9, 11, ... 13

5. 1, 2, 4, 8, ... 16

6. 1, 8, 27, 64, ... 125

7. 1, 3, 9, 27, 81, ... 243

8. 3, 6, 12, 24, 48, ... 96

9. 1, 4, 9, 16, 25, ... 36

10. 6, 7, 9, 12, 16, ... 21

Find the next shape in each of the following sequences.

11.

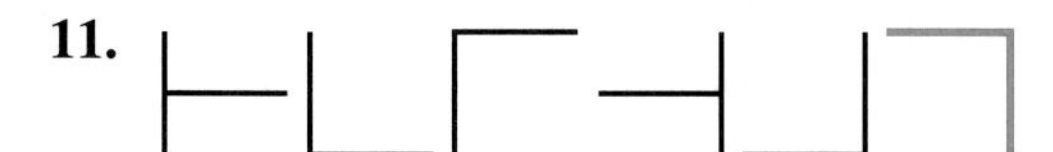

12.

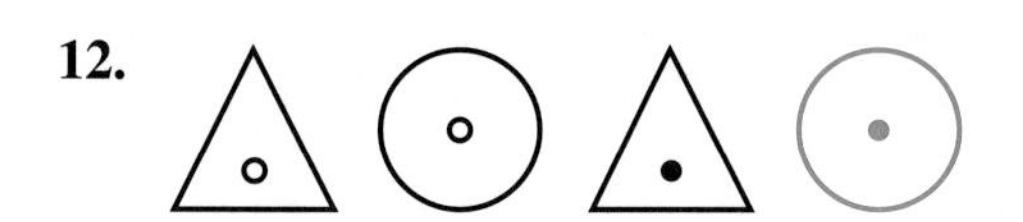

13.

14.

In each of the following, state whether or not the conclusion follows from the premises.

15. All animals are wild.
All lions are animals.
$\therefore$ All lions are wild.

Conclusion follows.

16. All students are hard workers.
All business majors are students.
$\therefore$ All business majors are hard workers.

Conclusion follows.

17. All teachers are serious.
All mathematicians are serious.
$\therefore$ All mathematicians are teachers.

Conclusion does not follow.

18. All boys ride bikes.
All Americans ride bikes.
$\therefore$ All Americans are boys.

Conclusion does not follow.

Solve the following problems.

19. In a given 30-day period, a man watched television 20 days and his wife watched television 25 days. If they watched television together 18 days, how many days did neither watch television?

3 days

20. In a class of 40 students, 21 students take both calculus and physics. If 30 students take calculus and 25 students take physics, how many do not take either calculus or physics?

6 students

21. Tom, Dick, Mary, and Joan all work for the same company. One is a secretary, one is a computer operator, one is a receptionist, and one is a mail clerk.
(a) Tom and Joan eat dinner with the computer operator.
(b) Dick and Mary carpool with the secretary.
(c) Mary works on the same floor as the computer operator and the mail clerk.
Who is the computer operator? Dick

22. Four cars—a Ford, a Buick, a Mercedes, and an Audi—are parked in a garage in four spaces.
(a) The Ford is in the last space.
(b) The Buick and Mercedes are next to each other.
(c) The Audi is next to the Ford but not next to the Buick.
Which car is in the first space? Buick

Answers to Selected Exercises

The solutions to selected odd-numbered exercises are given in the section beginning on page A-21.

In this section we provide the answers that we think most students will obtain when they work the exercises using the methods explained in the text. If your answer does not look exactly like the one given here, it is not necessarily wrong. In many cases there are equivalent forms of the answer that are correct. For example, if the answer section shows $\frac{3}{4}$ and your answer is 0.75, you have obtained the right answer but written it in a different (yet equivalent) form. Unless the directions specify otherwise, 0.75 is just as valid an answer as $\frac{3}{4}$.

In general, if your answer does not agree with the one given in the text, see whether it can be transformed into the other form. If it can, then it is the correct answer. If you still have doubts, talk with your instructor.

CHAPTER 1

SECTION 1.1 (page 5)

1. 2 1 **3.** 1 0 **5.** 8 2 **7.** 10, 678, 286 **9.** 60, 0, 502, 109 **11.** Answer varies. **13.** seventy-nine thousand, six hundred thirteen **15.** seven hundred twenty-five thousand, nine **17.** twenty-five million, seven hundred fifty-six thousand, six hundred sixty-five **19.** 83,135 **21.** 10,000,223 **23.** 3050 **25.** 13,112 **27.** 48,000,605 **29.** 800,621,020,215

SECTION 1.2 (page 13)

1. 49 **3.** 78 **5.** 889 **7.** 999 **9.** 7785 **11.** 899 **13.** 7676 **15.** 78,446 **17.** 15,928 **19.** 129,224 **21.** 143 **23.** 142 **25.** 121 **27.** 162 **29.** 102 **31.** 1781 **33.** 1154 **35.** 413 **37.** 1771 **39.** 1410 **41.** 9253 **43.** 11,624 **45.** 17,611 **47.** 15,954 **49.** 10,648 **51.** 16,858 **53.** 11,557 **55.** 12,078 **57.** 4250 **59.** 12,268 **61.** correct **63.** incorrect; should be 769 **65.** correct **67.** incorrect; should be 11,577 **69.** correct **71.** Answer varies. **73.** 35 miles **75.** 38 miles **77.** $90 **79.** 699 people **81.** 13,051 books **83.** 260 inches **85.** 708 feet

SECTION 1.3 (page 23)

1. 12 **3.** 51 **5.** 17 **7.** 213 **9.** 101 **11.** 6211 **13.** 3412 **15.** 2111 **17.** 13,160 **19.** 41,110 **21.** correct **23.** incorrect; should be 62 **25.** incorrect; should be 121 **27.** correct **29.** incorrect; should be 7222 **31.** 8 **33.** 29 **35.** 16 **37.** 61 **39.** 519 **41.** 9177 **43.** 7589 **45.** 8859 **47.** 3 **49.** 23 **51.** 1 **53.** 2833 **55.** 7775 **57.** 503 **59.** 156 **61.** 1942 **63.** 5687 **65.** 19,038 **67.** 15,778 **69.** 6584 **71.** correct **73.** correct **75.** correct **77.** correct **79.** Answer varies. **81.** 58 bushes **83.** 121 passengers **85.** $10,300 **87.** 312 miles **89.** 1329 students **91.** 9539 flags **93.** $144 **95.** $913 **97.** 758 people

SECTION 1.4 (page 33)

1. 20 **3.** 56 **5.** 0 **7.** 24 **9.** 36 **11.** 0 **13.** Answer varies. **15.** 270 **17.** 224 **19.** 2048 **21.** 1872 **23.** 8612 **25.** 10,084 **27.** 19,092 **29.** 258,447 **31.** 90 **33.** 320 **35.** 2220 **37.** 2000 **39.** 3750 **41.** 25,100 **43.** 270,000 **45.** 86,000,000 **47.** 48,500 **49.** 450,000 **51.** 1,940,000 **53.** 646 **55.** 2376 **57.** 3735 **59.** 2378 **61.** 6164 **63.** 18,192 **65.** 32,805 **67.** 14,564 **69.** 82,320 **71.** 183,996 **73.** 2,062,495 **75.** 66,005 **77.** 86,028 **79.** 19,422,180 **81.** 2,278,410 **83.** Answer varies. **85.** 24,000 pages **87.** 216 plants **89.** 418 miles **91.** $288 **93.** $4116 **95.** $3996 **97.** 50,568 **99.** 175 joggers **101.** 1058 calories **103.** $48

SECTION 1.5 (page 45)

1. $5\overline{)15}^{\,3}$ $\frac{15}{5} = 3$ **3.** $9\overline{)45}^{\,5}$ $45 \div 9 = 5$ **5.** $16 \div 2 = 8$ $\frac{16}{2} = 8$ **7.** 1 **9.** 5 **11.** meaningless **13.** 0 **15.** 0 **17.** meaningless **19.** 0 **21.** 8 **23.** 26 **25.** 47 **27.** 401 **29.** 627 R1 **31.** 1522 R5 **33.** 309 **35.** 1006 **37.** 5006 **39.** 811 R1 **41.** 2589 R2 **43.** 7324 R2 **45.** 3157 R2 **47.** 5841 **49.** 12,458 R3 **51.** 10,253 R5 **53.** 18,377 R6 **55.** correct **57.** incorrect; should be 1908 R1 **59.** incorrect; should be 670 R2 **61.** incorrect; should be 3568 R1 **63.** correct **65.** correct **67.** incorrect; should be 9628 R3 **69.** correct **71.** Answer varies. **73.** 156 cartons **75.** $14,441 **77.** $3250 **79.** 205 acres **81.** $1822

	2	3	5	10
83.	√	√	√	√
85.	√	X	X	X
87.	X	X	√	X
89.	X	√	X	X
91.	√	√	X	X
93.	X	X	X	X

95. $3045

SECTION 1.6 (page 53)

1. 32 **3.** 47 R10 **5.** 37 R7 **7.** 153 R7
9. 38 **11.** 236 R29 **13.** 2407 R1 **15.** 1239 R15
17. 522 R14 **19.** 9746 R1 **21.** 7746 R20
23. 3331 R82 **25.** 21 **27.** 1114 R196 **29.** 170
31. incorrect; should be 106 R7
33. incorrect; should be 658 **35.** incorrect should be 62
37. Answer varies. **39.** 25 hours **41.** $3906
43. $108 **45.** 1680 circuits **47.** $375

SECTION 1.7 (page 61)

1. 510 **3.** 1280 **5.** 7900 **7.** 86,800
9. 42,500 **11.** 8000 **13.** 15,800 **15.** 78,000
17. 6000 **19.** 53,000 **21.** 600,000
23. 9,000,000 **25.** 2370 2400 2000
27. 5390 5400 5000
29. 5050 5000 5000
31. 3130 3100 3000
33. 19,540 19,500 20,000
35. 26,290 26,300 26,000
37. 23,500 23,500 24,000
39. Answer varies. **41.** $40 + 60 + 90 + 60 = 250$; 250
43. $100 - 30 = 70$; 71 **45.** $80 \times 20 = 1600$; 1672
47. $800 + 800 + 300 + 700 = 2600$; 2635
49. $700 - 400 = 300$; 316
51. $300 \times 500 = 150{,}000$; 144,522
53. $8000 + 60 + 700 + 4000 = 12{,}760$; 12,605
55. $800 - 200 = 600$; 545 **57.** $600 \times 50 = 30{,}000$; 29,986 **59.** Answer varies.
61. 70,987,650; 70,990,000; 70,000,000
63. $5,465,485,400,000; $5,465,500,000,000; $5,465,000,000,000

SECTION 1.8 (page 67)

1. 2 **3.** 4 **5.** 12 **7.** 11 **9.** 2; 3; 9
11. 2; 6; 36 **13.** 2; 12; 144 **15.** 2; 15; 225
17. 400 **19.** 625 **21.** 1225 **23.** 1600
25. 2916 **27.** Answer varies. **29.** 65 **31.** 19
33. 5 **35.** 20 **37.** 24 **39.** 63 **41.** 118
43. 24 **45.** 35 **47.** 102 **49.** 10 **51.** 63
53. 33 **55.** 58 **57.** 7 **59.** 26 **61.** 36
63. 108 **65.** 19 **67.** 9 **69.** 44 **71.** 16
73. 8 **75.** 3 **77.** 7 **79.** 20 **81.** 7 **83.** 25
85. 16 **87.** 20 **89.** 209

SECTION 1.9 (page 75)

1. $600 + 200 + 50 = 850$; 836 members
3. $200 - 60 = 140$; 138 sold
5. $200 \times 20 = 4000$; 5664 kits
7. $3000 \div 700 \approx 4$; 4 toys
9. $8000 - 4000 = 4000$; 4174 people
11. $700 \times 4 = 2800$; $2700
13. $3000 \div 20 = 150$; 175 miles
15. $2000 - 500 - 300 - 300 - 200 - 200 = 500$; $193
17. $40{,}000 \times 100 = 4{,}000{,}000$; 6,011,280 square feet
19. $400 + 100 + 20 + 20 + 1 = 541$; 526 pounds
21. $(1000 \times 6) + (900 \times 20) = 24{,}000$; $20,961
23. $20 + 100 + 300 = 420$; 418 birds
25. Answer varies. **27.** Answer varies. **29.** $165
31. 2477 pounds **33.** $1800 **35.** 352 machines
37. 20 seats

CHAPTER 1 REVIEW EXERCISES (page 83)

1. 2 318 **2.** 56 478 **3.** 206 792
4. 1 768 710 618 **5.** seven hundred twenty-five
6. twelve thousand, four hundred twelve
7. three hundred nineteen thousand, two hundred fifteen
8. sixty-two million, five hundred thousand, five
9. 9150 **10.** 200,000,455 **11.** 82 **12.** 134
13. 4656 **14.** 15,657 **15.** 9179 **16.** 6979
17. 40,602 **18.** 49,855 **19.** 12 **20.** 32
21. 39 **22.** 184 **23.** 3803 **24.** 4327
25. 224 **26.** 25,866 **27.** 36 **28.** 0 **29.** 15
30. 81 **31.** 36 **32.** 36 **33.** 56 **34.** 81
35. 48 **36.** 45 **37.** 48 **38.** 8 **39.** 0
40. 64 **41.** 48 **42.** 0 **43.** 102 **44.** 528
45. 180 **46.** 89 **47.** 3834 **48.** 5467
49. 5396 **50.** 45,815 **51.** 14,518 **52.** 32,640
53. 465,525 **54.** 174,984 **55.** 612 **56.** 1872
57. 1176 **58.** 5100 **59.** 13,755 **60.** 30,184
61. 887,169 **62.** 500,856 **63.** $798 **64.** $312
65. $13,344 **66.** $1512 **67.** 17,500 **68.** 30,400
69. 194,400 **70.** 300,800 **71.** 128,000,000
72. 90,300,000 **73.** 6 **74.** 5 **75.** 6 **76.** 9
77. 9 **78.** 4 **79.** 9 **80.** 0 **81.** meaningless
82. 0 **83.** 8 **84.** 9 **85.** 81 **86.** 36
87. 6251 **88.** 352 **89.** 150 R4 **90.** 124 R25
91. 220 **92.** 18,600 **93.** 20,000 **94.** 70,000
95. 1500 1500 1000 **96.** 20,070 20,100 20,000
97. 98,200 98,200 98,000
98. 352,120 352,100 352,000 **99.** 5 **100.** 12
101. 9 **102.** 14 **103.** 3; 2; 8 **104.** 3; 5; 125
105. 5; 2; 32 **106.** 5; 4; 1024 **107.** 72 **108.** 4
109. 9 **110.** 4 **111.** 9 **112.** 6
113. $40 \times 10 = 400$; $420
114. $1000 \times 60 = 60{,}000$; 84,000 revolutions
115. $100 \times 6 = 600$; 720 cups
116. $8000 \times 40 = 320{,}000$; 320,000 nails
117. $2000 \times 10 = 20{,}000$; 24,000 hours
118. $80 \times 5 = 400$; 400 miles

119. $(30 \times 20) + (30 \times 40) = 1800$; $1550
120. $(60 \times 20) + (20 \times 10) = 1400$; $1024
121. $1300 - 1100 = 200$; $199
122. $400 - 100 = 300$; $247
123. $9000 \div 200 = 45$; 50 pounds
124. $30{,}000 \div 900 \approx 33$; 33 hours
125. $6000 \div 300 = 20$; 23 acres
126. $6000 \div 200 = 30$; 32 homes **127.** 320
128. 602 **129.** 115 **130.** 208 **131.** 1041
132. 1030 **133.** 32,062 **134.** 24,947 **135.** 3
136. 7 **137.** 93,635 **138.** 83,178 **139.** meaningless
140. 5 **141.** 13,800 **142.** 5501 **143.** 1,079,040
144. 115,713 **145.** 108 **146.** 207
147. two hundred eighty-six thousand, seven hundred fifty-three
148. one hundred eight thousand, two hundred ten
149. 7200 **150.** 500,000 **151.** 6 **152.** 10
153. $2128 **154.** $33,800 **155.** $2220
156. $15,782 **157.** 468 cards
158. 9000 subscriptions **159.** 52 miles
160. $27,940 **161.** $22,750 **162.** $3850
163. 3650 hours **164.** 2504 hours

CHAPTER 1 TEST (page 91)

1. five thousand, six hundred eighty
2. fifty-two thousand, eight **3.** 138,008 **4.** 16,486
5. 112,630 **6.** 3065 **7.** 3084 **8.** 120
9. 126,000 **10.** 1785 **11.** 4,450,743 **12.** 7747
13. meaningless **14.** 458 **15.** 160 **16.** 3570
17. 99,000 **18.** 51 **19.** 28
20. $500 + 500 + 500 + 400 - 800 = 1100$; $1140
21. $40{,}000 \div 300 \approx 133$; 123 days
22. $1000 - 700 - 200 - 70 = 30$; $165
23. $(100 \times 4) + (100 \times 4) = 800$; 1028 ovens
24. Answer varies. **25.** Answer varies.

CHAPTER 2

SECTION 2.1 (page 97)

1. $\frac{5}{8}$ **3.** $\frac{2}{3}$ **5.** $\frac{7}{5}$ **7.** $\frac{2}{11}$ **9.** $\frac{8}{25}$ **11.** $\frac{13}{71}$
13. 3 8 **15.** 12 7
17. proper: $\frac{1}{4}, \frac{3}{8}, \frac{7}{12}$ improper: $\frac{9}{7}, \frac{11}{4}, \frac{8}{8}$
19. proper: $\frac{3}{4}, \frac{9}{11}, \frac{7}{15}$ improper: $\frac{3}{2}, \frac{5}{5}, \frac{19}{18}$
21. Answer varies. **23.** 9 16 **25.** 18
27. 96 **29.** 7 **31.** 19

SECTION 2.2 (page 101)

1. $\frac{7}{4}$ **3.** $\frac{11}{3}$ **5.** $\frac{13}{4}$ **7.** $\frac{27}{4}$ **9.** $\frac{18}{11}$ **11.** $\frac{19}{3}$
13. $\frac{34}{3}$ **15.** $\frac{43}{4}$ **17.** $\frac{27}{8}$ **19.** $\frac{44}{5}$ **21.** $\frac{54}{11}$
23. $\frac{183}{8}$ **25.** $\frac{233}{13}$ **27.** $\frac{269}{15}$ **29.** $\frac{115}{18}$ **31.** $4\frac{1}{2}$
33. $2\frac{2}{3}$ **35.** 7 **37.** $3\frac{3}{8}$ **39.** $4\frac{3}{4}$ **41.** 9
43. $11\frac{3}{5}$ **45.** $5\frac{2}{9}$ **47.** $7\frac{1}{7}$ **49.** $16\frac{4}{5}$ **51.** $30\frac{3}{4}$
53. $26\frac{1}{7}$ **55.** Answer varies. **57.** $\frac{2041}{8}$ **59.** $\frac{1000}{3}$
61. 171 **63.** 13 **65.** 66 **67.** 49

SECTION 2.3 (page 107)

1. 1, 2, 3, 6 **3.** 1, 2, 4, 8 **5.** 1, 5, 25
7. 1, 2, 3, 6, 9, 18 **9.** 1, 2, 4, 5, 8, 10, 20, 40
11. 1, 2, 4, 8, 16, 32, 64 **13.** composite **15.** prime
17. composite **19.** prime **21.** prime
23. composite **25.** composite **27.** composite
29. 2^2 **31.** $3 \cdot 5$ **33.** 5^2 **35.** 2^5 **37.** $3 \cdot 13$
39. $2^3 \cdot 11$ **41.** $3 \cdot 5^2$ **43.** $2^2 \cdot 5^2$ **45.** 5^3
47. $3^2 \cdot 5^2$ **49.** $2^6 \cdot 5$ **51.** $2^3 \cdot 3^2 \cdot 5$
53. Answer varies. **55.** 64 **57.** 512 **59.** 1296
61. 108 **63.** 1125 **65.** 972 **67.** $2^3 \cdot 5 \cdot 7$
69. $2^6 \cdot 3 \cdot 5$ **71.** $2^6 \cdot 5^2$ **73.** 10 **75.** 700
77. 8 **79.** 27

SECTION 2.4 (page 115)

1. $\frac{1}{2}$ **3.** $\frac{3}{4}$ **5.** $\frac{5}{8}$ **7.** $\frac{6}{7}$ **9.** $\frac{9}{10}$ **11.** $\frac{6}{7}$
13. $\frac{4}{7}$ **15.** $\frac{1}{50}$ **17.** $\frac{8}{11}$ **19.** $\frac{5}{9}$
21. $\frac{2 \cdot 5}{2 \cdot 2 \cdot 2 \cdot 2} = \frac{5}{8}$ **23.** $\frac{5 \cdot 7}{2 \cdot 2 \cdot 2 \cdot 5} = \frac{7}{8}$
25. $\frac{3 \cdot 5 \cdot 5}{2 \cdot 3 \cdot 5 \cdot 5} = \frac{1}{2}$ **27.** $\frac{2 \cdot 2 \cdot 3 \cdot 3}{2 \cdot 2 \cdot 3} = 3$
29. $\frac{7 \cdot 11}{2 \cdot 2 \cdot 2 \cdot 3 \cdot 11} = \frac{7}{24}$ **31.** equivalent
33. not equivalent **35.** not equivalent **37.** equivalent
39. not equivalent **41.** equivalent
43. Answer varies. **45.** $\frac{7}{8}$ **47.** $\frac{8}{25}$
49. 1, 2, 3, 4, 6, 12 **51.** 1, 2, 4, 8, 16, 32, 64

SECTION 2.5 (page 123)

1. $\frac{3}{8}$ **3.** $\frac{1}{32}$ **5.** $\frac{9}{10}$ **7.** $\frac{3}{10}$ **9.** $\frac{5}{12}$ **11.** $\frac{9}{32}$
13. $\frac{1}{2}$ **15.** $\frac{13}{32}$ **17.** $\frac{21}{128}$ **19.** 6 **21.** 40
23. 25 **25.** 21 **27.** $31\frac{1}{2}$ **29.** 240 **31.** $213\frac{1}{3}$
33. 400 **35.** 810 **37.** $\frac{1}{2}$ square inch
39. 4 square yards **41.** $\frac{3}{10}$ square inch

43. Answer varies. **45.** 4 square inches
47. $1\frac{1}{3}$ square yards **49.** 1 square mile
51. 3360 lunches

SECTION 2.6 (page 129)

1. $\frac{5}{6}$ square yard **3.** $\frac{2}{3}$ square foot
5. 380 items are taxable **7.** \$2700 **9.** 375 students
11. 75 men **13.** \$32,000 **15.** \$6400 **17.** \$2000
19. Answer varies. **21.** \$75 **23.** 900 votes
25. $\frac{1}{32}$ of the estate **27.** 74 cartons

SECTION 2.7 (page 137)

1. $\frac{1}{2}$ **3.** $\frac{15}{32}$ **5.** $3\frac{1}{3}$ **7.** $\frac{5}{8}$ **9.** 4 **11.** $\frac{24}{25}$
13. 9 **15.** $22\frac{1}{2}$ **17.** $\frac{1}{14}$ **19.** $\frac{2}{9}$ acre
21. 35 shakers **23.** 88 dispensers **25.** 60 trips
27. 12 batches **29.** Answer varies.
31. 108 miles **33.** \$120,000 **35.** $\frac{15}{4}$ **37.** $\frac{46}{3}$
39. $\frac{1016}{9}$

SECTION 2.8 (page 145)

1. $7\frac{7}{8}$ **3.** $4\frac{1}{2}$ **5.** 4 **7.** $72\frac{1}{2}$ **9.** $49\frac{1}{2}$
11. 12 **13.** $1\frac{5}{21}$ **15.** $\frac{5}{6}$ **17.** $4\frac{4}{5}$ **19.** $\frac{2}{9}$
21. $\frac{3}{10}$ **23.** $\frac{17}{18}$ **25.** 13 yards **27.** 12 homes
29. $21\frac{7}{8}$ ounces **31.** 5000 dictionaries
33. 11 spacers **35.** 471 rolls **37.** 56 gallons
39. $\frac{1}{2}$ **41.** $\frac{5}{8}$ **43.** $\frac{9}{10}$

CHAPTER 2 REVIEW EXERCISES (page 151)

1. $\frac{3}{4}$ **2.** $\frac{5}{8}$ **3.** $\frac{1}{4}$
4. proper: $\frac{1}{2}, \frac{2}{3}, \frac{3}{8}$ improper: $\frac{2}{1}, \frac{6}{5}$
5. proper: $\frac{15}{16}, \frac{1}{8}$ improper: $\frac{6}{5}, \frac{16}{13}, \frac{5}{3}$ **6.** $\frac{43}{8}$ **7.** $\frac{181}{16}$
8. $9\frac{1}{2}$ **9.** $12\frac{3}{5}$ **10.** 1, 2, 3, 6
11. 1, 2, 3, 4, 6, 12 **12.** 1, 5, 11, 55
13. 1, 2, 3, 5, 6, 9, 10, 15, 18, 30, 45, 90 **14.** 5^2
15. $2^3 \cdot 3 \cdot 5$ **16.** $3^2 \cdot 5^2$ **17.** 36 **18.** 36
19. 1728 **20.** 2048 **21.** $\frac{3}{4}$ **22.** $\frac{7}{8}$ **23.** $\frac{35}{38}$
24. $\frac{5 \cdot 5}{2 \cdot 2 \cdot 3 \cdot 5}; \frac{5}{12}$ **25.** $\frac{2 \cdot 2 \cdot 2 \cdot 2 \cdot 2 \cdot 2 \cdot 2 \cdot 3}{2 \cdot 2 \cdot 2 \cdot 2 \cdot 2 \cdot 3}; 4$
26. equivalent **27.** equivalent **28.** $\frac{1}{2}$ **29.** $\frac{1}{3}$
30. $\frac{1}{7}$ **31.** $\frac{4}{21}$ **32.** 15 **33.** 625 **34.** $\frac{9}{16}$
35. $\frac{5}{3} = 1\frac{2}{3}$ **36.** $\frac{5}{2} = 2\frac{1}{2}$ **37.** 2 **38.** 16
39. 12 **40.** $\frac{8}{7} = 1\frac{1}{7}$ **41.** $\frac{2}{15}$ **42.** $\frac{4}{13}$
43. $\frac{15}{32}$ square yard **44.** $\frac{7}{12}$ square inch
45. 10 square feet **46.** 36 square yards
47. $3\frac{9}{16}$ **48.** $21\frac{3}{8}$ **49.** $5\frac{1}{6}$ **50.** $\frac{35}{64}$
51. 20 cans **52.** $\frac{2}{15}$ of the estate **53.** 36 pull cords
54. \$315 **55.** 30 pounds **56.** \$375
57. $\frac{5}{32}$ of the budget **58.** $\frac{1}{12}$ of the total **59.** $\frac{1}{4}$
60. $\frac{1}{6}$ **61.** $25\frac{5}{8}$ **62.** $28\frac{1}{2}$ **63.** $\frac{7}{48}$ **64.** $\frac{5}{32}$
65. 30 **66.** $2\frac{1}{6}$ **67.** $1\frac{3}{4}$ **68.** $46\frac{3}{4}$ **69.** $\frac{17}{3}$
70. $\frac{307}{8}$ **71.** $\frac{2 \cdot 2 \cdot 2}{2 \cdot 2 \cdot 3} = \frac{2}{3}$
72. $\frac{2 \cdot 2 \cdot 3 \cdot 3 \cdot 3}{2 \cdot 3 \cdot 5 \cdot 7} = \frac{18}{35}$ **73.** $\frac{1}{3}$ **74.** $\frac{3}{8}$ **75.** $\frac{2}{5}$
76. $\frac{1}{3}$ **77.** $152\frac{4}{9}$ ounces **78.** $144\frac{5}{8}$ quarts
79. $\frac{1}{2}$ square inch **80.** $\frac{3}{10}$ square meter

CHAPTER 2 TEST (page 157)

1. $\frac{3}{8}$ **2.** $\frac{5}{6}$ **3.** $\frac{5}{8}, \frac{7}{16}, \frac{2}{3}, \frac{3}{14}$
4. $\frac{23}{8}$ **5.** $20\frac{5}{6}$ **6.** 1, 3, 5, 15 **7.** $3^2 \cdot 5$
8. $2^5 \cdot 3$ **9.** $2^2 \cdot 5^3$ **10.** $\frac{5}{6}$ **11.** $\frac{2}{3}$
12. Answer varies. **13.** Answer varies.
14. $\frac{3}{8}$ **15.** 18 **16.** $\frac{1}{4}$ square inch
17. 483 students **18.** $\frac{7}{10}$ **19.** $15\frac{3}{4}$
20. 100 vehicles **21.** $17\frac{23}{32}$ **22.** $7\frac{17}{18}$
23. $4\frac{4}{15}$ **24.** $5\frac{1}{10}$ **25.** $30\frac{5}{8}$ grams

CUMULATIVE REVIEW 1–2 (page 159)

1. 6 3 **2.** 3 8 **3.** 146 **4.** 149,199
5. 1572 **6.** 3,221,821 **7.** 448 **8.** 105
9. 2,168,232 **10.** 440,300 **11.** 9 **12.** 7581
13. 4235 R2 **14.** 22 R26
15. 8630 8600 9000 **16.** 78,150 78,200 78,000
17. 7 **18.** 5 **19.** \$445 **20.** \$130
21. 47,000 hairs **22.** \$175
23. $\frac{7}{16}$ square foot **24.** 100 sports bottles
25. proper **26.** improper **27.** proper **28.** $\frac{3}{2}$
29. $\frac{67}{8}$ **30.** $5\frac{1}{2}$ **31.** $12\frac{7}{8}$ **32.** $2^3 \cdot 5$
33. $3 \cdot 5^2$ **34.** $2 \cdot 5^2 \cdot 7$ **35.** 36 **36.** 675
37. 640 **38.** $\frac{3}{8}$ **39.** $\frac{3}{7}$ **40.** $\frac{5}{9}$ **41.** $\frac{1}{2}$
42. 10 **43.** 25 **44.** $\frac{9}{16}$ **45.** $\frac{3}{10}$ **46.** $2\frac{2}{5}$

CHAPTER 3

SECTION 3.1 (page 165)

1. $\frac{2}{3}$ **3.** $\frac{7}{10}$ **5.** $\frac{1}{2}$ **7.** $\frac{1}{3}$ **9.** $\frac{5}{6}$ **11.** $\frac{13}{20}$
13. $\frac{10}{17}$ **15.** $\frac{3}{4}$ **17.** $\frac{11}{27}$ **19.** $\frac{1}{3}$ **21.** $\frac{8}{21}$
23. $\frac{3}{5}$ **25.** $\frac{2}{3}$ **27.** $\frac{1}{5}$ **29.** $\frac{1}{4}$ **31.** $\frac{1}{9}$ **33.** $\frac{3}{5}$
35. Answer varies. **37.** $\frac{1}{2}$ mile **39.** $\frac{3}{8}$ mile
41. $\frac{3}{4}$ acre **43.** $2 \cdot 5$ **45.** $2 \cdot 2 \cdot 5 \cdot 5$
47. $3 \cdot 5 \cdot 5$ **49.** $5 \cdot 5 \cdot 5$

SECTION 3.2 (page 173)

1. 12 **3.** 21 **5.** 72 **7.** 200 **9.** 180
11. 90 **13.** 120 **15.** 360 **17.** 120 **19.** 300
21. 108 **23.** 1350 **25.** $\frac{12}{24}$ **27.** $\frac{18}{24}$ **29.** $\frac{9}{24}$
31. 6 **33.** 36 **35.** 28 **37.** 55 **39.** 45
41. 72 **43.** 136 **45.** 96 **47.** 27
49. Answer varies. **51.** 7200 **53.** 10,584
55. $1\frac{2}{3}$ **57.** $1\frac{4}{5}$ **59.** $3\frac{6}{7}$

SECTION 3.3 (page 181)

1. $\frac{3}{5}$ **3.** $\frac{1}{2}$ **5.** $\frac{3}{4}$ **7.** $\frac{19}{22}$ **9.** $\frac{23}{36}$ **11.** $\frac{14}{15}$
13. $\frac{29}{36}$ **15.** $\frac{17}{20}$ **17.** $\frac{23}{30}$ **19.** $\frac{11}{12}$ **21.** $\frac{5}{6}$
23. $\frac{2}{3}$ **25.** $\frac{1}{2}$ **27.** $\frac{1}{6}$ **29.** $\frac{5}{24}$ **31.** $\frac{19}{45}$
33. $\frac{2}{15}$ **35.** $\frac{1}{36}$ **37.** $\frac{23}{24}$ ton **39.** $\frac{7}{12}$ acre
41. $\frac{41}{48}$ mile **43.** $\frac{3}{8}$ gallon **45.** Answer varies.
47. $\frac{1}{4}$ **49.** work and travel **51.** $\frac{7}{24}$ mile
53. $21\frac{1}{8}$ **55.** $13\frac{1}{2}$ **57.** $\frac{2}{5}$

SECTION 3.4 (page 189)

1. $17\frac{1}{2}$ **3.** $10\frac{3}{4}$ **5.** $15\frac{5}{8}$ **7.** $98\frac{2}{5}$ **9.** $20\frac{5}{14}$
11. $73\frac{39}{40}$ **13.** $42\frac{1}{8}$ **15.** $71\frac{3}{4}$ **17.** $67\frac{4}{15}$
19. $3\frac{1}{5}$ **21.** $4\frac{1}{4}$ **23.** $22\frac{7}{30}$ **25.** $10\frac{1}{8}$
27. $61\frac{23}{40}$ **29.** $6\frac{29}{48}$ **31.** $70\frac{1}{15}$ **33.** $372\frac{15}{16}$
35. $6\frac{4}{7}$ **37.** $216\frac{1}{4}$ **39.** $206\frac{1}{25}$ **41.** Answer varies.
43. $5\frac{11}{12}$ gallons **45.** $15\frac{7}{20}$ tons **47.** $22\frac{1}{2}$ hours
49. $3\frac{1}{4}$ yards **51.** $104\frac{5}{8}$ ft **53.** 50 cases
55. $14\frac{9}{20}$ in **57.** $1\frac{11}{24}$ ft **59.** 10 **61.** 23 **63.** 25

SECTION 3.5 (page 199)

1–11.

$2\frac{1}{6}$ $3\frac{1}{4}$ $3\frac{4}{5}$
$\frac{1}{4}$ $\frac{3}{8}$ $\frac{7}{9}$ $\frac{7}{8}$ $\frac{5}{4}$ $1\frac{7}{8}$ $\frac{7}{3}$ $\frac{11}{4}$ $\frac{7}{2}$
0 1 2 3 4
1. 2. 3. 10. 4. 12. 7. 5. 6. 11. 9. 8.

13. $<$ **15.** $<$ **17.** $<$ **19.** $<$
21. $<$ **23.** $>$ **25.** $\frac{4}{9}$ **27.** $\frac{49}{64}$ **29.** $\frac{8}{27}$
31. $\frac{125}{216}$ **33.** $\frac{81}{16} = 5\frac{1}{16}$ **35.** $\frac{1}{32}$
37. Answer varies. **39.** 8 **41.** 67 **43.** 1
45. $\frac{3}{16}$ **47.** $\frac{1}{4}$ **49.** $\frac{1}{3}$ **51.** $\frac{11}{16}$ **53.** $\frac{3}{8}$ **55.** $\frac{1}{4}$
57. $1\frac{1}{2}$ **59.** $\frac{1}{2}$ **61.** 3 **63.** $\frac{23}{112}$ **65.** $\frac{1}{4}$
67. $\frac{1}{10}$ **69.** eight thousand, four hundred thirty-six
71. four million, seventy-one thousand, two hundred eighty

CHAPTER 3 REVIEW EXERCISES (page 205)

1. $\frac{1}{2}$ **2.** $\frac{7}{8}$ **3.** $\frac{1}{3}$ **4.** $\frac{5}{14}$ **5.** $\frac{1}{5}$ **6.** $\frac{1}{4}$
7. $\frac{13}{31}$ **8.** $\frac{8}{27}$ **9.** $\frac{1}{2}$ of the lumber **10.** $\frac{1}{6}$ less

11. 40 **12.** 60 **13.** 60 **14.** 180 **15.** 120
16. 240 **17.** 8 **18.** 15 **19.** 10 **20.** 45
21. 63 **22.** 12 **23.** $\frac{5}{6}$ **24.** $\frac{7}{15}$ **25.** $\frac{7}{8}$ **26.** $\frac{15}{16}$
27. $\frac{11}{12}$ **28.** $\frac{29}{35}$ **29.** $\frac{31}{48}$ **30.** $\frac{1}{9}$ **31.** $\frac{7}{16}$ **32.** $\frac{7}{16}$
33. $\frac{5}{12}$ **34.** $\frac{17}{36}$ **35.** $\frac{23}{24}$ cubic yard
36. $\frac{109}{120}$ of the amount needed **37.** $11\frac{5}{6}$ **38.** $42\frac{1}{8}$
39. $96\frac{2}{7}$ **40.** $31\frac{43}{80}$ **41.** $5\frac{1}{6}$ **42.** $5\frac{7}{12}$
43. $17\frac{5}{6}$ **44.** $79\frac{7}{16}$ **45.** $2\frac{1}{12}$ gallons
46. $16\frac{7}{15}$ **47.** $15\frac{23}{24}$ pounds **48.** $4\frac{1}{16}$ acres
49.–52.

$\frac{3}{8}$ $\frac{7}{4}$ $2\frac{1}{5}$ $\frac{8}{3}$
0 1 2 3
49. 50.52.51.

53. < **54.** < **55.** <
56. > **57.** < **58.** > **59.** < **60.** >
61. $\frac{1}{4}$ **62.** $\frac{9}{16}$ **63.** $\frac{27}{125}$ **64.** $\frac{81}{4096}$ **65.** $\frac{2}{3}$
66. $5\frac{1}{3}$ **67.** $\frac{4}{9}$ **68.** 2 **69.** $\frac{3}{16}$ **70.** $1\frac{25}{64}$
71. $\frac{13}{15}$ **72.** $\frac{5}{8}$ **73.** $\frac{67}{86}$ **74.** $\frac{11}{16}$ **75.** $4\frac{5}{6}$
76. $24\frac{1}{4}$ **77.** $5\frac{3}{8}$ **78.** $11\frac{43}{80}$ **79.** $65\frac{5}{16}$ **80.** $\frac{8}{11}$
81. $\frac{1}{250}$ **82.** $\frac{1}{2}$ **83.** $\frac{2}{9}$ **84.** $\frac{11}{27}$ **85.** < **86.** <
87. < **88.** > **89.** 132 **90.** 420 **91.** 1008
92. 36 **93.** 108 **94.** 180 **95.** $56\frac{7}{8}$ feet **96.** $3\frac{13}{24}$

CHAPTER 3 TEST (page 209)

1. $\frac{1}{2}$ **2.** $\frac{4}{5}$ **3.** $\frac{1}{2}$ **4.** $\frac{1}{5}$ **5.** 16 **6.** 105
7. 420 **8.** $\frac{31}{48}$ **9.** $\frac{31}{35}$ **10.** $\frac{7}{18}$ **11.** $\frac{7}{40}$
12. $8\frac{5}{8}$ **13.** $2\frac{3}{4}$ **14.** $40\frac{29}{60}$ **15.** $\frac{5}{8}$
16. Answer varies. **17.** Answer varies.
18. $18\frac{1}{4}$ hours **19.** $35\frac{7}{8}$ gallons **20.** <
21. < **22.** $\frac{1}{3}$ **23.** $\frac{13}{48}$ **24.** $1\frac{1}{4}$ **25.** $1\frac{1}{2}$

CUMULATIVE REVIEW 1–3 (page 211)

1. 3 1 **2.** 3 8 **3.** 2850 2800 3000
4. 59,800 59,800 60,000
5. 10,000 + 300 + 50,000 + 50,000 = 110,300; 113,321
6. 20,000 − 10,000 = 10,000; 14,389
7. 1000 × 400 = 400,000; 528,360
8. 100,000 ÷ 40 = 2500; 3211 **9.** 26
10. 1,255,609 **11.** 921 **12.** 2,801,695 **13.** 192
14. 135 **15.** 216 **16.** 632 **17.** 209,322
18. 14,800 **19.** 158 **20.** 2693 R2 **21.** 32 R166
22. 64 feet **23.** 252 square feet **24.** 580 cans
25. 90,000 revolutions **26.** $\frac{3}{10}$ square inch
27. $4000 **28.** $18\frac{3}{8}$ cords **29.** $1454\frac{3}{8}$ feet
30. $2 \cdot 3 \cdot 5$ **31.** $2^4 \cdot 3^2$ **32.** $2 \cdot 5^3$ **33.** 200
34. 144 **35.** 432 **36.** 3 **37.** 8 **38.** 15
39. 7 **40.** 44 **41.** $\frac{1}{48}$ **42.** $\frac{9}{10}$ **43.** $1\frac{35}{192}$
44. proper **45.** proper **46.** improper **47.** $\frac{5}{8}$
48. $\frac{19}{25}$ **49.** $\frac{7}{20}$ **50.** $\frac{1}{2}$ **51.** $\frac{5}{22}$ **52.** $21\frac{1}{4}$
53. $1\frac{1}{8}$ **54.** $2\frac{3}{16}$ **55.** $13\frac{1}{2}$ **56.** $\frac{7}{9}$ **57.** $\frac{15}{16}$
58. $\frac{7}{36}$ **59.** $5\frac{7}{8}$ **60.** $26\frac{7}{24}$ **61.** $2\frac{5}{8}$ **62.** 150
63. 300 **64.** 144 **65.** 40 **66.** 77 **67.** 27
68. 60
69.–72.

$\frac{1}{9}$ $\frac{3}{4}$ $\frac{5}{3}$ $\frac{10}{3}$
0 1 2 3 4
70.69. 71. 72.

73. < **74.** < **75.** <

CHAPTER 4

SECTION 4.1 (page 221)

1. 7, 6, 3 **3.** 4, 1, 7 **5.** 4, 7, 0 **7.** 1, 6, 3
9. 1, 8, 9 **11.** 6, 2, 1
13. ones, tenths, hundredths
15. ones, tenths, hundredths, thousandths
17. tens, ones, tenths, hundredths, thousandths
19. $\frac{7}{10}$ **21.** $13\frac{2}{5}$ **23.** $\frac{7}{20}$ **25.** $\frac{33}{50}$ **27.** $10\frac{17}{100}$
29. $\frac{3}{50}$ **31.** $\frac{41}{200}$ **33.** $5\frac{1}{500}$ **35.** $\frac{343}{500}$
37. five tenths **39.** seventy-eight hundredths

41. one hundred five thousandths
43. twelve and four hundredths
45. one and seventy-five thousandths
47. 6.7 **49.** 0.32 **51.** 420.008 **53.** 0.0703
55. 75.030 **57.** Answer varies **59.** 3-C **61.** 4-A
63. eight thousand six and five thousand one ten-thousandths
65. $625\frac{1071}{2500}$ **67.** 8240, 8200, 8000
69. 19,710; 19,700; 20,000

SECTION 4.2 (page 229)

1. 16.9 **3.** 0.956 **5.** 0.80 **7.** 3.661 **9.** 794.0
11. 0.0980 **13.** 49 **15.** 9.09 **17.** 82.0002
19. \$0.42 **21.** \$1.22 **23.** \$0.30 **25.** \$17,250
27. \$310 **29.** \$379 **31.** Answer varies.
33. \$500 **35.** \$1.00 **37.** \$1000
39. 8000 + 6000 + 8000 = 22,000; 22,223
41. 80,000 + 100 + 800 = 80,900; 82,859

SECTION 4.3 (page 233)

1. 17.72 **3.** 240.742 **5.** 348.513 **7.** 11.98
9. 115.861 **11.** 59.323 **13.** 330.86895
15. Answer varies **17.** 40 + 20 + 8 = 68; 63.65
19. 400 + 1 + 20 = 421; 414.645
21. 60 + 500 + 6 = 566; 608.4363
23. 400 + 600 + 600 = 1600; 1586.308
25. \$300 + \$1 = \$301; \$311.09
27. 5 + 6 + 4 = 15 days; 14.49 days
29. \$7 + \$10 + \$1 = \$18; \$18.20
31. 8000 + 200 + 200 = 8400 miles; 7914.9 miles
33. 10 + 5 = 15 hours; 14.3 hours
35. \$400 + \$300 + \$700 = \$1400; \$1359.82
37. 90.3 miles **39.** \$39.78 **41.** 52.1 inches
43. 300 − 100 = 200; 197 **45.** 7000 − 100 = 6900; 6569

SECTION 4.4 (page 239)

1. 54.3 **3.** 32.566 **5.** 38.554 **7.** 20.104
9. 12.848 **11.** 89.7 **13.** 0.109 **15.** 0.91
17. 6.661 **19.** Answer varies.
21. \$20 − 7 = \$13; \$13.16 **23.** 9 − 4 = 5; 4.849
25. 2 − 2 = 0; 0.019 inches
27. 400 − 9 = 391; 375.194 liters
29. 0.2749 **31.** 5951.2858
33. 40 − 20 = 20 hours; 26.15 hours
35. \$20 − \$9 = \$11; \$10.88
37. 11 − 10 = 1 ounce; 0.65 ounce
39. 50,000 − 30,000 = 20,000 miles; 22,475.8 miles
41. 206.33869 **43.** 2.511317 **45.** \$75.53
47. 20 + 1 = 21; 21 − 9 = 12 gallons; 9.372 gallons
49. b = 1.39 cm **51.** k = 2.812 inches

53. 80 × 30 = 2400; 2324
55. 4000 × 200 = 800,000; 776,745

SECTION 4.5 (page 245)

1. 0.1344 **3.** 159.10 **5.** 155.844 **7.** 21.08667
9. 43.2 **11.** 0.432 **13.** 0.0432 **15.** 0.00432
17. 0.0000312 **19.** 0.000006 **21.** Answer varies.
23. 40 × 5 = 200; 190.08
25. 40 × 40 = 1600; 1558.2 **27.** 7 × 5 = 35; 30.038
29. 3 × 7 = 21; 19.24165 **31.** \$592.37 **33.** \$2.45
35. \$21.45 **37.** \$6461.00 **39.** \$347.52 **41.** \$76.50
43. \$121.30 **45.** 46.3 gallons **47.** \$388.34 **49.** \$4.09
51. $5\overline{)1000}$ (quotient 200); 190 R4 **53.** $20\overline{)20{,}000}$ (quotient 1000); 905 R15

SECTION 4.6 (page 255)

1. 3.9 **3.** 0.47 **5.** 400.2 **7.** 36 **9.** 10.90
11. 0.33 **13.** 25.3 **15.** 516.67 **17.** 24.291
19. 10,082.647 **21.** Answer varies.
23. unreasonable; 40 ÷ 8 = 5; $8\overline{)37.8}$ (quotient 4.725)
25. reasonable; 50 ÷ 50 = 1
27. unreasonable; 300 ÷ 5 = 60; $5.1\overline{)307.02}$ (quotient 60.2)
29. unreasonable; 9 ÷ 1 = 9; $1.25\overline{)9.3}$ (quotient 7.44)
31. \$1.00 **33.** \$19.46
35. \$0.30 **37.** \$5.89 per hour
39. 21.2 miles per gallon **41.** 28.0 ft **43.** 14.25
45. 73.4 **47.** 1.205 **49.** 0.334 **51.** \$0.03
53. \$237.25 **55.** $<$ **57.** $>$ **59.** $<$

SECTION 4.7 (page 263)

1. 0.5 **3.** 0.7 **5.** 0.167 **7.** 0.875 **9.** 0.667
11. 1.4 **13.** 3.571 **15.** 0.88 **17.** 0.688
19. 12.8 **21.** 0.010 **23.** 78.36
25. Answer varies. **27.** Answer varies.
29. $\frac{2}{5}$ **31.** $\frac{3}{8}$ **33.** $\frac{7}{20}$ **35.** 0.35 **37.** $\frac{1}{25}$
39. $\frac{3}{20}$ **41.** 0.2 **43.** $\frac{9}{100}$ **45.** Too much
47. shorter **49.** 0.9991 cm, 1.0007 cm **51.** More
53. 0.5399, 0.54, 0.5455 **55.** 5.0079, 5.79, 5.8, 5.804
57. 0.6009, 0.609, 0.628, 0.62812
59. 2.8902, 3.88, 4.876, 5.8751
61. 0.006, 0.043, $\frac{1}{20}$, 0.051 **63.** 0.37, $\frac{3}{8}$, $\frac{2}{5}$, 0.4001
65. 1.4 in **67.** 0.3 in **69.** 0.4 in
71. $\frac{6}{11}$, $\frac{5}{9}$, 0.571, $\frac{4}{7}$ **73.** 0.25, $\frac{4}{15}$, $\frac{3}{11}$, $\frac{1}{3}$
75. $\frac{1}{6}$, $\frac{3}{16}$, 0.188, $\frac{1}{5}$ **77.** $\frac{3}{4}$ **79.** $\frac{3}{4}$ **81.** $\frac{8}{11}$

CHAPTER 4 REVIEW EXERCISES (page 271)

1. 5, 8 **2.** 0, 2 **3.** 4, 2 **4.** 5, 9 **5.** 7, 6
6. $2\frac{3}{10}$ **7.** $\frac{3}{4}$ **8.** $4\frac{1}{20}$ **9.** $\frac{7}{8}$ **10.** $\frac{27}{1000}$
11. $1\frac{1}{625}$ **12.** eight tenths
13. four hundred and twenty-nine hundredths
14. twelve and seven thousandths
15. three hundred six ten-thousandths **16.** 8.3
17. 0.205 **18.** 70.0066 **19.** 0.30 **20.** 275.6
21. 72.79 **22.** 0.160 **23.** 0.091 **24.** 1.0
25. $15.83 **26.** $0.70 **27.** $17,625.79 **28.** $170
29. $39 **30.** $25 **31.** $10
32. 6 + 400 + 20 = 426; 444.86
33. 80 + 1 + 100 + 1 + 30 = 212; 233.515
34. 300 − 20 = 280; 290.7
35. 9 − 8 = 1; 1.2684
36. 13 − 10 = 3 hours; 2.75 hours
37. $200 + $40 = $240; $260.00
38. $2 + $5 + $20 = $27; $30 − $27 = $3; $4.14
39. 2 + 4 + 5 = 11 kilometers; 11.55 kilometers
40. 6 × 4 = 24; 22.7106
41. 40 × 3 = 120; 141.57
42. 0.0112 **43.** 0.000355
44. reasonable; 90 ÷ 10 = 9
45. unreasonable; 30 ÷ 3 = 10; $2.8\overline{)26.6}$ = 9.5
46. 14.467 **47.** 15.500 **48.** 0.4 **49.** $240
50. $1.38 **51.** 133 shares **52.** $3.12 **53.** 29.215
54. 10.15 **55.** 3.8 **56.** 0.64 **57.** 1.875
58. 0.111 **59.** 3.6008, 3.68, 3.806
60. 0.209, 0.2102, 0.215, 0.22
61. $\frac{1}{8}$, $\frac{3}{20}$, 0.159, 0.17 **62.** 404.865 **63.** 254.8
64. 3583.261 **65.** 29.0898 **66.** 0.03066
67. 9.4 **68.** 175.675 **69.** 9.04 **70.** 19.50
71. 8.19 **72.** 0.928 **73.** 35 **74.** 0.259
75. 0.3 **76.** $38.59 **77.** $4.51
78. 23.15 inches **79.** No **80.** $15.81
81. 1.8 meters **82.** 3.85 pounds **83.** $325

CHAPTER 4 Test (page 275)

1. $3\frac{4}{25}$ **2.** $\frac{1}{40}$ **3.** five and three ten-thousandths
4. seven hundred five thousandths **5.** 725.6
6. 0.630 **7.** $1.49 **8.** $7860
9. 8 + 80 + 40 = 128; 129.2028
10. 80 − 4 = 76; 75.498 **11.** 6 · 1 = 6; 6.948
12. 20 ÷ 5 = 4; 4.175 **13.** 839.762
14. 669.004 **15.** 0.0000483 **16.** 480
17. 3.5, 3.508, 3.51, 3.5108
18. 0.44, $\frac{9}{20}$, 0.4506, 0.451 **19.** 35.49 **20.** $462.87
21. Davida **22.** $5.35 **23.** 2.8 degrees
24. $4.55 per meter **25.** Answer varies.

CUMULATIVE REVIEW 1–4 (page 277)

1. 5, 9, 2 **2.** 0, 5, 8 **3.** 280,000 **4.** 0.391
5. $340 **6.** $0.85
7. 4000 + 600 + 9000 = 13,600; 13,339
8. 4 + 16 + 1 = 21; 20.683
9. 5000 − 2000 = 3000; 3209
10. 50 − 7 = 43; 44.506
11. 3000 × 200 = 600,000; 550,622
12. 7 × 7 = 49; 49.786
13. 100,000 ÷ 50 = 2000; 2690 **14.** 40 ÷ 8 = 5; 4.5
15. 9.671 **16.** $1\frac{4}{9}$ **17.** 73,225
18. $1\frac{2}{5}$ **19.** 4914 **20.** 93.603 **21.** 404 R3
22. $1\frac{17}{24}$ **23.** 233,728 **24.** 0.03264 **25.** 8
26. 45 **27.** $\frac{4}{31}$ **28.** $\frac{2}{3}$ **29.** 0.51 **30.** 22
31. 14 **32.** $\frac{1}{4}$ **33.** 20.81 **34.** 576 **35.** 14
36. $2^3 \cdot 5^2$ **37.** forty and thirty-five thousandths
38. 0.0306 **39.** $\frac{1}{8}$ **40.** $3\frac{2}{25}$ **41.** 2.6 **42.** 0.636
43. > **44.** 7.005, 7.5, 7.5005, 7.505
45. 0.8, 0.8015, $\frac{21}{25}$, $\frac{7}{8}$ **46.** $11.17 **47.** $3\frac{3}{8}$ inches
48. $144.05 **49.** 488 students **50.** $6\frac{5}{24}$ yards
51. 84 children **52.** $0.95 **53.** $194.50
54. $556 **55.** Too long.

CHAPTER 5

SECTION 5.1 (page 287)

1. $\frac{8}{9}$ **3.** $\frac{2}{1}$ **5.** $\frac{1}{3}$ **7.** $\frac{8}{5}$ **9.** $\frac{3}{8}$ **11.** $\frac{3}{5}$
13. $\frac{6}{5}$ **15.** $\frac{5}{6}$ **17.** $\frac{8}{5}$ **19.** $\frac{1}{12}$ **21.** $\frac{5}{16}$ **23.** $\frac{4}{1}$
25. $\frac{7}{2}$ **27.** $\frac{5}{4}$ **29.** $\frac{1}{100}$ **31.** Answer varies.
33. $\frac{2}{1}$ **35.** $\frac{3}{8}$ **37.** $\frac{7}{5}$ **39.** $\frac{6}{1}$ **41.** $\frac{38}{17}$ **43.** $\frac{1}{2}$
45. $\frac{34}{35}$

47. Answer varies. Some possibilities are 8 and 10, 12 and 15, 16 and 20.
49. 0.093 **51.** 1.025 **53.** 9.465

SECTION 5.2 (page 295)

1. $\frac{5 \text{ cups}}{3 \text{ people}}$ **3.** $\frac{3 \text{ feet}}{7 \text{ seconds}}$ **5.** $\frac{1 \text{ person}}{2 \text{ dresses}}$
7. $\frac{5 \text{ letters}}{1 \text{ minute}}$ **9.** $\frac{\$21}{2 \text{ visits}}$ **11.** $\frac{18 \text{ miles}}{1 \text{ gallon}}$
13. \$12 per hour or \$12/hour
15. 5 eggs per chicken or 5 eggs/chicken
17. 1.25 pounds/person **19.** \$103.30/day
21. 325.9; 21.0 **23.** 338.6; 20.9
25. 4 oz for \$0.89 **27.** 15 ounces for \$2.10
29. 18 ounces for \$1.41 **31.** Answer varies.
33. 1.75 pounds/week **35.** \$12.26/hour
37. \$11.50/share **39.** 0.25 crate/min or $\frac{1}{4}$ crate/min
4 min/crate
41. \$11.25/yard **43.** 11 **45.** 108 **47.** $3\frac{1}{2}$

SECTION 5.3 (page 301)

1. $\frac{\$9}{12 \text{ cans}} = \frac{\$18}{24 \text{ cans}}$ **3.** $\frac{20 \text{ dogs}}{45 \text{ cats}} = \frac{4 \text{ dogs}}{9 \text{ cats}}$
5. $\frac{120}{150} = \frac{8}{10}$ **7.** $\frac{2.2}{3.3} = \frac{3.2}{4.8}$ **9.** $\frac{1\frac{1}{2}}{8} = \frac{6}{32}$
11. true **13.** true **15.** false **17.** true
19. true **21.** false **23.** True **25.** False
27. False **29.** True **31.** False **33.** True
35. True **37.** Answer varies. **39.** False **41.** $\frac{2}{3}$
43. $1\frac{1}{4}$ **45.** $\frac{4}{7}$ **47.** $6\frac{1}{2}$

SECTION 5.4 (page 309)

1. 4 **3.** 2 **5.** 88 **7.** 40 **9.** 5 **11.** 10
13. 24.44 **15.** 50.4 **17.** 17.64
19. $\frac{6.67}{4} = \frac{5}{3}$ or $\frac{10}{6} = \frac{5}{3}$ or $\frac{10}{4} = \frac{7.5}{3}$ or $\frac{10}{4} = \frac{5}{2}$
21. 1 **23.** $3\frac{1}{2}$ **25.** False; $\frac{25 \text{ feet}}{18 \text{ sec}} = \frac{15 \text{ feet}}{10 \text{ sec}}$
27. True; $\frac{170 \text{ miles}}{6.8 \text{ gallons}} = \frac{330 \text{ miles}}{13.2 \text{ gallons}}$

SECTION 5.5 (page 313)

1. 22.5 hours **3.** \$35 **5.** 7 pounds **7.** \$153.45
9. 14 feet, 10 feet **11.** 3 runs **13.** 190 people
15. \$5.50 **17.** 625 stocks **19.** 4.6 hours
21. 4.06 meters **23.** 10.53 meters
25. Answer varies. **27.** $4\frac{3}{8}$ cups **29.** $7\frac{1}{2}$ tbsp
31. 4700 students **33.** 6 **35.** 2870
37. 0.0193

CHAPTER 5 REVIEW EXERCISES (page 321)

1. $\frac{3}{11}$ **2.** $\frac{19}{7}$ **3.** $\frac{3}{2}$ **4.** $\frac{9}{5}$ **5.** $\frac{2}{1}$ **6.** $\frac{2}{3}$
7. $\frac{1}{4}$ **8.** $\frac{3}{1}$ **9.** $\frac{3}{1}$ **10.** $\frac{3}{8}$ **11.** $\frac{5}{18}$ **12.** $\frac{24}{5}$
13. $\frac{10}{7}$ **14.** $\frac{7}{5}$ **15.** $\frac{5}{6}$ **16.** $\frac{5}{9}$
17. $\frac{\$5}{1 \text{ hour}}$ **18.** $\frac{145 \text{ miles}}{3 \text{ hours}}$ **19.** Answer varies.
20. Answer varies.
21. 0.2 pages/minute or $\frac{1}{5}$ page/minute; 5 minutes/page
22. \$8/hour; 0.125 hours/dollar or $\frac{1}{8}$ hour/dollar
23. 16 oz for \$2.80 **24.** 25 pounds for \$9.40
25. $\frac{5}{10} = \frac{20}{40}$ **26.** $\frac{7}{2} = \frac{35}{10}$ **27.** $\frac{1\frac{1}{4}}{5} = \frac{3}{12}$
28. true **29.** false **30.** false **31.** true
32. true **33.** true **34.** 1575 **35.** 20
36. 400 **37.** 12.5 **38.** 14.67 **39.** 8.17
40. 50.4 **41.** 0.57 **42.** 2.47
43. 27 cats **44.** 46 hits **45.** \$22.06
46. 3299 students **47.** 68 feet **48.** 14.7 milligrams
49. 22 hours **50.** 21 hours **51.** 105 **52.** 0
53. 128 **54.** 23.08 **55.** 6.5 **56.** 117.36
57. False **58.** False **59.** True **60.** $\frac{8}{5}$
61. $\frac{33}{80}$ **62.** $\frac{15}{4}$ **63.** $\frac{4}{1}$ **64.** $\frac{4}{5}$ **65.** $\frac{37}{7}$
66. $\frac{3}{8}$ **67.** $\frac{1}{12}$ **68.** $\frac{45}{13}$ **69.** 24,900 **70.** $\frac{8}{3}$
71. 75 feet for \$1.95 **72.** 762.5 feet
73. $\frac{1}{2}$ teaspoon or 0.5 teaspoon **74.** 21 points
75. Answer varies. **76.** 7.5 hours or $7\frac{1}{2}$ hours

CHAPTER 5 TEST (page 325)

1. $\frac{2}{3}$ **2.** $\frac{20 \text{ miles}}{1 \text{ gallon}}$ **3.** $\frac{\$1}{5 \text{ minutes}}$
4. $\frac{15}{4}$ **5.** $\frac{1}{80}$ **6.** $\frac{9}{2}$ **7.** 28 ounces for $3.15
8. Answer varies. **9.** False **10.** True **11.** 25
12. 2.67 **13.** 325 **14.** $10\frac{1}{2}$ **15.** 576 words
16. 6.4 hours **17.** 87 students **18.** Answer varies.
19. 23.8 grams **20.** 60 feet

CUMULATIVE REVIEW 1–5 (page 327)

1. 5 3 6 **2.** 9 0 5 **3.** 17,000 **4.** 0.05
5. $79 **6.** $2.56
7. $30 + 5000 + 400 = 5430$; 5585
8. $60 - 6 = 54$; 57.408
9. $5000 \times 800 = 4{,}000{,}000$; 3,791,664
10. $1 \times 18 = 18$; 17.4796
11. $50{,}000 \div 50 = 1000$; 907
12. $2000 \div 5 = 400$; 364
13. $6\frac{3}{5}$ **14.** 374,416 **15.** 29.34
16. 610 R 27 **17.** $\frac{1}{6}$ **18.** 68.381
19. $5\frac{14}{15}$ **20.** 35,852,728 **21.** 55.6 **22.** 2312
23. $13\frac{2}{5}$ **24.** 0.0076 **25.** 39 **26.** 18 **27.** 64
28. 0.95 **29.** one hundred five ten-thousandths
30. 60.071 **31.** 0.313 **32.** 4.778
33. 0.07, 0.0711, 0.7, 0.707 **34.** 0.305, $\frac{1}{3}, \frac{7}{20}, \frac{3}{8}$
35. $\frac{4}{1}$ **36.** $\frac{\$13}{2 \text{ hours}}$ **37.** $\frac{1}{12}$ **38.** $\frac{1}{3}$ **39.** $\frac{11}{5}$
40. 36 servings for $3.24 **41.** 21 **42.** 17.14
43. $11\frac{1}{4}$ **44.** 0.98 **45.** 250 pounds
46. 26.7 centimeters **47.** 8400 students **48.** $78
49. $\frac{2}{7}$ **50.** $31,500 **51.** $2\frac{9}{20}$ miles
52. 18.0 miles per gallon **53.** 200 residents
54. $1\frac{1}{4}$ teaspoons

CHAPTER 6

SECTION 6.1 (page 335)

1. 0.50 or 0.5 **3.** 0.18 **5.** 0.45 **7.** 1.40 or 1.4
9. 0.078 **11.** 1.00 or 1 **13.** 0.005 **15.** 0.0035
17. 60% **19.** 91% **21.** 7% **23.** 12.5%
25. 62.9% **27.** 200% **29.** 260% **31.** 3.12%
33. 416.2% **35.** 0.75% **37.** Answer varies
39. 0.07 **41.** 0.65 **43.** 8.6% **45.** 200%
47. 0.5% **49.** 1.536 **51.** 95%; 5%
53. 30%; 70% **55.** 75%; 25% **57.** 55%; 45%
59. 0.5 **61.** 0.75 **63.** 0.875 **65.** 0.8

SECTION 6.2 (page 345)

1. $\frac{1}{4}$ **3.** $\frac{1}{10}$ **5.** $\frac{17}{20}$ **7.** $\frac{3}{8}$ **9.** $\frac{1}{16}$ **11.** $\frac{1}{6}$
13. $\frac{1}{15}$ **15.** $\frac{1}{250}$ **17.** $1\frac{3}{10}$ **19.** $2\frac{1}{2}$ **21.** 25%
23. 30% **25.** 40% **27.** 37% **29.** 62.5%
31. 87.5% **33.** 52% **35.** 58% **37.** 5%
39. 83.3% **41.** 55.6% **43.** 14.3% **45.** $\frac{1}{10}$, 10%
47. $\frac{1}{8}$, 0.125 **49.** $\frac{1}{5}$, 20% **51.** 0.167, 16.7%
53. $\frac{1}{4}$, 25% **55.** $\frac{7}{8}$, 0.875 **57.** 0.667, 66.7%
59. 0.75, 75% **61.** 0.01, 1% **63.** 0.005, 0.5%
65. $2\frac{1}{2}$, 250% **67.** 3.25, 325% **69.** Answer varies.
71. $\frac{19}{25}$; 0.76; 76% **73.** $\frac{1}{5}$; 0.2; 20%
75. $\frac{3}{5}$; 0.6; 60% **77.** $\frac{9}{20}$; 0.45; 45%
79. $\frac{1}{10}$; 0.1; 10% **81.** $\frac{1}{4}$; 0.25; 25%
83. $\frac{2}{5}$; 0.4; 40% **85.** 40 **87.** 5 **89.** 10

SECTION 6.3 (page 353)

1. 80 **3.** 450 **5.** 40 **7.** 416.7 **9.** 50%
11. 300% **13.** 33.3% **15.** 26 **17.** 21.6
19. 13.2 **21.** Answer varies. **23.** 356 **25.** 0.4
27. 8.5 **29.** 12.25 **31.** 75% **33.** 57%
35. 87.5% **37.** 66.7%

SECTION 6.4 (page 357)

1. 25, unknown, 200 **3.** 81, unknown, 748
5. 15, 75, unknown **7.** 72, unknown, 18
9. unknown, 52, 26 **11.** unknown, 148, 88.8
13. 6.5, unknown, 27.17 **15.** .68, 487, unknown
17. Answer varies. **19.** p is unknown; 810; 640
21. p is unknown; 15; 0.90

23. 7; 1500; *a* is unknown **25.** 25; *b* is unknown; 240
27. 18; 110; *a* is unknown **29.** 32; 272; *a* is unknown
31. 16.8; *b* is unknown; 504

33. $\frac{7}{x} = \frac{5}{10}$, 14 **35.** $\frac{x}{36} = \frac{\frac{4}{3}}{12}$, 4

SECTION 6.5 (page 367)

1. 52 adults **3.** 109.2 meters **5.** 4.8 feet
7. 57 lamps **9.** 247.5 boxes **11.** $3.28
13. 1850 sales **15.** 42.625 pounds **17.** $21.60
19. 200 trucks **21.** 304 **23.** 420 envelopes
25. 680 books **27.** 2800 **29.** 50% **31.** 52%
33. 2% **35.** 1.5% **37.** 18.6% **39.** 9.2%
41. Answer varies. **43.** 54 children **45.** 836 drivers
47. $185,500 **49.** 220 students **51.** 15.5%
53. 1.5% **55.** 110% **57.** 248.0 million
59. $2,850,000,000 **61.** $825,000,000 **63.** $645
65. 2156 products **67.** 40.8425 **69.** 10.0936
71. 3600 **73.** 0.25

SECTION 6.6 (page 377)

1. $110 **3.** 765 fish **5.** 83.2 quarts **7.** 630 sales
9. 1029.2 meters **11.** $4.16 **13.** 200 apartments
15. 126 cookies **17.** 680 wires **19.** 1080 people
21. 300 gallons **23.** 50% **25.** 76% **27.** 125%
29. 1.5% **31.** 135% **33.** Answer varies.
35. 82.3 million homes **37.** 13 clients **39.** 4.9%
41. 700 gallons **43.** 478,175 Mustangs
45. $510,390 **47.** $11,900 **49.** 15, 375, 56.25
51. 36, unknown, 18 **53.** unknown, 830, 128.65

SECTION 6.7 (page 385)

1. $5, $105 **3.** 4%, $78 **5.** $21.90, $386.90
7. $12.10, $232.10 **9.** $16 **11.** 22%
13. $173.49 **15.** $4525 **17.** $25, $75
19. 25%, $189 **21.** $4.38, $13.12
23. $3.75, $33.75 **25.** Answer varies. **27.** $43.50
29. $31.50 **31.** 4% **33.** 32.5% **35.** 22.5%
37. $12 **39.** 6% **41.** $413; $767 **43.** 32.5%
45. $18.43 **47.** $4276.97 **49.** $11,943.70
51. 42 screwdrivers **53.** $2.88 **55.** 800
57. 32%

SECTION 6.8 (page 393)

1. $5 **3.** $144 **5.** $2.40 **7.** $810 **9.** $82
11. $72 **13.** $12 **15.** $37.50 **17.** $131.20
19. $18.75 **21.** $17.40 **23.** $338.33 **25.** $420
27. $685.10 **29.** $1725 **31.** $2535.75 **33.** $1869
35. $19,170.67 **37.** Answer varies.
39. $199.50 **41.** $780 **43.** $1025 **45.** $176.46
47. $159.50 **49.** $12,254.69 **51.** 6 **53.** 16
55. 25 **57.** 60

SECTION 6.9 (page 403)

1. $1262.50 **3.** $6205.20 **5.** $10,976.01
7. $2025.80 **9.** $2183.58 **11.** $4160.52
13. $24,983.54 **15.** $1372.80, $372.80
17. $2293.75, $1023.75 **19.** $2031.74, $551.74
21. $12,463.99, $4763.99
23. $2519.40, $2530.60, $2536.40
25. $28,344.80, $28,668.64, $28,839.36 **27.** quarterly
29. Answer varies. **31.** Answer varies.
33. **(a)** $2625 **(b)** $2668 **35.** $7845.55
37. **(a)** $11,901.75 **(b)** $4401.75
39. **(a)** $15,869; $16,084 **(b)** $215
41. **(a)** $40,223.50 **(b)** $10,223.50 **43.** 4.65
45. 106.3 **47.** 0.14

CHAPTER 6 REVIEW EXERCISES (page 411)

1. 0.5 **2.** 2.5 **3.** 0.137 **4.** 0.00085
5. 375% **6.** 2% **7.** 37.5% **8.** 0.2%
9. $\frac{6}{25}$ **10.** $\frac{1}{16}$ **11.** $3\frac{1}{4}$ **12.** $\frac{1}{4000}$ **13.** 80%
14. 62.5% **15.** 150% **16.** 0.25% **17.** 0.125
18. 12.5% **19.** $\frac{3}{20}$ **20.** 15% **21.** $2\frac{2}{5}$ **22.** 2.4
23. 1000 **24.** 96 **25.** 40; 150; 60
26. unknown; 90; 73 **27.** 46; 1040; unknown
28. 30; unknown; 418 **29.** unknown; 8; 3
30. 88; 1280; unknown **31.** 135 refrigerators
32. 1755 telephones **33.** 43.2 miles
34. 2.8 kilograms **35.** 500 tablets **36.** 2320 vials
37. 242 meters **38.** 17,000 cases **39.** 50%
40. 4.1% **41.** 9.5% **42.** 30.8% **43.** $8480
44. 4.4% **45.** $25.96 **46.** 80 dumpsters
47. 0.4% **48.** 175% **49.** 120 miles **50.** $143.75
51. $8.40; $218.40 **52.** $6.05; $116.05 **53.** $78
54. $4170 **55.** $10; $90 **56.** $146.50, $586
57. $16 **58.** $281.25 **59.** $3.50 **60.** $98.40
61. $397.25 **62.** $1203.60 **63.** $7063.79; 3513.79
64. $1609.08; $289.08 **65.** $5326.44; $1126.44
66. $18,991.35; $8491.35 **67.** 48 **68.** 1640
69. 23.28 meters **70.** 150% **71.** $0.51
72. 165 teachers **73.** 40% **74.** 249.4 liters
75. 0.25 **76.** 1 **77.** 400% **78.** 715%
79. 0.085 **80.** 62.1% **81.** 0.00375 **82.** 0.06%

83. 50% **84.** $\frac{19}{50}$ **85.** $\frac{7}{8}$ **86.** 37.5% **87.** $\frac{13}{40}$
88. 20% **89.** $\frac{1}{200}$ **90.** 175% **91.** $351.45
92. $1960.20 **93.** 90% **94.** **(a)** $15,835 **(b)** $3335
95. $3750 **96.** 8.1% **97.** 13.1% **98.** $7970
99. $7123 **100.** **(a)** $15,762.32 **(b)** $5962.32

CHAPTER 6 TEST (page 417)

1. 0.35 **2.** 0.0005 **3.** 2.00 or 2 **4.** 62.5%
5. 170% **6.** 80% **7.** $\frac{3}{8}$ **8.** $\frac{1}{400}$ **9.** 25%
10. 62.5% **11.** 250% **12.** 640 pens
13. 20% **14.** $16,800 **15.** $2854.20
16. $246.40 **17.** 35% **18.** Answer varies.
19. Answer varies. **20.** $3.84; $44.16
21. $68.25; $113.75 **22.** $516.95 **23.** $101.25
24. $2996
25. $3200 for 2 years at 8% quarterly; $110.19

CUMULATIVE REVIEW 1–6 (page 419)

1. 10,000 + 70 + 600 = 10,670; 10,524
2. 1 + 30 + 4 = 35; 34.738
3. 60,000 − 50,000 = 10,000; 9993
4. 6 − 3 = 3; 3.4945
5. 8000 × 600 = 4,800,000; 4,672,647
6. 30 × 9 = 270; 277.95
7. 40,000 ÷ 40 = 1000; 902
8. 2000 ÷ 8 = 250; 320 **9.** 7 ÷ 1 = 7; 8.45
11. 8 **12.** 39 **13.** 7900 **14.** 5,700,000
15. $375 **16.** $451.83 **17.** $1\frac{3}{8}$ **18.** $1\frac{1}{6}$
19. $13\frac{1}{15}$ **20.** $\frac{3}{8}$ **21.** $1\frac{5}{6}$ **22.** $8\frac{8}{15}$ **23.** $\frac{5}{8}$
24. $26\frac{5}{32}$ **25.** $28\frac{4}{5}$ **26.** $\frac{8}{9}$ **27.** 25
28. $\frac{11}{30}$ **29.** < **30.** < **31.** > **32.** $\frac{1}{24}$
33. $1\frac{1}{4}$ **34.** $\frac{1}{4}$ **35.** 0.4 **36.** 0.875 **37.** 0.85
38. 0.857 **39.** $\frac{3}{1}$ **40.** $\frac{3}{2}$ **41.** $\frac{1}{8}$ **42.** true
43. true **44.** 3 **45.** 3 **46.** 18 **47.** 30
48. 0.25 **49.** 1.397 **50.** 3.00 or 3 **51.** 0.0262
52. 68% **53.** 270% **54.** 2.3% **55.** $\frac{1}{25}$
56. $\frac{3}{8}$ **57.** $1\frac{1}{2}$ **58.** 87.5% or $87\frac{1}{2}$% **59.** 5%
60. 275% **61.** 280 folders **62.** $388.80
63. 180 cans **64.** 1700 miles **65.** 50% **66.** 40%
67. $2.16; $56.15 **68.** $27.44; $419.44
69. $731.10 **70.** $12,191.04 **71.** $53.20; $98.80
72. $53.66; $184.84 **73.** $842.52 **74.** $19,863.88
75. 28 watches **76.** 59.5 ounces **77.** 75%
78. 35% **79.** $48,000
80. **(a)** $7005.80 **(b)** $1255.80

CHAPTER 7

SECTION 7.1 (page 431)

1. 3 **3.** 2 **5.** 5280 **7.** 2000 **9.** 60
11. 2 **13.** 2 **15.** 80,000 to 90,000 pounds
17. 120 **19.** 112 **21.** 10 **23.** $1\frac{1}{2}$ or 1.5
25. $\frac{1}{4}$ or 0.25 **27.** $1\frac{1}{2}$ or 1.5 **29.** $2\frac{1}{2}$ or 2.5
31. $\frac{1}{2}$ day **33.** 5000 **35.** 17 **37.** 28 ounces
39. 216 **41.** 28 **43.** 518,400 **45.** 48,000
47. **(a)** pound/ounces **(b)** quarts/pints or pints/cups
(c) min/hr or sec/min **(d)** feet/inches
49. 174,240 **51.** 800 **53.** < **55.** > **57.** >

SECTION 7.2 (page 441)

1. 1000 **3.** $\frac{1}{1000}$ or 0.001 **5.** $\frac{1}{100}$ or 0.01
7. Answer varies—about 8 cm.
9. Answer varies—about 20 mm. **11.** cm **13.** m
15. km **17.** mm **19.** cm **21.** m
23. Answer varies. **25.** 700 cm **27.** 0.035 m
29. 9400 m **31.** 5.09 m **33.** 40 cm
35. 910 mm **37.** less; 18 cm or 0.18 m
39. 98,100,000 cm **41.** 0.0000056 km
43. $\frac{7}{8}$ **45.** $\frac{2}{25}$

SECTION 7.3 (page 449)

1. mL **3.** L **5.** kg **7.** g **9.** mL **11.** mg
13. L **15.** kg **17.** unreasonable
19. unreasonable **21.** reasonable **23.** unreasonable
25. Answer varies. **27.** Answer varies.
29. 15,000 mL **31.** 3 L **33.** 0.925 L
35. 0.008 L **37.** 4150 mL **39.** 8 kg **41.** 5200 g
43. 850 mg **45.** 30 g **47.** 0.598 g **49.** 0.06 L
51. 0.003 kg **53.** 990 mL **55.** 2000 mL
57. 0.95 kg **59.** greater; 5 mg or 0.005 g
61. 200 nickels **63.** 0.2 gram **65.** 7, 1, 8, 2

SECTION 7.4 (page 455)

1. $1.33 rounded to the nearest cent **3.** 9.75 kg
5. 25 servings **7.** 5.03 m **9.** 22.5 L
11. $358.50 **13.** $18 case **15.** 0.63 **17.** 5.733
19. 5.7225

SECTION 7.5 (page 461)

1. 21.8 yards **3.** 262.4 feet **5.** 4.8 m
7. 5.3 ounces **9.** 111.6 kg **11.** 30.3 quarts
13. 68.0 L **15.** 28°F **17.** 40°C **19.** 150°C
21. more; explanation varies. **23.** 16°C **25.** 40°C
27. 46°F **29.** 95°F **31.** 58°C **33.** 932°F
35. \$9.72 if gallons converted to liters; \$9.57 if liters converted to gallons **37.** 32 **39.** 81 **41.** 41

CHAPTER 7 REVIEW EXERCISES (page 467)

1. 16 **2.** 3 **3.** 2000 **4.** 24 **5.** 60 **6.** 2
7. 4 **8.** 5280 **9.** 12 **10.** 7 **11.** 60 **12.** 2
13. 48 **14.** 15,840 **15.** 4 **16.** 3
17. $2\frac{1}{2}$ or 2.5 **18.** $5\frac{1}{2}$ or 5.5
19. $\frac{3}{4}$ or 0.75 **20.** $2\frac{1}{4}$ or 2.25 **21.** 126,720
22. 128,000 **23.** 112 **24.** 345,600 **25.** mm
26. cm **27.** km **28.** m **29.** cm **30.** mm
31. 500 cm **32.** 12,000 m **33.** 8.5 cm
34. 3.7 m **35.** 6100 m **36.** 930 mm
37. mL **38.** L **39.** g **40.** kg **41.** L
42. mL **43.** mg **44.** g **45.** 5L
46. 8000 mL **47.** 4580 mg **48.** 700 g
49. 0.006g **50.** 0.035 L **51.** 31.5 L
52. 357 g **53.** 87.25 kg **54.** \$3.01
55. 10.9 yards **56.** 54.6 inches **57.** 67.0 miles
58. 1288 km **59.** 21.9 L **60.** 44.0 quarts
61. 0°C **62.** 100°C **63.** 37°C **64.** 20°C
65. 25°C **66.** 33°C **67.** 54°F **68.** 122°F
69. L **70.** kg **71.** cm **72.** mL **73.** mm
74. km **75.** g **76.** m **77.** mL **78.** g
79. mg **80.** L **81.** 27.5 mm
82. $\frac{3}{4}$ hour or 0.75 hour **83.** $7\frac{1}{2}$ feet or 7.5 feet
84. 130 cm **85.** 77°F **86.** 48 quarts
87. 0.7 g **88.** 810 mL **89.** 80 ounces
90. 60,000 g **91.** 1800 mL **92.** 30°C
93. 36 cm **94.** 0.055 L **95.** 1.26 m **96.** \$48.06
97. 91 meters **98.** 260 gallons

CHAPTER 7 TEST (page 471)

1. 42 **2.** 12 **3.** 9.5 or $9\frac{1}{2}$ **4.** 5.25 or $5\frac{1}{4}$
5. 128 **6.** 14,400 **7.** kg **8.** km **9.** mL
10. g **11.** cm **12.** mm **13.** L **14.** cm
15. 2.5 m **16.** 4600 m **17.** 0.5 cm **18.** 0.325 g
19. 16,000 mL **20.** 400 g **21.** 1055 cm
22. 0.095 L **23.** 6.32 kg **24.** 6.75 m **25.** 95°C
26. 0°C **27.** 5.5 m **28.** 32.4 kg
29. 33.2 ounces **30.** 10.3 km **31.** 23°C
32. 36°F **33.** Answer varies. **34.** Answer varies.

CUMULATIVE REVIEW 1–7 (page 473)

1. 100 + 3 + 70 = 173; 178.838
2. 30,000 − 800 = 29,200; 30,178
3. 90,000 × 200 = 18,000,000; 16,849,725
4. 60 × 5 = 300; 265.644
5. 20,000 ÷ 40 = 500; 530
6. 40 ÷ 8 = 5; 4.6 **7.** $5\frac{17}{24}$ **8.** 488,045 **9.** $5\frac{1}{3}$
10. 13.0 **11.** $\frac{8}{15}$ **12.** 7.0793 **13.** $\frac{5}{8}$
14. 307 R38 **15.** 0.00762 **16.** 5 **17.** 209
18. three hundred seven and nineteen hundredths
19. 0.0082 **20.** 0.083, 0.8033, 8.03, 8.3
21. 0.875 **22.** 87.5% or $87\frac{1}{2}$% **23.** $\frac{1}{25}$ **24.** 4%
25. $3\frac{1}{2}$ **26.** 3.5 **27.** $\frac{11}{1}$ **28.** $\frac{3}{8}$ **29.** $\frac{7}{4}$
30. 4 muffins for \$1.77 **31.** 12 **32.** 1.67 (rounded)
33. 5% **34.** 30 hours **35.** 72 inches
36. $1\frac{1}{4}$ or 1.25 hours **37.** 280 cm **38.** 0.065 g
39. 122.76 miles **40.** 10°C **41.** g **42.** cm
43. m **44.** kg **45.** mg **46.** mL **47.** 38°C
48. 12°C **49.** \$4.87; \$69.82 **50.** \$0.43
51. \$9.74 **52.** 93.6 km **53.** \$11,506.25
54. 88.6% **55.** 1 pound under the limit **56.** $\frac{1}{6}$ cup
57. \$56.70; \$132.30 **58.** 5.97 kg **59.** 120 rows
60. 2100 students

CHAPTER 8

SECTION 8.1 (page 483)

1. line named $\overleftrightarrow{CD}$ or $\overleftrightarrow{DC}$
3. line segment named $\overline{GF}$ or $\overline{FG}$ **5.** ray named $\overrightarrow{PQ}$
7. perpendicular **9.** parallel **11.** intersecting
13. $\angle AOS$ or $\angle SOA$ **15.** $\angle CRT$ or $\angle TRC$
17. $\angle AQC$ or $\angle CQA$ **19.** right (90°)
21. acute **23.** straight (180°)
25. Answer varies. **27.** true **29.** false
31. 144 **33.** 90 **35.** 68

SECTION 8.2 (page 489)

1. $\angle EOD$ and $\angle COD$; $\angle AOB$ and $\angle BOC$
3. $\angle HNE$ and $\angle ENF$; $\angle HNG$ and $\angle GNF$
$\angle HNE$ and $\angle HNG$; $\angle ENF$ and $\angle GNF$
5. 50° **7.** 7° **9.** 50° **11.** 84°
13. $\angle SON \cong \angle TOM$; $\angle TOS \cong \angle MON$
15. 63° **17.** 37° **19.** Answer varies.
21. $\angle ABF \cong \angle ECD$ Both are 138°.
$\angle ABC \cong \angle BCD$ Both are 42°.
23. 7 **25.** 14 **27.** 23

SECTION 8.3 (page 497)

1. $P = 28$ yd, $A = 48$ yd^2
3. $P = 3.6$ km, $A = 0.81$ km^2
5. $P = 22$ ft, $A = 10$ ft^2
7. $P = 29$ m, $A = 7$ m^2
9. $P = 196.2$ ft, $A = 1674.2$ ft^2
11. $P = 16$ in., $A = 16$ in.^{2}
13. $P = 38$ m, $A = 39$ m^2
15. $P = 98$ m, $A = 492$ m^2
17. $P = 78$ in., $A = 234$ in.^{2}
19. \$360 **21.** \$94.81 **23.** Answer varies.
25. Sketches vary. $A = 6528$ ft^2
27. 0.5 m **29.** 144 in. **31.** 75 yd

SECTION 8.4 (page 503)

1. 208 m **3.** 207.2 m **5.** 5.78 km **7.** 984 yd^2
9. 19.25 ft^2 **11.** 3099.6 cm^2 **13.** \$940.50
15. Answer varies. **17.** 3.02 m^2 **19.** 25,344 ft^2
21. 52 **23.** $13\frac{1}{3}$

SECTION 8.5 (page 509)

1. 27 yd **3.** 60 cm **5.** 1980 m^2 **7.** 302.46 cm^2
9. 198 m^2 **11.** 1196 m^2 **13.** 32° **15.** 48°
17. Answer varies. **19.** **(a)** 32 m^2 **(b)** 13.5 m^2
21. 50.24 **23.** 0.64

SECTION 8.6 (page 517)

1. 94 m **3.** 0.425 km
5. $C = 69.1$ ft, $A = 379.9$ ft^2
7. $C = 8.2$ m, $A = 5.3$ m^2
9. $C = 47.1$ cm, $A = 176.6$ cm^2
11. $C = 23.6$ m, $A = 44.2$ m^2
13. $C = 27.2$ km, $A = 58.7$ km^2 **15.** 219.8 cm
17. 785.0 ft **19.** 70,650 mi^2 **21.** Answer varies.
23. 57 cm^2 **25.** 197.8 cm^2 **27.** \$1170.33 (rounded)
29. 3.5 m **31.** 0.5 **33.** 2.75 **35.** 0.667
37. 10.5

SECTION 8.7 (page 527)

1. 550 cm^3 **3.** 150 in.^{3} **5.** 44,579.6 m^3
7. 3617.3 in.^{3} **9.** 471 ft^3 **11.** 410.3 cm^3
13. 418.7 m^3 **15.** 800 cm^3 **17.** 164.6 cm^3
19. Explanations vary. Correct answer is 192.3 cm^3.
21. 513 cm^3 **23.** 64 **25.** 4 **27.** 8

SECTION 8.8 (page 533)

1. 7 **3.** 8 **5.** 3.317 **7.** 2.828 **9.** 8.544
11. 10.050 **13.** 13.784 **15.** 13.229
17. $a^2 = 36$; $b^2 = 64$; $c^2 = 100$ side; $c = 10$; true
19. 39 ft **21.** 17 in. **23.** 6 in.
25. $\sqrt{73} = 8.544$ in. **27.** $\sqrt{65} = 8.062$ yd
29. $\sqrt{195} = 13.964$ cm **31.** $\sqrt{7.94} = 2.818$ m
33. $\sqrt{65.01} = 8.063$ cm **35.** $\sqrt{18.27} = 4.274$ km
37. $\sqrt{65} = 8.1$ ft **39.** $\sqrt{900} = 30$ m
41. Sketches vary. $\sqrt{135} = 11.6$ ft
43. Answer varies. **45.** $BC = 6.25$ ft; $BD = 3.75$ ft
47. 8 **49.** 18.4

SECTION 8.9 (page 541)

1. similar **3.** not similar **5.** similar
7. B and Q; C and R; A and P; AB and PQ; BC and QR; AC and PR
9. P and S; N and R; M and Q; MP and QS; MN and QR; NP and RS
11. $\frac{3}{2}; \frac{3}{2}; \frac{3}{2}$ **13.** $a = 5$ mm; $b = 3$ mm
15. $a = 6$ cm; $b = 15$ cm **17.** $x = 24.8$ m; $y = 15$ m
19. $n = 24$ ft **21.** Answer varies. **23.** $n = 110$ m
25. 50 m **27.** 8 **29.** 10 **31.** 15

CHAPTER 8 REVIEW EXERCISES (page 551)

1. line segment named $\overline{AB}$ or $\overline{BA}$
2. line named $\overleftrightarrow{CD}$ or $\overleftrightarrow{DC}$ **3.** ray named $\overrightarrow{OP}$
4. parallel **5.** perpendicular **6.** intersecting
7. acute **8.** obtuse **9.** straight, 180°
10. right, 90°
11. $\angle 1 \cong \angle 4$; $\angle 2 \cong \angle 5$; $\angle 3 \cong \angle 6$
$\angle 1$ and $\angle 2$; $\angle 4$ and $\angle 5$
$\angle 5$ and $\angle 1$; $\angle 4$ and $\angle 2$
12. $\angle 4 \cong \angle 7$; $\angle 3 \cong \angle 6$
$\angle 1$ and $\angle 2$; $\angle 3$ and $\angle 4$
$\angle 6$ and $\angle 7$; $\angle 6$ and $\angle 4$
$\angle 7$ and $\angle 3$
13. $\angle AOB$ and $\angle BOC$; $\angle BOC$ and $\angle COD$
$\angle COD$ and $\angle DOA$; $\angle DOA$ and $\angle AOB$
14. $\angle ERH$ and $\angle HRG$; $\angle HRG$ and $\angle GRF$
$\angle FRG$ and $\angle FRE$; $\angle FRE$ and $\angle ERH$
15. **(a)** 85° **(b)** 45° **(c)** 13°
16. **(a)** 171° **(b)** 40° **(c)** 90°
17. 4.84 m **18.** 128 in. **19.** 9.6 km
20. 61 ft **21.** 486 mm^2 **22.** 16.5 ft^2 or $16\frac{1}{2}$ ft^2
23. 39.7 m^2
24. $P = 50$ cm, $A = 140$ cm^2 **25.** $P = 102.1$ ft, $A = 567$ ft^2
26. $P = 200.2$ m, $A = 2074.0$ m^2 **27.** $P = 518$ cm, $A = 10{,}812$ cm^2

28. $P = 27.1$ m $A = 20.58$ m^2 **29.** $P = 20\frac{1}{4}$ ft or 20.25 ft $A = 14$ ft^2
30. 70° **31.** 24° **32.** 15.6 m
33. $11\frac{1}{2}$ ft or 11.5 ft
34. $C = 6.3$ cm $A = 3.1$ cm^2 **35.** $C = 109.3$ m $A = 950.7$ m^2
36. $C = 37.7$ in. $A = 113.0$ in.^{2} **37.** 20.3 m^2 **38.** 64 in.^{2}
39. 673 km^2 **40.** 1020 m^2 **41.** 229 ft^2
42. 132 ft^2 **43.** 5376 cm^2 **44.** 498.9 ft^2
45. 447.9 yd^2 **46.** 30 in.^{3} **47.** 96 cm^3
48. 45,000 mm^3 **49.** 267.9 m^3 **50.** 452.2 ft^3
51. 549.5 cm^3 **52.** 1808.6 m^3 **53.** 512.9 m^3
54. 16 yd^3 **55.** 2.646 **56.** 4.359 **57.** 4
58. 14 **59.** 7.616 **60.** 11 **61.** 10.247
62. 8.944 **63.** 17 in **64.** 7 cm **65.** 10.198 cm
66. 7.211 in **67.** 2.556 m **68.** 8.471 km
69. $y = 30$ ft $x = 34$ ft **70.** $y = 15$ m $x = 18$ m **71.** $x = 12$ mm $y = 7.5$ mm
72. $P = 18$ in. $A = 20.25$ in.^{2} or $20\frac{1}{4}$ in.^{2} **73.** $P = 10.3$ cm $A = 6.195$ cm^2
74. $C = 40.82$ m $A = 132.665$ m^2 **75.** $P = 54$ ft $A = 140$ ft^2
76. $P = 18.5$ yd or $18\frac{1}{2}$ yd $A = 15$ yd^2 **77.** $P = 7$ km $A = 1.96$ km^2
78. $C = 53.38$ m $A = 226.865$ m^2 **79.** $P = 78$ mm $A = 288$ mm^2
80. $P = 38$ mi $A = 58.5$ mi^2 **81.** parallel lines
82. line segment **83.** acute angle
84. intersecting lines **85.** right angle **86.** ray
87. straight angle **88.** obtuse angle
89. perpendicular lines **90.** 29° **91.** 95°
92. $P = 90$ m $A = 92$ m^2 **93.** $P = 282$ cm $A = 4190$ cm^2
94. 100.5 ft^3 **95.** 3.4 in.^{3} or $3\frac{3}{8}$ in.^{3} **96.** 7.4 m^3
97. 561 cm^3 **98.** 1271.7 cm^3 **99.** 1436.0 m^3
100. $x = 12.649$ km **101.** $\angle D = 72°$
102. $x = 12$ mm, $y = 14$ mm

CHAPTER 8 TEST (page 559)

1. e **2.** a **3.** f **4.** g **5.** d
6. Answers vary. **7.** 74° **8.** 105°
9. $\angle 1$ and $\angle 4$; $\angle 2$ and $\angle 5$; $\angle 3$ and $\angle 6$
10. $P = 23$ ft $A = 30$ ft^2 **11.** $P = 72$ mm $A = 324$ mm^2
12. $P = 26.2$ m $A = 33.12$ m^2 **13.** $P = 169$ cm $A = 1591$ cm^2
14. $P = 32.05$ m $A = 48$ m^2 **15.** $P = 37.8$ yd or $37\frac{4}{5}$ yd $A = 58.5$ yd^2
16. 107° **17.** 12.5 in. or $12\frac{1}{2}$ in. **18.** 5.7 km
19. 206.0 cm^2 **20.** 39.3 m^2 **21.** 6480 m^3
22. 33.5 ft^3 **23.** 5086.8 ft^3 **24.** 9.220 cm
25. $y = 12$ cm; $z = 6$ cm **26.** Answers vary.

CUMULATIVE REVIEW 1–8 (page 561)

1. $300 + 60{,}000 + 6000 = 66{,}300$; 64,574
2. $20 - 10 = 10$; 10.242
3. $4 \times 7 = 28$; 26.7472
4. $30{,}000 \times 70 = 2{,}100{,}000$; 2,074,587
5. $600 \div 3 = 200$; 200.8 **6.** $5000 \div 50 = 100$; 94
7. $2\frac{2}{9}$ **8.** 0.9132 **9.** 70 R79 **10.** $3\frac{23}{40}$
11. 0.000078 **12.** 12 **13.** 0.220 **14.** $4\frac{5}{12}$
15. 836.085 **16.** 9 **17.** 38
18. two hundred eight ten-thousandths
19. 665.51 **20.** 2.055; 2.5005; 2.505; 2.55
21. $\frac{1}{20}$ **22.** 5% **23.** 1.75 **24.** 175%
25. $\frac{2}{5}$ **26.** 0.4 **27.** $\frac{6}{1}$ **28.** $\frac{7}{3}$
29. 35 **30.** 100 **31.** 2.01 (rounded) **32.** 160%
33. $600 **34.** 300 seconds **35.** $2\frac{1}{2}$ or 2.5 pounds
36. 0.005 L **37.** 1250 g **38.** mm **39.** L
40. kg **41.** cm **42.** 0°C
43. $P = 11$ in. $A = 7$ in.^{2} **44.** $P = 5.9$ m $A = 1.575$ m^2 **45.** $C = 31.4$ ft $A = 78.5$ ft^2
46. $P = 76$ cm $A = 264$ cm^2 **47.** $P = 40.8$m $A = 95$ m^2 **48.** $P = 50$ yd $A = 142$ yd^2
49. $y = 24.2$ mm **50.** $x = 8.8$ ft **51.** 59%
52. 175 tissues for $1.89 **53.** 2255.3 cm^3
54. $15.40 **55.** $1\frac{1}{12}$ yard **56.** 42.9 pounds
57. 3.4 m **58.** 52 inches

CHAPTER 9

SECTION 9.1 (page 569)

1. +12 **3.** −12 **5.** +18,000 **7.** positive
9. negative **11.** neither **13.** negative
15.

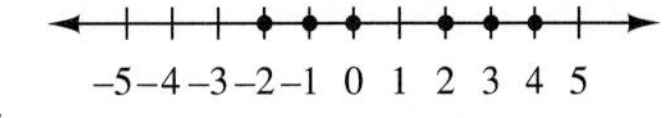

17.

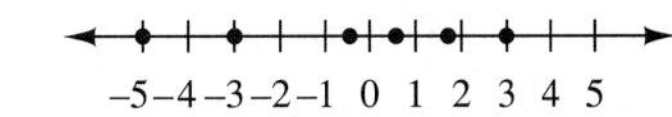

ANSWERS

19. 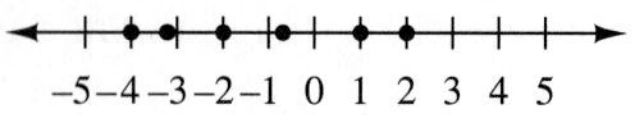 –5 –4 –3 –2 –1 0 1 2 3 4 5

21. 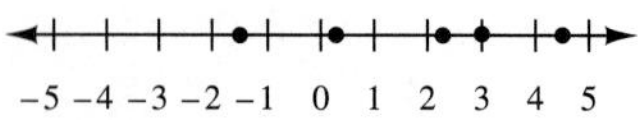 –5 –4 –3 –2 –1 0 1 2 3 4 5

23. 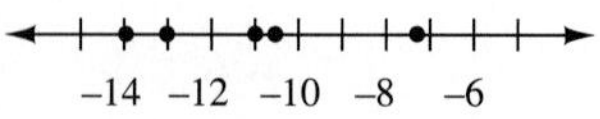 –14 –12 –10 –8 –6

25. $<$ **27.** $>$ **29.** $<$ **31.** $>$ **33.** $<$
35. $<$ **37.** $>$ **39.** $>$ **41.** 5 **43.** 5

45. 1 **47.** 251 **49.** 0 **51.** $\frac{1}{2}$ **53.** 9.5

55. 0.618 **57.** $\frac{3}{4}$ **59.** $-\frac{5}{2}$ **61.** -9 **63.** -4

65. -2 **67.** 54 **69.** 11 **71.** -163 **73.** $-\frac{4}{3}$

75. $4\frac{1}{2}$ **77.** -5.2 **79.** 1.4 **81.** Answer varies.

83. Answer varies. **85.** true **87.** true **89.** false

91. $\frac{19}{20}$ **93.** $\frac{7}{12}$

SECTION 9.2 (page 581)

1. 3

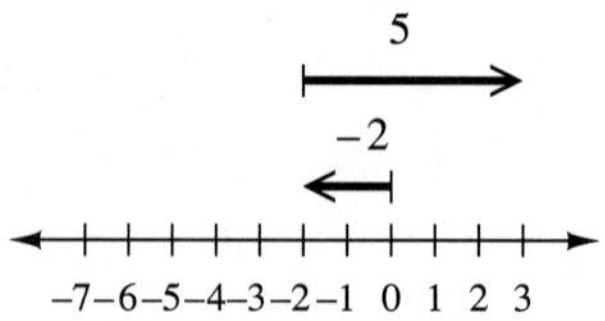

5, –2, –7 –6 –5 –4 –3 –2 –1 0 1 2 3

3. -7

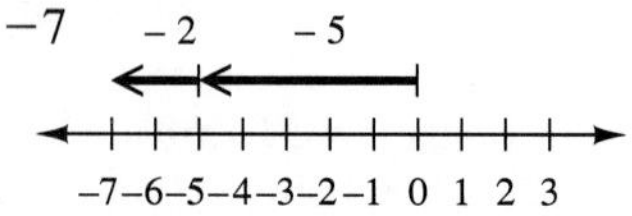

–2, –5, –7 –6 –5 –4 –3 –2 –1 0 1 2 3

5. -1

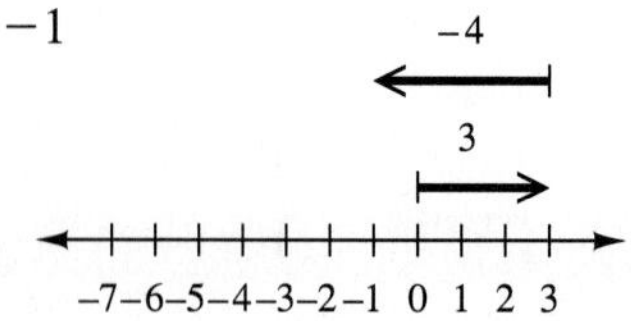

–4, 3, –7 –6 –5 –4 –3 –2 –1 0 1 2 3

7. -3 **9.** 7 **11.** -7 **13.** 1

15. $-8 + (-9) = -17$

17. $-\$52.50 + \$50 = -\$2.50$ **19.** -6.8

21. $\frac{1}{4}$ **23.** $-\frac{3}{10}$ **25.** $-\frac{26}{9}$ or $-2\frac{8}{9}$

27. Answer varies. **29.** -3 **31.** 9

33. $-\frac{1}{2}$ **35.** 6.2 **37.** 14 **39.** -2

41. -12 **43.** -25 **45.** -23 **47.** 5

49. 20 **51.** 11 **53.** $-\frac{3}{2}$ or $-1\frac{1}{2}$ **55.** -6

57. -3.6 **59.** Answer varies. **61.** -10

63. 12 **65.** -7 **67.** 5 **69.** -1

71. -4.4 **73.** -11 **75.** -10

77. $\frac{19}{4}$ or $4\frac{3}{4}$ **79.** 1058

81. 1513 **83.** 247 **85.** $\frac{3}{2}$ or $1\frac{1}{2}$

SECTION 9.3 (page 587)

1. -35 **3.** -27 **5.** -18 **7.** -50 **9.** -12

11. 32 **13.** 77 **15.** 133 **17.** 13 **19.** 0

21. 4 **23.** -4 **25.** $-\frac{1}{10}$ **27.** $\frac{14}{3}$ or $4\frac{2}{3}$

29. $-\frac{5}{6}$ **31.** $\frac{7}{4}$ or $1\frac{3}{4}$ **33.** -42.3 **35.** 6

37. -31.62 **39.** 4.5 **41.** -2 **43.** -5

45. -7 **47.** -14 **49.** 10 **51.** 4 **53.** -1

55. 191 **57.** $\frac{2}{3}$ **59.** $\frac{1}{3}$ **61.** -8

63. $-\frac{14}{3}$ or $-4\frac{2}{3}$ **65.** 4.73 **67.** -4.7

69. -5.3 **71.** 12 **73.** -0.36 **75.** $-\frac{7}{40}$

77. Answer varies. **79.** Answer varies. **81.** -2

83. -10 **85.** 9 **87.** 10

SECTION 9.4 (page 593)

1. 14 **3.** 0 **5.** 52 **7.** -39 **9.** -30

11. 17 **13.** 16 **15.** 30 **17.** 23 **19.** -43

21. 19 **23.** -3 **25.** 0 **27.** 47 **29.** 41

31. -2 **33.** 13 **35.** 126 **37.** 40 **39.** -20

41. 3 **43.** 8 **45.** 1.24 **47.** -1.47

49. -2.31 **51.** $-\frac{13}{10}$ or $-1\frac{3}{10}$ **53.** $\frac{5}{4}$ or $1\frac{1}{4}$

55. $\frac{4}{5}$ **57.** -1200 **59.** 1 **61.** Answer varies.

63. -27 **65.** 7 **67.** 6 **69.** $\frac{4}{3}$ or $1\frac{1}{3}$

71. $\frac{3}{2}$ or $1\frac{1}{2}$

SECTION 9.5 (page 599)

1. 28 **3.** -10 **5.** 8 **7.** -30 **9.** -8

11. 0 **13.** 2 **15.** -22 **17.** $-\frac{13}{8}$ or $-1\frac{5}{8}$

19. 30 **21.** 28 **23.** 140 **25.** $\frac{45}{2}$ or $22\frac{1}{2}$

27. 600 **29.** 318 **31.** 25.12 **33.** Answer varies.

35. -40 **37.** 113.04 **39.** -1 **41.** 1

SECTION 9.6 (page 607)

1. yes **3.** yes **5.** no **7.** 4 **9.** 30 **11.** 8
13. 10 **15.** −8 **17.** −6 **19.** 18 **21.** 9 **23.** −3
25. $\frac{7}{3}$ or $2\frac{1}{3}$ **27.** $10\frac{7}{8}$ **29.** $2\frac{1}{2}$ **31.** $3\frac{9}{10}$
33. 5.87 **35.** −1.51 **37.** 2 **39.** 4 **41.** −8
43. −6 **45.** 9 **47.** −7 **49.** 1.1 **51.** 34
53. 66 **55.** −36 **57.** −20 **59.** 4 **61.** 16
63. 1.3 **65.** −4.94 **67.** Answer varies. **69.** 19
71. 9 **73.** $\frac{8}{21}$ **75.** $-\frac{33}{13}$ or $-2\frac{7}{13}$ **77.** 1

SECTION 9.7 (page 615)

1. 1 **3.** 1 **5.** 2 **7.** −2 **9.** −4 **11.** 6
13. $6x + 24$ **15.** $7p - 56$ **17.** $-3m - 18$
19. $-2y + 6$ **21.** $-5z + 45$ **23.** $7m$ **25.** $8x$
27. $-2t$ **29.** $-4a$ **31.** 5 **33.** 9 **35.** −6
37. 1 **39.** 4 **41.** −2 **43.** 2 **45.** −1
47. 20 **49.** −9 **51.** −5 **53.** 8
55. Answer varies. **57.** −10 **59.** 2 **61.** 0
63. \$15.20 **65.** \$25.20

SECTION 9.8 (page 623)

1. $14 + x$ **3.** $-5 + x$ **5.** $x + 6$ **7.** $x - 9$
9. $x - 4$ **11.** $6x$ **13.** $3x$ **15.** $\frac{x}{2}$ **17.** $8 + 2x$
19. $7x - 10$ **21.** $2x + x$ **23.** Answer varies.
25. 7 **27.** −5 **29.** 20 **31.** 34 cm and 44 cm
33. 5 days **35.** 19 m **37.** 12 ft, 6 ft
39. 25 years old **41.** 227 m, 227 m, 252 m
43. 5 years **45.** \$2320 **47.** \$91.50

CHAPTER 9 REVIEW EXERCISES (page 631)

1. 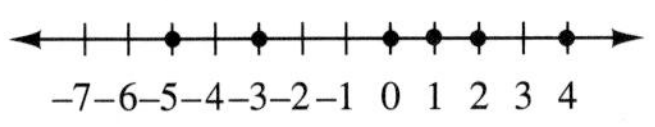

2. 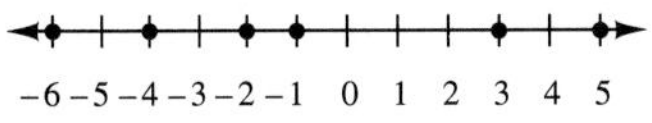

3. 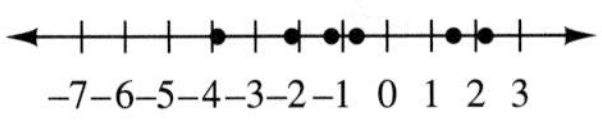

4. 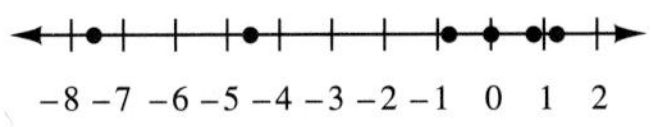

5. > **6.** < **7.** > **8.** < **9.** 8
10. 19 **11.** −7 **12.** −15 **13.** 2 **14.** 9
15. −19 **16.** −33 **17.** −7 **18.** −19
19. $\frac{3}{10}$ **20.** $-\frac{3}{8}$ **21.** −5.2 **22.** 0.3 **23.** −6
24. 14 **25.** $\frac{5}{8}$ **26.** −3.75 **27.** −6 **28.** −8
29. −7 **30.** −17 **31.** 11 **32.** 11 **33.** 13
34. −6 **35.** −24 **36.** −20 **37.** 15 **38.** 64
39. −8 **40.** −3 **41.** 5 **42.** 20 **43.** $-\frac{4}{7}$
44. $\frac{1}{6}$ **45.** 1.4 **46.** −14.79 **47.** −21
48. 23 **49.** −1 **50.** −2 **51.** −34 **52.** −32
53. −100 **54.** 117 **55.** −87.5 **56.** 1.328
57. 0 **58.** $\frac{1}{7}$ **59.** 27 **60.** −8 **61.** 0
62. 19 **63.** 33 **64.** −1 **65.** 35 **66.** 27
67. 9 **68.** 16 **69.** 1 **70.** 1
71. $-\frac{5}{4}$ or $-1\frac{1}{4}$ **72.** −8.05 **73.** −7 **74.** 8
75. 20 **76.** −55 **77.** 9 **78.** −4
79. $6r - 30$ **80.** $11p + 77$ **81.** $-9z + 27$
82. $-8x - 32$ **83.** $11r$ **84.** $-5z$ **85.** $-8p$
86. $2x$ **87.** −9 **88.** −5 **89.** −5 **90.** 7
91. −4 **92.** 2 **93.** $18 + x$ **94.** $\frac{1}{2}x$
95. $4x + 6$ **96.** $5x - 10$ **97.** −3 **98.** 7 days
99. 37 cm **100.** 21 and 30 years old **101.** 3
102. 40 **103.** −1 **104.** −7 **105.** −16
106. −9 **107.** 6 **108.** 5 **109.** 7 **110.** −19
111. 9 **112.** −4 **113.** 4 **114.** −10
115. −11 **116.** 3 **117.** 4 **118.** −1
119. −40 **120.** 3 **121.** 15 **122.** 13 inches
123. Pam weighs 137 pounds; Sam weighs 162 pounds.
124. 48 cm and 42 cm

CHAPTER 9 TEST (page 637)

1. 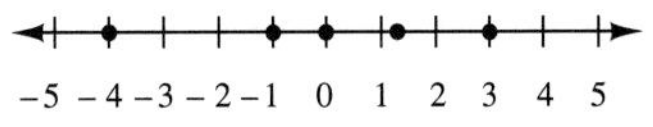

2. <, > **3.** 7, 15 **4.** −1
5. −13 **6.** 5.3 **7.** −7 **8.** 16 **9.** $\frac{1}{4}$
10. −32 **11.** 84 **12.** −25 **13.** 1
14. −21 **15.** −38 **16.** 29 **17.** Answer varies.
18. 110 **19.** 5 **20.** 31 **21.** −11 **22.** −15
23. 5 **24.** 2 **25.** −3 **26.** 10
27. 57 cm; 61 cm **28.** Answer varies.

CUMULATIVE REVIEW 1–9 (page 639)

1. $9 + 1 + 40 = 50$; 50.602
2. $6 \times 50 = 300$; 308.484
3. $40{,}000 \div 80 = 500$; 503 **4.** 8.906 **5.** $\frac{13}{20}$
6. 56.4 **7.** $\frac{2}{5}$ **8.** 534,072 **9.** $5\frac{1}{4}$ **10.** −5

11. 40 **12.** 4 **13.** -1.3 **14.** -5

15. $-\frac{1}{2}$ **16.** 5 **17.** 4 **18.** > **19.** <

20. < **21.** 0.06; $\frac{3}{50}$ **22.** 2.3; 230% **23.** $\frac{8}{3}$

24. 0.4 **25.** 233.33 (rounded) **26.** 45

27. 50 cars **28.** 4% **29.** 14 quarts **30.** 3 days

31. 3700 mL **32.** 0.4 m **33.** m **34.** mL

35. km **36.** g **37.** 100°C

38. $P = 9$ ft, $A = 5\frac{1}{16}$ or 5.0625 ft^2 **39.** $C = 28.26$ mm, $A = 63.585 \text{ mm}^2$

40. $P = 56$ mi, $A = 84 \text{ mi}^2$ **41.** $P = 6.45$ cm, $A = 2.185 \text{ cm}^2$

42. $P = 36$ ft, $A = 70 \text{ ft}^2$ **43.** $P = 188$ m, $A = 1636 \text{ m}^2$

44. $y = 13.2$ yd **45.** $x = 14$ in. **46.** -26

47. -5 **48.** 8 **49.** -4 **50.** -10

51. $350 for Reggie, $650 for Donald

52. 23 cm, 18 cm **53.** $25.52 **54.** $8\frac{1}{3}$ cups

55. 124% **56.** $4.30 **57.** 0.544 m^3 less

58. 11 feet **59.** 54 minutes **60.** Naomi's car

CHAPTER 10

SECTION 10.1 (page 647)

1. $32,000 **3.** $\frac{9800}{32000} = \frac{49}{160}$ **5.** $\frac{12100}{900} = \frac{121}{9}$

7. history **9.** $\frac{4000}{12000} = \frac{1}{3}$ **11.** $\frac{1800}{2000} = \frac{9}{10}$

13. $\frac{4000}{1000} = \frac{4}{1}$ **15.** $522,000 **17.** $174,000

19. $261,000 **21.** $3920 **23.** $2548 **25.** $1960

27. Answer varies. **29.** 90° **31.** 10%; 36°

33. 15% **35.** $630; 15%

37. **(a)** $200,000 **(b)** 22.5°; 72°; 108°; 90°; 67.5°

(c)

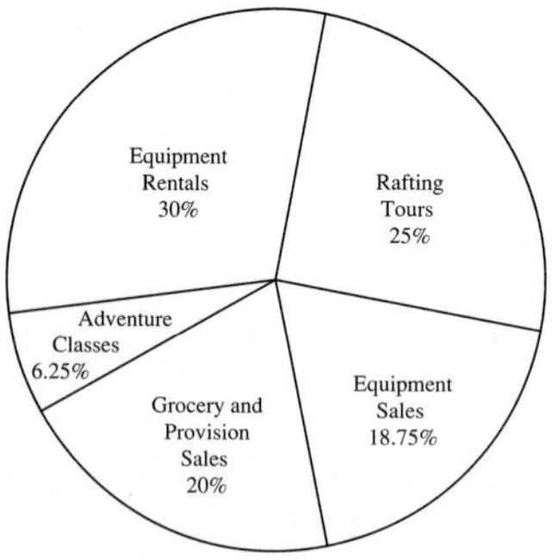

39. 30%; 108°
20%; 72°
15%; 54°
10%; 36°
5%; 18°
5%; 18°
15%; 54°

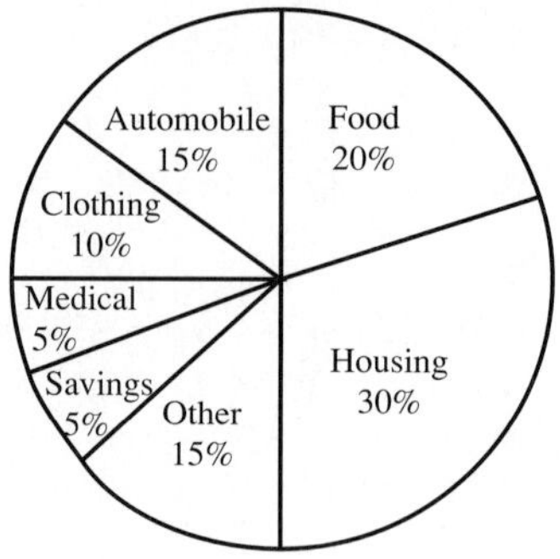

41. < **43.** > **45.** < **47.** >

SECTION 10.2 (page 657)

1. 3000 **3.** 2000 **5.** July 4 **7.** May; 10,000

9. 1500 **11.** 2500 **13.** 150,000 gallons **15.** 1991

17. 550,000 gallons **19.** April **21.** 300 **23.** 200

25. 3,000,000 **27.** 1,500,000 **29.** 3,500,000

31. Answer varies. **33.** $40,000 **35.** $25,000

37. $5000 **39.** $1848 **41.** $6552

SECTION 10.3 (page 663)

1. 60–65 years **3.** 13,000 members

5. 50,000 members **7.** $20,000–$25,000

9. 7 employees **11.** 28 employees

13. Answer varies. **15.** ||; 2 **17.** |||| |; 6

19. |; 1 **21.** ||; 2 **23.** |||; 3 **25.** ||||; 5

27. |||| |; 6 **29.** ||; 2 **31.** |||| |||| |; 11

33. |||| |||| |; 11 **35.** |||| |||; 8 **37.** ||||; 4

39. 25.6 **41.** 52.8 **43.** 39.4

SECTION 10.4 (page 671)

1. 13 years **3.** 60.3 **5.** $27,955 **7.** $58.24

9. $35,500 **11.** 6.7 **13.** 17.2 **15.** 16 messages

17. 516 customers **19.** 8 samples

21. 68 and 74 years **23.** Answer varies.

25. Answer varies. **27.** 2.6 **29.** 49

31. no mode

CHAPTER 10 REVIEW EXERCISES (page 681)

1. Lodging **2.** $\frac{400}{1700} = \frac{4}{17}$ **3.** $\frac{300}{1700} = \frac{3}{17}$

4. $\frac{280}{1700} = \frac{14}{85}$ **5.** $\frac{300}{160} = \frac{15}{8}$ **6.** $\frac{560}{400} = \frac{7}{5}$

7. $66,375 **8.** $13,275 **9.** $82,305 **10.** $26,550
11. $29,205 **12.** $47,790 **13.** March **14.** June
15. 8 million acre-feet **16.** 4 million acre-feet
17. 5 million acre-feet **18.** 2 million acre-feet
19. $25,000 **20.** $20,000 **21.** $35,000
22. $40,000 **23.** $30,000 **24.** $20,000 **25.** 28.5
26. 35.2 **27.** $51.10 **28.** 257.3 points
29. 39 customers **30.** $562 **31.** $64
32. 18 and 32 (bimodal) boats **33.** 36° **34.** 35%
35. 20% **36.** 25% **37.** 36%

38.

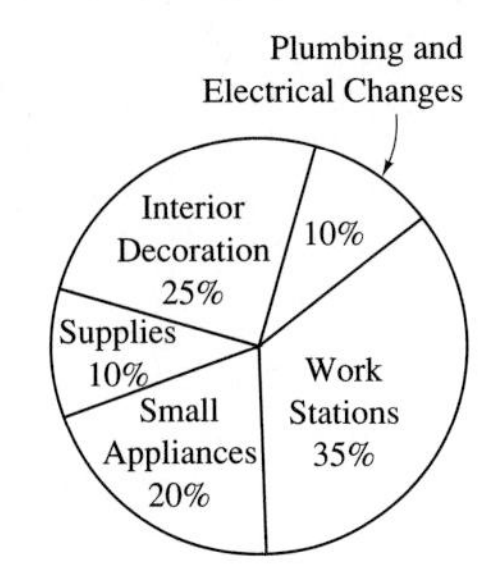

39. 50 years **40.** 144.7 **41.** 97
42. 31 and 43 (bimodal) living units **43.** 5.0 hours
44. 14.5 **45.** |||| ; 4 **46.** | ; 1 **47.** ~~||||~~ | ; 6
48. || ; 2 **49.** ~~||||~~ ~~||||~~ ||| ; 13 **50.** ~~||||~~ || ; 7
51. ~~||||~~ || ; 7

52.

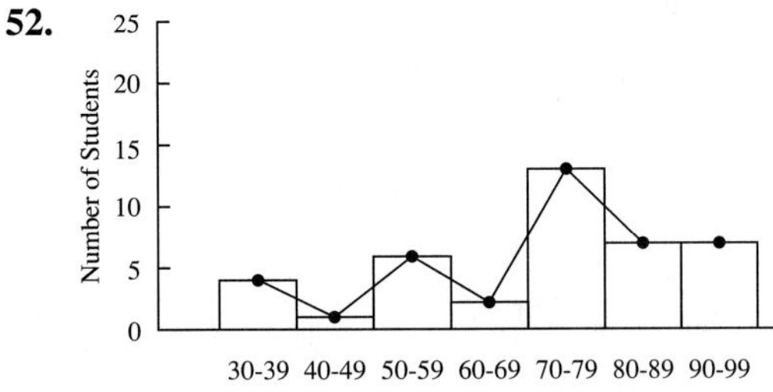

53. 32.3 **54.** $118.80

CHAPTER 10 TEST (page 687)

1. $1,232,000 **2.** $1,008,000 **3.** $1,680,000
4. $896,000 **5.** $112,000 **6.** $672,000
7. 72° **8.** 10% **9.** 108° **10.** 108° **11.** 36°

12.

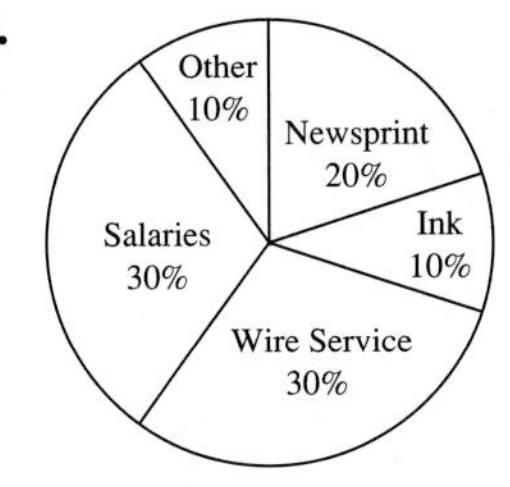

13. 3 **14.** 2 **15.** 4 **16.** 3 **17.** 3 **18.** 5

19.

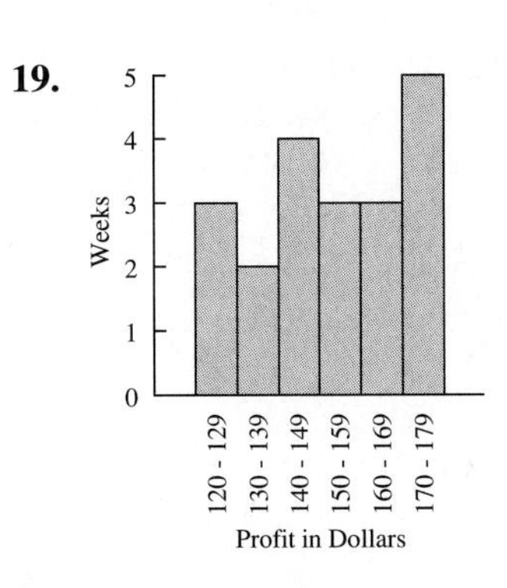

20. 75 **21.** 28.8 miles per gallon **22.** 478.9
23. Answer varies. **24.** Answer varies.
25. $11.30 **26.** 173.7 **27.** 31.5 calls
28. 9.3 liters **29.** 57 meters **30.** 103° and 104°

CUMULATIVE REVIEW 1–10 (page 691)

1. $65.74 **2.** $475 **3.** 75,700 **4.** 980,000
5. 30 **6.** 10 **7.** 144 **8.** 324
9. 60,000 + 200,000 + 90 + 20,000 = 280,090; 244,395
10. 3 + 800 + 40 + 50 + 8 = 901; 898.547
11. 300,000 − 100,000 = 200,000; 173,783
12. 900 − 60 = 840; 811.863
13. 7000 × 600 = 4,200,000; 4,485,640
14. 90 × 2 = 180; 148.0731
15. 40,000 ÷ 50 = 800; 856 **16.** 60 ÷ 4 = 15; 14.72
17. $1\frac{1}{2}$ **18.** $1\frac{7}{8}$ **19.** $11\frac{4}{15}$ **20.** $\frac{3}{8}$ **21.** $2\frac{7}{12}$
22. $30\frac{19}{20}$ **23.** $\frac{7}{10}$ **24.** $44\frac{2}{5}$ **25.** $8\frac{4}{5}$
26. $1\frac{1}{3}$ **27.** 18 **28.** $\frac{8}{21}$ **29.** $\frac{1}{12}$ **30.** $\frac{11}{16}$
31. 0.75 **32.** 0.625 **33.** 0.6 **34.** 0.55
35. 0.199, 0.207, 0.218, 0.22, 0.2215
36. 0.58, 0.608, $\frac{5}{8}$, 0.6319, $\frac{13}{20}$ **37.** $\frac{4}{1}$ **38.** $\frac{1}{8}$
39. true **40.** false **41.** 3 **42.** 4 **43.** 16.2
44. 52 **45.** 0.35 **46.** 2.5% **47.** 2.50 or 2.5
48. 0.0435 **49.** $\frac{1}{50}$ **50.** $\frac{5}{8}$ **51.** 15%
52. 325% **53.** 96 miles **54.** $48.60
55. 2800 people **56.** 320 **57.** 35% **58.** 25%
59. 3 **60.** 4 **61.** 120 **62.** 12,000
63. 100 cm **64.** 2.182 m **65.** 8300 mg
66. 0.23 kg **67.** 0.004 L **68.** 280 mL **69.** mL
70. g **71.** km **72.** kg **73.** 48.7 m^2
74. 38.7 cm^2 **75.** 38.3 ft^2 **76.** 132.7 cm^2

77. 882.5 cm^3 **78.** 99 m^3 **79.** 4m **80.** 17 cm
81. -15 **82.** 6.9 **83.** -30 **84.** 83.22
85. 5 **86.** -2.3 **87.** $x = 5$ **88.** $x = -6$
89. $x = 2$ **90.** $x = -7$ **91.** 27; 26; 19
92. 8.8; 8.6; 5.7 **93.** 12.5% **94.** $1124
95. 102 tanks **96.** 2166 employees **97.** $3697.50
98. $59,225.25 **99.** 17.5 L **100.** $3631.25

APPENDIX B: EXERCISES (page 707)

1. 37 **3.** 26 **5.** 16 **7.** 243 **9.** 36

11. ⊢ └ ┌ ⊣ ┘ ┐

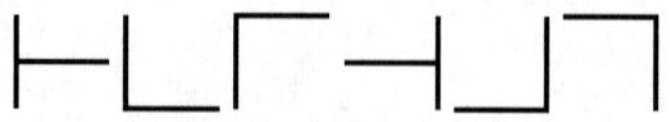

13. ⁄‾ _⁄ ‾\ _

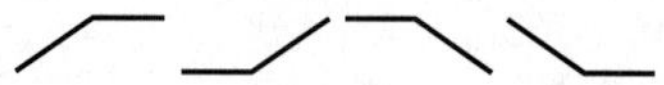

15. Conclusion follows. **17.** Conclusion does not follow.
19. 3 days **21.** Dick

Solutions to Selected Exercises

For the answers to all odd-numbered section exercises, all chapter review exercises, and all chapter tests, see the section beginning on page A-1.

If you would like to see more solutions, you may order the *Student's Solutions Manual for Intermediate Algebra* from your college bookstore. It contains solutions to the odd-numbered exercises that do not appear in this section, as well as solutions to all chapter review exercises and chapter tests.

As you are looking at these solutions, remember that many exercises and problems can be solved in a variety of ways. In this section we provide only one method of solving selected exercises from each section; space does not permit showing other methods that may be equally correct.

If you work the exercise differently but obtain the same answer, then *as long as your steps are mathematically valid,* your work is correct. Mathematical thinking is a creative process, and solving problems in more than one way is an example of this creativity.

CHAPTER 1

SECTION 1.1 (page 5)

1. 2: thousands; 1: tens
5. 8: millions; 2: thousands
9. 60: billions; 0: millions; 502: thousands; 109: ones
13. seventy-nine thousand, six hundred thirteen
17. twenty-five million, seven hundred fifty-six thousand, six hundred sixty-five
21. 10,000,223
25. 13,112

SECTION 1.2 (page 13)

1.
$$\begin{array}{r} 13 \\ +\ 36 \\ \hline 49 \end{array}$$
ones added
tens added

5.
$$\begin{array}{r} 317 \\ +\ 572 \\ \hline 889 \end{array}$$
ones added
tens added
hundreds added

9.
$$\begin{array}{r} 6310 \\ 252 \\ +\ 1223 \\ \hline 7785 \end{array}$$
ones added
tens added
hundreds added
thousands added

13.
$$\begin{array}{r} 1251 \\ 4311 \\ +\ 2114 \\ \hline 7676 \end{array}$$

17.
$$\begin{array}{r} 9213 \\ +\ 6715 \\ \hline 15{,}928 \end{array}$$

21.
$$\begin{array}{r} {\scriptstyle 1}\ \\ 78 \\ +\ 65 \\ \hline 143 \end{array}$$

25.
$$\begin{array}{r} {\scriptstyle 1}\ \\ 47 \\ +\ 74 \\ \hline 121 \end{array}$$

29.
$$\begin{array}{r} {\scriptstyle 1}\ \\ 73 \\ +\ 29 \\ \hline 102 \end{array}$$

33.
$$\begin{array}{r} {\scriptstyle 1}\ \\ 306 \\ +\ 848 \\ \hline 1154 \end{array}$$

37.
$$\begin{array}{r} {\scriptstyle 1}\ \ \\ 928 \\ +\ 843 \\ \hline 1771 \end{array}$$

41.
$$\begin{array}{r} {\scriptstyle 111}\ \\ 7968 \\ +\ 1285 \\ \hline 9253 \end{array}$$

45.
$$\begin{array}{r} {\scriptstyle 111}\ \\ 9625 \\ +\ 7986 \\ \hline 17{,}611 \end{array}$$

49.
$$\begin{array}{r} {\scriptstyle 2}\ \ \\ {\scriptstyle 1}18 \\ {\scriptstyle 1}708 \\ 9286 \\ +\ 636 \\ \hline 10{,}648 \end{array}$$

53.
$$\begin{array}{r} {\scriptstyle 1}\ \ \\ {\scriptstyle 1}321 \\ 9603 \\ 8 \\ 21 \\ +\ 1604 \\ \hline 11{,}557 \end{array}$$

57.
33
553
2 97
2772
437
63
+ 328
4250

61. 1021 correct
628
265
+ 128
1021

65. 5420 correct
4713
28
615
+ 64
5420

69. 14,332 correct
4 714
27
77
8 878
+ 636
14,332

73. Wilson and El Camino

Wilson to Southtown	11	miles
Southtown to El Camino	+ 24	miles
Wilson to El Camino	35	miles

Any other route is longer.

77.
$47 cost of the history book
+ $43 cost of the math book
$90 total cost for both books

81.
9 792 books from library
+ 3 259 books from book dealer
13,051 total books for sale

85.
286
308
+ 114
708 feet

SECTION 1.3 (page 23)

1. 27 − 15 = 12 Check 15 + 12 = 27 match

5. 77 − 60 = 17 Check 60 + 17 = 77 match

9. 552 − 451 = 101 Check 451 + 101 = 552 match

13. 5546 − 2134 = 3412 Check 2134 + 3412 = 5546 match

17. 24,392 − 11,232 = 13,160 Check 11,232 + 13,160 = 24,392 match

21. 47 − 32 = 15; 32 + 15 = 47 correct

25. 382 − 261 = 131; 261 + 131 = 392 incorrect; answer should be 121

29. 8643 − 1421 = 7212; 1421 + 7212 = 8633 incorrect; answer should be 7222

33.
11
6 1
− 3 2
2 9

37.
6 11
7 1 9
− 6 5 8
6 1

41.
7 15 11
9 8 6 1
− 6 8 4
9 1 7 7

45.
2 17 12 12 15
3 8, 3 3 5
− 2 9, 4 7 6
8, 8 5 9

49.
10
5 0
− 3 7
2 3

53.
3 10 3 11
4 0 4 1
− 1 2 0 8
2 8 3 3

57.
7 10
1 5 8 0
− 1 0 7 7
5 0 3

61.
9
5 10 11 10
6 0 2 0
− 4 0 7 8
1 9 4 2

65.
9
7 10 6 10 15
8 0, 7 0 5
− 6 1, 6 6 7
1 9, 0 3 8

69.
9 9
1 10 10 17 10
2 0, 0 8 0
− 1 3, 9 4 6
6 5 8 4

73. 1439 − 1169 = 270; 1169 + 270 = 1439 correct

77. 27,689 − 22,306 = 5 383; 22,306 + 5,383 = 27,689 correct

81.
98 rose bushes
− 40 sold
58 rose bushes left

85.

\$3900	Rent
\$2220	Food
\$1670	Fees
\$615	Books
+ \$1895	Other
\$10,300	Total student expenses

89.

5 11 15 (borrow marks)	
1 2, 6 2 5 (6, 2, 5 crossed out)	Students in fall
− 1 1, 2 9 6	students in spring
1 3 2 9	more students in fall

93.

\$1523	withheld from paycheck
− \$1379	owes for income tax
\$144	refund

97.

12,352	visitors on Tuesday
− 11,594	visitors on Monday
758	more visitors on Tuesday

SECTION 1.4 (page 33)

1. $2 \times 5 \times 2$

$2 \times 5 = 10 \rightarrow 10 \times 2 = 20$

5. $8 \cdot 9 \cdot 0$

$8 \cdot 9 = 72 \rightarrow 72 \cdot 0 = 0$

9. $(2)(3)(6)$

$(2)(3) = 6 \rightarrow 6(6) = 36$

17. $\begin{array}{r} ^{1} \\ 32 \\ \times\ 7 \\ \hline 224 \end{array}$

21. $\begin{array}{r} ^{1} \\ 624 \\ \times\ 3 \\ \hline 1872 \end{array}$

25. $\begin{array}{r} ^{2} \\ 2521 \\ \times\ 4 \\ \hline 10{,}084 \end{array}$

29. $\begin{array}{r} ^{4\,6\,1} \\ 36{,}921 \\ \times\ 7 \\ \hline 258{,}447 \end{array}$

33. $\begin{array}{r} 40 \\ \times\ 8 \\ \hline \end{array}$ First $\begin{array}{r} 4 \\ \times\ 8 \\ \hline 32 \end{array}$ $\begin{array}{r} 40 \\ \times\ 8 \\ \hline 320 \end{array}$ attach 1 zero.

37. $\begin{array}{r} 500 \\ \times\ 4 \\ \hline \end{array}$ First $\begin{array}{r} 5 \\ \times\ 4 \\ \hline 20 \end{array}$ $\begin{array}{r} 500 \\ \times\ 4 \\ \hline 2000 \end{array}$ attach 2 zeros.

41. $\begin{array}{r} 1255 \\ \times\ 20 \\ \hline \end{array}$ First $\begin{array}{r} 1255 \\ \times\ 2 \\ \hline 2510 \end{array}$ $\begin{array}{r} 1255 \\ \times\ 20 \\ \hline 25{,}100 \end{array}$ attach 1 zero.

45. $\begin{array}{r} 43{,}000 \\ \times\ 2\,000 \\ \hline \end{array}$ First $\begin{array}{r} 43 \\ \times\ 2 \\ \hline 86 \end{array}$ $\begin{array}{r} 43{,}000 \\ \times\ 2\,000 \\ \hline 86{,}000{,}000 \end{array}$ attach 6 zeros.

49. $\begin{array}{r} 500 \\ \times\ 900 \\ \hline \end{array}$ First $\begin{array}{r} 5 \\ \times\ 9 \\ \hline 45 \end{array}$ $\begin{array}{r} 500 \\ \times\ 900 \\ \hline 450{,}000 \end{array}$ attach 4 zeros.

53. $\begin{array}{r} 38 \\ \times\ 17 \\ \hline 266 \\ 38 \\ \hline 646 \end{array}$

57. $\begin{array}{r} 83 \\ \times\ 45 \\ \hline 415 \\ 332 \\ \hline 3735 \end{array}$

61. $\begin{array}{r} 67 \\ \times\ 92 \\ \hline 134 \\ 603 \\ \hline 6164 \end{array}$

65. $\begin{array}{r} 729 \\ \times\ 45 \\ \hline 3\,645 \\ 29\,16 \\ \hline 32{,}805 \end{array}$

69. $\begin{array}{r} 735 \\ \times\ 112 \\ \hline 1\,470 \\ 7\,35 \\ 73\,5 \\ \hline 82{,}320 \end{array}$

73. $\begin{array}{r} 7783 \\ \times\ 265 \\ \hline 38\,915 \\ 466\,98 \\ 1\,556\,6 \\ \hline 2{,}062{,}495 \end{array}$

77. $\begin{array}{r} 428 \\ \times\ 201 \\ \hline 428 \\ 85\,60 \\ \hline 86{,}028 \end{array}$

81. $\begin{array}{r} 2195 \\ \times\ 1038 \\ \hline 17\,560 \\ 65\,85 \\ 2\,195\,0 \\ \hline 2{,}278{,}410 \end{array}$

85.

800	pages in each volume
× 30	volumes
24,000	pages

89.

38	miles per gallon
× 11	gallons
418	miles

93. $\begin{array}{r} 84 \\ \times\ 49 \\ \hline 4116 \end{array}$

The total cost is \$4116

97. $97 \cdot 21 \cdot 43 \cdot 56 = 50{,}568$

101.

1406	calories in large meal
− 348	calories in small meal
1058	more calories in large meal

SOLUTIONS

SECTION 1.5 (page 45)

1. $\begin{array}{r} 3 \\ 5\overline{)15} \end{array}$ $\qquad \dfrac{15}{5} = 3$

5. $16 \div 2 = 8 \quad \dfrac{16}{2} = 8$

9. $\dfrac{10}{2} = 5$

13. $\dfrac{0}{4} = 0$

17. $0\overline{)15}$ meaningless

21. $\dfrac{8}{1} = 8$

25. $\begin{array}{r} 4\ 7 \\ 8\overline{)37^{5}6} \end{array}$
Check: $8 \times 47 = 376$

29. $\begin{array}{l} \ \ \ 6\ 2\ 7 \text{ R } 1 \\ 4\overline{)25^{1}0^{2}9} \end{array}$
Check: $4 \times 627 + 1 = 2508 + 1 = 2509$

33. $\begin{array}{r} 309 \\ 6\overline{)1854} \end{array}$
Check: $6 \times 309 = 1854$

37. $\begin{array}{r} 5\ 006 \\ 3\overline{)15{,}018} \end{array}$
Check: $3 \times 5006 = 15{,}018$

41. $\begin{array}{l} \ \ \ 2\ 5\ 8\ 9 \text{ R } 2 \\ 5\overline{)12{,}^{2}9^{4}4^{4}7} \end{array}$
Check: $5 \times 2589 + 2 = 12{,}945 + 2 = 12{,}947$

45. $\begin{array}{l} \ \ \ 3\ 1\ 5\ 7 \text{ R } 2 \\ 4\overline{)12{,}6^{2}3^{3}0} \end{array}$
Check: $4 \times 3157 + 2 = 12{,}628 + 2 = 12{,}630$

49. $\begin{array}{l} \ \ \ 1\ 2,\ 4\ 5\ 8 \text{ R } 3 \\ 6\overline{)7^{1}4{,}^{2}7^{3}5^{5}1} \end{array}$
Check: $6 \times 12{,}458 + 3 = 74{,}748 + 3 = 74{,}751$

53. $\begin{array}{l} \ \ \ 1\ 8,\ 3\ 7\ 7 \text{ R } 6 \\ 7\overline{)12^{5}8{,}^{2}6^{5}4^{5}5} \end{array}$
Check: $7 \times 18{,}377 + 6 = 128{,}639 + 6 = 128{,}645$

57. $3 \times 1908 + 2 = 5724 + 2 = 5726$ incorrect
$\begin{array}{l} \ \ \ 1\ 9\ 0\ 8 \text{ R } 1 \\ 3\overline{)5^{2}7\ 2^{2}5} \end{array}$

61. $6 \times 3568 + 2 = 21{,}408 + 2 = 21{,}410$ incorrect
$\begin{array}{l} \ \ \ 3,\ 5\ 6\ 8 \text{ R } 1 \\ 6\overline{)21{,}^{3}4^{4}0^{4}9} \end{array}$

65. $6 \times 11{,}523 + 2 = 69{,}138 + 2 = 69{,}140$ correct

69. $8 \times 27{,}822 = 222{,}576$ correct

73. $\begin{array}{r} 1\ 5\ 6 \\ 4\overline{)6^{2}2^{2}4} \end{array}$ cartons are needed.

77. $\begin{array}{r} 3\ 2\ 5\ 0 \\ 9\overline{)29{,}^{2}2^{4}5\ 0} \end{array}$ Each student received \$3250.

81. $\begin{array}{r} 1\ \ 8\ \ 2\ \ 2 \\ 58\overline{)105^{47}6^{12}7^{11}6} \end{array}$ each employee received.

85. The number 184 is
divisible by 2 since it ends in 4
not divisible by 3 since $1 + 8 + 4 = 13$
not divisible by 5 since it does not end in 0 or 5
not divisible by 10 since it does not end in 0

	2	3	5	10
184	√	×	×	×

89. The number 903 is
not divisible by 2 since it does not end in 0, 2, 4, 6 or 8
divisible by 3 since $9 + 3 = 12$
not divisible by 5 since it does not end in 0 or 5
not divisible by 10 since it does not end in 0

	2	3	5	10
903	×	√	×	×

93. The number 21,763 is
not divisible by 2 since it does not end in 0, 2, 4, 6 or 8
not divisible by 3 since $2 + 1 + 7 + 6 + 3 = 19$
not divisible by 5 since it does not end in 0 or 5
not divisible by 10 since it does not end in 0

	2	3	5	10
21,763	×	×	×	×

SECTION 1.6 (page 53)

1. $\begin{array}{r} 32 \\ 24\overline{)768} \\ 72 \\ 48 \\ 48 \\ 0 \end{array}$ Check $\begin{array}{r} 32 \\ \times\ 24 \\ \hline 128 \\ 64\ \\ \hline 768 \end{array}$ match

5. $\begin{array}{r} 37 \text{ R } 7 \\ 59\overline{)2190} \\ 177 \\ 420 \\ 413 \\ 7 \end{array}$ Check $\begin{array}{r} 37 \\ \times\ 59 \\ \hline 333 \\ 185\ \\ \hline 2183 \\ +\quad 7 \\ \hline 2190 \end{array}$ match

9. $\begin{array}{r} 38 \\ 58\overline{)2204} \\ 174 \\ 464 \\ 464 \\ 0 \end{array}$ Check $\begin{array}{r} 38 \\ \times\ 58 \\ \hline 304 \\ 190\ \\ \hline 2204 \end{array}$ match

13. $\begin{array}{r} 2\,407 \text{ R } 1 \\ 26\overline{)62{,}583} \\ 52 \\ 10\ 5 \\ 10\ 4 \\ 183 \\ 182 \\ 1 \end{array}$ Check $\begin{array}{r} 2407 \\ \times\quad 26 \\ \hline 14442 \\ 4814\ \\ \hline 62582 \\ +\quad 1 \\ \hline 62583 \end{array}$ match

17. $\begin{array}{r} 522 \text{ R } 14 \\ 46\overline{)24{,}026} \\ 23\ 0 \\ 1\ 02 \\ 92 \\ 106 \\ 92 \\ 14 \end{array}$ Check $\begin{array}{r} 522 \\ \times\quad 46 \\ \hline 3132 \\ 2088\ \\ \hline 24012 \\ +\quad 14 \\ \hline 24026 \end{array}$ match

21.

7 746 R 20
32)247,892
224
23 8
22 4
1 49
1 28
212
192
20

Check
7746
× 32
15492
23238
247872
+ 20
247892 match

25.

21
821)17,241
16 42
821
821
0

Check
21
× 821
21
42
168
17241 match

29.

170
900)153,000
90 0
63 00
63 00
00

Check
170
× 900
153000 match

33. $28 \times 658 + 9 = 18{,}424 + 9 = 18{,}433$ incorrect

658
28)18,424
16 8
1 62
1 40
224
224
0

41. Add.

$3200	amount borrowed
+ $706	interest
$3906	amount to be repaid

45. Multiply.

42	circuits in 1 hour
× 8	hours in a day
336	circuits in 1 day
× 5	days in work week
1680	total circuits

SECTION 1.7 (page 61)

1. 510 51|4 4 or less, leave original number. Change 4 to 0.
↑ tens place

5. 7900 78|62 5 or more, so add 1 to 8. Change 6 and 2 to 0.
↑ hundreds place

9. 42,500 42,4|95 5 or more, so add 1 to 4. Change 9 and 5 to 0.
↑ hundreds place

13. 15,800 15,7|58 5 or more, so add 1 to 7. Change 5 and 8 to 0.
↑ hundreds place

17. 6000 5|847 5 or more, so add 1 to 5. Change 8, 4 and 7 to 0.
↑ thousands place

21. 600,000 59|5,008 5 or more, so add 1 to 9, carry 1 to hundred thousand. Change 5 and 8 to 0.
↑ ten thousands place

25. ten 2370 236|5 5 or more, so add 1 to 6. Change 5 to 0.

hundred 2400 23|65 5 or more, so add 1 to 3. Change 6 and 5 to 0.

thousand 2000 2|365 4 or less, leave original number. Change 3, 6 and 5 to 0.

29. ten 5050 504|9 5 or more, so add 1 to 4. Change 9 to 0.

hundred 5000 50|49 4 or less, leave original number. Change 4 and 9 to 0.

thousand 5000 5|049 4 or less, leave original number. Change 4 and 9 to 0.

33. ten 19,540 19,53|9 5 or more, so add 1 to 3. Change 9 to 0.

hundred 19,500 19,5|39 4 or less, leave original number. Change 3 and 9 to 0.

thousand 20,000 19,|539 5 or more, so add 1 to 9. Carry 1 to ten thousands. Change 5, 3 and 9 to 0.

37. ten 23,500 23,50|2 4 or less, leave original number. Change 2 to 0.

hundred 23,500 23,5|02 4 or less, leave original number. Change 2 to 0.

thousand 24,000 23,|502 5 or more, so add 1 to 3. Change 5 and 2 to 0.

41. estimate
40
60
90
+ 60
250 exact: 250

45. estimate

$$\begin{array}{r} 80 \\ \times\ 20 \\ \hline 1600 \end{array}$$ exact: 1672

49. estimate

$$\begin{array}{r} 700 \\ -\ 400 \\ \hline 300 \end{array}$$ exact: 316

53. estimate

$$\begin{array}{r} 8000 \\ 60 \\ 700 \\ +\ 4000 \\ \hline 12{,}760 \end{array}$$ exact: 12,605

57. estimate

$$\begin{array}{r} 600 \\ \times\ 50 \\ \hline 30{,}000 \end{array}$$ exact: 29,986

61. ten

70,987,650 70,987,65|2 4 or less, leave original number. Change 2 to 0.

ten thousand

70,990,000 70,98|7,650 5 or more, so add 1 to 8. Change 7, 6 and 5 to 0.

ten million

70,000,000 7|0,987,652 4 or less, leave original number. Change 9, 8, 7, 6, 5 and 2 to 0.

SECTION 1.8 (page 67)

1. $2^2 = 4$, so $\sqrt{4} = 2$

5. $12^2 = 144$, so $\sqrt{144} = 12$

9. Exponent is 2, base is 3. $3^2 = 3 \times 3 = 9$

13. Exponent is 2, base is 12. $12^2 = 12 \times 12 = 144$

17. $20^2 = 20 \cdot 20 = 400$, so $\sqrt{400} = 20$

21. $35^2 = 35 \cdot 35 = 1225$, so $\sqrt{1225} = 35$

25. $54^2 = 54 \cdot 54 = 2916$, so $\sqrt{2916} = 54$

29. $8^2 + 4 - 3 = 64 + 4 - 3$
$= 68 - 3$
$= 65$

33. $15 \cdot 2 \div 6 = 30 \div 6$
$= 5$

37. $6 \cdot 2^2 + \frac{0}{6} = 6 \cdot 4 + \frac{0}{6}$
$= 24 + 0$
$= 24$

41. $3^3 \cdot 2^2 + (10 - 5) \cdot 2 = 3^3 \cdot 2^2 + 5 \cdot 2$
$= 27 \cdot 4 + 5 \cdot 2$
$= 108 + 10$
$= 118$

45. $5 \cdot 3 + 8 \cdot 3 - 4 = 15 + 24 - 4$
$= 39 - 4$
$= 35$

49. $8 + 10 \div 5 + \frac{0}{3} = 8 + 2 + 0$
$= 10 + 0$
$= 10$

53. $7 \cdot \sqrt{81} - 5 \cdot 6 = 7 \cdot 9 - 5 \cdot 6$
$= 63 - 30$
$= 33$

57. $4 \cdot \sqrt{49} - 7(5 - 2) = 4 \cdot \sqrt{49} - 7(3)$
$= 4 \cdot 7 - 7(3)$
$= 28 - 21$
$= 7$

61. $6^2 + 2^2 - 6 + 2 = 36 + 4 - 6 + 2$
$= 40 - 6 + 2$
$= 34 + 2$
$= 36$

65. $7 + 6 \div 3 + 5 \cdot 2 = 7 + 2 + 10$
$= 9 + 10$
$= 19$

69. $6^2 - 2^2 + 3 \cdot 4 = 36 - 4 + 3 \cdot 4$
$= 36 - 4 + 12$
$= 32 + 12$
$= 44$

73. $4 \cdot \sqrt{25} - 6 \cdot 2 = 4 \cdot 5 - 6 \cdot 2$
$= 20 - 12$
$= 8$

77. $7 \div 1 \cdot 8 \cdot 2 \div (21 - 5) = 7 \div 1 \cdot 8 \cdot 2 \div 16$
$= 7 \cdot 8 \cdot 2 \div 16$
$= 56 \cdot 2 \div 16$
$= 112 \div 16$
$= 7$

81. $4 \cdot \sqrt{16} - 3 \cdot \sqrt{9} = 4 \cdot 4 - 3 \cdot 3$
$= 16 - 9$
$= 7$

85. $8 \cdot 9 \div \sqrt{36} - 4 \div 2 + (14 - 8)$
$= 8 \cdot 9 \div \sqrt{36} - 4 \div 2 + 6$
$= 8 \cdot 9 \div 6 - 4 \div 2 + 6$
$= 72 \div 6 - 2 + 6$
$= 12 - 2 + 6$
$= 10 + 6$
$= 16$

89. $6 \cdot \sqrt{25} \cdot \sqrt{100} \div 3 \cdot \sqrt{4} + 9$
$= 6 \cdot 5 \cdot 10 \div 3 \cdot 2 + 9$
$= 30 \cdot 10 \div 3 \cdot 2 + 9$
$= 300 \div 3 \cdot 2 + 9$
$= 100 \cdot 2 + 9$
$= 200 + 9$
$= 209$

SECTION 1.9 (page 75)

1. *Step 1* The members in each division are given and the total number of members must be found

Step 2 Add the members in each division to get the total.

Step 3 A reasonable answer would be 850 since

$600 + 200 + 50 = 850$ (estimate)

Step 4 Find the actual total by adding all the members.

```
        836   check
        582
        208
      +  46
exact:  836   Total number of members.
```

The answer is reasonable.

5. *Step 1* The number of kits packaged in one hour is given and the total number of kits packaged in 24 hours must be found.

Step 2 Multiply the number of kits packaged in 1 hour by 24.

Step 3 A reasonable answer would be 4000 (estimate) (200 × 20)

Step 4 Multiply.

```
        236
      ×  24
exact: 5664   kits packaged in 24 hours
```

The answer is reasonable. Check

```
     236
24)5664
```

9. *Step* 1 The number of people at the lake on Friday is higher than that on Wednesday. The number on Friday and the amount higher is given. The number on Wednesday must be found.

Step 2 Subtract the number higher from the number on Friday.

Step 3 A reasonable answer would be 4000 (estimate) (8000 − 4000)

Step 4 Subtract.

```
        8392
      − 4218
exact:  4174   people at the lake on Wednesday
```

The answer is reasonable. Check:

```
  4218
+ 4174
  8392
```

13. *Step* 1 The total miles travelled was divided among 18 days. The miles traveled each day must be found.

Step 2 Divide the total miles by the number of days.

Step 3 A reasonable answer is 150 (estimate) (3000 ÷ 20)

Step 4 Divide.

```
            175
exact: 18)3150   175 miles were travelled each day.
```

The answer is reasonable. Check

```
   175
×   18
  3150
```

17. *Step 1* The square feet in one acre is given and the square feet in the given number of acres must be found.

Step 2 Multiply the square feet in one acre by the total acres.

Step 3 A reasonable answer is 4,000,000 (estimate) (40,000 × 100)

Step 4 Multiply.

```
          43,560
       ×     138
exact: 6,011,280   square feet in 138 acres.
```

The answer is reasonable. Check

```
        43,560
138)6,011,280
```

21. *Step* 1 The number and cost of wheelchairs and players is given and the total cost of all items must be found.

Step 2 First, find the cost of all wheelchairs and the cost of all players. Then add these costs to get the total cost.

Step 3 The cost of wheelchairs is about \$6000 (\$ 1000 × 6)
The cost of players is about \$18,000 (\$900 × 20)
Therefore, a reasonable answer would be \$24,000 (estimate)

Step 4 Cost of wheelchairs   Cost of players

```
  $1256          $895
×     6       ×    15
  $7536       $13,425
```

exact: Total cost = \$7536 + \$13,425 = \$20,961.
The answer is reasonable. Check

```
   $20,961
 − $13,425
    $7536

    1256          895
 6)7536     15)13,425
```

29. *Step* 1 The cost of packages of undershirts and socks is given and the total cost of a different number of these items must be found.

Step 2 First, find the number of packages of undershirts and socks required. Then find the cost of each item. Finally, add these costs to get the total cost.

Step 3 Packages of undershirts is 10 (30 ÷ 3)
Packages of socks is 3 (18 ÷ 6)
Cost of undershirts is about \$100 (\$10 × 10)
Cost of socks is about \$60 (\$20 × 3)
A reasonable answer is \$160 (estimate) (\$100 + \$60)

Step 4 Cost of undershirts   Cost of socks

```
    10          3
× $12       × $15
  $120        $45
```

exact: Total cost = \$120 + \$45 = \$165

The answer is reasonable. Check

$$\begin{array}{r} \$165 \\ -\ \$45 \\ \hline \$120 \end{array}$$

$$10\overline{)120} = 12 \qquad 3\overline{)45} = 15$$

33. *Step 1* The selling prices and the number of orange and lemon trees sold are given and the total sales must be found.

Step 2 First find the amount of sales of both orange trees and lemon trees. Finally, add these amounts to find the total sales.

Step 3 Sales of orange trees are \$800 (80 × \$10). Sales of lemon trees are \$1000 (50 × \$20). A reasonable answer for total sales is \$1800 (\$800 + \$1000).

Step 4 Sales of orange trees Sales of lemon trees

$$\begin{array}{r} 80 \\ \times\ \$10 \\ \hline \$800 \end{array} \qquad \begin{array}{r} 50 \\ \times\ \$20 \\ \hline \$1000 \end{array}$$

exact: Total sales = \$800 + \$1000 = \$1800

The answer is reasonable. Check

$$\begin{array}{r} \$1800 \\ -\ 1000 \\ \hline \$800 \end{array}$$

$$80\overline{)800} = 10 \qquad 50\overline{)1000} = 20$$

37. *Step 1* The total seating is given along with the information to find the number of seats on the main floor. The number of rows of seats in the balcony is given and the number of seats in each row of the balcony must be found.

Step 2 First the number of seats on the main floor must be found. Next, the number of seats on the main floor must be subtracted from the total number of seats to find the number of seats in the balcony. Finally the number of seats in the balcony must be divided by the number of rows of seats in the balcony.

Step 3 Number of seats on the main floor is 750 (30 × 25). Number of seats in balcony is about 500 (1300 − 800). A reasonable answer is about 20 (500 ÷ 25).

Step 4 Number of seats in balcony

$$\begin{array}{r} 1250 \\ -\ 750 \\ \hline 500 \end{array}$$

Number of seats in each row

exact: $25\overline{)500} = 20$

Number of seats in each row of the balcony is 20.

Answer is reasonable. Check

$$\begin{array}{r} 25 \\ \times\ 20 \\ \hline 500 \end{array} \qquad \begin{array}{r} 1250 \\ -\ 500 \\ \hline 750 \end{array} \qquad 30\overline{)750} = 25$$

CHAPTER 2

SECTION 2.1 (page 97)

1. $\frac{5}{8}$ (There are eight parts, and five are shaded.)

5. $\frac{7}{5}$ (The object is divided into five parts, and seven of these parts are shaded.)

9. $\frac{8}{25}$ (There are 25 students, and 8 are hearing-impaired.)

13. $\frac{3}{8}$ 3 ← numerator, 8 ← denominator

17. proper: $\frac{1}{4}, \frac{3}{8}, \frac{7}{12}$ improper: $\frac{9}{7}, \frac{11}{4}, \frac{8}{8}$

25. 2 × 3 × 3 = 6 × 3 = 18

29. 21 ÷ 3 = 7

SECTION 2.2 (page 101)

1. $1\frac{3}{4}$ $1 \cdot 4 = 4$ $4 + 3 = 7$ $1\frac{3}{4} = \frac{7}{4}$

5. $3\frac{1}{4}$ $3 \cdot 4 = 12$ $12 + 1 = 13$ $3\frac{1}{4} = \frac{13}{4}$

9. $1\frac{7}{11}$ $1 \cdot 11 = 11$ $11 + 7 = 18$ $1\frac{7}{11} = \frac{18}{11}$

13. $11\frac{1}{3}$ $11 \cdot 3 = 33$ $33 + 1 = 34$ $11\frac{1}{3} = \frac{34}{3}$

17. $3\frac{3}{8}$ $3 \cdot 8 = 24$ $24 + 3 = 27$ $3\frac{3}{8} = \frac{27}{8}$

21. $4\frac{10}{11}$ $4 \cdot 11 = 44$ $44 + 10 = 54$ $4\frac{10}{11} = \frac{54}{11}$

25. $17\frac{12}{13}$ $17 \cdot 13 = 221$ $221 + 12 = 233$ $17\frac{12}{13} = \frac{233}{13}$

29. $6\frac{7}{18}$ $6 \cdot 18 = 108$ $108 + 7 = 115$ $6\frac{7}{18} = \frac{115}{18}$

33. $3\overline{)8}$ = 2 R 2; $\begin{array}{r} 6 \\ \hline 2 \end{array}$ $\frac{8}{3} = 2\frac{2}{3}$

37. $8\overline{)27}$ = 3 R 3; $\begin{array}{r} 24 \\ \hline 3 \end{array}$ $\frac{27}{8} = 3\frac{3}{8}$

41.
```
    3
9)27
  27
   0
```
$\frac{27}{3} = 9$

45.
```
    5 R 2
9)47
  45
   2
```
$\frac{47}{9} = 5\frac{2}{9}$

49.
```
   16 R 4
5)84
  5
  34
  30
   4
```
$\frac{84}{5} = 16\frac{4}{5}$

53.
```
   26 R 1
7)183
  14
   43
   42
    1
```
$\frac{183}{7} = 26\frac{1}{7}$

57. $255 \cdot 8 = 2040 \quad 2040 + 1 = 2041 \quad 255\frac{1}{8} = \frac{2041}{8}$

61.
```
     171
15)2565
   15
   106
   105
    15
    15
     0
```
$\frac{2565}{15} = 171$

65. $15 \cdot 4 + 6 = 60 + 6$
$= 66$

SECTION 2.3 (page 107)

1. Factorizations of 6:

$1 \cdot 6 \quad 2 \cdot 3$

Factors of 6 are 1, 2, 3, 6.

5. Factorizations of 25:

$1 \cdot 25 \quad 5 \cdot 5$

Factors of 25 are 1, 5, 25.

9. Factorizations of 40:

$1 \cdot 40 \quad 2 \cdot 20 \quad 4 \cdot 10 \quad 5 \cdot 8$

Factors of 40 are 1, 2, 4, 5, 8, 10, 20, 40.

13. Because it can be divided by 2, 4 is composite.
17. Because it can be divided by 3, 9 is composite.
21. Because it can be divided by only itself and 1, 19 is prime.
25. Because it can be divided by 2 and 17, 34 is composite.

29.
```
2)4
2)2
  1
```
$4 = 2^2$

4 → ② ②

33.
```
5)25
5)5
  1
```
$25 = 5^2$

25 → ⑤ ⑤

37.
```
 3)39
13)13
    1
```
$39 = 3 \cdot 13$

39 → ③ ⑬

41.
```
3)75
5)25
 5)5
   1
```
$75 = 3 \cdot 5^2$

75 → ③ 25; 25 → ⑤ ⑤

45.
```
5)125
 5)25
  5)5
    1
```
$125 = 5^3$

125 → ⑤ 25; 25 → ⑤ ⑤

49.
```
2)320
 2)160
  2)80
   2)40
    2)20
     2)10
       5)5
         1
```
$320 = 2^6 \cdot 5$

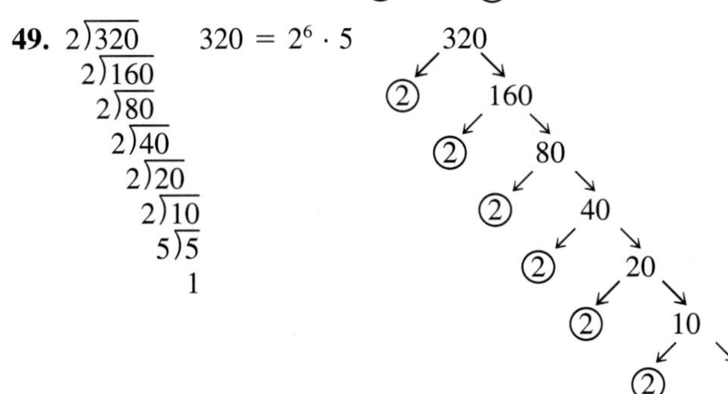

57. $8^3 = 8 \cdot 8 \cdot 8$
$= 512$

61. $2^2 \cdot 3^3 = 2 \cdot 2 \cdot 3 \cdot 3 \cdot 3$
$= 4 \cdot 27$
$= 108$

65. $3^5 \cdot 2^2 = 3 \cdot 3 \cdot 3 \cdot 3 \cdot 3 \cdot 2 \cdot 2$
$= 243 \cdot 4$
$= 972$

69.
```
2)960
 2)480
  2)240
   2)120
    2)60
     2)30
      3)15
       5)5
         1
```
$960 = 2^6 \cdot 3 \cdot 5$

73. $5 \cdot 1 \cdot 2 = 5 \cdot 2 = 10$
77. $24 \div 3 = 8$

SECTION 2.4 (page 115)

1. $\frac{10}{20} = \frac{10 \div 10}{20 \div 10} = \frac{1}{2}$

5. $\frac{30}{48} = \frac{30 \div 6}{48 \div 6} = \frac{5}{8}$

9. $\frac{63}{70} = \frac{63 \div 7}{70 \div 7} = \frac{9}{10}$

SOLUTIONS

13. $\frac{36}{63} = \frac{36 \div 9}{63 \div 9} = \frac{4}{7}$

17. $\frac{96}{132} = \frac{96 \div 12}{132 \div 12} = \frac{8}{11}$

21. $\frac{10}{16} = \frac{\overset{1}{\cancel{2}} \cdot 5}{\underset{1}{\cancel{2}} \cdot 2 \cdot 2 \cdot 2} = \frac{5}{8}$

25. $\frac{75}{150} = \frac{\overset{1}{\cancel{3}} \cdot \overset{1}{\cancel{5}} \cdot \overset{1}{\cancel{5}}}{2 \cdot \underset{1}{\cancel{3}} \cdot \underset{1}{\cancel{5}} \cdot \underset{1}{\cancel{5}}} = \frac{1}{2}$

29. $\frac{77}{264} = \frac{7 \cdot \overset{1}{\cancel{11}}}{2 \cdot 2 \cdot 2 \cdot 3 \cdot \underset{1}{\cancel{11}}} = \frac{7}{24}$

33. $\frac{12}{15} \times \frac{35}{45}$

$15 \cdot 35 = 525$

$12 \cdot 45 = 540$

not equivalent

The fractions are not equivalent.

37. $\frac{14}{16} \times \frac{35}{40}$

$16 \cdot 35 = 560$

$14 \cdot 40 = 560$

equivalent

The fractions are equivalent.

41. $\frac{25}{30} \times \frac{65}{78}$

$30 \cdot 65 = 1950$

$25 \cdot 78 = 1950$

equivalent

The fractions are equivalent.

45. $\frac{224}{256} = \frac{\overset{1}{\cancel{2}} \cdot \overset{1}{\cancel{2}} \cdot \overset{1}{\cancel{2}} \cdot \overset{1}{\cancel{2}} \cdot \overset{1}{\cancel{2}} \cdot 7}{\underset{1}{\cancel{2}} \cdot \underset{1}{\cancel{2}} \cdot \underset{1}{\cancel{2}} \cdot \underset{1}{\cancel{2}} \cdot \underset{1}{\cancel{2}} \cdot 2 \cdot 2 \cdot 2} = \frac{7}{8}$

49. Factorizations of 12:

$1 \cdot 12 \qquad 2 \cdot 6 \qquad 3 \cdot 4$

Factors of 12 are 1, 2, 3, 4, 6, 12

SECTION 2.5 (page 123)

1. $\frac{1}{2} \cdot \frac{3}{4} = \frac{1 \cdot 3}{2 \cdot 4} = \frac{3}{8}$

5. $\frac{3}{\underset{2}{\cancel{8}}} \cdot \frac{\overset{3}{\cancel{12}}}{5} = \frac{3 \cdot 3}{2 \cdot 5} = \frac{9}{10}$

9. $\frac{\overset{1}{\cancel{3}}}{\underset{2}{\cancel{4}}} \cdot \frac{5}{6} \cdot \frac{\overset{1}{\cancel{2}}}{\underset{1}{\cancel{3}}} = \frac{1 \cdot 5 \cdot 1}{2 \cdot 6 \cdot 1} = \frac{5}{12}$

13. $\frac{\overset{\overset{1}{\cancel{3}}}{\cancel{21}}}{\underset{\underset{2}{\cancel{6}}}{\cancel{30}}} \cdot \frac{\overset{1}{\cancel{5}}}{\underset{1}{\cancel{7}}} = \frac{1 \cdot 1}{2 \cdot 1} = \frac{1}{2}$

17. $\frac{\overset{1}{\cancel{16}}}{\underset{\underset{1}{\cancel{5}}}{\cancel{25}}} \cdot \frac{\overset{7}{\cancel{35}}}{\underset{2}{\cancel{32}}} \cdot \frac{\overset{3}{\cancel{15}}}{64} = \frac{1 \cdot 7 \cdot 3}{1 \cdot 2 \cdot 64} = \frac{21}{128}$

21. $72 \cdot \frac{5}{9} = \frac{\overset{8}{\cancel{72}}}{1} \cdot \frac{5}{\underset{1}{\cancel{9}}} = \frac{8 \cdot 5}{1 \cdot 1} = \frac{40}{1} = 40$

25. $42 \cdot \frac{7}{10} \cdot \frac{5}{7} = \frac{\overset{21}{\cancel{42}}}{1} \cdot \frac{\overset{1}{\cancel{7}}}{\underset{\underset{1}{\cancel{2}}}{\cancel{10}}} \cdot \frac{\overset{1}{\cancel{5}}}{\underset{1}{\cancel{7}}} = \frac{21 \cdot 1 \cdot 1}{1 \cdot 1} = \frac{21}{1} = 21$

29. $\frac{3}{5} \cdot 400 = \frac{3}{\underset{1}{\cancel{5}}} \cdot \frac{\overset{80}{\cancel{400}}}{1} = \frac{3 \cdot 80}{1 \cdot 1} = \frac{240}{1} = 240$

33. $\frac{28}{21} \cdot 640 \cdot \frac{15}{32} = \frac{\overset{4}{\cancel{28}}}{\underset{\underset{1}{\cancel{3}}}{\cancel{21}}} \cdot \frac{\overset{20}{\cancel{640}}}{1} \cdot \frac{\overset{5}{\cancel{15}}}{\underset{1}{\cancel{32}}} = \frac{4 \cdot 20 \cdot 5}{1 \cdot 1 \cdot 1}$

$= \frac{400}{1} = 400$

37. area = length · width

$= \frac{\overset{1}{\cancel{4}}}{\underset{1}{\cancel{5}}} \cdot \frac{\overset{1}{\cancel{5}}}{\underset{2}{\cancel{8}}}$

$= \frac{1}{2}$ square inch

41. area = length · width

$= \frac{\overset{1}{\cancel{7}}}{5} \cdot \frac{3}{\underset{2}{\cancel{14}}}$

$= \frac{3}{10}$ square inch

45. area = length · width

$= 16 \cdot \frac{1}{4}$

$= \frac{\overset{4}{\cancel{16}}}{1} \cdot \frac{1}{\underset{1}{\cancel{4}}}$

= 4 square inches

49. area = length · width

$= 2 \cdot \frac{1}{2}$

$= \frac{\overset{1}{\cancel{2}}}{1} \cdot \frac{1}{\underset{1}{\cancel{2}}}$

= 1 square mile

SECTION 2.6 (page 129)

1. Multiply the length by the width

$$\frac{5}{\underset{2}{\cancel{4}}} \cdot \frac{\overset{1}{\cancel{2}}}{3} = \frac{5}{6}$$

The area of the desk is $\frac{5}{6}$ square yard.

5. Multiply the total items by the fraction taxable.

$$\frac{1}{5} \cdot 1900 = \frac{1}{\underset{1}{\cancel{5}}} \cdot \frac{\overset{380}{\cancel{1900}}}{1} = \frac{380}{1} = 380$$

380 items are taxable.

9. Multiply the number of freshmen by the fraction receiving scholarships.

$$\frac{5}{24} \cdot 1800 = \frac{5}{\underset{1}{\cancel{24}}} \cdot \frac{\overset{75}{\cancel{1800}}}{1} = \frac{375}{1} = 375$$

375 students received scholarships.

13. Add the income for all twelve months to find the income for the year.

$$\begin{aligned} &2550 + 2375 + 2825 + 2520 + 2380 + 2765 \\ &\quad + 2660 + 1855 + 2280 + 3175 + 2810 \\ &\quad + 3805 = 32{,}000 \end{aligned}$$

The Gomes family had income of \$32,000 for the year.

17. Multiply the total income by the fraction saved.

$$\frac{1}{16} \cdot 32{,}000 = \frac{1}{\underset{1}{\cancel{16}}} \cdot \frac{\overset{2000}{\cancel{32{,}000}}}{1} = \frac{2000}{1} = 2000$$

The Gomes family saved \$2000 for the year.

21. Multiply the amount of earnings for an 8-hour day by a fraction having a numerator of the actual number of hours worked and a denominator of 8.

$$120 \cdot \frac{5}{8} = \frac{\overset{15}{\cancel{120}}}{1} \cdot \frac{5}{\underset{1}{\cancel{8}}} = 75$$

Kerry earned \$75 in 5 hours.

25. Multiply the remaining $\frac{1}{8}$ of the estate by the fraction going to the American Cancer Society.

$$\frac{1}{4} \cdot \frac{1}{8} = \frac{1}{32}$$

$\frac{1}{32}$ of the estate goes to the American Cancer Society.

SECTION 2.7 (page 137)

1. $\frac{1}{4} \div \frac{1}{2} = \frac{1}{\underset{2}{\cancel{4}}} \cdot \frac{\overset{1}{\cancel{2}}}{1} = \frac{1}{2}$

5. $\frac{5}{8} \div \frac{3}{16} = \frac{5}{\underset{1}{\cancel{8}}} \cdot \frac{\overset{2}{\cancel{16}}}{3} = \frac{10}{3} = 3\frac{1}{3}$

9. $\dfrac{\frac{7}{9}}{\frac{7}{36}} = \frac{7}{9} \div \frac{7}{36} = \frac{\overset{1}{\cancel{7}}}{\underset{1}{\cancel{9}}} \cdot \frac{\overset{4}{\cancel{36}}}{\underset{1}{\cancel{7}}} = \frac{4}{1} = 4$

13. $6 \div \frac{2}{3} = \frac{\overset{3}{\cancel{6}}}{1} \cdot \frac{3}{\underset{1}{\cancel{2}}} = \frac{9}{1} = 9$

17. $\dfrac{\frac{4}{7}}{8} = \frac{4}{7} \div \frac{8}{1} = \frac{\overset{1}{\cancel{4}}}{7} \cdot \frac{1}{\underset{2}{\cancel{8}}} = \frac{1}{14}$

21. Divide the amount of salt by the pounds in each shaker.

$$28 \div \frac{4}{5} = \frac{\overset{7}{\cancel{28}}}{1} \cdot \frac{5}{\underset{1}{\cancel{4}}} = \frac{35}{1} = 35$$

35 salt shakers can be filled.

25. Divide the total cords of wood by the amount per trip.

$$40 \div \frac{2}{3} = \frac{\overset{20}{\cancel{40}}}{1} \cdot \frac{3}{\underset{1}{\cancel{2}}} = \frac{60}{1} = 60$$

Pam has to make 60 trips.

33. First, divide the amount raised by the fraction of the total raised.

$$\$840{,}000 \div \frac{7}{8} = \frac{\overset{120{,}000}{\cancel{840{,}000}}}{1} \cdot \frac{8}{\underset{1}{\cancel{7}}} = \$960{,}000 \text{ needed}$$

Next, subtract the amount raised from the total funds needed.

$$\$960{,}000 - \$840{,}000 = \$120{,}000$$

The committee still needs \$120,000.

37. $15\frac{1}{3} \qquad 15 \cdot 3 = 45 \qquad 45 + 1 = 46 \qquad 15\frac{1}{3} = \frac{46}{3}$

SECTION 2.8 (page 145)

1. $2\frac{1}{4} \cdot 3\frac{1}{2} = \frac{9}{4} \cdot \frac{7}{2} = \frac{63}{8} = 7\frac{7}{8}$

5. $3\frac{1}{9} \cdot 1\frac{2}{7} = \frac{\overset{4}{\cancel{28}}}{\underset{1}{\cancel{9}}} \cdot \frac{\overset{1}{\cancel{9}}}{\underset{1}{\cancel{7}}} = \frac{4}{1} = 4$

9. $4\frac{1}{2} \cdot 2\frac{1}{5} \cdot 5 = \frac{9}{2} \cdot \frac{11}{\cancel{5}_1} \cdot \frac{\cancel{5}^1}{1} = 49\frac{1}{2}$

13. $3\frac{1}{4} \div 2\frac{5}{8} = \frac{13}{4} \div \frac{21}{8} = \frac{13}{\cancel{4}_1} \cdot \frac{\cancel{8}^2}{21} = \frac{26}{21} = 1\frac{5}{21}$

17. $6 \div 1\frac{1}{4} = \frac{6}{1} \div \frac{5}{4} = \frac{6}{1} \cdot \frac{4}{5} = \frac{24}{5} = 4\frac{4}{5}$

21. $1\frac{7}{8} \div 6\frac{1}{4} = \frac{15}{8} \div \frac{25}{4} = \frac{\cancel{15}^3}{\cancel{8}_2} \cdot \frac{\cancel{4}^1}{\cancel{25}_5} = \frac{3}{10}$

25. Multiply the number of dolls by the material per doll.

$$8 \cdot 1\frac{5}{8} = \frac{\cancel{8}^1}{1} \cdot \frac{13}{\cancel{8}_1} = \frac{13}{1} = 13$$

Shirley needs 13 yards of material.

29. Multiply the ounces of chemical by the gallons of water.

$$1\frac{3}{4} \cdot 12\frac{1}{2} = \frac{7}{4} \cdot \frac{25}{2} = \frac{175}{8} = 21\frac{7}{8}$$

$21\frac{7}{8}$ ounces of chemical are needed.

33. Divide the length of the tube by the length of each spacer.

$$5\frac{1}{2} \div \frac{1}{2} = \frac{11}{\cancel{2}_1} \cdot \frac{\cancel{2}^1}{1} = 11$$

11 spacers can be cut from the tube.

37. Divide the number of gallons in the tank by the fraction that it is full.

$$35 \div \frac{5}{8} = \frac{\cancel{35}^7}{1} \cdot \frac{8}{\cancel{5}_1} = 56$$

The tank will hold 56 gallons when it is full.

41. $\frac{25}{40} = \frac{5 \cdot \cancel{5}}{2 \cdot 2 \cdot 2 \cdot \cancel{5}} = \frac{5}{8}$

CHAPTER 3

SECTION 3.1 (page 165)

1. $\frac{1}{3} + \frac{1}{3} = \frac{1+1}{3} = \frac{2}{3}$

5. $\frac{1}{4} + \frac{1}{4} = \frac{1+1}{4} = \frac{2}{4} = \frac{1}{2}$

9. $\frac{7}{12} + \frac{3}{12} + \frac{7+3}{12} = \frac{10}{12} = \frac{5}{6}$

13. $\frac{3}{17} + \frac{2}{17} + \frac{5}{17} = \frac{3+2+5}{17} = \frac{10}{17}$

17. $\frac{2}{54} + \frac{8}{54} + \frac{12}{54} = \frac{2+8+12}{54} = \frac{22}{54} = \frac{11}{27}$

21. $\frac{16}{21} - \frac{8}{21} = \frac{16-8}{21} = \frac{8}{21}$

25. $\frac{14}{15} - \frac{4}{15} = \frac{14-4}{15} = \frac{10}{15} = \frac{2}{3}$

29. $\frac{43}{72} - \frac{25}{72} = \frac{43-25}{72} = \frac{18}{72} = \frac{1}{4}$

33. $\frac{746}{400} - \frac{506}{400} = \frac{746-506}{400} = \frac{240}{400} = \frac{3}{5}$

37. Add the distance jogged down the road to the distance jogged along the creek.

$$\frac{5}{12} + \frac{1}{12} = \frac{5+1}{12} = \frac{6}{12} = \frac{1}{2}$$

Maria jogged a total of $\frac{1}{2}$ mile.

41. Add the acres planted in the morning to the acres planted in the afternoon.

$$\frac{5}{12} + \frac{11}{12} = \frac{5+11}{12} = \frac{16}{12}$$

Then, subtract the acres destroyed by frost.

$$\frac{16}{12} - \frac{7}{12} = \frac{16-7}{12} = \frac{9}{12} = \frac{3}{4}$$

$\frac{3}{4}$ acre of seedlings remained.

45. $2\overline{)100}$, $2\overline{)50}$, $5\overline{)25}$, $5\overline{)5}$, 1 $100 = 2 \cdot 2 \cdot 5 \cdot 5$

100 → (2), 50; 50 → (2), 25; 25 → (5), (5)

49. $5\overline{)125}$, $5\overline{)25}$, $5\overline{)5}$, 1 $125 = 5 \cdot 5 \cdot 5$

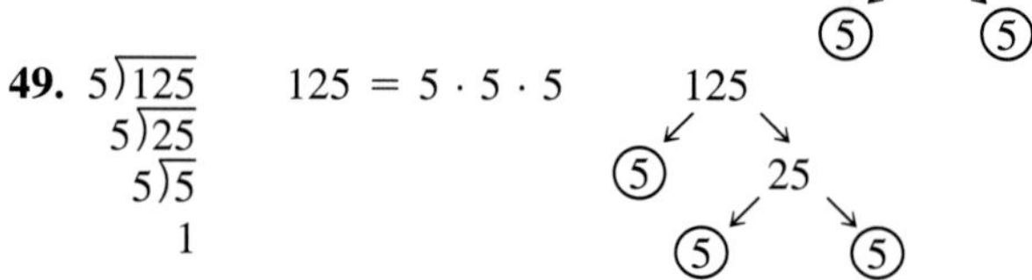

SECTION 3.2 (page 173)

1.

prime	2	3
6 =	2 ·	3
12 =	(2 · 2 ·)	(3)
LCM =	(2 · 2)	(3)

LCM = 2 · 2 · 3 = 12

5.

prime	2	3
18 =	2 ·	(3 · 3)
24 =	(2 · 2 · 2)	3
LCM =	(2 · 2 · 2)	(3 · 3)

LCM = 2 · 2 · 2 · 3 · 3 = 72

9.

prime	*2*	*3*	*5*
36 =	2 · 2 ·	3 · 3	
45 =		3 · 3 ·	5
LCM =	2 · 2	3 · 3	5

LCM = 2 · 2 · 3 · 3 · 5 = 180

13.

prime	*2*	*3*	*5*
15 =		3 ·	5
24 =	2 · 2 · 2 ·	3	
30 =	2 ·	3 ·	5
LCM =	2 · 2 · 2	3	5

LCM = 2 · 2 · 2 · 3 · 5 = 120

17.

prime	*2*	*3*	*5*
6 =	2 ·	3	
8 =	2 · 2 · 2		
10 =	2 ·		5
12 =	2 · 2 ·	3	
LCM =	2 · 2 · 2	3	5

LCM = 2 · 2 · 2 · 3 · 5 = 120

21.

prime	*2*	*3*
6 =	2 ·	3
9 =		3 · 3
27 =		3 · 3 · 3
36 =	2 · 2 ·	3 · 3
LCM =	2 · 2	3 · 3 · 3

LCM = 2 · 2 · 3 · 3 · 3 = 108

25. $\frac{1}{2} = \frac{1 \cdot 12}{2 \cdot 12} = \frac{12}{24}$

29. $\frac{3}{8} = \frac{3 \cdot 3}{8 \cdot 3} = \frac{9}{24}$

33. $\frac{9}{10} = \frac{9 \cdot 4}{10 \cdot 4} = \frac{36}{40}$

37. $\frac{5}{6} = \frac{5 \cdot 11}{6 \cdot 11} = \frac{55}{66}$

41. $\frac{9}{7} = \frac{9 \cdot 8}{7 \cdot 8} = \frac{72}{56}$

45. $\frac{8}{11} = \frac{8 \cdot 12}{11 \cdot 12} = \frac{96}{132}$

53.

prime	*2*	*3*	*7*
1512 =	2 · 2 · 2 ·	3 · 3 · 3 ·	7
392 =	2 · 2 · 2 ·		7 · 7
LCM =	2 · 2 · 2	3 · 3 · 3	7 · 7

LCM = 2 · 2 · 2 · 3 · 3 · 3 · 7 · 7 = 10,584

57. $5\overline{)9}$ = 1 R4 (subtract 5, remainder 4) $\frac{9}{5} = 1\frac{4}{5}$

SECTION 3.3 (page 181)

1. $\frac{2}{5} + \frac{1}{5} = \frac{2+1}{5} = \frac{3}{5}$

5. $\frac{9}{20} + \frac{3}{10} = \frac{9}{20} + \frac{6}{20} = \frac{9+6}{20} = \frac{15}{20} = \frac{3}{4}$

9. $\frac{2}{9} + \frac{5}{12} = \frac{8}{36} + \frac{15}{36} = \frac{8+15}{36} = \frac{23}{36}$

13. $\frac{1}{4} + \frac{2}{9} + \frac{1}{3} = \frac{9}{36} + \frac{8}{36} + \frac{12}{36} = \frac{9+8+12}{36} = \frac{29}{36}$

17. $\frac{4}{15} + \frac{1}{6} + \frac{1}{3} = \frac{8}{30} + \frac{5}{30} + \frac{10}{30} = \frac{8+5+10}{30} = \frac{23}{30}$

21. $\frac{8}{15} + \frac{3}{10} = \frac{16}{30} + \frac{9}{30} = \frac{16+9}{30} = \frac{25}{30} = \frac{5}{6}$

25. $\frac{2}{3} - \frac{1}{6} = \frac{4}{6} - \frac{1}{6} = \frac{4-1}{6} = \frac{3}{6} = \frac{1}{2}$

29. $\frac{7}{8} - \frac{2}{3} = \frac{21}{24} - \frac{16}{24} = \frac{21-16}{24} = \frac{5}{24}$

33. $\frac{4}{5} - \frac{2}{3} = \frac{12}{15} - \frac{10}{15} = \frac{12-10}{15} = \frac{2}{15}$

37. Add the tons of white sand, black sand and pea gravel.

$$\frac{1}{3} + \frac{3}{8} + \frac{1}{4} = \frac{8}{24} + \frac{9}{24} + \frac{6}{24} = \frac{23}{24}$$

$\frac{23}{24}$ ton of material was purchased.

41. Add the distances of each side of the field.

$$\frac{1}{6} + \frac{1}{4} + \frac{1}{16} + \frac{3}{8} = \frac{8}{48} + \frac{12}{48} + \frac{3}{48} + \frac{18}{48} = \frac{41}{48}$$

The total distance around the parcel of land is $\frac{41}{48}$ mile.

49. Change each fraction so that they have a common denominator.

$$\frac{1}{3} = \frac{8}{24} \quad \frac{1}{6} = \frac{4}{24} \quad \frac{1}{8} = \frac{3}{24} \quad \frac{1}{12} = \frac{2}{24} \quad \frac{7}{24} = \frac{7}{24}$$

The largest fraction is $\frac{8}{24}$. The greatest amount of time was spent in work and travel ($\frac{1}{3} = \frac{8}{24}$).

53. $6\frac{1}{2} \cdot 3\frac{1}{4} = \frac{13}{2} \cdot \frac{13}{4} = \frac{169}{8} = 21\frac{1}{8}$

57. $1\frac{1}{2} \div 3\frac{3}{4} = \frac{\overset{1}{\cancel{3}}}{\underset{1}{\cancel{2}}} \cdot \frac{\overset{2}{\cancel{4}}}{\underset{5}{\cancel{15}}} = \frac{2}{5}$

SOLUTIONS

SECTION 3.4 (page 189)

1. $$\begin{array}{r} 8\frac{3}{8} \\ +\ 9\frac{1}{8} \\ \hline 17\frac{4}{8} = 17\frac{1}{2} \end{array}$$

5. $$\begin{array}{rcl} \frac{3}{8} &=& \frac{3}{8} \\ +\ 15\frac{1}{4} &=& 15\frac{2}{8} \\ \hline && 15\frac{5}{8} \end{array}$$

9. $$\begin{array}{rcl} 14\frac{6}{7} &=& 14\frac{12}{14} \\ +\ 5\frac{1}{2} &=& 5\frac{7}{14} \\ \hline && 19\frac{19}{14} = 19 + 1\frac{5}{14} = 20\frac{5}{14} \end{array}$$

13. $$\begin{array}{rcl} 7\frac{1}{4} &=& 7\frac{2}{8} \\ 25\frac{3}{8} &=& 25\frac{3}{8} \\ +\ 9\frac{1}{2} &=& 9\frac{4}{8} \\ \hline && 41\frac{9}{8} = 41 + 1\frac{1}{8} = 42\frac{1}{8} \end{array}$$

17. $$\begin{array}{rcl} 27\frac{7}{15} &=& 27\frac{14}{30} \\ 30\frac{1}{2} &=& 30\frac{15}{30} \\ +\ 9\frac{3}{10} &=& 9\frac{9}{30} \\ \hline && 66\frac{38}{30} = 66 + 1\frac{8}{30} = 67\frac{8}{30} = 67\frac{4}{15} \end{array}$$

21. $$\begin{array}{rcl} 6\frac{7}{12} &=& 6\frac{7}{12} \\ -\ 2\frac{1}{3} &=& 2\frac{4}{12} \\ \hline && 4\frac{3}{12} = 4\frac{1}{4} \end{array}$$

25. $$\begin{array}{r} 19 \\ -\ 8\frac{7}{8} \\ \hline \end{array}$$

Borrow.

$$19 = 18 + 1 = 18 + \frac{8}{8} = 18\frac{8}{8}$$

Subtract. $$\begin{array}{r} 18\frac{8}{8} \\ -\ 8\frac{7}{8} \\ \hline 10\frac{1}{8} \end{array}$$

29. $$\begin{array}{rcl} 25\frac{13}{24} &=& 25\frac{26}{48} \\ -\ 18\frac{15}{16} &=& 18\frac{45}{48} \\ \hline \end{array}$$

Borrow.

$$25\frac{26}{48} = 24 + 1 + \frac{26}{48} = 24 + \frac{48}{48} + \frac{26}{48} = 24\frac{74}{48}$$

Subtract. $$\begin{array}{r} 24\frac{74}{48} \\ -\ 18\frac{45}{48} \\ \hline 6\frac{29}{48} \end{array}$$

33. $$\begin{array}{r} 429\frac{15}{16} \\ -\ 57 \\ \hline 372\frac{15}{16} \end{array}$$

37. $$\begin{array}{r} 415 \\ -\ 198\frac{3}{4} \\ \hline \end{array}$$

Borrow.

$$415 = 414 + 1 = 414 + \frac{4}{4} = 414\frac{4}{4}$$

Subtract. $$\begin{array}{r} 414\frac{4}{4} \\ -\ 198\frac{3}{4} \\ \hline 216\frac{1}{4} \end{array}$$

45. Add the tons recycled on Monday to the tons recycled on Tuesday.

$$\begin{array}{rcl} 5\frac{3}{4} &=& 5\frac{15}{20} \\ +\ 9\frac{3}{5} &=& 9\frac{12}{20} \\ \hline && 14\frac{27}{20} = 14 + 1\frac{7}{20} = 15\frac{7}{20} \end{array}$$

The total recycled on the two days was $15\frac{7}{20}$ tons.

49. First, add the yards of material used for all clothing.

$$\begin{array}{rcl} 3\frac{3}{4} &=& 3\frac{6}{8} \\ 4\frac{1}{8} &=& 4\frac{1}{8} \\ +\ 3\frac{7}{8} &=& 3\frac{7}{8} \\ \hline && 10\frac{14}{8} = 10 + 1\frac{6}{8} = 11\frac{6}{8} = 11\frac{3}{4} \end{array}$$

Next, subtract the yards of material used from the total yards of material bought.

$$\begin{array}{rcl} 15 & = & 14\frac{4}{4} \\ -\ 11\frac{3}{4} & = & 11\frac{3}{4} \\ \hline & & 3\frac{1}{4} \end{array}$$

Marv has $3\frac{1}{4}$ yards of material left.

53. Add the cases of each brand of oil.

$$\begin{array}{rcl} 16\frac{1}{2} & = & 16\frac{4}{8} \\ 12\frac{1}{8} & = & 12\frac{1}{8} \\ 8\frac{3}{4} & = & 8\frac{6}{8} \\ +\ 12\frac{5}{8} & = & 12\frac{5}{8} \\ \hline & & 48\frac{16}{8} = 48 + 2 = 50 \end{array}$$

Comet Auto Supply sold 50 cases of oil during the week.

57. First, add the length of the sections at each end.

$$\begin{array}{rcl} 5\frac{3}{4} & = & 5\frac{6}{8} \\ +\ 1\frac{1}{8} & = & 1\frac{1}{8} \\ \hline & & 6\frac{7}{8} \end{array}$$

Next, subtract this total from the total length of the object.

$$\begin{array}{rcccl} 8\frac{1}{3} & = & 8\frac{8}{24} & = & 7\frac{32}{24} \\ -\ 6\frac{7}{8} & = & 6\frac{21}{24} & = & 6\frac{21}{24} \\ \hline & & & & 1\frac{11}{24} \end{array}$$

The length of the section marked x is $1\frac{11}{24}$ feet.

61. $8 + 9 \div 3 + 6 \cdot 2 = 8 + 3 + 12$
$= 23$

SECTION 3.5 (page 199)

1., 5., and 9.

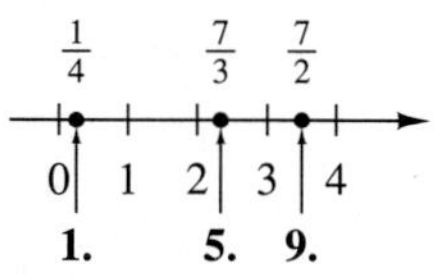

13. The least common multiple of 2 and 4 is 4.

$$\frac{1}{2} = \frac{2}{4} \text{ and } \frac{3}{4} = \frac{3}{4}; \text{ therefore, } \frac{1}{2} < \frac{3}{4}$$

17. The least common multiple of 8 and 12 is 24.

$$\frac{3}{8} = \frac{9}{24} \text{ and } \frac{5}{12} = \frac{10}{24}; \text{ therefore, } \frac{3}{8} < \frac{5}{12}$$

21. The least common multiple of 27 and 18 is 54.

$$\frac{19}{27} = \frac{38}{54} \text{ and } \frac{13}{18} = \frac{39}{54}; \text{ therefore, } \frac{19}{27} < \frac{13}{18}$$

25. $\left(\frac{2}{3}\right)^2 = \frac{2}{3} \cdot \frac{2}{3} = \frac{4}{9}$

29. $\left(\frac{2}{3}\right)^3 = \frac{2}{3} \cdot \frac{2}{3} \cdot \frac{2}{3} = \frac{8}{27}$

33. $\left(\frac{3}{2}\right)^4 = \frac{3}{2} \cdot \frac{3}{2} \cdot \frac{3}{2} \cdot \frac{3}{2} = \frac{81}{16} = 5\frac{1}{16}$

41. $8 \cdot 3^2 - \frac{10}{2} = 8 \cdot 9 - \frac{10}{2}$
$= 72 - 5$
$= 67$

45. $\left(\frac{3}{4}\right)^2 \cdot \left(\frac{1}{3}\right) = \frac{\overset{3}{\cancel{9}}}{16} \cdot \frac{1}{\underset{1}{\cancel{3}}}$
$= \frac{3}{16}$

49. $6 \cdot \left(\frac{2}{3}\right)^2 \cdot \left(\frac{1}{2}\right)^3 = \overset{2}{\cancel{6}} \cdot \frac{4}{\underset{3}{\cancel{9}}} \cdot \frac{1}{8}$
$= \frac{\overset{1}{\cancel{8}}}{3} \cdot \frac{1}{\underset{1}{\cancel{8}}}$
$= \frac{1}{3}$

53. $\frac{1}{2} + \left(\frac{1}{2}\right)^2 - \frac{3}{8} = \frac{1}{2} + \frac{1}{4} - \frac{3}{8}$
$= \frac{3}{4} - \frac{3}{8}$
$= \frac{3}{8}$

57. $\frac{9}{8} \div \left(\frac{2}{3} + \frac{1}{12}\right) = \frac{9}{8} \div \frac{9}{12}$
$= \frac{\overset{1}{\cancel{9}}}{\underset{2}{\cancel{8}}} \cdot \frac{\overset{3}{\cancel{12}}}{\underset{1}{\cancel{9}}}$
$= \frac{3}{2}$
$= 1\frac{1}{2}$

SOLUTIONS

61. $\frac{3}{8}\cdot\left(\frac{1}{4}+\frac{1}{2}\right)\cdot\frac{32}{3}=\frac{3}{8}\cdot\frac{\overset{1}{\cancel{3}}}{\underset{1}{\cancel{4}}}\cdot\frac{\overset{8}{\cancel{32}}}{\underset{1}{\cancel{3}}}$

$=\frac{3}{\underset{1}{\cancel{8}}}\cdot\overset{1}{\cancel{8}}$

$=3$

65. $\left(\frac{7}{8}-\frac{1}{4}\right)-\left(\frac{3}{4}\right)^2\cdot\frac{2}{3}=\frac{5}{8}-\left(\frac{3}{4}\right)^2\cdot\frac{2}{3}$

$=\frac{5}{8}-\frac{\overset{3}{\cancel{9}}}{\underset{8}{\cancel{16}}}\cdot\frac{\overset{1}{\cancel{2}}}{\underset{1}{\cancel{3}}}$

$=\frac{5}{8}-\frac{3}{8}$

$=\frac{2}{8}$

$=\frac{1}{4}$

69. eight thousand, four hundred thirty-six

CHAPTER 4

SECTION 4.1 (page 221)

1. 37.602
ones: 7
tenths: 6
tens: 3

5. 93.01472
thousandths: 4
ten-thousandths: 7
tenths: 0

9. 149.0832
hundreds: 1
hundredths: 8
ones: 9

13. 0.93
0: ones
9: tenths
3: hundredths

17. 60.372
6: tens
0: ones
3: tenths
7: hundredths
2: thousandths

21. $13.4 = 13\frac{4}{10} = 13\frac{2}{5}$ (lowest terms)

25. $0.66 = \frac{66}{100} = \frac{33}{50}$ (lowest terms)

29. $0.06 = \frac{6}{100} = \frac{3}{50}$ (lowest terms)

33. $5.002 = 5\frac{2}{1000} = 5\frac{1}{500}$ (lowest terms)

37. five tenths

41. one hundred five thousandths

45. one and seventy-five thousandths

49. 0.32

53. 0.0703

61. 1.006 which is part number 4-A.

65. $625.4284 = 625\frac{4284}{10{,}000} = 625\frac{1071}{2500}$

69. ten 19,710 19,70|5 5 or more, so add 1 to 0 in tens place. Change 5 to 0.

hundred 19,700 19,7|05 4 or less, leave 7 in hundreds place unchanged. Change 5 to 0.

thousand 20,000 19,|705 5 or more, so add 1 to 9. Carry 1 to ten-thousands place. Change 7 and 5 to 0.

SECTION 4.2 (page 229)

1. 16.9 16.8|974 (↑ tenths place) First digit cut is 5 or more. Round up by adding 1 tenth to the part you are keeping. (Drop digits to right of tenths.)

5. 0.80 0.79|9 (↑ hundredths place) First digit cut is 5 or more, so round up by adding 1 hundredth to the part you are keeping.

$$\begin{array}{r} 0.79 \\ +\ 0.01 \\ \hline 0.80 \end{array}$$

9. 794.0 793.9|88 (↑ tenths place) First digit cut is 5 or more so round up by adding 1 tenth to the part you are keeping.

$$\begin{array}{r} 739.9 \\ +\ 0.1 \\ \hline 740.0 \end{array}$$

13. 49 48.|512 (↑ ones place) First digit cut is 5 or more, so round up by adding 1 one to the part you are keeping.

17. 82.0002 82.0001|51 (↑ ten-thousandths place) First digit cut is 5 or more, so round up by adding 1 ten-thousandth to the part you are keeping.

21. \$1.22 \$1.22|25 (↑ cents place) First digit cut is 4 or less, so the part you are keeping stays the same.

25. \$17,250 \$17,249.|70 (↑ dollars place) First digit cut is 5 or more, so add \$1 to the part you are keeping.

$$\begin{array}{r} \$17{,}249 \\ +\qquad 1 \\ \hline \$17{,}250 \end{array}$$

29. \$379 \$378.|82 (↑ dollars place) First digit cut is 5 or more so add \$1 to the part you are keeping.

$$\begin{array}{r} \$378 \\ +\quad 1 \\ \hline \$379 \end{array}$$

33. $500 $499.|98 First digit cut is 5 or more, so add $1 to the part you are keeping. (dollars place)

$$\begin{array}{r} \$499 \\ +\quad 1 \\ \hline \$500 \end{array}$$

37. $1000 $999.|73 First digit cut is 5 or more, so add $1 to the part you are keeping. (dollars place)

$$\begin{array}{r} \$999 \\ +\quad 1 \\ \hline \$1000 \end{array}$$

41. Estimate:
80,000 + 100 + 800 = 80,900

Exact:
$$\begin{array}{r} 81{,}976 \\ 98 \\ +\quad 785 \\ \hline 82{,}859 \end{array}$$

SECTION 4.3 (page 233)

1.
$$\begin{array}{r} \overset{1\,2}{5.69} \\ 0.24 \\ +\ 11.79 \\ \hline 17.72 \end{array}$$

5.
$$\begin{array}{r} 8.763 \\ {}_{1}0.500 \\ +\ 339.250 \\ \hline 348.513 \end{array}$$

9.
$$\begin{array}{r} 14.230 \\ 8.000 \\ 74.630 \\ 18.715 \\ +\ 0.286 \\ \hline 115.861 \end{array}$$

13.
$$\begin{array}{r} {}_{2}39.76005 \\ 182.00000 \\ 4.79900 \\ 98.31000 \\ +\ 5.99990 \\ \hline 330.86895 \end{array}$$

17. estimate / exact
$$\begin{array}{r} 40 \\ 20 \\ +\ 8 \\ \hline 68 \end{array} \qquad \begin{array}{r} 37.25 \\ 18.90 \\ +\ 7.50 \\ \hline 63.65 \end{array}$$

21. estimate / exact
$$\begin{array}{r} 60 \\ 500 \\ +\ 6 \\ \hline 566 \end{array} \qquad \begin{array}{r} {}_{1}62.8173 \\ 539.9900 \\ +\ 5.6290 \\ \hline 608.4363 \end{array}$$

25. Add the amounts of her paycheck and her refund check.
estimate / exact
$$\begin{array}{r} \$300 \\ +\quad 1 \\ \hline \$301 \end{array} \qquad \begin{array}{r} \$310.14 \\ +\ 0.95 \\ \hline \$311.09 \end{array}$$

29. Add the cost of the muffins, croissants and cookie.
estimate / exact
$$\begin{array}{r} \$\ 7 \\ 10 \\ +\ 1 \\ \hline \$18 \end{array} \qquad \begin{array}{r} \$\ 7.42 \\ 10.09 \\ +\ 0.69 \\ \hline \$18.20 \end{array}$$

33. Add the hours worked on weekends (Saturday and Sunday).
estimate / exact
$$\begin{array}{r} 10 \\ +\ 5 \\ \hline 15 \text{ hours} \end{array} \qquad \begin{array}{r} 9.5 \\ +\ 4.8 \\ \hline 14.3 \text{ hours} \end{array}$$

37. Add the business miles only.
35.4 visit client
14.9 visit client
+ 40.0 business meeting
90.3 miles

41. Add the lengths of the four sides.
19.75 inches
19.75 inches
6.30 inches
6.30 inches
52.10 inches or 52.1 inches

45. Estimate:
7000 − 100 = 6900

Exact:
$$\begin{array}{r} 6708 \\ -\ 139 \\ \hline 6569 \end{array}$$

SECTION 4.4 (page 239)

1. 73.5 − 19.2 = 54.3 Check: 19.2 + 54.3 = 73.5

5. 58.254 − 19.700 = 38.554 Check: 19.700 + 38.554 = 58.254

9. 15.700 − 2.852 = 12.848 Check: 2.852 + 12.848 = 15.700

13. 0.400 − 0.291 = 0.109 Check: 0.291 + 0.109 = 0.400

17. 15.000 − 8.339 = 6.661 Check: 8.339 + 6.661 = 15.000

21. estimate / exact
$$\begin{array}{r} \$20 \\ -\ 7 \\ \hline \$13 \end{array} \qquad \begin{array}{r} \$19.74 \\ -\ 6.58 \\ \hline \$13.16 \end{array}$$

25. estimate / exact
$$\begin{array}{r} 2 \\ -\ 2 \\ \hline 0 \end{array} \qquad \begin{array}{r} 2.000 \\ -\ 1.981 \\ \hline 0.019 \text{ inch} \end{array}$$

29.
$$\begin{array}{r} 12.0000 \\ -\ 11.7251 \\ \hline 0.2749 \end{array}$$

33. Subtract the hours so far from the hours agreed.

estimate	exact
40	42.50
− 20	− 16.35
20 hours	26.15 hours

Tom must work 26.15 hours.

37. Subtract the weight of the lighter box from the weight of the heavier box.

estimate	exact
11	10.50
− 10	− 9.85
1 ounce	0.65 ounce

The difference in weight is 0.65 ounce.

41.
$$\begin{array}{r} 386.02100 \\ -\ 179.68231 \\ \hline 206.33869 \end{array}$$

45. Add the deposit to the balance, then subtract the checks and the service charge.

$$\begin{array}{rl} \$129.86 & \text{balance} \\ +\ 1749.82 & \text{deposit} \\ \hline \$1879.68 & \\ -\ 1802.15 & \text{checks} \\ \hline \$77.53 & \\ -\ 2.00 & \text{service charge} \\ \hline \$75.53 & \end{array}$$

Mitch has \$75.53 in his account at the end of the month.

49. Subtract the two inside measurements from the total length.

$$\begin{array}{rl} 3.00 & \text{total length} \\ -\ 0.91 & \text{left-most measurement} \\ \hline 2.09 & \\ -\ 0.70 & \text{middle measurement} \\ \hline 1.39 & \end{array}$$

b is 1.39 cm.

53.

estimate	exact
80	83
× 30	× 28
2400	664
	166
	2324

SECTION 4.5 (page 245)

1.
$$\begin{array}{rl} 0.042 & \leftarrow 3 \text{ decimal places} \\ \times\ 3.2 & \leftarrow 1 \text{ decimal place} \\ \hline 84 & \\ 126\ \ & \\ \hline 0.1344 & \leftarrow 4 \text{ decimal places in answer} \end{array}$$

5.
$$\begin{array}{rl} 234 & \leftarrow 0 \text{ decimal places} \\ \times\ 0.666 & \leftarrow 3 \text{ decimal places} \\ \hline 1404 & \\ 1404\ \ & \\ 1404\ \ \ \ & \\ \hline 155.844 & \leftarrow 3 \text{ decimal places in answer} \end{array}$$

9. $72 \times 0.6 = 43.2$

0 places + 1 place = 1 place

13. $0.72 \times 0.06 = 0.0432$

2 places + 2 places = 4 places

17. (0.006)(0.0052)

$$\begin{array}{rl} 0.0052 & \leftarrow 4 \text{ decimal places} \\ \times\ 0.006 & \leftarrow 3 \text{ decimal places} \\ \hline 0.0000312 & \leftarrow 7 \text{ decimal places in answer} \end{array}$$

Write four zeros in order to get 7 decimal places.

25.

estimate	exact	
40	37.1	←1 decimal place
× 40	× 42	←0 decimal place
1600	74 2	
	1484	
	1558.2	←1 decimal place in answer

29.

estimate	exact	
3	2.809	←3 decimal places
× 7	× 6.85	←2 decimal places
21	14045	
	2 2472	
	16 854	
	19.24165	←5 decimal places in answer

33. Multiply the cost of one meter of canvas by the number of meters needed.

$$\begin{array}{rl} \$4.09 & \\ \times\ 0.6 & \\ \hline \$2.454 & \text{rounds to } \$2.45 \end{array}$$

Sid will spend \$2.45 on the canvas.

37. Multiply the cost of the home by 0.07.

$$\begin{array}{r} \$92{,}300 \\ \times\ 0.07 \\ \hline \$6461.00 \end{array}$$

Ms. Rolack's fee was \$6461.

41. Multiply the number of sheets by the cost per sheet.

$$\begin{array}{rl} 5100 & \\ \times\ \$0.015 & \\ \hline \$76.500 & \text{or } \$76.50 \end{array}$$

The library will pay \$76.50 for the paper.

45. First multiply to find the total gallons of fertilizer used on the corn. Then subtract the result from 600 gallons, the amount originally in the tank.

$$\begin{array}{rl} 158.2 & \text{acres} \\ \times\ 3.5 & \text{gallons for each acre} \\ \hline 7910 & \\ 4746\ \ & \\ \hline 553.70 & \text{gallons used} \end{array}$$

$$\begin{array}{rl} 600.0 & \text{gallons in tank} \\ -\ 553.7 & \text{gallons used} \\ \hline 46.3 & \text{gallons left} \end{array}$$

There are 46.3 gallons of fertilizer left in the tank.

49. Multiply to find the cost of the rope, and multiply to find the cost of the wire. Add the results to find Barry's total purchases. Subtract the purchases from $15 (three $5 bills is $15).

Cost of rope	Cost of wire
16.5	1.05
× 0.47	× 3
$7.755	$3.15

rounds to $7.76

Purchases	Change
$7.76 rope	$15.00
+ 3.15 wire	− 10.91
$10.91	$ 4.09

Barry received $4.09 in change.

53. estimate

1000
20)20,000

exact

905 R15
21)19,020
18 9
120
105
15

SECTION 4.6 (page 255)

1.

3.9
7)27.3
21
6 3
6 3
0

5.

400.2
0.05)20.010
20 ↓
0 010
10
0

9.

10.904
15)163.570
15
13 5
13 5
070
60
10

rounds to 10.90 (You must write the zero in the hundredths place.)

13.

25.3
4.6)116.38
92
24 3
23 0
1 38
1 38
0

17.

24.2914
9.88)240.00 0000
197 6
42 40
39 52
2 880
1 976
90 40
88 92
1 480
988
4920
3952
968

rounds to 24.291

25. estimate:

1
50)50

The answer of 1.135 is reasonable.

29. estimate:

9
1)9

The answer of 0.744 is unreasonable.
Correct answer:

7.44
1.25)9.3000
8 75
550
500
500
500
0

33. Divide the balance by the number of months.

19.46
21)408.66
21
198
189
9 6
8 4
1 26
1 26
0

Aimee is paying $19.46 per month.

37. Divide the total earnings by the number of hours.

5.89
40)235.60
200
35 6
32 0
3 60
3 60
0

Darren earns $5.89 per hour.

SOLUTIONS

41. Find the average by adding the five lengths of U.S. athletes, then dividing by 5.

```
  29.200           28.04    rounds to 28.0
  27.040        5)140.235
  27.375          10
  28.020           40
+ 28.600           40
 140.235            0 23
                      20
                       3
```

The average long jump length is 28.0 feet.

45. $38.6 + 11.6 \cdot \underbrace{(13.4 - 10.4)}$ Work inside parentheses.

$38.6 + \underbrace{11.6 \cdot 3}$ Multiply next.

$38.6 + 34.8 = 73.4$

49. $33 - 3.2 \cdot \underbrace{(0.68 + 9)} - \underbrace{1.3^2}$ Parentheses first, then exponents.

$33 - \underbrace{3.2 \cdot 9.68} - 1.69$ Multiply next.

$\underbrace{33 - 30.976} - 1.69$ Subtract last.

$2.024 - 1.69 = 0.334$

53. Divide by 4 to find the quarterly installment, then add the $2.75 service fee.

```
   234.5        $234.50
 4)938.0      +    2.75
   8            $237.25
   13
   12
    18
    16
     2 0
     2 0
       0
```

Jenny's quarterly payment is $237.25.

57. The least common multiple of 6 and 9 is 18.

$$\frac{5}{6} = \frac{15}{18} \text{ and } \frac{7}{9} = \frac{14}{18} \text{ therefore } \frac{5}{6} > \frac{7}{9}$$

SECTION 4.7 (page 263)

1.

```
   0.5
 2)1.0
   1 0
     0
```

5.

```
   0.1666   rounds to 0.167
 6)1.0000
    6
    40
    36
     40
     36
      40
      36
       4
```

9.

```
   0.6666   rounds to 0.667
 3)2.0000
   1 8
     20
     18
      20
      18
       20
       18
        2
```

13. $3\frac{4}{7} = \frac{25}{7}$

```
    3.5714   rounds to 3.571
 7)25.0000
   21
    4 0
    3 5
      50
      49
       10
        7
       30
       28
        2
```

17.

```
     0.6875   rounds to 0.688
 16)11.0000
     9 6
     1 40
     1 28
       120
       112
        80
        80
         0
```

21.

```
     0.0101   rounds to 0.010
 99)1.0000
     99
      100
       99
        1
```

29. $0.4 = \frac{4}{10} = \frac{2}{5}$

33. $0.35 = \frac{35}{100} = \frac{7}{20}$

37. $0.04 = \frac{4}{100} = \frac{1}{25}$

41.

```
   0.2
 5)1.0
   1 0
     0
```

45. Write two zeros to the right of 0.5 so it has the same number of decimal places as 0.505. Then you can compare the numbers $0.505 > 0.500$
There was too much calcium in each capsule.

49. Write zeros so that all the numbers have four decimal places.
1.0100 > 1.0020 unacceptable
0.9991 > 0.9980 and 0.9991 < 1.0100 acceptable
1.0007 > 0.9980 and 1.0007 < 1.0100 acceptable
0.9900 < 0.9980 unacceptable
The lengths of 0.9991 cm and 1.0007 cm are acceptable.

53. 0.54 = 0.5400 = 5400 ten-thousandths
0.5455 = 5455 ten-thousandths ← 5455 is largest.
0.5399 = 5399 ten-thousandths ← 5399 is smallest.
(smallest) 0.5399 0.54 0.5455 (largest)

57. 0.628 = 0.62800 = 62800 hundred-thousandths
0.62812 = 62812 hundred-thousandths ← largest
0.609 = 0.60900 = 60900 hundred-thousandths
0.6009 = 0.60090 = 60090 hundred-thousandths ← smallest
(smallest) 0.6009 0.609 0.628 0.62812 (largest)

61. 0.043 = 43 thousandths
0.051 = 51 thousandths ← largest
0.006 = 6 thousandths ← smallest
$\frac{1}{20} = 0.05 = 0.050 = 50$ thousandths
(smallest) 0.006 0.043 $\frac{1}{20}$ 0.051 (largest)

65. $1\frac{7}{16} = \frac{23}{16}$

$16\overline{)23.00}$ = 1.43 rounds to 1.4
Length (a) is 1.4 inch

69. $8\overline{)3.00}$ = 0.37 rounds to 0.4
Length (e) is 0.4 inches

73. $\frac{3}{11} = 0.273 = 273$ thousandths (rounded)

$\frac{4}{15} = 0.267 = 267$ thousandths (rounded)

$0.25 = 0.250 = 250$ thousandths

$\frac{1}{3} = 0.333 = 333$ thousandths (rounded)

(smallest) 0.25 $\frac{4}{15}$ $\frac{3}{11}$ $\frac{1}{3}$ (largest)

77. $\frac{9}{12} = \frac{\cancel{3} \cdot 3}{2 \cdot 2 \cdot \cancel{3}} = \frac{3}{4}$

81. $\frac{96}{132} = \frac{\cancel{2} \cdot \cancel{2} \cdot 2 \cdot 2 \cdot 2 \cdot \cancel{3}}{\cancel{2} \cdot \cancel{2} \cdot \cancel{3} \cdot 11} = \frac{8}{11}$

CHAPTER 5

SECTION 5.1 (page 287)

1. $\frac{8}{9}$ 8 ← mentioned first; 9 ← mentioned second

5. $\frac{30 \text{ minutes}}{90 \text{ minutes}} = \frac{30}{90} = \frac{30 \div 30}{90 \div 30} = \frac{1}{3}$

9. $\frac{6 \text{ hours}}{16 \text{ hours}} = \frac{6}{16} = \frac{6 \div 2}{16 \div 2} = \frac{3}{8}$

13. $\frac{3}{2\frac{1}{2}} = \frac{\frac{3}{1}}{\frac{5}{2}} = \frac{3}{1} \div \frac{5}{2} = \frac{3}{1} \cdot \frac{2}{5} = \frac{6}{5}$

(Do not write the ratio as $1\frac{1}{5}$.)

17. 4 feet = 4 · 12 inches = 48 inches

$\frac{4 \text{ feet}}{30 \text{ inches}} = \frac{48 \text{ inches}}{30 \text{ inches}} = \frac{48}{30} = \frac{48 \div 6}{30 \div 6} = \frac{8}{5}$

(Do not write the ratio as $1\frac{3}{5}$.)

21. 2 days = 2 · 24 hours = 48 hours

$\frac{15 \text{ hours}}{2 \text{ days}} = \frac{15 \text{ hours}}{48 \text{ hours}} = \frac{15}{48} = \frac{15 \div 3}{48 \div 3} = \frac{5}{16}$

25. $\frac{35 \text{ years}}{10 \text{ years}} = \frac{35}{10} = \frac{35 \div 5}{10 \div 5} = \frac{7}{2}$

The ratio of Mr. Wilkins' age to his daughter's age is $\frac{7}{2}$.

29. $\frac{4}{400} = \frac{4 \div 4}{400 \div 4} = \frac{1}{100}$

The ratio of defective washers to the total number of washers is $\frac{1}{100}$.

33. $\frac{\text{taxes} \rightarrow \$400}{\text{transportation} \rightarrow \$200} = \frac{400 \div 200}{200 \div 200} = \frac{2}{1}$

37. $\frac{7 \text{ feet}}{5 \text{ feet}} = \frac{7}{5}$

The ratio of the length of the longest side to the length of the shortest side is $\frac{7}{5}$.

41. $\frac{9\frac{1}{2} \text{ inches}}{4\frac{1}{4} \text{ inches}} = \frac{9\frac{1}{2}}{4\frac{1}{4}} = \frac{\frac{19}{2}}{\frac{17}{4}} = \frac{19}{2} \div \frac{17}{4}$

$= \frac{19}{\cancel{2}} \cdot \frac{\cancel{4}}{17} = \frac{38}{17}$

The ratio of the length of the longest side to the length of the shortest side is $\frac{38}{17}$.

45. $59\frac{1}{2} \text{ days} \div 7 = \frac{\cancel{119}}{2} \cdot \frac{1}{\cancel{7}} = \frac{17}{2}$ weeks

$\frac{\frac{17}{2} \text{ weeks}}{8\frac{3}{4} \text{ weeks}} = \frac{\frac{17}{2}}{\frac{35}{4}} = \frac{17}{2} \div \frac{35}{4} = \frac{17}{\cancel{2}} \cdot \frac{\cancel{4}}{35} = \frac{34}{35}$

The ratio of $59\frac{1}{2}$ days to $8\frac{3}{4}$ weeks is $\frac{34}{35}$.

49. $7\overline{)0.6500}$ = 0.0928 rounds to 0.093

```
   0.0928
7)0.6500
  63
   20
   14
    60
    56
     4
```

53. $0.71\overline{)6.720000}$ = 9.4647 rounds to 9.465

```
         9.4647
0.71)6.720000
     6 39
       330
       284
        460
        426
         340
         284
          560
          497
           63
```

SECTION 5.2 (page 295)

1. $\frac{10 \text{ cups}}{6 \text{ people}} = \frac{10 \text{ cups} \div 2}{6 \text{ people} \div 2} = \frac{5 \text{ cups}}{3 \text{ people}}$

5. $\frac{14 \text{ people}}{28 \text{ dresses}} = \frac{1 \text{ person}}{2 \text{ dresses}}$

9. $\frac{\$63}{6 \text{ visits}} = \frac{\$21}{2 \text{ visits}}$

13. $\frac{\$60}{5 \text{ hours}} = \frac{\$12}{1 \text{ hour}} = \$12/\text{hour}$ Divide: $5\overline{)\$60}$ = 12

17. $\frac{7.5 \text{ pounds}}{6 \text{ people}} = \frac{1.25 \text{ pounds}}{1 \text{ person}} = 1.25 \text{ pounds/person}$

Divide: $6\overline{)7.50}$ = 1.25

21. Subtract to find miles traveled. Then divide miles by gallons to get miles per gallon. 27,758.2 − 27,432.3 = 325.9 and 325.9 ÷ 15.5 = 21.02 rounds to 21.0

Earl's car mileage was 21.0 miles per gallon.

25. 4 ounces: $\frac{\$0.89}{4 \text{ ounces}} = \$0.223$ per ounce

8 ounces: $\frac{\$2.13}{8 \text{ ounces}} = \$0.266$ per ounce

The best buy (lower cost per ounce) is 4 ounces for \$0.89.

29. 12 ounces: $\frac{\$1.09}{12 \text{ ounces}} = \$0.091$ per ounce

18 ounces: $\frac{\$1.41}{18 \text{ ounces}} = \$0.078$ per ounce ← lowest cost per ounce

28 ounces: $\frac{\$2.29}{28 \text{ ounces}} = \$0.082$ per ounce

40 ounces: $\frac{\$3.19}{40 \text{ ounces}} = \$0.080$ per ounce

The best buy is 18 ounces for \$1.41

33. $\frac{10.5 \text{ pounds}}{6 \text{ weeks}} = \frac{1.75 \text{ pounds}}{1 \text{ week}} = 1.75 \text{ pounds/week}$

Divide: $6\overline{)10.50}$ = 1.75

Her rate of loss was 1.75 pounds/week.

37. $\frac{\$1725}{150 \text{ shares}} = \frac{\$11.50}{1 \text{ share}} = \$11.50/\text{share}$

Divide: $150\overline{)1725.00}$ = 11.50

One share costs \$11.50

41. $\frac{\$51.75}{4.6 \text{ yards}} = \frac{\$11.25}{1 \text{ yard}} = \$11.25/\text{yard}$ Divide: $4.6\overline{)51.750}$ = 11.25

One yard of fabric costs \$11.25

45. $5\frac{2}{5} \cdot 20 = \frac{27}{\cancel{5}_1} \cdot \frac{\cancel{20}^4}{1} = \frac{108}{1} = 108$

SECTION 5.3 (page 301)

1. $\frac{\$9}{12 \text{ cans}} = \frac{\$18}{24 \text{ cans}}$

5. $\frac{120}{150} = \frac{8}{10}$ The common units (feet) cancel.

9. $\frac{1\frac{1}{2}}{8} = \frac{6}{32}$

13. $\frac{5}{8}$ is already in lowest terms

$\frac{25}{40} = \frac{25 \div 5}{40 \div 5} = \frac{5}{8}$ (lowest terms)

$\frac{5}{8} = \frac{5}{8}$ so the proportion is true

17. $\frac{42}{15} = \frac{42 \div 3}{15 \div 3} = \frac{14}{5}$ (lowest terms)

$\frac{28}{10} = \frac{28 \div 2}{10 \div 2} = \frac{14}{5}$ (lowest terms)

$\frac{14}{5} = \frac{14}{5}$ so the proportion is true

21. $\frac{7}{6}$ is already in lowest terms

$\frac{54}{48} = \frac{54 \div 6}{48 \div 6} = \frac{9}{8}$ (lowest terms)

$\frac{7}{6}$ does not equal $\frac{9}{8}$ so the proportion is false.

25. $\frac{20}{28} = \frac{12}{16}$

$28 \cdot 12 = 336$

$20 \cdot 16 = 320$

Cross products are not equal, so proportion is false.

29. $\dfrac{3.5}{4} = \dfrac{7}{8}$ $\quad 4 \cdot 7 = 28$ $\quad 3.5 \cdot 8 = 28$

Cross products are equal, so proportion is true.

33. $\dfrac{6}{3\frac{2}{3}} = \dfrac{18}{11}$ $\quad 3\dfrac{2}{3} \cdot 18 = \dfrac{11}{\cancel{3}_1} \cdot \dfrac{\cancel{18}^{6}}{1} = 66$ $\quad 6 \cdot 11 = 66$

Cross products are equal, so proportion is true.

37. (Answers may vary.)

$\dfrac{180 \text{ times}}{63 \text{ hits}} = \dfrac{140 \text{ times}}{49 \text{ hits}}$ $\quad 63 \cdot 140 = 8820$ $\quad 180 \cdot 49 = 8820$

Cross products are equal, so proportion is true. The coach is correct.

41. $\dfrac{16}{24} = \dfrac{16 \div 8}{24 \div 8} = \dfrac{2}{3}$

45. $\dfrac{36}{63} = \dfrac{36 \div 9}{63 \div 9} = \dfrac{4}{7}$

SECTION 5.4 (page 309)

1. $\dfrac{1}{3} = \dfrac{x}{12}$

$3 \cdot x = 1 \cdot 12$

$3 \cdot x = 12$

$\dfrac{\cancel{3}^{1} \cdot x}{\cancel{3}_{1}} = \dfrac{12}{3}$

$x = 4$

Check: $\dfrac{1}{3} = \dfrac{4}{12}$ $\quad 3 \cdot 4 = 12$ $\quad 1 \cdot 12 = 12$

Cross products are equal. Proportion is true.

5. $\dfrac{x}{11} = \dfrac{32}{4}$

$x \cdot 4 = 11 \cdot 32$

$x \cdot 4 = 352$

$\dfrac{x \cdot \cancel{4}^{1}}{\cancel{4}_{1}} = \dfrac{352}{4}$

$x = 88$

Check: $\dfrac{88}{11} = \dfrac{32}{4}$ $\quad 11 \cdot 32 = 352$ $\quad 88 \cdot 4 = 352$

Cross products are equal. Proportion is true.

9. $\dfrac{x}{25} = \dfrac{4}{20}$

$x \cdot 20 = 25 \cdot 4$

$x \cdot 20 = 100$

$\dfrac{x \cdot \cancel{20}^{1}}{\cancel{20}_{1}} = \dfrac{100}{20}$

$x = 5$

Check: $\dfrac{5}{25} = \dfrac{4}{20}$ $\quad 25 \cdot 4 = 100$ $\quad 5 \cdot 20 = 100$

Cross products are equal. Proportion is true.

13. $\dfrac{99}{55} = \dfrac{44}{x}$

$99 \cdot x = 55 \cdot 44$

$99 \cdot x = 2420$

$\dfrac{\cancel{99}^{1} \cdot x}{\cancel{99}_{1}} = \dfrac{2420}{99}$

$x = 24.44$ (rounded)

Check: $\dfrac{99}{55} = \dfrac{44}{24.44}$ $\quad 55 \cdot 44 = 2420$ $\quad 99 \cdot 24.44 = 2419.56$

Cross products are approximately equal (not exact due to rounding). Proportion is true.

17. $\dfrac{250}{24.8} = \dfrac{x}{1.75}$

$24.8 \cdot x = 250 \cdot 1.75$

$24.8 \cdot x = 437.5$

$\dfrac{\cancel{24.8}^{1} \cdot x}{\cancel{24.8}_{1}} = \dfrac{437.5}{24.8}$

$x = 17.64$ (rounded)

Check: $\dfrac{250}{24.8} = \dfrac{17.64}{1.75}$ $\quad 24.8 \cdot 17.64 = 437.472$ $\quad 250 \cdot 1.75 = 437.5$

Cross products are approximately equal (not exact due to rounding). Proportion is true.

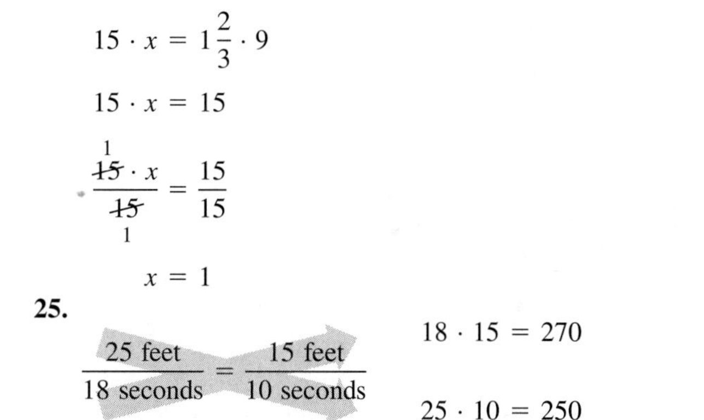

21. $\dfrac{15}{1\frac{2}{3}} = \dfrac{9}{x}$

$15 \cdot x = 1\dfrac{2}{3} \cdot 9$

$15 \cdot x = 15$

$\dfrac{\cancel{15}^{1} \cdot x}{\cancel{15}_{1}} = \dfrac{15}{15}$

$x = 1$

25. $\dfrac{25 \text{ feet}}{18 \text{ seconds}} = \dfrac{15 \text{ feet}}{10 \text{ seconds}}$ $\quad 18 \cdot 15 = 270$ $\quad 25 \cdot 10 = 250$

Cross products are not equal, so proportion is false.

SECTION 5.5 (page 313)

1. Let x represent the time to load 18 trucks.

$\dfrac{5 \text{ hours}}{4 \text{ trucks}} = \dfrac{x \text{ hours}}{18 \text{ trucks}}$ Both rates compare hours to trucks in the same order.

$\dfrac{5}{4} = \dfrac{x}{18}$

$4 \cdot x = 5 \cdot 18$

$4 \cdot x = 90$

$\dfrac{\cancel{4}^{1} \cdot x}{\cancel{4}_{1}} = \dfrac{90}{4}$

$x = 22.5$

It takes 22.5 hours to load 18 trucks.

5. Let x represent the pounds of seed needed for 4900 square feet.

$$\frac{5 \text{ pounds}}{3500 \text{ square feet}} = \frac{x \text{ pounds}}{4900 \text{ square feet}}$$

Both rates compare pounds to square feet in the same order.

$$\frac{5}{3500} = \frac{x}{4900}$$

$$3500 \cdot x = 5 \cdot 4900$$

$$3500 \cdot x = 24{,}500$$

$$\frac{\overset{1}{\cancel{3500}} \cdot x}{\underset{1}{\cancel{3500}}} = \frac{24{,}500}{3500}$$

$$x = 7$$

7 pounds of grass seed are needed.

9. Let x represent the actual length of the kitchen.

$$\frac{1 \text{ inch}}{4 \text{ feet}} = \frac{3.5 \text{ inches}}{x \text{ feet}}$$

Both ratios compare inches to feet in the same order.

$$1 \cdot x = 4 \cdot 3.5$$

$$1 \cdot x = 14$$

$$x = 14$$

The kitchen is 14 feet long.

Let x represent the actual width of the kitchen.

$$\frac{1 \text{ inch}}{4 \text{ feet}} = \frac{2.5 \text{ inches}}{x \text{ feet}}$$

Both ratios compare inches to feet in the same order.

$$1 \cdot x = 4 \cdot 2.5$$

$$1 \cdot x = 10$$

$$x = 10$$

The kitchen is 10 feet wide.

13. Let x represent the number of people who choose vanilla ice cream.

$$\frac{4 \text{ choose vanilla}}{5 \text{ people}} = \frac{x \text{ choose vanilla}}{238 \text{ people}}$$

Both ratios compare people who choose vanilla to the total group of people.

$$5 \cdot x = 4 \cdot 238$$

$$5 \cdot x = 952$$

$$\frac{\overset{1}{\cancel{5}} \cdot x}{\underset{1}{\cancel{5}}} = \frac{952}{5}$$

$$x = 190.4 \text{ rounds to } 190$$

You would expect 190 people to choose vanilla ice cream.

17. Let x represent the number of stocks that went up.

$$\frac{5 \text{ stocks up}}{6 \text{ stocks down}} = \frac{x \text{ stocks up}}{750 \text{ stocks down}}$$

Both ratios compare stocks going up to stocks going down in the same order.

$$\frac{5}{6} = \frac{x}{750}$$

$$6 \cdot x = 5 \cdot 750$$

$$6 \cdot x = 3750$$

$$\frac{\overset{1}{\cancel{6}} \cdot x}{\underset{1}{\cancel{6}}} = \frac{3750}{6}$$

$$x = 625$$

625 stocks went up.

21. Let x represent the width of the wing.

$$\frac{8 \text{ length}}{1 \text{ width}} = \frac{32.5 \text{ meters length}}{x \text{ meters width}}$$

Both ratios compare length to width in the same order.

$$\frac{8}{1} = \frac{32.5}{x}$$

$$8 \cdot x = 1 \cdot 32.5$$

$$8 \cdot x = 32.5$$

$$\frac{\overset{1}{\cancel{8}} \cdot x}{\underset{1}{\cancel{8}}} = \frac{32.5}{8}$$

$$x = 4.0625 \text{ rounds to } 4.06$$

The wing must be 4.06 meters wide.

29. Let x represent the amount of margarine for 15 servings.

$$\frac{6 \text{ tablespoons}}{12 \text{ servings}} = \frac{x \text{ tablespoons}}{15 \text{ servings}}$$

Both rates compare tablespoons to servings in the same order.

$$12 \cdot x = 6 \cdot 15$$

$$12 \cdot x = 90$$

$$\frac{\overset{1}{\cancel{12}} \cdot x}{\underset{1}{\cancel{12}}} = \frac{90}{12}$$

$$x = 7.5 \quad \text{or} \quad 7\frac{1}{2}$$

Use $7\frac{1}{2}$ tablespoons of margarine for 15 servings.

33. $0.06 \times 100 = 0.06 = 6$

Move the decimal point two places right.

37. $1.93 \div 100 = 01.93 = 0.0193$

Move the decimal point two places left.

CHAPTER 6

SECTION 6.1 (page 335)

1. 50% = 0̮5̮0. = 0.50 or 0.5 — Percent sign is dropped. Decimal is moved two places to the left.

5. 45% = 0̮4̮5. = 0.45

9. 7.8% = 0̮0̮7.8 = 0.078 — 0 is attached so the decimal point can be moved two places to the left.

13. 0.5% = 000.5 = 0.005 — Two zeros are attached.

17. 0.6 = 0.60% = 60% — Percent sign is attached. Decimal point is moved two places to the right.

21. 0.07 = 0.07% = 7%

25. 0.629 = 0.629% = 62.9%

29. 2.6 = 2.60 = 260% — Add 1 zero to right so that decimal point can be moved 2 places to the right.

33. 4.162 = 4.162% = 416.2%

41. 65% of the salespeople = 065. = 0.65

45. Success rate is 2 times = 2.00% = 200%

49. Blood pressure is 153.6% of normal = 153.6 = 1.536

53. 3 out of 10 shaded $= \frac{3}{10} = 0.3 = 30\%$ shaded

7 out of 10 unshaded $= \frac{7}{10} = 0.7 = 70\%$ unshaded

57. 55 out of 100 shaded $= \frac{55}{100} = 0.55 = 55\%$ shaded

45 out of 100 unshaded $= \frac{45}{100} = 0.45 = 45\%$ unshaded

61.
$$\begin{array}{r} 0.75 \\ 4\overline{)3.00} \\ \underline{2\,8} \\ 20 \\ \underline{20} \\ 0 \end{array}$$

65.
$$\begin{array}{r} 0.8 \\ 5\overline{)4.0} \\ \underline{4\,0} \\ 0 \end{array}$$

SECTION 6.2 (page 345)

1. $25\% = \frac{25}{100} = \frac{1}{4}$

5. $85\% = \frac{85}{100} = \frac{17}{20}$

9. $6.25\% = \frac{6.25}{100} = \frac{6.25 \cdot 100}{100 \cdot 100} = \frac{625}{10{,}000} = \frac{1}{16}$

13. $6\frac{2}{3}\% = \frac{6\frac{2}{3}}{100} = \frac{\frac{20}{3}}{100} = \frac{20}{3} \div \frac{100}{1} = \frac{\overset{1}{\cancel{20}}}{3} \cdot \frac{1}{\underset{5}{\cancel{100}}}$

$= \frac{1}{15}$

17. $130\% = \frac{130}{100} = \frac{13}{10}$ or $1\frac{3}{10}$

21. $\frac{1}{4} = \frac{p}{100}$

$4 \cdot p = 1 \cdot 100$

$4 \cdot p = 100$

$\frac{\overset{1}{\cancel{4}} \cdot p}{\underset{1}{\cancel{4}}} = \frac{100}{4}$

$p = 25$

$\frac{1}{4} = 25\%$

25. $\frac{2}{5} = \frac{p}{100}$

$5 \cdot p = 2 \cdot 100$

$5 \cdot p = 200$

$\frac{\overset{1}{\cancel{5}} \cdot p}{\underset{1}{\cancel{5}}} = \frac{200}{5}$

$p = 40$

$\frac{2}{5} = 40\%$

29. $\frac{5}{8} = \frac{p}{100}$

$8 \cdot p = 5 \cdot 100$

$8 \cdot p = 500$

$\frac{\overset{1}{\cancel{8}} \cdot p}{\underset{1}{\cancel{8}}} = \frac{500}{8}$

$p = 62.5$

$\frac{5}{8} = 62.5\%$

33. $\frac{13}{25} = \frac{p}{100}$

$25 \cdot p = 13 \cdot 100$

$25 \cdot p = 1300$

$\frac{\overset{1}{\cancel{25}} \cdot p}{\underset{1}{\cancel{25}}} = \frac{1300}{25}$

$p = 52$

$\frac{13}{25} = 52\%$

37. $\frac{1}{20} = \frac{p}{100}$

$20 \cdot p = 1 \cdot 100$

$20 \cdot p = 100$

$\frac{\overset{1}{\cancel{20}} \cdot p}{\underset{1}{\cancel{20}}} = \frac{100}{20}$

$p = 5$

$\frac{1}{20} = 5\%$

41. $\frac{5}{9} = \frac{p}{100}$

$9 \cdot p = 5 \cdot 100$

$9 \cdot p = 500$

$\frac{\overset{1}{\cancel{9}} \cdot p}{\underset{1}{\cancel{9}}} = \frac{500}{9}$

$p = 55.6$

$\frac{5}{9} = 55.6\%$

45. Decimal: 0.1 (given).

Fraction: $0.1 = \frac{1}{10}$

Percent: $0.1 = 0.10\% = 10\%$

49. Decimal: 0.2 (given).

Fraction: $0.2 = \frac{2}{10} = \frac{1}{5}$

Percent: $0.2 = 0.20\% = 20\%$

53. Decimal: 0.25 (given).

Fraction: $0.25 = \frac{25}{100} = \frac{1}{4}$

Percent: $0.25 = 0.25\% = 25\%$

57. Fraction: $\frac{2}{3}$ (given).

Decimal: $\frac{2}{3} = 0.667$ (rounded)

Percent: $\frac{2}{3} = 0.667 = 0.667\%$

$= 66.7\%$ (rounded)

61. Fraction: $\frac{1}{100}$ (given).

Decimal: $\frac{1}{100} = 0.01$

Percent: $\frac{1}{100} = 0.01 = 0.01\% = 1\%$

65. Decimal: 2.5 (given).

Fraction: $2.5 = 2\frac{5}{10} = 2\frac{1}{2}$

Percent: $2.5 = 2.50\% = 250\%$

73. Reduction is \$100 out of \$500.

Fraction: $\frac{100}{500} = \frac{1}{5}$

Decimal: $\frac{1}{5} = 0.2$

Percent: $0.2 = 20\%$

77. 22 out of 40 drive Cadillacs.
Therefore, $40 - 22 = 18$ drive Lincolns.

Fraction: $\frac{18}{40} = \frac{9}{20}$

Decimal: $\frac{9}{20} = 0.45$

Percent: $0.45 = 45\%$

81. 1050 students out of 4200 students use public transportation,

Fraction: $\frac{1050}{4200} = \frac{1}{4}$

Decimal: $\frac{1}{4} = 0.25$

Percent: $0.25 = 25\%$

85. $\frac{10}{5} = \frac{x}{20}$

$5 \cdot x = 10 \cdot 20$

$5 \cdot x = 200$

$\frac{\overset{1}{\cancel{5}} \cdot x}{\underset{1}{\cancel{5}}} = \frac{200}{5}$

$x = 40$

89. $\frac{42}{30} = \frac{14}{b}$

$42 \cdot b = 30 \cdot 14$

$42 \cdot b = 420$

$\frac{\overset{1}{\cancel{42}} \cdot b}{\underset{1}{\cancel{42}}} = \frac{420}{42}$

$b = 10$

SECTION 6.3 (page 353)

1. $\frac{20}{b} = \frac{25}{100}$

$\frac{20}{b} = \frac{1}{4}$

$b \cdot 1 = 20 \cdot 4$

$b = 80$

5. $\frac{16}{b} = \frac{40}{100}$

$\frac{16}{b} = \frac{2}{5}$

$b \cdot 2 = 16 \cdot 5$

$b \cdot 2 = 80$

$\frac{b \cdot \cancel{2}}{\cancel{2}} = \frac{80}{2}$

$b = 40$

9. $\frac{55}{110} = \frac{p}{100}$

$\frac{1}{2} = \frac{p}{100}$

$2 \cdot p = 100$

$\frac{\overset{1}{\cancel{2}} \cdot p}{\underset{1}{\cancel{2}}} = \frac{100}{2}$

$p = 50$

The percent is 50%.

13. $\frac{1.5}{4.5} = \frac{p}{100}$

$\frac{1}{3} = \frac{p}{100}$

$3 \cdot p = 100$

$\frac{\overset{1}{\cancel{3}} \cdot p}{\underset{1}{\cancel{3}}} = \frac{100}{1}$

$p = 33.3$ (rounded)

The percent is 33.3%.

17. $\frac{a}{144} = \frac{15}{100}$

$\frac{a}{144} = \frac{3}{20}$

$a \cdot 20 = 144 \cdot 3$

$a \cdot 20 = 432$

$\frac{a \cdot \overset{1}{\cancel{20}}}{\underset{1}{\cancel{20}}} = \frac{432}{20}$

$a = 21.6$

25. $\frac{20}{5000} = \frac{p}{100}$

$\frac{1}{250} = \frac{p}{100}$

$250 \cdot p = 100$

$\frac{\overset{1}{\cancel{250}} \cdot p}{\underset{1}{\cancel{250}}} = \frac{100}{250}$

$p = 0.4$

The percent is 0.4%.

29. $\frac{994.21}{8116} = \frac{p}{100}$

$8116 \cdot p = 994.21 \cdot 100$

$8116 \cdot p = 99421$

$\frac{\overset{1}{\cancel{8116}} \cdot p}{\underset{1}{\cancel{8116}}} = \frac{99421}{8116}$

$p = 12.25$

The percent is 12.25%.

33. $\frac{57}{100} = 0.57 = 57\%$

37. $\frac{2}{3} = 0.667 = 66.7\%$

SECTION 6.4 (page 357)

1. $\underbrace{25\%}_{p}$ of $\underbrace{\text{how much}}_{b}$ is $\underbrace{\$200}_{a}$

$p = 25$
$b =$ unknown
$a = 200$

5. $\underbrace{\text{What}}_{a}$ is $\underbrace{15\%}_{p}$ of $\underbrace{\$75}_{b}$? $p = 15$ $b = 75$ $a =$ unknown

9. $\underbrace{\text{26 lamps}}_{a}$ is $\underbrace{\text{what percent}}_{p}$ of $\underbrace{\text{52 lamps}}_{b}$?

$p =$ unknown
$b = 52$
$a = 26$

13. $\underbrace{\$27.17}_{a}$ is $\underbrace{6.5\%}_{p}$ of $\underbrace{\text{what number}}_{b}$?

$p = 6.5$
$b =$ unknown
$a = 27.17$

21. $p =$ unknown (what percent)
b = \$15 (cost of compact disc)
$a = \$0.90$ (sales tax)

25. $p = 25$ (25% of the bars)
$b =$ unknown (total number of bars)
$a = 240$ (candy bars sold)

29. $p = 32$ (32% of the people)
$b = 272$ (people tested)
$a =$ unknown (number having high cholesterol)

33. $\frac{7}{x} = \frac{5}{10}$ (proportion)

$5 \cdot x = 7 \cdot 10$

$5 \cdot x = 70$

$\frac{\overset{1}{\cancel{5}} \cdot x}{\underset{1}{\cancel{5}}} = \frac{70}{5}$

$x = 14$ (missing number)

SECTION 6.5 (page 367)

1. $0.10 \cdot 520 = 52$
$a = 52$ adults

5. $0.04 \cdot 120 = 4.8$
$a = 4.8$ feet

9. $0.225 \cdot 1100 = 247.5$
$a = 247.5$ boxes

13. $2.5 \cdot 740 = 1850$
$a = 1850$ sales

17. $0.009 \cdot 2400 = 21.6$
$a = \$21.60$

21. $p = 25 \quad a = 76$

$\frac{76}{b} = \frac{25}{100}$

$\frac{76}{b} = \frac{1}{4}$

$b = 304$

25. $p = 110 \quad a = 748$

$\frac{748}{b} = \frac{110}{100}$

$\frac{748}{b} = \frac{11}{10}$

$11b = 7480$

$\frac{\overset{1}{\cancel{11}}b}{\underset{1}{\cancel{11}}} = \frac{7480}{11}$

$b = 680$ books

29. $b = 64 \quad a = 32$

$\frac{32}{64} = \frac{p}{100}$

$64p = 3200$

$\frac{\overset{1}{\cancel{64}}p}{\underset{1}{\cancel{64}}} = \frac{3200}{64}$

$p = 50$

32 is 50% of 64.

33. $b = 400 \quad a = 8$

$\frac{8}{400} = \frac{p}{100}$

$400p = 800$

$\frac{\overset{1}{\cancel{400}}p}{\underset{1}{\cancel{400}}} = \frac{800}{400}$

$p = 2$

8 is 2% of 400.

37. $b = 172 \quad a = 32$

$\frac{32}{172} = \frac{p}{100}$

$172p = 3200$

$\frac{\overset{1}{\cancel{172}}p}{\underset{1}{\cancel{172}}} = \frac{3200}{172}$

$p = 18.60$ rounds to 18.6

18.6% of 172 is 32.

45. $p = 38$ (% wearing seat belts)
$b = 2200$ (total number of drivers)
$a =$ unknown (number wearing seat belts)

$\frac{a}{2200} = \frac{38}{100}$

$100a = 83600$

$\frac{\overset{1}{\cancel{100}}a}{\underset{1}{\cancel{100}}} = \frac{83600}{100}$

$a = 836$

836 drivers were wearing seat belts.

49. $p = 40$ (% of the applicants)
$b = 550$ (number of scholarship applicants)
$a =$ unknown (number receiving scholarships)

$\frac{a}{550} = \frac{40}{100}$

$100a = 22000$

$\frac{\overset{1}{\cancel{100}}a}{\underset{1}{\cancel{100}}} = \frac{22000}{100}$

$a = 220$

220 students will receive scholarships.

53. p = unknown (percent of gallons used)
b = 755,000 (number of gallons sold)
a = 11,700 (number of gallons used in steam irons)

$$\frac{11{,}700}{755{,}000} = \frac{p}{100}$$

$$755{,}000p = 1{,}170{,}000$$

$$\frac{\overset{1}{\cancel{755{,}000}}\,p}{\underset{1}{\cancel{755{,}000}}} = \frac{1{,}170{,}000}{755{,}000}$$

$$p = 1.54 \text{ rounds to } 1.5$$

1.5% of the water is used in steam irons.

57. p = 12.7 (percent of the total population)
b = unknown (total American population, in millions)
a = 31.5 (Americans who are 65 or older, in millions)

$$\frac{31.5}{b} = \frac{12.7}{100}$$

$$12.7b = 3150$$

$$\frac{\overset{1}{\cancel{12.7}}b}{\underset{1}{\cancel{12.7}}} = \frac{3150}{12.7}$$

$$b = 248.03 \text{ rounds to } 248.0$$

There are 248.0 million Americans in the U.S..

61. p = 11 (percent of total sales)
b = 7,500,000,000 (total sales)
a = unknown (sales of Post cereals)

$$\frac{a}{7{,}500{,}000{,}000} = \frac{11}{100}$$

$$100a = 82{,}500{,}000{,}000$$

$$\frac{\overset{1}{\cancel{100}}a}{\underset{1}{\cancel{100}}} = \frac{82{,}500{,}000{,}000}{100}$$

$$a = 825{,}000{,}000$$

$825,000,000 are the annual sales of Post cereals.

65. p = 86 (percent of products which failed)
b = 15,401 (number of new products)
a = unknown (number of products which failed)

$$\frac{a}{15{,}401} = \frac{86}{100}$$

$$100a = 1{,}324{,}486$$

$$\frac{\overset{1}{\cancel{100}}a}{\underset{1}{\cancel{100}}} = \frac{1{,}324{,}486}{100}$$

$$a = 13{,}244.86 \text{ rounds to } 13{,}245$$

15,401 − 13,245 = 2156
2156 products reached their objectives.

69.

```
     325.6
   × 0.031
     3 256
    9 7 68
   10.0 936
```

73.

```
         0.25
   688)172.00
       137 6
        34 40
        34 40
            0
```

SECTION 6.6 (page 377)

1. amount = percent · base
$a = 0.25 \cdot 440$ Write 25% as a decimal.
$a = 110$
25% of $440 is $110.

5. amount = percent · base
$a = 0.32 \cdot 260$
$a = 83.2$
32% of 260 quarts is 83.2 quarts.

9. amount = percent · base
$a = 0.124 \cdot 8300$
$a = 1029.2$
12.4% of 8300 meters is 1029.2 meters.

13. amount = percent · base
$30 = 0.15 \cdot b$

$$\frac{30}{0.15} = \frac{\overset{1}{\cancel{0.15}} \cdot b}{\underset{1}{\cancel{0.15}}}$$

$200 = b$
30 apartments is 15% of 200 apartments.

17. amount = percent · base
$238 = 0.35 \cdot b$

$$\frac{238}{0.35} = \frac{\overset{1}{\cancel{0.35}} \cdot b}{\underset{1}{\cancel{0.35}}}$$

$680 = b$
238 wires is 35% of 680 wires.

21. amount = percent · base
$3.75 = 0.0125 \cdot b$

$$\frac{3.75}{0.0125} = \frac{\overset{1}{\cancel{0.0125}} \cdot b}{\underset{1}{\cancel{0.0125}}}$$

$300 = b$
$1\frac{1}{4}$% of 300 gallons is 3.75 gallons.

25. amount = percent · base
$19 = p \cdot 25$

$$\frac{19}{25} = \frac{p \cdot \overset{1}{\cancel{25}}}{\underset{1}{\cancel{25}}}$$

$0.76 = p$ percent = 76%
19 videos is 76% of 25 videos.

29. amount = percent · base
$1.2 = p \cdot 80$

$$\frac{1.2}{80} = \frac{p \cdot \overset{1}{\cancel{80}}}{\underset{1}{\cancel{80}}}$$

$0.015 = p$ percent = 1.5%
1.5% of 80 is 1.2.

37. amount = percent · base
$a = 0.25 \cdot 52$
$a = 13$
There are 13 residential clients.

41. amount = percent · base
$175 = 0.25 \cdot b$

$$\frac{175}{0.25} = \frac{\overset{1}{\cancel{0.25}} \cdot b}{\underset{1}{\cancel{0.25}}}$$

$700 = b$
The tank contains 700 gallons when full.

45. amount = percent · base
$a = 0.325 \cdot 385{,}200$
$a = 125{,}190$
There was an increase in sales of $125,190.
Therefore, the volume of parts sales this year is $385,200 + $125,190 = $510,390

49. $\underbrace{\text{15\%}}_{p}$ of $\underbrace{\text{375 drums}}_{b}$ is $\underbrace{\text{56.25 drums}}_{a}$
$p = 15$ $b = 375$ $a = 56.25$

53. $\underbrace{\text{What percent}}_{p}$ of $\underbrace{\text{\$830}}_{b}$ is $\underbrace{\text{\$128.65?}}_{a}$
p = unknown $b = 830$ $a = 128.65$

SECTION 6.7 (page 385)

1. sales tax = cost of item · rate of tax
= $100 · 0.05
= $5
Sales tax is $5 and the total cost is $105 ($100 + $5).

5. sales tax = cost of item · rate of tax
= $365 · 0.06
= $21.90
Sales tax is $21.90 and the total cost is $386.90 ($365 + $21.90).

9. commission = rate of commission · sales
= 0.08 · $200
= $16
The amount of commission is $16.

13. commission = rate of commission · sales
= 0.03 · $5783
= $173.49
The amount of commission is $173.49.

17. discount = original price · rate of discount
= $100 · 0.25
= $25
The amount of discount is $25 and the sale price is $75 ($100 − $25).

21. discount = original price · rate of discount
= $17.50 · 0.25
= $4.38 (rounded)
The amount of discount is $4.38 and the sale price is $13.12 ($17.50 − $4.38).

29. discount = original price · rate of discount
= $45 · 0.30
= $13.50
The sale price of the shoe is $31.50 ($45 − $13.50).

33. amount of increase = 3286 − 2480
= 806
amount = percent · base
$806 = p \cdot 2480$

$$\frac{806}{2480} = \frac{p \cdot \overset{1}{\cancel{2480}}}{\underset{1}{\cancel{2480}}}$$

$0.325 = p$ percent = 32.5%
Student enrollment increased by 32.5%.

37. discount = original price · rate of discount
= $20 · 0.40
= $8
The sale price of the hair dryer is $12 ($20 − $8).

41. discount = original price · rate of discount
= $1180 · 0.35
= $413
The discount is $413, and the sale price is $767 ($1180 − $413).

45. discount = original price · rate of discount
= $18.50 · 0.06
= $1.11
The sale price is $17.39 ($18.50 − $1.11).
sales tax = cost of item · rate of tax
= $17.39 · 0.06
= $1.04 (rounded)
The total cost of the dictionary is $18.43 ($17.39 + $1.04).

49. discount = original price · rate of discount
= $13,905 · 0.18
= $2502.90
The sale price is $11,402.10 ($13,905 − $2502.90).
sales tax = cost of item · rate of tax
= $11,402.10 · 0.0475
= $541.60 (rounded)
The total cost of the boat is $11,943.70 ($11,402.10 + $541.60).

53. amount = percent · base
$a = 0.003 \cdot 960$
$a = 2.88$
0.3% of $960 is $2.88.

57. amount = percent · base
$147.2 = p \cdot 460$

$$\frac{147.2}{460} = \frac{p \cdot \overset{1}{\cancel{460}}}{\underset{1}{\cancel{460}}}$$

$0.32 = p$ percent = 32%
147.2 meters is 32% of 460 meters.

SECTION 6.8 (page 393)

1. $I = p \cdot r \cdot t$
$= 100 \cdot (0.05) \cdot 1$
$= 5$
The interest is \$5.

5. $I = p \cdot r \cdot t$
$= 60 \cdot (0.04) \cdot 1$
$= 2.4$
The interest is \$2.40.

9. $I = p \cdot r \cdot t$
$= 820 \cdot (0.04) \cdot 2.5$
$= 82$
The interest is \$82.

13. $I = p \cdot r \cdot t$
$= 300 \cdot (0.08) \cdot \frac{6}{12}$
$= 24 \cdot \frac{1}{2}$
$= 12$
The interest is \$12.

17. $I = p \cdot r \cdot t$
$= 820 \cdot (0.08) \cdot \frac{24}{12}$
$= 65.6 \cdot 2$
$= 131.2$
The interest is \$131.20.

21. $I = p \cdot r \cdot t$
$= 1160 \cdot (0.06) \cdot \frac{3}{12}$
$= 69.6 \cdot \frac{1}{4}$
$= 17.4$
The interest is \$17.40.

25. $I = p \cdot r \cdot t$
$= 400 \cdot (0.05) \cdot 1$
$= 20$
The interest is \$20.
amount due = principal + interest
= \$400 + \$20
= \$420
The total amount due is \$420.

29. $I = p \cdot r \cdot t$
$= 1500 \cdot (0.10) \cdot \frac{18}{12}$
$= 150 \cdot \frac{3}{2}$
= \$225
The interest is \$225.
amount due = principal + interest
= \$1500 + \$225
= \$1725
The total amount due is \$1725.

33. $I = p \cdot r \cdot t$
$= 1780 \cdot (0.10) \cdot \frac{6}{12}$
$= 178 \cdot \frac{1}{2}$
$= 89$
The interest is \$89.
amount due = principal + interest
= \$1780 + \$89
= \$1869
The total amount due is \$1869.

41. $I = p \cdot r \cdot t$
$= 6500 \cdot (0.08) \cdot \frac{18}{12}$
$= 520 \cdot \frac{3}{2}$
$= 780$
Joann will earn \$780 in interest.

45. $I = p \cdot r \cdot t$
$= 3850 \cdot (0.05) \cdot \frac{11}{12}$
$= 192.5 \cdot \frac{11}{12}$
= \$176.46 (rounded)
She will earn \$176.46 in interest.

49. $I = p \cdot r \cdot t$
$= 11{,}500 \cdot (0.0875) \cdot \frac{9}{12}$
$= 1006.25 \cdot \frac{3}{4}$
= 754.69 (rounded)
The total amount in the account will be \$12,254.69 (\$11,500 + \$754.69).

53. $\frac{4}{9} = \frac{4 \cdot 4}{9 \cdot 4} = \frac{16}{36}$

57. $\frac{15}{19} = \frac{15 \cdot 4}{19 \cdot 4} = \frac{60}{76}$

SECTION 6.9 (page 403)

1. 6% column, row 4 of the table gives 1.2625.
Multiply. $\$1000 \cdot 1.2625 = \$1262.50$

5. $4\frac{1}{2}\%$ (or 4.5%) column, row 6 of the table gives 1.3023.
Multiply. $\$8428.17 \cdot 1.3023 = \$10{,}976.01$ (rounded)

9. Look up 2.5% ($5\% \div 2$) and $2 \cdot 9 = 18$ time periods.
Find 1.5597.
Multiply. $\$1400 \cdot 1.5597 = \$2183.58$

13. Look up 2% ($8\% \div 4$) and $4 \cdot 3 = 12$ time periods.
Find 1.2682.
Multiply. $\$19{,}700 \cdot 1.2682 = \$24{,}983.54$

17. Look up 3% ($6\% \div 2$) and $2 \cdot 10 = 20$ time periods.
Find 1.8061.
Multiply. $\$1270 \cdot 1.8061 = \$2293.75$ compound amount
$\$2293.75 - \$1270 = \$1023.75$ compound interest

21. Look up 3.5% ($7\% \div 2$) and $2 \cdot 7 = 14$ time periods.
Find 1.6187.
Multiply. $\$7700 \cdot 1.6187 = \$12{,}463.99$ compound amount
$\$12{,}463.99 - \$7700 = \$4763.99$ compound interest

25. **(a)** annually 10%, 5 periods from the table is 1.6105
$\$17{,}600 \cdot 1.6105 = \$28{,}344.80$
(b) semiannually
5%, 10 periods from the table is 1.6289
$\$17{,}600 \cdot 1.6289 = \$28{,}668.64$
(c) quarterly 2.5%, 20 periods from the table is 1.6386
$\$17{,}600 \cdot 1.6386 = \$28{,}839.36$

33. **(a)** yearly 6%, 4 periods from the table is 1.2625
$\$10{,}000 \cdot 1.2625 = \$12{,}625$ compound amount
$\$12{,}625 - \$10{,}000 = \$2625$ compound interest
(b) semiannually
3%, 8 periods from the table is 1.2668
$\$10{,}000 \cdot 1.2668 = \$12{,}668$ compound amount
$\$12{,}668 - \$10{,}000 = \$2668$ compound interest

37. **(a)** Borrow \$7500 at 8% interest for 6 years.
8%, 6 periods from the table is 1.5869
$\$7500 \cdot 1.5869 = \$11{,}901.75$
The total amount to be repaid is \$11,901.75
(b) Amount of interest $= \$11{,}901.75 - \$7500$
$= \$4401.75$

41. **(a)** \$10,000 at 8% compounded quarterly for 2 years.
2%, 8 periods from the table is 1.1717
$\$10{,}000 \cdot 1.1717 = \$11{,}717$
\$31,717 (\$20,000 + \$11,717) at 8% compounded quarterly for 3 years.
2%, 12 periods from the table is 1.2682
$\$31{,}717 \cdot 1.2682 = \$40{,}223.50$ (rounded)
The total amount at the end of 5 years is \$40,223.50
(b) Interest earned
$= \$40{,}223.50 - \$10{,}000 - \$20{,}000$
$= \$10{,}223.50$

45. $19.3 + (6.7 - 5.2) \cdot 58 = 19.3 + 1.5 \cdot 58$
$= 19.3 + 87$
$= 106.3$

CHAPTER 7

SECTION 7.1 (page 431)

1. 1 yard = 3 feet

5. 1 mile = 5280 feet

9. 1 minute = 60 seconds

13. $8 \text{ quarts} = \frac{\overset{2}{\cancel{8}}\ \cancel{\text{quarts}}}{1} \cdot \frac{1 \text{ gallon}}{\underset{1}{\cancel{4}}\ \cancel{\text{quarts}}} = 2 \text{ gallons}$

17. $10 \text{ feet} = \frac{10\ \cancel{\text{feet}}}{1} \cdot \frac{12 \text{ inches}}{1\ \cancel{\text{foot}}} = 120 \text{ inches}$

21. $5 \text{ quarts} = \frac{5\ \cancel{\text{quarts}}}{1} \cdot \frac{2 \text{ pints}}{1\ \cancel{\text{quart}}} = 10 \text{ pints}$

25. $3 \text{ inches} = \frac{\overset{1}{\cancel{3}}\ \cancel{\text{inches}}}{1} \cdot \frac{1 \text{ foot}}{\underset{4}{\cancel{12}}\ \cancel{\text{inches}}} = \frac{1}{4} \text{ foot or } 0.25 \text{ foot}$

29. $5 \text{ cups} = \frac{5\ \cancel{\text{cups}}}{1} \cdot \frac{1 \text{ pint}}{2\ \cancel{\text{cups}}}$
$= \frac{5}{2} \text{ pints} = 2\frac{1}{2} \text{ pints or } 2.5 \text{ pints}$

33. $2\frac{1}{2} \text{ tons} = \frac{2\frac{1}{2}\ \cancel{\text{tons}}}{1} \cdot \frac{2000 \text{ pounds}}{1\ \cancel{\text{ton}}} = \frac{5}{\underset{1}{\cancel{2}}} \cdot \frac{\overset{1000}{\cancel{2000}}}{1} \text{pounds}$
$= 5000 \text{ pounds}$

37. $1\frac{3}{4} \text{ pounds} = \frac{1\frac{3}{4}\ \cancel{\text{pounds}}}{1} \cdot \frac{16 \text{ ounces}}{1\ \cancel{\text{pound}}}$
$= \frac{7}{\underset{1}{\cancel{4}}} \cdot \frac{\overset{4}{\cancel{16}}}{1} \text{ ounces}$
$= 28 \text{ ounces}$

41. $112 \text{ cups} = \frac{\overset{\overset{28}{\cancel{56}}}{\cancel{112}}\ \cancel{\text{cups}}}{1} \cdot \frac{1\ \cancel{\text{pint}}}{\underset{1}{\cancel{2}}\ \cancel{\text{cups}}} \cdot \frac{1 \text{ quart}}{\underset{1}{\cancel{2}}\ \cancel{\text{pints}}} = 28 \text{ quarts}$

45. $1\frac{1}{2} \text{ tons} = \frac{1\frac{1}{2}\ \cancel{\text{tons}}}{1} \cdot \frac{2000\ \cancel{\text{pounds}}}{1\ \cancel{\text{ton}}} \cdot \frac{16 \text{ ounces}}{1\ \cancel{\text{pound}}}$
$= \frac{3}{\underset{1}{\cancel{2}}} \cdot \frac{\overset{1000}{\cancel{2000}}}{1} \cdot \frac{16}{1} \text{ ounces}$
$= 48{,}000 \text{ ounces}$

49. $2\frac{3}{4} \text{ miles} = \frac{2\frac{3}{4}\ \cancel{\text{miles}}}{1} \cdot \frac{5280\ \cancel{\text{feet}}}{1\ \cancel{\text{mile}}} \cdot \frac{12 \text{ inches}}{1\ \cancel{\text{foot}}}$
$= \frac{11}{\underset{1}{\cancel{4}}} \cdot \frac{5280}{1} \cdot \frac{\overset{3}{\cancel{12}}}{1} \text{ inches}$
$= 174{,}240 \text{ inches}$

53. 2 weeks $= \frac{2 \text{ weeks}}{1} \cdot \frac{7 \text{ days}}{1 \text{ week}} = 14$ days

2 weeks < 15 days

57. 32 days $= \frac{32 \text{ days}}{1} \cdot \frac{1 \text{ week}}{7 \text{ days}} = \frac{32}{7}$ weeks $= 4\frac{4}{7}$ weeks

32 days > 4 weeks

SECTION 7.2 (page 441)

1. kilo means 1000, so 1 km = 1000 m

5. centi means $\frac{1}{100}$, so 1 cm $= \frac{1}{100}$ m or 0.01 m

9. (Answers may vary.) The width of a person's thumb is about 20 mm.

13. Ming-Na swam in the 200 m backstroke race.

17. An aspirin tablet is 10 mm across.

21. Dave's truck is 5 m long.

25. 7 m $= \frac{7 \text{ m}}{1} \cdot \frac{100 \text{ cm}}{1 \text{ m}} = 700$ cm

or:
From m to cm is two places right. 7 m = 7.00 cm = 700 cm

29. 9.4 km $= \frac{9.4 \text{ km}}{1} \cdot \frac{1000 \text{ m}}{1 \text{ km}} = 9400$ m

or:
From km to m is three places right. 9.4 km = 9.400 m = 9400 m

33. 400 mm $= \frac{\overset{40}{400 \text{ mm}}}{1} \cdot \frac{1 \text{ cm}}{\underset{1}{10 \text{ mm}}} = 40$ cm

or:
From mm to cm is one place left. 400 mm = 400. cm = 40 cm

37. From cm to m is two places left. 82 cm = 082. m = 0.82 m
82 cm is less than 1 m.
The difference in lengths is 0.18 m (1 m − 0.82 m) or 18 cm (100 cm − 82 cm).

41. From mm to km is six places left.
5.6 mm = 0000005.6 km = 0.0000056 km

45. $0.08 = \frac{8}{100} = \frac{8 \div 4}{100 \div 4} = \frac{2}{25}$

SECTION 7.3 (page 449)

1. The glass held 250 mL of water. (Liquids are measured in mL or L.)

5. Our labrador dog grew up to weigh 25 kg. (Weight is measured in mg, g, or kg.)

9. Andre donated 500 mL of blood today. (Blood is a liquid.)

13. The gas can for the lawn mower holds 4 L. (Gasoline is a liquid.)

17. Unreasonable; 4.1 liters is about 4 quarts (1 gallon).

21. Reasonable; 15 milliliters is about 3 teaspoons.

29. 15 L $= \frac{15 \text{ L}}{1} \cdot \frac{1000 \text{ mL}}{1 \text{ L}} = 15{,}000$ mL

or:
From L to mL is three places right.
15 L = 15.000 mL = 15,000 mL

33. 925 mL $= \frac{925 \text{ mL}}{1} \cdot \frac{1 \text{ L}}{1000 \text{ mL}} = 0.925$ L

or:
From mL to L is three places left. 925 mL = 0 925. L = 0.925 L

37. 4.15 L $= \frac{4.15 \text{ L}}{1} \cdot \frac{1000 \text{ mL}}{1 \text{ L}} = 4150$ mL

or:
From L to mL is three places right.
4.15 L = 4.150 mL = 4150 mL

41. 5.2 kg $= \frac{5.2 \text{ kg}}{1} \cdot \frac{1000 \text{ g}}{1 \text{ kg}} = 5200$ g

or:
From kg to g is three places right. 5.2 kg = 5.200 g = 5200 g

45. 30,000 mg $= \frac{30{,}000 \text{ mg}}{1} \cdot \frac{1 \text{ g}}{1000 \text{ mg}} = 30$ g

or:
From mg to g is three places left. 30,000 mg = 30000. g = 30 g

49. 60 mL $= \frac{60 \text{ mL}}{1} \cdot \frac{1 \text{ L}}{1000 \text{ mL}} = 0.06$ L

or:
From mL to L is three places left.
60 mL = 0060. L = 0.06 L

53. 0.99 L $= \frac{0.99 \text{ L}}{1} \cdot \frac{1000 \text{ mL}}{1 \text{ L}} = 990$ mL

or:
From L to mL is three places right. 0.99 L = 0.990 mL = 990 mL

57. From g to kg is three places left.
950 g = 0950 kg = 0.95 kg
The premature infant weighed 0.95 kg.

61. 1kg = 1.000 g = 1000 g
$\frac{1000 \text{ g}}{5 \text{ g}} = 200$; therefore, there are 200 nickels in 1 kg of nickels.

65. thousands: 7
hundredths: 1
thousandths: 8
hundreds: 2

SOLUTIONS

SECTION 7.4 (page 455)

1. Write 2 kg 50 g in terms of kilograms (the unit in the price).
2 kg → 2.00 kg
50 g → + 0.05 kg
2.05 kg

$\frac{\$0.65}{1 \cancel{kg}} \cdot \frac{2.05 \cancel{kg}}{1} = \$1.3325$ rounds to \$1.33

Pam will pay \$1.33 for the rice.

5. Write 4 L in terms of mL. 4 L = 4.000 mL = 4000 mL

$\frac{4000 \cancel{mL}}{1 \text{ jug}} \cdot \frac{1 \text{ serving}}{160 \cancel{mL}} = 25$ servings per jug

9. Write 750 mL in terms of L. 750 mL = 0 750. L = 0.75 L

$\frac{0.75 \text{ L}}{1 \cancel{day}} \cdot \frac{30 \cancel{days}}{1 \text{ month}} = 22.5$ L per month

The caretaker should order 22.5 L of chlorine.

13. \$16 case:
Capacity = 12 · 1 L = 12 L

Cost per liter $= \frac{\$16}{12 \text{ L}} = \$1.33$ per L

\$18 case:
400 mL = 0.4 L
Capacity = 36 · 0.4 L = 14.4 L

Cost per liter $= \frac{\$18}{14.4 \text{ L}} = \$1.25$ per L

The \$18 case is the better buy.

17.
```
   6.3 ← 1 decimal place
× 0.9 1 ← 2 decimal places
    6 3
 5 67
 5.73 3 ← 3 decimal places in answer
```

SECTION 7.5 (page 461)

1. The meters to yards value in the table is 1.09.
20 meters ≈ 20 · 1.09 ≈ 21.8 yards

5. The feet to meters value in the table is 0.30.
16 feet ≈ 16 · 0.30 ≈ 4.8 m

9. The pounds to kilograms value in the table is 0.45.
248 pounds ≈ 248 · 0.45 ≈ 111.6 kg

13. The gallons to liters value in the table is 3.78.
18 gallons ≈ 18 · 3.78 ≈ 68.04 rounds to 68.0 L

17. 40°C is the more reasonable temperature because normal body temperature is about 37°C.

25. $C = \frac{5(F - 32)}{9}$

$= \frac{5(104 - 32)}{9}$

$= \frac{5(72)}{9}$

$= \frac{5(\cancel{72}^{8})}{\cancel{9}_{1}}$

= 40

Thus, 104°F = 40°C

29. $F = \frac{9 \cdot C}{5} + 32$

$= \frac{9 \cdot 35}{5} + 32$

$= \frac{9 \cdot \cancel{35}^{7}}{\cancel{5}_{1}} + 32$

= 63 + 32

= 95

Thus, 35°C = 95°F

33. $F = \frac{9 \cdot C}{5} + 32$

$= \frac{9 \cdot 500}{5} + 32$

$= \frac{9 \cdot \cancel{500}^{100}}{\cancel{5}_{1}} + 32$

= 900 + 32

= 932

The kiln temperature reaches 932°F

37. $\underbrace{2 \cdot 8}_{16} + \underbrace{2 \cdot 8}_{16} = 32$ Multiply before adding.

41. $\underbrace{(5^2)}_{25} + \underbrace{(4^2)}_{16} = 41$ Do exponents before adding.

CHAPTER 8

SECTION 8.1 (page 483)

1. This is line $\overleftrightarrow{CD}$ or $\overleftrightarrow{DC}$. A line is a straight row of points that goes on forever in both directions.

5. This is ray $\overrightarrow{PQ}$. A ray is a part of a line that has only one endpoint and goes on forever in one direction.

9. These are parallel lines. Parallel lines are lines that never cross.

13. $\angle AOS$ or $\angle SOA$ The middle letter, *0,* identifies the vertex.

17. $\angle AQC$ or $\angle CQA$ The middle letter, *Q,* identifies the vertex.

21. This is an acute angle. Acute angles measure between 0° and 90°.

29. False. $\angle USQ$ is a straight angle and so is $\angle PQR$, therefore each measures 180°.

33. (180 − 75) − 15 = 105 − 15 = 90 Do what is inside parentheses first.

SECTION 8.2 (page 489)

1. $\angle EOD$ and $\angle COD$ are complementary angles because $75° + 15° = 90°$
$\angle AOB$ and $\angle BOC$ are complementary angles because $25° + 65° = 90°$

5. The complement of 40° is 50°, because $90° - 40° = 50°$

9. The supplement of 130° is 50°, because $180° - 130° = 50°$

13. If two angles are vertical angles, they are congruent. $\angle SON \cong \angle TOM$ and $\angle TOS \cong \angle MON$

17. $\angle EOF = 37°$
Since $\angle EOF$ and $\angle AOH$ are vertical angles, $\angle EOF \cong \angle AOH$

21. $\angle ABF \cong \angle ECD$. Both are 138°, since $180° - 42° = 138°$. $\angle ABC \cong \angle BCD$, both are 42°.

25. $2 \cdot 3 + 2 \cdot 4 = 6 + 8$
$= 14$

SECTION 8.3 (page 497)

1. $P = 2 \cdot 8 \text{ yd} + 2 \cdot 6 \text{ yd} = 16 \text{ yd} + 12 \text{ yd} = 28 \text{ yd}$
$A = 8 \text{ yd} \cdot 6 \text{ yd} = 48 \text{ yd}^2$ (square units for area)

5. $P = 2 \cdot 10 \text{ ft} + 2 \cdot 1 \text{ ft} = 20 \text{ ft} + 2 \text{ ft} = 22 \text{ ft}$
$A = 10 \text{ ft} \cdot 1 \text{ ft} = 10 \text{ ft}^2$ (square units for area)

9. $P = 2 \cdot 76.1 \text{ ft} + 2 \cdot 22 \text{ ft} = 152.2 \text{ ft} + 44 \text{ ft} = 196.2 \text{ ft}$
$A = 76.1 \text{ ft} \cdot 22 \text{ ft} = 1674.2 \text{ ft}^2$ (square units for area)

13.

7m, 3m, 12m, 9m, 2m; 7m, 3m, 9m, 2m

$P = 12 \text{ m} + 7 \text{ m} + 3 \text{ m} + 5 \text{ m} + 9 \text{ m} + 2 \text{ m}$
$= 38 \text{ m}$
$A = 7 \text{ m} \cdot 3 \text{ m} + 9 \text{ m} \cdot 2 \text{ m} = 21 \text{ m}^2 + 18 \text{ m}^2$
$= 39 \text{ m}^2$

17.

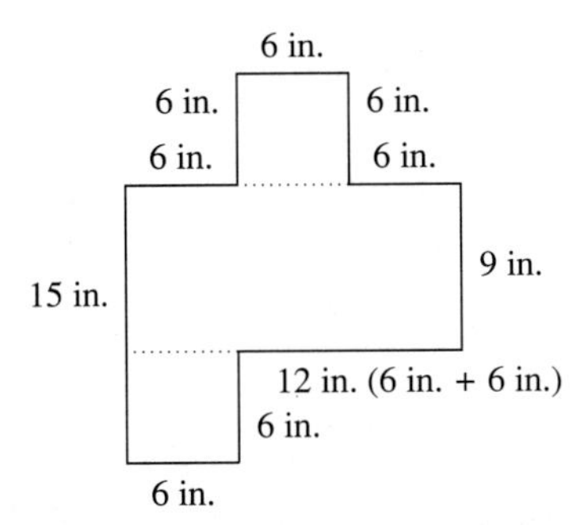

$P = 15 \text{ in.} + 6 \text{ in.} + 6 \text{ in.} + 6 \text{ in.} + 6 \text{ in.} + 6 \text{ in.} + 9 \text{ in.} + 12 \text{ in.} + 6 \text{ in.} + 6 \text{ in.} = 78 \text{ in.}$
$A = 6 \text{ in.} \cdot 6 \text{ in.} + 18 \text{ in.} \cdot 9 \text{ in.} + 6 \text{ in.} \cdot 6 \text{ in.}$
$= 36 \text{ in.}^2 + 162 \text{ in.}^2 + 36 \text{ in.}^2$
$= 234 \text{ in.}^2$

21. $P = 2 \cdot 4.4 \text{ m} + 2 \cdot 5.1 \text{ m} = 8.8 \text{ m} + 10.2 \text{ m} = 19 \text{ m}$

$$\text{Cost} = \frac{19 \cancel{\text{m}}}{1} \cdot \frac{\$4.99}{1 \cancel{\text{m}}} = \$94.81$$

Tyra will have to spend $94.81 for the strip.

25. Length of inner rectangle $= 172 \text{ ft} - 12 \text{ ft} - 12 \text{ ft} = 148 \text{ ft}$
Width of inner rectangle $= 124 \text{ ft} - 12 \text{ ft} - 12 \text{ ft} = 100 \text{ ft}$
Shaded area $= 172 \text{ ft} \cdot 124 \text{ ft} - 148 \text{ ft} \cdot 100 \text{ ft}$
$= 21{,}328 \text{ ft}^2 - 14{,}800 \text{ ft}^2$
$= 6528 \text{ ft}^2$

The area of land that cannot be built on is 6528 ft^2.

29. $12 \text{ ft} = \frac{12 \cancel{\text{ft}}}{1} \cdot \frac{12 \text{ in.}}{1 \cancel{\text{ft}}} = 144 \text{ in.}$

SECTION 8.4 (page 503)

1. $P = 46 \text{ m} + 58 \text{ m} + 46 \text{ m} + 58 \text{ m} = 208 \text{ m}$

5. $P = 3 \text{ km} + 0.8 \text{ km} + 0.95 \text{ km} + 1.03 \text{ km}$
$= 5.78 \text{ km}$

9. $A = 5.5 \text{ ft} \cdot 3.5 \text{ ft} = 19.25 \text{ ft}^2$ (square units for area)

13. $A = \frac{1}{2} \cdot 45 \text{ ft} \cdot (80 \text{ ft} + 110 \text{ ft}) = \frac{1}{2} \cdot 45 \text{ ft} \cdot (190 \text{ ft})$
$= 4275 \text{ ft}^2$

$$\text{Cost} = \frac{4275 \cancel{\text{ft}^2}}{1} \cdot \frac{\$0.22}{1 \cancel{\text{ft}^2}} = \$940.50$$

It will cost $940.50 to put sod on the yard.

17. $A = 1.3 \text{ m} \cdot 0.8 \text{ m} + 1.8 \text{ m} \cdot 1.1 \text{ m}$
$= 1.04 \text{ m}^2 + 1.98 \text{ m}^2$
$= 3.02 \text{ m}^2$

21. $6\frac{1}{2} \cdot 8 = \frac{13}{\cancel{2}_1} \cdot \frac{\cancel{8}^4}{1} = \frac{52}{1} = 52$

SECTION 8.5 (page 509)

1. $P = 9 \text{ yd} + 7 \text{ yd} + 11 \text{ yd} = 27 \text{ yd}$

5. $A = \frac{1}{2} \cdot 60 \text{ m} \cdot 66 \text{ m} = 1980 \text{ m}^2$ (square units for area)

9. $A = 12 \text{ m} \cdot 12 \text{ m} + \frac{1}{2} \cdot 12 \text{ m} \cdot 9 \text{ m} = 144 \text{ m}^2 + 54 \text{ m}^2 = 198 \text{ m}^2$

13. *Step 1* Add the two angles given. $90° + 58° = 148°$
Step 2 Subtract the sum from 180°.
$180° - 148° = 32°$
The missing angle is 32°.

21.
$$\begin{array}{r} 3.14 \\ \times\ 16 \\ \hline 18\ 84 \\ 31\ 4 \\ \hline 50.24 \end{array}$$

SOLUTIONS

SECTION 8.6 (page 517)

1. $d = 2 \cdot r = 2 \cdot 47\ m = 94$ m

5. $C = 2 \cdot \pi \cdot r = 2 \cdot 3.14 \cdot 11$ ft $= 69.08$ ft ≈ 69.1 ft

$A = \pi \cdot r^2 = 3.14 \cdot 11$ ft $\cdot$ 11 ft $= 379.94\ \text{ft}^2$

$\approx 379.9\ \text{ft}^2$ (square units for area)

9. $C = \pi \cdot d = 3.14 \cdot 15$ cm $= 47.1$ cm

$r = \dfrac{d}{2} = \dfrac{15 \text{ cm}}{2} = 7.5$ cm

$A = 3.14 \cdot 7.5$ cm $\cdot$ 7.5 cm $= 176.625\ \text{cm}^2$

$\approx 176.6\ \text{cm}^2$ (square units for area)

13. $C = 3.14 \cdot 8.65$ km $= 27.161$ cm ≈ 27.2 cm

$r = \dfrac{8.65 \text{ km}}{2} = 4.325$ km

$A = 3.14 \cdot 4.325$ km $\cdot$ 4.325 km $\approx 58.7\ \text{km}^2$

17. $C = 3.14 \cdot 250$ ft $= 785.0$ ft

The circumference of the dome is 785.0 ft.

25. Area of large circle $= 3.14 \cdot 12$ cm $\cdot$ 12 cm

$= 452.16\ \text{cm}^2$

Area of small circle $= 3.14 \cdot 9$ cm $\cdot$ 9 cm

$= 254.34\ \text{cm}^2$

Shaded area $= 452.16\ \text{cm}^2 - 254.34\ \text{cm}^2 \approx 197.8\ \text{cm}^2$

29. 22 m $= 2 \cdot 3.14 \cdot r$

$\dfrac{22 \text{ m}}{6.28} = \dfrac{\overset{1}{\cancel{6.28}} \cdot r}{\underset{1}{\cancel{6.28}}}$

$3.5 \text{ m} \approx r$

The radius of the pool is about 3.5 m.

33. $2\frac{3}{4} = \frac{11}{4}$

$$\begin{array}{r} 2.75 \\ 4\overline{)11.00} \\ \underline{8} \\ 3\,0 \\ \underline{2\,8} \\ 20 \\ \underline{20} \\ 0 \end{array}$$

37. $10\frac{1}{2} = \frac{21}{2}$

$$\begin{array}{r} 10.5 \\ 2\overline{)21.0} \\ \underline{20} \\ 1\,0 \\ \underline{1\,0} \\ 0 \end{array}$$

SECTION 8.7 (page 527)

1. $V = l \cdot w \cdot h = 12.5$ cm $\cdot$ 4 cm $\cdot$ 11 cm $= 550\ \text{cm}^3$ (cubic units for volume)

5. $V = \dfrac{4}{3} \cdot \pi \cdot r \cdot r \cdot r$

$= \dfrac{4 \cdot 3.14 \cdot 22 \text{ m} \cdot 22 \text{ m} \cdot 22 \text{ m}}{3}$

$\approx 44{,}579.6\ \text{m}^3$

9. $V = \pi \cdot r \cdot r \cdot h = 3.14 \cdot 5$ ft $\cdot$ 5 ft $\cdot$ 6 ft $= 471\ \text{ft}^3$

13. $V = \dfrac{\pi \cdot r \cdot r \cdot h}{3} = \dfrac{3.14 \cdot 5 \text{ m} \cdot 5 \text{ m} \cdot 16 \text{ m}}{3}$

$\approx 418.7\ \text{m}^3$

17. $r = \dfrac{d}{2} = \dfrac{6.8 \text{ cm}}{2} = 3.4$ cm

$V = \dfrac{4 \cdot 3.14 \cdot 3.4 \text{ cm} \cdot 3.4 \text{ cm} \cdot 3.4 \text{ cm}}{3} \approx 164.6\ \text{cm}^3$

The volume of the ball is about $164.6\ \text{cm}^3$.

21. Volume of base $= 11$ cm $\cdot$ 9 cm $\cdot$ 3 cm $= 297\ \text{cm}^3$

Volume of column $= 9$ cm $\cdot$ 2 cm $\cdot$ 12 cm $= 216\ \text{cm}^3$

Volume of entire object $= 297\ \text{cm}^3 + 216\ \text{cm}^3$

$= 513\ \text{cm}^3$

25. $\sqrt{16} = 4$, since $4 \cdot 4 = 16$

SECTION 8.8 (page 533)

1. $\sqrt{49} = 7$, since $7 \cdot 7 = 49$

5. Using a calculator, $\sqrt{11} = 3.3166 \approx 3.317$

9. Using a calculator, $\sqrt{73} = 8.5440 \approx 8.544$

13. Using a calculator, $\sqrt{190} = 13.7840 \approx 13.784$

17. $a^2 = 36$, $b^2 = 64$, $c^2 = 100$

Since $36 + 64 = 100$, the Pythagorean Theorem is true.

21. hypotenuse $= \sqrt{(\text{leg})^2 + (\text{leg})^2}$

$= \sqrt{8^2 + 15^2}$

$= \sqrt{64 + 225}$

$= \sqrt{289}$

$= 17$

The hypotenuse is 17 in. long.

25. hypotenuse $= \sqrt{(\text{leg})^2 + (\text{leg})^2}$

$= \sqrt{8^2 + 3^2}$

$= \sqrt{64 + 9}$

$= \sqrt{73}$

≈ 8.544

The hypotenuse is about 8.544 in. long.

29. leg $= \sqrt{(\text{hypotenuse})^2 - (\text{leg})^2}$

$= \sqrt{22^2 - 17^2}$

$= \sqrt{484 - 289}$

$= \sqrt{195}$

≈ 13.964

The leg is about 13.964 cm long.

33. leg $= \sqrt{(\text{hypotenuse})^2 - (\text{leg})^2}$

$= \sqrt{11.5^2 - 8.2^2}$

$= \sqrt{132.25 - 67.24}$

$= \sqrt{65.01}$

≈ 8.063

The leg is about 8.063 cm long.

37. hypotenuse $= \sqrt{(\text{leg})^2 + (\text{leg})^2}$
$= \sqrt{4^2 + 7^2}$
$= \sqrt{16 + 49}$
$= \sqrt{65}$
≈ 8.1
The length of the loading dock is about 8.1 ft.

41. leg $= \sqrt{(\text{hypotenuse})^2 - (\text{leg})^2}$
$= \sqrt{12^2 - 3^2}$
$= \sqrt{144 - 9}$
$= \sqrt{135}$
≈ 11.6
The ladder will reach about 11.6 ft high on the building.

45. length $AC = \sqrt{12^2 + 9^2}$
$= \sqrt{144 + 81}$
$= \sqrt{225}$
$= 15$ ft
length $BC = 15 \text{ ft} - 8.75 \text{ ft} = 6.25$ ft
length $BD = \sqrt{6.25^2 - 5^2}$
$= \sqrt{39.0625 - 25}$
$= \sqrt{14.0625}$
$= 3.75$ ft

49. $\dfrac{x}{9.2} = \dfrac{15.6}{7.8}$

$x \cdot 7.8 = 9.2 \cdot 15.6$

$x \cdot 7.8 = 143.52$

$\dfrac{x \cdot \overset{1}{\cancel{7.8}}}{\underset{1}{\cancel{7.8}}} = \dfrac{143.52}{7.8}$

$x = 18.4$

SECTION 8.9 (page 541)

1. Similar because the triangles have the same shape.
5. Similar because the triangles have the same shape.
9. Corresponding angles: P and S
N and R
M and Q
Corresponding sides:
MP and QS (the longest side in each triangle)
MN and QR (the shortest side in each triangle)
NP and RS

13. $\dfrac{a}{10} = \dfrac{6}{12}$ $\qquad$ $\dfrac{b}{6} = \dfrac{6}{12}$

$a \cdot 12 = 10 \cdot 6$ $\qquad$ $b \cdot 12 = 6 \cdot 6$

$\dfrac{a \cdot \overset{1}{\cancel{12}}}{\underset{1}{\cancel{12}}} = \dfrac{60}{12}$ $\qquad$ $\dfrac{b \cdot \overset{1}{\cancel{12}}}{\underset{1}{\cancel{12}}} = \dfrac{36}{12}$

$a = 5$ mm $\qquad$ $b = 3$ mm

17. $\dfrac{x}{18.6} = \dfrac{28}{21}$ $\qquad$ $\dfrac{y}{20} = \dfrac{21}{28}$

$x \cdot 21 = 18.6 \cdot 28$ $\qquad$ $y \cdot 28 = 20 \cdot 21$

$\dfrac{x \cdot \overset{1}{\cancel{21}}}{\underset{1}{\cancel{21}}} = \dfrac{520.8}{21}$ $\qquad$ $\dfrac{y \cdot \overset{1}{\cancel{28}}}{\underset{1}{\cancel{28}}} = \dfrac{420}{28}$

$x = 24.8$ m $\qquad$ $y = 15$ m

25. $\dfrac{x}{120} = \dfrac{100}{100 + 140}$

$\dfrac{x}{120} = \dfrac{100}{240}$

$x \cdot 240 = 120 \cdot 100$

$\dfrac{x \cdot \overset{1}{\cancel{240}}}{\underset{1}{\cancel{240}}} = \dfrac{12000}{240}$

$x = 50$ m

29. $13 - 2 \cdot 5 + 7 = 13 - 10 + 7 = 3 + 7 = 10$
Do multiplication first, then add and subtract from left to right.

CHAPTER 9

SECTION 9.1 (page 569)

1. $+12$ Above zero is positive.
5. $+18{,}000$ Above sea level is positive.
9. negative
13. negative
17.

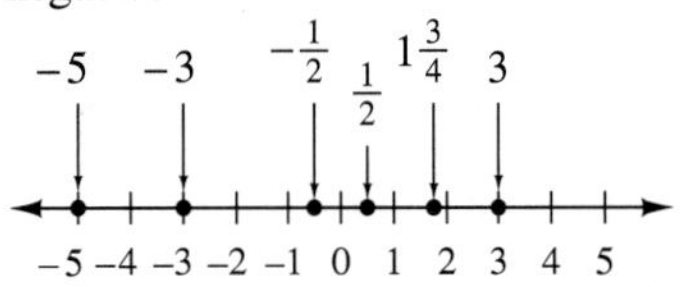

21.

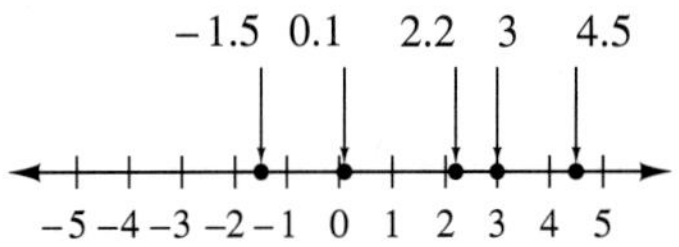

25. Since 9 is to the left of 14 on a number line, 9 is less than 14. $9 < 14$
29. Since -6 is to the left of 3 on a number line, -6 is less than 3. $-6 < 3$
33. Since -11 is to the left of -2 on a number line, -11 is less than -2. $-11 < -2$
37. Since 2 is to the right of -1 on a number line, 2 is greater than -1. $2 > -1$
41. $|5| = 5$
45. $|-1| = 1$
49. $|0| = 0$
53. $|-9.5| = 9.5$
57. $\left|\dfrac{3}{4}\right| = \dfrac{3}{4}$
61. $-|-9| = -9$ because $|-9|$ is 9 and $-(9)$ is -9.

65. The opposite of 2 is -2.
69. The opposite of -11 is 11 (positive 11).
73. The opposite of $\frac{4}{3}$ is $-\frac{4}{3}$.
77. The opposite of 5.2 is -5.2.
85. True. $|-5| = 5$, therefore $|-5| > 0$.
89. False. $-|-4| = -4$ and $-|-7| = -7$; -4 is to the right of -7 on the number line, therefore, $-4 > -7$ and $-|-4| < -|-7|$ is false.
93. $\frac{5}{6} - \frac{1}{4} = \frac{10}{12} - \frac{3}{12} = \frac{7}{12}$

SECTION 9.2 (page 581)

1. $-2 + 5 = 3$
See graph in the answer section.
5. $3 + (-4) = -1$
See graph in the answer section.
9. $|-1| = 1 \quad |8| = 8 \quad 8 - 1 = 7$
The positive number 8 has the larger absolute value, so the answer is positive.

$$-1 + 8 = 7$$

13. $|6| = 6 \quad |-5| = 5 \quad 6 - 5 = 1$
The positive number 6 has the larger absolute value, so the answer is positive.

$$6 + (-5) = 1$$

17. The overdrawn amount is negative ($-\$52.50$) and the deposit is positive.

$$-\$52.50 + \$50 = -\$2.50$$

21. $\left|-\frac{1}{2}\right| = \frac{1}{2} \quad \left|\frac{3}{4}\right| = \frac{3}{4} \quad \frac{3}{4} - \frac{1}{2} = \frac{3}{4} - \frac{2}{4} = \frac{1}{4}$
The positive number $\frac{3}{4}$ has the larger absolute value, so the answer is positive.

$$-\frac{1}{2} + \frac{3}{4} = \frac{1}{4}$$

25. $\left|-\frac{7}{3}\right| = \frac{7}{3} \quad \left|-\frac{5}{9}\right| = \frac{5}{9} \quad \frac{7}{3} + \frac{5}{9} = \frac{21}{9} + \frac{5}{9} = \frac{26}{9}$
Write a negative sign in front of the sum, since both numbers are negative.

$$-\frac{7}{3} + \left(-\frac{5}{9}\right) = -\frac{26}{9} \quad \text{or} \quad -2\frac{8}{9}$$

29. The additive inverse of 3 is -3 (change sign).
33. The additive inverse of $\frac{1}{2}$ is $-\frac{1}{2}$ (change sign).
37. $19 - 5 = 19 + (-5)$ change 5 to its opposite (-5) and add
$= 14$
41. $7 - 19 = 7 + (-19)$ change 19 to its opposite (-19) and add
$= -12$
45. $-9 - 14 = -9 + (-14)$ change 14 to its opposite (-14) and add
$= -23$
49. $6 - (-14) = 6 + (+14)$ change (-14) to its opposite $(+14)$ and add
$= 20$
53. $-\frac{7}{10} - \frac{4}{5} = -\frac{7}{10} + \left(-\frac{4}{5}\right)$ change $\frac{4}{5}$ to its opposite $\left(-\frac{4}{5}\right)$ and add

$$= -\frac{7}{10} + \left(-\frac{8}{10}\right)$$
$$= -\frac{15}{10} = -\frac{3}{2} \quad \text{or} \quad -1\frac{1}{2}$$

57. $-6.4 - (-2.8) = -6.4 + (+2.8)$ change (-2.8) to its opposite $(+2.8)$ and add
$= -3.6$
61. $-2 + (-11) - (-3) = -13 - (-3)$
$= -13 + 3$
$= -10$

65. $-12 - (-3) - (-2) = -12 + 3 - (-2)$
$= -9 - (-2)$
$= -9 + 2$
$= -7$
69. $\frac{1}{2} - \frac{2}{3} + \left(-\frac{5}{6}\right) = \frac{1}{2} + \left(-\frac{2}{3}\right) + \left(-\frac{5}{6}\right)$

$$= \frac{3}{6} + \left(-\frac{4}{6}\right) + \left(-\frac{5}{6}\right)$$
$$= -\frac{1}{6} + \left(-\frac{5}{6}\right)$$
$$= -\frac{6}{6} = -1$$

73. $-2 + (-11) + |-2| = -13 + |-2|$
$= -13 + 2$
$= -11$
77. $2\frac{1}{2} + 3\frac{1}{4} - \left(-1\frac{3}{8}\right) - 2\frac{3}{8} = 5\frac{3}{4} - \left(-1\frac{3}{8}\right) - 2\frac{3}{8}$

$$= 5\frac{3}{4} + 1\frac{3}{8} - 2\frac{3}{8}$$
$$= 5\frac{6}{8} + 1\frac{3}{8} - 2\frac{3}{8}$$
$$= 6\frac{9}{8} - 2\frac{3}{8}$$
$$= 4\frac{6}{8} = 4\frac{3}{4}$$

81.
```
      7 1.2 0 ← 2 decimal place
   × 2 1.2 5 ← 2 decimal places
    3 5 6 0 0
   1 4 2 4 0
    7 1 2 0
 1 4 2 4 0
 1 5 1 3.0 0 0 0 ← 4 decimal places in answer
```
1513.0000 or 1513

85. $\frac{7}{9} \div \frac{14}{27} = \frac{\overset{1}{\cancel{7}}}{\underset{1}{\cancel{9}}} \cdot \frac{\overset{3}{\cancel{27}}}{\underset{2}{\cancel{14}}} = \frac{3}{2}$ or $1\frac{1}{2}$

SECTION 9.3 (page 587)

1. $-5 \cdot 7 = -35$ (different signs, product is negative)
5. $3 \cdot (-6) = -18$ (different signs, product is negative)
9. $-1 \cdot 12 = -12$ (different signs, product is negative)
13. $11 \cdot 7 = 77$ (same signs, product is positive)
17. $-13 \cdot (-1) = 13$ (same signs, product is positive)

21. $-\frac{1}{2} \cdot (-8) = -\frac{1}{\cancel{2}_1} \cdot \left(\frac{-\cancel{8}^4}{1}\right) = \frac{4}{1} = 4$

25. $\frac{\cancel{3}^1}{5} \cdot \left(-\frac{1}{\cancel{6}_2}\right) = -\frac{1}{10}$

29. $-\frac{\cancel{7}^1}{\cancel{15}_3} \cdot \frac{\cancel{25}^5}{\cancel{14}_2} = -\frac{5}{6}$

33. $9 \cdot (-4.7) = -42.3$
37. $-6.2 \cdot (5.1) = -31.62$
41. $\frac{-14}{7} = -2$ (different signs, quotient is negative)
45. $\frac{-28}{4} = -7$ (different signs, quotient is negative)
49. $\frac{-20}{-2} = 10$ (same signs, quotient is positive)
53. $\frac{-18}{18} = -1$ (different signs, quotient is negative)

57. $\frac{-\frac{5}{7}}{-\frac{15}{14}} = -\frac{\cancel{5}^1}{\cancel{7}_1} \cdot \left(-\frac{\cancel{14}^2}{\cancel{15}_3}\right) = \frac{2}{3}$

61. $5 \div \left(-\frac{5}{8}\right) = \frac{\cancel{5}^1}{1} \cdot \left(-\frac{8}{\cancel{5}_1}\right) = -8$

65. $\frac{-18.92}{-4} = 4.73$

69. $\frac{45.58}{-8.6} = -5.3$

73. $(-0.6)(-0.2)(-3) = (0.12)(-3) = -0.36$

81. $-36 \div (-2) \div (-3) \div (-3) \div (-1)$ Divide from left to right.
$18 \div (-3) \div (-3) \div (-1)$
$-6 \div (-3) \div (-1)$
$2 \div (-1) = -2$

85. $8 + 4 \cdot 2 \div 8$ Multiply first.
$8 + 8 \div 8$ Divide next.
$8 + 1 = 9$ Add.

SECTION 9.4 (page 593)

1. $20 \div 5 + 10$ Divide first.
$4 + 10 = 14$ Add.

5. $6^2 + 4^2$ Exponents first.
$36 + 16 = 52$ Add.

9. $(-2)^5 + 2$ Exponents first.
$-32 + 2 = -30$ Add.

13. $2 - (-5) + 3^2$ Exponents first.
$2 - (-5) + 9$ Add and subtract from left to right.
$2 + (+5) + 9$
$7 + 9 = 16$

17. $3 + 5 \cdot (6 - 2)$ Parentheses first.
$3 + 5 \cdot (4)$ Multiply next.
$3 + 20 = 23$ Add.

21. $-6 + (-5) \cdot (9 - 14)$ Parentheses first.
$-6 + (-5) \cdot (-5)$ Multiply next.
$-6 + 25 = 19$ Add.

25. $9 \div (-3)^2 + (-1)$ Exponents first.
$9 \div 9 + (-1)$ Division next.
$1 + (-1) = 0$ Add.

29. $(-2) \cdot (-7) + 3 \cdot 9$ Multiply from left to right.
$14 + 27 = 41$ Add.

33. $2 \cdot 5 - 3 \cdot 4 + 5 \cdot 3$ Multiply from left to right.
$10 - 12 + 15$ Add and subtract from left to right.
$-2 + 15 = 13$

37. $-12 \cdot (-1) + 5^2 - (-3)$ Exponents first.
$-12 \cdot (-1) + 25 - (-3)$ Multiply next.
$12 + 25 - (-3)$ Add and subtract from left to right.
$37 - (-3) = 40$

41. $3^2 \cdot (2 - 5) \div (4 + 5) - (-6)$ Parentheses first.
$3^2 \cdot (-3) \div (9) - (-6)$ Exponents next.
$9 \cdot (-3) \div (9) - (-6)$ Multiply and divide from left to right.
$-27 \div (9) - (-6)$
$-3 - (-6) = 3$

45. $(-0.3)^2 + (-0.5)^2 + 0.9$ Exponents first.
$0.09 + 0.25 + 0.9$ Add from left to right.
$0.34 + 0.9 = 1.24$

49. $(0.5)^2 \cdot (-8) - (0.31)$ Exponents first.
$0.25 \cdot (-8) - (0.31)$ Multiply next.
$-2 - (0.31) = -2.31$

53. $\left(-\frac{1}{2}\right)^2 - \left(\frac{3}{4} - \frac{7}{4}\right)$ Parentheses first.
$\left(-\frac{1}{2}\right)^2 - \left(-\frac{4}{4}\right)$ Exponents next.
$\frac{1}{4} - \left(-\frac{4}{4}\right) = \frac{5}{4}$ or $1\frac{1}{4}$

57. $5^2 \cdot (9 - 11) \cdot (-3) \cdot (-2)^3$ Parentheses first.
$5^2 \cdot (-2) \cdot (-3) \cdot (-2)^3$ Exponents next.
$25 \cdot (-2) \cdot (-3) \cdot (-8)$ Multiply from left to right.
$-50 \cdot (-3) \cdot (-8)$
$150 \cdot (-8) = -1200$

65. $-7 \cdot \left(6 - \frac{5}{\cancel{8}_1} \cdot \cancel{24}^3 + \cancel{3}^1 \cdot \frac{8}{\cancel{3}_1}\right)$ Work inside parentheses; do multiplications first.
$-7 \cdot (6 - 15 + 8)$ Add and subtract inside parentheses.
$-7 \cdot (-9 + 8)$
$-7 \cdot (-1) = 7$

69. $\frac{\cancel{6}^2}{\cancel{7}_1} \cdot \frac{\cancel{14}^2}{\cancel{9}_3} = \frac{4}{3}$ or $1\frac{1}{3}$

SOLUTIONS

SECTION 9.5 (page 599)

1. $2r + 4s = 2(2) + 4(6)$
$= 4 + 24$
$= 28$

5. $2r + 4s = 2(-4) + 4(4)$
$= -8 + 16$
$= 8$

9. $2r + 4s = 2(0) + 4(-2)$
$= 0 + (-8)$
$= -8$

13. $6k + 2s = 6(1) + 2(-2)$
$= 6 + (-4)$
$= 2$

17. $-m - 3n = -\left(\frac{1}{2}\right) - 3\left(\frac{3}{8}\right)$
$= -\frac{1}{2} - \frac{9}{8}$
$= -\frac{4}{8} + \left(-\frac{9}{8}\right) = -\frac{13}{8}$ or $-1\frac{5}{8}$

21. $P = 2l + 2w$
$= 2(9) + 2(5)$
$= 18 + 10$
$= 28$

25. $A = \frac{1}{2}bh$
$= \frac{1}{2}(15)(3)$
$= \frac{1}{2}(45)$
$= \frac{45}{2}$ or $22\frac{1}{2}$

29. $d = rt$
$= (53)(6)$
$= 318$

37. $V = \frac{4\pi r^3}{3}$
$= \frac{4(3.14)(3)^3}{3}$
$= \frac{4(3.14)(27)}{3}$
$= \frac{12.56(27)}{3}$
$= \frac{339.12}{3} = 113.04$

41. $-\frac{\overset{1}{\cancel{4}}}{\underset{1}{\cancel{3}}} \cdot \left(-\frac{\overset{1}{\cancel{3}}}{\underset{1}{\cancel{4}}}\right) = 1$ (same signs, product is positive)

SECTION 9.6 (page 607)

1. $x + 7 = 11$
$4 + 7 = 11$
$11 = 11$ true
Yes, 4 is a solution of the equation.

5. $2z - 1 = -15$
$2(-8) - 1 = -15$
$-16 - 1 = -15$
$-17 = -15$ false
No, -8 is not a solution of the equation.

9. $k + 10 = 40$
$k + 10 - 10 = 40 - 10$
$k = 30$
Check: $30 + 10 = 40$
$40 = 40$ true
The solution is 30.

13. $8 = r - 2$
$8 + 2 = r - 2 + 2$
$10 = r$
Check: $8 = 10 - 2$
$8 = 8$ true
The solution is 10.

17. $7 = r + 13$
$7 - 13 = r + 13 - 13$
$-6 = r$
Check: $7 = -6 + 13$
$7 = 7$ true
The solution is -6.

21. $-8 + x = 1$
$-8 + 8 + x = 1 + 8$
$x = 9$
Check: $-8 + 9 = 1$
$1 = 1$ true
The solution is 9.

25. $d + \frac{2}{3} = 3$
$d + \frac{2}{3} - \frac{2}{3} = 3 - \frac{2}{3}$
$d = \frac{7}{3}$ or $2\frac{1}{3}$
Check: $\frac{7}{3} + \frac{2}{3} = 3$
$\frac{9}{3} = 3$
$3 = 3$ true
The solution is $\frac{7}{3}$.

29. $k - 2 = \frac{1}{2}$
$k - 2 + 2 = \frac{1}{2} + 2$
$k = \frac{5}{2}$ or $2\frac{1}{2}$
Check: $\frac{5}{2} - 2 = \frac{1}{2}$
$\frac{1}{2} = \frac{1}{2}$ true
The solution is $\frac{5}{2}$.

33. $x - 0.8 = 5.07$
$x - 0.8 + 0.8 = 5.07 + 0.8$
$x = 5.87$
Check: $5.87 - 0.8 = 5.07$
$5.07 = 5.07$ true
The solution is 5.87.

37. $6z = 12$
$\frac{\overset{1}{\cancel{6}} \cdot z}{\underset{1}{\cancel{6}}} = \frac{12}{6}$
$z = 2$
Check: $6(2) = 12$
$12 = 12$ true
The solution is 2.

41. $3y = -24$

$\frac{\cancel{3}^{1} \cdot y}{\cancel{3}_{1}} = \frac{-24}{3}$

$y = -8$

The solution is -8.

Check:
$3(-8) = -24$
$-24 = -24$ true

45. $-36 = -4p$

$\frac{-36}{-4} = \frac{-\cancel{4}^{1} \cdot p}{-\cancel{4}_{1}}$

$9 = p$

The solution is 9.

Check:
$-36 = -4(9)$
$-36 = -36$ true

49. $-8.4p = -9.24$

$\frac{\cancel{-8.4}^{1} \cdot p}{\cancel{-8.4}_{1}} = \frac{-9.24}{-8.4}$

$p = 1.1$

The solution is 1.1

Check:
$-8.4(1.1) = -9.24$
$-9.24 = -9.24$ true

53. $11 = \frac{a}{6}$

$6 \cdot 11 = \frac{\cancel{6}^{1}}{1} \cdot \frac{a}{\cancel{6}_{1}}$

$66 = a$

The solution is 66.

Check:
$11 = \frac{66}{6}$
$11 = 11$ true

57. $-\frac{2}{5}p = 8$

$-\frac{\cancel{5}^{1}}{\cancel{2}_{1}} \cdot -\frac{\cancel{2}^{1}}{\cancel{5}_{1}}p = -\frac{5}{\cancel{2}_{1}} \cdot \cancel{8}^{4}$

$p = -20$

The solution is -20.

Check:
$-\frac{2}{5}(-20) = 8$
$8 = 8$ true

61. $\frac{3}{8}x = 6$

$\frac{\cancel{8}^{1}}{\cancel{3}_{1}} \cdot \frac{\cancel{3}^{1}}{\cancel{8}_{1}}x = \frac{8}{\cancel{3}_{1}} \cdot \cancel{6}^{2}$

$x = 16$

The solution is 16.

Check:
$\frac{3}{8}(16) = 6$
$6 = 6$ true

65. $\frac{z}{-3.8} = 1.3$

$\frac{\cancel{-3.8}^{1}}{1} \cdot \frac{z}{\cancel{-3.8}_{1}} = (-3.8)(1.3)$

$z = -4.94$

The solution is -4.94.

Check:
$\frac{-4.94}{-3.8} = 1.3$
$1.3 = 1.3$
true

69. $x - 17 = 5 - 3$
$x - 17 = 2$
$x - 17 + 17 = 2 + 17$
$x = 19$ The solution is 19.

73. $\frac{7}{2}x = \frac{4}{3}$

$\frac{\cancel{2}^{1}}{\cancel{7}_{1}} \cdot \frac{\cancel{7}^{1}}{\cancel{2}_{1}}x = \frac{2}{7} \cdot \frac{4}{3}$

$x = \frac{8}{21}$ The solution is $\frac{8}{21}$.

77. $\frac{1}{2} + \frac{\cancel{3}^{1}}{\cancel{4}_{1}} \cdot \frac{\cancel{8}^{2}}{\cancel{9}_{3}} - \frac{1}{6}$ Multiply first.

$\frac{1}{2} + \frac{2}{3} - \frac{1}{6}$ Add and subtract from left to right.

$\frac{7}{6} - \frac{1}{6}$

$\frac{6}{6} = 1$

SECTION 9.7 (page 615)

1. $7p + 5 = 12$
$7p + 5 - 5 = 12 - 5$ Subtract 5 from both sides.
$7p = 7$

$\frac{\cancel{7}^{1} \cdot p}{\cancel{7}_{1}} = \frac{7}{7}$ Divide both sides by 7.

$p = 1$

Check:
$7(1) + 5 = 12$
$7 + 5 = 12$
$12 = 12$ true The solution is 1.

5. $-3m + 1 = -5$
$-3m + 1 - 1 = -5 - 1$ Subtract 1 from both sides.
$-3m = -6$

$\frac{\cancel{-3}^{1} \cdot m}{\cancel{-3}_{1}} = \frac{-6}{-3}$ Divide both sides by -3.

$m = 2$

Check:
$-3(2) + 1 = -5$
$-6 + 1 = -5$
$-5 = -5$ true The solution is 2.

9. $-5x - 4 = 16$
$-5x - 4 + 4 = 16 + 4$ Add 4 to both sides.
$-5x = 20$

$\frac{\cancel{-5}^{1} \cdot x}{\cancel{-5}_{1}} = \frac{20}{-5}$ Divide both sides by -5.

$x = -4$

Check:
$-5(-4) - 4 = 16$
$20 - 4 = 16$
$16 = 16$ true The solution is -4.

13. $6(x + 4) = 6 \cdot x + 6 \cdot 4 = 6x + 24$

17. $-3(m + 6) = -3 \cdot m + (-3) \cdot 6 = -3m - 18$

21. $-5(z - 9) = -5 \cdot z - (-5) \cdot 9 = -5z - (-45)$
$= -5z + 45$

25. $10x - 2x = (10 - 2)x = 8x$

29. $-5a + a = (-5 + 1)a = -4a$

33. $45 = 13m - 8m$ Combine like terms.
$45 = 5m$

$$\frac{45}{5} = \frac{\cancel{5} \cdot m}{\cancel{5}}$$ Divide both sides by 5.

$9 = m$

Check:
$45 = 13(9) - 8(9)$
$45 = 117 - 72$
$45 = 45$ true The solution is 9.

37. $-12 = 6y - 18y$ Combine like terms.
$-12 = -12y$

$$\frac{-12}{-12} = \frac{\cancel{-12} \cdot y}{\cancel{-12}}$$ Divide both sides by -12.

$1 = y$

Check:
$-12 = 6(1) - 18(1)$
$-12 = 6 - 18$
$-12 = -12$ true The solution is 1.

41. $9 + 7z = 9z + 13$
$9 + 7z - 9z = 9z + 13 - 9z$ Subtract $9z$ from both sides.
$9 - 2z = 13$
$9 - 2z - 9 = 13 - 9$ Subtract 9 from both sides.
$-2z = 4$

$$\frac{\cancel{-2} \cdot z}{\cancel{-2}} = \frac{4}{-2}$$ Divide both sides by -2.

$z = -2$

Check:
$9 + 7(-2) = 9(-2) + 13$
$9 + (-14) = -18 + 13$
$-5 = -5$ true The solution is -2.

45. $-3.6m + 1 = 2.4m + 7$
$-3.6m + 1 - 2.4m = 2.4m + 7 - 2.4m$
Subtract $2.4m$ from both sides.
$-6m + 1 = 7$
$-6m + 1 - 1 = 7 - 1$ Subtract 1 from both sides.
$-6m = 6$

$$\frac{\cancel{-6} \cdot m}{\cancel{-6}} = \frac{6}{-6}$$ Divide both sides by -6.

$m = -1$

Check:
$-3.6(-1) + 1 = 2.4(-1) + 7$
$3.6 + 1 = -2.4 + 7$
$4.6 = 4.6$ true The solution is -1.

49. $-10 = 2(y + 4)$ Use distributive property.
$-10 = 2y + 8$
$-10 - 8 = 2y + 8 - 8$ Subtract 8 from both sides.
$-18 = 2y$

$$\frac{-18}{2} = \frac{\cancel{2} \cdot y}{\cancel{2}}$$ Divide both sides by 2.

$-9 = y$

Check:
$-10 = 2(-9 + 4)$
$-10 = 2(-5)$
$-10 = -10$ true The solution is -9.

53. $6(x - 1) = 42$ Use distributive property.
$6x - 6 = 42$
$6x - 6 + 6 = 42 + 6$ Add 6 to both sides.
$6x = 48$

$$\frac{\cancel{6} \cdot x}{\cancel{6}} = \frac{48}{6}$$ Divide both sides by 6.

$x = 8$

Check:
$6(8 - 1) = 42$
$6(7) = 42$
$42 = 42$ true The solution is 8.

57. $30 - 40 = -2x + 7x - 4x$
$-10 = (-2 + 7 - 4)x$
$-10 = 1x$
$-10 = x$ The solution is -10.

61. $3 + 4r = 10 - 7$
$3 + 4r = 3$
$3 + 4r - 3 = 3 - 3$ Subtract 3 from both sides.
$4r = 0$

$$\frac{\cancel{4} \cdot r}{\cancel{4}} = \frac{0}{4}$$ Divide both sides by 4.

$r = 0$ The solution is 0.

65. Sales tax $= (0.06)(\$420)$
$= \$25.20$

SECTION 9.8 (page 623)

1. $14 + x$
5. $x + 6$
9. $x - 4$
13. $3x$
17. $8 + 2x$
21. $2x + x$

25. Let x represent the unknown number.

$\underbrace{\text{four times a number}}_{4x}\ \underbrace{\text{decreased by 2}}_{-2}\ \underbrace{\text{result is}}_{=}\ 26$

$$4x - 2 = 26$$
$$4x - 2 + 2 = 26 + 2$$
$$4x = 28$$
$$\frac{\cancel{4} \cdot x}{\cancel{4}} = \frac{28}{4}$$
$$x = 7$$

The number is 7.
Check: 4 times 7 is 28
28 decreased by 2 is 26 true

29. Let x represent the unknown number.

$\underbrace{\text{half a number}}_{\frac{1}{2}x}\ \underbrace{\text{is added to}}_{+}\ \underbrace{\text{twice the number}}_{2x}\ \underbrace{\text{answer is}}_{=}\ 50$

$$2\frac{1}{2}x = 50$$
$$\frac{5}{2}x = 50$$
$$\frac{\cancel{2}}{\cancel{5}} \cdot \frac{\cancel{5}}{\cancel{2}}x = \frac{2}{\cancel{5}} \cdot \cancel{50}^{10}$$
$$x = 20$$

The number is 20.
Check: half of 20 is 10
twice 20 is 40
10 added to 40 is 50 true

33. Let x represent the number of days Kerwin rented the saw.

$\underbrace{\text{one-time \$9 fee}}_{9}\ \underbrace{\text{added to}}_{+}\ \underbrace{\text{\$16 per day}}_{16 \cdot x}\ \underbrace{\text{total is}}_{=}\ \$89$

$$9 + 16x = 89$$
$$9 + 16x - 9 = 89 - 9$$
$$16x = 80$$
$$\frac{\cancel{16} \cdot x}{\cancel{16}} = \frac{80}{16}$$
$$x = 5$$

Kerwin rented the saw for 5 days.
Check: \$16 per day for 5 days is \$80
\$80 plus \$9 one-time fee is \$89 true

37. Let x represent the width of the rectangle. The length will be twice x, which is $2x$.

$$P = 2 \cdot \text{length} + 2 \cdot \text{width}$$
$$36 = 2 \cdot 2x + 2 \cdot x$$
$$36 = 4x + 2x$$
$$36 = 6x$$
$$\frac{36}{6} = \frac{\cancel{6} \cdot x}{\cancel{6}}$$
$$6 = x$$

Width $= x$ so width is 6 ft.
Length $= 2x$ so length is 2(6) or 12 ft.
Check: $36 = 2(12) + 2(6)$
$36 = 24 + 12$
$36 = 36$ true

41. Let x represent the length of each equal part. The length of the third part is $x + 25$

Total length is 706

$$x + x + x + 25 = 706$$
$$3x + 25 = 706$$
$$3x + 25 - 25 = 706 - 25$$
$$3x = 681$$
$$\frac{\cancel{3} \cdot x}{\cancel{3}} = \frac{681}{3}$$
$$x = 227$$

Two parts are each 227 m long.
The third part is 227 + 25 or 252 m long.
Check: 227 + 227 + 252 = 706 true

45. amount = percent · base
$a = 0.80 \cdot 2900$
$a = 2320$
80% of \$2900 is \$2320.

CHAPTER 10

SECTION 10.1 (page 647)

1. The total cost of adding the art studio is \$12,100 + \$9800 + \$2000 + \$900 + \$1800 + \$3000 + \$2400 = \$32,000

5. $\dfrac{\text{carpentry}}{\text{window coverings}} = \dfrac{12100}{900} = \dfrac{121}{9}$

9. $\dfrac{\text{business majors}}{\text{total students}} = \dfrac{4000}{12000} = \dfrac{1}{3}$

13. $\dfrac{\text{business majors}}{\text{science majors}} = \dfrac{4000}{1000} = \dfrac{4}{1}$

17. Total cost = \$1,740,000
Cost of doors and thresholds = 10% of total cost
= 0.10 · \$1,740,000
= \$174,000

21. Total income = \$19,600
Budget for supplies = 20% of total income
= 0.20 · \$19,600
= \$3920

25. Total income = \$19,600
Budget for national dues = 10% of total income
= 0.10 · \$19,600
= \$1960

29. 25% of total is rent.
number of degrees = 25% · 360°
= 0.25 · 360°
= 90°

33. percent of total $= \frac{630}{4200} = 0.15 = 15\%$

37. **(a)** Total sales
= \$12,500 + \$40,000 + \$60,000
+ \$50,000 + \$37,500 = \$200,000

(b) Adventure classes = \$12,500
percent of total $= \frac{12{,}500}{200{,}000} = 0.0625 = 6.25\%$
number of degrees = 0.0625 · 360° = 22.5°
Grocery and provision sales = \$40,000
percent of total $= \frac{40{,}000}{200{,}000} = 0.2 = 20\%$
number of degrees = 0.2 · 360° = 72°
Equipment rentals = \$60,000
percent of total $= \frac{60{,}000}{200{,}000} = 0.3 = 30\%$
number of degrees = 0.3 · 360° = 108°
Rafting tours = \$50,000
percent of total $= \frac{50{,}000}{200{,}000} = 0.25 = 25\%$
number of degrees = 0.25 · 360° = 90°
Equipment sales = \$37,500
percent of total $= \frac{37{,}500}{200{,}000} = 0.1875 = 18.75\%$
number of degrees = 0.1875 · 360° = 67.5°

(c) See graph in the answer section.

41. $\frac{4}{5} = \frac{32}{40}$ and $\frac{7}{8} = \frac{35}{40}$; therefore, $\frac{4}{5} < \frac{7}{8}$

45. 25.9% < 26.01%

SECTION 10.2 (page 657)

1. The attendance on July 5 was 3000.

5. The greatest attendance was on July 4.

9. Unemployed workers in February of 1995 = 7000
Unemployed workers in February of 1994 = 5500

7000 − 5500 = 1500

There were 1500 more workers unemployed in February of 1995.

13. 150,000 gallons of supreme unleaded gasoline were sold in 1991.

17. 700,000 gallons of supreme unleaded gasoline were sold in 1995. 150,000 gallons of supreme unleaded gasoline were sold in 1991.

700,000 − 150,000 = 550,000

There was an increase of 550,000 gallons of supreme unleaded gasoline sales.

21. There were 300 burglaries in June.

25. Chain Store A sold 3,000,000 CD's in 1995.

29. Chain Store B sold 3,500,000 CD's in 1994.

33. The total sales in 1995 was \$40,000.

37. The profit in 1994 was \$5000

41. Amount going to county general fund = 39% of \$16,800
= 0.39 · \$16,800
= \$6552

SECTION 10.3 (page 663)

1. The age group of 60–65 years has the greatest number of members.

5. Total members in 40–60 years age group
= 10,000 + 15,000 + 11,000 + 14,000
= 50,000

9. There are 7 employees who earn \$15–\$20 thousand.

17.

number of sets	tally	frequency
140–149	𝍸 I	6

21.

temperature	tally	frequency
70°–74°	II	2

25.

temperature	tally	frequency
90°–94°	𝍸	5

29.

temperature	tally	frequency
110°–114°	II	2

33.

new homes sold	tally	frequency
11–15	𝍸 𝍸 I	11

37.

new homes sold	tally	frequency
31–35	IIII	4

41. (8 · 6) + (3 · 8) ÷ 5 = 48 + 24 ÷ 5
= 48 + 4.8
= 52.8

SECTION 10.4 (page 671)

1. Mean $= \frac{4 + 7 + 15 + 18 + 21}{5} = \frac{65}{5} = 13$ years

5. Mean
$= \frac{21{,}900 + 22{,}850 + 24{,}930 + 29{,}710 + 28{,}340 + 40{,}000}{6}$
$= \frac{167{,}730}{6}$
= \$27,955

9.

Policy Amount	*Number of Policies Sold*	*Product*
$10,000	6	10,000 · 6 = 60,000
$20,000	24	20,000 · 24 = 480,000
$25,000	12	25,000 · 12 = 300,000
$30,000	8	30,000 · 8 = 240,000
$50,000	5	50,000 · 5 = 250,000
$100,000	3	100,000 · 3 = 300,000
$250,000	2	250,000 · 2 = 500,000
Totals	60	2,130,000

$$\text{Weighted mean} = \frac{2{,}130{,}000}{60} = \$35{,}500$$

13.

Value	*Frequency*	*Product*
12	4	12 · 4 = 48
13	2	13 · 2 = 26
15	5	15 · 5 = 75
19	3	19 · 3 = 57
22	1	22 · 1 = 22
23	5	23 · 5 = 115
Totals	20	343

$$\text{Weighted mean} = \frac{343}{20} = 17.15 = 17.2 \text{ (rounded)}$$

17. 328 420 516 592 715
↑ (516)

The median is 516 customers.

21. 74, 68, 68, 68, 75, 75, 74, 74, 70
(arrows mark each 68 and each 74)

Because both 68 and 74 occur three times, each is a mode. This list is bimodal.

29. 21, 32, 38, 46, 49, 53, 58, 97, 97
↑ (49)
median is the middle number

SOLUTIONS

Index

W

Y

Z